LIST OF CACTACEAE NAMES

(1950–1990)

LIST OF CACTACEAE NAMES

from

REPERTORIUM PLANTARUM SUCCULENTARUM

(1950–1990)

Edited by

URS EGGLI
Städtische Sukkulenten-Sammlung, Zürich

NIGEL TAYLOR
Royal Botanic Gardens, Kew

First published 1991

General editor of series J.M. Lock

ISBN 0 947643 37 0

Cover design by Media Resources, RBG, Kew.

Produced from laser-printed copy supplied by
Städtische Sukkulenten-Sammlung, Zürich

Printed and Bound in Great Britain by
Whitstable Litho Ltd., Whitstable, Kent.

Introduction

For more than 40 years, IOS has been publishing the annual ***Repertorium Plantarum Succulentarum*** (RPS), listing new names of succulent plants. It might seem superfluous to have such a publication, giving information similar to that in Index Kewensis, but the continuing success RPS enjoys clearly shows the need for its existence. In its earlier years, the annual publication was a considerable advantage over the 5-yearly supplements of Index Kewensis. In more modern times, the inclusion of information additional to that given in the now annual Index Kewensis (infraspecific nomenclature, typification details [since vol. 33]) makes RPS an important source of data.
With these advantages in mind, the computerization of all the nomenclatural data included in the last 41 volumes of RPS has been undertaken at the Municipal Succulent Collection, Zürich (ZSS), and the 41-year name-list of Cactaceae is the first family checklist to be published from this database.
A tribute has to be paid to the former editors and compilers of RPS who – without the modern bibliographical aids and computer databases available nowadays – managed to produce a nearly complete coverage of the nomenclature of succulent plants from 1950 onwards (1: H. M. Roan & A. J. A. Uitewaal; 2-5: H. M. Roan & G. D. Rowley; 6-12: G. D. Rowley; 13-26: G. D. Rowley & L. E. Newton; 27-29: G. D. Rowley, L. E. Newton & D. R. Hunt; 30-32: G. D. Rowley, L. E. Newton & N. P. Taylor; 33+: U. Eggli & N. P. Taylor). The rapid computerization of this data would have been impossible without their work. Although some omissions and errors have been detected during this phase, RPS is to the best of our knowledge still the most complete nomenclatural list in this specialized field.
It is the hope of the IOS that this checklist, giving an as complete overview as possible of the name changes for cacti during the past 41 years, will be of use to all who work with this fascinating family. Some of the readers would probably have liked to see more information included in the list, especially a preferred classification of some sort. While this idea has considerable appeal, it is thought that a mere checklist giving nomenclatural facts (or at the most some interpretations of such facts) is preferable at present, instead of including taxonomic decisions which are always subjective.

Notes on the List of Cactaceae Names (1950-1990) from the Repertorium Plantarum Succulentarum Database

This alphabetic list of names of *Cactaceae* published between 1950 and 1990 is an extract from the database of *Repertorium Plantarum Succulentarum* (RPS) maintained at the Municipal Succulent Collection Zürich. The database contains nomenclatural and taxonomic information published in serialized form in the yearly issues of *Repertorium Plantarum Succulentarum*, initiated and currently published by the *International Organization for Succulent Plant Study* (IOS).

The data from the yearly issues of RPS was input into the database directly without going back to the original references. In many cases, however, corrections have been made when mistakes have been noted, or when the contents of the database have shown that the information as published in RPS is incorrect. All such cases have been labelled accordingly. The data from the RPS issues has been supplemented with information from other sources. It has become evident that a (fortunately small) number of taxa has been omitted from earlier issues of RPS.

The following notes are of importance to understand the scope of this extract from the RPS database:

— *Nomina nuda* which are indicated globally for whole works in individual issues of RPS have normally been included in the database, when the relevant works have been available, with the exception of the following works: Haage, W. & O. Sadovsky (1958): Kakteen-Sterne (names of hybrids); Ito, Y. (1954): Cacti – Cultivation and information (many nude names for suprageneric taxa); Ito, Y. (1962): Gardening of Cacti (includes hundreds of new cultivars and many new names and combinations, but all apparently without basionym or latin diagnosis (G. Rowley, in litt. 27. 4. 1990); Johnson, H. (1955): Succulent Parade for the Indoor Gardener (trade catalogue); and Winter, H. (1955 [undated]): Kakteen – Cacti – Cactées (trade catalogue, with many provisional names by F. Ritter).

— Entries which have been corrected from the published versions are labelled accordingly. In many cases, names have been reported repeatedly in different issues of RPS; such entries have been combined into one entry without further comment.

— In comparison with the published version of RPS, additional typification information has been included. Typification data has been included in RPS since vol. 33, but this information is also available in the database for a number of taxa covered in earlier issues. In addition to collector's name and number and herbarium acronyms, this list contains also the country and state / province / department where the type was collected, if this information is recorded in the database. All data from the register of type specimens of *Cactaceae* in Swiss herbaria (cf. U. Eggli, Trop. subtrop. Pfl.-welt, 59, 1987) has been incorporated in this list (where applicable), including some corrections.

It is hoped that no names included in the published versions of RPS have been omitted inadvertently from this database. Despite careful checking, no guarantee can be given, and the compilers at ZSS would be grateful if notified of any omissions or errors.

It should be noted that inclusion of a name as valid, invalid or illegitimate is not a final decision but is based on available evidence. *No name given here in this extract is accepted by the compilers and editors of RPS (in the sense of ICBN Art. 34.1).*

Otherwise, the conventions used in recent issues of RPS are also in effect for this extract from the RPS database (see below).

Acknowledgements are due to all those who have been involved in the preparation of the computerized RPS database. Dr R. Baumberger (Zürich) compiled all the *Cactaceae* names published 1960 - 1980; Dr B. Burr (Bonn) compiled all names in RPS 1-7. The remainder of the published RPS names have been inserted into the database by ZSS staff, mostly by the editor of this compilation. Funding for part of the data entry of *Cactaceae* names has been made available through Dr D. R. Hunt, RBG Kew, via the CITES *Cactaceae* name checklist project.
Various IOS members have helped with the corrections of preliminary print-outs for this checklist. The editors are particularly indebted to Dr D. R. Hunt B. E. Leuenberger, S. Arroyo Leuenberger and D. C. Zappi have kindly helped with the translations of the introduction and these notes.

Conventions Used:

— Validly published names are given in **bold face** type, accompanied by an indication of the nomenclatural type (name or specimen dependent on rank), followed by the herbarium acronyms of the herbaria where the holotype and possible isotypes are said to be deposited (first acronym for holotype), according to Index Herbariorum, ed. 7 and its supplements published in Taxon. Invalid, illegitimate, or incorrect names are given in *italic* type face. In each case, a full bibliographic reference is given. For new combinations, the basionym is also listed. For invalid or illegitimate names, the articles of the ICBN which have been contravened are indicated in brackets.

— Abbreviations for periodicals are those suggested in 'A Bibliography of Succulent Plant Periodicals' (U. Eggli in Bradleya 3: 103-119, 1985, and supplements in l.c. 5: 101-104, 1987) for specialized periodicals, or those of Bibliographia Periodicorum Huntianum for other periodicals.

Vorwort

Seit mehr als 40 Jahren publiziert die IOS jährlich das *Repertorium Plantarum Succulentarum* (RPS), das eine Liste der neuen Namen sukkulenter Pflanzen darstellt. Eine solche Publikation könnte auf den ersten Blick überflüssig erscheinen, enthält sie doch ähnliche Informationen wie der Index Kewensis. Der fortdauernde Erfolg, dessen sich das RPS erfreut, zeigt jedoch klar seine Existenzberechtigung. In den früheren Jahres seines Bestehens lag der Hauptvorteil gegenüber den 5-Jahres-Supplementen des Index Kewensis in seiner jährlichen Erscheinungsweise. In den letzten Jahren wurde das RPS vor allem deshalb zu einer wertvollen Informationsquelle, weil im Vergleich zum jährlich erscheinenden Kew Index zusätzliche Angaben enthalten sind (infraspezifische Nomenklatur, Typifikations-Angaben [ab Band 33]).
Mit Blick auf diese Vorteile wurden sämtliche bereits erschienenen 41 Bände des Repertoriums an der Städtischen Sukkulenten-Sammlung Zürich (ZSS) in eine Computer-Datenbank übertragen. Die 41 Jahre umfassende Namensliste der *Cactaceae* ist nun die erste aus dieser Datenbank hervorgehende 'Checkliste'.
Eine ganze Anzahl von Redaktoren waren seit 1950 für die Zusammenstellung des RPS verantwortlich – und dies in den früheren Jahren ohne die heute erhältlichen Computer-Datenbanken. Ohne deren Arbeit würden wir heute nicht über diese fast vollständige Übersicht über die Nomenklatur der sukkulenten Pflanzen verfügen (1: H. M. Roan & A. J. A. Uitewaal; 2-5: H. M. Roan & G. D. Rowley; 6-12: G. D. Rowley; 13-26: G. D. Rowley & L. E. Newton; 27-29: G. D. Rowley, L. E. Newton & D. R. Hunt; 30-32: G. D. Rowley, L. E. Newton & N. P. Taylor; 33+: U. Eggli & N. P. Taylor). Die rasche Computerisierung der RPS-Daten wäre ohne die von diesen Redaktoren geleistete Vorarbeit nicht möglich gewesen. Obwohl während dieser Arbeit eine Reihe von Fehlern und Auslassungen entdeckt wurden, handelt es sich unseres Wissens beim RPS noch immer um die umfangreichste nomenklatorische Liste in diesem Spezialgebiet.
Die Herausgeber und die IOS hofft, dass die vorliegende Liste, welche einen Überblick über die Namensänderungen bei den Kakteen während der vergangenen 41 Jahren gibt, für alle Bearbeiter dieser Familie ein brauchbares Werkzeug liefert. Einige Leser würden sich vielleicht zusätzliche Informationen wünschen, z. Bsp. eine bevorzugte Klassifikation. Diese Idee hat sicher ihre Vorzüge, aber es ist unsere Überzeugung, dass eine 'nackteÿ Checkliste mit nomenklatorischen Fakten (oder einigen Interpretationen solcher Fakten) einer Liste mit taxonomischen Entscheidungen (welche immer bis zu einem gewissen Grad subjektiv sind) vorzuziehen ist.

Bemerkungen zur Liste der Cactaceae-Namen (1950-1990) aus der Datenbank des Repertorium Plantarum Succulentarum

Diese alphabetische Liste der zwischen 1950 und 1990 publizierten *Cactaceae*-Namen stellt einen Auszug aus der *Repertorium Plantarum Succulentarum* (RPS) -Datenbank dar, welche an der Städtischen Sukkulenten-Sammlung Zürich aufgebaut wurde. Diese Datenbank enthält die nomenklatorische und taxonomische Information, die in Zeitschriftenform in den jährlichen Ausgaben des von der *Internationalen Organisation für Sukkulenten-Forschung* (IOS) begonnenen und weiterhin publizierten *Repertorium Plantarum Succulentarum* aufgelistet wurden.

Die Angaben aus den jährlichen RPS-Ausgaben wurden direkt in die Datenbank übernommen, ohne zur ursprünglichen Veröffentlichung zurückzugehen. In vielen Fällen konnten jedoch Fehler berichtigt werden, wenn sich durch den Inhalt der wachsenden Datenbank zeigte, dass die im RPS veröffentlichte Information fehlerhaft gewesen ist. Diese Fälle wurden entsprechend gekennzeichnet. In einigen Fällen konnten die Angaben aus dem RPS durch andere Quellen vervollständigt werden. Es hat sich gezeigt, dass eine (glücklicherweise kleine) Anzahl von Taxa bisher übersehen worden ist.

Die folgenden Bemerkungen sind für das Verständnis des Umfanges dieses Auszugs aus der RPS-Datenbank wichtig:

— *Nomina nuda* wurden in einzelnen RPS-Heften oft für ganze Veröffentlichungen global zitiert; diese Namen wurden jedoch einzeln in die Datenkbank eingefügt, mit Ausnahme der folgenden uns nicht vorliegenden Werke: Haage, W. & O. Sadovsky (1958): Kakteen-Sterne (Namen für Hybriden); Ito, Y. (1954): Cacti – Cultivation and information (viele *nomina nuda* für supragenerische Taxa); Ito, Y. (1962): Gardening of Cacti (enthält Hunderte von neuen Kultivaren und viele neue Namen und Kombinationen, aber alle offenbar ohne Angabe von Basionym bzw. ohne lateinische Diagnose (G. Rowley, in litt. 27. 4. 1990); Johnson, H. (1955): Succulent Parade for the Indoor Gardener (Handels-Katalog); und Winter, H. (1955 [undatiert]): Kakteen – Cacti – Cactées Handelskatalog mit vielen provisorischen Namen von F. Ritter).

— Einträge, die im Vergleich zur RPS-Veröffentlichung abweichen, wurden entsprechend gekennzeichnet. In mehreren Fällen wurde ein Name in verschiedenen RPS-Ausgaben mehrfach gennant; hier wurden solche Mehrfachnennungen ohne weiteren Hinweis zu einem einzigen Eintrag zusammengefasst.

— Im Vergleich zur publizierten Version des RPS enthält die Datenbank und damit auch diese Checkliste zusätzliche Typusangaben. Typusangaben wurden im RPS seit Band 33 gemacht, aber diese Angaben sind in der Datenbank auch für eine Anzahl von früher publizierten Taxa vorhanden. Zusätzlich zu Sammlername und -nummer sowie den Abkürzungen der Herbarien enthält diese Checkliste auch Angaben über die geographische Herkunft des Typus' (Land, Staat / Provinz / Departement), sofern diese Angaben in der Datenbank vorhanden sind. Alle relevanten Angaben aus aus dem Register der Typus-Belege von *Cactaceae* in Schweizer Herbarien (cf. U. Eggli, Trop. subtrop. Pfl.-welt, 59, 1987) wurden ebenfalls in die Datenbank übernommen und wenn nötig korrigiert.

Es wurde alles unternommen, um keine in der publizierten Version des RPS vorhandenen Namen versehentlich nicht in die Datenbank aufzunehmen. Trotz sorgfältiger Kontrolle kann jedoch diesbezüglich keine Garantie gegeben werden, und die Redaktion in der Städtischen Sukkulenten-Sammlung ist dankbar für alle Hinweise auf Auslassungen oder Fehler.

Es muss darauf hingewiesen werden, dass die Aufnahme eines Namens als gültig, ungültig oder illegitim keine endgültige Entscheidung darstellt, sondern den heutigen Stand unseres Wissens wiedergibt. *Keiner der in dieser Liste aufgeführten Namen wird von den Redaktoren und Mitarbeitern des RPS im Sinne von ICBN Art. 34.1 akzeptiert.*

Im Weiteren gelten dieselben Konventionen wie in den neueren Ausgaben des RPS (vgl. auch unten).

Ein Dank geht an alle welche bei der Erstellung der Computer-Datenbank der RPS-Daten mitgewirkt haben. Dr. R. Baumberger (Zürich) hat alle zwischen 1960 und 1980 erfassten *Cactaceae*-Namen erfasst; Dr. B. Burr (Bonn) erfasste die verbleibenden Namen aus den RPS-Bänden 1 bis 7. Die übrigen der in RPS nachgewiesenen Namen wurden durch Mitarbeiter der Sukkulenten-Sammlung Zürich (ZSS) erfasst, hauptsächlich durch den Redaktor dieses Auszugs. Finanzielle Unterstützung für die Erfassung der Kakteennamen wurde durch Dr. D. R. Hunt, RBG Kew durch das CITES *Cactaceae* name checklist project zugänglich gemacht.
Verschiedene IOS-Mitglieder halfen beim Korrigieren der vorläufigen Versionen dieser Checkliste. Die Redaktoren bedanken sich vor allem bei Dr. D. R. Hunt, sowie für die Übersetzungen der Vorworte und Bemerkungen bei B. E. Leuenberger, S. Arroyo Leuenberger und D. C. Zappi.

Benutzte Konventionen:

— Gültig publizierte Namen werden in **Fettschrift** wiedergegeben, zusammen mit einer Angabe des nomenklatorischen Typus' (Name oder Belegexemplar, je nach Rangstufe), gefolgt von den Abkürzungen derjenigen Herbarien, wo Holotypus und allfällige Isotypen hinterlegt sind (erste Abkürzung für den Holotypus; die Abkürzungen richten sich nach Index Herbariorum, ed. 7, und den in Taxon publizierten Nachträgen). Ungültige, illegitime oder unkorrekte Namen werden in *Kursivschrift* wiedergegeben. In allen Fällen folgt auf den Namen das ausführliche bibliographische Zitat. Für neue Kombinationen wird zudem das Basionym angegeben, während für ungültige oder illegitime Namen in Klammern diejenigen ICBN-Artikel zitiert werden, die übertreten worden sind.

— Die Zeitschriften-Abkürzungen richten sich nach 'A Bibliography of Succulent Plant Periodicals' (U. Eggli in Bradleya 3: 103-119, 1985, und Nachträge in l.c. 5: 101-104, 1987) für die spezialisierten Sukkulenten- Zeitschriften, sowie nach der Bibliographie Periodicorum Huntianum für die übrigen Zeitschriften.

Introducción

Hace más de 40 años que la IOS publica anualmente el *Repertorium Plantarum Succulentarum* (RPS) enumerando los nuevos nombres de suculentas publicados durante el correspondiente año. Puede esta publicación parecer superflua, dado que existe una información similar en el Index Kewensis, pero el éxito continuo del RPS indica claramente su necesidad. En los primeros años, la publicación anual fue un gran aporte sobre los suplementos penta-anuales del Index Kewensis. En épocas más recientes, la inclusión de información adicional en comparación con el Index Kewensis (taxa infraespecíficas, detalles de tipificación [desde el vol. 33]) hacen del RPS una fuente importante de datos.

Con esta idea en mente en la Colección Municipal de Suculentas de Zurich (ZSS), se ha efectuado la computación de todos los datos nomenclaturales incluidos en los últimos 40 volúmenes del RPS. La lista completa de los 41 años de nombres de *Cactaceae* es la primer lista de nombres a publicar de una familia, basada en este banco de datos.

Deben mencionarse a los primeros editores del RPS quienes – sin la ayuda de los modernos métodos bibliográficos con que hoy contamos y sin un banco de datos computado – se ingeniaron para producir un panorama casi completo de la nomenclatura de plantas suculentas (1: H. M. Roan & A. J. A. Uitewaal; 2-5: H. M. Roan & G. D. Rowley; 6-12: G. D. Rowley; 13-26: G. D. Rowley & L. E. Newton; 27-29: G. D. Rowley, L. E. Newton & D. R. Hunt; 30-32: G. D. Rowley, L. E. Newton & N. P. Taylor; 33+: U. Eggli & N. P. Taylor). La rápida computación de estos datos hubiera sido imposible sin ese trabajo. No obstante algunas omisiones y errores han sido detectadas durante esta fase, el RPS sigue siendo la lista nomenclatural más completa en esta especialidad.

Es la esperanza de la IOS que esta lista dé una visión completa de los cambios de nombres ocurridos en cactus en los últimos 41 años, siendo útil para todos los que trabajan en esta fascinante familia. Posiblemente a algunos de los lectores les gustaría ver más información que la incluida en la lista, especialmente la referida a una u otra clasificación elegida. Si bien esa idea parece atractiva, se cree preferible por el momento una lista dando unicamente casos nomenclaturales, (o a lo sumo algunas interpretaciones de esos casos) en vez de incluir decisiones taxonómicas, las cuales muchas veces tendrían que ser tomadas sobre bases inseguras o evidencias incompletas.

Notas sobre la lista taxonómica de los nombres de Cactaceae publicada entre 1950 y 1990 extracto del banco de datos del *Repertorium Plantarum Succulentarum*

La lista taxonómica de los nombres de *Cactaceae* publicada entre 1950 y 1990 es un extracto del banco de datos del *Repertorium Plantarum Succulentarum* (RPS), datos que son mantenidos en la Colección Municipal de Suculentas de Zurich. El banco de datos contiene información nomenclatural y taxonómica publicada en los primeros volúmenes de la serie del *Repertorium Plantarum Succulentarum*, iniciado y posteriormente publicado por la *Organización Internacional para el Estudio de Plantas Suculentas* (IOS).
Los datos de los volúmenes anuales del RPS han sido puesto en el banco de datos "como son", sin revisar la versión original. Sin embargo, en muchos casos, se han hecho correcciones cuando el error era evidente o cuando el contenido del banco de datos mostraba que la información publicada en el RPS era incorrecta. Todos esos casos han sido señalados.
Los datos de los volúmenes del RPS han sido complementados con informaciones de otros lados. Resulta evidente que un número (afortunadamente pequeño) de taxa han sido omitidas en el RPS.
Las siguientes notas son importantes para entender el concepto de ese extracto tomado del banco de datos del RPS:

— *Nomina nuda* cuando so indicados en forma global para todo el trabajo en volúmenes individuales del RPS, han sido normalmente incluidos en el banco de datos, en los casos que el trabajo relevante ha sido accesible, con la excepción de las siguientes obras: Haage, W. & O. Sadovsky (1958): Kakteen-Sterne (nombres de híbridos); Ito, Y. (1954): Cacti – Cultivation and Information (muchos *nomina nuda* para taxa supragenéricos); Ito, Y. (1962): Gardening of Cacti (incluye cientos de nuevos cultivares y muchos nuevos nombres y combinaciones, pero aparentemente sin basyónimos ni diagnsosis latinas [G. Rowley in litt. 27. 4. 1990]; Johnson, H. (1955): Succulent Parade for the Indoor Gardener (catálogo comercial); y Winter, H. [1955 [undated]: Kakteen – Cacti – Cactées (catálogo comercial, con muchos nombres provisionales de F. Ritter).

— Entradas que han sido corregidas en comparación con la versión impresa son señaladas. En muchos casos, nombres han sido repetidamente mencionados en diferentes volúmenes del RPS; esas entradas han sido combinadas en una sola entrada sin comentarios adicionales.

— En comparación con la versión impresa del RPS, información sobre tipificación está incluida para taxa adicionales. Datos de tipificación han sido incluidos en la impresión del RPS solamente desde el volúmen 33, pero esa información está también accesible en el banco de datos de un número de taxa incluídas en volúmenes anteriores. En esta lista al nombre y número del colector y sigla del herbario, si la información fue recopilada en el banco de datos, se ha agregado tambien el país y la provincia y/o departamento donde el especimen tipo fue colectado. Todos los datos registrados de los especímenes tipo de *Cactaceae* de los herbarios Suizos (cf. U. Eggli, Trop. subtrop. Pfl.-welt, 59, 1987), incluyendo algunas correcciones han sido incorporados (cuando corresponde) en esta lista.

Es de esperar que ningun nombre incluido en la versión impresa del RPS haya sido inadvertidamente omitido del banco de datos. No obstante el cuidadoso control realizado, no se pueden dar garantías, por lo cual los recopiladores de la ZSS agradecen sinceramente toda comunicación sobre cualquier posible omisión o error.

Debe ser advertido que la inclusión de un taxon como válido, inválido o ilegítimo no es una decisión final, pero sí basada en las evidencias accesibles. *Ningún nombre dado como aceptado en este extracto es aceptado por el compilador y editor del RPS o de esa lista en el sentido del ICBN art. 34.1.*

Las convenciónes usadas en los últimos volúmenes del RPS estan también en efecto en ese extracto del banco de datos del RPS (ver abajo).

Se agradece a todos aquellos que han estado involucrados en la preparación de la computación del banco de datos del RPS. Dr. R. Baumberger (Zurich) ha compilado los nombres de *Cactaceae* publicados de 1960-1980; Dr. B. Burr (Bonn) ha compilado los nombres del RPS 1-7. Los restantes nombres publicados en RPS han sido incorporados en el banco de datos por miembros del ZSS, principalmente por el editor de esta recopilación. Parte de los fondos para la entrada de nombres de *Cactaceae* en el banco de datos, han sido recibidos a través del Dr. D. Hunt, RBG Kew, del proyecto de la lista de nombres de *Cactaceae* de la CITES.
Varios miembros de la IOS han ayudado con las correcciones preliminares para esta lista. Los editores agradecen especialmente al Dr. D. R. Hunt. Tambien agradecen a B. E. Leuenberger, S. Arroyo Leuenberger y D. C. Zappi.

Convenciones usadas en la lista:

— Los nombres válidos son dados en **letra gruesa** acompañados por una indicación del tipo nomenclatural (nombre o espécimen, según el rango), seguido de la sigla del herbario donde el holotipo y posible isotipo han sido depositados (la primera sigla es del holotipo), de acuerdo al Index Herbariorum, ed. 7 y suplementos publicados en Taxon. Nombres inválidos, ilegítimos o incorrectos son dados en *itálica*. En ambos casos se da una referencia bibliográfica completa. Para las nuevas combinaciones los basónimos están también enumerados. Para nombres inválidos o ilegítimos, los artículos del ICBN involucrados son mencionados entre paréntesis.
Los recopiladores quieren señalar que no se acepta ningún nombre que pueda ser inadvertidamente revalidado en este volúmen del RPS.

— Las abreviaturas usadas para las publicaciones periódicas son las sugeridas por 'A Bibliography of Succulent Plant Periodicals' (U. Eggli en Bradleya 3: 103-119, 1985, y suplementos, lc.4 55: 101-104, 1987) para revistas especializadas, para las otras revistas la 'Bibliographia Periodicorum Huntianum'.

Introdução

Por mais de 40 anos, a IOS vem publicando anualmente o *Repertorium Plantarum Succulentarum* (RPS), com finalidade de listar nomes novos de plantas suculentas. Tal publicação pode parecer supérflua, por apresentar o mesmo tipo de informação que está contida no Index Kewensis, porém o sucesso contínuo do RPS prova claramente a necessidade de sua existência. Nos seus primeiros anos, o RPS, devido ao seu caráter anual obteve considerável vantagem sobre os suplementos quinqueniais de Index Kewensis. Atualmente, através da inclusão de informações adicionais sobre nomenclatura infraespecífica e detalhes sobre tipificação (a partir do vol. 33), o RPS tranformou-se numa fonte de dados mais completa.

Tendo em vista essas vantagens, a computadorização de todos os dados de nomenclatura incluídos nos últimos 41 volumes do RPS, foi realizada na Coleção Municipal de Suculentas, Zurique, (ZSS), e a lista nomenclatural de 41 anos da família *Cactaceae* é o primeiro "checklist" a ser publicado a partir desse banco de dados.

Devemos aque prestar uma homenagem aos primeiros editores / compiladores do RPS, os quais, sem os modernos auxílios bibliográficos e bancos de dados computadorizados, conseguiram obter uma cobertura quase completa da nomenclatura de plantas suculentas, a partir de 1950 (1: H. M. Roan & A. J. A. Uitewaal; 2-5: H. M. Roan & G. D. Rowley; 6-12: G. D. Rowley; 13-26: G. D. Rowley & L. E. Newton; 27-29: G. D. Rowley, L. E. Newton & D. R. Hunt; 30-32: G. D. Rowley, L. E Newton & N. P. Taylor; 33+: U. Eggli & N. P. Taylor). A computadorização rápida desses dados teria sido impossível sem o seu trabalho. Embora alguns erros e omissões tenham sido detectadas durante essa fase, RPS trata-se da mais completa lista de nomenclatura existente nesse campo especializado.

A IOS espera que este "checklist", que fornece uma visão das mais completas no que concerne às modificações de nomenclatura das *Cactaceae* durante os últimos 41 anos, seja útil para todos aqueles que trabalham com essa família fascinante. É provável que alguns dos leitores esperassem encontrar mais informações anexas à presente lista, especialmente a inclusão de um sistema de classificação. Essa idéia, apesar de atraente, foi deixada de lado por tratar-se de um mero "checklist" aonde constam questões nomenclaturais (ou até algumas interpretações de tais questões), no qual não procede incluir decisões taxonômicas, que são sempre subjetivas.

Notas na Lista Nomenclatural de Cactaceae (1950-1990) do Banco de Dados do Repertorium Plantarum Suculentarum.

Esta lista alfabética de nomes de *Cactaceae* publicados entre 1950 e 1990 foi extraída do banco de dados do *Repertorium Plantarum Suculentarum* (RPS), mantido na Coleção Municipal de Suculentas, Zurique. O banco de dados contém a informação nomenclatural e taxonômica publicada de maneira seriada a partir dos fascículos anuais do *Repertorium Plantarum Suculentarum*, iniciado e publicado até o presente pela *Organização Internacional para Estudo das Plantas Suculentas* (IOS).
Os dados incluídos nos fascículos anuais do RPS foi introduzido no banco de dados diretamente, sem retornar às referências originais. Em muitos casos, correções foram feitas quando os erros foram percebidos, ou quando o conteúdo do banco de dados demonstrou que a informação contida no RPS estava incorreta. Todos esses casos foram indicados adequadamente. Os dados dos fascículos do RPS foram suplementados com informação proveniente de outras fontes. Evidentemente, alguns (poucos) táxons foram omitidos pelos fascículos anteriores do RPS.

As notas subsequentes são importantes para que se compreenda a abrangência deste extrato do banco de dados do RPS:

— Os *nomina nuda* que foram indicados de forma global para um trabalho inteiro em fascículos do RPS foram, de modo geral, incluídos no banco de dados, com excessão daqueles que caíram em uso generalizado, e com excessão daqueles das seguintes trabalhos: Haage, W. & O. Sadovsky (1958): Kakteen-Sterne (nomes de híbridos); Ito, Y. (1954): Cacti - Cultivation and information, Tokyo (muitos *nomina nuda* para táxons supragénericos); Ito, Y. (1962): Gardening of Cacti (inclui centenas de novos cultivares e muitos nomes novos e combinações, mas todos aparentemente sem basônimo ou diagnoses latinas (G. Rowley, in litt. 27. 4. 1990); Johnson, H. (1955): Succulent Parade for the Indoor Gardener (catálogo de venda); e Winter, H. (1955 [sem data]): Kakteen - Cacti - Cactées (catálogo de venda, com muitos nomes provisórios de F. Ritter).

— Entradas que foram corrigidas a partir das versões publicadas foram indicadas de maneira adequada. Em vários casos, nomes que foram compilados repetidamente em diferentes fascículos do RPS foram combinados em uma só entrada sem maiores comentários.

— Informação adicional referente à tipificação foi incluída. Desde o vol. 33 o RPS já incluía informação sobre tipificação dos táxons, porém o banco de dados foi adicionado desse tipo de informação de maneira retroativa. Em adição ao nome e número de coletor e sigla do herbário, esta lista contém também o país, estado / província / departamento de origem do táxon, sempre e quando esta informação se encontrar disponível no banco de dados. Todos os dados de registro de espécimes-tipo de *Cactaceae* em herbários suíços (cf. U. Eggli, Trop. subtrop. Pfl.-welt, 59, 1987) foi incorporada nesta lista, quando aplicável, incluindo algumas correções.

Espera-se que nenhum nome incluído nas versões publicadas de RPS tenha sido omitido inadvertidamente por este banco de dados. Apesar de ter sido conferida cuidadosamente, nenhuma garantia pode ser oferecida, e os compiladores da ZSS estão prontos a receber comunicações de erros ou omissões que possam ocorrer.

Deve ficar claro que a inclusão de um nome como válido, inválido ou ilegítimo não se trata de uma decisão final, sendo porém baseada nas evidências disponíveis. Nenhum nome dado neste extrato é aceito pelos compiladores e editores do RPS (de acordo com o ICBN Art. 34.1), por ter sido o seu local de publicação válida creditado no presente.
Finalmente, as convenções utilizadas nos fascículos mais recentes do RPS são seguidas neste extrato do banco de dados do RPS (veja abaixo). Agradecemos a todos aqueles que estiveram envolvidos na organização do banco de dados computadorizado do RPS. Ao Dr. R. Baumberger (Zurique), que compilou todos os nomes de Cactaceae publicados entre 1960 e 1980; ao Dr B. Burr (Bonn), compilador dos nomes incluídos nos RPS 1-7. Os nomes remanescentes publicados no RPS foram incluídos no banco de dados pelos pesquisadores do ZSS, a maior parte pelo editor desta compilação. O financiamento de parte da inclusão de nomes de *Cactaceae* foi possibilitado graças ao Dr. D.R. Hunt, RBG Kew, através do projeto da CITES que visa produzir um "checklist" da família *Cactaceae.*
Vários membros da IOS prestaram auxílio na correção das provas preliminares deste checklist. Os editores agradecem particularmente ao Dr. D. R. Hunt, e também a B. E. Leuenberger, S. Arroyo Leuenberger e D. C. Zappi pelas traduções das notas introdutorias.

Convenções utilizadas:

— Nomes válidos publicados são citados em **negrito**, acompanhados por uma indicação do tipo nomenclatural (nome ou espécime, dependendo do nível), seguido pela sigla do herbário onde o holótipo e possíveis isótipos são declarados como depositados (primeira sigla para o holótipo), de acordo com o Index Herbariorum, ed. 7 e seus suplementos, como publicados em Taxon. Nomes inválidos, ilegítimos ou incorretos são citados em *itálico.* Em cada caso, a referência bibliográfica completa é dada. Para as novas combinações, o basônimo também é citado. Para nomes inválidos ou ilegítimos, os artigos do ICBN que sofreram contravenção são indicados entre parênteses.

— As abreviações para os periódicos seguem aquelas sugeridas em "A Bibliography of Succulent Plant Periodicals" (U. Eggli in Bradleya 3: 103-119, 1985, e suplementos in ibid. 5: 101-104, 1987) para periódicos especializados, ou aquelas de Bibliographia Periodicorum Huntianum para outros periódicos.

Cactaceae (1950 - 1990)

Acanthanthus Ito, The Cactaceae, 354, 1981. *Nom. illeg.* (Art. 63.1). [Based on *Acantholobivia* Backeberg.] [RPS -]
Acanthocalycium subgen. *Amblyophori* Y. Ito, Bull. Takarazuka Insectarium 71: 13-20, 1950. *Nom. inval.* (Art. 36.1). [RPS 3]
Acanthocalycium subgen. *Goniophori* Y. Ito, Bull. Takarazuka Insectarium 71: 13-20, 1950. *Nom. inval.* (Art. 36.1). [RPS 3]
Acanthocalycium andreaeanum (Backeberg) Donald, Ashingtonia 1(11): 125, 1975. *Nom. inval.*, based on *Neochilenia andreaeana, nom. inval.* (Art. 9.5, 37.1). [Erroneously included as valid in RPS 26.] [RPS 26]
Acanthocalycium aurantiacum Rausch, Kakt. and. Sukk. 21(3): 45, 1970. Typus: *Rausch* 148 (ZSS). [First invalidly published in l.c. 19(5): 92, ill., 1968.] [Argentina: Catamarca] [RPS 21]
Acanthocalycium brevispinum Ritter, Taxon 13: 144, 1964. [RPS 15]
Acanthocalycium catamarcense Ritter, Kakt. Südamer. 2: 465, 1980. [RPS 31]
Acanthocalycium ferrarii Rausch, Succulenta 55(5): 81-82, ills., 1976. Typus: *Rausch* 572 (ZSS). [See l.c. 56(2): 30, 1977, for location of type.] [Argentina: Tucuman] [RPS 27]
Acanthocalycium formosanum var. **albispinum** (F. A. C. Weber) Y. Ito, Explan. Diagr. Austroechinocactinae, 145, 1957. Basionym: *Echinopsis formosana* var. *albispina*. [RPS 8]
Acanthocalycium formosanum var. **crassispinum** (Monville) Y. Ito, Explan. Diagr. Austroechinocactinae, 145, 1957. Basionym: *Echinocactus formosanus* var. *crassispinus*. [RPS 8]
Acanthocalycium formosanum var. **gilliesii** (Salm-Dyck) Y. Ito, Explan. Diagr. Austroechinocactinae, 145, 1957. Basionym: *Echinopsis formosana* var. *gilliesii*. [RPS 8]
Acanthocalycium formosanum var. **laevior** (Monville) Y. Ito, Explan. Diagr. Austroechinocactinae, 145, 1957. Basionym: *Echinopsis formosana* var. *laevior*. [RPS 8]
Acanthocalycium formosanum var. **rubrispinum** (Monville) Y. Ito, Explan. Diagr. Austroechinocactinae, 145, 1957. Basionym: *Echinopsis formosana* var. *rubrispina*. [RPS 8]
Acanthocalycium formosanum var. **spinosior** (Salm-Dyck) Y. Ito, Explan. Diagr. Austroechinocactinae, 145, 1957. Basionym: *Echinopsis formosana* var. *spinosior*. [RPS 8]
Acanthocalycium glaucum Ritter, Taxon 13: 143-144, 1964. [RPS 15]
Acanthocalycium griseum Backeberg, Descr. Cact. Nov. 3: 5, 1963. *Nom. inval.* (Art. 9.5). [Erroneously included as valid in RPS 14.] [RPS 14]
Acanthocalycium spiniflorum fa. *klimpelianum* (Weidlich & Werdermann) Donald, Ashingtonia 1: 116, 1975. *Nom. inval.* (Art. 33.2), based on *Echinopsis klimpeliana*. [RPS -]
Acanthocalycium spiniflorum fa. *peitscherianum* (Backeberg) Donald, Ashingtonia 1: 116, 1975. *Nom. inval.* (Art. 33.2), based on *Acanthocalycium peitscherianum*. [RPS -]
Acanthocalycium spiniflorum fa. *violaceum* (Werdermann) Donald, Ashingtonia 1: 116, 1975. *Nom. inval.* (Art. 33.2), based on *Echinopsis violacea*. [RPS -]
Acanthocalycium spiniflorum var. *klimpelianum* (Weidlich & Werdermann) Donald, Ashingtonia 1: 116, 1975. *Nom. inval.* (Art. 33.2), based on *Echinopsis klimpeliana*. [RPS -]
Acanthocalycium thionanthum var. *aurantiacum* (Rausch) Donald, Ashingtonia 1: 124, 1975. *Nom. inval.* (Art. 33.2), based on *Acanthocalycium aurantiacum*. [RPS -]
Acanthocalycium thionanthum var. *brevispinum* (Ritter) Donald, Ashingtonia 1: 124, 1975. *Nom. inval.* (Art. 33.2), based on *Acanthocalycium brevispinum*. [RPS -]
Acanthocalycium thionanthum var. *glaucum* (Ritter) Donald, Ashingtonia 1: 124, 1975. *Nom. inval.* (Art. 33.2), based on *Acanthocalycium glaucum*. [RPS -]
Acanthocalycium thionanthum var. *variflorum* (Backeberg) Donald, Ashingtonia 1: 124, 1975. *Nom. inval.*, based on *Acanthocalycium variflorum, nom. inval.* (Art. 9.5). [Sphalm. 'variiflorum'.] [RPS -]
Acanthocalycium variflorum Backeberg, Kakt.-Lex., 57, 1966. *Nom. inval.* (Art. 9.5). [Sphalm. 'variiflorum'.] [RPS 17]
Acanthocerei Buxbaum, Madroño 14(6): 182, 1958. Typus: *Acanthocereus*. [Published as 'linea'.] [RPS 9]
Acanthocereus chiapensis Bravo, Cact. Suc. Mex. 17: 117-118, 1972. [RPS 23]
Acanthocereus griseus Backeberg, Cactus (Paris) No. 85: 103-106, 1965. *Nom. inval.* (Art. 37.1). [Erroneously included as valid in RPS 16 and 17. Publication repeated in Kakt.-Lex., 58, 1966.] [RPS 16]
Acanthocereus princeps (Pfeiffer) Backeberg, Descr. Cact. Nov. 3: 5, 1963. Basionym: *Cereus princeps*. [RPS 14]
Acanthocereus tetragonus var. **micracanthus** Dugand, Phytologia 13: 383, 1966. [RPS 17]
Acantholobivia euanthema (Backeberg) Y. Ito, Explan. Diagr. Austroechinocactinae, 131, 1957. Basionym: *Lobivia euanthema*. [RPS 8]
Acantholobivia haagei (Fric & Schelle) Y. Ito, Explan. Diagr. Austroechinocactinae, 131,

1957. Basionym: *Rebutia haagei*. [Validity of basionym not assessed.] [RPS 8]

Acantholobivia incuiensis (Rauh & Backeberg) Rauh & Backeberg, in Rauh, Beitr. Kenntn. peruan. Kakt.veg., 471, 1958. Basionym: *Lobivia incuiensis*. [RPS 10]

Acanthopetalus Y. Ito, Explan. Diagr. Austroechinocactinae, 141, 292, 1957. *Nom. illeg*. (Art. 63.1). [Based on the same type as *Setiechinopsis* (Backeberg) Backeberg.] [RPS 8]

Acanthopetalus mirabilis (Spegazzini) Y. Ito, Explan. Diagr. Austroechinocactinae, 143, 1957. *Nom. illeg*. (Art. 63.1), based on *Setiechinopsis mirabilis*. [RPS 8]

Acanthorhipsalis brevispina F. Ritter, Kakt. Südamer. 4: 1260, fig. 114 (p. 1529), 1981. *Nom. inval*. (Art. 37.1). [RPS 32]

Acanthorhipsalis houlletiana (Lemaire) Volgin, Byull. Mosk. Obsch. Ispyt. Prir. Biol. 87(3): 83, 1982. Basionym: *Rhipsalis houlletiana*. [RPS 33]

Acanthorhipsalis incachacana (Cardenas) Volgin, Byull. Mosk. Obsch. Ispyt. Prir. Biol. 87(3): 83, 1982. Basionym: *Rhipsalis incachacana*. [RPS 33]

Acanthorhipsalis incahuasina Cardenas, Cactus (Paris) 7(34): 127, ill., 1952. Typus: *Cardenas* 4857 (BOLV). [Through a printer's mistake, the name *A. paranganiensis* was substituted for the correct epithet; this was later corrected and pcorrection slip was supplied to cover the mistake.] [Bolivia: Cordillera] [RPS 3]

Acanthorhipsalis monacantha var. **samaipatana** (Cardenas) Backeberg, Die Cact. 2: 701, 1959. Basionym: *Rhipsalis monacantha* var. *samaipatana*. [RPS 10]

Acanthorhipsalis paranganiensis Cardenas, Cactus (Paris) 6(34): 126-127, ill., 1952. [RPS 3]

Acanthorhipsalis Kimnach, Cact. Succ. J. (US) 55(4): 177-181, 1983. *Nom. illeg*. (Art. 64.1). [*Non Acanthorhipsalis* Britton & Rose 1923; later corrected to *Lymanbensonia*.] [RPS 34]

Akersia Buining, Succulenta 1961: 25-27, 1961. Typus: *A. roseiflora*. [RPS 12]

Akersia roseiflora Buin., Succulenta 1961: 25, 1960. [RPS 12]

Ancistrocactus crassihamatus (F. A. C. Weber) L. Benson, Cact. Succ. J. (US) 41: 188, 1969. *Nom. inval*., based on *Echinocactus crassihamatus, nom. inval*. (Art. 35). [RPS 20]

Ancistrocactus scheeri fa. **brevihamatus** (Engelmann) Krainz, Kat. ZSS ed. 2, 35, 1967. Basionym: *Echinocactus brevihamatus*. [RPS 18]

Ancistrocactus scheeri fa. **megarhizus** (Rose) Krainz, Kat. ZSS ed. 2, 35, 1967. Basionym: *Echinocactus megarhizus*. [RPS 18]

Ancistrocactus uncinatus (Galeotti) L. Benson, Cact. Succ. J. (US) 41(4): 188, 1969. Basionym: *Echinocactus uncinatus*. [RPS 20]

Ancistrocactus uncinatus var. **wrightii** (Engelmann) L. Benson, Cact. Succ. J. (US) 41(4): 188, 1969. Basionym: *Echinocactus uncinatus* var. *wrightii*. [RPS 20]

Anisocereus foetidus (MacDougall & Miranda) W. T. Marshall, Saguaroland Bull. 11: 35, 1957. Basionym: *Pterocereus foetidus*. [RPS 8]

Aparadoa van Vliet, Succulenta 65(4): 86, 1986. *Nom. inval*. (Art. 36.1, 37.1). [A substitute for *Brasiliparodia*.] [RPS 37]

× **Aporberocereus** Rowley, Nation. Cact. Succ. J. 37(2): 46, 1982. [= *Aporocactus* × *Weberocereus*.] [RPS 33]

× **Aporechinopsis** Rowley, Name that Succulent, 113, 1980. [= *Aporocactus* × *Echinopsis*.] [RPS 31]

× **Aporepiphyllum** Rowley, Name that Succulent, 114, 1980. [= *Aporocactus* × *Epiphyllum*.] [RPS 31]

× **Aporepiphyllum freiburgensis** (Weingart) Rowley, Nation. Cact. Succ. J. 37(2): 46, 1982. Basionym: *Cereus freiburgensis*. [Sphalm. 'frieburgensis'; = *Aporocactus flagelliformis* × *Epiphyllum ?crenatum*.] [RPS 33]

× **Aporochia** Rowley, Epiphytes 4(13): 12, 1972. [= *Aporocactus* × *Nopalxochia*.] [RPS 31]

× **Aporoheliocereus mallisonii** (Otto & Dietrich) P. Heath, Taxon 38(2): 281, 1989. Basionym: *Cereus mallisonii*. [RPS 40]

× **Aporoheliocereus mallisonii** fa. **smithii** (Pfeiffer) P. Heath, Taxon 38(2): 281, 1989. Basionym: *Cereus smithii*. [RPS 40]

× *Aporoheliocereus smithianus* (Sweet) P. Heath, Epiphytes 7(28): 92, 1983. *Nom. inval*. (Art. 32D Ex. 1). [Based on *Cereus smithianus* Sweet, *nom. inval*., cf. Heath in Taxon 38(2): 281, 1989.] [RPS 34]

× *Aporoheliocereus smithianus* nv. *mallisonii* (Otto & Dietrich) P. Heath, Epiphytes 7(28): 92, 1983. *Nom. inval*. (Art. 43.1), based on *Cereus mallisonii*. [RPS 34]

× **Aporoheliocereus smithii** (Pfeiffer) Rowley, Nation. Cact. Succ. J. 37(2): 46, 1982. Basionym: *Cereus smithii*. [= *Aporocactus flagelliformis* × *Heliocereus speciosus*.] [RPS 33]

× **Aporoheliochia** P. Heath, Epiphytes 8(30): 40, 1984. [*Aporocactus* × *Heliocereus* × *Nopalxochia*.] [RPS 35]

× **Aporophyllum freiburgensis** (Weingart) P. Heath, Epiphytes 8(30): 39, 1984. Basionym: *Cereus freiburgensis*. [sphalm. 'frieburgensis'] [RPS 35]

Arequipa australis Ritter, Kakt. Südamer. 3: 1123, fig. 28 (p. 1131), 1980. [RPS 31]

Arequipa clavata (Söhrens) Y. Ito, Explan. Diagr. Austroechinocactinae, 156, 1957. Basionym: *Echinocactus clavatus*. [RPS 8]

Arequipa erectocylindrica Rauh & Backeberg, Descr. Cact. Nov. 1: 18, 1957. Typus: *Rauh* K142a (1956) (ZSS). [Dated 1956, published Jan. 1957.] [Peru: Arequipa] [RPS 7]

Arequipa haynii (Otto) Krainz, Die Kakt. Lief. 25: C Vb, 1963. Basionym: *Echinocactus haynii*. [RPS 14]

Arequipa mirabilis (Buining) Backeberg, Kakt.-Lex. 62, 1966. Basionym: *Matucana mirabilis*. [RPS 17]

Arequipa rettigii var. **borealis** Ritter, Kakt. Südamer. 4: 1369, 1981. Typus: *Ritter* 127b (U). [Type cited for U l.c. 1: iii, 1979.]

[Chile: Arequipa] [RPS 32]

Arequipa rettigii var. **erectocylindrica** (Rauh & Backeberg) Krainz, Die Kakt. Lief. 36/37: C Vb, 1967. Basionym: *Arequipa erectocylindrica.* [RPS 18]

Arequipa soehrensii Backeberg, Die Cact. 2: 1054-1055, 1959. *Nom. inval.* (Art. 37.1). [= *Arequipiopsis soehrensii* Kreuzinger *nom. nud.* Erroneously included as valid in RPS 10.] [RPS 10]

Arequipa spinosissima Ritter, Taxon 13(3): 115, 1964. Typus: *Ritter* 196 (U, ZSS). [First mentioned (with Krainz & Ritter as authors) in Backeberg, Die Cact. 2: 1053, 1959 (cf. RPS 10).] [Peru: Arequipa] [RPS 15]

Arequipa varicolor (Backeberg) Backeberg, Cactus (Paris) No. 37: 217, 1953. Basionym: *Oreocereus varicolor.* [Sphalm. 'variicolor'.] [RPS 4]

Ariocarpus subgen. **Neogomesia** (Castañeda) Buxbaum, in Krainz, Die Kakt. C VIIIb, 1963. Basionym: *Neogomesia.* [Sphalm. 'Neogomezia'.] [RPS 14]

Ariocarpus agavoides (Castañeda) Anderson, Amer. J. Bot. 49: 615, 1962. Basionym: *Neogomesia agavoides.* [RPS 13]

Ariocarpus fissuratus var. **hintonii** Stuppy & N. P. Taylor, Bradleya 7: 84-88, ills., 1989. Typus: *Hinton* s.n. (K). [Mexico: San Luis Potosi] [RPS 40]

Ariocarpus kotschoubeyanus var. *elephantidens* Skarupke, Stachelpost 9(43): 33-34, ills., 1973. *Nom. inval.* (Art. 37.1). [RPS 23]

Ariocarpus retusus var. **furfuraceus** (Watson) Frank, in Krainz, Die Kakt. 63, CVIIIb, 1975. Basionym: *Mammillaria furfuracea.* [RPS 26]

Ariocarpus trigonus var. **elongatus** (Salm-Dyck) Backeberg, Die Cact. 5: 3087, 1961. Basionym: *Anhalonium elongatum.* [RPS 12]

Ariocarpus trigonus var. **minor** Voldan, Kaktusy 12(1): 3-6, ills., 1976. [Probably invalid, as no type is explicitly cited, although W is stated to be the herbarium where the type has been deposited.] [RPS 27]

Armatocereus arboreus Rauh & Backeberg, Descr. Cact. Nov. 1: 13, 1957. [Dated 1956, published Jan. 1957.] [RPS 7]

Armatocereus arduus Ritter, Kakt. Südamer. 4: 1271, 1981. Typus: *Ritter* 1060 (U, ZSS). [Holotype cited for U l.c. 1: iii, 1979.] [Peru: La Libertad] [RPS 32]

Armatocereus armatus Johnson, Catalogue Johnson, [unpaged], 1954. *Nom. inval.* (Art. 36.1). [RPS -]

Armatocereus balsasensis Ritter, Kakt. Südamer. 4: 1271-1272, 1981. Typus: *Ritter* 273a (U, ZSS). [Holotype cited for U l.c. 1: iii, 1979.] [Peru: Cajamarca] [RPS 32]

Armatocereus brevispinus J. E. Madsen, Fl. Ecuador 35: 9-10, 1989. Typus: *Madsen* 75910 (AAU). [Ecuador: Prov. Loja] [RPS 40]

Armatocereus cartwrightianus var. **longispinus** Backeberg, Descr. Cact. Nov. 1: 14, 1957. [Dated 1956, published Jan. 1957.] [RPS 7]

Armatocereus churinensis Rauh & Backeberg, in Backeberg, Descr. Cact. Nov. [1:] 13, 1957. [Dated 1956, published Jan. 1957.] [RPS 7]

Armatocereus ghiesbreghtii (Schumann) Ritter, Backeberg's Descr. & Erört. taxon. nomenkl. Fragen, [unpaged], 1961. Basionym: *Cereus ghiesbreghtii.* [RPS 15]

Armatocereus ghiesbreghtii var. **oligogonus** (Rauh & Backeberg) Ritter, Kakt. Südamer. 4: 1275, 1981. Basionym: *Armatocereus oligogonus.* [RPS 32]

Armatocereus humilis (Britton & Rose) Backeberg, Die Cact. 2: 905, 1959. Basionym: *Lemaireocereus humilis.* [RPS 10]

Armatocereus mataranus Ritter, Succulenta 46(2): 23, ill., 1967. Typus: *Ritter* 672 (U, ZSS). [First invalidly published (without type) in l.c. 45: 117, 1966.] [Peru] [RPS 18]

Armatocereus mataranus var. **ancashensis** Ritter, Kakt. Südamer. 4: 1277, 1981. Typus: *Ritter* 672a (U). [Peru: Ancash] [RPS 32]

Armatocereus oligogonus Rauh & Backeberg, Descr. Cact. Nov. [1:] 13, 1957. [Dated 1956, published Jan. 1957.] [RPS 7]

Armatocereus procerus Rauh & Backeberg, Cactus (Paris) No. 51: 95, 1956. Typus: *Rauh* K32 (1954) (ZSS [iso]). [Repeated in Descr. Cact. Nov. 1: 13 (dated 1956, published 1957); emendated diagnosis see Kakt. and. Sukk. 8(6): 82-87, 1957.] [Peru] [RPS 7]

Armatocereus procerus var. **armatus** Ritter, Kakt. Südamer. 4: 1277, 1981. Typus: *Ritter* 131 (U, ZSS). [Holotype cited for U l.c. 1: iii, 1979.] [Peru: Ancash] [RPS 32]

Armatocereus rauhii Backeberg, Descr. Cact. Nov. 1: 13, 1957. Typus: *Rauh* K127 (1954) (ZSS). [Dated 1956, published 1957.] [Peru: Jaen] [RPS 7]

Armatocereus riomajensis Rauh & Backeberg, Descr. Cact. Nov. 1: 13, 1957. Typus: *Rauh* K152a (1956) (ZSS [iso]). [Dated 1956, published 1957.] [Peru] [RPS 7]

Armatocereus rupicola Ritter, Kakt. Südamer. 4: 1278, 1981. Typus: *Ritter* 1318 (U). [Peru: Cajamarca] [RPS 32]

Arrojadoa albiflora Buining & Brederoo, Succulenta 54(2): 21-27, ills., 1975. Typus: *Horst et Uebelmann* HU 401 (U, ZSS [type number]). [Brazil: Bahia] [RPS 26]

Arrojadoa aureispina Buining & Brederoo, Kakt. and. Sukk. 23(4): 95-98, ills., 1972. Typus: *Horst et Uebelmann* HU 154 (1966) (U, ZSS). [Brazil: Bahia] [RPS 23]

Arrojadoa aureispina var. **anguinea** P. J. Braun & E. Esteves Pereira, Cact. Succ. J. (US) 60(4): 174-180, ills., SEM-ills., 1988. Typus: *Braun* 80a (ZSS). [Brazil: Bahia] [RPS 39]

Arrojadoa aureispina var. **guanambensis** P. J. Braun & Heimen, Kakt. and. Sukk. 31(11): 334-337, 1980. Typus: *Braun et Heimen* 80 (). [Brazil: Bahia] [RPS 31]

Arrojadoa beateae P. J. Braun & E. Esteves Pereira, Kakt. and. Sukk. 40(10): 250-256, ills., SEM-ills., 1989. Typus: *Esteves Pereira* 261 (UFG, ZSS). [Brazil: Minas Gerais]

[RPS 40]

Arrojadoa canudosensis Buining & Brederoo, Cact. Succ. J. (US) 44: 111-113, 1972. [RPS 23]

Arrojadoa dinae Buining & Brederoo, Kakt. and. Sukk. 24: 99-101, 1973. [RPS 24]

Arrojadoa eriocaulis Buining & Brederoo, Kakt. and. Sukk. 24(11): 241-244, ills., 1973. Typus: *Horst et Uebelmann* HU 349 (1972) (U, ZSS). [Brazil: Minas Gerais] [RPS 24]

Arrojadoa eriocaulis var. **albicoronata** van Heek et al., Kakt. and. Sukk. 33(11): 224-227, ills., 1982. Typus: *van Heek et al.* s.n. (KOELN [Succulentarium]). [The type may be van Heek 81/113.] [RPS 34]

Arrojadoa horstiana P. J. Braun & Heimen, Kakt. and. Sukk. 32(8): 186-190, 1981. [RPS 32]

Arrojadoa multiflora Ritter, Kakt. Südamer. 1: 89-90, fig. 62-63 (pp. 296-297), col. fig. 2 (p. 275), 1979. Typus: *Ritter* 1243 (ZSS). [Brazil: Bahia] [RPS -]

Arrojadoa penicillata var. **decumbens** Backeberg & Voll, Arq. Jard. Bot. Rio de Janeiro 9: 164-166, ill. (p. 159), 1950. [Volume for 1949, published 1950.] [RPS -]

Arrojadoa penicillata var. **spinosior** Brederoo & Theunissen, Succulenta 59(1): 20-27, 1980. [RPS 31]

Arrojadoa polyantha (Werdermann) D. Hunt, Bradleya 5: 92, 1987. Basionym: *Cephalocereus polyanthus*. [RPS 38]

Arrojadoa rhodantha ssp. **canudosensis** (Buining & Brederoo) P. J. Braun, Bradleya 6: 90, 1988. Basionym: *Arrojadoa canudosensis*. [Erroneously included at the rank of variety in RPS 39.] [RPS 39]

Arrojadoa rhodantha ssp. **reflexa** P. J. Braun, Kakt. and. Sukk. 35(2): 34-38, ills., 1984. *Braun* 68 (KOELN [Succulentarium]). [Brazil] [RPS 35]

Arrojadoa rhodantha var. **occibahiensis** P. J. Braun, Kakt. and. Sukk. 36(6): 114-115, ills., 1985. Typus: *Horst et Uebelmann* HU 616 (ZSS). [Brazil: Bahia] [RPS 36]

Arrojadoa rhodantha var. **theunisseniana** (Buining & Brederoo) P. J. Braun, Bradleya 6: 90, 1988. Basionym: *Arrojadoa theunisseniana*. [RPS 39]

Arrojadoa theunisseniana Buining & Brederoo, in Krainz, Die Kakt. Lief. 52: C Vb, 1973. [RPS 24]

Arthrocereus subgen. **Cutakia** Backeberg, Cact. Succ. J. (US) 22(5): 153, 1950. Typus: *Arthrocereus mello-barretoi* Backeberg & Voll. [RPS 1]

Arthrocereus subgen. *Euarthrocereus* Backeberg, Cact. Succ. J. (US) 22(5): 153, 1950. *Nom. inval.* (Art. 21.3). [Typus: *Cereus microsphaericus* K. Schumann.] [RPS 1]

Arthrocereus subgen. **Praearthrocereus** Buxbaum, in Krainz, Die Kakt. Lief. 48/49: C Va (Jan. 1972), 1972. Typus: *A. rondonianus* Backeberg. [RPS 23]

Arthrocereus subgen. **Pygmaeocereus** (Backeberg & Johnson) Buxbaum, in Krainz, Die Kakt. Lief. 48/49: C Va (Jan. 1972), 1972. Basionym: *Pygmaeocereus*. [P. bylesianus Andreae & Backeberg] [RPS 23]

Arthrocereus bylesianus (Johnson & Backeberg) Buxbaum, Kakt. and. Sukk. 20: 97, 1969. Basionym: *Pygmaeocereus bylesianus*. [RPS 20]

Arthrocereus densiaculeatus (Backeberg) Krainz, Kat. ZSS ed. 2, 36, 1967. *Nom. inval.*, based on *Pygmaeocereus densiaculeatus*, *nom. inval.* (Art. 9.5). [Erroneously included as valid in RPS 18] [RPS 18]

Arthrocereus itabiriticola P. J. Braun, Kakt. and. Sukk. 37(11): 234-237, ills., SEM-ills., 1986. Typus: *Horst et Uebelmann* HU 330 (ZSS). [Brazil: Minas Gerais] [RPS 37]

Arthrocereus melanurus (Schumann) Diers et al., Kakt. and. Sukk. 38(12): 314, 1987. Basionym: *Cereus melanurus*. [RPS 38]

Arthrocereus melanurus var. **estevesii** Diers & P. J. Braun, Kakt. and. Sukk. 39(5): 100-105, ills., SEM-ills., 1988. Typus: *Horst, Braun et Esteves* 1 (KOELN [Succulentarium], ZSS). [Brazil: Minas Gerais] [RPS 39]

Arthrocereus mello-barretoi Backeberg & Voll, Arq. Jard. Bot. Rio de Janeiro 9: 157-162, ills., 1950. Typus: *Mello Barreto* s.n. (RB 65.044). [Volume for 1949, published 1950.] [Brazil: Minas Gerais] [RPS -]

Arthrocereus odorus Ritter, Kakt. Südamer. 1: 225, 1979. Typus: *Ritter* 1354 (U). [Brazil: Minas Gerais] [RPS -]

Arthrocereus rondonianus Backeberg & Voll, Cact. Succ. J. (US) 23(4): 120, 1951. [First mentioned (without Latin diagnosis) in Backeberg, Blätt. Kakt.-forsch. 1935(4).] [RPS 3]

Arthrocereus rowleyanus (Backeberg) Buxbaum, Kakt. and. Sukk. 20: 97, 1969. *Nom. inval.*, based on *Pygmaeocereus rowleyanus*, *nom. inval.* (Art. 9.5). [RPS 20]

Arthrocereus spinosissimus (Buining & Brederoo) Ritter, Kakt. Südamer. 1: 244, 1979. Basionym: *Eriocereus spinosissimus*. [RPS 30]

Astrophytum subgen. *Euastrophytum* Backeberg, Cact. Succ. J. (US) 22(5): 154, 1950. *Nom. inval.* (Art. 21.3). [Typus: *Astrophytum myriostigma* Lemaire.] [RPS 1]

Astrophytum subgen. **Neoastrophytum** Backeberg, Cact. Succ. J. (US) 22(5): 154, 1950. Typus: *Echinocactus asterias* Zuccarini. [RPS 1]

Astrophytum asterias cv. **Super Kabuto** R. Mayer, Succulenta 69(7-8): 1068, ill., 1990. [RPS 41]

Astrophytum asterias var. *nudum* Y. Ito, The Cactaceae, 508, ill. (p. 509), 1981. *Nom. inval.* (Art. 37.1). [RPS -]

Astrophytum asterias var. *pubesente* Y. Ito, The Cactaceae, 508, ill. (p. 509), 1981. *Nom. inval.* (Art. 37.1). [Spelling of the varietal epithet exactly as given in the protologue.] [RPS -]

Astrophytum asterias var. *striatipetalum* Okumura & Y. Ito, The Cactaceae, 508, ill. (p. 509), 1981. *Nom. inval.* (Art. 37.1). [RPS -]

Astrophytum capricorne fa. **aureum** (Möller) Krainz, Kat. ZSS ed. 2, 36, 1967. Basionym:

Echinocactus capricornis var. *aureus*. [RPS 18]

Astrophytum capricorne fa. **crassispinum** (Möller) Krainz, Kat. ZSS ed. 2, 36, 1967. Basionym: *Echinocactus capricornis* var. *crassispinus*. [RPS 18]

Astrophytum capricorne fa. **minus** (Runge & Quehl) Krainz, Kat. ZSS ed. 2, 37, 1967. Basionym: *Echinocactus capricornis* var. *minor*. [RPS 18]

Astrophytum capricorne fa. **niveum** (Kayser) Krainz, Kat. ZSS ed. 2, 37, 1967. Basionym: *Echinocactus capricornis* var. *niveus*. [RPS 18]

Astrophytum capricorne var. *crassispinum-nudum* Y. Ito, The Cactaceae, 514, ill. (p. 515), 1981. *Nom. inval.* (Art. 37.1). [Varietal epithet given as 'crassispinum nudum' in the protologue.] [RPS -]

Astrophytum capricorne var. *majus* Y. Ito, Cacti, 110, 1952. *Nom. inval.* (Art. 36.1). [RPS 4]

Astrophytum columnare (Schumann) Sadkovsky & Schütz, Die Gattung Astrophytum 159, 1979. Basionym: *Echinocactus myriostigma* var. *columnaris*. [RPS 30]

Astrophytum crassispinum (Möller) W. Haage & Sadovsky, Kakt. and. Sukk. 9: 81, 1958. Basionym: *Echinocactus capricornis* var. *crassispinus*. [First mentioned (as *nom. nud.*) in l.c. 8: 137, 1957.] [RPS 9]

Astrophytum myriostigma fa. **nudum** (Meyer) Krainz, Kat. ZSS ed. 2, 37, 1967. Basionym: *Echinocactus myriostigma* var. *nuda*. [RPS 18]

Astrophytum myriostigma fa. **quadricostatum** (Möller) Krainz, Kat. ZSS ed. 2, 37, 1967. Basionym: *Echinocactus myriostigma* ssp. *quadricostatus*. [RPS 18]

Astrophytum myriostigma fa. *rotundum* Sadovsky, Friciana 4(27): 6, 1964. *Nom. inval.* (Art. 37.1). [Sphalm. 'rotunda'. The entry in RPS 15 for this taxon is incorrect.] [RPS 15]

Astrophytum myriostigma fa. **tulense** (Kayser) Krainz, Kat. ZSS ed. 2, 37, 1967. Basionym: *Astrophytum myriostigma* ssp. *tulense*. [RPS 18]

Astrophytum myriostigma subvar. *glabrum* Backeberg, Die Cact. 5: 2656, 1960. *Nom. inval.* (Art. 37.1). [Erroneously recorded as valid in RPS 12.] [RPS 12]

Astrophytum myriostigma subvar. **nudum** (R. Meyer) Backeberg, Die Cact. 5: 2662, 1961. Basionym: *Echinocactus myriostigma* var. *nuda*. [RPS 12]

Astrophytum myriostigma subvar. **tulense** (Kayser) Backeberg, Die Cact. 5: 2660, 1961. Basionym: *Astrophytum myriostigma* ssp. *tulense*. [RPS 12]

Astrophytum myriostigma var. *caespitosum* Y. Ito, The Cactaceae, 510-511, ill., 1981. *Nom. inval.* (Art. 37.1). [Sphalm. 'caespitoum'.] [RPS -]

Astrophytum myriostigma var. **coahuilense** (Möller) Y. Ito, Cacti, 109, 1952. [Given as *comb. nud.* in RPS 4. Other sources give Borg as combining author.] [RPS 4]

Astrophytum myriostigma var. *coahuilense-pubesente* Y. Ito, The Cactaceae, 510, ill. (p. 511), 1981. *Nom. inval.* (Art. 37.1). [Epithet given as 'coahuilense pubesente' in the protologue.] [RPS -]

Astrophytum myriostigma var. *columnare-octogonum* Y. Ito, The Cactaceae, 512, ill., 1981. *Nom. inval.* (Art. 37.1). [Varietal epithet given as 'columare octogonum' (sic !) in the protologue.] [RPS -]

Astrophytum myriostigma var. **columnaris** (Schumann) Y. Ito, Cacti, 109, 1952. [Given as *comb. nud.* in RPS 4.] [RPS 4]

Astrophytum myriostigma var. **nudum** (R. Meyer) Y. Ito, Cacti, 109, 1952. [Given as *comb. nud.* in RPS 4.] [RPS 4]

Astrophytum myriostigma var. *nudum-octogonum* Y. Ito, The Cactaceae, 512, 1981. *Nom. inval.* (Art. 37.1). [Varietal epithet given as 'nudum octogonum' in the protologue.] [RPS -]

Astrophytum myriostigma var. *nudum-pubesente* Y. Ito, The Cactaceae, 512, ill., 1981. *Nom. inval.* (Art. 37.1). [Varietal epithet given as 'nudum pubescente' in the protologue.] [RPS -]

Astrophytum myriostigma var. **quadricostatum** (Möller) Y. Ito, Cacti, 110, 1952. [Given as *comb. nud.* in RPS 4. Other sources give Baum as the combining author.] [RPS 4]

Astrophytum myriostigma var. *quadricostatum-pubesente* Y. Ito, The Cactaceae, 510, ill. (p. 511), 1981. *Nom. inval.* (Art. 37.1). [Epithet given as 'quadricostatum pubesente' in the protologue.] [RPS -]

Astrophytum myriostigma var. *strongylogonum* Backeberg, Die Cact. 5: 2661, 1961. *Nom. inval.* (Art. 37.1). [Erroneously included as valid in RPS 12.] [RPS 12]

Astrophytum myriostigma var. *tulense* (Kayser) Y. Ito, The Cactaceae, 511, 1981. *Nom. inval.* (Art. 33.2), based on *Astrophytum myriostigma* ssp. *tulense*. [RPS -]

Astrophytum niveum (Kayser) W. Haage & Sadovsky, Kakt. and. Sukk. 9: 81, 1958. [First mentioned (as *nom. nud.*) in l.c. 8: 137, 1957.] [RPS 9]

Astrophytum ornatum fa. **glabrescens** (F. A. C. Weber) Krainz, Die Kakt. Lief. 31/32: C VIe, 1965. Basionym: *Astrophytum glabrescens*. [RPS 16]

Astrophytum ornatum fa. **mirbelii** (Lemaire) Krainz, Die Kakt. Lief. 31/32: C VIe, 1965. Basionym: *Echinocactus mirbelii*. [RPS 16]

Astrophytum ornatum var. **glabrescens** (F. A. C. Weber *ex* Schumann) Backeberg, Die Cact. 5: 2664, 1961. Basionym: *Echinocactus ornatus* var. *glabrescens*. [First published (as nude combination) by Ito, Cacti, 110, 1952 (cf. RPS 4).] [RPS 12]

Astrophytum ornatum var. *pubesente* Y. Ito, The Cactaceae, 513, ill., 1981. *Nom. inval.* (Art. 37.1). [Spelling of the varietal epithet exactly as given in the protologue.] [RPS -]

Astrophytum tulense (Kayser) Sadovsky & Schütz, Die Gattung Astrophytum, 159, 1979. Basionym: *Astrophytum myriostigma* ssp. *tulense*. [RPS 30]

Austrocactus coxii (Schumann) Backeberg, Die

Cact. 3: 1562, 1959. Basionym: *Echinocactus coxii*. [RPS 10]

Austrocactus hibernus Ritter, Sukkulentenkunde 7/8: 34-36, ill., 1963. Typus: *Ritter* 226 (U 116 932B, ZSS). [Chile] [RPS 14]

Austrocactus philippii (Regel & Schmidt) Buxbaum & Ritter, Sukkulentenkunde 7/8: 11, 35, 1963. Basionym: *Cereus philippii*. [RPS 14]

Austrocactus spiniflorus (Philippi) Ritter, Sukkulentenkunde 7/8: 35, 1963. Basionym: *Opuntia spiniflora*. [RPS 14]

Austrocephalocereus subgen. **Espostoopsis** (Buxbaum) Buxbaum, in Krainz, Die Kakt. Lief. 44/45: C Va (Oct. 1970), 1970. Basionym: *Espostoopsis*. [RPS 21]

Austrocephalocereus albicephalus Buining & Brederoo, Kakt. and. Sukk. 24(4): 73-75, ills., 1973. Typus: *Horst et Uebelmann* HU 348 (1968) (U, ZSS). [Brazil: Minas Gerais] [RPS 24]

Austrocephalocereus dolichospermaticus Buining & Brederoo, Kakt. and. Sukk. 25(4): 76-79, ills., 1974. Typus: *Horst et Uebelmann* HU 395 (1972) (U, ZSS). [Brazil: Bahia] [RPS 25]

Austrocephalocereus dybowskii (Roland-Gosselin) Backeberg, Cact. Succ. J. (US) 23: 149, 1951. Basionym: *Cereus dybowskii*. [Repeated by Backeberg in Die Cact. 4: 2498, 1960.] [RPS 3]

Austrocephalocereus estevesii Buining & Brederoo, Cact. Succ. J. (US) 47(6): 267-271, ills., 1975. Typus: *Horst et Uebelmann* HU 432 (1974) (U, ZSS). [Brazil: Goiás] [RPS 27]

Austrocephalocereus estevesii ssp. **grandiflorus** Diers & E. Esteves Pereira, Kakt. and. Sukk. 40(4): 77-81, ills., SEM-ills., 1989. Typus: *Esteves Pereira* 149 (KOELN [Succulentarium ?]). [Brazil: Goiás] [RPS 40]

Austrocephalocereus estevesii ssp. **insigniflorus** Diers & E. Esteves Pereira, Kakt. and. Sukk. 39(11): 264-268, ills., SEM-ills., 1988. Typus: *Esteves Pereira* 122 (KOELN). [Brazil: Minas Gerais] [RPS 39]

Austrocephalocereus fluminensis (Miquel) Buxbaum, Beitr. Biol. Pflanz. 44: 240, 1968. Basionym: *Cereus fluminensis*. [RPS 19]

Austrocephalocereus lehmannianus (Werdermann) Backeberg, Cact. Succ. J. (US) 23: 149, 1951. Basionym: *Cephalocereus lehmannianus*. [Repeated by Backeberg in Die Cact. 4: 2499, 1960.] [RPS 3]

Austrocephalocereus salvadorensis (Werdermann) Buxbaum, in Krainz, Die Kakt. Lief. 33, 1966. Basionym: *Pilocereus salvadorensis*. [RPS 17]

Austrocylindropuntia chuquisacana (Cardenas) Ritter, Kakt. Südamer. 2: 485-486, 1980. Basionym: *Opuntia chuquisacana*. [RPS 31]

Austrocylindropuntia clavarioides var. **ruiz-lealii** (Castellanos) Backeberg, Die Cact. 1: 161, 1958. Basionym: *Opuntia ruiz-lealii*. [RPS 9]

Austrocylindropuntia colubrina (Castellanos) Backeberg, Die Cact. 1: 156, 1958. Basionym: *Opuntia colubrina*. [RPS 9]

Austrocylindropuntia floccosa (Salm-Dyck) Ritter, Kakt. Südam. 4: 1244-1246, 1981. Basionym: *Opuntia floccosa*. [RPS 32]

Austrocylindropuntia haematacantha Backeberg, Cact. Succ. J. (US) 23(1): 13, 1951. Basionym: *Cylindropuntia haematacantha*. [RPS 3]

Austrocylindropuntia humahuacana (Backeberg) Backeberg, Cact. Succ. J. (US) 23(1): 13, 1951. Basionym: *Cylindropuntia humahuacana*. [RPS 3]

Austrocylindropuntia hypsophila (Spegazzini) Backeberg, Cact. Succ. J. (US) 23(1): 13, 1951. Basionym: *Opuntia hypsophila*. [The name given as basionym was itself a combination, and is here corrected.] [RPS 3]

Austrocylindropuntia inarmata Backeberg, Die Cact. 6: 3578, 1962. *Nom. inval.* (Art. 9.5). [Erroneously included as valid in RPS 13.] [RPS 13]

Austrocylindropuntia intermedia Rauh & Backeberg, Descr. Cact. Nov. 1: 6, 1957. [Dated 1956, published 1957.] [RPS 7]

Austrocylindropuntia ipatiana (Cardenas) Backeberg, Die Cact. 1: 153, 1958. Basionym: *Opuntia ipatiana*. [RPS 9]

Austrocylindropuntia lagopus (Schumann) Ritter, Kakt. Südamer. 4: 1242-1243, 1981. Basionym: *Opuntia lagopus*. [RPS 32]

Austrocylindropuntia lagopus fa. **rauhii** (Backeberg) Ritter, Kakt. Südamer. 4: 1243-1244, 1981. Basionym: *Tephrocactus rauhii*. [RPS 32]

Austrocylindropuntia lauliacoana Ritter, Kakt. Südamer. 4: 1247-1248, 1981. Typus: *Ritter* 1449 (U). [Holotype cited for U l.c. 1: iii, 1979.] [Peru] [RPS 32]

Austrocylindropuntia machacana Ritter, Kakt. Südamer. 4: 1246-1247, 1981. Typus: *Ritter* 701 (U). [Holotype cited for U l.c. 1: iii, 1979.] [Peru: Ancash] [RPS 32]

Austrocylindropuntia maldonadensis (Arechavelata) Backeberg, Cact. Succ. J. (US) 23(1): 13, 1951. Basionym: *Cylindropuntia maldonadensis*. [RPS 3]

Austrocylindropuntia malyana (Rausch) Ritter, Kakt. Südamer. 4: 1244, 1981. Basionym: *Tephrocactus malyanus*. [RPS 32]

Austrocylindropuntia miquelii var. **filesii** Backeberg, Descr. Cact. Nov. 1: 6, 1957. [Dated 1956, published 1957.] [RPS 7]

Austrocylindropuntia salmiana var. **albiflora** (Schumann) Backeberg, Die Cact. 1: 157, 1958. Basionym: *Opuntia albiflora*. [RPS 9]

Austrocylindropuntia salmiana var. **spegazzinii** (F. A. C. Weber) Backeberg, Die Cact. 1: 156, 1958. Basionym: *Opuntia spegazzinii*. [RPS 9]

Austrocylindropuntia schickendantzii (F. A. C. Weber) Backeberg, Cact. Succ. J. (US) 23(1): 13, 1951. Basionym: *Opuntia schickendantzii*. [The name given as basionym is itself a combination; this has here been corrected to the correct basionym.] [RPS 3]

Austrocylindropuntia shaferi (Britton & Rose) Backeberg, Cact. Succ. J. (US) 23(1): 14,

1951. Basionym: *Opuntia shaferi*. [The name given as basionym was itself a combination, and is here corrected.] [RPS 3]

Austrocylindropuntia shaferi var. **humahuacana** (Backeberg) Kiesling, Cact. Succ. J. Gr. Brit. 42(4): 110, 1980. Basionym: *Cylindropuntia humahuacana*. [RPS 31]

Austrocylindropuntia spegazzinii (F. A. C. Weber) Backeberg, Cact. Succ. J. (US) 23(1): 14, 1951. Basionym: *Opuntia spegazzinii*. [The name given as basionym is itself a combination, and is here corrected.] [RPS 3]

Austrocylindropuntia steiniana Backeberg, Descr. Cact. Nov. 1: 6, 1957. [Dated 1956, published 1957.] [RPS 7]

Austrocylindropuntia tephrocactoides Rauh & Backeberg, Descr. Cact. Nov. 1: 6, 1957. [Dated 1956, published 1957.] [RPS 7]

Austrocylindropuntia verschaffeltii var. **hypsophila** (Spegazzini) Backeberg, Die Cact. 1: 149, 1958. Basionym: *Opuntia hypsophila*. [RPS 9]

Austrocylindropuntia verschaffeltii var. **longispina** Backeberg, Descr. Cact. Nov. 1: 6, 1957. [Dated 1956, published 1957.] [RPS 7]

Austrocylindropuntia vestita var. **chuquisacana** (Cardenas) Backeberg, Die Cact. 1: 151, 1958. Basionym: *Opuntia chuquisacana*. [RPS 9]

Austrocylindropuntia vestita var. *intermedia* Backeberg, Die Cact. 6. 3581, 1962. *Nom. inval.* (Art. 9.5). [Erroneously included as valid in RPS 13.] [RPS 13]

Austrocylindropuntia vestita var. **maior** Backeberg, Cact. & Succ. J. (US) 23: 14, 1951. [RPS -]

Austrocylindropuntia vestita var. **major** Backeberg, Cact. Succ. J. (US) 23(1): 14, 1951. [RPS 3]

Austrocylindropuntia vestita var. **shaferi** (Britton & Rose) Ritter, Kakt. Südamer. 2: 380-382, 1980. Basionym: *Opuntia shaferi*. [RPS 31]

Austrocylindropuntia weingartiana (Backeberg) Backeberg, Cact. Succ. J. (US) 23(1): 14, 1951. Basionym: *Cylindropuntia weingartiana*. [RPS 3]

Aylostera albiflora (Ritter) Backeberg, Descr. Cact. Nov. 3: 5, 1963. Basionym: *Rebutia albiflora*. [RPS 14]

Aylostera albipilosa (Ritter) Backeberg, Descr. Cact. Nov. 3: 5, 1963. Basionym: *Rebutia albipilosa*. [RPS 14]

Aylostera camargoensis (Rausch) Rausch, Lobivia 85, 129, 1987. *Nom. inval.* (Art. 34.1), based on *Rebutia camargoensis*. [Name erroneously applied to *Rebutia camargoensis* in incorrect basionym citation for combination under *Lobivia*.] [RPS -]

Aylostera huasiensis (Rausch) Rausch, Lobivia 85, 13, 1987. *Nom. inval.* (Art. 34.1), based on *Rebutia huasiensis*. [Name erroneously applied to *Aylostera huasiensis* in incorrect basionym citation for combination under *Lobivia*.] [RPS -]

Aylostera krugeri Cardenas, Cactus (Paris) No. 57: 260-261, 1957. [Sphalm. 'krugerii', sometimes corrected to 'kruegeri'.] [RPS 8]

Aylostera muscula (Ritter & Thiele) Backeberg, Descr. Cact. Nov. 3: 5, 1963. Basionym: *Rebutia muscula*. [RPS 14]

Aylostera narvaecensis Cardenas, Cact. Succ. J. (US) 43: 245-246, 1971. [Sphalm. 'narvaecense'.] [RPS 22]

Aylostera pseudodeminuta var. **albiseta** Backeberg, Cact. Succ. J. (US) 23(3): 82, 1951. [RPS 3]

Aylostera pseudodeminuta var. **grandiflora** Backeberg, Cact. Succ. J. (US) 23(3): 82, 1951. [RPS 3]

Aylostera pseudodeminuta var. **schneideriana** Backeberg, Cact. Succ. J. (US) 23(3): 82, 1951. [RPS 3]

Aylostera pseudominuscula (Backeberg) Y. Ito, Cacti, 60, 1952. [Given as *comb. nud.* in RPS 4.] [RPS 4]

Aylostera pulvinosa (Ritter & Buining) Backeberg, Descr. Cact. Nov. 3: 5, 1963. Basionym: *Rebutia pulvinosa*. [RPS 14]

Aylostera spegazziniana var. **atroviridis** Backeberg, Cact. Succ. J. (US) 23(3): 82, 1951. [RPS 3]

Aylostera steinmannii (Solms-Laubach) Backeberg, Die Cact. 3: 1528, 1959. Basionym: *Echinocactus steinmannii*. [RPS 10]

Aylostera tuberosa (Ritter) Backeberg, Descr. Cact. Nov. 3: 5, 1963. Basionym: *Rebutia tuberosa*. [RPS 14]

Aylostera walteriana (Backeberg) Y. Ito, Explan. Diagr. Austroechinocactinae, 129, 1957. Basionym: *Rebutia walteriana*. [RPS 8]

Aylostera zavaletae Cardenas, Kakt and. Sukk. 16: 177-178, 1965. [RPS 16]

Aylostera zecheri (Rausch) Rausch, Lobivia 85, 13, 1987. *Nom. inval.* (Art. 34.1), based on *Rebutia zecheri*. [Name erroneously applied to *Rebutia zecheri* in incorrect basionym citation for combination under *Lobivia*.] [RPS -]

Azureocereus hertlingianus (Backeberg) Rauh, Cactus (Paris) No. 51: 94, 1956. *Nom. inval.* (Art. 33.2), based on *Clistanthocereus hertlingianus*. [RPS 7]

Azureocereus viridis Rauh & Backeberg, Descr. Cact. Nov. [1:] 14, 1957. [Dated 1956, published 1957.] [RPS 7]

Backebergia Bravo, Anales Inst. Biol. UNAM 24: 215, 232, 1953. Typus: *Backebergia chrysomalla*. [Publication year variously cited as 1953 (e.g. RPS 4) or 1954.] [RPS 4]

Backebergia chrysomalla (Schumann) Bravo, Anales Inst. Biol. UNAM 24: 215, 232, 1953. Basionym: *Pilocereus chrysomallus*. [Sphalm. 'chrysomallus'.] [RPS 4]

Backebergia militaris (Audot) Sanchez-Mejorada, Cact. Succ. J. (US) 45: 171-174, 1973. Basionym: *Cereus militaris*. [RPS 24]

Binghamia acanthura (Vaupel) Borg, Cacti, ed. 2, 185, 1951. Basionym: *Cereus acanthurus*. [RPS 9]

Binghamia eriotricha (Werdermann & Backeberg) Borg, Cacti, ed. 2, 185, 1951. Basionym: *Cereus eriotrichus*. [RPS 9]

Blossfeldia atroviridis Ritter, Succulenta 44(2):

23, 1965. Typus: *Ritter* 748 (ZSS [type collection]). [No herbarium cited for the type.] [Bolivia: Cochabamba] [RPS 16]

Blossfeldia atroviridis var. **intermedia** Ritter, Kakt. Südamer. 2: 551, 1980. Typus: *Ritter* 748a (U, ZSS). [Holotype cited for U l.c. 1: iii, 1979.] [Bolivia: Oropeza] [RPS 31]

Blossfeldia campaniflora Backeberg, Cact. Succ. J. Gr. Brit. 21: 32, 1959. *Nom. inval.* (Art. 37.1). [Erroneously included as valid in RPS 10.] [RPS 10]

Blossfeldia fechseri Backeberg, Die Cact. 6: 3909, 1962. *Nom. inval.* (Art. 37.1). [Erroneously included as valid in RPS 13.] [RPS 13]

Blossfeldia liliputana fa. *campaniflora* (Backeberg) Krainz, Kat. ZSS ed. 2, 39, 1967. *Nom. inval.*, based on *Blossfeldia campaniflora, nom. inval.* (Art. 37.1). [Erroneously included as valid in RPS 18.] [RPS 18]

Blossfeldia liliputana fa. *fechseri* (Backeberg) Krainz, Kat. ZSS ed. 2, 39, 1967. *Nom. inval.*, based on *Blossfeldia fechseri, nom. inval.* (Art. 37.1). [Erroneously included as valid 66.] [RPS 18]

Blossfeldia liliputana var. **atroviridis** (Ritter) Krainz, Die Kakt. Lief. 63: C VIe, 1975. Basionym: *Blossfeldia atroviridis*. [RPS 26]

Blossfeldia liliputana var. **caineana** Cardenas, Cactus 82: 53, 1964. [RPS 15]

Blossfeldia liliputana var. *campaniflora* (Backeberg) Krainz, Die Kakt., part 63, C VIe (Oct. 1975), 1975. *Nom. inval.*, based on *Blossfeldia campaniflora, nom. inval.* (Art. 37.1). [Erroneously included as valid in RPS 26.] [RPS 26]

Blossfeldia liliputana var. *fechseri* (Backeberg) Krainz, Die Kakt. Lief. 63: C VIe, 1975. *Nom. inval.*, based on *Blossfeldia fechseri, nom. inval.* (Art. 37.1). [Erroneously included as valid in RPS 26. Combination repeated by F. Ritter, Kakt. Südamer. 2: 435, 1980 (RPS 31).] [RPS 26]

Blossfeldia liliputana var. **formosa** Ritter, Kakt. Südamer. 2: 435, 1980. [RPS 31]

Blossfeldia minima Ritter, Kakt. Südamer. 2: 552-553, 1980. Typus: *Ritter* 750 (U, ZSS). [Holotype cited for U l.c. 1: iii, 1979.] [Bolivia: Uzurduy] [RPS 31]

Blossfeldia pedicellata Ritter, Succulenta 44(2): 23, 1965. Typus: *Ritter* 749 (ZSS [type collection]). [No herbarium cited for the type.] [Bolivia: Chuquisaca] [RPS 16]

Bolivicereus Cardenas, Cact. Succ. J. (US) 23(3): 91, 1951. Typus: *Bolivicereus samaipatanus*. [RPS 2]

Bolivicereus brevicaulis Ritter, Kakt. Südamer. 2: 703, 1980. [RPS 31]

Bolivicereus croceus Ritter, Kakt. Südamer. 2: 704-705, 1980. [RPS 31]

Bolivicereus pisacensis Knize, Biota 7(57): 252, 1969. [RPS 20]

Bolivicereus rufus Ritter, Kakt. Südamer. 2: 703-704, 1980. [RPS 31]

Bolivicereus samaipatanus Cardenas, Cact. Succ. J. (US) 23(3): 91-93, 1951. [RPS 2]

Bolivicereus samaipatanus var. **divimiseratus** Cardenas, Nation. Cact. Succ. J. 6(1): 9, 1951. [RPS 2]

Bolivicereus samaipatanus var. **multiflorus** Cardenas, Cact. Succ. J. (US) 23(3): 93-94, 1951. [RPS 2]

Bolivicereus serpens (Humboldt, Bonpland & Kunth) Backeberg, Die Cact. 6: 3676, 1962. Basionym: *Cactus serpens*. [RPS 13]

Bolivicereus soukupii Knize, Biota 7(57): 252-253, 1969. [Sphalm. 'soukupi'. Entry repeated in RPS 25.] [RPS 20]

Bolivicereus tenuiserpens (Rauh & Backeberg) Backeberg, Die Cact. 6: 3677, 1962. Basionym: *Cleistocactus tenuiserpens*. [RPS 13]

Borzicactella F. Ritter, Kakt. Südamer. 4: 1385-1387, 1981. Typus: *B. tenuiserpens*. [= *Borzicactella* Johnson (*nom. nud.*)] [RPS 32]

Borzicactella serpens (Kunth) Ritter, Kakt. Südamer. 4: 1387, 1981. Basionym: *Cactus serpens*. [RPS 32]

Borzicactella tenuiserpens (Backeberg) Ritter, Kakt. Südamer. 4: 1387, 1981. Basionym: *Cleistocactus tenuiserpens*. [RPS 32]

Borzicactinae Buxbaum, Madroño 14(6): 191, 1958. Typus: *Borzicactus*. [RPS 9]

Borzicactus subgen. **Bolivicereus** (Cardenas) Buxbaum, in Krainz, Die Kakt. Lief. 55-56: C Vb (Dec. 1973), 1974. Basionym: *Bolivicereus*. [Dated Dec. 1973, published early in 1974.] [RPS 25]

Borzicactus aequatorialis Backeberg, Descr. Cact. Nov. [1:] 17, 1957. [Dated 1956, published 1957.] [RPS 7]

Borzicactus aurantiacus (Vaupel) Kimnach & P. C. Hutchison, Cact. Succ. J. (US) 29(2): 46-51, 1957. Basionym: *Echinocactus aurantiacus*. [RPS 8]

Borzicactus aurantiacus var. **calvescens** (Kimnach & P. C. Hutchison) Donald, Nation. Cact. Succ. J. 26: 10, 1971. Basionym: *Borzicactus calvescens*. [RPS 22]

Borzicactus aurantiacus var. **megalanthus** (Ritter) Donald, Nation. Cact. Succ. J. 26: 10, 1971. Basionym: *Matucana megalantha*. [RPS 22]

Borzicactus aureiflorus (Ritter) Donald, Nation. Cact. Succ. J. 26. 10, 1971. Basionym: *Matucana aureiflora*. [RPS 22]

Borzicactus aureispinus (Ritter) Rowley, Rep. Pl. Succ. 24: 6, 1973. *Nom. inval.* (Art. 43.1 ?), based on *Winteria aureispina*, incorrect name (Art. 11.3). [Erroneously included as valid in RPS 24.] [RPS 24]

Borzicactus cajamarcensis Ritter, Kakt. Südamer. 4: 1375, 1981. Typus: *Ritter* 679 (U, ZSS). [Holotype cited for U l.c. 1: iii, 1979.] [Peru: Cajamarca] [RPS 32]

Borzicactus calocephalus (Skarupke) Donald, Ashingtonia Species Cat. Cact. [part 12], unnumbered page, 1976. *Nom. inval.*, based on *Matucana calocephala, nom. inval.* (Art. 37.1). [Distributed as centre-page insert with Ashingtonia 2(4).] [RPS 27]

Borzicactus calvescens Kimnach & P. C. Hutchison, Cact. Succ. J. (US) 29(4): 111-115, 1957. Typus: *West* 8176 (UC, US, ZSS). [Peru: La Libertad] [RPS 8]

Borzicactus calviflorus Ritter, Taxon 13: 118, 1964. [RPS 15]

Borzicactus celsianus (Lemaire) Kimnach, Cact. & Succ. J. (US) 32: 59, 1960. Basionym: *Pilocereus celsianus*. [RPS 11]
Borzicactus celsianus var. **trollii** (Kupper) Rowley, Excelsa 12:35, 1986. Basionym: *Cereus trollii*. [RPS 37]
Borzicactus doelzianus (Backeberg) Kimnach, Cact. & Succ. J. (US) 32: 59, 1960. Basionym: *Morawetzia doelziana*. [RPS 11]
Borzicactus fieldianus var. **tessellatus** (Akers & Buining) Krainz, Kat. ZSS ed. 2, 39, 1967. Basionym: *Borzicactus tessellatus*. [RPS 18]
Borzicactus formosus (Ritter) Donald, Nation. Cact. Succ. J. 26: 10, 1971. Basionym: *Matucana formosa*. [RPS 22]
Borzicactus fossulatus (Labouret) Kimnach, Cact. & Succ. J. (US) 32: 59, 1960. Basionym: *Pilocereus fossulatus*. [RPS 11]
Borzicactus fruticosus (Ritter) Donald, Nation. Cact. Succ. J. 26(1): 10, 1971. Basionym: *Matucana fruticosa*. [RPS 22]
Borzicactus gracilis (Akers & Buining) Buxbaum & Krainz, in Krainz: Die Kakt. C Vb, 1962. Basionym: *Maritimocereus gracilis*. [RPS 13]
Borzicactus gracilis var. **aticensis** (Rauh & Backeberg) Krainz, Die Kakt. C Vb, 1962. Basionym: *Loxanthocereus aticensis*. [RPS 13]
Borzicactus gracilis var. **camanaensis** (Rauh & Backeberg) Krainz, Die Kakt. C Vb, 1962. Basionym: *Loxanthocereus camanaensis*. [RPS 13]
Borzicactus haynei (Otto) Kimnach, Cact. Succ. J. (US) 32: 92, 1960. Basionym: *Echinocactus haynii*. [RPS 11]
Borzicactus haynei var. **atrispinus** (Rauh & Backeberg) Donald, Nation. Cact. Succ. J. 25: 111, 1970. Basionym: *Matucana hystrix* var. *atrispina*. [RPS 21]
Borzicactus haynei var. **blancii** (Backeberg) Donald, Nation. Cact. Succ. J. 25: 72, 1970. [Given as "comb. illeg. (Art. 60) in RPS 21; nomenclatural status not assessed.] [RPS 21]
Borzicactus haynei var. **breviflora** (Rauh & Backeberg) Donald, Nation. Cact. Succ. J. 25: 72, 1970. Basionym: *Matucana breviflora*. [RPS 21]
Borzicactus haynei var. **hystrix** (Rauh & Backeberg) Donald, Nation. Cact. Succ. J. 25: 72, 1970. [Given as "comb. illeg. (Art. 60) in RPS 21; nomenclatural status not assessed.] [RPS 21]
Borzicactus haynei var. **perplexa** (Backeberg) Donald, Nation. Cact. Succ. J. 25: 111, 1970. Basionym: *Matucana herzogiana* var. *perplexa*. [RPS 21]
Borzicactus hempelianus (Gürke) Donald, Nation. Cact. Succ. J. 25: 71, 1970. Basionym: *Echinocactus hempelianus*. [RPS 24]
Borzicactus hempelianus var. **rettigii** (Quehl) Donald, Nation. Cact. Succ. J. 25: 71, 1970. Basionym: *Echinocactus rettigii*. [RPS 24]
Borzicactus hempelianus var. **spinosissimus** (Ritter) Donald, Nation. Cact. Succ. J. 25: 71, 1970. Basionym: *Arequipa spinosissima*. [RPS 24]
Borzicactus hempelianus var. **weingartianus** (Backeberg) Donald, Nation. Cact. Succ. J. 25: 71, 1970. Basionym: *Arequipa weingartiana*. [RPS 24]
Borzicactus hendriksenianus (Backeberg) Kimnach, Cact. Succ. J. (US) 32: 92, 1960. Basionym: *Oreocereus hendriksenianus*. [RPS 11]
Borzicactus huagalensis Donald & Lau, Nation. Cact. Succ. J. 25(2): 33, 1970. [RPS 21]
Borzicactus icosagonus fa. **aurantiaciflorus** (Backeberg) Krainz, Die Kakt. Lief. 15: C Vb, 1960. Basionym: *Seticereus icosagonus* var. *aurantiaciflorus*. [RPS 11]
Borzicactus intertextus (Ritter) Donald, Nation. Cact. Succ. J. 26(1): 10, 1971. Basionym: *Matucana intertexta*. [RPS 22]
Borzicactus intertextus var. **celendinensis** (Ritter) Donald, Nation. Cact. Succ. J. 26(1): 10, 1971. Basionym: *Matucana celendinensis*. [RPS 22]
Borzicactus keller-badensis (Backeberg & Krainz) Krainz, Kat. ZSS ed. 2, 39, 1967. Basionym: *Loxanthocereus keller-badensis*. [RPS 18]
Borzicactus krahnii Donald, Cact. Succ. J. (US) 51(2): 52-55, 1979. [RPS 30]
Borzicactus leucotrichus (Philippi) Kimnach, Cact. Succ. J. (US) 32: 93, 1960. Basionym: *Echinocactus leucotrichus*. [RPS 11]
Borzicactus madisoniorum P. C. Hutchison, Cact. Succ. J. (US) 35(6): 167-172, 1963. [RPS 14]
Borzicactus madisoniorum var. **pujupatii** Donald & Lau, Nation. Cact. Succ. J. 26(3): 71, 1971. [RPS 22]
Borzicactus mirabilis (Buining) Donald, Nation. Cact. Succ. J. 25: 111, 1970. Basionym: *Matucana mirabilis*. [RPS 21]
Borzicactus myriacanthus (Vaupel) Donald, Ashingtonia 1(9): 104, 1974. Basionym: *Echinocactus myriacanthus*. [RPS 25]
Borzicactus neoroezlii Ritter, Kakt. and. Sukk. 12: 54, 1961. Typus: *Ritter* 301 (U, ZSS). [Peru] [RPS 12]
Borzicactus oreodoxus (Ritter) Donald, Nation. Cact. Succ. J. 26: 10, 1971. Basionym: *Eomatucana oreodoxa*. [RPS 22]
Borzicactus paucicostatus (Ritter) Donald, Nation. Cact. Succ. J. 26: 10, 1971. Basionym: *Matucana paucicostata*. [RPS 22]
Borzicactus paucicostatus fa. **robustispinus** Donald & Lau, Nation. Cact. Succ. J. 26: 71-73, 1971. [RPS 22]
Borzicactus pisacensis (Knize) Rowley, Rep. Pl. Succ. 25: 5, 1976. Basionym: *Bolivicereus pisacensis*. [RPS 25]
Borzicactus piscoensis (Rauh & Backeberg) Rauh & Backeberg, in Rauh, Beitr. Kenntn. peruan. Kakt.veg., 291, 1958. Basionym: *Loxanthocereus piscoensis*. [RPS 10]
Borzicactus pseudothelegonus (Rauh & Backeberg) Backeberg, Descr. Cact. Nov. 3: 5, 1963. *Nom. inval.*, based on *Cereus pseudothelegonus, nom. inval.* (Art. 9.5, 34.1). [Erroneously included as valid in RPS 14.] [RPS 14]
Borzicactus purpureus Ritter, Kakt. Südamer. 4: 1383, 1981. [RPS 32]

Borzicactus ritteri (Buining) Donald, Nation. Cact. Succ. J. 26(1): 10, 1971. Basionym: *Matucana ritteri*. [RPS 22]
Borzicactus samaipatanus (Cardenas) Kimnach, Cact. Succ. J. (US) 32: 93, 1960. Basionym: *Bolivicereus samaipatanus*. [RPS 11]
Borzicactus samaipatanus var. **divimiseratus** (Cardenas) Krainz, Die Kakt. C Vb, 1962. Basionym: *Bolivicereus samaipatanus* var. *divimiseratus*. [RPS 13]
Borzicactus samaipatanus var. **multiflorus** (Cardenas) Krainz, Die Kakt. C Vb, 1962. Basionym: *Bolivicereus samaipatanus* var. *multiflorus*. [RPS 13]
Borzicactus sammensis Ritter, Taxon 13: 118, 1964. Typus: *Ritter* 304 (ZSS). [Holotype not located at ZSS, but several later preparations from living material of the type collection are present.] [Peru: Dept. Ancash] [RPS 15]
Borzicactus sepium var. **morleyanus** (Britton & Rose) Krainz, Die Kakt. CVb, 1971. Basionym: *Borzicactus morleyanus*. [RPS 22]
Borzicactus serpens (Humboldt, Bonpland & Kunth) Kimnach, Cact. Succ. J. (US) 32: 95, 1960. Basionym: *Cactus serpens*. [RPS 11]
Borzicactus sextonianus (Backeberg) Kimnach, Cact. & Succ. J. (US) 32: 95, 1960. Basionym: *Erdisia sextoniana*. [RPS 11]
Borzicactus soukupii (Knize) Rowley, Rep. Pl. Succ. 25: 6, 1976. Basionym: *Bolivicereus soukupii*. [RPS 25]
Borzicactus sulcifer (Rauh & Backeberg) Kimnach, Cact. Succ. J. (US) 32: 95, 1960. Basionym: *Loxanthocereus sulcifer*. [RPS 11]
Borzicactus tenuiserpens (Rauh & Backeberg) Kimnach, Cact. Succ. J. (US) 32: 95, 1960. Basionym: *Cleistocactus tenuiserpens*. [RPS 11]
Borzicactus tessellatus Akers & Buining, Succulenta 1954(6): 81-83, 1954. Typus: *Anonymus* s.n. (DS). [Peru: Dept. Lima] [RPS 5]
Borzicactus trollii (Kupper) Kimnach, Cact. Succ. J. (US) 32: 96, 1960. Basionym: *Cereus trollii*. [RPS 11]
Borzicactus tuberculatus Donald, Cact. Succ. J. (US) 51(2): 55, 1979. [RPS 30]
Borzicactus variabilis (Rauh & Backeberg) Donald, Nation. Cact. Succ. J. 25: 72, 1970. Basionym: *Matucana variabilis*. [RPS 21]
Borzicactus weberbaueri (Vaupel) Donald, Nation. Cact. Succ. J. 26(1): 10, 1971. Basionym: *Echinocactus weberbaueri*. [RPS 22]
Borzicactus weberbaueri var. **flammeus** Donald, Ashingtonia 1(9): 100-102, 1974. [RPS 25]
Borzicactus weberbaueri var. **myriacanthus** (Vaupel) Donald, Nation. Cact. Succ. J. 26(1): 10, 1971. Basionym: *Echinocactus myriacanthus*. [RPS 22]
× **Borzimoza** Rowley, Name that Succulent, 116, 1980. [= *Borzicactus* × *Denmoza*.] [RPS 31]
× **Borzinopsis** Rowley, Nation. Cact. Succ. J. 37(2): 47, 1982. [= *Borzicactus* × *Echinopsis*.] [RPS 33]
× **Borzipostoa** Rowley, Nation. Cact. Succ. J. 37(2): 47, 1982. [= *Borzicactus* × *Espostoa*.] [RPS 33]
× **Borzipostoa mirabilis** (Rauh & Backeberg) Rowley, Nation. Cact. Succ. J. 37(2): 47, 1982. Basionym: *Neobinghamia mirabilis*. [= *Borzicactus icosagonus* × *Espostoa lanata*.] [RPS 33]
× **Borziroya** Rowley, Nation. Cact. Succ. J. 37(2): 47, 1982. [= *Borzicactus* × *Oroya*.] [RPS 33]
Brasilicactus elachisanthus (F. A. C. Weber) Backeberg, Die Cact. 3: 1578, 1959. Basionym: *Echinocactus elachisanthus*. [RPS 10]
Brasilicactus graessneri var. **albisetus** Cullmann, Kakt. and. Sukk. 6(1): 105, 1955. [Sphalm. 'albiseta'.] [RPS 6]
Brasilicereus breviflorus Ritter, Kakt. Südamer. 1: 228, 1979. [RPS 30]
Brasilicereus markgrafii Backeberg & Voll, Arq. Jard. Bot. Rio de Janeiro 9: 154-157, ill., 1950. Typus: *Markgraf et al.* s.n. (RB 65.043). [Volume for 1949, published 1950.] [Brazil: Minas Gerais] [RPS -]
Brasiliopuntia neoargentina Backeberg, Descr. Cact. Nov. [1:] 9, 1957. [Dated 1956, published 1957.] [RPS 7]
Brasiliopuntia schulzii (Castellanos & Lelong) Backeberg, Die Cact. 1: 373, 1958. Basionym: *Opuntia schulzii*. [RPS 9]
Brasiliopuntia subacarpa Rizzini & Mattos-F., Rev. Brasil. Biol. 46(2): 324, 1986. Typus: *Rizzini et Mattos-Filho* 38 (RB). [RPS 37]
Brasiliparodia Ritter, Kakt. Südamer. 1: 144-149, 1979. Typus: *Parodia buenekeri* Buin. [RPS 30]
Brasiliparodia alacriportana (Backeberg & Voll) Ritter, Kakt. Südamer. 1: 149, 1979. Basionym: *Parodia alacriportana*. [RPS 30]
Brasiliparodia brevihamata (W. Haage) Ritter, Kakt. Südamer. 1: 151, 1979. Basionym: *Parodia brevihamata*. [RPS 30]
Brasiliparodia brevihamata fa. *conjungens* Ritter, Kakt. Südamer. 1: 151, 1979. *Nom. inval*. (Art. 37.1). [Erroneously included as valid in RPS 30: 4, corrected in RPS 37: 20.] [RPS 30]
Brasiliparodia brevihamata var. *mollispina* Ritter, Kakt. Südamer. 1: 151, 1979. *Nom. inval*. (Art. 37.1). [Erroneously included as valid in RPS 30: 4, corrected in RPS 37: 20.] [RPS 30]
Brasiliparodia buenekeri (Buining) Ritter, Kakt. Südamer. 1: 149, 1979. Basionym: *Parodia buenekeri*. [RPS 30]
Brasiliparodia buenekeri fa. *conjungens* Ritter, Kakt. Südamer. 1: 150, 1979. *Nom. inval*. (Art. 37.1). [Erroneously included as valid in RPS 30: 4, corrected in RPS 37: 20.] [RPS 30]
Brasiliparodia buenekeri var. *intermedia* Ritter, Kakt. Südamer. 1: 151, 1979. *Nom. inval*. (Art. 37.1). [Erroneously included as valid in RPS 30: 4, corrected in RPS 37: 20.] [RPS 30]
Brasiliparodia catarinensis Ritter, Kakt.

Südamer. 1: 152, 1979. *Nom. inval.* (Art. 37.1). [Erroneously included as valid in RPS 30: 4, corrected in RPS 37: 20.] [RPS 30]

Brasiliparodia rechensis (Buining) Ritter, Kakt. Südamer. 1: 149, 1979. Basionym: *Notocactus rechensis.* [RPS 30]

Browningia ser. *Acutisquamae* Buxbaum, in Krainz, Die Kakt. C IV/1 (Nov. 1965), 1965. *Nom. inval.* (Art. 22). [B. candelaris (Meyen) Britton & Rose (*typus generis*).] [RPS 16]

Browningia ser. **Rotundisquamae** Buxbaum, in Krainz, Die Kakt. C IV/1 (Nov. 1965), 1965. Typus: *B. altissima* (Ritter) Buxbaum. [RPS 16]

Browningia subgen. **Azureocereus** (Akers & Johnson) Buxbaum, in Krainz, Die Kakt. C IV/1 (Nov. 1965), 1965. Basionym: *Azureocereus.* [RPS 16]

Browningia albiceps Ritter, Kakt. Südamer. 4: 1322, 1981. Typus: *Ritter* 1319 (U). [Peru: Cajamarca] [RPS 32]

Browningia altissima (Ritter) Buxbaum, in Krainz, Die Kakt. C IV/1, 1965. Basionym: *Gymnanthocereus altissimus.* [RPS 16]

Browningia amstutziae (Rauh & Backeberg) Hutchison *ex* Buxbaum, in Krainz, Die Kakt. Lief. 31/32: C IV/1, 1965. Basionym: *Gymnocereus amstutziae.* [RPS 16]

Browningia ciliisquama Ritter *ex* Buxbaum, in Krainz, Die Kakt. Lief. 31/32: C IV/1 (Nov. 1965), 1965. *Nom. inval.*, based on *Azureocereus ciliisquamus, nom. inval.* (Art. nom. nud.). [RPS 16]

Browningia columnaris Ritter, Kakt. Südamer. 4: 1323-1324, 1981. Typus: *Ritter* 1294 (U). [Peru: Ayacucho] [RPS 32]

Browningia hertlingiana (Backeberg) Buxbaum, in Krainz, Die Kakt. C IV/1, 1965. Basionym: *Clistanthocereus hertlingianus.* [RPS 16]

Browningia icaensis Ritter, Kakt. Südamer. 4: 1320-1322, 1981. Typus: *Ritter* 193 (U). [Peru: Ica] [RPS 32]

Browningia pilleifera (Ritter) P. C. Hutchison, Cact. Succ. J. (US) 40: 23-25, 1968. Basionym: *Gymnanthocereus pilleifer.* [RPS 19]

Browningia riosaniensis (Backeberg) Rowley, Nation. Cact. Succ. J. 37(2): 48, 1982. Basionym: *Rauhocereus riosaniensis.* [RPS 33]

Browningia viridis (Rauh & Backeberg) Buxbaum, in Krainz, Die Kakt. C IV/1, 1965. Basionym: *Azureocereus viridis.* [RPS 16]

Browningieae Buxbaum, in Krainz, Die Kakt. C IV/1 (July 1966), 1966. Typus: *Browningia* Britton & Rose. [RPS 17]

Buiningia Buxbaum, in Krainz, Die Kakt. Lief. 46/47: C IV (June 1971), 1971. Typus: *B. brevicylindrica.* [RPS 22]

Buiningia aurea (Ritter) Buxbaum, in Krainz, Die Kakt., Lief. 46/47: C IV (June 1971), 1971. Basionym: *Coleocephalocereus aureus.* [RPS -]

Buiningia brevicylindrica Buining, in Krainz, Die Kakt., Lief. 46/47 (1. 6. 1971), CIV, unpaged, 1971. Typus: *Horst et Uebelmann* HU 167 (ZSS, ZSS). [Localization of holotype unclear; holotype specimen at ZSS is labelled "Buining 710 715").] [Brazil: Minas Gerais] [RPS 22]

Buiningia brevicylindrica var. **elongata** Buining, in Krainz, Die Kakt. 46/47, 1971. Typus: *Horst et Uebelmann* HU 271 (1968) (ZSS). [There is some doubt whether the specimen labelled as holotype at ZSS is in fact the holotype.] [Brazil: Minas Gerais] [RPS 22]

Buiningia brevicylindrica var. **longispina** Buining, in Krainz, Die Kakt. CIV, 1971. Typus: *Horst et Uebelmann* HU 167a (1966) (ZSS, ZSS). [There is some doubt whether the specimen labelled as holotype at ZSS is in fact the holotype.] [Brazil: Minas Gerais] [RPS 22]

Buiningia purpurea Buining & Brederoo, Kakt. and. Sukk. 24(6): 121-123, 1973. Typus: *Horst et Uebelmann* HU 359 (1971) (U, ZSS). [Brazil: Minas Gerais] [RPS 24]

Buiningia roseiflora Uebelmann, Feld-Nummern, 11, 1972. *Nom. inval.* (Art. 36.1, 37.1). [Published as provisional name without description.] [RPS -]

Cactara Boom, Succulenta 1957: 121-124, 1957. *Nom. inval.* (Art. H6). [Published as provisional name..] [Suggested as collective generic name for hybrid cactus cultivars involving more than one genus.] [RPS 8]

Cactus matanzanus (León) Borg, Cacti, ed. 2, 340, 1951. Basionym: *Melocactus matanzanus.* [RPS 9]

Cactus microcephalus (Miquel) Borg, Cacti, ed. 2, 340, 1951. Basionym: *Melocactus microcephalus.* [RPS 9]

Calymmanthium F. Ritter, Kakt. and. Sukk. 13: 25, figs. 1-3, 1962. Typus: *C. substerile.* [First mentioned as nude name in Backeberg, Die Cact. 2: 886-887, 984, 986, 1959 (cf. RPS 10).] [RPS 13]

Calymmanthium fertile Ritter, Kakt. Südamer. 4: 1266-1267, fig. 1129-1130, 1981. [This seems to be the valid publication place for this taxon. First invalidly published (without type) in Succulenta 45: 117, 135, ill., 1966 (this citation erroneously given as valid publication in RPS 17).] [RPS -]

Calymmanthium substerile Ritter, Kakt. and. Sukk. 13: 25, 1962. Typus: *Ritter* 315 (U 097 800; ZSS). [First mentioned as nude name in Backeberg, Die Cact. 2: 887, 1959 (cf. RPS 10).] [Peru: Jaen] [RPS 13]

Carnegiea subgen. *Eucarnegiea* Backeberg, Cact. Succ. J. (US) 22(5): 154, 1950. *Nom. inval.* (Art. 21.3). [Typus: *Cereus giganteus* Engelmann..] [RPS 1]

Carnegiea subgen. **Rooksbya** Backeberg, Cact. Succ. J. (US) 22(5): 154, 1950. Typus: *Cereus euphorbioides* Haworth. [RPS 1]

Carnegiea polylopha (De Candolle) D. Hunt, Bradleya 6: 100, 1988. Basionym: *Cereus polylophus.* [RPS 39]

Castellanosia Cardenas, Cact. Succ. J. (US) 23(3): 90, 1951. Typus: *Castellanosia caineana.* [RPS 2]

Castellanosia caineana Cardenas, Cact. Succ. J. (US) 23(3): 90-91, 1951. [RPS 2]

× **Cephalepiphyllum** Mottram, Contr. New Class. Cact. Fam., 26, 1990. [=

Cephalocereus × *Epiphyllum*.] [RPS 41]

Cephalocereinae Buxbaum, Bot. Stud. 12: 92, 1961. Typus: *Cephalocereus* Pfeiffer. [RPS 12]

Cephalocereus subgen. **Neodawsonia** (Backeberg) Bravo, Cact. Suc. Mex. 19: 47, 1974. Basionym: *Neodawsonia*. [RPS 25]

Cephalocereus subgen. **Pilosocereus** (Byles & Rowley) Bravo, Cact. Suc. Mex. 19: 47, 1974. Basionym: *Pilosocereus*. [RPS 25]

Cephalocereus atroviridis (Backeberg) Borg, Cacti, ed. 2, 149, 1951. Basionym: *Cereus atroviridis*. [RPS 9]

Cephalocereus backebergii (Weingart) Borg, Cacti, ed. 2, 149, 1951. Basionym: *Cereus backebergii*. [Repeated by Buxbaum, Bot. Stud. 12: 101, 1961 (cf. RPS 12).] [RPS 9]

Cephalocereus bradei (Backeberg & Voll) Dawson, Los Angeles County Mus. Contr. Sci. 10: 6, 1957. Basionym: *Pilocereus bradei*. [RPS 8]

Cephalocereus claroviridis (Backeberg) Borg, Cacti, ed. 2, 149, 1951. Basionym: *Cereus claroviridis*. [RPS 9]

Cephalocereus cuyabensis (Backeberg) Dawson, Los Angeles County Mus. Contr. Sci. 10: 1, 1957. Basionym: *Pilocereus cuyabensis*. [RPS 8]

Cephalocereus fricii (Backeberg) Borg, Cacti, ed. 2, 149, 1951. Basionym: *Cereus fricii*. [RPS 9]

Cephalocereus fulviceps (F. A. C. Weber) Moore, Baileya 19: 165, 1975. Basionym: *Pilocereus fulviceps*. [RPS 26]

Cephalocereus guerreronis (Backeberg) Buxbaum, Bot. Stud. 12: 101, 1961. Basionym: *Pilocereus guerreronis*. [RPS 12]

Cephalocereus hapalacanthus (Werdermann) Dawson, Los Angeles County Mus. Contr. Sci. 10: 6, 1957. Basionym: *Pilocereus hapalacanthus*. [RPS 8]

Cephalocereus machrisii Dawson, Los Angeles County Mus. Contr. Sci. 10: 8, 1957. [RPS 8]

Cephalocereus militaris (Audot) Moore, Baileya 19: 165, 1975. Basionym: *Cereus militaris*. [RPS 26]

Cephalocereus minensis (Werdermann) Dawson, Los Angeles County Mus. Contr. Sci. 10: 6, 1957. Basionym: *Cereus minensis*. [RPS 8]

Cephalocereus moritzianus var. **backebergii** (Weingart) Krainz, Kat. ZSS ed. 2, 41, 1967. Basionym: *Cereus backebergii*. [RPS 18]

Cephalocereus palmeri var. **sartorianus** (Rose) Krainz, Kat. ZSS ed. 2, 41, 1967. Basionym: *Cephalocereus sartorianus*. [RPS 18]

Cephalocereus perlucens (Schumann) Borg, Cacti, ed. 2, 144, 1951. Basionym: *Cereus perlucens*. [RPS 9]

Cephalocereus tehuacanus (Weingart) Borg, Cacti, ed. 2, 150, 1951. Basionym: *Pilocereus tehuacanus*. [RPS 9]

Cephalocereus tetetzo var. **nudus** (Dawson) Dawson, Desert Pl. Life 24: 53-57, 1952. Basionym: *Cephalocereus nudus*. [RPS 4]

Cephalocleistocactus Ritter, Succulenta 1959(8): 107-111, 1959. Typus: *C. chrysocephalus*. [RPS 10]

Cephalocleistocactus chrysocephalus Ritter, Succulenta 1959(8): 107-111, ills., 1959. Typus: *Ritter* 326 (?, ZSS). [ZSS has several preparations from the type collection.] [Bolivia: La Paz] [RPS 10]

Cephalocleistocactus pallidus Backeberg, Kakt.-Lex. 79, 1966. Typus: *Ritter* 324 p.p. (?, ZSS). [ZSS has authentic material from the type collection.] [See Eggli in Bradleya 3: 102, 1985, on the typification of this taxon. *Cleistocactus palhuayensis* Ritter and *C. palhuayensis* var. *camachoensis* are based on the same collection.] [Bolivia: Muñecas] [RPS 17]

Cephalocleistocactus ritteri (Backeberg) Backeberg, Die Cact. 6: 3691, 1962. Basionym: *Cleistocactus ritteri*. [RPS 13]

Cephalocleistocactus schattatianus Backeberg, Descr. Cact. Nov. 3: 5, 1963. *Nom. inval.* (Art. 9.5). [Erroneously included as valid in RPS 14. Based on cultivated material without origin in the collection of Marnier-Lapostolle (cf. RPS 11).] [RPS 14]

Cephalophorus chrysacanthus (F. A. C. Weber) Boom, Succulenta 46: 107, 1967. Basionym: *Pilocereus chrysacanthus*. [Given as "comb. illeg." in RPS 18, validity not assessed.] [RPS 18]

Cephalophorus palmeri (Rose) Boom, Succulenta 46: 107, 1967. Basionym: *Cephalocereus palmeri*. [Given as "comb. illeg." in RPS 18, validity not assessed.] [RPS 18]

× **Cepheliocereus** Rowley, Name that Succulent, 118, 1980. [= *Cephalocereus* × *Heliocereus*.] [RPS 31]

Cerei Mottram, Contr. New Class. Cact. Fam., 4, 10, 1990. Typus: *Cereus* Miller. [Validated by reference to *Cereeae* Salm-Dyck 1845 (as 'Cereastreae', corrected), published at the rank of *linea*.] [RPS 41]

× **Cerephyllum** Mottram, Contr. New Class. Cact. Fam., 26, 1990. [= *Cereus* × *Epiphyllum*.] [RPS 41]

Cereus cv. **Rettigsche Hybride** Backeberg, Cactus (Paris) No. 48: 13-14, ill., 1956. [RPS 7]

Cereus subgen. **Brasilicereus** (Backeberg) P. J. Braun, Bradleya 6: 87, 1988. Basionym: *Brasilicereus*. [RPS 39]

Cereus subgen. **Ebneria** (Backeberg) D. Hunt, Bradleya 6: 100, 1988. Basionym: *Monvillea* subgen. *Ebneria*. [RPS 39]

Cereus subgen. *Eucereus* Backeberg, Cact. Succ. J. (US) 22(5): 154, 1950. *Nom. inval.* (Art. 21.3). [Typus: *Cactus hexagonus* Linné..] [RPS 1]

Cereus subgen. **Neocereus** Backeberg, Cact. Succ. J. (US) 22(5): 154, 1950. Typus: *Cereus huntingtonianus* Weingart. [RPS 1]

Cereus adelmarii (Rizzini & Mattos-F.) P. J. Braun, Bradleya 6: 86, 1988. Basionym: *Monvillea adelmarii*. [Sphalm. 'adelmari'.] [RPS 39]

Cereus aethiops var. **landbeckii** (Philippi) Backeberg, Die Cact. 4: 2333, 1960. Basionym: *Cereus landbeckii*. [RPS 11]

Cereus aethiops var. **melanacanthus** (Schumann) Backeberg, Die Cact. 4: 2333,

1960. Basionym: *Cereus coerulescens* var. *melanacanthus*. [RPS 11]

Cereus alacriportanus var. **bageanus** (Ritter) P. J. Braun, Bradleya 6: 86, 1988. Basionym: *Piptanthocereus bageanus*. [RPS 39]

Cereus alticostatus (Ritter) P. J. Braun, Bradleya 6: 86, 1988. Basionym: *Monvillea alticostata*. [RPS 39]

Cereus bicolor Rizzini & Mattos-F., Rev. Brasil. Biol. 45(3): 307, 1985. Typus: *Rizzini, Mattos-F. et Saddi* s.n. (RB). [Brazil] [RPS 36]

Cereus braunii Cardenas, Succulenta 1956(1): 2-5, 1956. [RPS 7]

Cereus calcirupicola (Ritter) Rizzini, Rev. Brasil. Biol. 46: 782, 1987. Basionym: *Piptanthocereus calcirupicola*. [Vol. for 1986, publ. 1987.] [RPS 38]

Cereus calcirupicola var. **albicans** Rizzini, Rev. Brasil. Biol. 46: 784, 1987. *Rizzini et Mattos-Filho* s.n. (RB). [Type not found, N. P. Taylor, in litt. 1989.] [Vol. for 1986, publ. 1987.] [Brazil] [RPS 38]

Cereus calcirupicola var. **cabralensis** (Ritter) P. J. Braun, Bradleya 6: 86, 1988. Basionym: *Piptanthocereus cabralensis*. [RPS 39]

Cereus calcirupicola var. **cipoensis** (Ritter) P. J. Braun, Bradleya 6: 86, 1988. Basionym: *Piptanthocereus cipoensis*. [RPS 39]

Cereus calcirupicola var. **pluricostatus** (Ritter) Rizzini, Rev. Brasil. Biol. 46: 784, 1987. Basionym: *Piptanthocereus calcirupicola* var. *pluricostatus*. [Vol. for 1986, publ. 1987. Basionym incorrectly given in RPS 38.] [RPS 38]

Cereus campinensis (Backeberg & Voll) P. J. Braun, Bradleya 6: 86, 1988. Basionym: *Pilocereus campinensis*. [RPS 39]

Cereus campinensis var. **piedadensis** (Ritter) P. J. Braun, Bradleya 6: 86, 1988. Basionym: *Monvillea piedadensis*. [RPS 39]

Cereus chachapoyensis Ochoa & Backeberg, in Backeberg, Die Cact. 2: 859, 1959. *Nom. inval.* (Art. 34.1, 37.1). [Erroneously included as valid in RPS 10.] [RPS 10]

Cereus cochabambensis Cardenas, Cact. Succ. J. (US) 42: 30-31, 1970. [RPS 21]

Cereus cochabambensis var. **longicarpus** Cardenas, Cact. Succ. J. (US) 42: 184, 1970. [Sphalm. 'longicarpa'.] [RPS 21]

Cereus comarapanus Cardenas, Succulenta 1956(1): 3-6, 1956. [RPS 7]

Cereus crassisepalus Buining & Brederoo, in Krainz, Die Kakt. 53: CIVb, 1973. Typus: *Horst et Uebelmann* HU 169 (1966) (U, ZSS). [Brazil: Minas Gerais] [RPS 24]

Cereus deflexispinus Rauh & Backeberg, Descr. Cact. Nov. [1:] 14, 1957. [Dated 1956, published 1957. Validity in view of *Cereus deflexispinus* Monville not assessed.] [RPS 7]

Cereus eriophorus var. **fragrans** (Small) L. Benson, Cact. Succ. J. (US) 41: 126, 1969. Basionym: *Harrisia fragrans*. [RPS 20]

Cereus goiasensis (Ritter) P. J. Braun, Bradleya 6: 86, 1988. Basionym: *Piptanthocereus goiasensis*. [RPS 39]

Cereus gracilis var. **aboriginum** (Small) L. Benson, Cact. Succ. J. (US) 41: 126, 1969. Basionym: *Harrisia aboriginum*. [RPS 20]

Cereus gracilis var. **simpsonii** (Small) L. Benson, Cact. Succ. J. (US) 41: 126, 1969. Basionym: *Harrisia simpsonii*. [RPS 20]

Cereus grandiflorus var. **armatus** (Schumann) L. Benson, Cact. Succ. J. (US) 41: 126, 1969. Basionym: *Cereus nycticalus* var. *armatus*. [RPS 20]

Cereus huilunchu Cardenas, Succulenta 1951(4): 49-52, 1951. [RPS 2]

Cereus lanosus (Ritter) P. J. Braun, Bradleya 6: 86, 1988. Basionym: *Piptanthocereus lanosus*. [RPS 39]

Cereus markgrafii (Backeberg & Voll) P. J. Braun, Bradleya 6: 87, 1988. Basionym: *Brasilicereus markgrafii*. [RPS 39]

Cereus neonesioticus (Ritter) P. J. Braun, Bradleya 6: 86, 1988. Basionym: *Piptanthocereus neonesioticus*. [RPS 39]

Cereus neonesioticus var. **interior** (Ritter) P. J. Braun, Bradleya 6: 86, 1988. Basionym: *Piptanthocereus neonesioticus* var. *interior*. [RPS 39]

Cereus neotetragonus Backeberg, Die Cact. 4: 2363, 1960. *Nom. inval.* (Art. 37.1), based on *Cactus tetragonus*. [Erroneously included as valid in RPS 11. *Nom. nov. pro Cactus tetragonus sensu* Willdenow *non* Linné, therefore to be treated as new species (cf. Art. 33, Note 1).] [RPS 11]

Cereus phaeacanthus var. **breviflorus** (Ritter) P. J. Braun, Bradleya 6: 87, 1988. Basionym: *Brasilicereus breviflorus*. [RPS 39]

Cereus pseudothelegonus Rauh & Backeberg, in Backeberg, Die Cact. 2: 1136, 1959. *Nom. inval.* (Art. 9.5, 34.1). [Erroneously included as valid in RPS 10.] [RPS 10]

Cereus ridleii Dardano de Andrade Lima, in Backeberg, Die Cact. 4: 2352, 1960. [RPS 11]

Cereus robinii (Lemaire) L. Benson, Cact. Succ. J. (US) 41: 126, 1969. Basionym: *Pilocereus robinii*. [RPS 20]

Cereus robinii var. **deeringii** (Small) L. Benson, Cact. Succ. J. (US) 41: 126, 1969. Basionym: *Cephalocereus deeringii*. [RPS 20]

Cereus saddianus (Rizzini & Mattos-F.) P. J. Braun, Bradleya 6: 86, 1988. Basionym: *Monvillea saddiana*. [RPS 39]

Cereus sericifer (Ritter) P. J. Braun, Bradleya 6: 86, 1988. Basionym: *Piptanthocereus sericifer*. [RPS 39]

Cereus tacuaralensis Cardenas, Cactus (Paris) No. 80/81: 19, 1964. [RPS 15]

Cereus uruguayanus Kiesling, Darwinion 24: 448, 1982. [RPS -]

Cereus vargasianus Cardenas, Succulenta 1951(3): 33-36, 1951. [RPS 2]

× **Cerevillea** Rowley, Nation. Cact. Succ. J. 37(2): 48, 1982. [= *Cereus* × *Monvillea*.] [RPS 33]

× **Chamaebivia** cv. **Silandra** Timmermans, Succulenta 46: 51-54, 1967. [= *Chamaecereus silvestrii* × *Lobivia densispina*.] [RPS 18]

Chamaecereus cv. **John Pilbeam** Mottram, Kakt. Sukk. Ohra 21: 25, 1979. [= *Chamaecereus silvestrii* × ?.] [RPS 30]

Chamaecereus cv. **Wotan** Mottram, Kakt. Sukk. Ohra 21: 26, 1979. [= *Chamaecereus silvestrii* × *Echinopsis rowleyi*.] [RPS 30]

Chamaecereus cv. **Yellow Bird** Mottram, Kakt. Sukk. Ohra 21: 26, 1979. [= *Chamaecereus silvestrii* × *Lobivia melanea* Fric (*nom. nud.*).] [RPS 30]

× **Chamaelobivia** Y. Ito, Bull. Takarazuka Insectarium 71: 13-20, 1950. [Republished by Y. Ito in Cacti, 64, 1952, and in Explan. Diag. Austroechinocactinae, 110, 289, 1957 (with 'type' designation); and by Timmermans in Succulenta 1963: 38, 1963.] [RPS 3]

× **Chamaelobivia** cv. **Domino** Timmermans, Succulenta 1963: 38, 1963. [= *Chamaecereus silvestrii* × *Lobivia densispina*.] [RPS 14]

Chamaelobivia matuzakii Y. Ito, Explan. Diag. Austroechinocactinae, 112, 289, 1957. [RPS 8]

Chamaelobivia tanahashii Y. Ito, Cacti, 64, 1952. *Nom. inval.* (Art. 36.1). [Sphalm. 'tananhashii'.] [RPS 4]

Chamaelobivia tanahashii var. **notatiflora** Y. Ito, Explan. Diag. Austroechinocactinae, 111, 289, 1957. [RPS 8]

Chamaelobivia tanahasii var. *notatiflora* Y. Ito, Cacti, 64, 1952. *Nom. inval.* (Art. 36.1). [RPS 4]

Chamaelobivia tanahasii var. *sanguiniflora* Y. Ito, Cacti, 64, 1952. *Nom. inval.* (Art. 36.1). [RPS 4]

Chamaelobivia tudae Y. Ito, Explan. Diag. Austroechinocactinae, 112, 289, 1957. [First invalidly (Art. 36.1) published in Y. Ito, Cacti, 64, 1952 (cf. RPS 4).] [RPS 8]

Chamaelobivia tudae var. **cariniflora** Y. Ito, Explan. Diag. Austroechinocactinae, 112, 290, 1957. [RPS 8]

Chamaelobivia tudae var. **chrysantha** Y. Ito, Explan. Diag. Austroechinocactinae, 113, 290, 1957. [RPS 8]

Chamaelobivia tudae var. **cinnabarina** Y. Ito, Explan. Diag. Austroechinocactinae, 112, 290, 1957. [RPS 8]

Chamaelobivia tudae var. **croceantha** Y. Ito, Explan. Diag. Austroechinocactinae, 112, 290, 1957. [RPS 8]

Chamaelobivia tudae var. **rubriflora** Y. Ito, Explan. Diag. Austroechinocactinae, 112, 290, 1957. [RPS 8]

Chamaelobivia tudae var. **salmonea** Y. Ito, Explan. Diag. Austroechinocactinae, 112, 290, 1957. [RPS 8]

Chamaelobivia tudae var. **streptantha** Y. Ito, Explan. Diag. Austroechinocactinae, 113, 290, 1957. [RPS 8]

Chamaelobivia tudae var. **striatipetala** Y. Ito, Explan. Diag. Austroechinocactinae, 113, 290, 1957. [Sphalm. 'striasipetala'.] [RPS 8]

Chiapasia nelsonii var. **hondurensis** (Kimnach) Backeberg, Kakt.-Lex., 84, 1966. Basionym: *Disocactus nelsonii* var. *hondurensis*. [RPS 17]

Chichipia Backeberg, in Marnier-Lapostolle, Liste Cact. Jard. Bot. Les Cèdres, 12, 1950. *Nom. illeg.* (Art. 63). [The type species is also the type of *Polaskia* Backeberg 1949.] [RPS 7]

Chileniopsis villosa var. **nigra** (A. Berger) Y. Ito, Cacti, 81, 1952. [Given as *comb. nud.* in RPS 4.] [RPS 4]

Chileorebutia Fric *ex* Ritter, Cactus (Paris) No. 65: 191-194, 1959. *Nom. illeg.* (Art. 63.1). [First published in a supplement to l.c. no. 64.] [Published as a provisional name by Fric in Kreuzinger, Sukk.-Verz., 27, 1935, and validated (incl. Latin diagnosis) by Ritter l.c. The generic name is, however, based on the same type as *Thelocephala* Ito 1957, and therefore illegitimate.] [RPS 10]

Chileorebutia aerocarpa Ritter, Cactus (Paris) No. 64: Suppl.: 1, 1959. *Nom. inval.* (Art. 43.1). [RPS 10]

Chileorebutia aricensis Ritter, Katalog H. Winter, [unpaged], 1959. *Nom. inval.* (Art. 36.1, 37.1). [Published as provisional name without description, year of first usage not established.] [RPS -]

Chileorebutia duripulpa Ritter, Taxon 12: 123, 1963. Incorrect name (Art. 11.3). [RPS 14]

Chileorebutia esmeraldana Ritter, Taxon 12: 123, 1963. Incorrect name (Art. 11.3). [RPS 14]

Chileorebutia fulva Ritter, Cactus (Paris) No. 66: 10, 1960. *Nom. inval.* (Art. 36.1). [Given as *nom. nud.*] [RPS -]

Chileorebutia glabrescens Ritter, Cactus (Paris) No. 64: Suppl.: 1, 1959. *Nom. inval.* (Art. 43.1). [RPS 10]

Chileorebutia iquiquensis Ritter, Taxon 12: 32, 1963. Incorrect name (Art. 11.3). [RPS -]

Chileorebutia krausii Ritter, Cactus (Paris) 15(66): 16-18, ill., 1960. Based on *Ritter* 502. Incorrect name (Art. 11.3). [First mentioned in l.c. 15(63): Suppl.: [5], 1959 (*nom. inval.* Art 37.1; supplement erroneously labelled as supplement to no. 64). Published as *nom. nov.* for *Echinocactus odieri* var. *mebbesii* sensu Schumann 1898. The spelling 'krausii' is correct, as the taxon is named for Peter Kraus.] [Chile: Dept. Copiapo] [RPS 10]

Chileorebutia malleolata Ritter, Taxon 12(3): 123, 1963. Based on *Ritter* 517. Incorrect name (Art. 11.3). [Chile] [RPS 14]

Chileorebutia malleolata var. *solitaria* Ritter, Taxon 12: 123, 1963. Incorrect name (Art. 11.3). [RPS 14]

Chileorebutia napina (Philippi) Ritter, Cactus (Paris) No. 64: Suppl.: [unpaged], 1959. *Nom. inval.* (Art. 43.1), based on *Echinocactus napinus*. [The name given as basionym was itself a combination, and is here corrected.] [RPS 10]

Chileorebutia odieri (Salm-Dyck) Ritter, Cactus (Paris) No. 64: Suppl.: [unpaged], 1959. *Nom. inval.* (Art. 43.1), based on *Echinocactus odieri*. [RPS 10]

Chileorebutia reichei (Schumann) Ritter, Cactus (Paris) No. 64: Suppl.: [unpaged], 1959. *Nom. inval.* (Art. 43.1), based on *Echinocactus reichei*. [RPS 10]

Chileorebutia saxifraga Ritter, Cactus (Paris) 15(66): 10, 1960. *Nom. inval.* (Art. 36.1, 37.1). [Again mentioned as *nomen nudum* in Backeberg, Die Cact. 6: 3777, 1962.] [RPS -]

Chilita subgen. **Acentracantha** Buxbaum, Sukkulentenkunde 5: 25, 1954. Typus: *Chilita lasiacantha* (Engelmann) Orcutt. [RPS 5]

Chilita subgen. **Euancistracantha** Buxbaum, Sukkulentenkunde 5: 18, 1954. Typus: *Chilita bombycina* (Quehl) Orcutt. [RPS 5]

Chilita subgen. **Euebnerella** Buxbaum, Sukkulentenkunde 5: 20, 1954. Typus: *Chilita wildii* (Dietrich) Orcutt. [RPS 5]

Chilita subgen. **Procochemiea** Buxbaum, Sukkulentenkunde 5: 15, 1954. Typus: *Cochemiea sheldonii* (Britton & Rose) Orcutt. [RPS 5]

Chilita subgen. **Rectochilita** Buxbaum, Sukkulentenkunde 5: 23, 1954. Typus: *Chilita multiceps* (Salm-Dyck) Orcutt. [RPS 5]

Chilita alamensis (Craig) Buxbaum, Sukkulentenkunde 5: 16, 1954. *Nom. inval.* (Art. 33.2), based on *Mammillaria alamensis*. [RPS 5]

Chilita ancistroides (Lemaire) Buxbaum, Sukkulentenkunde 5: 21, 1954. *Nom. inval.* (Art. 33.2), based on *Mammillaria ancistroides*. [RPS 5]

Chilita angelensis (Craig) Buxbaum, Sukkulentenkunde 5: 18, 1954. *Nom. inval.* (Art. 33.2), based on *Mammillaria angelensis*. [RPS 5]

Chilita aureilanata (Bödeker) Buxbaum, Sukkulentenkunde 5: 25, 1954. *Nom. inval.* (Art. 33.2), based on *Mammillaria aureilanata*. [RPS 5]

Chilita aurihamata (Bödeker) Buxbaum, Sukkulentenkunde 5: 21, 1954. *Nom. inval.* (Art. 33.2), based on *Mammillaria aurihamata*. [RPS 5]

Chilita blossfeldiana (Bödeker) Buxbaum, Sukkulentenkunde 5: 17, 1954. *Nom. inval.* (Art. 33.2), based on *Mammillaria blossfeldiana*. [RPS 5]

Chilita boolii (Lindsay) Buxbaum, Sukkulentenkunde 5: 17, 1954. *Nom. inval.* (Art. 33.2), based on *Mammillaria boolii*. [RPS 5]

Chilita capensis (Gates) Buxbaum, Sukkulentenkunde 5: 17, 1954. *Nom. inval.* (Art. 33.2), based on *Mammillaria capensis*. [RPS 5]

Chilita colonensis (Craig) Buxbaum, Sukkulentenkunde 5: 16, 1954. *Nom. inval.* (Art. 33.2), based on *Mammillaria colonensis*. [RPS 5]

Chilita criniformis (De Candolle) Buxbaum, Sukkulentenkunde 5: 21, 1954. *Nom. inval.* (Art. 33.2), based on *Mammillaria criniformis*. [RPS 5]

Chilita crinita (De Candolle) Buxbaum, Sukkulentenkunde 5: 21, 1954. *Nom. inval.* (Art. 33.2), based on *Mammillaria crinita*. [RPS 5]

Chilita dioica (Brandegee) Buxbaum, Sukkulentenkunde 5: 17, 1954. *Nom. inval.* (Art. 33.2), based on *Mammillaria dioica*. [RPS 5]

Chilita erectohamata (Bödeker) Buxbaum, Sukkulentenkunde 5: 21, 1954. *Nom. inval.* (Art. 33.2), based on *Mammillaria erectohamata*. [RPS 5]

Chilita erythrosperma (Bödeker) Buxbaum, Sukkulentenkunde 5: 21, 1954. *Nom. inval.* (Art. 33.2), based on *Mammillaria erythrosperma*. [RPS 5]

Chilita estanzuelensis (Möller) Buxbaum, Sukkulentenkunde 5: 25, 1954. *Nom. inval.* (Art. 33.2), based on *Mammillaria estanzuelensis*. [Author of the basionym given as 'Berger' by Buxbaum.] [RPS 5]

Chilita fasciculata (Engelmann) Buxbaum, Sukkulentenkunde 5: 15, 1954. *Nom. inval.* (Art. 33.2), based on *Mammillaria fasciculata*. [RPS 5]

Chilita gasseriana (Bödeker) Buxbaum, Sukkulentenkunde 5: 19, 1954. *Nom. inval.* (Art. 33.2), based on *Mammillaria gasseriana*. [RPS 5]

Chilita gilensis (Bödeker) Buxbaum, Sukkulentenkunde 5: 21, 1954. *Nom. inval.* (Art. 33.2), based on *Mammillaria gilensis*. [RPS 5]

Chilita gueldemanniana (Backeberg) Buxbaum, Sukkulentenkunde 5: 15, 1954. *Nom. inval.* (Art. 33.2), based on *Mammillaria gueldemanniana*. [RPS 5]

Chilita haehneliana (Bödeker) Buxbaum, Sukkulentenkunde 5: 21, 1954. *Nom. inval.* (Art. 33.2), based on *Mammillaria haehneliana*. [RPS 5]

Chilita herrerae (Werdermann) Buxbaum, Sukkulentenkunde 5: 25, 1954. *Nom. inval.* (Art. 33.2), based on *Mammillaria herrerae*. [RPS 5]

Chilita humboldtii (Ehrenberg) Buxbaum, Sukkulentenkunde 5: 25, 1954. *Nom. inval.* (Art. 33.2), based on *Mammillaria humboldtii*. [RPS 5]

Chilita hutchisoniana (Gates) Buxbaum, Sukkulentenkunde 5: 17, 1954. *Nom. inval.* (Art. 33.2), based on *Mammillaria hutchisoniana*. [RPS 5]

Chilita icamolensis (Bödeker) Buxbaum, Sukkulentenkunde 5: 21, 1954. *Nom. inval.* (Art. 33.2), based on *Mammillaria icamolensis*. [RPS 5]

Chilita inaiae (Craig) Buxbaum, Sukkulentenkunde 5: 15, 1954. *Nom. inval.* (Art. 33.2), based on *Mammillaria inaiae*. [RPS 5]

Chilita insularis (Gates) Buxbaum, Sukkulentenkunde 5: 17, 1954. *Nom. inval.* (Art. 33.2), based on *Mammillaria insularis*. [RPS 5]

Chilita knebeliana (Bödeker) Buxbaum, Sukkulentenkunde 5: 21, 1954. *Nom. inval.* (Art. 33.2), based on *Mammillaria knebeliana*. [RPS 5]

Chilita lengdobleriana (Bödeker) Buxbaum, Sukkulentenkunde 5: 25, 1954. *Nom. inval.* (Art. 33.2), based on *Mammillaria lengdobleriana*. [RPS 5]

Chilita magallanii (Schmoll *ex* Craig) Buxbaum, Sukkulentenkunde 5: 25, 1954. *Nom. inval.* (Art. 33.2), based on *Mammillaria magallanii*. [RPS 5]

Chilita microcarpa (Engelmann) Buxbaum, Sukkulentenkunde 5: 12, 1954. *Nom. inval.* (Art. 33.2), based on *Mammillaria microcarpa*. [The name may also be invalid (Art. 34) because the basionym name can be interpreted as having been a provisional name only.] [RPS 5]

Chilita moelleriana (Bödeker) Buxbaum, Sukkulentenkunde 5: 19, 1954. *Nom. inval.* (Art. 33.2), based on *Mammillaria moelleriana*. [RPS 5]

Chilita monancistra (Berge *ex* Salm-Dyck) Buxbaum, Sukkulentenkunde 5: 21, 1954. *Nom. inval* (Art. 33.2), based on *Mammillaria monancistra*. [RPS 5]

Chilita painteri (Rose) Buxbaum, Sukkulentenkunde 5: 19, 1954. *Nom. inval.* (Art. 33.2), based on *Mammillaria painteri*. [RPS 5]

Chilita phitauiana (Baxter) Buxbaum, Sukkulentenkunde 5: 18, 1954. *Nom. inval.* (Art. 33.2), based on *Mammillaria phitauiana*. [RPS 5]

Chilita pilispina (J. A. Purpus) Buxbaum, Sukkulentenkunde 5: 23, 1954. *Nom. inval.* (Art. 33.2), based on *Mammillaria pilispina*. [RPS 5]

Chilita posseltiana (Bödeker) Buxbaum, Sukkulentenkunde 5: 19, 1954. *Nom. inval.* (Art. 33.2), based on *Mammillaria posseltiana*. [RPS 5]

Chilita pubispina (Bödeker) Buxbaum, Sukkulentenkunde 5: 21, 1954. *Nom. inval.* (Art. 33.2), based on *Mammillaria pubispina*. [RPS 5]

Chilita pygmaea (Britton & Rose) Buxbaum, Sukkulentenkunde 5: 21, 1954. *Nom. inval.* (Art. 33.2), based on *Neomammillaria pygmaea*. [RPS 5]

Chilita rettigiana (Bödeker) Buxbaum, Sukkulentenkunde 5: 19, 1954. *Nom. inval.* (Art. 33.2), based on *Mammillaria rettigiana*. [RPS 5]

Chilita sanluisensis (Shurly) Buxbaum, Sukkulentenkunde 5: 23, 1954. *Nom. inval.* (Art. 33.2), based on *Mammillaria sanluisensis*. [RPS 5]

Chilita schieliana (Schick) Buxbaum, Sukkulentenkunde 5: 23, 1954. *Nom. inval.* (Art. 33.2), based on *Mammillaria schieliana*. [RPS 5]

Chilita sinistrohamata (Bödeker) Buxbaum, Sukkulentenkunde 5: 19, 1954. *Nom. inval.* (Art. 33.2), based on *Mammillaria sinistrohamata*. [RPS 5]

Chilita tacubayensis (Fedde) Buxbaum, Sukkulentenkunde 5: 19, 1954. *Nom. inval.* (Art. 33.2), based on *Mammillaria tacubayensis*. [RPS 5]

Chilita trichacantha (Schumann) Buxbaum, Sukkulentenkunde 5: 21, 1954. *Nom. inval.* (Art. 33.2), based on *Mammillaria trichacantha*. [RPS 5]

Chilita unihamata (Bödeker) Buxbaum, Sukkulentenkunde 5: 19, 1954. *Nom. inval.* (Art. 33.2), based on *Mammillaria unihamata*. [RPS 5]

Chilita viereckii (Bödeker) Buxbaum, Sukkulentenkunde 5: 23, 1954. *Nom. inval.* (Art. 33.2), based on *Mammillaria viereckii*. [RPS 5]

Chilita weingartiana (Bödeker) Buxbaum, Sukkulentenkunde 5: 19, 1954. *Nom. inval.* (Art. 33.2), based on *Mammillaria weingartiana*. [RPS 5]

Chilita yaquensis (Craig) Buxbaum, Sukkulentenkunde 5: 15, 1954. *Nom. inval.* (Art. 33.2), based on *Mammillaria yaquensis*. [RPS 5]

Chilita zeilmanniana (Bödeker) Buxbaum, Sukkulentenkunde 5: 21, 1954. *Nom. inval.* (Art. 33.2), based on *Mammillaria zeilmanniana*. [RPS 5]

Chimerophora Ito, The Cactaceae, 658, 1981. *Nom. inval.* (Art. 37.1). [Spelling as given in the original.] [RPS -]

Chimerophora cosmphora Y. Ito, The Cactaceae, 658, 1981. *Nom. inval.* (Art. 36.1, 37.1). [Spelling as given in the original source.] [RPS -]

Chimerophora pseudotubiflora Y. Ito, The Cactaceae, 658, 1981. *Nom. inval.* (Art. 36.1, 37.1). [RPS -]

Chrysocactus Y. Ito, Bull. Takarazuka Insectarium 71: 13-20, 1950. *Nom. inval.* (Art. 36.1). [RPS 3]

Cinnabarinea Fric *ex* Ritter, Kakt. Südamer. 2: 633-634, 1980. Typus: *Echinocactus cinnabarinus*. [RPS 31]

Cinnabarinea acanthoplegma (Backeberg) Ritter, Kakt. Südamer. 2: 635-636, 1980. *Nom. inval.*, based on *Pseudolobivia acanthoplegma*, *nom. inval.* (Art. 37.1). [Erroneously included as valid in RPS 31.] [RPS 31]

Cinnabarinea acanthoplegma var. *leucosiphus* (Cardenas) Ritter, Kakt. Südamer. 6: 636, 1980. *Nom. inval.* (Art. 43.1), based on *Lobivia taratensis* var. *leucosiphus*. [Erroneously included as valid in RPS 31.] [RPS 31]

Cinnabarinea boedekeriana (Harden) Ritter, Kakt. Südamer. 2: 635, 1980. Basionym: *Echinopsis boedekeriana*. [RPS 31]

Cinnabarinea cinnabarina (Hooker) Fric *ex* Ritter, Kakt. Südamer. 2: 633, 1980. Basionym: *Echinocactus cinnabarinus*. [RPS 31]

Cinnabarinea neocinnabarina (Backeberg) Ritter, Kakt. Südamer. 2: 636, 1980. *Nom. inval.*, based on *Lobivia neocinnabarina*, *nom. inval.* (Art. 9.5). [Erroneously included as valid in RPS 31.] [RPS 31]

Cinnabarinea oligotricha (Cardenas) Ritter, Kakt. Südamer. 2: 636, 1980. Basionym: *Lobivia oligotricha*. [RPS 31]

Cinnabarinea prestoana (Cardenas) Ritter, Kakt. Südamer. 2: 637, 1980. Basionym: *Lobivia prestoana*. [RPS 31]

Cinnabarinea pseudocinnabarina (Backeberg) Ritter, Kakt. Südamer. 2: 636, 1980. *Nom. inval.*, based on *Lobivia pseudocinnabarina*, *nom. inval.* (Art. 9.5). [Erroneously included as invalid in RPS 31.] [RPS 31]

Cinnabarinea pseudocinnabarina var. *microthelis* Ritter, Kakt. Südamer. 2: 636-637, 1980. *Nom. inval.* (Art. 43.1). [Erroneously included as valid in RPS 31.] [RPS 31]

Cinnabarinea purpurea (Donald & Lau) Ritter, Kakt. Südamer. 2: 637, 1980. Basionym:

Weingartia purpurea. [RPS 31]
Cinnabarinea torotorensis (Cardenas) Ritter, Kakt. Südamer. 2: 637, 1980. Basionym: *Weingartia torotorensis*. [RPS 31]
Cinnabarinea walterspielii (Bödeker) Ritter, Kakt. Südamer. 2: 634-635, 1980. Basionym: *Lobivia walterspielii*. [RPS 31]
Cinnabarinea walterspielii var. **sanguiniflora** Ritter, Kakt. Südamer. 2: 635, 1980. [RPS 31]
Cinnabarinea zudanensis (Cardenas) Ritter, Kakt. Südamer. 2: 637, 1980. Basionym: *Lobivia zudanensis*. [RPS 31]
Cipocereus Ritter, Kakt. Südamer. 1: 49, 54, 1979. Typus: *C. pleurocarpus*. [RPS 30]
Cipocereus minensis (Werdermann) Ritter, Kakt. Südamer. 1: 57, 1979. Basionym: *Cereus minensis*. [RPS 30]
Cipocereus pleurocarpus Ritter, Kakt. Südamer. 1: 54, 1979. [RPS 30]
× **Cleistoborzicactus** Rowley, Nation. Cact. Succ. J. 37(2): 48, 1982. [= *Cleistocactus* × *Borzicactus*.] [RPS 33]
Cleistocactus cv. **Jupiter** Mottram, Cact. Succ. J. (US) 61(4): 154, ills., 1989. [= *Cleistocactus samaipatanus* × *C. winteri*.] [RPS 40]
Cleistocactus cv. **Stuart** Mottram, Cact. Succ. J. (US) 61(4): 156, ill. (p. 155), 1989. [= *Cleistocactus baumannii* ssp. *santacruzensis* × *C. brookei*.] [RPS 40]
Cleistocactus subgen. **Annemarnieria** Buxbaum, Cactus (Paris) No. 51: 87-91, 1956. Typus: *C. smaragdiflorus* (F. A. C. Weber) Britton & Rose. [RPS 7]
Cleistocactus subgen. *Eucleistocactus* Buxbaum, Cactus (Paris) No. 51: 87-91, 1956. *Nom. illeg.* (Art. 21.3). [RPS 7]
Cleistocactus grex **Brookmann** Mottram, Cact. Succ. J. (US) 61(4): 156, 1989. [Typus: *Cleistocactus* cv. Stuart.] [RPS 40]
Cleistocactus acanthurus (Vaupel) D. Hunt, Bradleya 5: 92, 1987. Basionym: *Cereus acanthurus*. [RPS 38]
Cleistocactus angosturensis Cardenas, Cact. Succ. J. (US) 28(2): 60-61, 1956. [RPS 7]
Cleistocactus areolatus var. **herzogianus** (Backeberg) Backeberg, Die Cact. 2: 1005, 1959. Basionym: *Cleistocactus herzogianus*. [RPS 10]
Cleistocactus aureispinus (Ritter) D. Hunt, Bradleya 5: 92, 1987. *Nom. illeg.* (Art. 64), based on *Winteria aureispina*, incorrect name (Art. 11.3). [*Non C. aureispinus* Fric 1928; = *C. winteri* D. Hunt, Bradleya 6: 100, 1988.] [RPS 38]
Cleistocactus ayopayanus Cardenas, Cact. Succ. J. (US) 28(2): 58-59, 1956. [RPS 7]
Cleistocactus azerensis Cardenas, Cact. & Succ. J. (US) 33: 74, 1961. [RPS 12]
Cleistocactus baumannii ssp. **santacruzensis** (Backeberg) Mottram, Cact. Succ. J. (US) 61(4): 156, in adnot, 1989. Basionym: *Cleistocactus santacruzensis*. [as 'santa-cruzensis'.] [RPS 40]
Cleistocactus brevispinus Ritter, Kakt. Südamer. 4: 1360-1361, 1981. Typus: *Ritter* 1297 (U). [Peru: Apurimac] [RPS 32]
Cleistocactus brookei Cardenas, Cact. Succ. J. (US) 24(5): 144-146, 1952. [RPS 3]
Cleistocactus brookei var. **flavispinus** Ritter, Kakt. Südamer. 2: 680, 1980. [RPS 31]
Cleistocactus bruneispinus Backeberg, Die Cact. 2: 1001, 1959. Basionym: *Cereus baumannii* var. *colubrinus*. [*Nom. nov.*] [RPS 10]
Cleistocactus buchtienii var. **flavispinus** Cardenas, Cact. Succ. J. (US) 24(6): 182, 1952. [RPS 3]
Cleistocactus candelilla Cardenas, Cact. Succ. J. (US) 24(5): 146-147, 1952. [RPS 3]
Cleistocactus candelilla var. **pojoensis** Cardenas, Cact. Succ. J. (US) 24(5): 147-148, 1952. [RPS 3]
Cleistocactus capadalensis Ritter, Kakt. Südamer. 2: 677, 1980. [RPS 31]
Cleistocactus chacoanus Ritter, Kakt. Südamer. 2: 671-672, 1980. Typus: *Ritter* 841 (U, ZSS). [Bolivia: Gran Chaco] [RPS 31]
Cleistocactus chacoanus var. **santacruzensis** Ritter, Kakt. Südamer. 2: 672, 1980. Typus: *Ritter* 356 (U, ZSS). [Based on the same type collection as *Cleistocactus santacruzensis* Backeberg, and therefore incorrectly designated as illegitimate by Mottram in Cact. Succ. J. (US) 61: 156, 1989.] [Bolivia: Santa Cruz] [RPS 31]
Cleistocactus clavicaulis Cardenas, Cactus (Paris) No. 80/81: 20, 1964. [RPS 15]
Cleistocactus compactus Backeberg, Descr. Cact. Nov. [1:] 17, 1957. [Dated 1956, published 1957.] [RPS 7]
Cleistocactus crassicaulis Cardenas, Cact. Succ. J. (US) 33: 77, 1961. [RPS 12]
Cleistocactus crassicaulis var. **paucispinus** Ritter, Kakt. Südamer. 2: 674-675, 1980. [RPS 31]
Cleistocactus crassiserpens Rauh & Backeberg, Descr. Cact. Nov. [1:] 17, 1957. [Dated 1956, published 1957.] [RPS 7]
Cleistocactus croceiflorus Ritter, Kakt. Südamer. 1: 272, 1979. [RPS 30]
Cleistocactus dependens Cardenas, Cact. Succ. J. (US) 24(5): 143-144, 1952. [RPS 3]
Cleistocactus ferrarii Kiesling, Hickenia 2(7): 37, ill., 1984. *Kiesling* 1722 (SI). [Argentina] [RPS 35]
Cleistocactus fieldianus (Britton & Rose) D. Hunt, Bradleya 5: 92, 1987. Basionym: *Borzicactus fieldianus*. [RPS 38]
Cleistocactus flavescens fa. **wendlandiorum** (Backeberg) Krainz, Kat. ZSS ed. 2, 43, 1967. Basionym: *Cleistocactus wendlandiorum*. [RPS 18]
Cleistocactus flavispinus (Schumann) Backeberg, Die Cact. 2: 1000, 1959. Basionym: *Cereus baumannii* var. *flavispinus*. [RPS 10]
Cleistocactus fossulatus Mottram, Chileans 13(43): 30, 1985. Typus: *Ritter* 100 (K, ZSS [type collection]). [Holotype specimen ex cult. Mottram 4.1.] [Publication date according to Index Kewensis, alternatively cited as 1986. syn. *Oreocereus fossulatus* Backeberg (1934) pro parte excl. typ., non *Pilocereus fossulatus* Labouret. First invalidly published (Art. 36) in l.c. 12(42): 138, 1984 (RPS 35: 3).] [Bolivia: Dept. La

Paz] [RPS 37]

Cleistocactus fossulatus var. *rubrispinus* Mottram, Chileans 12(42): 142, 1984. *Nom. inval.* (Art. 43). [Based on *Oreocereus fossulatus* Ritter *non* (Labouret) Backeberg var. rubrispinus Ritter, *nom. inval.* (Art. 43).] [RPS 35]

Cleistocactus fusiflorus Cardenas, Cactus (Paris) No. 57: 252-254, 1957. [RPS 8]

Cleistocactus glaucus Ritter, Taxon 13(3): 114, 1964. Typus: *Ritter* 112 (U 117 611, ZSS). [Holotype cited for U l.c. 12(1): 28.] [Bolivia: Murillo] [RPS 15]

Cleistocactus glaucus var. **plurispinus** Ritter, Taxon 13: 114, 1964. [RPS 15]

Cleistocactus granjaensis Ritter, Kakt. Südamer. 2: 687, 1980. Typus: *Ritter* 106 p.p. (U, ZSS). [Holotype cited for U l.c. 1: iii, 1979.] [Bolivia: Murillo] [RPS 31]

Cleistocactus grossei Weingart *ex* Backeberg, Descr. Cact. Nov. [1:] 17, 1957. [Dated 1956, published 1957.] [RPS 7]

Cleistocactus hildegardiae Ritter, Kakt. Südamer. 2: 685, fig. 13 (p. 710), 1980. Typus: *Ritter* 1126 (U). [Bolivia: Prov. Mendez] [RPS 31]

Cleistocactus hildewinterae Backeberg, Kakt.-Lex., 87, 1966. *Nom. inval.* (Art. 36.1, 37.1). [Only mentioned as provisional and ascribed to F. Ritter.] [RPS -]

Cleistocactus hildewinterae var. *flavispinus* Backeberg, Kakt.-Lex., 87, 1966. *Nom. inval.* (Art. 36, 37, 43). [Only mentioned as provisional and ascribed to F. Ritter.] [RPS -]

Cleistocactus horstii P. J. Braun, Kakt. and. Sukk. 33(10): 204-209, ills., 1982. Typus: *Horst et Uebelmann* HU 373 (1974) (KOELN [Succulentarium], ZSS). [Brazil: Mato Grosso] [RPS 33]

Cleistocactus ianthinus Cardenas, Cact. Succ. J. (US) 28(2): 57-58, 1956. [RPS 7]

Cleistocactus jugatiflorus Backeberg, Die Cact. 6: 3685, 1962. *Nom. inval.* (Art. 37.1). [Erroneously included as valid in RPS 13.] [RPS 13]

Cleistocactus jujuyensis var. **fulvus** Ritter, Kakt. Südamer. 2: 480, 1980. [RPS 31]

Cleistocactus laniceps var. **plurispinus** Ritter, Kakt. Südamer. 2: 673, 1980. [RPS 31]

Cleistocactus leonensis J. E. Madsen, Fl. Ecuador 35: 18-19, ills., 1989. Typus: *Madsen* 61087 (AAU, QCA). [Ecuador: Prov. Azuay] [RPS 40]

Cleistocactus luribayensis Cardenas, Cact. Succ. J. (US) 28(2): 59-60, 1956. [RPS 7]

Cleistocactus mendozae Cardenas, Cact. Succ. J. (US) 35: 202, 1963. [RPS 14]

Cleistocactus micropetalus Ritter, Kakt. Südamer. 2: 675, 1980. Typus: *Ritter* 830 (U, ZSS). [Holotype cited for U l.c. 1: iii, 1979.] [Bolivia: Avilez] [RPS 31]

Cleistocactus morawetzianus var. **pycnacanthus** Rauh & Backeberg, Descr. Cact. Nov. [1:] 17, 1957. Typus: *Rauh* K75 (1954) (ZSS, ZSS). [Dated 1956, published 1957.] [Peru] [RPS 7]

Cleistocactus muyurinensis Ritter, Taxon 13(3): 114, 1964. Typus: *Ritter* 821 (U, ZSS). [Holotype cited for U l.c. 12(1): 28.] [Bolivia: Santa Cruz] [RPS 15]

Cleistocactus neoroezlii (Ritter) Buxbaum, in Krainz, Die Kakt. 57, CVb, 1974. Basionym: *Borzicactus neoroezlii*. [RPS 25]

Cleistocactus nivosus Borg, Cacti, ed. 2, 195, 1951. [RPS 3]

Cleistocactus orthogonus Cardenas, Cactus (Paris) No. 64: 161, 1959. [RPS 10]

Cleistocactus palhuayensis Ritter & Shahori, in Ritter, Kakt. Südamer. 2: 688-689, 1980. Typus: *Ritter* 324 p.p. (U, ZSS). [Holotype cited for U l.c. 1: iii, 1979.] [*Cephalocleistocactus pallidus* Backeberg is based on another individual of the type collection.] [Bolivia: Muñecas] [RPS 31]

Cleistocactus palhuayensis var. **camachoensis** Ritter & Shahori, in Ritter, Kakt. Südamer. 2: 688-689, 1980. [RPS 31]

Cleistocactus paraguariensis Ritter, Kakt. Südamer. 1: 232, 1979. [RPS 30]

Cleistocactus parapetiensis Cardenas, Cact. Succ. J. (US) 24(6): 182-183, 1952. [RPS 3]

Cleistocactus parviflorus var. **aiquilensis** Ritter, Kakt. and. Sukk. 14(6): 102-104, ill., 1963. Typus: *Ritter* 359 (U 117 874B, ZSS). [Type erroneously cited as Ritter 539 (instead of Ritter 359).] [Bolivia: Cochabamba] [RPS 14]

Cleistocactus parviflorus var. **comarapanus** Ritter, Kakt. Südamer. 2: 683-684, 1980. Typus: *Ritter* 358 (U, ZSS). [Holotype cited for U l.c. 1: iii, 1979.] [Bolivia: Valle Grande] [RPS 31]

Cleistocactus parviflorus var. **herzogianus** (Backeberg) Backeberg, Descr. Cact. Nov. 3: 5, 1963. Basionym: *Cleistocactus herzogianus*. [RPS 14]

Cleistocactus piraymirensis Cardenas, Cact. Succ. J. (US) 33: 78, 1961. [RPS 12]

Cleistocactus pojoensis (Cardenas) Backeberg, Die Cact. 2: 1003, 1959. Basionym: *Cleistocactus candelilla* var. *pojoensis*. [RPS 10]

Cleistocactus pungens Ritter, Taxon 13: 115, 1964. [RPS 15]

Cleistocactus pycnacanthus (Rauh & Backeberg) Backeberg, Kakt.-Lex. 88, 1966. Basionym: *Cleistocactus morawetzianus* var. *pycnacanthus*. [RPS 17]

Cleistocactus reae Cardenas, Cactus (Paris) No. 57: 251-252, 1957. [RPS 8]

Cleistocactus ressinianus Cardenas, Cact. Succ. J. (US) 28(2): 55-56, 1956. [RPS 7]

Cleistocactus ritteri Backeberg, Kakt. and. Sukk. 10(11): 163, 1959. Typus: *Ritter* 325 (?, ZSS). [The type was cited to be a living plant in the collection of Karius, Muggensturm; but preserved material of the type collection was at the time already deposited at ZSS.] [Bolivia: Yungas] [RPS 10]

Cleistocactus rojoi Cardenas, Cact. Succ. J. (US) 28(2): 56-57, 1956. [RPS 7]

Cleistocactus samaipatanus (Cardenas) D. Hunt, Bradleya 5: 92, 1987. Basionym: *Bolivicereus samaipatanus*. [RPS 38]

Cleistocactus santacruzensis Backeberg, Das Kakteenlexikon, 89, 1966. *Ritter* 356 (unde

?). [Bolivia: Santa Cruz] [RPS -]

Cleistocactus sepium var. **morleyanus** (Britton & Rose) J. E. Madsen, Fl. Ecuador 35: 21, 1989. Basionym: *Borzicactus morleyanus*. [RPS 40]

Cleistocactus sepium var. **ventimigliae** (Riccobono) J. E. Madsen, Fl. Ecuador 35: 22, 1989. Basionym: *Borzicactus ventimigliae*. [RPS 40]

Cleistocactus smaragdiflorus fa. **rojoi** (Cardenas) Ritter, Kakt. Südamer. 2: 678, 1980. Basionym: *Cleistocactus rojoi*. [RPS 31]

Cleistocactus smaragdiflorus var. *gracilior* Backeberg, Kakt.-Lex. 89, 1966. *Nom. inval.* (Art. 37.1). [Erroneously included as valid in RPS 17; and ommitted from the list published by Eggli in Bradleya 3, 1985.] [RPS 17]

Cleistocactus strausii var. **fricii** (Dörfl.) Backeberg, Die Cact. 2: 1015, 1959. Basionym: *Pilocereus strausii* var. *fricii*. [RPS 10]

Cleistocactus sucrensis Cardenas, Cact. Succ. J. (US) 24(5): 148-149, 1952. [RPS 3]

Cleistocactus tarijensis Cardenas, Cact. Succ. J. (US) 28(2): 54-55, 1956. [RPS 7]

Cleistocactus tenuiserpens Rauh & Backeberg, in Backeberg, Descr. Cact. Nov. [1:] 17, 1957. [Dated 1956, published 1957.] [RPS 7]

Cleistocactus tupizensis var. **sucrensis** (Cardenas) Backeberg, Die Cact. 2: 1012, 1959. Basionym: *Cleistocactus sucrensis*. [RPS 10]

Cleistocactus vallegrandensis Cardenas, Cact. Succ. J. (US) 33: 75, 1961. [RPS 12]

Cleistocactus varispinus Ritter, Taxon 13(3): 114, 1964. Typus: *Ritter* 108 (U 117 850B, ZSS). [Holotype cited for U l.c. 12(1): 28.] [Sphalm. 'variispinus'.] [Bolivia: La Paz] [RPS 15]

Cleistocactus villaazulensis Ritter, Kakt. Südamer. 4: 1359-1360, 1981. Typus: *Ritter* 1296 (U, ZSS [iso?]). [Holotype cited for U l.c. 1: iii, 1979.] [Peru: Huancavelica] [RPS 32]

Cleistocactus villamontesii Cardenas, Cact. Succ. J. (US) 33: 76, 1961. [RPS 12]

Cleistocactus villamontesii var. *longiflorior* Backeberg, Kakt.-Lex. 90, 1966. *Nom. inval.* (Art. 37.1). [Erroneously included as valid in RPS 17.] [RPS 17]

Cleistocactus viridialabastri Cardenas, Cact. Succ. J. (US) 35: 201, 1963. [RPS 14]

Cleistocactus viridiflorus Backeberg, Descr. Cact. Nov. 3: 5, 1963. Typus: *Ritter* 323 (ZSS [type number]). [Based on a living plant (cult. St. Pie) of Ritter 323; believed to be validly published as Ritter had previously deposited herbarium material.] [Bolivia: Muñecas] [RPS 14]

Cleistocactus vulpis-cauda Ritter & Cullmann, Kakt. and. Sukk. 13(3): 38-40, ills., 1962. Typus: *Ritter* 847 (ZSS, ZSS). [Holotype cited for ZSS, but not found; several isotypes present.] [Bolivia: Tomina] [RPS 13]

Cleistocactus wendlandiorum Backeberg, Kakt. and. Sukk. 6(2): 113-117, ills., 1955. [Based on cultivated material, no type indicated.] [RPS 6]

×**Cleistopsis** Strigl, Kakt. and. Sukk. 30(9): 226-227, 1979. [= *Cleistocactus* × *Echinopsis*.] [RPS 30]

Clistanthocereus calviflorus (Ritter) Backeberg, Kakt.-Lex., 91, 1966. Basionym: *Borzicactus calviflorus*. [RPS 17]

Clistanthocereus samnensis (Ritter) Backeberg, Kakt.-Lex., 91, 1966. Basionym: *Borzicactus samnensis*. [RPS 17]

Clistanthocereus tessellatus (Akers & Buining) Backeberg, Die Cact. 2: 939, 1959. Basionym: *Borzicactus tessellatus*. [RPS 10]

Cochiseia Earle, Saguaroland Bull. 30: 61, 64-66, 72, 1976. Typus: *Cochiseia robbinsorum* Earle. [RPS 27]

Cochiseia robbinsorum Earle, Saguaroland Bull. 30: 61, 64-66, 1976. Typus: *Robbins s.n.* (ARIZ). [USA: Arizona] [RPS 27]

Coleocephalocereus subgen. **Buiningia** (Buxbaum) P. J. Braun, Bradleya 6: 92, 1988. Basionym: *Buiningia*. [RPS 39]

Coleocephalocereus subgen. **Lagenopsis** Buxbaum, in Krainz, Die Kakt. Lief. 48/49: C IVb, 1972. Typus: *Cereus luetzelburgii*. [RPS 23]

Coleocephalocereus albicephalus (Buining & Brederoo) F. Brandt, Kakt. Orch.-Rundschau 6(5): 124, 1981. Basionym: *Austrocephalocereus albicephalus*. [RPS 32]

Coleocephalocereus aureispinus Buining & Brederoo, Kakt. and. Sukk. 25(4): 73-75, ills., 1974. Typus: *Horst et Uebelmann* HU 391 (U, ZSS). [Brazil: Bahia] [RPS 25]

Coleocephalocereus aureus Ritter, Kakt. and. Sukk. 19(8): 158-160, ill., 1968. Typus: *Ritter* 1341 (U, ZSS [seeds only]). [Brazil: Minas Gerais] [RPS 19]

Coleocephalocereus braunii Diers & E. Esteves Pereira, Kakt. and. Sukk. 36(2): 28-35, ills., 1985. Typus: *P. J. Braun* 470 (KOELN [Succulentarium], ZSS). [Brazil] [RPS 36]

Coleocephalocereus brevicylindricus (Buining) Ritter, Kakt. Südamer. 1: 122, 1979. Basionym: *Buiningia brevicylindrica*. [RPS 30]

Coleocephalocereus brevicylindricus var. **elongatus** (Buining) Ritter, Kakt. Südamer. 1: 122, 1979. Basionym: *Buiningia brevicylindrica* var. *elongata*. [RPS 30]

Coleocephalocereus brevicylindricus var. **longispinus** (Buining) Ritter, Kakt. Südamer. 1: 122, 1979. Basionym: *Buiningia brevicylindrica* var. *longispina*. [RPS 30]

Coleocephalocereus buxbaumianus Buining, Succulenta 53: 28-33, 1974. [RPS 25]

Coleocephalocereus decumbens Ritter, Kakt. and. Sukk. 19: 160-161, 1968. [RPS 19]

Coleocephalocereus diersianus P. J. Braun & E. Esteves Pereira, Kakt. and. Sukk. 39(3): 48-53, ills., SEM-ills., 1988. Typus: *Horst, Braun et Esteves* 3 (KOELN [Succulentarium]). [Brazil: Minas Gerais] [RPS 39]

Coleocephalocereus dybowskii (Roland-Gosselin) F. Brandt, Kakt. Orch.-Rundschau 6(5): 124, 1981. Basionym: *Cereus dybowskii*. [RPS 32]

Coleocephalocereus elongatus (Buining) P. J. Braun, Bradleya 6: 92, 1988. Basionym: *Buiningia brevicylindrica* var. *elongata*. [RPS 39]

Coleocephalocereus estevesii Diers, Kakt. and. Sukk. 29(9): 201-205, 1978. [RPS 29]

Coleocephalocereus flavisetus Ritter, Kakt. Südamer. 1: 127, 1979. [RPS 30]

Coleocephalocereus fluminensis var. **braamhaarii** P. J. Braun, Kakt. and. Sukk. 33(6): 118-121, ills., 1982. Typus: *Braamhaar* 1/1980 (KOELN [Succulentarium]). [Brazil] [RPS 33]

Coleocephalocereus goebelianus (Vaupel) Buining, Kakt. and. Sukk. 21: 202-206, 1970. [First mentioned invalidly (Art. 33.2) by Backeberg (and ascribed to Ritter) in Kakt.-Lex., 92, 1966.] [RPS 21]

Coleocephalocereus lehmannianus (Werdermann) F. Brandt, Kakt. Orch.-Rundschau 6(5): 124, 1981. *Nom. inval.* (Art. 33.2), based on *Cephalocereus lehmannianus*. [RPS 32]

Coleocephalocereus luetzelburgii (Vaupel) Buxbaum, in Krainz, Die Kakt. Lief. 48-49: C IVb, 1972. Basionym: *Cereus luetzelburgii*. [RPS 23]

Coleocephalocereus minensis (Werdermann) F. Brandt, Succulenta 60(5): 117-118, 1981. Basionym: *Cereus minensis*. [RPS 32]

Coleocephalocereus pachystele Ritter, Kakt. and. Sukk. 19: 157, 1968. [*Spec. nov. pro Cephalocereus purpureus* Werdermann (1933), *non* Gürke (1908).] [RPS 19]

Coleocephalocereus paulensis Ritter, Kakt. and. Sukk. 19: 161-162, 1968. [RPS 19]

Coleocephalocereus pleurocarpus (Ritter) F. Brandt, Succulenta 60(5): 117, 1981. Basionym: *Cipocereus pleurocarpus*. [RPS 32]

Coleocephalocereus pluricostatus Buining & Brederoo, in Krainz, Die Kakt. Lief. 46/47: C IVb, 4 pp. of text, ills., 1971. Typus: *Horst et Uebelmann* HU 245 (1968) (ZSS, ZSS). [Brazil: Minas Gerais] [RPS 22]

Coleocephalocereus purpureus (Buining & Brederoo) Ritter, Kakt. Südamer. 1: 128, 1979. Basionym: *Buiningia purpurea*. [RPS 30]

Copiapoa subgen. **Pilocopiapoa** (Ritter) Ritter, Kakt. Südamer. 3: 1046-1053, 1981. Basionym: *Pilocopiapoa*. [RPS 31]

Copiapoa alticostata Ritter, Taxon 12(1): 29, 1963. [RPS 14]

Copiapoa applanata Backeberg, Die Cact. 3: 1913, 1959. *Nom. inval.* (Art. 37.1). [Erroneously included as valid in RPS 10.] [RPS 10]

Copiapoa atacamensis Middleditch, Chileans 11(37): 20-21, 1980. Based on *Rose* 19410. Incorrect name (Art. 57.1). [Chile] [RPS 31]

Copiapoa atacamensis var. **calderana** (Ritter) A. E. Hoffmann, Cact. Fl. Chil., 98-99, ill., 1989. Basionym: *Copiapoa calderana*. [RPS 40]

Copiapoa barquitensis Ritter, Cactus (Paris) 15(66): 19, 1960. *Nom. inval.* (Art. 36.1, 37.1). [RPS -]

Copiapoa boliviana (Pfeiffer) Ritter, Kakt. Südamer. 3: 1089-1090, 1980. Basionym: *Echinocactus bolivianus*. [RPS 31]

Copiapoa bridgesii (Pfeiffer) Backeberg, Die Cact. 3: 1909, 1959. Basionym: *Echinocactus bridgesii*. [RPS 10]

Copiapoa brunnescens Backeberg, Die Cact. 3: 1901, 1959. *Nom. inval.* (Art. 36.1). [Published as provisional name.] [RPS 10]

Copiapoa calderana Ritter, Cactus (Paris) No. 65: 197, 1959. [RPS 10]

Copiapoa calderana var. **spinosior** Ritter, Kakt. Südamer. 3: 1082, 1980. [RPS 31]

Copiapoa carrizalensis Ritter, Cactus (Paris) No. 63: 139, 1959. [RPS 10]

Copiapoa carrizalensis var. **gigantea** Ritter, Taxon 12(1): 29, 1963. [RPS 14]

Copiapoa castanea Ritter, in Backeberg, Die Cact. 6: 3820, 1962. *Nom. inval.* (Art. 36.1/37.1). [RPS 13]

Copiapoa chanaralensis Ritter, in Backeberg, Die Cact. 6: 3820, 1962. *Nom. inval.* (Art. 36.1, 37.1). [Later validly described by Ritter as *C. chaniaralensis*.] [RPS 13]

Copiapoa chaniaralensis Ritter, Kakt. Südamer. 3: 1063-1064, 1980. [First mentioned as *C. chanaralensis* (*nom. nud.*) in Backeberg, Die Cact. 6: 3820, 1962. Probably correctable to 'chanaralensis' under Art. 73.6.] [RPS 31]

Copiapoa cinerascens var. **grandiflora** (Ritter) A. E. Hoffmann, Cact. Fl. Chil., 100-101, ill., 1989. Basionym: *Copiapoa grandiflora*. [RPS 40]

Copiapoa cinerascens var. **intermedia** Ritter, Kakt. Südamer. 3: 1084, 1980. Typus: *Ritter* 216 (ZSS [status ?]). [Published as a new name for *Copiapoa applanata* Backeberg.] [Chile] [RPS 31]

Copiapoa cinerea var. **albispina** Ritter, Taxon 12: 30, 1963. [RPS 14]

Copiapoa cinerea var. **columna-alba** (Ritter) Backeberg, Die Cact. 6: 3820, 1962. Basionym: *Copiapoa columna-alba*. [RPS 13]

Copiapoa cinerea var. **dealbata** (Ritter) Backeberg, Die Cact. 6: 3823, 1962. Basionym: *Copiapoa dealbata*. [RPS 13]

Copiapoa cinerea var. **eremophila** (Ritter) A. E. Hoffmann, Cact. Fl. Chil., 106, 1989. Basionym: *Copiapoa eremophila*. [RPS 40]

Copiapoa cinerea var. **gigantea** (Backeberg) N. P. Taylor, Bradleya 5: 92, 1987. Basionym: *Copiapoa gigantea*. [RPS 38]

Copiapoa cinerea var. **haseltoniana** (Backeberg) N. P. Taylor, Cact. Succ. J. Gr. Brit. 43(2-3): 53, 1981. Basionym: *Copiapoa haseltoniana*. [RPS 32]

Copiapoa cinerea var. **tenebrosa** (Ritter) A. E. Hoffmann, Cact. Fl. Chil., 106, ill. (p. 107), 1989. Basionym: *Copiapoa tenebrosa*. [RPS 40]

Copiapoa columna-alba Ritter, Cactus (Paris) No. 65: 199, 1959. [RPS 10]

Copiapoa columna-alba var. **nuda** Ritter, Kakt. Südamer. 3: 1095, 1980. [RPS 31]

Copiapoa conglomerata (Philippi) Lembcke, Kakt. and. Sukk. 17: 29-31, 1966. Basionym: *Echinocactus conglomeratus*. [RPS 17]

Copiapoa coquimbana var. **alticostata** (Ritter)

A. E. Hoffmann, Cact. Fl. Chil., 108, 1989. Basionym: *Copiapoa alticostata.* [RPS 40]

Copiapoa coquimbana var. **armata** Ritter, Kakt. Südamer. 3: 1075, 1980. [RPS 31]

Copiapoa coquimbana var. **fiedleriana** (Schumann) A. E. Hoffmann, Cact. Fl. Chil., 108, ill. (p. 109), 1989. Basionym: *Echinocactus fiedlerianus.* [RPS 40]

Copiapoa coquimbana var. **pendulina** (Ritter) A. E. Hoffmann, Cact. Fl. Chil., 108, 1989. Basionym: *Copiapoa pendulina.* [RPS 40]

Copiapoa coquimbana var. **pseudocoquimbana** (Ritter) A. E. Hoffmann, Cact. Fl. Chil., 108, 1989. Basionym: *Copiapoa pseudocoquimbana.* [RPS 40]

Copiapoa coquimbana var. **vallenarensis** (Ritter) A. E. Hoffmann, Cact. Fl. Chil., 108, ill. (p. 111), 1989. Basionym: *Copiapoa vallenarensis.* [RPS 40]

Copiapoa coquimbana var. **wagenknechtii** Ritter, Taxon 12(1): 30, 1963. Typus: *Ritter* 718 (). [Chile] [RPS 14]

Copiapoa cuprea Ritter, Cactus (Paris) No. 63: 136, 1959. [RPS 10]

Copiapoa cupreata (Poselger) Backeberg, Die Cact. 3: 1920, 1959. Basionym: *Echinocactus cupreatus.* [RPS 10]

Copiapoa dealbata Ritter, Cactus (Paris) 14(63): 137-139, ills., 1959. Typus: *Ritter* 509 (ZSS, U, ZSS). [No herbarium has been cited for the holotype, and it is unclear, which specimen is the actual holotype.] [Chile] [RPS 10]

Copiapoa dealbata fa. **gigantea** (Ritter) A. E. Hoffmann, Cact. Fl. Chil., 112, ill. (p. 113), 1989. Basionym: *Copiapoa carrizalensis* var. *gigantea.* [RPS 40]

Copiapoa dealbata var. **carrizalensis** (Ritter) A. E. Hoffmann, Cact. Fl. Chil., 112, 1989. Basionym: *Copiapoa carrizalensis.* [RPS 40]

Copiapoa desertorum Ritter, Kakt. Südamer. 3: 1060, 1980. [First mentioned (as *nom. nud.*) in Backeberg, Die Cact. 6: 3826, 1962.] [RPS 31]

Copiapoa desertorum var. **hornilloensis** (Ritter) A. E. Hoffmann, Cact. Fl. Chil., 114, 1989. Basionym: *Copiapoa hornilloensis.* [RPS 40]

Copiapoa desertorum var. **rubriflora** (Ritter) A. E. Hoffmann, Cact. Fl. Chil., 114, 1989. Basionym: *Copiapoa rubriflora.* [RPS 40]

Copiapoa desertorum var. *rupestris* (Ritter) A. E. Hoffmann, Cact. Fl. Chil., 114, ill. (p. 115), 1989. Incorrect name (Art. 57), based on *Copiapoa rupestris.* [RPS 40]

Copiapoa dura Ritter, Taxon 12(1): 31, 1963. Typus: *Ritter* 546 (U, ZSS [seeds only]). [Holotype cited for U l.c. p. 28.] [First mentioned in Backeberg, Die Cact. 6: 3827, 1962 (as *nom. nud.*).] [Chile] [RPS 14]

Copiapoa echinata Ritter, Cactus (Paris) 14(63): 133-134, 1959. Typus: *Ritter* 506 (ZSS [not found], ZSS [seeds only]). [Chile] [RPS 10]

Copiapoa echinata fa. **pulla** Ritter, Kakt. Südamer. 3: 1081, 1980. [RPS 31]

Copiapoa echinata var. **borealis** Ritter, Cactus (Paris) 14(63): 133-134, ill., 1959. Typus: *Ritter* 506a (ZSS [not found], ZSS [seeds only]). [Chile] [RPS -]

Copiapoa echinoides var. **cuprea** (Ritter) A. E. Hoffmann, Cact. Fl. Chil., 116, ill. (p. 117), 1989. Basionym: *Copiapoa cuprea.* [RPS 40]

Copiapoa eremophila Ritter, Kakt. Südamer. 3: 1104-1105, 1980. Typus: *Ritter* 476 (U, ZSS [seeds only]). [Holotype cited for U l.c. 1: iii, 1979.] [First mentioned in Backeberg, Die Cact. 6: 3828, 1962 (as *nom. inval.*).] [Chile] [RPS 31]

Copiapoa esmeraldana Ritter, Kakt. Südamer. 3: 1064-1065, 1980. [RPS 31]

Copiapoa ferox Lembcke & Backeberg, in Backeberg, Die Cact. 3: 1922, 1959. *Nom. inval.* (Art. 37.1). [Erroneously included as valid in RPS 10.] [RPS 10]

Copiapoa gigantea var. **haseltoniana** (Backeberg) Ritter, Kakt. Südamer. 3: 1101-1102, 1980. Basionym: *Copiapoa haseltoniana.* [RPS 31]

Copiapoa grandiflora Ritter, Taxon 12(1): 30, 1963. Typus: *Ritter* 523 (U, ZSS [seeds only]). [First mentioned in Backeberg, Die Cact. 6: 3828, 1962 (as *nom. nud.*).] [Chile] [RPS 14]

Copiapoa haseltoniana Backeberg, Descr. Cact. Nov. [1:] 33, 1957. [RPS 7]

Copiapoa hornilloensis Ritter, Kakt. Südamer. 3: 1060, 1980. [RPS 31]

Copiapoa humilis (Philippi) P. C. Hutchison, Cact. Succ. J. (US) 25: 34, 1953. Basionym: *Echinocactus humilis.* [RPS 4]

Copiapoa humilis var. **esmeraldana** (Ritter) A. E. Hoffmann, Cact. Fl. Chil., 118, ill. (p. 119), 1989. Basionym: *Copiapoa esmeraldana.* [RPS 40]

Copiapoa humilis var. **longispina** (Ritter) A. E. Hoffmann, Cact. Fl. Chil., 118, ill. (p. 119), 1989. Basionym: *Copiapoa longispina.* [RPS 40]

Copiapoa humilis var. *paposoensis* (Ritter) A. E. Hoffmann, Cact. Fl. Chil., 118, 1989. *Nom. inval.*, based on *Copiapoa paposoensis, nom. inval.* (Art. 37). [RPS 40]

Copiapoa humilis var. **taltalensis** (Werdermann) A. E. Hoffmann, Cact. Fl. Chil., 118, ill. (p. 119), 1989. Basionym: *Echinocactus taltalensis.* [RPS 40]

Copiapoa hypogaea Ritter, Cactus (Paris) 15(66): 19-20, 1960. Typus: *Ritter* 261 (ZSS). [Type cited for Z, later corrected to ZSS.] [Chile: Chañaral] [RPS 11]

Copiapoa hypogaea var. **barquitensis** Ritter, Kakt. Südamer. 3: 1086, 1980. Typus: *Ritter* 654 (U, ZSS [seeds only]). [Holotype cited for U l.c. 1: iii, 1979.] [Chile] [RPS 31]

Copiapoa hypogaea var. **laui** (Diers) A. E. Hoffmann, Cact. Fl. Chil., 120, ill. (p. 121), 1989. Basionym: *Copiapoa laui.* [Sphalm. 'lauii'.] [RPS 40]

Copiapoa intermedia Ritter, in Backeberg, Die Cact. 6: 3832, 1962. *Nom. inval.* (Art. 36.1/37.1). [RPS 13]

Copiapoa krainziana Ritter, Taxon 12(1): 30-31, 1963. Typus: *Ritter* 210 (ZSS, ZSS). [First mentioned in Backeberg, Die Cact. 6: 3834, 1962 (as *nom. nud.*).] [Chile: Taltal] [RPS 14]

Copiapoa krainziana var. **scopulina** Ritter, Taxon 12(1): 30-31, 1963. Typus: *Ritter* 209 (U, U, ZSS). [Chile: Taltal] [RPS 14]

Copiapoa laui Diers, Kakt. and. Sukk. 31(12): 362-365, ills., 1980. Typus: *Lau* 891 (KOELN [Succulentarium], ZSS [type number]). [Sphalm. 'lauii'.] [Chile: Antofagasta] [RPS 31]

Copiapoa lembckei Backeberg, Die Cact. 3: 1922, 1959. *Nom. inval.* (Art. 37.1). [Erroneously included as valid in RPS 10.] [RPS 10]

Copiapoa longispina Ritter, Taxon 12(1): 31, 1963. [RPS 14]

Copiapoa longistaminea Ritter, Taxon 12(1): 31-32, 1963. Typus: *Ritter* 531 (U, ZSS [seeds only]). [Chile] [RPS 14]

Copiapoa marginata var. **bridgesii** (Pfeiffer) A. E. Hoffmann, Cact. Fl. Chil., 124, ill. (p. 125), 1989. Basionym: *Echinocactus bridgesii*. [RPS 40]

Copiapoa megarhiza var. **echinata** (Ritter) A. E. Hoffmann, Cact. Fl. Chil., 126, ill. (p. 127), 1989. Basionym: *Copiapoa echinata*. [RPS 40]

Copiapoa megarhiza var. **microrhiza** Ritter, Kakt. Südamer. 3: 1081, 1980. [RPS 31]

Copiapoa melanohystrix Ritter, Kakt. Südamer. 3: 1096-1097, 1980. [RPS 31]

Copiapoa microsperma Ritter, Cactus (Paris) 15(66): 23, 1959. *Nom. inval.* (Art. 36.1, 37.1). [RPS -]

Copiapoa mollicula Ritter, Taxon 12(1): 30, 1963. Typus: *Ritter* 525 (U, ZSS [seeds only]). [Chile] [RPS 14]

Copiapoa montana Ritter, Cactus (Paris) 15(66): 21-22, 1960. Typus: *Ritter* 211a (). [Appears to based on 2 syntypes, *Ritter 211a* and *Ritter 522*, but '522' is a re-numbering for '211a' (cf. Kakt. Südamer. 3: 1088).] [Chile] [RPS 11]

Copiapoa olivana Ritter, Kakt. Südamer. 3: 1088, 1980. [RPS 31]

Copiapoa paposoensis Ritter, Kakt. Südamer. 3: 1068, 1980. *Nom. inval.* (Art. 37). [RPS 31]

Copiapoa pendulina Ritter, Cactus (Paris) 14(63): 134-135, ill., 1959. Typus: *Ritter* 504 (ZSS [not found], ZSS [seeds only]). [Chile] [RPS 10]

Copiapoa pepiniana var. **fiedleriana** (Schumann) Backeberg, Die Cact. 3: 1919, 1959. Basionym: *Echinocactus fiedlerianus*. [RPS 10]

Copiapoa pseudocoquimbana Ritter, Taxon 12(1): 30, 1963. Typus: *Ritter* 1086 (). [Chile: Dept. La Serena] [RPS 14]

Copiapoa pseudocoquimbana var. **chaniarensis** Ritter, Kakt. Südamer. 3: 1077, 1980. [RPS 31]

Copiapoa pseudocoquimbana var. **domeykoensis** Ritter, Kakt. Südamer. 3: 1077, 1980. [RPS 31]

Copiapoa pseudocoquimbana var. **vulgata** Ritter, Taxon 12(1): 30, 1963. Typus: *Ritter* 230 (U 145 255B, ZSS). [Chile] [RPS 14]

Copiapoa rarissima Ritter, Kakt. Südamer. 3: 1088-1089, 1980. [RPS 31]

Copiapoa rubriflora Ritter, Taxon 12(1): 31, 1963. Typus: *Ritter* 211 (?, ZSS). [Chile] [RPS 14]

Copiapoa rupestris Ritter, Taxon 12(1): 31, 1963. Typus: *Ritter* 528 (U, ZSS). [Chile] [RPS 14]

Copiapoa scopulina Ritter, in Kat. H. Winter, [unpaged], 1957. *Nom. inval.* (Art. 36.1, 37.1). [Published as provisional name; year of first usage not established.] [RPS -]

Copiapoa serpentisulcata Ritter, Cactus (Paris) 15(66): 22-23, 1960. Typus: *Ritter* 246 (ZSS, ZSS). [Type cited for Z, later corrected to ZSS.] [Chile] [RPS 11]

Copiapoa serpentisulcata var. **castanea** Ritter, Kakt. Südamer. 3: 1093-1094, 1980. [RPS 31]

Copiapoa solaris (Ritter) Ritter, Kakt. Südamer. 3: 1047, 1980. Basionym: *Pilocopiapoa solaris*. [RPS 31]

Copiapoa streptocaulon (Hooker) Ritter, Kakt. and. Sukk. 12: 4, 1961. Basionym: *Echinocactus streptocaulon*. [RPS 12]

Copiapoa subnuda Ritter, in Backeberg, Die Cact. 6: 3827, 1962. *Nom. inval.* (Art. 36.1/37.1). [RPS 13]

Copiapoa tenebrosa Ritter, Kakt. Südamer. 3: 1098-1099, 1980. [RPS 31]

Copiapoa tenuissima Ritter, Taxon 12(1): 31, 1963. *Nom. inval.* (Art. 37.1). [First mentioned in Backeberg, Die Cact. 6: 3840, 1962 (as *nom. nud.*). Based on 2 syntypes, *Ritter* 539 (U !) and *Ritter* 540. Erroneously included as valid in RPS 14.] [RPS 14]

Copiapoa tocopillana Ritter, Kakt. Südamer. 3: 1072-1073, 1980. [RPS 31]

Copiapoa totoralensis Ritter, Cactus (Paris) 15(66): 23-24, ill., 1960. Typus: *Ritter* 512 (ZSS). [Type erroneously cited for Z instead of ZSS.] [Chile] [RPS 11]

Copiapoa vallenarensis Ritter, Kakt. Südamer. 3: 1077-1078, 1980. [RPS 31]

Copiapoa varispinata Ritter, Kakt. Südamer. 3: 1070, 1980. [Sphalm. 'varispinata'] [RPS 31]

Copiapoa wagenknechtii Ritter, in Backeberg, Die Cact. 6: 3841, 1962. *Nom. inval.* (Art. 36.1/37.1). [RPS 13]

Corryocactinae Buxbaum, in Krainz, Die Kakt. C IV a, 1964. Typus: *Corryocactus*. [RPS 15]

Corryocactus acervatus Ritter, Kakt. Südamer. 4: 1287, 1981. Typus: *Ritter* 558 (U). [Peru: Arequipa] [RPS 32]

Corryocactus apiciflorus (Vaupel) P. C. Hutchison, Sukkulentenkunde 7/8: 9, 1963. Basionym: *Cereus apiciflorus*. [RPS 14]

Corryocactus aureus (Meyen) P. C. Hutchison, Sukkulentenkunde 7/8: 9, 1963. Basionym: *Cactus aureus*. [RPS 14]

Corryocactus ayacuchoensis Rauh & Backeberg, in Backeberg, Descr. Cact. Nov. [1:] 12, 1957. [Dated 1956, published 1957.] [RPS 7]

Corryocactus ayacuchoensis var. **leucacanthus** Rauh & Backeberg, in Backeberg, Descr. Cact. Nov. [1:] 12, 1957. [Dated 1956, published 1957.] [RPS 7]

Corryocactus ayopayanus Cardenas, Rev. Agric. Cochabamba 7(7): 21, 1952. [RPS 3]

Corryocactus brachycladus Ritter, Kakt.

Südamer. 4: 1289, 1981. Typus: *Ritter* 657 (U). [Peru: Ancash] [RPS 32]

Corryocactus brevispinus Rauh & Backeberg, in Backeberg, Descr. Cact. Nov. [1:] 12, 1957. [Dated 1956, published 1957.] [RPS 7]

Corryocactus brevistylus var. **puquiensis** (Backeberg) Ritter, Kakt. Südamer. 4: 1279, 1981. Basionym: *Corryocactus puquiensis.* [RPS 32]

Corryocactus chachapoyensis Ochoa & Backeberg, in Backeberg, Die Cact. 2: 859, 1959. *Nom. inval.*, based on *Cereus chachapoyensis, nom. inval.* (Art. 34.1, 37.1). [RPS 10]

Corryocactus charazanensis Cardenas, Cactus (Paris) No. 57: 247-249, 1957. [RPS 8]

Corryocactus chavinilloensis Ritter, Kakt. Südamer. 4: 1290, 1981. Typus: *Ritter* 689 (U). [Peru: Huanuco] [RPS 32]

Corryocactus cuajonesensis Ritter, Kakt. Südamer. 4: 1287, 1981. Typus: *Ritter* 638 (U). [Peru: Moquegua] [RPS 32]

Corryocactus erectus (Backeberg) Ritter, Kakt. Südamer. 4: 1281-1282, 1981. Basionym: *Erdisia erecta.* [RPS 32]

Corryocactus gracilis Ritter, Kakt. Südamer. 4: 1282-1283, 1981. Typus: *Ritter* 1299 (U). [Peru: Ayacucho] [RPS 32]

Corryocactus heteracanthus Backeberg, Descr. Cact. Nov. [1:] 12, 1957. [Dated 1956, published 1957.] [RPS 7]

Corryocactus huincoensis Ritter, Kakt. Südamer. 4: 1288-1289, 1981. Typus: *Ritter* 1070 (U). [Peru: Lima] [RPS 32]

Corryocactus krausii Backeberg, Descr. Cact. Nov. [1:] 12, 1957. [Dated 1956, published 1957.] [RPS 7]

Corryocactus matucanensis Ritter, Kakt. Südamer. 4: 1287-1288, 1981. Typus: *Ritter* 629 (U). [Peru: Lima] [RPS 32]

Corryocactus maximus (Backeberg) P. C. Hutchison, Sukkulentenkunde 7/8: 9, 1963. Basionym: *Erdisia maxima.* [RPS 14]

Corryocactus megarhizus Ritter, Kakt. Südamer. 4: 1286-1287, 1981. Typus: *Ritter* 559 (U). [Peru: Ayacucho] [RPS 32]

Corryocactus melaleucus Ritter, Kakt. Südamer. 4: 1289, 1981. Typus: *Ritter* 687 (U). [Peru: Lima] [RPS 32]

Corryocactus melanotrichus var. **caulescens** Cardenas, Rev. Agric. Cochabamba 7(7): 20, 1952. [RPS 3]

Corryocactus odoratus Ritter, Kakt. Südamer. 4: 1285-1286, 1981. Typus: *Ritter* 1301 (U). [Peru: Huancavelica] [RPS 32]

Corryocactus otuyensis Cardenas, Cactus (Paris) No. 78: 87, 1963. [RPS 14]

Corryocactus pachycladus Rauh & Backeberg, in Backeberg, Descr. Cact. Nov. [1:] 12, 1957. [Dated 1956, published 1957.] [RPS 7]

Corryocactus perezianus Cardenas, Rev. Agric. Cochabamba 7(7): 22, 1952. [RPS 3]

Corryocactus pilispinus Ritter, Kakt. Südamer. 4: 1291, 1981. Typus: *Ritter* 1071 (U). [Peru: La Libertad] [RPS 32]

Corryocactus prostratus Ritter, Kakt. Südamer. 4: 1283-1284, 1981. Typus: *Ritter* 180 (U, ZSS). [Holotype cited for U l.c. 1: iii, 1979.] [Peru: Arequipa] [RPS 32]

Corryocactus pulquinensis Cardenas, Nation. Cact. Succ. J. 12(4): 84, 1957. [RPS 8]

Corryocactus puquiensis Rauh & Backeberg, in Backeberg, Descr. Cact. Nov. [1:] 12, 1957. Typus: *Rauh* K48 (1954) (ZSS). [Dated 1956, published 1957.] [Peru] [RPS 7]

Corryocactus pyroporphyranthus Ritter, Kakt. Südamer. 4: 1284-1285, 1981. Typus: *Ritter* 1481 (U). [Peru] [RPS 32]

Corryocactus quadrangularis (Rauh & Backeberg) Ritter, Backebergs Descr. & Erört. taxon. Fragen, [unpaged], 1958. Basionym: *Erdisia quadrangularis.* [RPS 15]

Corryocactus quivillanus Ritter, Kakt. Südamer. 4: 1290-1291, 1981. Typus: *Ritter* 688 (U). [Peru: Huanuco] [RPS 32]

Corryocactus serpens Ritter, Kakt. Südamer. 4: 1285, 1981. Typus: *Ritter* 1302 (U). [Peru: Huancavelica] [RPS 32]

Corryocactus solitarius Ritter, Kakt. Südamer. 4: 1289-1290, 1981. Typus: *Ritter* 634 (U). [Peru: La Libertad] [RPS 32]

Corryocactus spiniflorus (Philippi) P. C. Hutchison, Sukkulentenkunde 7/8: 9, 1963. Basionym: *Opuntia spiniflora.* [RPS 14]

Corryocactus squarrosus (Vaupel) P. C. Hutchison, Sukkulentenkunde 7/8: 9, 1963. Basionym: *Cereus squarrosus.* [RPS 14]

Corryocactus tarijensis Cardenas, Rev. Agric. Cochabamba 7(7): 23, 1952. [RPS 3]

Corryocactus tenuiculus (Backeberg) P. C. Hutchison, Sukkulentenkunde 7/8: 9, 1963. Basionym: *Erdisia tenuicula.* [RPS 14]

Corynopuntia planibulbispina Backeberg, Die Cact. 6: 3603, 1962. *Nom. inval.* (Art. 9.5). [Erroneously included as valid in RPS 13.] [RPS 13]

Corynopuntia reflexispina (Wiggins & Rollason) Backeberg, Die Cact. 1: 365, 1958. Basionym: *Opuntia reflexispina.* [RPS 9]

Corynopuntia stanlyi var. **kunzei** (Rose) Backeberg, Die Cact. 1: 360, 1958. Basionym: *Opuntia kunzei.* [Validity of name questionable because the basionym of the binomial may be invalid under Art. 34.] [RPS 9]

Corynopuntia stanlyi var. **parishii** (Orcutt) Backeberg, Die Cact. 1: 361, 1958. Basionym: *Opuntia parishii.* [Validity of name questionable because the basionym of the binomial may be invalid under Art. 34.] [RPS 9]

Corynopuntia stanlyi var. **wrightiana** (Baxter) Backeberg, Die Cact. 1: 360, 1958. Basionym: *Grusonia wrightiana.* [Validity of name questionable because the basionym of the binomial may be invalid under Art. 34.] [RPS 9]

Corynopuntia vilis var. *bernhardinii* Hildmann *ex* Backeberg, Die Cact. 6: 3601, 1962. *Nom. inval.* (Art. 33.2, 37.1). [Erroneously included as valid in RPS 13. Given as new combination, but based on a nude name.] [RPS 13]

Coryphantha sect. **Escobaria** (Britton & Rose) H. E. Moore, Baileya 20: 29, 1976.

Basionym: *Escobaria*. [RPS 27]

Coryphantha sect. *Euescobaria* (Buxbaum) Moran, Gentes Herbar. 8(4): 318, 1953. *Nom. inval*. (Art. 21.3). [RPS 4]

Coryphantha sect. **Lepidocoryphantha** (Backeberg) Moran, Gentes Herbar. 8(4): 318, 1953. Basionym: *Lepidocoryphantha*. [RPS 4]

Coryphantha sect. **Pseudocoryphantha** (Buxbaum) Moran, Gentes Herbar. 8(4): 318, 1953. Basionym: *Escobaria* subgen. *Pseudocoryphantha*. [RPS 4]

Coryphantha albicolumnaria (Hester) D. Zimmerman, Cact. Succ. J. (US) 44: 157-158, 1972. Basionym: *Escobaria albicolumnaria*. [Treated as invalid by N. Taylor, Cact. Succ. J. Gr. Brit. 40: 35, 1978, as the exact page reference of the basionym is not cited. This is here treated as bibliographical error.] [RPS 23]

Coryphantha alversonii var. *exaltissima* Wiegand & Backeberg, Die Cact. 5: 3001, 1961. *Nom. inval*. (Art. 37.1). [Erroneously included as valid in RPS 12.] [RPS 12]

Coryphantha bernalensis Bremer, Cact. Succ. J. (US) 56(4): 165-166, ills., 1984. *Bremer* 1076-8 (ASU). [RPS 35]

Coryphantha bussleri (Mundt) Scheinvar, Phytologia 49(4): 313, 1981. Basionym: *Mammillaria bussleri*. [RPS 32]

Coryphantha calipensis Bravo, Cact. Suc. Mex. 9: 79-80, 1964. [RPS 15]

Coryphantha chlorantha var. **deserti** (Engelmann) Backeberg, Die Cact. 5: 3003, 1961. Basionym: *Mammillaria deserti*. [RPS 12]

Coryphantha clava var. **schlechtendalii** (Ehrenberg) Backeberg, Die Cact. 5: 3040, 1961. Basionym: *Mammillaria schlechtendalii*. [RPS 12]

Coryphantha clavata var. **ancistracantha** (Lemaire) Backeberg, Die Cact. 5: 2995, 1961. Basionym: *Mammillaria ancistracantha*. [RPS 12]

Coryphantha clavata var. **radicantissima** (Quehl) Heinrich, In Backeberg, Die Cact. 5: 2994-2995, 1961. Basionym: *Mammillaria radicantissima*. [RPS 12]

Coryphantha cornifera var. **echinus** (Engelmann) L. Benson, Cact. Succ. J. (US) 41: 189, 1969. Basionym: *Mammillaria echinus*. [RPS 20]

Coryphantha cuencamensis Bremer, Cact. Succ. J. (US) 52(4): 183-184, 1980. [RPS 31]

Coryphantha dasyacantha var. **varicolor** (Tiegel) L. Benson, Cact. Succ. J. (US) 41: 189, 1969. Basionym: *Coryphantha varicolor*. [RPS 20]

Coryphantha delicata Bremer, Cact. Succ. J. (US): 51(2): 76-77, 1979. [RPS 30]

Coryphantha duncanii (Hester) L. Benson, Cact. Succ. J. (US) 41: 189, 1969. Basionym: *Escobesseya duncanii*. [RPS 20]

Coryphantha elephantidens var. **barciae** Bremer, Cact. Suc. Mex. 22(3): 64-66, 1977. [RPS 28]

Coryphantha elephantidens var. *roseiflora* Y. Ito, The Cactaceae, 545, ill., 1981. *Nom. inval*. (Art. 37.1). [RPS -]

Coryphantha garessii Bremer, Cact. Succ. J. (US) 52(2): 82-83, 1980. [RPS 31]

Coryphantha gracilis Bremer & Lau, Cact. Succ. J. (US) 49(2): 71-73, ills., 1977. Typus: *Lau* 645 (1972) (UNM, ZSS). [Mexico: Chihuahua] [RPS 28]

Coryphantha grandis Bremer, Cact. Succ. J. (US): 50(3): 134-135, 1978. [RPS 29]

Coryphantha grata Bremer, Cact. Succ. J. (US) 53(6): 276-277, 1981. [RPS 32]

Coryphantha greenwoodii Bravo, Cact. Suc. Mex. 15: 27-29, 45-46, 1970. [RPS 21]

Coryphantha henricksonii (Glass & Foster) Glass & Foster, Cact. Succ. J. (US): 51(3): 125, 1979. Basionym: *Escobaria henricksonii*. [RPS 30]

Coryphantha indensis Bremer, Cact. Suc. Mex. 22(4): 73, 75-77, 1977. [RPS 28]

Coryphantha jalpanensis Buchenau, Cact. Suc. Mex. 10: 25, 36-39, 48, 1965. [RPS 16]

Coryphantha laredoi Glass & Foster, Cact. Succ. J. (US): 50(5): 235-236, 1978. Typus: *Glass et Foster* 3761 (POM). [Mexico: Coahuila] [RPS 29]

Coryphantha laui Bremer, Cact. Succ. J. (US) 51: 278-279, 1979. Typus: *Bremer* 476-3 (ASU). [Mexico: Coahuila] [RPS 30]

Coryphantha macromeris var. **runyonii** (Britton & Rose) L. Benson, Cact. Succ. J. (US) 41: 188, 1969. Basionym: *Coryphantha runyonii*. [RPS 20]

Coryphantha maliterrarum Bremer, Cact. Succ. J. (US) 56(2): 71-72, ills., 1984. Typus: *Bremer* 1076-7 (ASU). [RPS 35]

Coryphantha melleospina Bravo, Anales Inst. Biol. UNAM 25(1-2): 525-526, ill., 1954. Typus: *Bravo Hollis* s.n. (MEXU). [Mexico: Oaxaca] [RPS 6]

Coryphantha missouriensis var. **caespitosa** (Engelmann) L. Benson, Cact. Succ. J. (US) 41: 189, 1969. Basionym: *Mammillaria similis* var. *caespitosa*. [RPS 20]

Coryphantha missouriensis var. **marstonii** (Clover) L. Benson, Cacti of Arizona, ed. 3, 26, 204, 1969. Basionym: *Coryphantha marstonii*. [RPS 20]

Coryphantha missouriensis var. **robustior** (Engelmann) L. Benson, Cact. Succ. J. (US) 41: 190, 1969. Basionym: *Mammillaria similis* var. *robustior*. [RPS 20]

Coryphantha muehlenpfordtii var. **robustispina** (Schott) W. T. Marshall, Arizona's Cactuses, ed. 2, 94, 1953. Basionym: *Mammillaria robustispina*. [RPS 4]

Coryphantha neglecta Bremer, Cact. Suc. Mex. 24(1): 3-5, 1979. [RPS 30]

Coryphantha neoscheeri Backeberg, In Backeberg, Die Cact. 5: 3051, 1961. [*Nom. nov. pro Mammillaria scheeri* Mühlenpfordt 1847 *non* Mühlenpfordt 1845. Validity and legitimacy not assessed.] [RPS 12]

Coryphantha orcuttii (Bödeker) D. A. Zimmerman, Cact. Succ. J. (US) 44: 156, 1972. Basionym: *Escobaria orcuttii*. [RPS 23]

Coryphantha organensis D. A. Zimmerman, Cact. Succ. J. (US) 44: 114-117, ills., 1972. Typus: *Zimmerman* 1535 (SNM). [USA:

New Mexico] [RPS 23]

Coryphantha ottoniana (Pfeiffer) Y. Ito, The Cactaceae, 553, 1981. *Nom. inval.* (Art. 33.2). [Based on *Echinocactus ottonianus* Poselger 1853 according to Ito.] [RPS -]

Coryphantha poselgeriana var. **saltillensis** (Poselger) Bremer, Cact. Suc. Mex. 22(1): 16, 1977. Basionym: *Echinocactus saltillensis.* [RPS 28]

Coryphantha poselgeriana var. **valida** Heinrich, in Backeberg, Die Cact. 5: 3050, 1961. Basionym: *Mammillaria valida.* [*Nom. nov. pro Mammillaria valida* J. A. Purpus 1911 (*non M. valida* Engelmann).] [RPS 12]

Coryphantha potosiana (Jacobi) Glass & Foster, Cact. Succ. J. (US) 43: 7, 1971. Basionym: *Mammillaria potosiana.* [First mentioned (without basionym) in l.c. 42: 265, 1970.] [RPS 21]

Coryphantha pseudoradians Bravo, Anales Inst. Biol. UNAM 25(1-2): 527-528, ill., 1954. Typus: *Bravo Hollis* s.n. (MEXU). [Mexico: Oaxaca] [RPS 6]

Coryphantha pulleineana (Backeberg) Glass, Cact. Suc. Mex. 13: 34-35, 42-43, 1968. Basionym: *Neolloydia pulleineana.* [RPS 19]

Coryphantha pusilliflora Bremer, Cact. Succ. J. (US) 54(3): 133-134, 144, ills., 1982. Typus: *Bremer 477-2* (ARIZ). [RPS 33]

Coryphantha radians var. *echina* (Schumann) Y. Ito, Cacti, 115, 1952. *Nom. inval.* (Art. 33.2). [RPS 4]

Coryphantha radians var. **impexicoma** (Schumann) Y. Ito, Cacti, 115, 1952. [Given as *comb. nud.* in RPS 4.] [RPS 4]

Coryphantha radians var. **pectinoides** (J. Coulter) Bravo, Cact. Suc. Mex. 27(1): 17, 1982. Basionym: *Cactus radians* var. *pectinoides.* [RPS 33]

Coryphantha radians var. **pseudoradians** (Bravo) Bravo, Cact. Suc. Mex. 27(1): 17, 1982. Basionym: *Coryphantha pseudoradians.* [RPS 33]

Coryphantha radians var. **sulcata** (Coulter) Y. Ito, Cacti, 115, 1952. [Given as *comb. nud.* in RPS 4.] [RPS 4]

Coryphantha recurvispina (De Vriese) Bremer, Cact. Suc. Mex. 21(1): 12, 1976. Basionym: *Mammillaria recurvispina.* [RPS 27]

Coryphantha retusa var. **melleospina** (Bravo) Bravo, Cact. Suc. Mex. 27(1): 17, 1982. Basionym: *Coryphantha melleospina.* [RPS 33]

Coryphantha retusa var. *pallidispina* Backeberg, In Backeberg, Die Cact. 6: 3874, 1962. *Nom. inval.* (Art. 37.1). [Erroneously included as valid in RPS 13.] [RPS 13]

Coryphantha rhaphidacantha var. **ancistracantha** (Schumann) Y. Ito, Cacti, 116, 1952. [Given as *comb. nud.* in RPS 4.] [RPS 4]

Coryphantha robbinsorum (Earle) A. Zimmerman, Cact. Succ. J. (US): 50(6): 294, 1978. Basionym: *Cochiseia robbinsorum.* [RPS 29]

Coryphantha roseana (Bödeker) Moran, Gentes Herbar. 8(4): 318, 1953. Basionym: *Echinocactus roseanus.* [RPS 4]

Coryphantha salm-dyckiana var. **brunnea** (Salm-Dyck) Unger, Kakt. and. Sukk. 37(5): 86, 1986. Basionym: *Mammillaria salm-dyckiana* var. *brunnea.* [RPS 37]

Coryphantha scheeri (Kuntze) L. Benson, Cact. Succ. J. (US) 41: 234, 1969. Basionym: *Mammillaria scheeri.* [RPS 20]

Coryphantha scheeri var. **robustispina** (Schott) L. Benson, Cacti of Arizona, ed. 3, 25, 195, 1969. Basionym: *Mammillaria robustispina.* [RPS 20]

Coryphantha scheeri var. **uncinata** L. Benson, Cact. Succ. J. (US) 41: 234, 1969. [RPS 20]

Coryphantha scheeri var. **valida** (Engelmann) L. Benson, The Cacti of Arizona ed. 3, 25, 195, 1969. Basionym: *Mammillaria scheeri* var. *valida.* [RPS 20]

Coryphantha sneedii var. **leei** (Rose *ex* Bödeker) L. Benson, Cact. Succ. J. (US) 41: 189, 1969. Basionym: *Escobaria leei.* [RPS 20]

Coryphantha strobiliformis (Poselger) Moran, Gentes Herbar. 8(4): 318, 1953. Basionym: *Echinocactus strobiliformis.* [Author citation sometimes given as 'Orcutt'.] [RPS 4]

Coryphantha strobiliformis var. **durispina** (Quehl) L. Benson, Cact. Succ. J. (US) 41: 189, 1969. Basionym: *Mammillaria strobiliformis.* [RPS 20]

Coryphantha strobiliformis var. **orcuttii** (Bödeker) L. Benson, Cacti of Arizona, ed. 3, 26, 204, 1969. Basionym: *Escobaria orcuttii.* [The name given as basionym is incorrect, and is here corrected.] [RPS 20]

Coryphantha sulcata var. **nickelsiae** (Brandegee) L. Benson, Cact. Succ. J. (US) 41: 188, 1969. Basionym: *Mammillaria nickelsiae.* [RPS 20]

Coryphantha tripugionacantha Lau, Cact. Suc. Mex. 33(1): 1, 20-24, ills., SEM-ills., 1988. Typus: *Lau* 1464 (MEXU). [The type is cited as 'Lau 1469' but according to the list of field numbers, this applies to a *Nyctocereus*, whereas Lau 1464 has the correct data associated with the new taxon.] [Mexico: Zacatecas] [RPS 39]

Coryphantha valida (Purpus) Bremer, Cact. Suc. Mex. 22(1): 16-17, 1977. Basionym: *Mammillaria valida.* [RPS 28]

Coryphantha vivipara fa. **sonorensis** P. Fischer, Cact. Succ. J. (US) 52(1): 27-28, 1980. [RPS 31]

Coryphantha vivipara var. **aggregata** (Engelmann) W. T. Marshall, Arizona's Cact., 93-94, 1950. Basionym: *Mammillaria aggregata.* [= Desert Bot. Gard. Ariz. Sci. Bull. 1: 93, 1950.] [RPS 2]

Coryphantha vivipara var. **alversonii** (Coulter) L. Benson, Cacti of Arizona, ed. 3, 26, 200, 1969. Basionym: *Cactus radiosus* var. *alversonii.* [RPS 20]

Coryphantha vivipara var. **arizonica** (Engelmann) W. T. Marshall, Arizona's Cactuses, 94, 1950. Basionym: *Mammillaria arizonica.* [RPS 2]

Coryphantha vivipara var. **bisbeeana** (Orcutt) L. Benson, Cacti of Arizona, ed. 3, 25, 197,

1969. Basionym: *Coryphantha bisbeeana*. [RPS 20]

Coryphantha vivipara var. **buoflama** P. Fischer, Cact. Succ. J. (US) 52(1): 27-28, 1980. [RPS 31]

Coryphantha vivipara var. **deserti** (Engelmann) W. T. Marshall, Arizona's Cactuses, 94, 1950. Basionym: *Mammillaria deserti*. [RPS 2]

Coryphantha vivipara var. **kaibabensis** P. Fischer, Cact. Succ. J. (US) 51(6): 286-287, 1979. [RPS 30]

Coryphantha vivipara var. **neomexicana** (Engelmann) Backeberg, Die Cact. 5: 2999, 1961. Basionym: *Mammillaria vivipara* var. *neomexicana*. [Sphalm. 'neo-mexicana'. Based on *Mammillaria vivipara radiosa neo-mexicana* Engelmann (validity of basionym not assessed).] [RPS 12]

Coryphantha vivipara var. **radiosa** (Engelmann) Backeberg, Die Cact. 5: 2998, 1961. Basionym: *Mammillaria radiosa*. [RPS 12]

Coryphantha vivipara var. **rosea** (Clokey) L. Benson, Cacti of Arizona, ed. 3, 26, 200, 1969. Basionym: *Coryphantha rosea*. [RPS 20]

Coryphantha wohlschlageri Holzeis, Kakt. and. Sukk. 41(3): 50-52, ills., 1990. Typus: *Wohlschlager* 223 (WU). [Mexico: San Luis Potosi] [RPS 41]

Coryphanthae Buxbaum, Österr. Bot. Zeitschr. 98(1/2): 94-95, 1951. *Nom. inval.* (Art. 36.1). [Published as 'linea'.] [RPS -]

× **Cosmopsis** Y. Ito, Shaboten 94: 23, 1977. [= *Cosmantha* (auct. ?) × *Echinopsis*.] [RPS 28]

× **Cosmopsis** cv. **Shitanryu** Y. Ito, Shaboten 94: 23, 1977. [= *Cosmantha grandiflora* × *Echinopsis kermesina*; also spelled 'Shitanruyu'.] [RPS 28]

Cryptocereus Alexander, Cact. Succ. J. (US) 22(6): 164-165, 1950. Typus: *C. anthonyanus*. [RPS 1]

Cryptocereus anthonyanus Alexander, Cact. Succ. J. (US) 22(6): 163-166, ill., 1950. Typus: *MacDougall* s.n. (NY, ZSS). [Mexico: Chiapas] [RPS 1]

Cryptocereus imitans (Kimnach & P. C. Hutchison) Backeberg, Die Cact. 2: 734, 1959. Basionym: *Werckleocereus imitans*. [RPS 10]

Cryptocereus rosei (Kimnach) Backeberg, Descr. Cact. Nov. 3: 5, 1963. Basionym: *Eccremocactus rosei*. [Repeated in Kakt.-Lex., 109, 1966.] [RPS 14]

Cullmannia C. Distefano, Kakt. and. Sukk. 7(1): 8-9, 1956. Typus: *Cullmannia viperina*. [RPS 7]

Cullmannia viperina (F. A. C. Weber) C. Distefano, Kakt. and. Sukk. 7(1): 9-10, 1956. Basionym: *Cereus viperinus*. [RPS 7]

Cumarinia (Knuth) Buxbaum, Österr. Bot. Zeitschr. 98(1-2): 60-61, 1951. Basionym: *Coryphantha* subgen. *Cumarinia*. [RPS 2]

Cumarinia odorata (Bödeker) Hort., Österr. Bot. Zeitschr. 98(1/2): 87, sub fig. 20, 1951. Basionym: *Coryphantha odorata*. [Citation frequently given as l.c., 61.] [RPS -]

Cumulopuntia Ritter, Kakt. Südamer. 2: 399-400, 1980. Typus: *Opuntia ignescens*. [RPS 31]

Cumulopuntia alboareolata (Ritter *ex* Backeberg) Ritter, Kakt. Südamer. 4: 1249-1250, 1981. Basionym: *Tephrocactus alboareolatus*. [RPS 32]

Cumulopuntia berteri (Colla) Ritter, Kakt. Südamer. 3: 885, 1980. [RPS 31]

Cumulopuntia boliviana (Salm-Dyck) Ritter, Kakt. Südamer. 2: 492-493, 1980. Basionym: *Opuntia boliviana*. [RPS 31]

Cumulopuntia crassicylindrica (Backeberg) Ritter, Kakt. Südamer. 4: 1254, 1981. Basionym: *Tephrocactus crassicylindricus*. [RPS 32]

Cumulopuntia echinacea (Ritter) Ritter, Kakt. Südamer. 3: 884, 1980. Basionym: *Tephrocactus echinaceus*. [RPS 31]

Cumulopuntia famatinensis Ritter, Kakt. Südamer. 2: 400-401, 1980. [RPS 31]

Cumulopuntia frigida Ritter, Kakt. Südamer. 2: 493-494, 1980. [RPS 31]

Cumulopuntia galerasensis Ritter, Kakt. Südamer. 4: 1249, 1981. Typus: *Ritter* 1045 (U). [Peru: Ayacucho] [RPS 32]

Cumulopuntia hystrix Ritter, Kakt. Südamer. 3: 883-884, 1980. [RPS 31]

Cumulopuntia ignescens (Vaupel) Ritter, Kakt. Südamer. 3: 880-882, 1980. Basionym: *Opuntia ignescens*. [RPS 31]

Cumulopuntia ignota (Britton & Rose) Ritter, Kakt. Südamer. 4: 1250-1251, 1981. Basionym: *Opuntia ignota*. [RPS 32]

Cumulopuntia kuehnrichiana (Werdermann & Backeberg) Ritter, Kakt. Südamer. 4: 1253-1254, 1981. *Nom. inval.* (Art. 33.2), based on *Opuntia kuehnrichiana*. [RPS 32]

Cumulopuntia multiareolata (Ritter) Ritter, Kakt. Südamer. 4: 1252-1253, 1981. Basionym: *Tephrocactus multiareolatus*. [RPS 32]

Cumulopuntia pampana Ritter, Kakt. Südamer. 2: 402-403, 1980. [RPS 31]

Cumulopuntia pentlandii (Salm-Dyck) Ritter, Kakt. Südamer. 2: 488-490, 1980. Basionym: *Opuntia pentlandii*. [RPS 31]

Cumulopuntia pentlandii var. **dactylifera** (Vaupel) Ritter, Kakt. Südamer. 2: 490-491, 1980. Basionym: *Opuntia dactylifera*. [RPS 31]

Cumulopuntia pyrrhacantha (Schumann) Ritter, Kakt. Südamer. 4: 1249, 1981. Basionym: *Opuntia pyrrhacantha*. [RPS 32]

Cumulopuntia rauppiana (Schumann) Ritter, Kakt. Südamer. 4: 1252, 1981. Basionym: *Opuntia rauppiana*. [RPS 32]

Cumulopuntia rossiana (Heinrich & Backeberg) Ritter, Kakt. Südamer. 2: 486-488, 1980. Basionym: *Tephrocactus pentlandii* var. *rossianus*. [RPS 31]

Cumulopuntia subterranea (R. Fries) Ritter, Kakt. Südamer. 2: 401-402, 1980. Basionym: *Opuntia subterranea*. [RPS 31]

Cumulopuntia ticnamarensis Ritter, Kakt. Südamer. 3: 885, 1980. [RPS 31]

Cumulopuntia tortispina Ritter, Kakt. Südamer. 3: 882-883, 1980. [RPS 31]

Cumulopuntia tubercularis Ritter, Kakt. Südamer. 3: 888, 1980. [RPS 31]

Cumulopuntia tumida Ritter, Kakt. Südamer. 4: 1254-1255, 1981. Typus: *Ritter* 1324 (U). [Peru: Arequipa] [RPS 32]

Cumulopuntia unguispina (Backeberg) Ritter, Kakt. Südamer. 4: 1251, 1981. Basionym: *Tephrocactus unguispinus*. [RPS 32]

Cumulopuntia unguispina var. **major** Ritter, Kakt. Südamer. 4: 1251-1252, 1981. Typus: *Ritter* 1077 (U). [Peru: Arequipa] [RPS 32]

× **Cylindrantha** Ito, The Cactaceae, 643, 1981. [= *Cylindrolobivia* × *Cosmantha*. The validity of the name depends on the status of the generic names given as parents.] [RPS -]

× **Cylindrocalycium** Ito, The Cactaceae, 644, 1981. [= *Cylindrolobivia* × *Acanthocalycium*. The validity of the name depends on the status of the first-named parent.] [RPS -]

Cylindrolobivia Ito, The Cactaceae, 338, 1981. *Nom. inval.* (Art. 36.1, 37.1). [Said to have been published first in Ito, Illust. Flow. Cacti, 145, 1967. Status and validity not assessed. The name is used again (as latter homonym) in The Cactaceae, 640 for a hybrid genus "Cylindrolobivia × Lobivopsis".] [RPS -]

Cylindrolobivia pseudoschickendantzii Y. Ito, The Cactaceae, 643, 1981. *Nom. inval.* (Art. 37.1, 43.1?). [RPS -]

Cylindrolobivia vatteri Y. Ito, The Cactaceae, 340, 1981. *Nom. inval.* (Art. 37.1). [Sphalm. 'vateri' and attributed to 'Ritter & Ito'.] [RPS -]

Cylindrolobivia vatteri var. *sulphurea* Y. Ito, The Cactaceae, 340, 1981. *Nom. inval.* (Art. 43.1). [Sphalm. 'vateri'.] [RPS -]

× **Cylindropsis** Ito, The Cactaceae, 639, 1981. [= *Cylindrolobivia* × *Echinopsis*. The validity of the name depends on the status of the first-named parent.] [RPS -]

Cylindropuntia abyssi (Hester) Backeberg, Die Cact. 1: 184, 1958. Basionym: *Opuntia abyssi*. [RPS 9]

Cylindropuntia acanthocarpa var. **ramosa** (Peebles) Backeberg, Die Cact. 1: 181, 1958. Basionym: *Opuntia acanthocarpa* var. *ramosa*. [RPS 9]

Cylindropuntia acanthocarpa var. **thornberi** (Thornber & Bonker) Backeberg, Die Cact. 1: 184, 1958. Basionym: *Opuntia thornberi*. [RPS 9]

Cylindropuntia alamosensis (Britton & Rose) Backeberg, Die Cact. 1: 177, 1958. Basionym: *Opuntia alamosensis*. [RPS 9]

Cylindropuntia arbuscula var. **congesta** (Griffiths) Backeberg, Die Cact. 1: 174, 1958. Basionym: *Opuntia congesta*. [RPS 9]

Cylindropuntia bigelowii var. **hoffmannii** (Wolf) Backeberg, Kakt.-Lex., 111, 1966. Basionym: *Opuntia bigelowii* var. *hoffmannii*. [RPS 17]

Cylindropuntia brevispina (Gates) Backeberg, Die Cact. 1: 205, 1958. Basionym: *Opuntia brevispina*. [RPS 9]

Cylindropuntia brittonii (G. Ortega) Backeberg, Die Cact. 1: 171, 1958. Basionym: *Opuntia brittonii*. [RPS 9]

Cylindropuntia burrageana (Britton & Rose) Backeberg, Die Cact. 1: 205, 1958. Basionym: *Opuntia burrageana*. [RPS 9]

Cylindropuntia clavarioides var. **ruiz-lealii** (Castellanos) Krainz, Die Kakteen Lief. 33, 1966. Basionym: *Opuntia ruiz-lealii*. [RPS 17]

Cylindropuntia densiaculeata Backeberg, Descr. Cact. Nov. [1:] 6, 1957. [Dated 1956, published 1957.] [RPS 7]

Cylindropuntia fulgida var. **mamillata** (Schott) Backeberg, Die Cact. 1: 204, 1958. Basionym: *Opuntia mamillata*. [RPS 9]

Cylindropuntia hualpaensis (Hester) Backeberg, Die Cact. 1: 178, 1958. Basionym: *Opuntia hualpaensis*. [RPS 9]

Cylindropuntia hystrix (Grisebach) Areces, Ciencias ser. 10 Bot. 15: 4, 1976. Basionym: *Opuntia hystrix*. [Publication date according to Index Kewensis; variously also cited as 1979.] [RPS 30]

Cylindropuntia imbricata var. **argentea** (Anthony) Backeberg, Die Cact. 1: 195, 1958. Basionym: *Opuntia imbricata* var. *argentea*. [RPS 9]

Cylindropuntia intermedia (Rauh & Backeberg) Rauh & Backeberg, in Rauh, Beitr. Kenntn. peruan. Kakt.veg., 194, 1958. *Nom. inval.* (Art. 33.2), based on *Austrocylindropuntia intermedia*. [Basionym inferred by inference.] [RPS 10]

Cylindropuntia leptocaulis var. *glauca* Backeberg, Die Cact. 6: 3583, 1962. *Nom. inval.* (Art. 9.5). [Erroneously included as valid in RPS 13.] [RPS 13]

Cylindropuntia leptocaulis var. *tenuispina* Backeberg, Die Cact. 6: 3584, 1962. *Nom. inval.* (Art. 9.5). [Erroneously included as valid in RPS 13.] [RPS 13]

Cylindropuntia metuenda (Pittier) Backeberg, Die Cact. 1: 174, 1958. Basionym: *Opuntia metuenda*. [Treated as valid in RPS 9, but probably invalid because of incomplete basionym citation.] [RPS 9]

Cylindropuntia multigeniculata (Clokey) Backeberg, Die Cact. 1: 186, 1958. Basionym: *Opuntia multigeniculata*. [RPS 9]

Cylindropuntia × **munzii** (Wolf) Backeberg, Kakt.-Lex., 113, 1966. Basionym: *Opuntia* × *munzii*. [RPS 17]

Cylindropuntia recondita var. **perrita** (Griffiths) Backeberg, Die Cact. 1: 178, 1958. Basionym: *Opuntia perrita*. [RPS 9]

Cylindropuntia rosarica (Lindsay) Backeberg, Die Cact. 1: 185, 1958. Basionym: *Opuntia rosarica*. [RPS 9]

Cylindropuntia rosea (De Candolle) Backeberg, Die Cact. 1: 197, 1958. Basionym: *Opuntia rosea*. [RPS 9]

Cylindropuntia rosea var. *atrorosea* Backeberg, Descr. Cact. Nov. 3: 5, 1963. *Nom. inval.* (Art. 9.5). [Erroneously included as valid in RPS 14.] [RPS 14]

Cylindropuntia tephrocactoides (Rauh & Backeberg) Rauh & Backeberg, in Rauh, Beitr. Kenntn. peruan. Kakt.veg., 194, 1958. *Nom. inval.* (Art. 33.2), based on *Austrocylindropuntia tephrocactoides*. [RPS 10]

Cylindropuntia tesajo var. **cineracea** (Wiggins) Backeberg, Die Cact. 1: 173, 1958.

Basionym: *Opuntia cineracea.* [RPS 9]
Cylindropuntia tunicata var. **aricensis** Ritter, Kakt. Südamer. 3: 890-892, 1980. [RPS 31]
Cylindropuntia tunicata var. **chilensis** Ritter, Kakt. Südamer. 3: 889-890, 1980. [RPS 31]
Cylindropuntia verschaffeltii fa. **longispina** (Backeberg) Krainz, Die Kakt. Lief. 33, 1966. Basionym: *Austrocylindropuntia verschaffeltii* var. *longispina.* [RPS 17]
Cylindropuntia whipplei var. **enodis** (Peebles) Backeberg, Die Cact. 1: 180, 1958. Basionym: *Opuntia whipplei* var. *enodis.* [RPS 9]
Cylindropuntia wigginsii (L. Benson) Robinson, Phytologia 26: 175, 1973. Basionym: *Opuntia wigginsii.* [RPS 24]
Cylindrorebutia einsteinii (Fric) Subik & Pazout, Succulenta 49(3): 36, ills. (p. 37), 1970. *Nom. inval.* (Art. 33.2), based on *Rebutia einsteinii.* [RPS -]
Cylindrorebutia einsteinii var. steineckei Fric *ex* Subik & Pazout, Succulenta 49(3): 36, 1970. *Nom. inval.* (Art. 43.1, 33.2). [Several names by Fric and Kreuzinger or Backeberg are cited (without page reference) as syonyms.] [RPS -]
Cylindrorebutia karreri Fric *ex* Subik & Pazout, Succulenta 49(3): 39, ill., 1970. *Nom. inval.* (Art. 36.1, 37.1). [Various names are cited (without page references) as synonyms, and the name is ascribed to Fric.] [RPS -]
Cylindrorebutia nicolai Fric *ex* Subik & Pazout, Succulenta 49(3): 31, ill. (p. 40), 1970. *Nom. inval.* (Art. 36.1, 37.1). [Various names are cited (without page references) as synonyms, and the name is ascribed to Fric.] [RPS -]
Cylindrorebutia rubriviridis (Fric *ex* Backeberg) Subik & Pazout, Succulenta 49(3): 38, ill., 1970. Basionym: *Mediolobivia schmiedcheniana* var. *rubriviridis.* [Sphalm. 'rubriviride'. The Backeberg name is cited, besides other names, in the synonymy and is taken to represent the basionym.] [RPS -]
× **Cylindrosia** Ito, Shaboten 91: 25, 1976. [= *Cylindrorebutia* × *Soehrensia.* Publication repeated in Ito, The Cactaceae, 643, 1981.] [RPS 27]
× **Cylindrosia** cv. **Saigenryu** Y. Ito, Shaboten 91: 25, 1976. [= *Cylindrorebutia huascha* var. *roseiflora* × *Soehrensia bruchii.*] [RPS 27]
Dactylanthocactus Y. Ito, Bull. Takarazuka Insectarium 71: 13-20, 1950. *Nom. illeg.* (Art. 10). [Republished in Explan. Diag. Austroechinocactinae, 225, 294, 1957.] [RPS 3]
Dactylanthocactus graessneri (Schumann) Y. Ito, Explan. Diagr. Austroechinocactinae, 225, 1957. *Nom. inval.* (Art. 43.1), based on *Echinocactus graessneri.* [RPS 8]
Delaetia Backeberg, Die Cact. 6: 3788, 1962. *Nom. inval.* (Art. 37.1). [Invalid, because the type species is not validly published.] [RPS 13]
Delaetia woutersiana Backeberg, In Backeberg, Die Cact. 6: 3788, 1962. *Nom. inval.* (Art. 37.1). [Erroneously included as valid in RPS 13.] [RPS 13]
Denmoza rhodacantha var. **coccinea** (Monville) Y. Ito, Explan. Diagr. Austroechinocactinae, 154, 1957. Basionym: *Echinopsis rhodacantha* var. *coccinea.* [RPS 8]
Denmoza rhodacantha var. **gracilior** (Labouret) Y. Ito, Explan. Diagr. Austroechinocactinae, 154, 1957. Basionym: *Echinopsis rhodacantha* var. *gracilior.* [RPS 8]
Digitorebutia cv. **A. V. Fric** Donald & Cullmann, Cact. Succ. J. Gr. Brit. 19: 35, 40, 1957. [*Rebutia fricii* Hort.] [RPS 8]
Digitorebutia brachyantha var. **ritteri** (Wessner) Donald, Nation. Cact. Succ. J. 12(2): 11, 1957. Basionym: *Lobivia ritteri.* [RPS 8]
Digitorebutia canacruzensis (Rausch) Rausch, Lobivia 85, 56, 1987. *Nom. inval.* (Art. 34.1), based on *Rebutia canacruzensis.* [Name erroneously applied to *Rebutia canacruzensis* in incorrect basionym citation for combination under *Lobivia.*] [RPS -]
Digitorebutia carmeniana (Rausch) Rausch, Lobivia 85, 100, 1987. *Nom. inval.* (Art. 34.1), based on *Rebutia carmeniana.* [Name erroneously applied to *Rebutia carmeniana* in incorrect basionym citation for combination under *Lobivia.*] [RPS -]
Digitorebutia christinae (Rausch) Rausch, Lobivia 85, 129, 1987. *Nom. inval.* (Art. 34.1), based on *Rebutia christinae.* [Name erroneously applied to *Rebutia christinae* in incorrect basionym citation for combination under *Lobivia.*] [RPS -]
Digitorebutia cincinnata (Rausch) Rausch, Lobivia 85, 129, 1987. *Nom. inval.* (Art. 34.1), based on *Rebutia cincinnata.* [Name erroneously applied to *Rebutia cincinnata* in incorrect basionym citation for combination under *Lobivia.*] [RPS -]
Digitorebutia costata var. **eucaliptana** (Backeberg) Donald, Nation. Cact. Succ. J. 12(2): 11, 1957. Basionym: *Lobivia eucaliptana.* [RPS 8]
Digitorebutia diersiana (Rausch) Rausch, Lobivia 85, 116, 1987. *Nom. inval.* (Art. 34.1), based on *Rebutia diersiana.* [Name erroneously applied to *Rebutia diersiana* in incorrect basionym citation for combination under *Lobivia.*] [RPS -]
Digitorebutia diersiana var. *minor* (Rausch) Rausch, Lobivia 85, 116, 1987. *Nom. inval.* (Art. 34.1, 43.1), based on *Rebutia diersiana* var. *minor.* [Name erroneously applied to *Rebutia diersiana* var. *minor* in incorrect basionym citation for combination under *Lobivia.*] [RPS -]
Digitorebutia eos (Rausch) Rausch, Lobivia 85, 57, 1987. *Nom. inval.* (Art. 33.2). [Name erroneously applied to *Rebutia eos* in incorrect basionym citation for combination under *Lobivia.*] [RPS -]
Digitorebutia euanthema var. *longispina* (Backeberg) Donald, Nation. Cact. Succ. J. 12(2): 11, 1957. *Nom. inval.*, based on *Pygmaeolobivia pygmaea* var. *longispina, nom. inval.* (Art. 36.1). [RPS 8]
Digitorebutia friedrichiana (Rausch) Rausch,

Lobivia 85, 116, 1987. *Nom. inval.* (Art. 34.1), based on *Rebutia friedrichiana*. [Name erroneously applied to *Rebutia friedrichiana* in incorrect basionym citation for combination under *Lobivia*.] [RPS -]

Digitorebutia haagei var. atrovirens (Backeberg) Donald, Nation. Cact. Succ. J. 12(2): 11, 1957. Basionym: *Lobivia atrovirens*. [RPS 8]

Digitorebutia haagei var. digitiformis (Backeberg) Donald, Nation. Cact. Succ. J. 12(2): 11, 1957. Basionym: *Lobivia digitiformis*. [RPS 8]

Digitorebutia haagei var. orurensis (Backeberg) Donald, Nation. Cact. Succ. J. 12(2): 11, 1957. Basionym: *Lobivia orurensis*. [RPS 8]

Digitorebutia haagei var. pectinata (Backeberg) Donald, Nation. Cact. Succ. J. 12(2): 11, 1957. Basionym: *Lobivia pectinata*. [RPS 8]

Digitorebutia iscayachensis (Rausch) Rausch, Lobivia 85, 116, 1987. *Nom. inval.* (Art. 34.1), based on *Rebutia iscayachensis*. [Name erroneously applied to *Rebutia iscayachensis* in incorrect basionym citation for combination under *Lobivia*.] [RPS -]

Digitorebutia mudanensis (Rausch) Rausch, Lobivia 85, 57, 1987. *Nom. inval.* (Art. 34.1), based on *Rebutia mudanensis*. [Name erroneously applied to *Rebutia mudanensis* in incorrect basionym citation for combination under *Lobivia*.] [RPS -]

Digitorebutia nazarenoensis Rausch, Succulenta 58(8): 186, ill. (p. 185), 1979. Typus: *Rausch* 484 (ZSS, ZSS). [Argentina: Santa Victoria] [RPS 30]

Digitorebutia nigricans (Wessner) Buining, Succulenta 1957: 53, 1957. Basionym: *Lobivia nigricans*. [RPS -]

Digitorebutia pallida (Rausch) Rausch, Lobivia 85, 57, 1987. *Nom. inval.* (Art. 34.1), based on *Rebutia pallida*. [Name erroneously applied to *Rebutia pallida* in incorrect basionym citation for combination under *Lobivia*.] [RPS -]

Digitorebutia raulii (Rausch) Rausch, Lobivia 85, 13, 1987. *Nom. inval.* (Art. 34.1), based on *Rebutia raulii*. [Name erroneously applied to *Rebutia raulii* in incorrect basionym citation for combination under *Lobivia*.] [RPS -]

Digitorebutia rauschii (Zecher) Rausch, Lobivia 85, 129, 1987. *Nom. inval.* (Art. 34.1), based on *Rebutia rauschii*. [Name erroneously applied to *Rebutia rauschii* in incorrect basionym citation for combination under *Lobivia*.] [RPS -]

Digitorebutia yuquinensis (Rausch) Rausch, Lobivia 85, 13, 1987. *Nom. inval.* (Art. 34.1), based on *Rebutia yuquinensis*. [Name erroneously applied to *Digitorebutia yuquinensis* in incorrect basionym citation for combination under *Lobivia*.] [RPS -]

Diploperianthium Ritter, in Backeberg, Die Cact. 1: 72, in clav, 1958. *Nom. inval.* (Art. 36.1). [Casually mentioned.] [RPS 9]

× **Disberocereus** E. Meier, Kakt. and. Sukk. 41(4): 80, 1990. [= *Disocactus* × *Weberocereus*.] [RPS 41]

Discocactus albispinus Buining & Brederoo, Cact. Succ. J. (US) 46(6): 252-257, ills., 1974. Typus: *Horst et Uebelmann* HU 390 (U, ZSS). [Brazil: Bahia] [RPS 25]

Discocactus araneispinus Buining & Brederoo *ex* Theunissen, Succulenta 56: 258-259, 1977. Typus: *Horst et Uebelmann* HU 440 (1972) (U, ZSS). [Again reported as new species in Buining, Gattung Discocactus, 39-46, ills., 1980.] [Brazil: Bahia] [RPS 28]

Discocactus boliviensis Backeberg *ex* Theunissen, Succulenta 56: 258, 1977. [First invalidly published by Backeberg in Descr. Cact. Nov. 3: 5, 1963 (cf. RPS 14, erroneously reported as valid publication).] [RPS 28]

Discocactus boomianus Buining & Brederoo, Succulenta 50(2): 26-29, ills., 1971. Typus: *Horst et Uebelmann* HU 222 (1967) (U, ZSS). [The collection at U is not labelled as holotype.] [Brazil: Bahia] [RPS 22]

Discocactus buenekeri Abraham, Kakt. and. Sukk. 38(11): 282-285, ills., 1987. Typus: *Büneker* in *Abraham* 27 (KOELN). [Detailed locality information deposited with the holotype] [Brazil: Bahia] [RPS 38]

Discocactus cangaensis Diers & E. Esteves Pereira, Cact. Succ. J. (US) 52(3): 107-111, 1980. [RPS 31]

Discocactus caracolensis Uebelmann, Feld-Nummern, [unpaged], 1972. *Nom. inval.* (Art. 36.1, 37.1). [Published as provisional name without description.] [RPS -]

Discocactus catingicola Buining & Brederoo, Kakt. and. Sukk. 25: 265-267, 1974. [RPS 25]

Discocactus cephaliaciculosus Buining & Brederoo, Kakt. and. Sukk. 26(5): 97-100, ills., 1975. *Nom. inval.* (Art. 37.1). [Based on 2 syntypes, *Horst & Uebelmann HU* 430 and *Horst & Uebelmann HU 431*. Erroneously included as valid in RPS 26.] [RPS 26]

Discocactus cipolandensis Uebelmann, Feld-Nummern, 7, 1972. *Nom. inval.* (Art. 36.1, 37.1). [Published as provisional name without description, also applied to *Horst & Uebelmann* 198a = *D. semicampaniflorus*.] [RPS -]

Discocactus conorhizus Uebelmann, Feld-Nummern, [unpaged], 1972. *Nom. inval.* (Art. 36.1, 37.1). [Published as provisional name without description.] [RPS -]

Discocactus corumbensis Uebelmann, Feld-Nummern, [unpaged], 1972. *Nom. inval.* (Art. 36.1, 37.1). [Published as provisional name without description.] [RPS -]

Discocactus crystallophilus Diers & E. Esteves Pereira, Kakt. and. Sukk. 32(11): 258-262, 1981. [RPS 32]

Discocactus diersianus E. Esteves Pereira, Cact. Succ. J. (US) 51(4): 179-183, 1979. [RPS 30]

Discocactus estevesii Diers, Cact. Succ. J. (US): 50(2): 83-85, 1978. [RPS 29]

Discocactus ferricola Buining & Brederoo, Kakt. and. Sukk. 26(1): 2-5, ills., 1975. Typus: *Horst et Uebelmann* HU 195 (1971) (U, ZSS). [Brazil: Mato Grosso] [RPS 26]

Discocactus flavispinus Buining & Brederoo *in* Theunissen, Succulenta 56: 259, 1977. Typus: *Horst et Uebelmann* HU 326a (1974) (U). [Again reported as new species in Buining, Gattung Discocactus, 157-161, ills., 1980.] [Brazil: Mato Grosso] [RPS 28]

Discocactus goianus Diers & E.Esteves Pereira, Kakt. and. Sukk. 31(3): 73-79, 1980. [RPS 31]

Discocactus griseus Buining & Brederoo, Succulenta 54(10): 185-190, ills., 1975. Typus: *Horst et Uebelmann* HU 343 (U, ZSS). [Brazil: Minas Gerais] [RPS 26]

Discocactus grossoanus Uebelmann, Feld-Nummern, [unpaged], 1972. *Nom. inval.* (Art. 36.1, 37.1). [Published as provisional name without description; sphalm. 'grossoana'.] [RPS -]

Discocactus hartmannii var. bonitoensis (Buining & Brederoo) P. J. Braun, 25 Jahre HU Horst-Uebelmann, [19], 1984. Basionym: *Discocactus hartmannii* ssp. *bonitoensis*. [Combination repeated in Kakt. and. Sukk. 36(2): 23, 1985.] [RPS 35]

Discocactus hartmannii var. magnimammus (Buining & Brederoo) P. J. Braun, 25 Jahre HU Horst-Uebelmann, [27], 1984. Basionym: *Discocactus magnimammus*. [Combination repeated in Kakt. and. Sukk. 36(2): 23, 1985.] [RPS 35]

Discocactus hartmannii var. **mamillosus** (Buining & Brederoo) P. J. Braun, 25 Jahre HU Horst-Uebelmann, [19], 1984. Basionym: *Discocactus mamillosus*. [Combination repeated in Kakt. and. Sukk. 36(2): 23, 1985.] [RPS 35]

Discocactus hartmannii var. **patulifolius** (Buining & Brederoo) P. J. Braun, 25 Jahre HU Horst-Uebelmann, [18], 1984. Basionym: *Discocactus patulifolius*. [Combination repeated in Kakt. and. Sukk. 36(2): 23, 1985.] [RPS 35]

Discocactus horstii Buining & Brederoo, in Krainz, Die Kakt. Lief. 52: C VIf, 3 pp., ill., 1973. Typus: *Horst et Uebelmann* HU 360 (U, ZSS). [Brazil: Minas Gerais] [RPS 24]

Discocactus iguatemiensis Uebelmann, Feld-Nummern, [unpaged], 1972. *Nom. inval.* (Art. 36.1, 37.1). [Published as provisional name without description.] [RPS -]

Discocactus latispinus Buining & Brederoo *in* Theunissen, Succulenta 56(11): 259, 1977. Typus: *Horst et Uebelmann* HU 146 (U, ZSS). [Again reported as new species in Buining, Gattung Discocactus, 95-100, ills., 1980.] [Brazil: Minas Gerais] [RPS 28]

Discocactus lindaianus Diers & E. Esteves Pereira, Cact. Succ. J. Amer. 53(2): 56-60, 1981. [RPS 32]

Discocactus magnimammus Buining & Brederoo, Kakt. and. Sukk. 25(11): 242-245, ills., 1974. Typus: *Horst et Uebelmann* HU 324 (1972) (U, ZSS). [Brazil: Mato Grosso] [RPS 25]

Discocactus magnimammus ssp. **bonitoensis** Buining & Brederoo *in* Theunissen, Succulenta 56: 259-260, 1977. Typus: *Horst et Uebelmann* HU 193 (1974) (U). [Again reported as new subspecies in Buining, Gattung Discocactus, 204-208, ills., 1980.] [Brazil: Mato Grosso] [RPS 28]

Discocactus mamillosus Buining & Brederoo, Kakt. and. Sukk. 25(10): 217-219, ills., 1974. Typus: *Horst et Uebelmann* HU 191 (1972) (U, ZSS). [Brazil: Mato Grosso] [RPS 25]

Discocactus manecoensis Uebelmann, Feld-Nummern, [unpaged], 1972. *Nom. inval.* (Art. 36.1, 37.1). [Published as provisional name without description.] [RPS -]

Discocactus melanochlorus Buining & Brederoo *in* Theunissen, Succulenta 56: 260, 1977. Typus: *Horst et Uebelmann* HU 453 (U). [Again reported as new species in Buining, Gattung Discocactus, 135-140, ills., 1980.] [Brazil: Mato Grosso do Sul] [RPS 28]

Discocactus multicolorispinus P. J. Braun & Brederoo, Kakt. and. Sukk. 32(3): 54-59, 1981. [RPS 32]

Discocactus nigrisaetosus Buining & Brederoo *in* Theunissen, Succulenta 56: 260, 1977. Typus: *Horst et Uebelmann* HU 448 (1974) (U, ZSS). [Again reported as new species in Buining, Gattung Discocactus, 129-134, ills., 1980.] [Brazil: Bahia] [RPS 28]

Discocactus pachythele Buining & Brederoo, Cact. Succ. J. (US) 47(4): 163-166, ills., 1975. Typus: *Horst et Uebelmann* HU 198 (1972) (U, ZSS [type number]). [Brazil: Mato Grosso] [RPS 27]

Discocactus paranaensis Backeberg, Die Cact. 4: 2628-2629, 1960. *Nom. inval.* (Art. 37.1). [Erroneously included as valid in RPS 11.] [RPS 11]

Discocactus patulifolius Buining & Brederoo, Kakt. and. Sukk. 25(9): 195-197, ills., 1974. Typus: *Horst et Uebelmann* HU 190 (1972) (U, ZSS). [Brazil: Mato Grosso] [RPS 25]

Discocactus prominentigibbus Diers & E. Esteves Pereira, Kakt. and. Sukk. 39(1): 14-19, ills., SEM-ills., 1988. Typus: *Esteves Pereira* 151 (KOELN [Succulentarium]). [Brazil: Goiás] [RPS 39]

Discocactus pseudolatispinus Diers & E. Esteves Pereira, Kakt. and. Sukk. 38(10): 242-247, ills., SEM-ills., 1987. Typus: *Esteves Pereira* 111 (KOELN). [Detailed locality information deposited with the holotype] [Brazil: Minas Gerais] [RPS 38]

Discocactus pugionacanthus Buining & Brederoo *in* Theunissen, Succulenta 56: 260-261, 1977. Typus: *Horst et Uebelmann* HU 462 (1974) (U). [Again reported as new species in Buining, Gattung Discocactus, 61-65, ills., 1980.] [Brazil: Minas Gerais] [RPS 28]

Discocactus pulvinicapitatus Buining & Brederoo *in* Theunissen et al., Die Gatt. Discocactus, 100-105, ills., 1980. Typus: *Horst et Uebelmann* HU 425 (U, ZSS [type number]). [Brazil: Minas Gerais] [RPS -]

Discocactus rapirhizus Buining & Brederoo, Ashingtonia 2(3): 44-47, ills., 1975. Typus: *Horst et Uebelmann* HU 200 (1974) (U, ZSS). [Brazil: Goiás] [RPS 26]

Discocactus semicampaniflorus Buining & Brederoo, Cact. Succ. J. (US) 47(3): 122-123, ill., 1975. Typus: *Horst et Uebelmann* HU

198a (1972) (U, ZSS). [Publication date according to Index Kewensis; variously also cited as 1976.] [Brazil: Mato Grosso] [RPS 27]

Discocactus silicicola Buining & Brederoo, Cact. Succ. J. (US) 47(5): 214-217, ills., 1975. Typus: *Horst et Uebelmann* HU 325 (1972) (U, ZSS). [Brazil: Mato Grosso] [RPS 27]

Discocactus silvaticus Buining & Brederoo *in* Theunissen, Succulenta 56: 261, 1977. Typus: *Horst et Uebelmann* HU 455 (1974) (U, ZSS). [Again reported as new species in Buining, Gattung Discocactus, 140-145, ills., 1980.] [Brazil: Mato Grosso] [RPS 28]

Discocactus spinosior Buining & Brederoo *in* Theunissen, Succulenta 56: 261, 1977. Typus: *Horst et Uebelmann* HU 205a (1974) (U, ZSS). [Again reported as new species in Buining, Gattung Discocactus, 146-150, ills., 1980.] [Brazil: Bahia] [RPS 28]

Discocactus squamibaccatus Buining & Brederoo *in* Theunissen, Succulenta 56: 261-262, 1977. Typus: *Horst et Uebelmann* HU 428 (1974) (U). [Sphalm. 'sguamibaccatus'. Again reported as new species in Buining, Gattung Discocactus, 183-188, ills., 1980.] [Brazil: Goiás] [RPS 28]

Discocactus subterraneo-proliferans Diers & E. Esteves Pereira, Kakt. and. Sukk. 31(9): 266-271, 1980. [RPS 31]

Discocactus subviridigriseus Buining & Brederoo *in* Theunissen, Succulenta 56: 262, 1977. Typus: *Horst et Uebelmann* HU 438 (1974) (U, ZSS). [Again reported as new species in Buining, Gattung Discocactus, 115-122, ills., 1980.] [Brazil: Bahia] [RPS 28]

Discocactus woutersianus Brederoo & van den Broek, Succulenta 59(9): 197-203, 1980. [J. Riha (Kaktusy 26(3): 59, 1990) reports that this taxon is the hybrid *Discocactus horstii* × *'D. insignis'* [= *D. pseudoinsignis*] on the basis of controlled crossing in cultivation.] [RPS 31]

Discocactus zehntneri fa. **albispinus** (Buining & Brederoo) Riha, Kaktusy 19(2): 26, 1983. Basionym: *Discocactus albispinus*. [RPS 34]

Discocactus zehntneri var. **albispinus** (Buining & Brederoo) P. J. Braun, Succulenta 69(10): 215, 1990. Basionym: *Discocactus albispinus*. [RPS 41]

Discocactus zehntneri var. **araneispinus** (Buining & Brederoo) P. J. Braun, Succulenta 69(10): 215, 1990. Basionym: *Discocactus araneispinus*. [RPS 41]

Discocactus zehntneri var. **boomianus** (Buining & Brederoo) P. J. Braun, Succulenta 69(10): 218, 1990. Basionym: *Discocactus boomianus*. [RPS 41]

Discocactus zehntneri var. **horstiorum** P. J. Braun, Succulenta 69(10): 218, ill. (p. 220), 1990. Typus: *Horst* 667 (ZSS, B, K, KOELN). [Brazil: Bahia] [RPS 41]

× **Disheliocereus** Rowley, Nation. Cact. Succ. J. 37(2): 48, 1982. [*Disocactus* × *Heliocereus*.] [RPS 33]

Disocactinae Buxbaum, Madroño 14(6): 183, 1958. Typus: *Disocactus*. [RPS 9]

Disocactus cv. **Frühlingsanfang** Petersen, Kakt. and. Sukk. 36(2): 25, ill., 1985. [= *D. macranthus* (*Lau* 1263) × *D. nelsonii*.] [RPS 36]

Disocactus subgen. **Pseudorhipsalis** (Britton & Rose) Kimnach, Cact. Succ. J. (US) 51(4): 170, 1979. Basionym: *Pseudorhipsalis*. [RPS 31]

Disocactus acuminatus (Cufodontis) Kimnach, Cact & Succ. J. (US) 33: 14, 1961. Basionym: *Pseudorhipsalis acuminata*. [RPS 12]

Disocactus alatus (Swartz) Kimnach, Cact. & Succ. J. (US) 33: 14, 1961. Basionym: *Cactus alatus*. [RPS 12]

Disocactus amazonicus (Schumann) D. Hunt, Cact. Succ. J. Gr. Britt. 44(1): 2, 1982. Basionym: *Wittia amazonica*. [RPS 33]

Disocactus himantocladus (Gosselin) Kimnach, Cact. Succ. J. (US) 33: 14, 1961. Basionym: *Rhipsalis himantoclada*. [RPS 12]

Disocactus horichii Kimnach, Cact. Succ. J. (US) 51(4): 168-170, ills., 1979. Typus: *Horich* s.n. (HNT, UC, US, ZSS [type number]). [Costa Rica] [RPS 30]

Disocactus kimnachii Rowley, Brit. Cact. Succ. J. 5(3): 84, 1987. Basionym: *Nopalxochia horichii*. [*Nom. nov.* needed to avoid conflict with *Disocactus horichii* Kimnach 1979.] [RPS 38]

Disocactus lankesteri Kimnach, Cact. Succ. J. (US) 51(4): 166-168, 1979. [RPS 30]

Disocactus macranthus var. *glaucocladus* Ewald, Epiphytes 13(52): 106-108, ill., 1989. *Nom. inval.* (Art. 37.1). [Location of type given, but type collection not cited clearly.] [RPS 40]

Disocactus nelsonii var. **hondurensis** Kimnach, Cact. Succ. J. (US) 37: 29, 31-33, 1965. [RPS 16]

Disocactus quezaltecus (Standley & Steyermark) Kimnach, Cact. Succ. J. (US) 31: 137, 1959. Basionym: *Bonifazia quezalteca*. [RPS 10]

Disocactus ramulosus (Salm-Dyck) Kimnach, Cact. Succ. J. (US) 33: 14, 1961. Basionym: *Cereus ramulosus*. [RPS 12]

Disocactus ramulosus var. **angustissimus** (F. A. C. Weber) Kimnach, Cact. Succ. J. (US) 58(2): 67, 1987. Basionym: *Rhipsalis angustissima*. [RPS 38]

× **Disochia** Rowley, Nation. Cact. Succ. J. 37(2): 48, 1982. [= *Disocactus* × *Nopalxochia*.] [RPS 33]

× **Disochia** cv. **Märzsonne** Petersen, Kakt. and. Sukk. 36(2): 25, ill., 1985. [= *Disocactus macranthus* (*Lau* 1263) × *Nopalxochia phyllanthoides*.] [RPS 36]

× **Disophyllum** cv. **Frühlingsahnen** Petersen, Kakt. and. Sukk. 41(10): 238-239, ill., 1990. [*Disocactus macranthus* fa. × ?] [RPS 41]

× **Disophyllum** cv. **Frühlingselfe** Petersen, Kakt. and. Sukk. 41(10): 238-239, ill., 1990. [*Disocactus macranthus* fa. × ?] [RPS 41]

× **Disophyllum** cv. **Frühlingsgold** Petersen, Kakt. and. Sukk. 41(10): 238-239, ill., 1990. [*Disocactus macranthus* fa. × ?] [RPS 41]

× **Disophyllum** cv. **Frühlingspracht** Petersen,

Kakt. and. Sukk. 41(10): 238-239, ill., 1990. [*Disocactus macranthus* fa. × ?] [RPS 41]
× **Disophyllum** cv. **Frühlingsreigen** Petersen, Kakt. and. Sukk. 41(10): 238-239, ill., 1990. [*Disocactus macranthus* fa. × ?] [RPS 41]
× **Disophyllum** cv. **Frühlingsstern** Petersen, Kakt. and. Sukk. 41(10): 238-239, ill., 1990. [*Disocactus macranthus* fa. × ?] [RPS 41]
× **Disophyllum** cv. **Frühlingstraum** Petersen, Kakt. and. Sukk. 41(10): 238-239, ill., 1990. [*Disocactus macranthus* fa. × ?] [RPS 41]
× **Disoselenicereus** E. Meier, Kakt. and. Sukk. 41(4): 80, 1990. [= *Disocactus* × *Selenicereus*.] [RPS 41]
Dolichothele albescens (Tiegel) Backeberg, Cact. Succ. J. (US) 23(5): 152, 1951. [RPS 3]
Dolichothele aylostera (Werdermann) Backeberg, Cact. Succ. J. (US) 23(5): 152, 1951. [RPS 3]
Dolichothele balsasoides (Craig) Backeberg, Die Cact. 5: 3527, 1961. Basionym: *Mammillaria balsasoides*. [RPS 12]
Dolichothele baumii (Bödeker) Werdermann & Buxbaum, Österr. Bot. Zeitschr. 98 (1-2): 506-508, 1951. Basionym: *Mammillaria baumii*. [Also published by Backeberg in Cact. Succ. J. (US) 23(5): 152, c. Oct. 1951 (may have priority over Buxbaum's combination, whose publication date is not know with certainty (K copy stamped 11. 12. 1951).] [RPS 2]
Dolichothele beneckei (Ehrenberg) Backeberg, Die Cact. 5: 3524, 1961. Basionym: *Mammillaria beneckei*. [RPS 12]
Dolichothele longimamma fa. **gigantothele** (Berg) Krainz, Die Kakt. Lief. 57: C VIIIc, 1974. Basionym: *Mammillaria longimamma* var. *gigantothele*. [RPS 25]
Dolichothele longimamma fa. **globosa** (Linke) Krainz, Die Kakt. Lief. 57: C VIIIc, 1974. Basionym: *Mammillaria globosa*. [RPS 25]
Dolichothele longimamma ssp. **uberiformis** (Zuccarini) Krainz, Die Kakt. Lief. 57: C VIIIc, 1974. Basionym: *Mammillaria uberiformis*. [RPS 25]
Dolichothele nelsonii (Britton & Rose) Backeberg, Die Cact. 5: 3525, 1961. Basionym: *Neomammillaria nelsonii*. [RPS 12]
Dolichothele surculosa (Bödeker) Backeberg, Cact. Succ. J. (US) 23(5): 152, 1951. [RPS 3]
Dolichothele zephyranthoides (Scheidweiler) Backeberg, Die Cact. 5: 3528, 1961. Basionym: *Mammillaria zephyranthoides*. [RPS 12]
Dracocactus Y. Ito, Bull. Takarazuka Insectarium 71: 13-20, 1950. *Nom. inval.* (Art. 36.1). [RPS 3]
Ebnerella F. Buxbaum, Oesterr. Bot. Zeitschr. 98(1-2): 88, 1951. Typus: *Mammillaria wildii* Dietrich. [RPS 2]
Ebnerella subgen. **Archiebnerella** Buxbaum, Österr. Bot. Zeitschr. 98(1-2): 91, 1951. Typus: *Mammillaria zephyranthoides* Scheidweiler. [RPS 2]
Ebnerella angelensis (Craig) Buxbaum, Österr. Bot. Zeitschr. 98(1-2): 89, 1951. Basionym: *Mammillaria angelensis*. [RPS 2]
Ebnerella armillata (Brandegee) Buxbaum, Österr. Bot. Zeitschr. 98(1-2): 89, 1951. Basionym: *Mammillaria armillata*. [RPS 2]
Ebnerella aureilanata (Backeberg) Buxbaum, Österr. Bot. Zeitschr. 98(1-2): 89, 1951. Basionym: *Mammillaria aureilanata*. [RPS 2]
Ebnerella aurihamata (Bödeker) Buxbaum, Österr. Bot. Zeitschr. 98(1-2): 89, 1951. Basionym: *Mammillaria aurihamata*. [RPS 2]
Ebnerella barbata (Engelmann) Buxbaum, Österr. Bot. Zeitschr. 98(1-2): 89, 1951. Basionym: *Mammillaria barbata*. [RPS 2]
Ebnerella blossfeldiana (Bödeker) Buxbaum, Österr. Bot. Zeitschr. 98(1-2): 89, 1951. Basionym: *Mammillaria blossfeldiana*. [RPS 2]
Ebnerella bocasana (Poselger) Buxbaum, Österr. Bot. Zeitschr. 98(1-2): 89, 1951. Basionym: *Mammillaria bocasana*. [RPS 2]
Ebnerella boedekeriana (Quehl) Buxbaum, Österr. Bot. Zeitschr. 98(1-2): 89, 1951. Basionym: *Mammillaria boedekeriana*. [RPS 2]
Ebnerella bombycina (Quehl) Buxbaum, Österr. Bot. Zeitschr. 98(1-2): 89, 1951. Basionym: *Mammillaria bombycina*. [RPS 2]
Ebnerella bullardiana (Gates) Buxbaum, Österr. Bot. Zeitschr. 98(1-2): 89, 1951. Basionym: *Mammillaria bullardiana*. [RPS 2]
Ebnerella capensis (Gates) Buxbaum, Österr. Bot. Zeitschr. 98(1-2) 89, 1951. Basionym: *Mammillaria capensis*. [RPS 2]
Ebnerella carretii (Rebut) Buxbaum, Österr. Bot. Zeitschr. 98(1-2): 89, 1951. Basionym: *Mammillaria carretii*. [RPS 2]
Ebnerella crinita (de Candolle) Buxbaum, Österr. Bot. Zeitschr. 98(1-2): 89, 1951. Basionym: *Mammillaria crinita*. [RPS 2]
Ebnerella denudata (Engelmann) Buxbaum, Österr. Bot. Zeitschr. 98(1-2): 89, 1951. Basionym: *Mammillaria denudata*. [RPS 2]
Ebnerella dioica (Brandegee) Buxbaum, Österr. Bot. Zeitschr. 98(1-2): 89, 1951. Basionym: *Mammillaria dioica*. [RPS 2]
Ebnerella dumetorum (Purpus) Buxbaum, Österr. Bot. Zeitschr. 98(1-2): 89, 1951. Basionym: *Mammillaria dumetorum*. [RPS 2]
Ebnerella erectohamata (Bödeker) Buxbaum, Österr. Bot. Zeitschr. 98(1-2): 89, 1951. Basionym: *Mammillaria erectohamata*. [RPS 2]
Ebnerella erythrosperma (Bödeker) Buxbaum, Österr. Bot. Zeitschr. 98(1-2): 89, 1951. Basionym: *Mammillaria erythrosperma*. [RPS 2]
Ebnerella fasciculata (Engelmann) Buxbaum, Österr. Bot. Zeitschr. 98(1-2): 89, 1951. Basionym: *Mammillaria fasciculata*. [RPS 2]
Ebnerella fraileana (Britton & Rose) Buxbaum, Österr. Bot. Zeitschr. 98(1-2): 89, 1951. Basionym: *Mammillaria fraileana*. [RPS 2]
Ebnerella gasseriana (Bödeker) Buxbaum, Österr. Bot. Zeitschr. 98(1-2): 89, 1951. Basionym: *Mammillaria gasseriana*. [RPS 2]

Ebnerella gilensis (Bödeker) Buxbaum, Österr. Bot. Zeitschr. 98(1-2): 89, 1951. Basionym: *Mammillaria gilensis*. [RPS 2]

Ebnerella glochidiata (Martius) Buxbaum, Österr. Bot. Zeitschr. 98(1-2): 89, 1951. Basionym: *Mammillaria glochidiata*. [RPS 2]

Ebnerella goodridgei (Scheer) Buxbaum, Österr. Bot. Zeitschr. 98(1-2): 89, 1951. Basionym: *Mammillaria goodridgei*. [RPS 2]

Ebnerella guirocobensis (Craig) Buxbaum, Österr. Bot. Zeitschr. 98(1-2): 89, 1951. Basionym: *Mammillaria guirocobensis*. [RPS 2]

Ebnerella haehneliana (Bödeker) Buxbaum, Österr. Bot. Zeitschr. 98(1-2): 89, 1951. Basionym: *Mammillaria haehneliana*. [RPS 2]

Ebnerella humboldtii (Ehrenberg) Buxbaum, Österr. Bot. Zeitschr. 98(1-2): 89, 1951. Basionym: *Mammillaria humboldtii*. [RPS 2]

Ebnerella hutchisoniana (Gates) Buxbaum, Österr. Bot. Zeitschr. 98(1-2): 89, 1951. Basionym: *Mammillaria hutchisoniana*. [RPS 2]

Ebnerella icamolensis (Bödeker) Buxbaum, Österr. Bot. Zeitschr. 98(1-2): 89, 1951. Basionym: *Mammillaria icamolensis*. [RPS 2]

Ebnerella inaiae (Craig) Buxbaum, Österr. Bot. Zeitschr. 98(1-2): 89, 1951. Basionym: *Mammillaria inaiae*. [RPS 2]

Ebnerella insularis (Gates) Buxbaum, Österr. Bot. Zeitschr. 98(1-2): 89, 1951. Basionym: *Mammillaria insularis*. [RPS 2]

Ebnerella jaliscana (Britton & Rose) Buxbaum, Österr. Bot. Zeitschr. 98(1-2): 89, 1951. Basionym: *Mammillaria jaliscana*. [RPS 2]

Ebnerella knebeliana (Bödeker) Buxbaum, Österr. Bot. Zeitschr. 98(1-2): 89, 1951. Basionym: *Mammillaria knebeliana*. [RPS 2]

Ebnerella kunzeana (Bödeker & Quehl) Buxbaum, Österr. Bot. Zeitschr. 98(1-2): 89, 1951. Basionym: *Mammillaria kunzeana*. [RPS 2]

Ebnerella lasiacantha (Engelmann) Buxbaum, Österr. Bot. Zeitschr. 98(1-2): 89, 1951. Basionym: *Mammillaria lasiacantha*. [RPS 2]

Ebnerella longicoma (Britton & Rose) Buxbaum, Österr. Bot. Zeitschr. 98(1-2): 89, 1951. Basionym: *Mammillaria longicoma*. [RPS 2]

Ebnerella magallanii (Schmoll) Buxbaum, Österr. Bot. Zeitschr. 98(1-2): 89, 1951. Basionym: *Mammillaria magallanii*. [RPS 2]

Ebnerella mainae (Brandegee) Buxbaum, Österr. Bot. Zeitschr. 98(1-2): 89, 1951. Basionym: *Mammillaria mainae*. [RPS 2]

Ebnerella mazatlanensis (Schumann) Buxbaum, Österr. Bot. Zeitschr. 98(1-2): 89, 1951. Basionym: *Mammillaria mazatlanensis*. [RPS 2]

Ebnerella mercadensis (Patoni) Buxbaum, Österr. Bot. Zeitschr. 98(1-2): 89, 1951. Basionym: *Mammillaria mercadensis*. [RPS 2]

Ebnerella microcarpa (Engelmann) Buxbaum, Österr. Bot. Zeitschr. 98(1-2): 89, 1951. Basionym: *Mammillaria microcarpa*. [Validity of this name depending on the status of the basionym which might be invalid under Art. 34.] [RPS 2]

Ebnerella moelleriana (Bödeker) Buxbaum, Österr. Bot. Zeitschr. 98(1-2): 89, 1951. Basionym: *Mammillaria moelleriana*. [RPS 2]

Ebnerella monancistra (A. Berger) Buxbaum, Österr. Bot. Zeitschr. 98(1-2): 90, 1951. Basionym: *Mammillaria monancistra*. [RPS 2]

Ebnerella multiceps (Salm-Dyck) Buxbaum, Österr. Bot. Zeitschr. 98(1-2): 90, 1951. Basionym: *Mammillaria multiceps*. [RPS 2]

Ebnerella multiformis (Britton & Rose) Buxbaum, Österr. Bot. Zeitschr. 98(1-2): 90, 1951. Basionym: *Mammillaria multiformis*. [RPS 2]

Ebnerella multihamata (Bödeker) Buxbaum, Österr. Bot. Zeitschr. 98(1-2): 90, 1951. Basionym: *Mammillaria multihamata* var. *fittkaui*. [RPS 2]

Ebnerella nunezii (Britton & Rose) Buxbaum, Österr. Bot. Zeitschr. 98(1-2): 90, 1951. Basionym: *Neomammillaria nunezii*. [Basionym erroneously cited as *Mammillaria nunezii*.] [RPS 2]

Ebnerella occidentalis (Britton & Rose) Buxbaum, Österr. Bot. Zeitschr. 98(1-2): 90, 1951. Basionym: *Neomammillaria occidentalis*. [Basionym erroneously given as *Mammillaria occidentalis*.] [RPS 2]

Ebnerella oliviae (Orcutt) Buxbaum, Österr. Bot. Zeitschr. 98(1-2): 90, 1951. Basionym: *Mammillaria oliviae*. [RPS 2]

Ebnerella painteri (Rose) Buxbaum, Österr. Bot. Zeitschr. 98(1-2): 90, 1951. Basionym: *Mammillaria painteri*. [RPS 2]

Ebnerella phitauiana (Baxter) Buxbaum, Österr. Bot. Zeitschr. 98(1-2): 90, 1951. Basionym: *Mammillaria phitauiana*. [RPS 2]

Ebnerella plumosa (F. A. C. Weber) Buxbaum, Österr. Bot. Zeitschr. 98(1-2): 90, 1951. Basionym: *Mammillaria plumosa*. [RPS 2]

Ebnerella posseltiana (Bödeker) Buxbaum, Österr. Bot. Zeitschr. 98(1-2): 90, 1951. Basionym: *Mammillaria posseltiana*. [RPS 2]

Ebnerella prolifera (Miller) Buxbaum, Österr. Bot. Zeitschr. 98(1-2): 90, 1951. Basionym: *Mammillaria prolifera*. [RPS 2]

Ebnerella pubispina (Bödeker) Buxbaum, Österr. Bot. Zeitschr. 98(1-2): 90, 1951. Basionym: *Mammillaria pubispina*. [RPS 2]

Ebnerella pygmaea (Britton & Rose) Buxbaum, Österr. Bot. Zeitschr. 98(1-2): 90, 1951. Basionym: *Neomammillaria pygmaea*. [Basionym erroneously cited as *Mammillaria pygmaea*.] [RPS 2]

Ebnerella rekoi (Britton & Rose) Buxbaum, Österr. Bot. Zeitschr. 98(1-2): 90, 1951. Basionym: *Neomammillaria rekoi*. [Basionym erroneously given as

Mammillaria rekoi.] [RPS 2]

Ebnerella rettigiana (Bödeker) Buxbaum, Österr. Bot. Zeitschr. 98(1-2): 90, 1951. Basionym: *Mammillaria rettigiana.* [RPS 2]

Ebnerella scheidweileriana (Otto) Buxbaum, Österr. Bot. Zeitschr. 98(1-2): 90, 1951. Basionym: *Mammillaria scheidweileriana.* [RPS 2]

Ebnerella schelhasei (Pfeiffer) Buxbaum, Österr. Bot. Zeitschr. 98(1-2): 90, 1951. Basionym: *Mammillaria schelhasei.* [RPS 2]

Ebnerella schiedeana (Ehrenberg) Buxbaum, Österr. Bot. Zeitschr. 98(1-2): 90, 1951. Basionym: *Mammillaria schiedeana.* [RPS 2]

Ebnerella seideliana (Quehl) Buxbaum, Österr. Bot. Zeitschr. 98(1-2): 90, 1951. Basionym: *Mammillaria seideliana.* [RPS 2]

Ebnerella sheldonii (Britton & Rose) Buxbaum, Österr. Bot. Zeitschr. 98(1-2): 90, 1951. Basionym: *Neomammillaria sheldonii.* [Basionym erroneously given as *Mammillaria sheldonii.*] [RPS 2]

Ebnerella sinistrohamata (Bödeker) Buxbaum, Österr. Bot. Zeitschr. 98(1-2): 90, 1951. Basionym: *Mammillaria sinistrohamata.* [RPS 2]

Ebnerella solisii (Britton & Rose) Buxbaum, Österr. Bot. Zeitschr. 98(1-2): 90, 1951. Basionym: *Neomammillaria solisii.* [Basionym erroneously cited as *Mammillaria solisii.*] [RPS 2]

Ebnerella sphacelata (Martius) Buxbaum, Österr. Bot. Zeitschr. 98(1-2): 90, 1951. Basionym: *Mammillaria sphacelata.* [RPS 2]

Ebnerella surculosa (Bödeker) Buxbaum, Österr. Bot. Zeitschr. 98(1-2): 90, 1951. Basionym: *Mammillaria surculosa.* [RPS 2]

Ebnerella surculosa (Fric) Buxbaum, Österr. Bot. Zeitschr. 98(1-2): 89, 1951. Basionym: *Mammillaria esshaussieri.* [Sphalm. 'esshaussierii'.] [RPS 2]

Ebnerella swinglei (Britton & Rose) Buxbaum, Österr. Bot. Zeitschr. 98(1-2): 90, 1951. Basionym: *Neomammillaria swinglei.* [Basionym erroneously given as *Mammillaria swinglei.*] [RPS 2]

Ebnerella tacubayensis (Fedde) Buxbaum, Österr. Bot. Zeitschr. 98(1-2): 90, 1951. Basionym: *Mammillaria tacubayensis.* [RPS 2]

Ebnerella trichacantha (Schumann) Buxbaum, Österr. Bot. Zeitschr. 98(1-2): 90, 1951. Basionym: *Mammillaria trichacantha.* [RPS 2]

Ebnerella unihamata (Bödeker) Buxbaum, Österr. Bot. Zeitschr. 98(1-2): 90, 1951. Basionym: *Mammillaria unihamata.* [RPS 2]

Ebnerella verhaertiana (Bödeker) Buxbaum, Österr. Bot. Zeitschr. 98(1-2): 90, 1951. Basionym: *Mammillaria verhaertiana.* [RPS 2]

Ebnerella viereckii (Bödeker) Buxbaum, Österr. Bot. Zeitschr. 98(1-2): 90, 1951. Basionym: *Mammillaria viereckii.* [RPS 2]

Ebnerella weingartiana (Bödeker) Buxbaum, Österr. Bot. Zeitschr. 98(1-2): 90, 1951. Basionym: *Mammillaria weingartiana.* [RPS 2]

Ebnerella wilcoxii (Toumey *ex* Schumann) Buxbaum, Österr. Bot. Zeitschr. 98(1-2): 90, 1951. Basionym: *Mammillaria* × *wilcoxii.* [RPS 2]

Ebnerella wildii (Dietrich) Buxbaum, Österr. Bot. Zeitschr. 98(1-2): 90, 1951. Basionym: *Mammillaria wildii.* [RPS 2]

Ebnerella wrightii (Engelmann) Buxbaum, Österr. Bot. Zeitschr. 98(1-2): 90, 1951. Basionym: *Mammillaria wrightii.* [RPS 2]

Ebnerella yaquensis (Craig) Buxbaum, Österr. Bot. Zeitschr. 98(1-2): 90, 1951. Basionym: *Mammillaria yaquensis.* [RPS 2]

Ebnerella zeilmanniana (Bödeker) Buxbaum, Österr. Bot. Zeitschr. 98(1-2): 90, 1951. Basionym: *Mammillaria zeilmanniana.* [RPS 2]

Ebnerella zephyranthoides (Scheidweiler) Buxbaum, Österr. Bot. Zeitschr. 98(1-2): 90, 1951. Basionym: *Mammillaria zephyranthoides.* [RPS 2]

Eccremocactus imitans (Kimnach & P. C. Hutchison) Kimnach, Cact. Succ. J. (US) 34: 82, 1962. Basionym: *Werckleocereus imitans.* [RPS 13]

Eccremocactus rosei Kimnach, Cact. Succ. J. (US) 34: 78-82, 1962. [RPS 13]

× *Echinaporus* Ito, The Cactaceae, 649-650, ill., 1981. *Nom. inval.* (Art. H6.2). [= *Echinopsis* × *Aporocactus*; the parentage may be doubted from Ito's illustration.] [RPS -]

× **Echinobergia** Mottram, Contr. New Class. Cact. Fam., 35, 1990. [Validity depends on the application of the name *Echinofossulocactus.*] [= *Echinofossulocactus* × *Leuchtenbergia.*] [RPS 41]

× **Echinobivia** Rowley, Nation. Cact. Succ. J. 21: 82, 1966. [= *Echinopsis* × *Lobivia.*] [RPS 17]

× *Echinobivia* cv. *Andenken an A. V. Fric* Kilian, Stachelpost 6: 282, 1970. *Nom. inval.* (Art. 30 [ICBNCP]). [RPS 22]

× *Echinobivia* cv. *Andenken an Dr. Paul Schmidt* Kilian, Stachelpost 7: 371, 427, 1971. *Nom. inval.* (Art. 30 [ICBNCP]). [RPS 22]

× **Echinobivia** cv. **Aubergine** Kilian, Stachelpost 7: 373, 427, 1971. [RPS 22]

× **Echinobivia** cv. **Leibnitz** G. Unger, Kakt. and. Sukk. 36(11): 242-243, ills., 1985. [= *Lobivia backebergii* var. *hertrichiana* × *Echinopsis eyriesii.*] [RPS 36]

Echinocacti Mottram, Contr. New Class. Cact. Fam., 4, 10, 1990. *Nom. inval.* (Art. 36.1). [Published at the rank of *linea.*] [RPS 41]

Echinocactus acunensis W. T. Marshall, Saguaroland Bull. 7: 33-34, 1953. [RPS -]

Echinocactus erectocentrus var. *pallidus* (Backeberg) Weniger, Cacti of the Southwest, 90-91, pl. 24, 1970. *Nom. inval.*, based on *Echinomastus pallidus, nom. inval.* (Art. 36.1, 37.1). [Not dated.] [RPS -]

Echinocactus flavidispinus (Backeberg) Weniger, Cacti of the Southwest, 87, 1970. *Nom. inval.* (Art. 33.2), based on *Thelocactus bicolor* var. *flavidispinus.* [Not dated. The name given as basionym is itself a combination (here corrected).] [RPS -]

Echinocactus grusonii cv. *L. J. van Veen* Janse, Succulenta 56: 41-43, 1977. *Nom. inval.* (Art. 30 [ICBNCP]). [= *Echinocactus grusonii subinermis* hort.] [RPS 28]

Echinocactus horizonthalonius var. **nicholii** L. Benson, Cacti of Arizona, ed. 3, 23, 1969. [RPS 20]

Echinocactus ingens var. **grandis** (Rose) Krainz, Kat. ZSS ed. 2, 49, 1967. Basionym: *Echinocactus grandis*. [RPS 18]

Echinocactus ingens var. **palmeri** (Rose) Krainz, Kat. ZSS ed. 2, 49, 1967. Basionym: *Echinocactus palmeri*. [RPS 18]

Echinocactus mariposensis (Hester) Weniger, Cacti of the Southwest, 92, 1970. Basionym: *Echinomastus mariposensis*. [Not dated.] [RPS -]

Echinocactus mesae-verdae (Boissevain) L. Benson, Leaflets West. Bot. 6: 163, 1951. Basionym: *Coloradoa mesae-verdae*. [RPS 3]

Echinocactus platyacanthus fa. **grandis** (Rose) Bravo, Cact. Suc. Mex. 25(3): 64, 1980. Basionym: *Echinocactus grandis*. [RPS 31]

Echinocactus platyacanthus fa. **visnaga** (Hooker) Bravo, Cact. Suc. Mex. 25(3): 65, 1980. Basionym: *Echinocactus visnaga*. [RPS 31]

Echinocactus tobuschii (W. T. Marshall) Weniger, Cacti of the Southwest, 78, 1970. *Nom. inval.* (Art. 33.2), based on *Ancistrocactus tobuschii*. [RPS -]

Echinocereeae Buxbaum, Madroño 14(6): 193, 1958. Typus: *Echinocereus*. [Sphalm. 'Echinocereae'.] [RPS 9]

Echinocereus sect. **Erecti** (Schumann) Bravo, Cact. Suc. Mex. 27(1): 16, 1982. Basionym: *Echinocereus* ser. *Erecti*. [Sphalm. 'Erecta'. Lectotypus: *E. fendleri* (Engelmann) Rümpler. The lectotypi fication is unnecessary, as the basionym has already been typified by Buxbaum in 1974] [RPS 33]

Echinocereus sect. **Morangaya** (Rowley) N. P. Taylor, Genus Echinocereus, 31, 1985. Basionym: *Morangaya*. [RPS 36]

Echinocereus sect. **Pulchellus** N. P. Taylor, Genus Echinocereus, 140, 1985. Typus: *E. pulchellus* (C. Martius) Schumann. [RPS 36]

Echinocereus sect. **Reichenbachii** N. P. Taylor, Genus Echinocereus, 105, 1985. Typus: *E. reichenbachii* (Terscheck *ex* Walpers) Hort. F. A. Haage. [RPS 36]

Echinocereus sect. **Triglochidiata** Bravo, Cact. Suc. Mex. 18: 108-109, 1973. Typus: *E. triglochidiatus* Engelmann. Spelling altered to 'Triglochidiatus' by N. Taylor, Genus Echinocereus, 58, 1985.. [RPS 24]

Echinocereus sect. **Wilcoxia** (Britton & Rose) N. P. Taylor, Genus Echinocereus, 134, 1985. Basionym: *Wilcoxia*. [RPS 36]

Echinocereus ser. **Compacti** H. Kunzmann, Kakt. and. Sukk. 36(4): 76, 1985. Typus: *E. triglochidiatus* Engelmann. [RPS 36]

Echinocereus ser. *Decalophi* (Salm-Dyck) H. Kunzmann, Kakt. and. Sukk. 36(4): 74, 1985. *Nom. inval.* (Art. 33.2). [Based on *Cereus* [infrageneric group] *Decalophi* Salm-Dyck. Lectotype *E. stramineus* of Kunzmann unnecessarily supersedes choice of *E. fendleri* by Buxbaum, 1974.] [RPS 36]

Echinocereus ser. **Fasciculatae** Bravo, Cact. Suc. Mex. 27(1): 16, 1982. Typus: *Echinocereus fasciculatus* Benson. The validity of the name is questionable and depends on the validity of its type (see comment there).. [RPS 33]

Echinocereus ser. *Pectinati* (Salm-Dyck) H. Kunzmann, Kakt. and. Sukk. 36(4): 74, 1985. *Nom. inval.* (Art. 33.2). [Based on *Cereus* [infrageneric group] *Pectinati* Salm-Dyck.] [RPS 36]

Echinocereus ser. **Penicillati** H. Kunzmann, Kakt. and. Sukk. 36(4): 78, 1985. Typus: *E. longisetus* (Engelmann) Lemaire. [RPS 36]

Echinocereus ser. *Viridiflorae* Bravo, Cact. Suc. Mex. 27(1): 16, 1982. *Nom. inval.* (Art. 32.1, 22.1). [Typus: *E. viridiflorus (typus generis)*.] [RPS 33]

Echinocereus subser. **Breviseti** H. Kunzmann, Kakt. and. Sukk. 36(4): 79, 1985. Typus: *E. nivosus* Glass & Foster. [RPS 36]

Echinocereus subser. *Brevispini* H. Kunzmann, Kakt. and. Sukk. 36(4): 79, 1985. *Nom. illeg.* (Art. 63.1). [Based on *E. pectinatus* (Scheidweiler) Engelmann, the type of *E.* subser. *Pectinati* (Salm-Dyck) Schumann 1898.] [RPS 36]

Echinocereus subser. **Echinacanthi** H. Kunzmann, Kakt. and. Sukk. 36(4): 79, 1985. Typus: *E. grandis* Britton & Rose. [RPS 36]

Echinocereus subser. **Engelmannianae** H. Kunzmann, Kakt. and. Sukk. 36(4): 78, 1985. Typus: *E. engelmannii* (Parry *ex* Engelmann) Lemaire. [RPS 36]

Echinocereus subser. *Fendlerianae* H. Kunzmann, Kakt. and. Sukk. 36(4): 77, 1985. *Nom. illeg.* (Art. 63.1). [Based on *E. fendleri* (Engelmann) Rümpler, which was selected as lectotype of *E.* subser. *Decalophi* (Salm-Dyck) Schumann by Buxbaum in 1974.] [RPS 36]

Echinocereus subser. *Matthesianae* H. Kunzmann, Kakt. and. Sukk. 36(4): 75, 1985. *Nom. inval.* (Art. 37). [Based on *E. matthesianus* Backeberg, *nom. inval.* (Arts. 9.5, 37).] [RPS 36]

Echinocereus subser. **Nigrispini** H. Kunzmann, Kakt. and. Sukk. 36(4): 80, 1985. Typus: *E. palmeri* Britton & Rose. [RPS 36]

Echinocereus subser. *Pectinispini* H. Kunzmann, Kakt. and. Sukk. 36(4): 80, 1985. *Nom. inval.* (Art. 22.1, 32.1b). [Based on *E. viridiflorus* Engelmann (*typus generis*).] [RPS 36]

Echinocereus subser. **Peniculispini** H. Kunzmann, Kakt. and. Sukk. 36(4): 80, 1985. Typus: *E. russsanthus* Weniger. [RPS 36]

Echinocereus subser. **Roseiani** H. Kunzmann, Kakt. and. Sukk. 36(4): 76, 1985. Typus: *E. rosei* Wooton & Standley. [RPS 36]

Echinocereus subser. **Straminei** H. Kunzmann, Kakt. and. Sukk. 36(4): 77, 1985. Typus: *E. stramineus* (Engelmann) Rümpler. [RPS 36]

Echinocereus subser. **Triglochidiati** H. Kunzmann, Kakt. and. Sukk. 36(4): 76, 1985. Typus: *E. triglochidiatus* Engelmann. Can also be considered to be a *stat. nov.* for *E.*

sect. *Triglochidiata* H. Bravo-H. (1973) with corrected orthography. [RPS 36]

Echinocereus adustus var. **schwarzii** (Lau) N. P. Taylor, Kew Mag. 2(2): 268, 1985. Basionym: *Echinocereus schwarzii.* [RPS 36]

Echinocereus albatus Backeberg, Die Cact. 4: 2007-2008, fig. 1910; l.c. 6: 3847, fig. 3485 (1962), 1960. *Nom. inval.* (Art. 37.1). [Erroneously included as valid in RPS 11.] [RPS 11]

Echinocereus baileyi var. *caespiticus* Backeberg, Die Cactaceae 4: 2011, 1960. *Nom. inval.* (Art. 37.1). [Erroneously included as valid in RPS 11.] [RPS 11]

Echinocereus berlandieri var. **angusticeps** (Clover) L. Benson, Cact. Succ. J. (US) 48(2): 59, 1976. Basionym: *Echinocereus angusticeps.* [RPS 27]

Echinocereus berlandieri var. **papillosus** (A. Linke *ex* Rümpler) L. Benson, Cact. Succ. J. (US) 48(2): 59, 1976. Basionym: *Echinocereus papillosus.* [RPS 27]

Echinocereus blanckii var. **angusticeps** (Clover) L. Benson, in Lundell et al., Flora of Texas 2: 260, 1969. Basionym: *Echinocereus angusticeps.* [RPS 21]

Echinocereus blanckii var. **berlandieri** (Engelmann) Backeberg, Die Cact. 4: 1999, 1997, 1960. Basionym: *Cereus berlandieri.* [Basionym given on p. 1997.] [RPS -]

Echinocereus blanckii var. **papillosus** (A. Linke) L. Benson, Cact. Succ. J. (US) 41: 126, 1969. Basionym: *Echinocereus papillosus.* [RPS 20]

Echinocereus bristolii var. **pseudopectinatus** N. P. Taylor, Genus Echinocereus, 120, 1985. Typus: *A. Lau* 607 (K). [Ex cult. N. P. Taylor.] [Mexico] [RPS 36]

Echinocereus caespitosus var. *perbellus* (Britton & Rose) Weniger, Cacti of the Southwest, 23, 1970. *Nom. inval.* (Art. 33.2), based on *Echinocereus perbellus.* [Not dated.] [RPS -]

Echinocereus caespitosus var. *purpureus* (Lahman) Weniger, Cacti of the Southwest, 23, 1970. *Nom. inval.* (Art. 33.2), based on *Echinocereus purpureus.* [Not dated.] [RPS -]

Echinocereus chisoensis var. **fobeanus** (Oehme) N. P. Taylor, Kew Mag. 2(2): 261, 1985. Basionym: *Echinocereus fobeanus.* [RPS 36]

Echinocereus chloranthus var. **cylindricus** (Engelmann) N. P. Taylor, Kew Mag. 1(4): 169, 1984. Basionym: *Cereus viridiflorus* var. *cylindricus.* [RPS 35]

Echinocereus chloranthus var. *flavispinus* Y. Ito, Cacti, 133, 1952. *Nom. inval.* (Art. 36.1). [RPS 4]

Echinocereus chloranthus var. **neocapillus** Weniger, Cact. Succ. J. (US) 41: 39-41, 1969. [See N. Taylor, Genus Echinocereus, 103, 1985, on the typification of the taxon and a possible choice of a lectotype by Benson 1982.] [RPS 20]

Echinocereus chloranthus var. **russanthus** (Weniger) Lamb *ex* Rowley, Rep. Pl. Succ. 23: 7, 1974. Basionym: *Echinocereus russanthus.* [First invalidly published in Monthly Notes Exotic Coll. 1972: 62.] [RPS 23]

Echinocereus cinerascens var. **ehrenbergii** (Pfeiffer) Bravo, Cact. Suc. Mex. 19: 47, 1974. Basionym: *Cereus ehrenbergii.* [RPS 25]

Echinocereus cinerascens var. **septentrionalis** N. P. Taylor, Bradleya 6: 68-69, ill. (p. 84), 1988. Typus: *Taylor* 281 (MEXU, K). [Mexico: San Luis Potosi] [RPS 39]

Echinocereus cinerascens var. **tulensis** (Bravo) N. P. Taylor, Bradleya 6: 69, ill. (p. 84), 1988. Basionym: *Echinocereus tulensis.* [RPS 39]

Echinocereus coccineus var. **arizonicus** (Rose) Ferguson, Cact. Succ. J. (US) 61(5): 221, 1989. Basionym: *Echinocereus arizonicus.* [RPS 40]

Echinocereus coccineus var. *conoideus* (Engelmann & Bigelow) Weniger, Cacti of the Southwest, 42, 1970. *Nom. inval.* (Art. 33.2), based on *Cereus conoideus.* [Erroneously ascribed to Engelmann.] [RPS -]

Echinocereus coccineus var. **gurneyi** (L. Benson) Heil & Brack, Cact. Succ. J. (US) 60(1): 26, 1988. Basionym: *Echinocereus triglochidiatus* var. *gurneyi.* [RPS 39]

Echinocereus coccineus var. **kunzei** (Gürke) Backeberg, Die Cact. 4: 2070, 1960. Basionym: *Echinocereus kunzei.* [RPS 11]

Echinocereus coccineus var. **paucispinus** (Engelmann) Ferguson, Cact. Succ. J. (US) 61(5): 222, 1989. Basionym: *Cereus paucispinus.* [RPS 40]

Echinocereus dasyacanthus var. **ctenoides** (Engelmann) Backeberg, Die Cact. 4: 2021, 1960. Basionym: *Cereus ctenoides.* [Repeated (*nom. inval.*, Art. 33) by Weniger, Cacti of the Southwest, 25, 1970.] [RPS 11]

Echinocereus dasyacanthus var. *hildmannii* (Arendt) Weniger, Cacti of the Southwest, 33, 1970. *Nom. inval.* (Art. 33.2), based on *Echinocereus hildmannii.* [Sphalm. 'hildmanii'. Erroneously attributed to Arendt.] [RPS -]

Echinocereus delaetii var. **freudenbergeri** (G. R. W. Frank) N. P. Taylor, Genus Echinocereus, 97, 1985. Basionym: *Echinocereus freudenbergeri.* [RPS 36]

Echinocereus engelmannii var. **acicularis** L. Benson, Cacti of Arizona, ed. 3, 22, 138-139, fig. 3.19, 1969. Typus: *Benson* 16616 (POM 311313). [USA: Arizona] [RPS 20]

Echinocereus engelmannii var. **armatus** L. Benson, Cact. Succ. J. (US) 41: 33, 1969. Typus: *Benson* 14767 (POM 284927 (2 sheets)). [USA: California] [RPS 20]

Echinocereus engelmannii var. **howei** L. Benson, Cact. Succ. J. (US) 46: 80, 1974. Typus: *Howe* 4570 (POM 317886). [USA: California] [RPS 25]

Echinocereus engelmannii var. **purpureus** L. Benson, Cact. Succ. J. (US) 41: 126-127, 1969. Typus: *Benson* 13637 (POM 285578). [USA: Utah] [RPS 20]

Echinocereus enneacanthus fa. **brevispinus** W. O. Moore, Brittonia 19: 93-94, fig. 34, 1967. Typus: *Clover* s.n. (MICH). [Ex cult. *Moore*

508.] [USA: Texas] [RPS 18]

Echinocereus enneacanthus fa. **intermedius** W. O. Moore, Brittonia 19: 93, 1967. Typus: *Clover in Moore* 431 (MICH). [USA: Texas] [RPS 18]

Echinocereus enneacanthus var. **brevispinus** (W. O. Moore) L. Benson, Cact. Succ. J. (US) 41: 127, 1969. Basionym: *Echinocereus enneacanthus* fa. *brevispinus.* [RPS 20]

Echinocereus enneacanthus var. **conglomeratus** (Förster) L. Benson, Cact. Succ. J. (US) 46: 80, 1974. Basionym: *Echinocereus conglomeratus.* [RPS 25]

Echinocereus enneacanthus var. **dubius** (Engelmann) L. Benson, Cact. Succ. J. (US) 41: 127, 1969. Basionym: *Cereus dubius.* [RPS 20]

Echinocereus enneacanthus var. *major* hort. *ex* Borg, Cacti, 223, 1951. *Nom. inval.* (Art. 36.1). [RPS -]

Echinocereus enneacanthus var. **stramineus** (Engelmann) L. Benson, Cact. Succ. J. (US) 41: 127, 1969. Basionym: *Cereus stramineus.* [RPS 20]

Echinocereus fasciculatus (Engelmann) L. Benson, The Cacti of Arizona ed. 3, 21, 132, ills., 1969. Basionym: *Mammillaria fasciculata.* [The validity of this name is questionable due to the fact that the basionym as cited by Benson was a provisional name only. This also affects any subordinate combinations.] [RPS 20]

Echinocereus fasciculatus var. **bonkerae** (Thornber & Bonker) L. Benson, Cacti of Arizona, ed. 3, 22, 136, ills., 1969. Basionym: *Echinocereus bonkerae.* [May be invalid, depending on the validity of *Echinocereus fasciculatus.*] [RPS 20]

Echinocereus fasciculatus var. **boyce-thompsonii** (Orcutt) L. Benson, Cacti of Arizona, ed. 3, 21, 132, ills., 1969. Basionym: *Echinocereus boyce-thompsonii.* [May be invalid, depending on the validity of *Echinocereus fasciculatus.*] [RPS 20]

Echinocereus fendleri var. **albiflorus** (Weingart) Backeberg, Die Cact. 4: 2047, 1960. Basionym: *Echinocereus albiflorus.* [RPS 11]

Echinocereus fendleri var. **fasciculatus** (Engelmann) N. P. Taylor, Kew Mag. 2(2): 252, 1985. Basionym: *Mammillaria fasciculata.* [RPS 36]

Echinocereus fendleri var. **kuenzleri** (Castetter et al.) L. Benson, Cacti U.S. & Canada, 942, 1982. Basionym: *Echinocereus kuenzleri.* [RPS 33]

Echinocereus fendleri var. **ledingii** (Peebles) N. P. Taylor, Kew Mag. 2(2): 253, 1985. Basionym: *Echinocereus ledingii.* [RPS 36]

Echinocereus ferreirianus H. Gates, Saguaroland Bull. 7(1): 8-11, ills., 1953. Typus: *Gates* s.n. (DS). [Mexico: Baja California Norte] [RPS 4]

Echinocereus ferreirianus var. **lindsayi** (Meyran) N. P. Taylor, Genus Echinocereus, 46, 1985. Basionym: *Echinocereus lindsayi.* [RPS 36]

Echinocereus freudenbergeri G. R. W. Frank, Kakt. and. Sukk. 32(5): 102-105, ills., 1981. Typus: *Freudenberger* s.n. (ZSS). [Mexico: Coahuila] [RPS 32]

Echinocereus galtieri Rebut *ex* Borg, Cacti, ed. 2, 228, in synon, 1951. *Nom. inval.* (Art. 36.1). [RPS -]

Echinocereus knippelianus var. **kruegeri** Glass & Foster, Cact. Succ. J. (US): 50(2): 79-80, ills., 1978. Typus: *Glass et Foster* 3902 (POM, ZSS). [Mexico: Nuevo Leon] [RPS 29]

Echinocereus knippelianus var. **reyesii** Lau, Cact. Succ. J. (US) 52(6): 264-265, 1980. Typus: *Lau* 1237A (POM). [Mexico: Nuevo Leon] [RPS 31]

Echinocereus kuenzleri Castetter et al., Cact. Succ. J. (US) 48(2): 77-78, figs. 1-2, 1976. Typus: *Kuenzler* 3585 (UNM 55571). [USA: New Mexico] [RPS 27]

Echinocereus laui G. R. W. Frank, Kakt. and. Sukk. 29(4): 74-77, ills., 1978. Typus: *Lau* 780 (ZSS). [Sphalm. 'lauii'.] [Mexico: Sonora] [RPS 29]

Echinocereus leucanthus N. P. Taylor, Genus Echinocereus, 136, 1985. [*Nom. nov. pro Wilcoxia albiflora* Backeberg, *non Echinocereus albiflorus* Weingart. Type as for *Wilcoxia albiflora* Backeberg.] [RPS 36]

Echinocereus lindsayi Meyran, Cact. Suc. Mex. 20: 77, 79-83, ills., 1976. Typus: *Meyran et al. in Sanchez-Mejorada* 2424 (MEXU). [Probably already published 1975.] [Mexico: Baja California Norte] [RPS 27]

Echinocereus longisetus var. *albatus* (Backeberg) W. Sterk, Succulenta 59(1): 12-13, ill., 1980. *Nom. inval.*, based on *Echinocereus albatus, nom. inval.* (Art. 37.1). [RPS -]

Echinocereus longisetus var. **delaetii** (M. Gürke) N. P. Taylor, Bradleya 6: 79, 1988. Basionym: *Cephalocereus delaetii.* [RPS 39]

Echinocereus longispinus Lahman, Cact. Succ. J. (US) 22: 128, 1950. Typus: *Lahman* s.n. (MO [lecto], NY [iso]). [Lectotype selected from various isotypes by Benson, 1982.] [First published invalidly (Art. 36.1) in l.c. 7: 135, 1936.] [USA: Oklahoma] [RPS 9]

Echinocereus mariae Backeberg, Nation. Cact. Succ. J. 20: 19, 1965. Based on *Polaski* s.n.. *Nom. inval.* (Art. 9.5). [Erroneously included as valid in RPS 16., and incorrectly treated as valid by Eggli, Bradleya 3: 102, 1985.] [?] [RPS 16]

Echinocereus maritimus var. **hancockii** (E. Dawson) N. P. Taylor, Genus Echinocereus, 44, 1985. Basionym: *Echinocereus hancockii.* [RPS 36]

Echinocereus marksianus Backeberg, Kakt.-Lex., 124, 1966. *Nom. inval.* (Art. 37.1). [RPS -]

Echinocereus matthesianus Backeberg, Descr. Cact. Nov. 3: 6, 1963. *Nom. inval.* (Art. 9.5, 37.1). [Also in Kakt.-Lex., fig. 90, 1966.] [RPS 14]

Echinocereus matudae Bravo, Anales Inst. Biol. UNAM 31: 119-121, ill., 1960. Typus: *Matuda* s.n. (MEXU?). [No herbarium cited.] [No specimen is actually cited as type, but a single collection is mentioned which is taken as representing the type.]

[Mexico: Chihuahua] [RPS -]

Echinocereus melanacanthus (Engelmann) W. H. Earle, Cacti Southwest, 61, 1963. *Nom. illeg.* (Art. 63.1), based on *Cereus coccineus* var. *melanacanthus*. [Incorrectly used; *E. coccineus* 1848 appears in synonymy and has priority.] [RPS -]

Echinocereus metornii G. R. W. Frank, Kakt. and. Sukk. 41(10): 210-218, ills., SEM-ills., 1990. Typus: *Metorn* 49 (ZSS). [Mexico: Coahuila] [RPS 41]

Echinocereus mombergerianus G. R. W. Frank, Kakt. and. Sukk. 41(11): 261, 1990. Typus: *Momberger* 1 (ZSS). [First published (with unclear typification and therefore invalidly [Art. 37.1]) in l.c. 40(11): 272-277, ills., 1989 (cf. RPS 40).] [Mexico: Baja California del Norte] [RPS 41]

Echinocereus morricalii Riha, Kaktusy 11: 75, 78, 1975. [RPS 26]

Echinocereus nicholii (L. Benson) Parfitt, Phytologia 63(3): 157-158, 1987. Basionym: *Echinocereus engelmannii* var. *nicholii*. [RPS 38]

Echinocereus nivosus Glass & Foster, Cact. Succ. J. (US) 50(1): 18-19, ills., 1978. Typus: *Glass et Foster* 3764 (POM). [Mexico: Coahuila] [RPS 29]

Echinocereus oklahomensis Lahman, Cact. Succ. J. (US) 22: 128, 1950. Typus: *Lahman* s.n. (MO, US). [First published invalidly (Art. 36.1) in l.c. 6: 141, ill., 1935.] [USA: Oklahoma] [RPS 9]

Echinocereus pamanesiorum Lau, Cact. Succ. Mex. 26(2): 25, 36-41, ills., 1981. Typus: *Lau* 1247 (MEXU, ZSS). [Mexico: Zacatecas] [RPS 32]

Echinocereus parkeri N. P. Taylor, Bradleya 6: 73, ills. (pp. 73-74), 1988. Typus: *Hansen et al.* 3863 (MEXU). [Mexico: Nuevo Leon] [RPS 39]

Echinocereus parkeri var. **gonzalezii** N. P. Taylor, Bradleya 6: 74, ills. (p. 74, 84), 1988. Typus: *Gonzalez Medrano* 8515 (MEXU). [Mexico: Tamaulipas] [RPS 39]

Echinocereus pectinatus fa. **castaneus** (Engelmann) Krainz, Kat. ZSS ed. 2, 50, 1967. Basionym: *Cereus caespitosus* var. *castaneus*. [RPS 18]

Echinocereus pectinatus fa. **rigidissimus** (Engelmann) Krainz, Kat. ZSS ed. 2, 50, 1967. Basionym: *Cereus pectinatus* var. *rigidissimus*. [First mentioned (attributed to Engelmann) by Schelle, Handb. Kakt.-kult., 131, 1907.] [RPS 18]

Echinocereus pectinatus var. **bristolii** (W. T. Marshall) W. T. Marshall, Saguaroland Bull. 10(7): 81-82, 1956. Basionym: *Echinocereus bristolii*. [RPS 7]

Echinocereus pectinatus var. **dasyacanthus** (Engelmann) N. P. Taylor, Kew Mag. 1(4): 179, 1984. Basionym: *Echinocereus dasyacanthus*. [First published invalidly (Art. 33.2) and attributed to L. Benson by W. H. Earle, Saguaroland Bull. 25: 80, 1971.] [RPS 35]

Echinocereus pectinatus var. **minor** (Engelmann) L. Benson, Cact. Succ. J. (US) 40: 119-127, 1968. Basionym: *Cereus dasyacanthus* var. *minor*. [RPS 19]

Echinocereus pectinatus var. **rubispinus** G. R. W. Frank & Lau, Kakt. and. Sukk. 33(2): 32-35, ills., 1982. Typus: *Lau* 88 (ZSS). [Mexico: Chihuahua] [RPS 33]

Echinocereus pectinatus var. **wenigeri** L. Benson, Cact. Succ. J. (US) 40: 119-127, fig. 3, 1968. Typus: *Benson* 16521 (POM 311338). [USA: Texas] [RPS 19]

Echinocereus pentalophus var. **ehrenbergii** (Pfeiffer) Backeberg, Die Cact. 4: 2003, 1960. Basionym: *Cereus ehrenbergii*. [RPS 11]

Echinocereus pentalophus var. **leonensis** (Mathsson) N. P. Taylor, Genus Echinocereus, 78, 1985. Basionym: *Echinocereus leonensis*. [RPS 36]

Echinocereus polyacanthus var. **densus** (Regel) N. P. Taylor, Kew Mag. 1(4): 159, 1984. Basionym: *Echinopsis valida* var. *densa*. [RPS 35]

Echinocereus polyacanthus var. *galtieri* Rebut *ex* Borg, Cacti, ed. 2, 228, 1951. *Nom. inval.* (Art. 36.1). [RPS -]

Echinocereus polyacanthus var. **huitcholensis** (F. A. C. Weber) N. P. Taylor, Bradleya 6: 82, 1988. Basionym: *Cereus huitcholensis*. [RPS 39]

Echinocereus polyacanthus var. *longispinus* hort. *ex* Borg, Cacti, ed. 2, 228, 1951. *Nom. inval.* (Art. 36.1). [RPS -]

Echinocereus polyacanthus var. *neomexicanus* (Standley) Weniger, Cacti of the Southwest, 44, 1970. *Nom. inval.* (Art. 33.2), based on *Echinocereus neomexicanus*. [RPS -]

Echinocereus polyacanthus var. *nigrispinus* hort. *ex* Borg, Cacti, ed. 2, 228, 1951. *Nom. inval.* (Art. 36.1). [RPS -]

Echinocereus polyacanthus var. **pacificus** (Engelmann) N. P. Taylor, Kew Mag. 1(4): 160, 1984. Basionym: *Cereus phoeniceus* var. *pacificus*. [RPS 35]

Echinocereus polyacanthus var. *rosei* (Wooton & Standley) Weniger, Cacti of the Southwest, 43, 1970. *Nom. inval.* (Art. 33.2), based on *Echinocereus rosei*. [RPS -]

Echinocereus primolanatus F. Schwarz *ex* N. P. Taylor, Genus Echinocereus, 130, 1985. Typus: *Anonymus* s.n. (K). [Ex cult. D. Parker 1985.] [Syn. *E. primolanatus* F. Schwarz *ex* Backeberg, Die Cact. 4: 2043, 1960, *nom. inval.* (Arts. 37, 9.5).] [Mexico: Coahuila] [RPS 36]

Echinocereus pseudopectinatus (N. P. Taylor) N. P. Taylor, Bradleya 7: 74, 1989. Basionym: *Echinocereus bristolii* var. *pseudopectinatus*. [RPS 40]

Echinocereus pulchellus var. **sharpii** N. P. Taylor, Bradleya 7: 75-77, ills., 1989. Typus: *Gonzalez G.* s.n. (K). [Ex cult. N. P. Taylor 1988.] [Mexico: Nuevo Leon] [RPS 40]

Echinocereus pulchellus var. **weinbergii** (Weingart) N. P. Taylor, Kew Mag. 2(2): 272, 1985. Basionym: *Echinocereus weinbergii*. [RPS 36]

Echinocereus rayonesensis N. P. Taylor, Bradleya 6: 75-76, ill., 1988. Typus: *Lau* 1101 (K). [Ex cult. R. Mottram.] [Mexico: Nuevo Leon] [RPS 39]

Echinocereus reichenbachii var. **albertii** L. Benson, Cact. Succ. J. (US) 41: 127, 1969. Typus: *Albert et Benson* 16550 (POM 317080). [USA: Texas] [RPS 20]

Echinocereus reichenbachii var. **albispinus** (Lahmann) L. Benson, Cact. Succ. J. (US) 41: 127, 1969. Basionym: *Echinocereus albispinus*. [Incorrectly used name, as *Echinocereus baileyi* [var. *baileyi*] is cited in synonymy.] [RPS 20]

Echinocereus reichenbachii var. **armatus** (Poselger) N. P. Taylor, Genus Echinocereus, 133, 1985. Basionym: *Cereus pectinatus* var. *armatus*. [RPS 36]

Echinocereus reichenbachii var. **chisoensis** (W. T. Marshall) L. Benson, Cact. Succ. J. (US) 41: 127, 1969. Basionym: *Echinocereus chisoensis*. [Sphalm. 'chisosensis'.] [RPS 20]

Echinocereus reichenbachii var. **fitchii** (Britton & Rose) L. Benson, Cact. Succ. J. (US) 41: 127, 1969. Basionym: *Echinocereus fitchii*. [RPS 20]

Echinocereus reichenbachii var. **perbellus** (Britton & Rose) L. Benson, Cact. Succ. J. (US) 41: 24, 1969. Basionym: *Echinocereus perbellus*. [RPS 20]

Echinocereus rigidissimus var. **rubispinus** (G. R. W. Frank) N. P. Taylor, Kew Mag. 1(4): 175, 1984. Basionym: *Echinocereus pectinatus* var. *rubispinus*. [RPS 35]

Echinocereus roetteri var. **lloydii** (Britton & Rose) Backeberg, Die Cact. 4: 2027, 1960. Basionym: *Echinocereus lloydii*. [RPS 11]

Echinocereus russanthus Weniger, Cact. Succ. J. (US) 41: 41-42, fig. 5, 1969. Typus: *Weniger* 712 (UNM). [USA: Texas] [RPS 20]

Echinocereus scheeri var. **gentryi** (Clover) N. P. Taylor, Kew Mag. 1(4): 154, 1984. Basionym: *Echinocereus gentryi*. [RPS 35]

Echinocereus scheeri var. **gentryi** cv. **Cucumis** (Werdermann) N. P. Taylor, Kew Mag. 1(4): 155, 1984. Basionym: *Echinocereus cucumis*. [RPS 35]

Echinocereus scheeri var. **koehresianus** G. R. W. Frank, Kakt. and. Sukk. 39(8): 186-189, ills., SEM-ills., 1988. Typus: *Lau* 1143 (ZSS). [Mexico: Sinaloa] [RPS 39]

Echinocereus scheeri var. **obscuriensis** Lau, Kakt. and. Sukk. 40(2): 34-36, ills., SEM-ills., 1989. Typus: *Lau* 91 (ZSS). [Mexico: frontier Sonora - Chihuahua] [RPS 40]

Echinocereus schereri G. R. W. Frank, Kakt. and. Sukk. 41(8): 154-159, ills., SEM-ills., 1990. Typus: *Scherer* 123 (ZSS). [Mexico: Durango] [RPS 41]

Echinocereus schmollii (Weingart) N. P. Taylor, Genus Echinocereus, 140, 1985. Basionym: *Cereus schmollii*. [RPS 36]

Echinocereus schwarzii Lau, Cact. Succ. J. (US) 54(1): 27-29, ills., 1982. Typus: *Lau* 1305 (POM). [Mexico: Durango] [RPS 33]

Echinocereus sciurus var. **floresii** (Backeberg) N. P. Taylor, Genus Echinocereus, 115, 1985. Basionym: *Echinocereus floresii*. [RPS 36]

Echinocereus spinigemmatus Lau, Kakt. and. Sukk. 35(11): 249-250, ills., (12): 281, ill. (additional information), 1984. Typus: *Lau* 1246 (ZSS). [Mexico: Jalisco] [RPS 35]

Echinocereus stoloniferus var. **tayopensis** (W. T. Marshall) N. P. Taylor, Kew Mag. 2(2): 258, 1985. Basionym: *Echinocereus tayopensis*. [RPS 36]

Echinocereus stramineus var. **conglomeratus** (Förster) Bravo, Cact. Suc. Mex. 19: 47, 1974. Basionym: *Echinocereus conglomeratus*. [RPS 25]

Echinocereus stramineus var. *major* hort. *ex* BOrg, Cacti, 222, 1951. *Nom. inval.* (Art. 33.1). [RPS -]

Echinocereus stramineus var. **occidentalis** N. P. Taylor, Bradleya 6: 70-71, ill. (pp. 70, 84), 1988. Typus: *Taylor* 240B (MEXU, K). [Mexico: Durango] [RPS 39]

Echinocereus subinermis fa. **luteus** (Britton & Rose) Krainz, Kat. ZSS ed. 2, 51, 1967. Basionym: *Echinocereus luteus*. [RPS 18]

Echinocereus subinermis var. **aculeatus** G. Unger, Kakt. and. Sukk. 35(7): 164, ills., 1984. Typus: *Reppenhagen et Unger* s.n. (ZSS). [Mexico: Chihuahua] [RPS 35]

Echinocereus subinermis var. **luteus** (Britton & Rose) Backeberg, Die Cact. 4: 1994, 1960. Basionym: *Echinocereus luteus*. [RPS 11]

Echinocereus subinermis var. **ochoterenae** (J. Gonzalez Ortega) G. Unger, Kakt. and. Sukk. 35(7): 164, 1984. Basionym: *Echinocereus ochoterenae*. [RPS 35]

Echinocereus subterraneus Backeberg, Die Cact. 4: 2012, 2014, 1960. *Nom. inval.* (Art. 37.1). [Erroneously included as valid in RPS 11.] [RPS 11]

Echinocereus tayopensis W. T. Marshall, Saguaroland Bull. 10(7): 78-80, ills., 1956. Typus: *Gold et Sanchez-Mejorada* s.n. (DES). [Mexico: Sonora] [RPS 7]

Echinocereus triglochidiatus fa. *inermis* (Schumann) Ferguson, Cact. Succ. J. (US) 61(5): 219, 1989. *Nom. inval.*, based on *Echinocereus phoeniceus* var. *inermis*, *nom. inval.* (Art. 43.1). [RPS 40]

Echinocereus triglochidiatus var. **acifer** (Otto) Bravo, Cact. Suc. Mex. 23(3): 66, 1978. Basionym: *Cereus acifer*. [RPS 29]

Echinocereus triglochidiatus var. **arizonicus** (Rose *ex* Orcutt) L. Benson, Cacti of Arizona, ed. 3, 21, 129, ill., 1969. Basionym: *Echinocereus arizonicus*. [RPS 20]

Echinocereus triglochidiatus var. **gurneyi** L. Benson, Cact. Succ. J. (US) 41: 126, 1969. Typus: *Correll et Benson* 16488 (POM 317078). [USA: Texas] [RPS 20]

Echinocereus triglochidiatus var. **inermis** (Schumann) Arp, Cact. Succ. J. (US) 45: 132 (19 May), 1973. Basionym: *Echinocereus phoeniceus* var. *inermis*. [Repeated by Rowley in Rep. Pl. Succ. 22: 8, 1973 (23 May)] [RPS 22]

Echinocereus triglochidiatus var. **multicostatus** (Schumann) W. T. Marshall, Saguaroland Bull. 7(7): 67, 1953. Basionym: *Echinocereus leeanus* var. *multicostatus*. [RPS 4]

Echinocereus triglochidiatus var. **pacificus** (Engelmann) Bravo, Cact. Suc. Mex. 23(3): 66, 1978. Basionym: *Cereus phoeniceus* var. *pacificus*. [RPS 29]

Echinocereus triglochidiatus var. **rosei** (Wooton & Standley) W. T. Marshall, Arizona's Cact., ed. 1, 63, 1950. Basionym: *Echinocereus rosei*. [Repeated by Roan & Rowley in Repert. Pl.Succ. 2: 2, 1952.] [RPS 2]

Echinocereus tulensis Bravo, Cact. Suc. Mex. 18: 110-111, 1973. Typus: *Sanchez-Mejorada* 2085 (MEXU). [Fig. 63, l.c., is excluded from the concept of the taxon by N. P. Taylor in Bradleya 6: 69, 1988.] [Mexico: San Luis Potosi] [RPS 24]

Echinocereus vatteri B. Botzenhart, Kakt. and. Sukk. 19(7): front cover, 1968. *Nom. inval.* (Art. 36, 37). [RPS -]

Echinocereus viereckii var. **morricalii** (Riha) N. P. Taylor, Genus Echinocereus, 93, 1985. Basionym: *Echinocereus morricalii*. [RPS 36]

Echinocereus viridiflorus fa. **chloranthus** (Engelmann) Krainz, Kat. ZSS ed. 2, 51, 1967. Basionym: *Cereus chloranthus*. [RPS 18]

Echinocereus viridiflorus fa. **davisii** (Houghton) Krainz, Kat. ZSS ed. 2, 51, 1967. Basionym: *Echinocereus davisii*. [RPS 18]

Echinocereus viridiflorus var. **chloranthus** (Engelmann) Backeberg, Die Cact. 4: 2015, 1960. Basionym: *Cereus chloranthus*. [RPS 11]

Echinocereus viridiflorus var. **correllii** L. Benson, Cact. Succ. J. (US) 41: 128, 1969. Typus: *Correll et Benson* 16485 (POM 317079). [USA: Texas] [RPS 20]

Echinocereus viridiflorus var. *intermedius* Backeberg, Die Cact. 4: 2015, 1960. *Nom. inval.* (Art. 37.1). [Erroneously included as valid in RPS 11.] [RPS 11]

Echinocereus viridiflorus var. *standleyi* (Britton & Rose) Weniger, Cacti of the Southwest, 15, 1970. *Nom. inval.* (Art. 33.2), based on *Echinocereus standleyi*. [Not dated. Attributed to Orcutt.] [RPS -]

×*Echinocylindra* Ito, The Cactaceae, 644, 1981. *Nom. inval.* (Art. H6.2). [= *Echinopsis* × *Cylindrolobivia* (*nom. inval.* ?).] [RPS -]

Echinofossulocactus acroacanthus (Stieber) Rowley, Rep. Pl. Succ. 23: 7, 1974. Basionym: *Echinocactus acroacanthus*. [RPS 23]

Echinofossulocactus arrigens fa. **xiphacanthus** (Miquel) Krainz, Kat. ZSS ed. 2, 52, 1967. Basionym: *Echinocactus xiphacanthus*. [RPS 18]

Echinofossulocactus caespitosus Backeberg, Cact. Succ. J. Gr. Brit. 12(4): 81, 1950. [RPS 1]

Echinofossulocactus caespitosus var. *gracilispinus* Bravo, Cact. Suc. Mex. 14: 34, 1969. *Nom. inval.* (Art. 33.2). [Sphalm. 'gracilispisnus'.] [RPS 20]

Echinofossulocactus densispinus Tiegel *ex* Pechanek, Friciana 32: 7-9, 1965. [RPS 16]

Echinofossulocactus erectocentrus Backeberg, Die Cact. 2772-2773, 1961. *Nom. inval.* (Art. 37.1). [Erroneously included as valid in RPS 12.] [RPS 12]

Echinofossulocactus flexispinus (Salm-Dyck) Bravo, Cact. Suc. Mex. 14: 44, 1969. Basionym: *Echinocactus flexispinus*. [RPS 20]

Echinofossulocactus multiareolatus Bravo, Anales Inst. Biol. UNAM 30: 59-61, 1959. [RPS 10]

Echinofossulocactus pentacanthus var. **david-boudetianus** Bravo, Cact. Suc. Mex. 14: 36, 1969. [RPS 20]

Echinofossulocactus phyllacanthus var. **hookeri** (Mühlenpfordt) Bravo, Cact. Suc. Mex. 14: 19, 1969. Basionym: *Echinocactus hookeri*. [RPS 20]

Echinofossulocactus rosasianus Whitm. *ex* Pechanek, Friciana 32: 7-8, 1965. [RPS 16]

Echinofossulocactus sulphureus (Dietrich) Meyran, Cact. Suc. Mex. 22(2): 36-40, 1977. Basionym: *Echinocactus sulphureus*. [First published as nude combination by Y. Ito, Cacti, 101, 1952 (cf. RPS 4).] [RPS 28]

Echinofossulocactus tegelbergii Schütz, Kaktusy 72(8): 93-94, 1972. *Nom. inval.* (Art. 37.1). [RPS 23]

Echinofossulocactus tellii Hort. *ex* Y. Ito, Cacti, 102, 1952. *Nom. inval.* (Art. 36.1). [Given as *comb. nud.* in RPS 4.] [RPS 4]

Echinofossulocactus tetraxiphus var. *longiflorus* Bravo, Cact. Suc. Mex. 14: 84, 1969. *Nom. inval.* (Art. 36.1,37.1). [RPS 20]

Echinofossulocactus tricuspidatus var. *longispinus* Bravo, Cact. Suc. Mex. 14: 20, 1969. *Nom. inval.* (Art. 36.1, 37.1). [RPS 20]

Echinofossulocactus vaupelianus var. *rectispinus* Bravo, Cact. Suc. Mex. 14: 65, 1969. *Nom. inval.* (Art. 33.2), based on *Stenocactus rectispinus*. [RPS 20]

Echinofossulocactus zacatecasensis var. *moranensis* Bravo, Cact. Suc. Mex. 14: 67, 1969. *Nom. inval.* (Art. 36.1, 37.1). [RPS 20]

Echinolobivia Y. Ito, Bull. Takarazuka Insectarium 71: 13-20, 1950. *Nom. inval.* (Art. 36.1). [RPS 3]

Echinomastus acunensis W. T. Marshall, Saguaroland Bull. 7(3): 33-34, 1953. [See also *Echinocactus acunensis* W. T. Marshall.] [RPS 4]

Echinomastus centralis (Rose) Y. Ito, Cacti, 102, 1952. [Given as *comb. nud.* in RPS 4.] [RPS 4]

Echinomastus erectocentrus var. **acunensis** (W. T. Marshall) Bravo, Cact. Suc. Mex. 25(3): 65, 1980. Basionym: *Echinomastus acunensis*. [RPS 31]

Echinomastus intertextus var. **dasyacanthus** (Engelmann) Backeberg, Die Cact. 5: 2832, 1961. Basionym: *Echinocactus intertextus* var. *dasyacanthus*. [RPS 12]

Echinomastus johnsonii var. **lutescens** (Parish) Wiggins, in Shreve & Wiggins, Veg. Fl. Sonoran Des. 2: 1011, 1964. Basionym: *Echinocactus johnsonii* var. *lutescens*. [RPS 18]

Echinomastus kakui Backeberg, Descr. Cact. Nov. 3: 6, 1963. *Nom. inval.* (Art. 36.1,37.1). [RPS 14]

Echinomastus laui G. Frank & Zecher, Cact. Succ. J. (US) 50(4): 188-189, ills., 1978. Typus: *Zecher* 729 (ZSS). [Mexico: San

Luis Potosi] [RPS 29]
Echinomastus mapimiensis Backeberg, Cact. Succ. J. Gr. Brit. 15: 67-68, 1953. [RPS 4]
Echinomastus pallidus Backeberg, Die Cact. 5: 2826-2827, ill., 1960. *Nom. inval.* (Art. 36.1, 37.1). [Given as provisional name.] [RPS -]
Echinomastus unguispinus var. *crassihamatus* Kaku & Y. Ito, The Cactaceae, 480, 1981. *Nom. inval.* (Art. 37.1). [RPS -]
Echinomastus unguispinus var. **durangensis** (Runge) Bravo, Cact. Suc. Mex. 25: 65, 1980. Basionym: *Echinocactus durangensis.* [RPS 31]
Echinomastus unguispinus var. **laui** (Frank & Zecher) Glass & Foster, Cact. Succ. J. (US) 51(3): 124, ill., 1978. Basionym: *Echinomastus laui.* [RPS 30]
Echinomastus unguispinus var. **minimus** Lau, Kakt. and. Sukk. 34(9): 204-207, ills., 1983. Typus: *Lau* 1236 (ZSS). [Mexico: Durango] [RPS 34]
Echinomastus warnockii (L. Benson) Glass & Foster, Cact. Succ. J. (US) 47(5): 218-223, 1975. Basionym: *Neolloydia warnockii.* [Publication date sometimes incorrectly given as 1986.] [RPS 27]
× **Echinoparodia** Mottram, Contr. New Class. Cact. Fam., 36, 1990. [= *Echinopsis* × *Parodia.*] [RPS 41]
Echinopsideae Friedrich & Rowley, Rep. Pl. Succ. 25: 6, 1976. *Nom. illeg.* (Art. 63.1). [RPS 25]
Echinopsidinae Friedrich & Rowley, Rep. Pl. Succ. 25: 6, 1976. Typus: *Echinopsis.* [RPS 25]
Echinopsis cv. **Apricot** Kilian, Stachelpost 8: 125, 1972. [Parentage not given.] [RPS 23]
Echinopsis cv. **Apricot Delight** B. Braun, Stachelpost 8: 124-125, 1972. [Parentage not given.] [RPS 23]
Echinopsis cv. **Aurora** B. Braun, Stachelpost 8: 124, 1972. [Parentage not given.] [RPS 23]
Echinopsis cv. **Azalee** Kilian, Stachelpost 7: 372, 427, 1971. [Referred to × *Echinobivia* in RPS 22.] [RPS 22]
Echinopsis cv. **Daphne** Kilian, Stachelpost 7: 372-373, 427, 1971. [Referred to × *Echinobivia* in RPS 22.] [RPS 22]
Echinopsis cv. **Dr. Stauch** Strigl, Kakt. and. Sukk. 26: 101-102, ill. (front cover), 1975. [= *Trichocereus purpureopilosus* × *Echinopsis* hybr.] [RPS 26]
Echinopsis cv. **Gerritse** Timmermans, Succulenta 46: 51-53, 1967. [= *Echinopsis eyriesii* × (*Echinocereus viridiflorus* (sic !) × *Lobivia densispina*).] [RPS 18]
Echinopsis cv. **Heidelberg** Kilian, Stachelpost 7: 404, 428, 1971. [Referred to × *Echinobivia* in RPS 22.] [RPS 22]
Echinopsis cv. **Maya** Kilian, Stachelpost 7: 299, 1971. [Referred to × *Echinobivia* in RPS 22.] [RPS 22]
Echinopsis cv. **Meyrl** Kilian, Stachelpost 7: 404, 428, 1971. [Referred to × *Echinobivia* in RPS 22.] [RPS 22]
Echinopsis cv. **Morgenzauber** Kilian, Stachelpost 7: 403, 427, 1971. [Referred to × *Echinobivia* in RPS 22.] [RPS 22]
Echinopsis cv. **Nürnberg** Kilian, Stachelpost 6: 282-283, 1970. [Referred to × *Echinobivia* in RPS 22.] [RPS 22]
Echinopsis cv. **Niederrhein** Kilian, Stachelpost 6: 281-282, 1970. [Referred to × *Echinobivia* in RPS 22.] [RPS 22]
Echinopsis cv. **Orange Glory** Kilian, Stachelpost 7: 404, 428, 1971. [Referred to × *Echinobivia* in RPS 22.] [RPS 22]
Echinopsis cv. **Rosy Star** Kilian, Stachelpost 6: 282, 1970. [Referred to × *Echinobivia* in RPS 22.] [RPS 22]
Echinopsis cv. **Schachenfeuer** H. Theobald, Kakt. and. Sukk. 30(3): 56-57, ill., 1982. [= *Echinopsis oxygona* × *E. eyriesii* × *E. aurea* × *E. densispina.*] [RPS 33]
Echinopsis cv. **Whitestone Peach** Mottram, Brit. Cact. Succ. J. 6(3): 78, ill., 1988. [= (*Echinopsis chamaecereus* × *Echinopsis* sp.) × *Echinopsis* 'Stars and Stripes'.] [RPS 39]
Echinopsis cv. **Whitestone Pink** Mottram, Brit. Cact. Succ. J. 6(3): 77-78, ill., 1988. [= (*Echinopsis chamaecereus* × *Echinopsis* sp.) × *Echinopsis* 'Stars and Stripes'.] [RPS 39]
Echinopsis subgen. *Ancistrechinopsis* Y. Ito, Bull. Takarazuka Insectarium 71: 13-20, 1950. *Nom. inval.* (Art. 36.1). [RPS 3]
Echinopsis subgen. *Gladiechinopsis* Y. Ito, Bull. Takarazuka Insectarium 71: 13-20, 1950. *Nom. inval.* (Art. 36.1). [RPS 3]
Echinopsis subgen. *Lamprechinopsis* Y. Ito, Bull. Takarazuka Insectarium 71: 13-20, 1950. *Nom. inval.* (Art. 36.1). [RPS 3]
Echinopsis adolfofriedrichii G. Moser, Nation. Cact. Succ. J. 37(2): 39-40, ills., 1982. Typus: *Moser* 946 (K). [RPS 33]
Echinopsis albispinosa var. **fuauxiana** Backeberg, Descr. Cact. Nov. [1:] 28, 1957. [Dated 1956, published 1957.] [RPS 7]
Echinopsis amblayensis (Rausch) Friedrich *ex* Rowley, Kakt. and. Sukk. 25(4): 82, 1974. Basionym: *Lobivia amblayensis.* [RPS 25]
Echinopsis amblayensis var. **albispina** (Rausch) Friedrich, Kakt. and. Sukk. 25: 82, 1974. Basionym: *Lobivia amblayensis* var. *albispina.* [RPS 25]
Echinopsis ancistrophora ssp. **arachnacantha** (Buining & Ritter) Rausch, Lobivia 3: 140, 1976. Basionym: *Lobivia arachnacantha.* [RPS 27]
Echinopsis ancistrophora ssp. **cardenasiana** (Rausch) Rausch, Lobivia 3: 140, 1976. Basionym: *Lobivia cardenasiana.* [RPS 27]
Echinopsis ancistrophora ssp. **pojoensis** (Rausch) Rausch, Lobivia 3: 140, 1976. Basionym: *Lobivia pojoensis.* [RPS 27]
Echinopsis ancistrophora var. **densiseta** (Rausch) Rausch, Lobivia 3: 140, 1976. Basionym: *Lobivia arachnacantha* var. *densiseta.* [RPS 27]
Echinopsis ancistrophora var. **grandiflora** (Rausch) Rausch, Lobivia 3: 140, 1976. Basionym: *Lobivia pojoensis* var. *grandiflora.* [RPS 27]
Echinopsis ancistrophora var. **hamatacantha** (Backeberg) Rausch, Lobivia 3: 140, 1976. Basionym: *Echinopsis hamatacantha.* [Erroneously included as invalid in RPS 27.] [RPS 27]
Echinopsis ancistrophora var. **kratochviliana**

(Backeberg) Rausch, Lobivia 3: 140, 1976. Basionym: *Echinopsis kratochviliana.* [RPS 27]

Echinopsis ancistrophora var. **megalocephala** (Rausch) Rausch, Lobivia 3: 140, 1976. Basionym: *Echinopsis rauschii* var. *megalocephala.* [RPS 27]

Echinopsis ancistrophora var. **polyancistra** (Backeberg) Rausch, Lobivia 3: 140, 1976. Basionym: *Echinopsis polyancistra.* [RPS 27]

Echinopsis ancistrophora var. **sulphurea** (Vasquez) Rausch, Lobivia 3: 140, 1976. Basionym: *Lobivia arachnacantha* var. *sulphurea.* [RPS 27]

Echinopsis ancistrophora var. **torrecillasensis** (Cardenas) Rausch, Lobivia 3: 140, 1976. Basionym: *Echinopsis torrecillasensis.* [RPS 27]

Echinopsis ancistrophora var. *vallegrandensis* Rausch, Lobivia 3: 140, 1976. *Nom. inval.* (Art. 33). [RPS 27]

Echinopsis angelesiae (Kiesling) Rowley, Rep. Succ. Pl. 29: 5, 1980. Basionym: *Trichocereus angelesiae.* [Sphalm. 'angelesii'.] [RPS 29]

Echinopsis antezanae (Cardenas) Friedrich & Rowley, IOS Bull. 3(3): 94, 1974. Basionym: *Trichocereus antezanae.* [RPS 25]

Echinopsis arachnacantha (Buining & Ritter) Friedrich, Kakt. and. Sukk. 25: 82, 1974. Basionym: *Lobivia arachnacantha.* [RPS 25]

Echinopsis arachnacantha var. **densiseta** (Rausch) Friedrich, Kakt. and. Sukk. 25: 82, 1974. Basionym: *Lobivia arachnacantha* var. *densiseta.* [RPS 25]

Echinopsis arachnacantha var. **sulphurea** (Vasquez) Rowley, Rep. Pl. Succ. 25: 6, 1976. Basionym: *Lobivia arachnacantha* var. *sulphurea.* [RPS 25]

Echinopsis arachnacantha var. **torrecillasensis** (Cardenas) Friedrich, Kakt. and. Sukk. 25: 82, 1974. Basionym: *Echinopsis torrecillasensis.* [RPS 25]

Echinopsis arachnacantha var. **vallegrandensis** Rausch, Kakt. and. Sukk. 26(1): 1-2, ill., 1975. Typus: *Rausch* 184 (ZSS). [Type cited for W, but placed in ZSS.] [Bolivia: Santa Cruz] [RPS 26]

Echinopsis arebaloi Cardenas, Cact. Succ. J. (US) 28(3): 73-74, 1956. [RPS 7]

Echinopsis atacamensis (Philippi) Friedrich & Rowley, IOS Bull. 3(3): 94, 1974. Basionym: *Cereus atacamensis.* [RPS 25]

Echinopsis aurantiaca (Rausch) Friedrich & Rowley, IOS Bull. 1974: 94, 1974. Basionym: *Acanthocalycium aurantiacum.* [RPS 25]

Echinopsis aurea var. **albiflora** (Rausch) Ullmann, Kaktusy 26(1): 7, 1990. Basionym: *Lobivia aurea* var. *albiflora.* [RPS 41]

Echinopsis aurea var. **callochrysea** (Ritter) Ullmann, Kaktusy 26(1): 7, 1990. Basionym: *Hymenorebutia aurea* var. *callochrysea.* [RPS 41]

Echinopsis aurea var. **catamarcensis** (Ritter) Ullmann, Kaktusy 26(1): 7, 1990. Basionym: *Hymenorebutia aurea* var. *catamarcensis.* [RPS 41]

Echinopsis aurea var. **depressicostata** (Ritter) Ullmann, Kaktusy 26(1): 7, 1990. Basionym: *Hymenorebutia aurea* var. *depressicostata.* [RPS 41]

Echinopsis aurea var. **dobeana** (Dölz) Ullmann, Kaktusy 26(1): 7, 1990. Basionym: *Lobivia dobeana.* [RPS 41]

Echinopsis aurea var. **fallax** (Oehme) Ullmann, Kaktusy 26(1): 7, 1990. Basionym: *Lobivia fallax.* [RPS 41]

Echinopsis aurea var. **lariojensis** (Ritter) Ullmann, Kaktusy 26(1): 7, 1990. Basionym: *Hymenorebutia aurea* var. *lariojensis.* [RPS 41]

Echinopsis aurea var. **quinesensis** Rausch, Kakt. and. Sukk. 21(3): 45, 1970. Typus: *Rausch* 112 (ZSS). [Type cited for W, but placed in ZSS.] [First published invalidly in Kakt. and. Sukk. 16(11): 214, 1965, and in l.c. 17: 107, 1966.] [Argentina: San Luis] [RPS 21]

Echinopsis aurea var. *shaferi* (Britton & Rose) Rausch, Kakt. and. Sukk. 16(11): 214, 1965. *Nom. inval.* (Art. 33.2), based on *Lobivia shaferi.* [RPS 16]

Echinopsis aurea var. **sierragrandensis** (Rausch) Ullmann, Kaktusy 26(1): 7, 1990. Basionym: *Lobivia aurea* var. *sierragrandensis.* [First invalidly published (Art. 33.2) by Lambert in Succulenta 67(9): 182, 1988 (cf. RPS 39).] [RPS 41]

Echinopsis aurea var. **tortuosa** (Rausch) Ullmann, Kaktusy 26(1): 7, 1990. Basionym: *Lobivia aurea* var. *tortuosa.* [RPS 41]

Echinopsis ayopayana Ritter & Rausch, Succulenta 47: 85-86, 1968. [RPS 19]

Echinopsis berlingii Y. Ito, Explan. Diag. Austroechinocactinae, 64, 285, 1957. [First used as a *nomen nudum* by Backeberg in 1936.] [RPS 8]

Echinopsis bertramiana (Backeberg) Friedrich & Rowley, IOS Bull. 3(3): 94, 1974. Basionym: *Trichocereus bertramianus.* [RPS 25]

Echinopsis blossfeldiana Robl., in Backeberg, Die Cact. 2: 1303, 1959. *Nom. inval.* (Art. 36.1, 37.1). [Reported as name of uncertain application and status, dating probably from 1948.] [RPS 10]

Echinopsis boyuibensis Ritter, Succulenta 1965: 25-26, 1965. [RPS 16]

Echinopsis brasiliensis Fric *ex* Pazout, Friciana 17: 3, 1963. [RPS 14]

Echinopsis brevispina (Ritter) Friedrich & Rowley, IOS Bull. 1974: 94, 1974. Basionym: *Acanthocalycium brevispinum.* [RPS 25]

Echinopsis bruchii (Britton & Rose) Friedrich & Glätzle, Bradleya 1:96, 1983. Basionym: *Lobivia bruchii.* [RPS 34]

Echinopsis cabrerae (Kiesling) Rowley, Rep. Pl. Succ. 27: 5, 1979. Basionym: *Trichocereus cabrerae.* [RPS 27]

Echinopsis cajasensis Ritter, Kakt. Südamer. 2: 630, 1980. [RPS 31]

Echinopsis calliantholilacina Cardenas, Cactus (Paris) No. 85: 110-111, 1965. [RPS 16]

Echinopsis callichroma Cardenas, Kakt. and. Sukk. 16: 49-50, 1965. [RPS 16]

Echinopsis calochlora var. *albispina* Backeberg, Die Cact. 6: 3719, 1962. *Nom. inval.* (Art. 37.1). [Erroneously included as valid in RPS 13.] [RPS 13]

Echinopsis calorubra Cardenas, Nation. Cact. Succ. J. 12: 62, 1957. [RPS 8]

Echinopsis camarguensis (Cardenas) Friedrich & Rowley, IOS Bull. 3(3): 94, 1974. Basionym: *Trichocereus camarguensis.* [RPS 25]

Echinopsis candicans (Gillies *ex* Salm-Dyck) D. Hunt, Bradleya 5: 92, 1987. Basionym: *Cereus candicans.* [Backeberg (Die Cact.) reports this name with the author "F. A. C. Weber".] [RPS 38]

Echinopsis candicans var. *gladiata* (Lemaire) Friedrich & Rowley, IOS Bull. 3(3): 94, 1974. *Nom. inval.* (Art. 43.1), based on *Cereus gladiatus.* [RPS 25]

Echinopsis candicans var. *tenuispina* (Pfeiffer) Friedrich & Rowley, IOS Bull. 3(3): 94, 1974. *Nom. inval.* (Art. 43.1), based on *Cereus candicans* var. *tenuispinus.* [RPS 25]

Echinopsis cardenasiana (Rausch) Friedrich, Kakt. and. Sukk. 25: 82, 1974. Basionym: *Lobivia cardenasiana.* [RPS 25]

Echinopsis carmineoflora (Hoffmann & Backeberg) Friedrich, Kakt. and. Sukk. 25: 82, 1974. *Nom. inval.*, based on *Pseudolobivia carmineoflora, nom. inval.* (Art. 37.1). [Erroneously included as valid in RPS 25.] [RPS 25]

Echinopsis cephalomacrostibas (Werdermann & Backeberg) Friedrich & Rowley, IOS Bull. 3(3): 94, 1974. Basionym: *Cereus cephalomacrostibas.* [RPS 25]

Echinopsis cerdana Cardenas, Cactus (Paris) No. 65: 177, 1959. [RPS 10]

Echinopsis chacoana var. **spinosior** Ritter, Succulenta 1965: 26, 1965. [RPS 16]

Echinopsis chalaensis (Rauh & Backeberg) Friedrich & Rowley, IOS Bull. 3(3): 94, 1974. Basionym: *Trichocereus chalaensis.* [RPS 25]

Echinopsis chamaecereus Friedrich & Glätzle, Bradleya 1: 96, 1983. Basionym: *Cereus silvestrii.* [*Non Echinopsis silvestrii* Spegazzini.] [RPS 34]

Echinopsis chiloensis (Colla) Friedrich & Rowley, IOS Bull. 3(3): 94, 1974. Basionym: *Cactus chiloensis.* [RPS 25]

Echinopsis chiloensis var. **eburnea** (Philippi) Friedrich & Rowley, IOS Bull. 3(3): 94, 1974. Basionym: *Eulychnia eburnea.* [RPS 25]

Echinopsis cochabambensis Backeberg, Descr. Cact. Nov. [1:] 28, 1957. [Dated 1956, published 1957.] [RPS 7]

Echinopsis comarapana Cardenas, Nation. Cact. Succ. J. 12: 61, 1957. [RPS 8]

Echinopsis conaconensis (Cardenas) Friedrich & Rowley, IOS Bull. 3(3): 94, 1974. Basionym: *Trichocereus conaconensis.* [RPS 25]

Echinopsis coquimbana (Molina) Friedrich & Rowley, IOS Bull. 3(3): 94, 1974. Basionym: *Cactus coquimbanus.* [Combination repeated in Rep. Pl. Succ. 25: 7, 1976.] [RPS 25]

Echinopsis coquimbana A. E. Hoffmann, Cact. Fl. Chile, 86, 1989. *Nom. inval.* (Art. 36.1). [Based on *Trichocereus coquimbanus sensu* Britton & Rose, *non Cactus coquimbanus* Molina, *nec Echinopsis coquimbana* (Molina) Friedrich & Rowley.] [RPS 40]

Echinopsis coronata Cardenas, Nation. Cact. Succ. J. 12: 63, 1957. [RPS 8]

Echinopsis cotacajesi Cardenas, Cact. Succ. J. (US) 42: 184, 1970. [RPS 21]

Echinopsis courantii (Schumann) Friedrich & Rowley, IOS Bull. 3(3): 95, 1974. Basionym: *Cereus candicans* var. *courantii.* [RPS 25]

Echinopsis crassicaulis (Backeberg ex Kiesling) Friedrich & Glätzle, Bradleya 1:96, 1983. Basionym: *Lobivia crassicaulis.* [RPS 34]

Echinopsis cuzcoensis (Britton & Rose) Friedrich & Rowley, IOS Bull 3(3): 95, 1974. Basionym: *Trichocereus cuzcoensis.* [RPS 25]

Echinopsis cylindracea (Backeberg) Friedrich, Kakt. and. Sukk. 25: 82, 1974. Basionym: *Lobivia cylindracea.* [RPS 25]

Echinopsis dehrenbergii var. **blossfeldii** Backeberg, Descr. Cact. Nov. [1:] 27, 1957. [Dated 1956, published 1957.] [RPS 7]

Echinopsis deserticola (Werdermann) Friedrich & Rowley, IOS Bull. 3(3): 95, 1974. Basionym: *Cereus deserticola.* [RPS 25]

Echinopsis deserticola var. **fulvilana** (Ritter) A. E. Hoffmann, Cact. Fl. Chil., 86-87, ill., 1989. Basionym: *Trichocereus fulvilanus.* [RPS 40]

Echinopsis elongata (Backeberg) Friedrich, Kakt. and. Sukk. 25: 82, 1974. Basionym: *Lobivia elongata.* [Repeated by Friedrich & Glätzle, Bradleya 1: 96, 1983.] [RPS 25]

Echinopsis escayachensis (Cardenas) Friedrich & Rowley, IOS Bull. 3(3): 95, 1974. Basionym: *Trichocereus escayachensis.* [RPS 25]

Echinopsis fabrisii (Kiesling) Rowley, Rep. Pl. Succ. 27: 5, 1979. Basionym: *Trichocereus fabrisii.* [RPS 27]

Echinopsis fallax (Oehme) Friedrich, Kakt. and. Sukk. 25: 80, 82, 1974. Basionym: *Lobivia fallax.* [RPS 25]

Echinopsis fallax var. **albiflora** (Rausch) Lambert, Succulenta 67(9): 182, 1988. Basionym: *Lobivia aurea* var. *albiflora.* [RPS 39]

Echinopsis fallax var. **callochrysea** (Ritter) Lambert, Succulenta 67(9): 182, 1988. Basionym: *Hymenorebutia aurea* var. *callochrysea.* [RPS 39]

Echinopsis fallax var. **catamarcensis** (Ritter) Lambert, Succulenta 67(9): 182, 1988. Basionym: *Hymenorebutia aurea* var. *catamarcensis.* [RPS 39]

Echinopsis fallax var. **depressicostata** (Ritter) Lambert, Succulenta 67(9): 182, 1988. Basionym: *Hymenorebutia aurea* var. *depressicostata.* [RPS 39]

Echinopsis fallax var. **shaferi** (Britton & Rose)

Lambert, Succulenta 67(9): 182, 1988. Basionym: *Lobivia shaferi*. [RPS 39]

Echinopsis fallax var. *tortuosa* (Rausch) Lambert, Succulenta 67(9): 182, 1988. *Nom. inval.* (Art. 33.2), based on *Lobivia aurea* var. *tortuosa*. [RPS 39]

Echinopsis frankii (Bozsing) Friedrich *ex* Donald, Ashingtonia Species Cat. Cact. [part 3], unnumbered page, 1974. *Nom. inval.* (Art. 33.2). [Published as "nov. comb. in press".] [RPS -]

Echinopsis fricii (Rausch) Friedrich, Kakt. and. Sukk. 25: 83, 1974. Basionym: *Lobivia fricii*. [RPS 25]

Echinopsis friedrichii Rowley, IOS Bull. 3(3): 95, 1974. Basionym: *Trichocereus shaferi*. [*Nom. nov., non Echinopsis shaferi* Britton & Rose.] [RPS 25]

Echinopsis fulvilana (Ritter) Friedrich & Rowley, IOS Bull. 3(3): 95, 1974. Basionym: *Trichocereus fulvilanus*. [RPS 25]

Echinopsis gladispina Y. Ito, Explan. Diagr. Austroechinocactinae, 68, 1957. [*Nom. nov. pro Echinopsis campylacantha* R. Meyer 1898 and *E. salpingophora* Preinreich 1898, cf. RPS 8. Validity not assessed.] [RPS 8]

Echinopsis glauca (Ritter) Friedrich & Rowley, IOS Bull. 3(3): 95, 1974. Basionym: *Trichocereus glaucus*. [RPS 25]

Echinopsis glauca fa. **pendens** (Ritter) Friedrich & Rowley, IOS Bull. 3(3): 95, 1974. Basionym: *Trichocereus glaucus* fa. *pendens*. [RPS 25]

Echinopsis glaucina Friedrich & Rowley, IOS Bull. 3(3): 95, 1974. Basionym: *Acanthocalycium glaucum*. [*Nom. nov., non Echinopsis glauca* (Ritter) Friedrich & Rowley.] [RPS 25]

Echinopsis grandis (Britton & Rose) Friedrich & Rowley, IOS Bull. 1974: 95, 1974. Basionym: *Lobivia grandis*. [RPS 25]

Echinopsis hammerschmidii Cardenas, Cact. Succ. J. (US) 28(3): 72-73, 1956. [RPS 7]

Echinopsis herbasii Cardenas, Cact. Succ. J. (US) 28(4): 111-112, 1956. [RPS 7]

Echinopsis herzogiana (Cardenas) Friedrich & Rowley, IOS Bull. 3(3): 95, 1974. Basionym: *Trichocereus herzogianus*. [RPS 25]

Echinopsis herzogiana var. **totorensis** (Cardenas) Friedrich & Rowley, IOS Bull. 3(3): 95, 1974. Basionym: *Trichocereus herzogianus* var. *totorensis*. [RPS 25]

Echinopsis hualfinensis (Rausch) Friedrich & Glätzle, Bradleya 1: 96, 1983. Basionym: *Lobivia hualfinensis*. [RPS 34]

Echinopsis huascha (F. A. C. Weber) Friedrich & Rowley, IOS Bull. 1974: 95, 1974. Basionym: *Cereus huascha*. [RPS 25]

Echinopsis huascha var. **auricolor** (Backeberg) Friedrich & Rowley, IOS Bull. 1974: 95, 1974. Basionym: *Trichocereus auricolor*. [RPS 25]

Echinopsis huascha var. **rubriflora** (F. A. C. Weber) Friedrich & Rowley, IOS Bull. 1974: 95, 1974. Basionym: *Cereus huascha* var. *rubriflorus*. [RPS 25]

Echinopsis hystrichoides Ritter, Kakt. Südamer. 2: 626, 1980. [RPS 31]

Echinopsis ibicuatensis Cardenas, Cact. Succ. J. (US) 28(3): 74-75, 1956. [RPS 7]

Echinopsis ingens (Backeberg) Friedrich & Rowley, IOS Bull. 3: 96, 1974. Basionym: *Soehrensia ingens*. [Probably only validated in Rep. Pl. Succ. 25: 8, 1976.] [RPS 25]

Echinopsis kermesina (Krainz) Krainz, Die Kakt. C Va, 1961. Basionym: *Pseudolobivia kermesina*. [Repeated invalidly (Art. 33.2) by Y. Ito in Full Bloom of Cactus Flowers, 12, 1962.] [RPS 12]

Echinopsis kermesina var. *cylindrica* Y. Ito, The Cactaceae, 265, 1981. *Nom. inval.* (Art. 37.1). [RPS -]

Echinopsis kladiwaiana Rausch, Kakt. and. Sukk. 23(10): 264-265, ill., 1972. Typus: *Rausch* 447 (W [not found], ZSS). [Bolivia] [RPS 23]

Echinopsis klingleriana Cardenas, Cactus (Paris) No. 85: 109, 1965. [RPS 16]

Echinopsis knuthiana (Backeberg) Friedrich & Rowley, IOS Bull. 1974: 96, 1974. Basionym: *Trichocereus knuthianus*. [RPS 25]

Echinopsis kuehnrichii (Fric) Friedrich & Glätzle, Bradleya 1: 96, 1983. Basionym: *Lobivia kuehnrichii*. [RPS 34]

Echinopsis lageniformis (Förster) Friedrich & Rowley, IOS Bull. 1974: 96, 1974. Basionym: *Cereus lageniformis*. [RPS 25]

Echinopsis lamprochlora (Lemaire) F.A.C. Weber ex Friedrich & Glätzle, Bradleya 1: 96, 1983. Basionym: *Cereus lamprochlorus*. [RPS 34]

Echinopsis lecoriensis Cardenas, Cact. & Succ. J. (US) 35: 158, 1963. [RPS 14]

Echinopsis leucantha var. **volliana** Backeberg, Descr. Cact. Nov. [1:] 28, 1957. [Dated 1956, published 1957.] [RPS 7]

Echinopsis leucomalla (Wessner) Friedrich, Kakt. and. Sukk. 25: 82, 1974. [RPS 25]

Echinopsis litoralis (Johnson) Friedrich & Rowley, IOS Bull. 3(3): 96, 1974. Basionym: *Cereus litoralis*. [RPS 25]

Echinopsis luteiflora (Backeberg) Friedrich *ex* Donald, Ashingtonia Species Cat. Cact. [part 3], unnumbered page, 1974. *Nom. inval.* (Art. 33.2). [Published as "nov. comb. in press".] [RPS -]

Echinopsis macrogona (Salm-Dyck) Friedrich & Rowley, IOS Bull. 3(3): 96, 1974. Basionym: *Cereus macrogonus*. [RPS 25]

Echinopsis mamillosa var. **flexilis** Rausch, Kakt. and. Sukk. 28(10): 235, ill., 1977. Typus: *Rausch* 510 (ZSS). [Argentina: Salta] [RPS 28]

Echinopsis mamillosa var. **kermesina** (Krainz) Friedrich, Kakt. and. Sukk. 22: 45, 1971. Basionym: *Pseudolobivia kermesina*. [RPS 22]

Echinopsis mamillosa var. **orozasana** Ritter, Succulenta 1965: 25, 1965. [RPS 16]

Echinopsis mamillosa var. **ritteri** (Bödeker) Ritter, Succulenta 1965: 24, 1965. Basionym: *Echinopsis ritteri*. [RPS 16]

Echinopsis mamillosa var. **tamboensis** Ritter, Succulenta 1965: 25, 1965. [RPS 16]

Echinopsis manguinii (Backeberg) Friedrich &

Rowley, IOS Bull. 3(3): 96, 1974. Basionym: *Trichocereus manguinii.* [RPS 25]

Echinopsis mataranensis Cardenas, Cact. Succ. J. (US) 42: 184, 1970. [RPS 21]

Echinopsis narvaecensis (Cardenas) Friedrich & Rowley, IOS Bull. 3(3): 96, 1974. Basionym: *Trichocereus narvaecensis.* [RPS 25]

Echinopsis nealeana (Backeberg) Friedrich, Kakt. and. Sukk. 25: 82, 1974. Basionym: *Lobivia nealeana.* [RPS 25]

Echinopsis nigripilis (Philippi) Friedrich & Rowley, IOS Bull. 3(3): 96, 1974. Basionym: *Cereus nigripilis.* [RPS 25]

Echinopsis obrepanda var. **aguilarii** (Vasquez) Rausch, Lobivia 1: 52, 1975. Basionym: *Lobivia aguilarii.* [RPS 26]

Echinopsis obrepanda var. **calorubra** (Cardenas) Rausch, Lobivia 1: 52, 1975. Basionym: *Echinopsis calorubra.* [RPS 26]

Echinopsis obrepanda var. **fiebrigii** (Gürke) Friedrich, Kakt. and. Sukk. 25: 82, 1974. Basionym: *Echinopsis fiebrigii.* [RPS 25]

Echinopsis obrepanda var. **mizquensis** (Rausch) Rausch, Lobivia 1: 52, 1975. Basionym: *Lobivia mizquensis.* [RPS 26]

Echinopsis orurensis (Cardenas) Friedrich & Rowley, IOS Bull. 3(3): 96, 1974. Basionym: *Trichocereus orurensis.* [RPS 25]

Echinopsis orurensis var. **albiflora** (Cardenas) Friedrich & Rowley, IOS Bull. 3(3): 96, 1974. Basionym: *Trichocereus orurensis* var. *albiflorus.* [RPS 25]

Echinopsis oxygona fa. **brevispina** Ritter, Kakt. Südamer. 1: 239, 1979. [RPS 30]

Echinopsis pachanoi (Britton & Rose) Friedrich & Rowley, IOS Bull. 3(3): 96, 1974. Basionym: *Trichocereus pachanoi.* [RPS 25]

Echinopsis pamparuizii Cardenas, Cact. Succ. J. (US) 42: 32-34, 1970. [RPS 21]

Echinopsis paraguayensis Mundt *ex* Ritter, Kakt. Südamer. 1: 264, 1979. [Derived from a *nomen nudum* by Mundt.] [RPS 30]

Echinopsis pasacana (F. A. C. Weber) Friedrich & Rowley, IOS Bull. 3(3): 96, 1974. Basionym: *Pilocereus pasacana.* [RPS 25]

Echinopsis pecheretiana (Backeberg) Friedrich & Rowley, IOS Bull. 3(3): 96, 1974. Basionym: *Helianthocereus pecheretianus.* [RPS 25]

Echinopsis peitscheriana (Backeberg) Friedrich & Rowley, IOS Bull. 3(3): 96, 1974. Basionym: *Acanthocalycium peitscherianum.* [RPS 25]

Echinopsis pelecyogona Y. Ito, The Cactaceae, 635, 1981. *Nom. inval.* (Art. 37.1). [RPS -]

Echinopsis pelecyrhachis var. **lobivioides** (Backeberg) Friedrich, Kakt. and. Sukk. 25: 82, 1974. Basionym: *Echinopsis lobivioides.* [RPS 25]

Echinopsis pereziensis Cardenas, Cactus (Paris) No. 78: 88, 1963. [RPS 14]

Echinopsis peruviana (Britton & Rose) Friedrich & Rowley, IOS Bull. 3(3): 97, 1974. Basionym: *Trichocereus peruvianus.* [RPS 25]

Echinopsis poco (Backeberg) Friedrich & Rowley, IOS Bull. 3(3): 97, 1974. Basionym: *Trichocereus poco.* [RPS 25]

Echinopsis poco var. **albiflora** (Cardenas) Friedrich & Rowley, IOS Bull. 3(3): 97, 1974. Basionym: *Trichocereus poco* var. *albiflorus.* [RPS 25]

Echinopsis poco var. **friciana** (Cardenas) Friedrich & Rowley, Rep. Pl. Succ. 25: 9, 1976. Basionym: *Trichocereus poco* var. *fricianus.* [RPS 25]

Echinopsis pojoensis Cardenas, Cactus (Paris) No. 64: 165, 1959. [RPS 10]

Echinopsis pseudocachensis (Backeberg) Friedrich, Kakt. and. Sukk. 25: 82, 1974. Basionym: *Lobivia pseudocachensis.* [RPS 25]

Echinopsis pseudocandicans (Backeberg ex Kiesling) Friedrich & Glätzle, Bradleya 1: 96, 1983. *Nom. inval.*, based on *Trichocereus pseudocandicans, nom. inval.* (Art. 37.1). [Erroneously included as valid in RPS 34. The name cited as basionym is itself a 'combination' of an invalid name.] [RPS 34]

Echinopsis pseudomamillosa Cardenas, Cactus (Paris) No. 64: 164, 1959. [Sphalm. 'pseudomammillosa'.] [RPS 10]

Echinopsis puquiensis (Rauh & Backeberg) Friedrich & Rowley, IOS Bull. 3(3): 97, 1974. Basionym: *Trichocereus puquiensis.* [RPS 25]

Echinopsis purpureopilosa (Weingart) Friedrich & Rowley, IOS Bull. 3(3): 97, 1974. Basionym: *Trichocereus purpureopilosus.* [RPS 25]

Echinopsis quinesensis (Rausch) Friedrich & Glätzle, Bradleya 1: 96, 1983. Basionym: *Echinopsis aurea* var. *quinesensis.* [RPS 34]

Echinopsis randallii (Cardenas) Friedrich & Rowley, IOS Bull. 1974: 95, 1974. Basionym: *Trichocereus randallii.* [RPS 25]

Echinopsis rauschii Friedrich, Kakt. and. Sukk. 1974: 83, 1974. Basionym: *Lobivia pojoensis.* [*Nom. nov.* for *Lobivia pojoensis* Rausch (*non Echinopsis pojoensis* Cardenas).] [RPS 25]

Echinopsis rauschii var. **grandiflora** (Rausch) Friedrich, Kakt. and. Sukk. 25: 83, 1974. Basionym: *Lobivia pojoensis* var. *grandiflora.* [RPS 25]

Echinopsis rauschii var. **megalocephala** Rausch, Kakt. and. Sukk. 25(11): 241-242, ill., 1974. Typus: *Rausch* 272 (ZSS). [Holotype cited for W but placed in ZSS.] [Bolivia: Cochabamba] [RPS 25]

Echinopsis rebutioides (Backeberg) Friedrich, Kakt. and. Sukk. 25: 82, 1974. Basionym: *Lobivia rebutioides.* [RPS 25]

Echinopsis rhodotricha var. **brevispina** Ritter, Kakt. Südamer. 1: 263, 1979. [RPS 30]

Echinopsis rhodotricha var. **chacoana** (Schütz) Ritter, Kakt. Südamer. 1: 263, 1979. Basionym: *Echinopsis chacoana.* [RPS 30]

Echinopsis rhodotricha var. **spinosior** (Ritter) Ritter, Kakt. Südamer. 1: 262, 1979. Basionym: *Echinopsis chacoana* var. *spinosior.* [RPS 30]

Echinopsis riviere-de-caraltii Cardenas, Cact. Succ. J. (US) 43: 242-243, 1971. *Cardenas* 6327 (BOLV). [Bolivia: Boeto] [RPS 22]

Echinopsis rojasii Cardenas, Rev. Agric.

Cochabamba 7(6): 31, 1951. [RPS 2]

Echinopsis rojasii var. **albiflora** Cardenas, Rev. Agric. Cochabamba 7(6): 33, 1951. [RPS 2]

Echinopsis roseo-lilacina Cardenas, Cactus (Paris) No. 57: 254-255, 1957. [RPS 8]

Echinopsis rowleyi Friedrich, IOS Bull. 3(3): 97, 1974. Basionym: *Lobivia grandiflora*. [*Nom. nov., non Echinopsis grandiflora* Linke.] [RPS 25]

Echinopsis rubinghiana (Backeberg) Friedrich & Rowley, IOS Bull. 3(3): 97, 1974. *Nom. inval.*, based on *Trichocereus rubinghianus, nom. inval.* (Art. 9.5). [Erroneously included as valid in RPS 25.] [RPS 25]

Echinopsis santaensis (Rauh & Backeberg) Friedrich & Rowley, IOS Bull. 3(3): 97, 1974. Basionym: *Trichocereus santaensis*. [RPS 25]

Echinopsis santiaguensis (Spegazzini) Friedrich & Rowley, IOS Bull. 3(3): 97, 1974. Basionym: *Cereus santiaguensis*. [RPS 25]

Echinopsis schieliana (Backeberg) D. Hunt, Bradleya 5: 92, 1987. Basionym: *Lobivia schieliana*. [RPS 38]

Echinopsis schoenii (Rauh & Backeberg) Friedrich & Rowley, IOS Bull. 3(3): 97, 1974. Basionym: *Trichocereus schoenii*. [RPS 25]

Echinopsis silvatica Ritter, Succulenta 1965: 24, 1965. [RPS 16]

Echinopsis skottsbergii (Backeberg) Friedrich & Rowley, IOS Bull. 3(3): 97, 1974. Basionym: *Trichocereus skottsbergii*. [RPS 25]

Echinopsis skottsbergii var. **breviata** (Backeberg) Friedrich & Rowly, IOS Bull. 3(3): 98, 1974. Basionym: *Trichocereus skottsbergii* var. *breviatus*. [RPS 25]

Echinopsis spachiana (Lemaire) Friedrich & Rowley, IOS Bull. 3(3): 98, 1974. Basionym: *Cereus spachianus*. [RPS 25]

Echinopsis spinibarbis (Otto *ex* Pfeiffer) A. E. Hoffmann, Cact. Fl. Chil., 90-91, ill., 1989. Basionym: *Cereus spinibarbis*. [RPS 40]

Echinopsis strigosa (Salm-Dyck) Friedrich & Rowley, IOS Bull. 3(3): 98, 1974. Basionym: *Cereus strigosus*. [RPS 25]

Echinopsis subdenudata Cardenas, Cact. Succ. J. (US) 28(3): 71-72, 1956. [Sphalm. 'subdenudatus'.] [RPS 7]

Echinopsis sucrensis Cardenas, Cact. Succ. J. (US) 35: 200, 1963. [RPS 14]

Echinopsis tacaquirensis (Vaupel) Friedrich & Rowley, IOS Bull. 3(3): 98, 1974. Basionym: *Cereus tacaquirensis*. [RPS 25]

Echinopsis tapecuana Ritter, Succulenta 1965: 24, 1965. [RPS 16]

Echinopsis taquimbalensis (Cardenas) Friedrich & Rowley, IOS Bull. 3(3): 98, 1974. Basionym: *Trichocereus taquimbalensis*. [RPS 25]

Echinopsis taquimbalensis var. **wilkeae** (Backeberg) Friedrich & Rowley, IOS Bull. 3(3): 98, 1974. Basionym: *Trichocereus taquimbalensis* var. *wilkeae*. [RPS 25]

Echinopsis taratensis (Cardenas) Friedrich & Rowley, IOS Bull. 3(3): 98, 1974. Basionym: *Trichocereus taratensis*. [RPS 25]

Echinopsis tarijensis (Vaupel) Friedrich & Rowley, IOS Bull. 3(3): 98, 1974. Basionym: *Cereus tarijensis*. [RPS 25]

Echinopsis tarmaensis (Rauh & Backeberg) Friedrich & Rowley, IOS Bull. 3(3): 98, 1974. Basionym: *Trichocereus tarmaensis*. [RPS 25]

Echinopsis tegeleriana (Backeberg) D. Hunt, Bradleya 5: 92, 1987. Basionym: *Lobivia tegeleriana*. [RPS 38]

Echinopsis terscheckii (Parmentier) Friedrich & Rowley, IOS Bull. 3(3): 98, 1974. Basionym: *Cereus terscheckii*. [RPS 25]

Echinopsis terscheckii var. **montana** (Backeberg) Friedrich & Rowley, Rep. Pl. Succ. 25: 10, 1976. Basionym: *Trichocereus terscheckii* var. *montanus*. [RPS 25]

Echinopsis thelegona (F. A. C. Weber) Friedrich & Rowley, IOS Bull. 3(3): 98, 1974. Basionym: *Cereus thelegonus*. [RPS 25]

Echinopsis thelegonoides (Spegazzini) Friedrich & Rowley, IOS Bull. 3(3): 98, 1974. Basionym: *Cereus thelegonoides*. [RPS 25]

Echinopsis toralapana Cardenas, Cactus 82: 41, 1964. [RPS 15]

Echinopsis torrecillasensis Cardenas, Cact. Succ. J. (US) 28(4): 110-111, ills., 1956. Typus: *Cardenas* 5060 (BOLV, ZSS [authentic material ?]). [Bolivia: Florida] [RPS 7]

Echinopsis trichosa (Cardenas) Friedrich & Rowley, IOS Bull. 3(3): 98, 1974. Basionym: *Trichocereus trichosus*. [RPS 25]

Echinopsis tucumanensis Y. Ito, Explan. Diag. Austroechinocactinae, 64, 285, 1957. [First used as a nude name by Fric in Price-List for 1928, [2] (sphalm. 'tucumanense'), and again by Backeberg in 1936 or 1937.] [RPS 8]

Echinopsis tulhuayacensis (Ochoa) Friedrich & Rowley, IOS Bull. 3(3): 98, 1974. Basionym: *Trichocereus tulhuayacensis*. [RPS 25]

Echinopsis tunariensis (Cardenas) Friedrich & Rowley, IOS Bull. 3(3): 98, 1974. Basionym: *Trichocereus tunariensis*. [RPS 25]

Echinopsis uebelmanniana (Lembcke & Backeberg) A. E. Hoffmann, Cact. Fl. Chil., 92-93, ill., 1989. *Nom. inval.*, based on *Soehrensia uebelmanniana, nom. inval.* (Art. 37.1). [RPS 40]

Echinopsis uyupampensis (Backeberg) Friedrich & Rowley, IOS Bull. 3(3): 99, 1974. Basionym: *Trichocereus uyupampensis*. [RPS 25]

Echinopsis vallegrandensis Cardenas, Cactus (Paris) No. 64: 163, 1959. [Sphalm. 'vellegradensis'. Although this spelling is used consistently, it is here corrected as it is an obvious error for this taxon from Valle Grande.] [RPS 10]

Echinopsis vasquezii (Rausch) Rowley, Rep. Pl. Succ. 24: 6, 1975. Basionym: *Trichocereus vasquezii*. [Repeated by Rowley in Ashingtonia 2(4): centre inset (inval. Art. 33.2) and in Rep. Pl. Succ. 27: 5, 1979.] [RPS 24]

Echinopsis volliana (Backeberg) Friedrich & Rowley, IOS Bull. 3(3): 99, 1974.

Basionym: *Trichocereus vollianus*. [RPS 25]

Echinopsis volliana var. **rubrispina** (Backeberg) Friedrich & Rowley, IOS Bull. 3(3): 99, 1974. Basionym: *Trichocereus vollianus* var. *rubrispinus*. [RPS 25]

Echinopsis walteri (Kiesling) Friedrich & Glätzle, Bradleya 1: 96, 1983. Basionym: *Lobivia walteri*. [RPS 34]

Echinopsis werdermanniana (Backeberg) Friedrich & Rowley, IOS Bull. 3(3): 99, 1974. Basionym: *Trichocereus werdermannianus*. [RPS 25]

Echinopsis werdermannii Fric *ex* Fleischer, Friciana 1(7): 1, 1962. [RPS 13]

Echinopsis wilkeae (Backeberg) Friedrich *ex* Donald, Ashingtonia Species Cat. Cact. [part 4], unnumbered page, 1974. *Nom. inval.* (Art. 33.2). [Published as "nov. comb. in press".] [RPS -]

Echinopsis wilkeae var. *carminata* (Backeberg) Friedrich *ex* Donald, Ashingtonia Species Cat. Cact. [part 4], unnumbered page, 1974. *Nom. inval.* (Art. 43.1). [Published as "nov. comb. in press".] [RPS -]

Echinopsis xiphacantha var. **brevispina** (R. Meyer) Y. Ito, Explan. Diagr. Austroechinocactinae, 66, 1957. Basionym: *Echinopsis campylacantha* var. *brevispina*. [RPS 8]

Echinopsis xiphacantha var. **leucantha** (Labouret) Y. Ito, Explan. Diagr. Austroechinocactinae, 66, 1957. Basionym: *Echinopsis campylacantha* var. *leucantha*. [RPS 8]

Echinopsis xiphacantha var. **longispina** (R. Meyer) Y. Ito, Explan. Diagr. Austroechinocactinae, 66, 1957. Basionym: *Echinopsis campylacantha* var. *longispina*. [RPS 8]

Echinopsis yungasensis Ritter, Kakt. Südamer. 2: 631, 1980. Typus: *Ritter* 331 (U, ZSS). [Bolivia: Sud-Yungas] [RPS 31]

Eomatucana Ritter, Kakt. and. Sukk. 16: 230, 1965. Typus: *E. oreodoxa*. [RPS 16]

Eomatucana madisoniorum (P. C. Hutchison) Ritter, Kakt. Südamer. 4: 1487, 1981. Basionym: *Borzicactus madisoniorum*. [RPS 32]

Eomatucana oreodoxa Ritter, Kakt. and. Sukk. 16: 230, 1965. [RPS 16]

Epiphyllanthus opuntioides (Löfgren & Dusen) Moran, Gent. Herb. 8(4): 338, 1953. Basionym: *Epiphyllum opuntioides*. [RPS 4]

Epiphyllopsis gaertneri var. *serrata* (Lindinger) Backeberg, Die Cact. 2: 724, 1959. *Nom. inval.*, based on *Rhipsalidopsis serrata, nom. inval.* (Art. 36.1). [Erroneously included as valid in RPS 10.] [RPS 10]

Epiphyllopsis gaertneri var. **tiburtii** Backeberg & Voll, Arq. Jard. Bot. Rio de Janeiro 9:149-151, ill., 1950. Typus: *Tiburtius* s.n. (RB 65.041). [Vol. for 1949, published 1950.] [Brazil: Paraná] [RPS -]

Epiphyllum cv. **Clarence Wright** P. C. Hutchison, Cact. Succ. J. (US) 50(5): 238, 1978. [= *Epiphyllum* 'Reward' × *E.* 'Discovery'.] [RPS 29]

Epiphyllum cv. **Fanfare** Rowley, Ashingtonia 1: 19-20, ill., 1973. [Published as 'Epicactus'.] [RPS 24]

Epiphyllum chrysocardium Alexander, Cact. Succ. J. (US) 28(1): 3-6, ill., 1956. Typus: *MacDougall* A198 (NY, G, ZSS). [Mexico: Chiapas] [RPS 7]

Epiphyllum crenatum var. **kimnachii** Bravo, Anales Inst. Biol. UNAM 35: 77-80, ills., 1965. Typus: *MacDougall* s.n. (MEXU?, UC). [The name appears to be based on several syntypes, but the citation of a single collection in the introduction to the description is interpreted as being sufficient as type citation under the Code.] [Mexico: Chiapas] [RPS 17]

Epiphyllum eichlamii (Weingart) L. Williams, Fieldiana 29: 378, 1962. Basionym: *Phyllocactus eichlamii*. [RPS 18]

Epiphyllum floribundum Kimnach, Cact. Succ. J. (US) 62(2): 83-85, ills., 1990. Typus: *Mathias* s.n. (HNT, MO, UCLA, US). [Peru: Dept. Loreto] [RPS 41]

Epiphyllum gigas Woodson & Cutak, Ann. Missouri Bot. Gard. 45(1): 87, 1958. [RPS 9]

Epiphyllum laui Kimnach, Cact. Succ. J. (US) 62(3): 148-151, ills., 1990. Typus: *Lau* 1319 (HNT, CAS, MEXU). [Mexico: Chiapas] [RPS 41]

Epiphyllum macrocarpum (F. A. C. Weber) Backeberg, Die Cact. 2: 754, 1959. Basionym: *Phyllocactus macrocarpus*. [RPS 10]

Epiphyllum oxypetalum var. **purpusii** (Weingart) Backeberg, Die Cact. 2: 747, 1959. Basionym: *Phyllocactus purpusii*. [RPS 10]

Epiphyllum phyllanthus var. **boliviense** (F. A. C. Weber) Backeberg, Die Cact. 2: 746, 1959. Basionym: *Phyllocactus phyllanthus* var. *boliviensis*. [RPS 10]

Epiphyllum phyllanthus var. **columbiense** (F. A. C. Weber) Backeberg, Die Cact. 2: 746, 1959. Basionym: *Phyllocactus phyllanthus* var. *columbiensis*. [RPS 10]

Epiphyllum phyllanthus var. **guatemalense** (Britton & Rose) Kimnach, Cact. Succ. J. (US) 36: 110, 1964. Basionym: *Epiphyllum guatemalense*. [RPS 15]

Epiphyllum phyllanthus var. **hookeri** (Haworth) Kimnach, Cact. Succ. J. (US) 36: 113, 1964. Basionym: *Epiphyllum hookeri*. [RPS 15]

Epiphyllum phyllanthus var. **paraguayense** (F. A. C. Weber) Backeberg, Die Cact. 2: 746, 1959. Basionym: *Phyllocactus phyllanthus* var. *paraguayensis*. [RPS 10]

Epiphyllum phyllanthus var. **pittieri** (F. A. C. Weber) Kimnach, Cact. Succ. J. (US) 36: 115, 1964. Basionym: *Phyllocactus pittieri*. [RPS 15]

Epiphyllum phyllanthus var. **rubrocoronatum** Kimnach, Cact. Succ. J. (US) 36: 110, 1964. [RPS 15]

Epiphyllum quezaltecum (Standley & Steyermark) L. Williams, Fieldiana 29: 378, 1962. Basionym: *Bonifazia quezalteca*. [RPS 18]

Epiphyllum rubrocoronatum (Kimnach) Dodson & Gentry, Selbyana 2: 31, 1977. Basionym: *Epiphyllum phyllanthus* var.

rubrocoronatum. [RPS -]

Epiphyllum ruestii (Weingart) Backeberg, Die Cact. 2: 751, 1959. Basionym: *Phyllocactus ruestii.* [RPS 10]

Epiphyllum steyermarkii Croizat, Phytologia 28(1): 17-20, ills., 1974. Typus: *Steyermark* 108741 (VEN). [Venezuela: Est. Miranda] [RPS 26]

Epiphyllum thomasianum var. **costaricense** (F. A. C. Weber) Kimnach, Cact. Succ. J. (US) 37: 168, 1965. Basionym: *Phyllocactus costaricensis.* [RPS 16]

Epithelantha bokei L. Benson, Cact. Succ. J. (US) 41: 185-186, 1969. [RPS 20]

Epithelantha densispina Bravo, Anales Inst. Biol. UNAM 22(1): 19-21, ill., 1951. [RPS 2]

Epithelantha micromeris fa. **elongata** (Backeberg) Bravo, Cact. Suc. Mex. 25(3): 65, 1980. Basionym: *Epithelantha pachyrhiza* var. *elongata.* [RPS 31]

Epithelantha micromeris var. **bokei** (L. Benson) Glass & Foster, Cact. Succ. J. (US) 50(4): 185, 1978. Basionym: *Epithelantha bokei.* [RPS 29]

Epithelantha micromeris var. **densispina** (Bravo) Backeberg, Cactus (Paris) No. 39: 31; No. 40: 57, 1954. Basionym: *Epithelantha densispina.* [RPS 5]

Epithelantha micromeris var. **polycephala** (Backeberg) Glass & Foster, Cact. Succ. J. (US) 50(4): 187, 1978. Basionym: *Epithelantha polycephala.* [RPS 29]

Epithelantha micromeris var. **rufispina** (Bravo) Backeberg, Cactus (Paris) No. 39: 31; No. 40: 57, 1954. Basionym: *Epithelantha rufispina.* [RPS 5]

Epithelantha micromeris var. **unguispina** (Bödeker) Backeberg, Cactus (Paris) No. 39: 31; No. 40: 58, 1954. Basionym: *Mammillaria micromeris* var. *unguispina.* [RPS 5]

Epithelantha pachyrhiza (W. T. Marshall) Backeberg, Cactus (Paris) No. 39: 31; No. 40: 60, 1954. Basionym: *Epithelantha micromeris* var. *pachyrhiza.* [Name first used by Bravo Hollis and ascribed to Marshall in Anales Inst. Biol. UNAM 22(1): 16, 1951.] [RPS 5]

Epithelantha pachyrhiza var. **elongata** Backeberg, Cactus (Paris) 9(39): 32, (40): 62, 1954. [RPS 5]

Epithelantha polycephala Backeberg, Cactus (Paris) No. 39: 32, No. 40: 62, 1954. [RPS 5]

Epithelantha rufispina Bravo, Anales Inst. Biol. UNAM 22(1): 22-23, ill., 1951. [RPS 2]

Epithelantha spinosior C. Schmoll, Anales Inst. Biol. Mexico 22(1): 11-14, 1951. [RPS 2]

× **Epixochia amarantina** (Regel *pro var.*) Rowley, in Backeberg, Die Cact. 6: 3556, 1962. Basionym: *Phyllocactus crenatus* var. *amarantinus.* [= *Epiphyllum crenatum* × *Nopalxochia phyllanthoides.*] [RPS 13]

× **Epixochia splendens** (Regel *pro var.*) Rowley, in Backeberg, Die Cact. 6: 3556, 1962. Basionym: *Phyllocactus crenatus* var. *splendens.* [= *Epiphyllum crenatum* × *Nopalxochia ackermannii.*] [RPS 13]

Erdisia aureispina Backeberg & Jacobsen, in Backeberg, Descr. Cact. Nov. [1:] 12, 1957. [Dated 1956, published 1957.] [RPS 7]

Erdisia fortalezensis Ritter, Taxon 13: 116, 1964. [RPS 15]

Erdisia quadrangularis Rauh & Backeberg, in Backeberg, Descr. Cact. Nov. [1:] 12-13, 1957. Typus: *Rauh* K120 (1956) (ZSS). [Dated 1956, published 1957.] [Peru] [RPS 7]

Erdisia ruthae Johnson, in Backeberg, Die Cact. 6: 3662, 1962. *Nom. inval.* (Art. 36.1/37.1). [RPS 13]

Erdisia tenuicula Backeberg, Descr. Cact. Nov. [1:] 12, 1957. [Dated 1956, published 1957.] [RPS 7]

Eriocactus ampliocostatus Ritter, Kakt. Südamer. 1: 253, 1979. Incorrect name (Art. 11.3). [RPS 30]

Eriocactus claviceps Ritter, Succulenta 1966: 113, 115-116, 1966. Incorrect name (Art. 11.3). [RPS 17]

Eriocactus grossei var. *aureispinus* Ritter, Kakt. Südamer. 1: 253, 1979. Incorrect name (Art. 11.3). [RPS 30]

Eriocactus leninghausii fa. *apelii* (Heinrich) Backeberg, Die Cact. 3: 1680, 1959. Incorrect name, based on *Eriocephala leninghausii* fa. *apelii,* incorrect name (Art. 11.3). [RPS 10]

Eriocactus leninghausii var. *minor* Ritter, Kakt. Südamer. 1: 156, 1979. Incorrect name (Art. 11.3). [RPS 30]

Eriocactus magnificus Ritter, Succulenta 45(4): 50, 1966. Incorrect name (Art. 11.3). [RPS 17]

Eriocactus nigrispinus (Schumann) Ritter, Kakt. Südamer. 1: 256, 1979. Incorrect name (Art. 11.3), based on *Echinocactus nigrispinus.* [RPS 30]

Eriocactus warasii Ritter, Bradea 1: 353-355, 1973. Incorrect name (Art. 11.3). [RPS 24]

Eriocephala grossei (Schumann) Y. Ito, Explan. Diagr. Austroechinocactinae, 252, 1957. Incorrect name (Art. 11.3), based on *Echinocactus grossei.* [Sphalm. 'Eriocephalus' fide RPS 8.] [RPS 8]

Eriocephala leninghausii (Haage) Y. Ito, Explan. Diagr. Austroechinocactinae, 253, 1957. Incorrect name (Art. 11.3), based on *Pilocereus leninghausii.* [Sphalm. 'Eriocephalus' fide RPS 8. The name given as basionym was itself a combination, and is here corrected.] [RPS 8]

Eriocephala schumanniana var. *longispina* (Haage jr.) Y. Ito, Explan. Diagr. Austroechinocactinae, 252, 1957. Incorrect name (Art. 11.3), based on *Echinocactus schumannianus* var. *longispinus.* [Sphalm. 'Eriocephalus schumannianus var. longispinus' fide RPS 8.] [RPS 8]

Eriocephala schumanniana var. *nigrispina* (Haage jr.) Y. Ito, Explan. Diagr. Austroechinocactinae, 252, 1957. Incorrect name (Art. 11.3), based on *Echinocactus schumannianus* var. *nigrispinus.* [Sphalm. 'Eriocephalus schumannianus var. nigrispinus' fide RPS 8.] [RPS 8]

Eriocereus arendtii (Schumann) Ritter, Kakt. Südamer. 1: 242, 1979. Basionym: *Cereus*

arendtii. [RPS 30]
Eriocereus jusbertii cv. **'Drawert'** Täuber, Succulenta 63(4): 94-96, ills., 1984. [RPS 35]
Eriocereus polyacanthus Ritter, Kakt. Südamer. 2: 436, 1980. Typus: *Ritter* 413 (U, ZSS). [Argentina: Catamarca] [RPS 31]
Eriocereus pomanensis var. **uruguayensis** (Osten) Backeberg, Die Cact. 4: 2095, 1960. Basionym: *Harrisia tortuosa* var. *uruguayensis*. [RPS 11]
Eriocereus regelii (Weingart) Backeberg, Die Cact. 4: 2093, 1960. Basionym: *Cereus regelii*. [RPS 11]
Eriocereus spinosissimus Buining et al., Kakt. and. Sukk. 28(3): 49-51, 1977. [RPS 28]
Eriocereus tarijensis Ritter, Kakt. Südamer. 2: 557-558, 1980. [RPS 31]
Eriosyce algarrobensis Ritter, Kakt. Südamer. 3: 914, 1980. [RPS 31]
Eriosyce ausseliana Ritter, Kakt. Südamer. 3: 913, 1980. Based on *Ritter* 254. *Nom. inval.* (Art. 36.1). [Given as provisional name. Later (i.e. field number list in l.c., 4) also treated as variety of *E. ihotzkyana*.] [Chile: La Serena] [RPS -]
Eriosyce ceratistes var. **celsii** (Labouret) Y. Ito, Explan. Diagr. Austroechinocactinae, 135, 1957. Basionym: *Echinocactus ceratistes* var. *celsii*. [RPS 8]
Eriosyce ceratistes var. **combarbalensis** Backeberg, Descr. Cact. Nov. [1:] 33, 1957. [Dated 1956, published 1957.] [RPS 7]
Eriosyce ceratistes var. **coquimbensis** Backeberg, Descr. Cact. Nov. [1:] 32, 1957. [Dated 1956, published 1957.] [RPS 7]
Eriosyce ceratistes var. **jorgensis** Backeberg, Descr. Cact. Nov. [1:] 33, 1957. [Dated 1956, published 1957.] [RPS 7]
Eriosyce ceratistes var. **melanacantha** (Labouret) Y. Ito, Explan. Diagr. Austroechinocactinae, 134, 1957. Basionym: *Echinocactus ceratistes* var. *melanacanthus*. [RPS 8]
Eriosyce ceratistes var. **mollesensis** Backeberg, Descr. Cact. Nov. [1]: 32, 1957. Typus: *Jiles* 5 (S). [Dated 1956, published 1957. Typification resolved by Backeberg in Die Cact. 3: 1880, 1959.] [Chile: Coquimbo] [RPS 7]
Eriosyce ceratistes var. **vallenarensis** Backeberg, Descr. Cact. Nov. [1]: 32, 1957. Typus: *Kraus* 17 (). [Dated 1956, published 1957. Typification resolved by Backeberg in Die Cact. 3: 1881, 1959.] [Chile: Atacama] [RPS 7]
Eriosyce ceratistes var. **zorillaensis** Backeberg, Descr. Cact. Nov. [1]: 32, 1957. [Dated 1956, published 1957.] [RPS 7]
Eriosyce ihotzkyana Ritter, Kakt. Südamer. 3: 912-913, 1980. Typus: *Ritter* 253 (ZSS). [None of the several specimens at ZSS is labelled as holotype.] [Chile] [RPS 31]
Eriosyce lapampaensis Ritter, Kakt. Südamer. 3: 914-915, 1980. Typus: *Ritter* 255 (ZSS). [None of the several specimens at ZSS is labelled as holotype.] [Chile: Atacama] [RPS 31]
Eriosyce megacarpa Ritter, Kakt. Südamer. 3: 917-918, 1980. Typus: *Ritter* 514 (U, ZSS). [Holotype cited for U l.c. 1: iii, 1979.] [Chile: Chañaral] [RPS 31]
Eriosyce rodentiophila Ritter, Kakt. Südamer. 3: 916-917, 1980. Typus: *Ritter* 264 (U, ZSS). [Holotype cited for U l.c. 1: iii, 1979.] [Chile: Chañaral] [RPS 31]
Eriosyce rodentiophila var. **lanata** Ritter, Kakt. Südamer. 3: 917, 1980. [RPS 31]
Eriosyce sandillon var. **algarrobensis** (Ritter) A. E. Hoffmann, Cact. Fl. Chil., 146, ill. (p. 147), 1989. Basionym: *Eriosyce algarrobensis*. [*E. ceratistes*, cited in the synonymy of the species, would have priority and should have been used for the species.] [RPS 40]
Eriosyce sandillon var. *mollesensis* (Backeberg) A. E. Hoffmann, Cact. Fl. Chil., 144, ill. (p. 145), 1989. *Nom. inval.* (Art. 33.2), based on *Eriosyce ceratistes* var. *mollesensis*. [Sphalm. 'mollensis'; and with incorrect basionym citation. Moreover, *E. ceratistes*, cited in the synonymy of the species, would have priority and should have been used for the species.] [RPS 40]
Eriosyce sandillon var. *vallenarensis* (Backeberg) A. E. Hoffmann, Cact. Fl. Chil., 144, 1989. *Nom. inval.* (Art. 33.2), based on *Eriosyce ceratistes* var. *vallenarensis*. [With incorrect basionym citation. Moreover, *E. ceratistes*, cited in the synonymy of the species, would have priority and should have been used for the species.] [RPS 40]
Eriosyce spinibarbis Ritter, Kakt. Südamer. 3: 915, 1980. [RPS 31]
Erythrorhipsalis burchellii (Britton & Rose) Volgin, Vestn. Moskovsk. Univ. Ser. 16, Biol. 36(3): 19, 1981. Basionym: *Rhipsalis burchellii*. [RPS 36]
Erythrorhipsalis campos-portoana (Loefgren) Volgin, Vestn. Moskovsk. Univ. Ser. 16, Biol. 36(3): 19, 1981. Basionym: *Rhipsalis campos-portoana*. [RPS 36]
Erythrorhipsalis cereuscula (Haworth) Volgin, Vestn. Moskovsk. Univ. Ser. 16, Biol. 36(3): 19, 1981. Basionym: *Rhipsalis cereuscula*. [RPS 36]
Erythrorhipsalis cribrata (Lemaire) Volgin, Vestn. Moskovsk. Univ. Ser. 16, Biol. 36(3): 19, 1981. Basionym: *Hariota cribrata*. [RPS 36]
Escobaria sect. **Acharagma** N. P. Taylor, Kakt. and. Sukk. 34(8): 185, 1983. Typus: *Echinocactus roseanus* Bödeker. [RPS 34]
Escobaria sect. **Neobesseya** (Britton & Rose) N. P. Taylor, Kakt. and. Sukk. 34(7): 155, 1983. Basionym: *Neobesseya*. [RPS 34]
Escobaria sect. **Pleurantha** N. P. Taylor, Kakt. and. Sukk. 34(5): 123, 1983. Typus: *Escobaria chihuahuensis* Britton & Rose. [RPS 34]
Escobaria subgen. *Euescobaria* Buxbaum, Österr. Bot. Zeitschr. 98(1-2): 78, 1951. *Nom. inval.* (Art. 21.3). [Type not cited.] [RPS 2]
Escobaria subgen. **Neobesseya** (Britton & Rose) V. John & Riha, Kaktusy 17(3): 63, 1981. Basionym: *Neobesseya*. [RPS 32]
Escobaria subgen. **Ortegocactus** (Alexander) V.

John & Riha, Kaktusy 17(2): 42, 1981. Basionym: *Ortegocactus*. [RPS 32]

Escobaria subgen. **Primibaria** V. John & Riha, Kaktusy 17(2): 41, 1981. Typus: *Escobaria roseana* (Bödeker) Buxbaum. [RPS 32]

Escobaria subgen. **Protomammillaria** V. John & Riha, Kaktusy 17(3): 65, 1981. Typus: *Escobaria henricksonii* Glass & Foster. [RPS 32]

Escobaria subgen. **Pseudocoryphantha** Buxbaum, Österr. Bot. Zeitschr. 98(1-2): 78, 1951. Typus: *Escobaria chlorantha* (Engelmann) Buxbaum. [RPS 2]

Escobaria aggregata (Engelmann) Buxbaum, Österr. Bot. Zeitschr. 98(1-2): 78, 1951. Basionym: *Mammillaria aggregata*. [Basionym erroneously given as *Coryphantha aggregata*.] [RPS 2]

Escobaria aguirreana (Glass & Foster) N. P. Taylor, Kakt. and. Sukk. 34(8): 185, 1983. Basionym: *Gymnocactus aguirreanus*. [RPS 34]

Escobaria arizonica (Engelmann) Buxbaum, Österr. Bot. Zeitschr. 98(1-2): 78, 1951. Basionym: *Mammillaria arizonica*. [Basionym erroneously given as *Coryphantha arizonica*.] [RPS 2]

Escobaria asperispina (Bödeker) D. Hunt, Cact. Succ. J. Gr. Brit. 40(1): 13, 1978. Basionym: *Coryphantha asperispina*. [RPS 29]

Escobaria chaffeyi fa. **viridiflora** (Fric) Riha, Kaktusy 22(2): 25-27, ill., 1986. Basionym: *Fobea viridiflora*. [RPS 37]

Escobaria chlorantha (Engelmann) Buxbaum, Österr. Bot. Zeitschr. 98(1-2): 78, 1951. Basionym: *Mammillaria chlorantha*. [Basionym erroneously given as *Coryphantha chlorantha*.] [RPS 2]

Escobaria cubensis (Britton & Rose) D. Hunt, Cact. Succ. J. Gr. Brit. 40(1): 13, 1978. Basionym: *Coryphantha cubensis*. [RPS 29]

Escobaria dasyacantha var. **chaffeyi** (Britton & Rose) N. P. Taylor, Kakt. and. Sukk. 34(7): 157, 1983. Basionym: *Escobaria chaffeyi*. [RPS 34]

Escobaria dasyacantha var. **duncanii** (Hester) N. P. Taylor, Kakt. and. Sukk. 34(?): 157, 1983. Basionym: *Escobesseya duncanii*. [RPS 34]

Escobaria dasyacantha var. **varicolor** (Tiegel) D. Hunt, Cact. Succ. J. Gr. Brit. 40(1): 13, 1983. Basionym: *Coryphantha varicolor*. [RPS 29]

Escobaria deserti (Engelmann) Buxbaum, Österr. Bot. Zeitschr. 98(1-2): 78, 1951. Basionym: *Mammillaria deserti*. [Basionym erroneously given as *Coryphantha deserti*.] [RPS 2]

Escobaria duncanii (Hester) Backeberg, Die Cact. 5: 2966, 1961. Basionym: *Escobesseya duncanii*. [RPS 12]

Escobaria emskoetteriana (Quehl) Backeberg, Die Cact. 5: 2958. Cact, 1961. Basionym: *Mammillaria emskoetteriana*. [Also attributed to Borg, Cacti, 304, 1937 (not assessed).] [RPS 12]

Escobaria guadalupensis Brack & Heil, Cact. Succ. J. (US) 58(4): 165-167, ills., 1986. Typus: *Anonymus* (SJNM 2774). [RPS 37]

Escobaria henricksonii Glass & Foster, Cact. Succ. J. (US) 49(5): 195-196, 1977. Typus: *Henrickson* 7744 (POM). [Mexico: Chihuahua] [RPS 28]

Escobaria hesteri (Y. Wright) Buxbaum, Österr. Bot. Zeitschr. 98(1-2): 78, 1951. Basionym: *Coryphantha hesteri*. [RPS 2]

Escobaria laredoi (Glass & Foster) N. P. Taylor, Cact. Succ. J. Gr. Brit. 41(1): 20, 1979. Basionym: *Coryphantha laredoi*. [RPS 30]

Escobaria minima (Baird) D. Hunt, Cact. Succ. J. Gr. Brit. 40(2): 30, 1978. Basionym: *Coryphantha minima*. [RPS 29]

Escobaria missouriensis (Sweet) D. Hunt, Cact. Succ. J. Gr. Brit. 40(1): 13, 1978. Basionym: *Mammillaria missouriensis*. [RPS 29]

Escobaria missouriensis var. **asperispina** (Bödeker) N. P. Taylor, Kakt. and Sukk. 34(8): 185, 1983. Basionym: *Coryphantha asperispina*. [RPS 34]

Escobaria missouriensis var. **caespitosa** (Engelmann) D. Hunt, Cact. Succ. J. Gr. Brit. 40(1): 13, 1978. Basionym: *Mammillaria similis* var. *caespitosa*. [RPS 29]

Escobaria missouriensis var. **marstonii** (Clover) D. Hunt, Cact. Succ. J. Gr. Brit. 40(1): 13, 1978. Basionym: *Coryphantha marstonii*. [RPS 29]

Escobaria missouriensis var. **robustior** (Engelmann) D. Hunt, Cact. Succ. J. Gr. Brit. 40(1): 13, 1978. Basionym: *Mammillaria similis* var. *robustior*. [RPS 29]

Escobaria missouriensis var. **similis** (Engelmann) N. P. Taylor, Kakt. and Sukk. 34(8): 184, 1983. Basionym: *Mammillaria similis*. [RPS 34]

Escobaria nellieae (Croizat) Backeberg, Die Cact. 5: 2967, 1961. Basionym: *Coryphantha nellieae*. [RPS 12]

Escobaria neomexicana (Engelmann) Buxbaum, Österr. Bot. Zeitschr. 98(1-2): 78, 1951. Basionym: *Mammillaria neomexicana*. [Sphalm. 'neo-mexicana'. Basionym erroneously given as *Coryphantha neomexicana*.] [RPS 2]

Escobaria oklahomensis (Lahmann) Buxbaum, Österr. Bot. Zeitschr. 98(1-2): 78, 1951. Basionym: *Coryphantha oklahomensis*. [RPS 2]

Escobaria orcuttii var. **koenigii** Castetter, Pierce & Schwerin, Cact. Succ. J. (US) 47(2): 68-69, 1976. Typus: *Castetter* 961 (UNM). [USA: New Mexico] [RPS 27]

Escobaria orcuttii var. **macraxina** Castetter, Pierce & Schwerin, Cact. Succ. J. (US) 47(2): 67, ill., 1975. Typus: *Heil* 4287 (UNM). [USA: New Mexico] [RPS 27]

Escobaria organensis (D. Zimmerman) Castetter, Pierce & Schwerin, Cact. Succ. J. (US) 47(2): 60, 1975. Basionym: *Coryphantha organensis*. [RPS 27]

Escobaria radiosa (Engelmann) G. Frank, Kakt. and. Sukk. 11: 157, 1960. *Nom. inval.* (Art. 33.2, 34.1), based on *Mammillaria radiosa*. [RPS -]

Escobaria rigida Backeberg, Die Cact. 5: 3969-2971, pl. 220, 1961. *Nom. inval.* (Art. 37.1). [Erroneously omitted from RPS so far.] [RPS -]

Escobaria robbinsorum (Earle) D. Hunt, Cact. Succ. J. Gr. Brit. 40(1): 13, 1978. Basionym: *Cochiseia robbinsorum*. [RPS 29]

Escobaria roseana (Bödeker) Buxbaum, Österr. Bot. Zeit. 98: 1-2, 1951. Basionym: *Echinocactus roseanus*. [Attributed to Schmoll. Repeated by Backeberg in Cact. Succ. J. (US) 23(5): 151, 1952. Incorrectly treated as invalid combination in RPS 3.] [RPS 3]

Escobaria sandbergii Castetter, Pierce & Schwerin, Cact. Succ. J. (US) 47(2): 62-64, 1975. Typus: *Pierce* 3409 (UNM 38739). [USA: New Mexico] [RPS 27]

Escobaria sneedii var. **leei** (Rose *ex* Bödeker) D. Hunt, Cact. Succ. J. Gr. Brit. 40(2): 30, 1978. Basionym: *Escobaria leei*. [RPS 29]

Escobaria strobiliformis var. **durispina** (Quehl) Bravo, Cact. Suc. Mex. 27(1): 17, 1982. Basionym: *Mammillaria strobiliformis* var. *durispina*. [RPS 33]

Escobaria tuberculosa var. **pubescens** (Quehl) Y. Ito, Cacti, 113, 1952. [Given as *comb. nud.* in RPS 4.] [RPS 4]

Escobaria tuberculosa var. **varicolor** (Tiegel) Brack & Heil, Cact. Succ. J. (US) 60(1): 17, 1988. Basionym: *Coryphantha varicolor*. [RPS 39]

Escobaria villardii Castetter, Pierce & Schwerin, Cact. Succ. J. (US) 47(2): 64-66, 1975. Typus: *Reaves* 3984 (UNM 50789). [USA: New Mexico] [RPS 27]

Escobaria vivipara (Nuttall) Buxbaum, Österr. Bot. Zeitschr. 98(1-2): 78, 1951. Basionym: *Cactus viviparus*. [Basionym erroneously given as *Coryphantha vivipara*.] [RPS 2]

Escobaria vivipara var. **alversonii** (Coulter) D. Hunt, Cact. Succ. J. Gr. Brit. 40(1): 13, 1978. Basionym: *Cactus radiosus* var. *alversonii*. [RPS 29]

Escobaria vivipara var. **arizonica** (Engelmann) D. Hunt, Cact. Succ. J. Gr. Brit. 40(1): 13, 1978. Basionym: *Mammillaria arizonica*. [RPS 29]

Escobaria vivipara var. **bisbeeana** (Orcutt) D. Hunt, Cact. Succ. J. Gr. Brit. 40(1): 13, 1978. Basionym: *Coryphantha bisbeeana*. [RPS 29]

Escobaria vivipara var. **buoflama** (P. Fischer) N. P. Taylor, Kakt. and. Sukk. 34(6): 140, 1983. Basionym: *Coryphantha vivipara* var. *buoflama*. [RPS 34]

Escobaria vivipara var. **deserti** (Engelmann) D. Hunt, Cact. Succ. J. Gr. Brit. 40(1): 13, 1978. Basionym: *Mammillaria deserti*. [RPS 29]

Escobaria vivipara var. **kaibabensis** (P. Fischer) N. P. Taylor, Kakt. and Sukk. 34(6): 139, 1983. Basionym: *Coryphantha vivipara* var. *kaibabensis*. [RPS 34]

Escobaria vivipara var. **neomexicana** (Engelmann) Buxbaum, in Krainz, Die Kakt. Lief. 50-51: C VIIIc, 1973. Basionym: *Mammillaria vivipara* var. *neomexicana*. [Sphalm. 'neo-mexicana'.] [RPS 24]

Escobaria vivipara var. **radiosa** (Engelmann) D. Hunt, Cact. Succ. J. Gr. Brit. 40(1): 13, 1978. Basionym: *Mammillaria radiosa*. [RPS 29]

Escobaria vivipara var. **rosea** (Clokey) D. Hunt, Cact. Succ. J. Gr. Brit. 40(1): 13, 1978. Basionym: *Coryphantha rosea*. [RPS 29]

Escobaria zilziana (Bödeker) Backeberg, Die Cact. 5: 2957, 1961. Basionym: *Coryphantha zilziana*. [*Non* Backeberg 1962, cf. N. P. Taylor in Kakt. and. Sukk. 34(7): 155, 1983.] [RPS 12]

Escobesseya albicolumnaria Hester *ex* L. Benson, in Lundell, Fl. Texas 2: 311, 1969. *Nom. inval.* (Art. 33.2, 43.1), based on *Escobaria albicolumnaria*. [RPS -]

Escontria lepidantha (Eichlam) Buxbaum, Bot. Stud. 12: 45, 99, 1961. Basionym: *Cereus lepidanthus*. [RPS 12]

× **Espostingia** Rowley, Nation. Cact. Succ. J. 37(2): 48, 1982. [*Espostoa* × *Browningia*.] [RPS 33]

Espostoa subgen. **Facheiroa** (Britton & Rose) Buxbaum, Österr. Bot. Zeitschr. 106: 155, 1959. Basionym: *Facheiroa*. [RPS 10]

Espostoa baumannii Knize, Biota 7: 263, 1969. Typus: *Knize* 300 (not indicated). [sphalm. 'baumanni'. Published either Dec. 1968 or early in 1969.] [RPS 36]

Espostoa baumannii var. **arborescens** Knize, Biota 7: 263, 1969. Typus: *Knize* 299 (not indicated). [Sphalm. 'baumanni'. Published either in Dec. 1968 or early in 1969.] [RPS 36]

Espostoa blossfeldiorum (Werdermann) Buxbaum, in Krainz, Die Kakt. Lief. 33: [unpaged], 1966. Basionym: *Cephalocereus blossfeldiorum*. [First published as nude combination in Österr. Bot. Zeitschr. 106: 155, 1959 (cf. RPS 10).] [RPS 17]

Espostoa calva Ritter, Kakt. Südamer. 4: 1432, 1981. Typus: *Ritter* 1314 (U). [Peru: Amazonas] [RPS 32]

Espostoa frutescens J. E. Madsen, Fl. Ecuador 35: 36-37, 1989. Typus: *Madsen* 61064 (AAU, QCA). [Ecuador: Prov. El Oro / Azuay] [RPS 40]

Espostoa guentheri (Kupper) Buxbaum, Österr. Bot. Zeitschr. 106: 155, 1959. *Nom. inval.* (Art. 33.2), based on *Cephalocereus guentheri*. [RPS 10]

Espostoa haagei (Rümpler) Ritter, Backeberg's Descr. & Erört. taxon. Fragen, [unpaged], 1958. [RPS 15]

Espostoa haagei var. *samnensis* Ritter, Kakt. Südamer. 4: 1435, 1981. Based on *Ritter* 144a. *Nom. inval.* (Art. 36.1). [Described provisionally.] [Peru: La Libertad] [RPS -]

Espostoa huanucoensis Ritter, Kakt. Südamer. 4: 1435-1436, 1981. Typus: *Ritter* 665 (U, ZSS [seeds only]). [Holotype cited for U l.c. 1: iii, 1979.] [First mentioned (as invalid name) in Backeberg, Die Cact. 4: 2540, 1960 (cf. RPS 11, where the name is reported as valid).] [Peru: Huanuco] [RPS 32]

Espostoa hylaea Ritter, Taxon 13: 143, 1964. [RPS 15]

Espostoa lanata var. **floridaensis** Ritter, Kakt. Südamer. 4: 1442-143, 1981. Typus: *Ritter* 281f (U). [Peru: Cajamarca] [RPS 32]

Espostoa lanata var. **sericata** (Backeberg) Backeberg, Die Cact. 4: 2529, 1960. Basionym: *Cereus sericatus*. [RPS 11]

Espostoa lanianuligera Ritter, Kakt. Südamer.

4: 1443-1444, 1981. Typus: *Ritter* 660 (U, ZSS). [Holotype cited for U l.c. 1: iii, 1979.] [Peru: Cajamarca] [RPS 32]

Espostoa laticornua Rauh & Backeberg, in Backeberg, Descr. Cact. Nov. [1:] 35, 1957. Typus: *Rauh* s.n. (ZSS). [Dated 1956, published 1957.] [Peru] [RPS 7]

Espostoa laticornua var. **atroviolacea** Rauh & Backeberg, in Backeberg, Descr. Cact. Nov. [1:] 35, 1957. Typus: *Rauh* K88 (1954) (ZSS). [Dated 1956, published 1957.] [Peru] [RPS 7]

Espostoa laticornua var. **rubens** Rauh & Backeberg, in Backeberg, Descr. Cact. Nov. [1:] 35, 1957. Typus: *Rauh* K129 (1954) (ZSS). [Dated 1956, published 1957. See W. Rauh, Beitr. Kennt. peruan. Kakt.-veg., 527, 1958, for details about typification as collection number, etc.] [Peru] [RPS 7]

Espostoa melanostele fa. **inermis** (Backeberg) Krainz, Die Kakt. Lief. 26: C Va, 1964. Basionym: *Pseudoespostoa melanostele* var. *inermis*. [RPS 15]

Espostoa mirabilis Ritter, Taxon 13(4): 143, 1964. Typus: *Ritter* 670 (U, ZSS [seeds only]). [Type cited for U l.c. 12: 28, 1963.] [First mentioned without Latin diagnosis in Backeberg, Die Cact. 4: 2540, 1960.] [Peru: Cajamarca] [RPS 15]

Espostoa mirabilis var. **primigena** Ritter, Taxon 13(4): 143, 1964. Typus: *Ritter* 1061 (U, ZSS). [Type cited for U l.c. 12: 28, 1963.] [Peru: Cajamarca] [RPS 15]

Espostoa nana Ritter, Taxon 13(4): 143, 1964. Typus: *Ritter* 166 (U, ZSS). [Type cited for U l.c. 12: 28, 1963.] [Peru: Ancash] [RPS 15]

Espostoa procera Rauh & Backeberg, Cactus (Paris) No. 49: 28-29, 1956. [Repeated in Backeberg, Descr. Cact. Nov. [1:] 35, 1957 (dated 1956).] [RPS 7]

Espostoa ritteri Buining, Succulenta 39(3): 25-27, ills., 1960. Typus: *Ritter* 274 (ZSS). [Peru] [RPS 11]

Espostoa ruficeps Ritter, Kakt. Südamer. 4: 1448-1450, 1981. Typus: *Ritter* 573a (U, ZSS [seeds only]). [Holotype cited for U l.c. 1: iii, 1979.] [Peru: Huari] [RPS 32]

Espostoa senilis (Ritter) N. P. Taylor, Cact. Succ. J. Gr. Brit. 40(2): 54, 1978. Basionym: *Thrixanthocereus senilis*. [RPS 29]

Espostoa superba Ritter, Kakt. and. Sukk. 11(6): 85-86, ill., 1960. Typus: *Ritter* 572 (U, ZSS). [The type was cited for ZSS, and a specimen labelled as "holotype" is present there; another specimen labelled as "holotype" in Ritter's handwriting is, however, also present at U.] [Peru: Jaen] [RPS 11]

Espostoa ulei (Gürke) Buxbaum, Österr. Bot. Zeitschr. 106: 155, 1959. *Nom. inval.* (Art. 33.2), based on *Cephalocereus ulei*. [RPS 10]

× **Espostocactus** Mottram, Contr. New Class. Cact. Fam., 38, 1990. [= *Espostoa* × *Cleistocactus*.] [RPS 41]

Espostoopsis F. Buxbaum, in Krainz, Die Kakt. C Va (July 1968), 1968. Typus: *E. dybowskii*. [First mentioned as manuscript name in Katalog ZSS ed. 2, 58, 1967 (*nom. nud.*).] [RPS 19]

Espostoopsis dybowskii (Gosselin) Buxbaum, in Krainz, Die Kakt. CVa, 1968. Basionym: *Cereus dybowskii*. [RPS 19]

Eulychnia acida var. **elata** Ritter, Kakt. Südamer. 3: 896, 1980. [RPS 31]

Eulychnia acida var. **procumbens** Ritter, Kakt. Südamer. 3: 895-896, 1980. [RPS 31]

Eulychnia aricensis Ritter, Taxon 13(3): 115, 1964. Typus: *Ritter* 197 p.p. (ZSS). [Chile] [RPS 15]

Eulychnia barquitensis Ritter, Kakt. Südamer. 3: 899, 1980. Typus: *Ritter* 215 (U, ZSS). [Type cited for U l.c. 1: iii, 1979.] [Chile: Chañaral] [RPS 31]

Eulychnia breviflora var. **taltalensis** Ritter, Kakt. Südamer. 3: 898-899, 1980. Typus: *Ritter* 214 (U, ZSS [seeds only]). [Type cited for U l.c. 1: iii, 1979.] [Chile] [RPS 31]

Eulychnia breviflora var. **tenuis** Ritter, Kakt. Südamer. 3: 898, 1980. Typus: *Ritter* 215a (U, ZSS [seeds only]). [Type cited for U l.c. 1: iii, 1979.] [Chile] [RPS 31]

Eulychnia cephalophora Ritter, Katalog H. Winter, [unpaged], 1957. [Published as provisional name.] [RPS -]

Eulychnia iquiquensis var. **pullilana** Ritter, Kakt. Südamer. 3: 901, 1980. [RPS 31]

Eulychnia longispina (Salm-Dyck) Ritter, Kakt. and. Sukk. 16: 212, 1965. Basionym: *Cereus longispinus*. [RPS 16]

Eulychnia morromorenoensis Ritter, Kakt. Südamer. 3: 901-902, 1980. Typus: *Ritter* 202a p.p. (ZSS). [Chile] [RPS 31]

Eulychnia procumbens Backeberg, Descr. Cact. Nov. 3: 6, 1963. *Nom. inval.* (Art. 9.5). [RPS 14]

Eulychnia ritteri Cullmann, Kakt. and. Sukk. 9(8): 121-122, 1958. Typus: *Ritter* 276 (ZSS). [Peru: Arequipa] [RPS 9]

Eulychnia saint-pieana Ritter, Taxon 13(3): 115, 1964. Typus: *Ritter* 479a (U, ZSS [seeds only]). [Type cited for U l.c. 12: 28, 1963.] [Chile: Atacama] [RPS 15]

Eulychnia saint-pieana var. **barquitensis** (Ritter) A. E. Hoffmann, Cact. Fl. Chil., 142, 1989. Basionym: *Eulychnia barquitensis*. [RPS 40]

Facheiroa subgen. **Zehntnerella** (Britton & Rose) P. J. Braun & E. Esteves Pereira, Kakt. and. Sukk. 37(3): 56, 1986. Basionym: *Zehntnerella*. [RPS 37]

Facheiroa cephaliomelana Buining & Brederoo, Kakt. and. Sukk. 26(6): 121-124, ills., 1975. Typus: *Horst et Uebelmann* HU 447 (1974) (U, ZSS). [Brazil: Bahia] [RPS 26]

Facheiroa chaetacantha (Ritter) P. J. Braun & E. Esteves Pereira, Kakt. and. Sukk. 40(8): 202, 1989. Basionym: *Zehntnerella chaetacantha*. [RPS 40]

Facheiroa chaetacantha var. **montealtoi** (Ritter) P. J. Braun & E. Esteves Pereira, Kakt. and. Sukk. 40(8): 202, 1989. Basionym: *Zehntnerella chaetacantha* var. *montealtoi*. [RPS 40]

Facheiroa estevesii P. J. Braun, Kakt. and. Sukk. 37(4): 74-79, ills., SEM-ills., 1986. Typus: *Esteves Pereira* 186 (KOELN). [Brazil]

[RPS 37]

Facheiroa pilosa Ritter, Kakt. Südamer. 1: 219-220, 1979. Typus: *Ritter* 1000 (ZSS [not received]). [Brazil: Minas Gerais] [RPS 30]

Facheiroa squamosa (Gürke) P. J. Braun & E. Esteves Pereira, Kakt. and. Sukk. 40(8): 199, 1989. Basionym: *Cereus squamosus*. [RPS 40]

Facheiroa squamosa var. **polygona** (Ritter) P. J. Braun & E. Esteves Pereira, Kakt. and. Sukk. 40(8): 202, 1989. Basionym: *Zehntnerella polygona*. [RPS 40]

Facheiroa tenebrosa P. J. Braun & E. Esteves Pereira, Kakt. and. Sukk. 39(6): 126-131, ills., SEM-ill., 1988. Typus: *Esteves Pereira* 183 (KOELN [Succulentarium], ZSS). [Brazil: Bahia] [RPS 39]

× **Ferobergia** Glass, Cact. Succ. J. (US) 38: 177-178, 1966. [*Ferocactus* × *Leuchtenbergia*.] [RPS 17]

× **Ferobergia** cv. **Gil Tegelberg** Glass, Cact. Succ. J. (US) 38: 149, 177-178, 1966. [= *Ferocactus acanthodes* × *Leuchtenbergia principis*.] [RPS 17]

× **Ferobergia** cv. **Violet** G. Unger, Kakt. and. Sukk. 35(3): 68-72, ills., 1984. [= *Leuchtenbergia principis* × *Ferocactus fordii*.] [RPS 35]

Ferocacti Buxbaum, Madroño 14(6): 198, 1958. Typus: *Ferocactus*. [Published as 'linea'.] [RPS 9]

Ferocactinae Buxbaum, Madroño 14(6): 197, 1958. Typus: *Ferocactus*. [RPS 9]

Ferocactus sect. **Bisnaga** (Orcutt) N. P. Taylor & J. Y. Clark, Bradleya 1: 6, 1983. Basionym: *Bisnaga*. [RPS 34]

Ferocactus sect. **Sclerocactus** (Britton & Rose) N. P. Taylor, Cact. Succ. J. Gr. Brit. 41(4): 90, 1979. Basionym: *Sclerocactus*. [RPS 30]

Ferocactus subgen. **Ancistrocactus** (Schumann) N. P. Taylor, Cact. Succ. J. Gr. Brit. 41(4): 90, 1979. Basionym: *Echinocactus* subgen. *Ancistrocactus*. [RPS 30]

Ferocactus subgen. **Hamatacanthus** Bravo, Cact. Suc. Mex. 21: 66, 1976. Typus: *F. hamatacanthus* Britton & Rose. [RPS 27]

Ferocactus subgen. **Pennisquama** Buxbaum, Beitr. Biol. Pfl. 41: 156, 1965. Typus: *F. flavovirens* Britton & Rose. [RPS 17]

Ferocactus subgen. **Stenocactus** (Schumann) N. P. Taylor, Cact. Succ. J. Gr. Brit. 42(4): 108, 1980. Basionym: *Echinocactus* subgen. *Stenocactus*. [RPS 31]

Ferocactus acanthodes var. **eastwoodiae** L. Benson, The Cacti of Arizona ed. 3, 23, 1969. [RPS 20]

Ferocactus acanthodes var. **lecontei** (Engelmann) Lindsay, Cact. Succ. J. (US) 27(6): 169, 1955. Basionym: *Echinocactus lecontei*. [RPS 6]

Ferocactus acanthodes var. **tortulispinus** (Gates) Lindsay, Cact. Succ. J. (US) 27(6): 168-169, 1955. Basionym: *Ferocactus tortulispinus*. [Sphalm. 'tortulospinus'.] [RPS 6]

Ferocactus bicolor (Galeotti *ex* Pfeiffer) N. P. Taylor, Cact. Succ. J. Gr. Brit. 41: 30, 1979. Basionym: *Echinocactus bicolor*. [RPS 30]

Ferocactus bicolor var. **bolaensis** (Runge) N. P. Taylor, Cact. Succ. J. Gr. Brit. 41(2): 30, 1979. Basionym: *Echinocactus bolaensis*. [RPS 30]

Ferocactus bicolor var. **flavidispinus** (Backeberg) N. P. Taylor, Cact. Succ. J. Gr. Brit. 41: 30, 1979. Basionym: *Thelocactus bicolor* var. *flavidispinus*. [RPS 30]

Ferocactus bicolor var. **schwarzii** (Backeberg) N. P. Taylor, Cact. Succ. J. Gr. Brit. 41: 30, 1979. Basionym: *Thelocactus schwarzii*. [RPS 30]

Ferocactus californicus (Labouret) Y. Ito, Cacti, 103, 1952. Basionym: *Echinocactus californicus*. [RPS 4]

Ferocactus coptonogonus (Lemaire) N. P. Taylor, Cact. Succ. J. Gr. Brit. 42(4): 108, 1980. Basionym: *Echinocactus coptonogonus*. [RPS 31]

Ferocactus coulteri var. *rufispinus* Y. Ito, The Cactaceae, 499, 1981. *Nom. inval.* (Art. 37.1, 43.1?). [RPS -]

Ferocactus crispatus (De Candolle) N. P. Taylor, Cact. Succ. J. Gr. Brit. 42(4): 108, 1980. Basionym: *Echinocactus crispatus*. [RPS 31]

Ferocactus cylindraceus var. *chrysacanthus* (Hort. *ex* Schelle) Y. Ito, The Cactaceae, 490, 1981. *Nom. inval.* (Art. 33.2), based on *Echinocactus cylindraceus* var. *chrysacanthus*. [RPS -]

Ferocactus cylindraceus var. **eastwoodiae** (L. Benson) N. P. Taylor, Bradleya 2: 33, 1984. Basionym: *Ferocactus acanthodes* var. *eastwoodiae*. [RPS 35]

Ferocactus cylindraceus var. **lecontei** (Engelmann) Bravo, Cact. Suc. Mex. 25(3): 65, 1980. Basionym: *Echinocactus lecontei*. [RPS 31]

Ferocactus cylindraceus var. **tortulispinus** (H. Gates) Bravo, Cact. Suc. Mex. 25(3): 65, 1980. Basionym: *Ferocactus tortulispinus*. [RPS 31]

Ferocactus diguetii var. **carmenensis** Lindsay, Cact. Succ. J. (US) 27(6): 167-168, 1955. [RPS 6]

Ferocactus eastwoodiae (L. Benson) L. Benson, Cacti U.S. & Can. 969, 1982. Basionym: *Ferocactus acanthodes* var. *eastwoodiae*. [RPS 33]

Ferocactus echidne var. **victoriensis** (Rose) Lindsay, Cact. Succ. J. (US) 27(6): 168, 1955. Basionym: *Echinocactus victoriensis*. [RPS 6]

Ferocactus emoryi var. **rectispinus** (Engelmann) N. P. Taylor, Bradleya 2: 37, 1984. Basionym: *Echinocactus emoryi* var. *rectispinus*. [RPS 35]

Ferocactus fordii var. **grandiflorus** Lindsay, Cact. Succ. J. (US) 27(6): 164-165, 1955. [RPS 6]

Ferocactus gatesii Lindsay, Cact. Succ. J. (US) 27(5): 150-151, 1955. [RPS 6]

Ferocactus glaucus (Schumann) N. P. Taylor, Cact. Succ. J. Gr. Brit. 41(4): 90, 1979. Basionym: *Echinocactus glaucus*. [RPS 30]

Ferocactus gracilis var. **coloratus** (Gates) Lindsay, Cact. Succ. J. (US) 27(6): 169, 1955. Basionym: *Ferocactus coloratus*. [RPS 6]

Ferocactus hamatacanthus var. **crassispinus** (Engelmann) L. Benson, Cact. Succ. J. (US) 46: 80, 1974. Basionym: *Echinocactus hamatacanthus* var. *crassispinus*. [RPS 25]
Ferocactus hamatacanthus var. **sinuatus** (Dietrich) L. Benson, Cact. Succ. J. (US) 41: 128, 1969. Basionym: *Echinocactus sinuatus*. [RPS 20]
Ferocactus hastifer (Werdermann & Bödeker) N. P. Taylor, Cact. Succ. J. Gr. Brit. 41(4): 90, 1979. Basionym: *Echinocactus hastifer*. [RPS 30]
Ferocactus heterochromus (F. A. C. Weber) N. P. Taylor, Cact. Succ. J. Gr. Brit. 41(4): 90, 1979. Basionym: *Echinocactus heterochromus*. [RPS 30]
Ferocactus histrix (De Candolle) Lindsay, Cact. Succ. J. (US) 27(6): 171-173, 1955. Basionym: *Echinocactus histrix*. [RPS 6]
Ferocactus latispinus var. **flavispinus** (F. A. C. Weber) Y. Ito, Cacti, 105, 1952. Basionym: *Echinocactus latispinus* var. *flavispinus*. [RPS 4]
Ferocactus latispinus var. **greenwoodii** (C. Glass) N. P. Taylor, Bradleya 2: 27, 1984. Basionym: *Ferocactus recurvus* var. *greenwoodii*. [RPS 35]
Ferocactus latispinus var. **spiralis** (Karwinsky *ex* Pfeiffer) N. P. Taylor, Bradleya 2: 26, 1984. Basionym: *Echinocactus spiralis*. [RPS 35]
Ferocactus leucacanthus (Zuccarini) N. P. Taylor, Cact. Succ. J. Gr. Brit. 41(4): 90, 1979. Basionym: *Echinocactus leucacanthus*. [RPS 30]
Ferocactus lindsayi Bravo, Cact. Suc. Mex. 11: 9-12, 1966. [RPS 17]
Ferocactus macrodiscus var. *decolor* (Monville) Y. Ito, The Cactaceae, 497, 1981. *Nom. inval.* (Art. 33.2), based on *Echinocactus macrodiscus* var. *decolor*. [RPS -]
Ferocactus macrodiscus var. **septentrionalis** Meyran, Cact. Suc. Mex. 32(3): 51-54, ills., 1987. Typus: *G. A. Navarro* s.n. (MEXU). [Mexico: Guanajuato] [RPS 38]
Ferocactus mathssonii (Berge *ex* Schumann) N. P. Taylor, Cact. Succ. J. Gr. Brit. 41(4): 91, 1979. Basionym: *Echinocactus mathssonii*. [RPS 30]
Ferocactus mesae-verdae (Boissevain & Davidson) N. P. Taylor, Cact. Succ. J. Gr. Brit. 41(4): 90, 1979. Basionym: *Coloradoa mesae-verdae*. [RPS 30]
Ferocactus neohaematacanthus Y. Ito, The Cactaceae, 499, 1981. *Nom. inval.* (Art. 33.2). [*Nom. nov. inval. pro Echinocactus electracanthus* var. *haematacanthus* Salm-Dyck 1850.] [RPS -]
Ferocactus parviflorus (Clover & Jotter) N. P. Taylor, Cact. Succ. J. Gr. Brit. 41(4): 90, 1979. Basionym: *Sclerocactus parviflorus*. [RPS 30]
Ferocactus peninsulae var. **santa-maria** (Britton & Rose) N. P. Taylor, Bradleya 2: 30, 1984. Basionym: *Ferocactus santa-maria*. [RPS 35]
Ferocactus peninsulae var. **townsendianus** (Britton & Rose) N. P. Taylor, Bradleya 2: 29, 1984. Basionym: *Ferocactus townsendianus*. [RPS 35]
Ferocactus peninsulae var. **viscainensis** (Gates) Lindsay, Cact. Succ. J. (US) 27(6): 169-170, 1955. Basionym: *Ferocactus viscainensis*. [RPS 6]
Ferocactus pfeifferi (Zuccarini) Backeberg, Die Cact. 5: 2742, 1961. *Nom. inval.* (Art. 33.2, 43), based on *Echinocactus pfeifferi*. [RPS -]
Ferocactus phyllacanthus (Dietrich & Otto) N. P. Taylor, Cact. Succ. J. Gr. Brit. 42(4): 108, 1980. Basionym: *Echinocactus phyllacanthus*. [RPS 31]
Ferocactus piliferus (Lemaire *ex* Ehrenberg) Unger, Kakt. and. Sukk. 37(2): 45, 1986. Basionym: *Echinocactus piliferus*. [RPS 37]
Ferocactus piliferus fa. **flavispinus** (*hort. ex* Schelle) Unger, Kakt. and. Sukk. 37(2): 45, 1986. Basionym: *Echinocactus pilosus* fa. *flavispinus*. [RPS 37]
Ferocactus piliferus var. **stainesii** (Salm-Dyck) Unger, Kakt. and. Sukk. 37(2): 45, 1986. Basionym: *Echinocactus pilosus* var. *stainesii*. [RPS 37]
Ferocactus polyancistrus (Engelmann & Bigelow) N. P. Taylor, Cact. Succ. J. Gr. Brit. 41(4): 90, 1979. Basionym: *Echinocactus polyancistrus*. [RPS 30]
Ferocactus pottsii (Salm-Dyck) Backeberg, Die Cact. 5: 2738, 1961. Basionym: *Echinocactus pottsii*. [RPS 12]
Ferocactus pubispinus (Engelmann) N. P. Taylor, Cact. Succ. J. Gr. Brit. 41(4): 90, 1979. Basionym: *Echinocactus pubispinus*. [RPS 30]
Ferocactus recurvus (Miller) Lindsay, Cact. Succ. J. (US) 27(6): 173-174, 1955. Basionym: *Cactus recurvus*. [First mentioned by Y. Ito in Cacti, 105, 1952 (cf. RPS 6).] [RPS 6]
Ferocactus recurvus var. **greenwoodii** Glass, Cact. Succ. J. (US) 40: 158-161, 1968. [RPS 19]
Ferocactus reppenhagenii Unger, Kakt. and. Sukk. 25(3): 50-54, ills., 1974. Typus: *Reppenhagen* s.n. (ZSS). [Mexico: Michoacan] [RPS 25]
Ferocactus scheeri (Salm-Dyck) N. P. Taylor, Cact. Succ. J. Gr. Brit. 41(4): 90, 1979. Basionym: *Echinocactus scheeri*. [RPS 30]
Ferocactus schwarzii Lindsay, Cact. Succ. J. (US) 27(3): 70-72, 1955. [RPS 6]
Ferocactus setispinus (Engelmann) L. Benson, Cact. Succ. J. (US) 41: 128, 1969. Basionym: *Echinocactus setispinus*. [RPS 20]
Ferocactus spinosior (Engelmann) N. P. Taylor, Cact. Succ. J. Gr. Brit. 41(4): 90, 1979. Basionym: *Echinocactus whipplei* var. *spinosior*. [RPS 30]
Ferocactus stainesii var. **haematacanthus** (Salm-Dyck) Backeberg, Die Cact. 5: 2700, 1961. Basionym: *Echinocactus electracanthus* var. *haematacanthus*. [RPS 12]
Ferocactus stainesii var. **pilosus** (Galeotti) Backeberg, Die Cact. 5: 2702, 1961. Basionym: *Echinocactus pilosus*. [RPS 12]
Ferocactus stainesii var. **pringlei** (Coulter) Backeberg, Die Cact. 5: 2701, 1961.

Basionym: *Echinocactus pilosus* var. *pringlei*. [RPS 12]

Ferocactus tobuschii (W. T. Marshall) N. P. Taylor, Cact. Succ. J. Gr. Brit. 41(4): 90, 1979. Basionym: *Mammillaria tobuschii*. [RPS 30]

Ferocactus townsendianus var. **santa-maria** (Britton & Rose) Lindsay, Cact. Succ. J. (US) 27(6): 170, 1955. Basionym: *Ferocactus santa-maria*. [RPS 6]

Ferocactus uncinatus var. **wrightii** (Engelmann) N. P. Taylor, Cact. Succ. J. Gr. Brit. 41(2): 31, 1979. Basionym: *Echinocactus uncinatus* var. *wrightii*. [RPS 30]

Ferocactus vaupelianus (Werdermann) N. P. Taylor, Cact. Succ. J. Gr. Brit. 42(4): 108, 1980. Basionym: *Echinocactus vaupelianus*. [RPS 31]

Ferocactus victoriensis (Rose) Backeberg, Die Cact. 5: 2728, 1961. Basionym: *Echinocactus victoriensis*. [RPS 12]

Ferocactus viridescens var. **littoralis** Lindsay, Cact. Succ. J. (US) 36: 8-10, 1964. [RPS 15]

Ferocactus whipplei (Engelmann & Bigelow) N. P. Taylor, Cact. Succ. J. Gr. Brit. 41(4): 90, 1979. Basionym: *Echinocactus whipplei*. [RPS 30]

Ferocactus wislizeni var. **albispinus** (Toumey) Y. Ito, Cacti, 105, 1952. Basionym: *Echinocactus wislizeni* var. *albispinus*. [RPS 4]

Ferocactus wislizeni var. *falconeri* (Orcutt) Y. Ito, The Cactaceae, 495, 1981. *Nom. inval.* (Art. 33.2), based on *Echinocactus falconeri*. [Sphalm. 'flalconeri'.] [RPS -]

Ferocactus wislizeni var. **herrerae** (J. Gonzalez Ortega) N. P. Taylor, Bradleya 2: 34, 1984. Basionym: *Ferocactus herrerae*. [RPS 35]

Ferocactus wislizeni var. **phoeniceus** (Kunze) Y. Ito, Cacti, 105, 1952. [Basionym information not available.] [RPS 4]

Ferocactus wislizeni var. *roseus* Y. Ito, The Cactaceae, 495, 1981. *Nom. inval.* (Art. 37.1). [Sphalm. 'roseus'.] [RPS -]

Ferocactus wislizeni var. **tiburonensis** Lindsay, Cact. Succ. J. (US) 27(6): 166-167, 1955. [RPS 6]

Ferocactus wrightiae (L. Benson) N. P. Taylor, Cact. Succ. J. Gr. Brit. 41(4): 90, 1979. Basionym: *Sclerocactus wrightiae*. [RPS 30]

× **Ferofossulocactus** Rowley, Name that Succulent, 124, 1980. [= *Ferocactus* × *Echinofossulocactus*.] [RPS 31]

× **Ferofossulocactus** cv. **Leibnitz** G. Unger, Kakt. and. Sukk. 36(11): 243, ill., 1985. [= *Ferocactus macrodiscus* × *Echinofossulocactus coptonogonus*.] [RPS 36]

Floresia Krainz & Ritter, in Backeberg, Die Cact. 1: 54, 73, in clav, 1958. *Nom. inval.* (Art. 36.1). [RPS 9]

Floribunda Ritter, Kakt. Südamer. 1: 58, 1979. Typus: *F. pusilliflora*. [RPS 30]

Floribunda pusilliflora Ritter, Kakt. Südamer. 1: 58, 1979. [RPS 30]

Frailea alacriportana Backeberg & Voll, Arq. Jard. Bot. Rio de Janeiro 9: 173-174, ill., 1950. *Kennicke* s.n. (?RB). [Based on 2 syntypes, both probably a living plant, cited as "Viveiro 15.830" and no. 14.114, respectively.] [Volume for 1949, published 1950.] [Brazil: Rio Grande do Sul] [RPS -]

Frailea albiareolata Buining & Brederoo, in Krainz, Die Kakt. Lief. 53: C VIe, 1973. [RPS 24]

Frailea albicolumnaris Ritter, Succulenta 49: 124, 184-185, 1970. [In RPS 30: 6, 1979, erroneously reported has having been published in Kakt. Südamer. 1, 1979. Corrected in RPS 36: 4.] [RPS 21]

Frailea albifusca Ritter, Succulenta 49: 124, 184-185, 1970. [In RPS 30: 6, 1979, erroneously reported has having been published in Kakt. Südamer. 1, 1979. Corrected in RPS 36: 4.] [RPS 21]

Frailea asperispina Ritter, Succulenta 49: 124, 184-185, 1970. [In RPS 30: 6, 1979, erroneously reported has having been published in Kakt. Südamer. 1, 1979. Corrected in RPS 36: 4.] [RPS 21]

Frailea asterioides var. **backebergii** Ritter, Kakt. Südamer. 1: 210, 1979. [RPS 30]

Frailea asterioides var. **harmoniana** Ritter, Kakt. Südamer. 1: 211, 1979. [RPS 30]

Frailea aureinitens Buining & Brederoo, Succulenta 55(4): 61-65, ills., 1976. Typus: *Horst et Uebelmann* HU 178 (U, ZSS). [Brazil: Rio Grande do Sul] [RPS 27]

Frailea aureispina Ritter, Succulenta 49: 124, 184-185, 1970. [In RPS 30: 6, 1979, erroneously reported has having been published in Kakt. Südamer. 1, 1979. Corrected in RPS 36: 4.] [RPS 21]

Frailea aureispina var. **pallidior** Ritter, Succulenta 49: 125, 184-185, 1970. [In RPS 30: 6, 1979, erroneously reported has having been published in Kakt. Südamer. 1, 1979. Corrected in RPS 36: 4.] [RPS 21]

Frailea bueneckeri Abraham, Succulenta 69(3): 64-68, ills., 1990. Typus: *Büneker* 27 (KOELN [Succulentarium]). [Brazil: Rio Grande do Sul] [RPS 41]

Frailea carminifilamentosa Kilian, in Backeberg, Descr. Cact. Nov. 3: 6, 1963. *Nom. inval.* (Art. 37.1). [RPS 14]

Frailea carminifilamentosa var. *winkelmanniana* Kilian *ex* Backeberg, Kakt.-Lex., 159, 1966. *Nom. inval.* (Art. 37.1, 43.1). [RPS -]

Frailea cataphracta var. **duchii** Moser, Nation. Cact. Succ. J. 32(4): 83-84, ills., 1977. [RPS 28]

Frailea cataphracta var. **tuyensis** Buining & Moser, Succulenta 50(4): 64-65, (7): 135, 1971. Typus: *Friedrich* s.n. (ZSS). [See H. Krainz, Die Kakt. Lief. 48/49: C VIe, 1972, for details on the type collection.] [Paraguay] [RPS 22]

Frailea cataphractoides Backeberg, Kakt.-Lex., 159, 1966. *Nom. inval.* (Art. 9.5). [Erroneously included as valid in RPS 17.] [RPS 17]

Frailea chiquitana Cardenas, Nation. Cact. Succ. J. 6(1): 8-9, 1951. [RPS 2]

Frailea chrysacantha Hrabe, Kaktusy 65: 131, 1965. *Nom. inval.* (Art. 37.1, 9.5). [Republished in Succulenta 47: 140-141, 1968. Based on a nude name by Fric.

Erroneously reported as valid in RPS 16.] [RPS 16]

Frailea concepcionensis Buining & Moser, Succulenta 50(3): 49-50, (7): 135, 1971. Typus: *Friedrich* s.n. (ZSS). [See H. Krainz, Die Kakt. Lief. 48/49: C VIe, 1972, for details on the type collection.] [Paraguay] [RPS 22]

Frailea curvispina Buining & Brederoo, in Krainz, Die Kakt. Lief. 50-51: C VIe, 1972. [Publication date variously also cited as 1973.] [RPS 24]

Frailea deminuta Buining & Brederoo, in Krainz, Die Kakt. Lief. 54: C VIe, 1973. [RPS 24]

Frailea friedrichii Buining & Moser, Succulenta 50(2): 25, (7): 135, 1971. Typus: *Friedrich* s.n. (ZSS). [See H. Krainz, Die Kakt. Lief. 48/49: C VIe, 1972, for details on the type collection.] [Paraguay] [RPS 22]

Frailea fulviseta Buining & Brederoo, Kakt. and Sukk. 24: 170-172, 1973. [RPS 24]

Frailea grahliana var. **rubrispina** (Schelle) Y. Ito, Explan. Diagr. Austroechinocactinae, 233, 1957. Basionym: *Echinocactus grahlianus* var. *rubrispinus*. [RPS 8]

Frailea hlineckyana Cervinka, Frailea 1971(2): 13, 1971. [This may not be the original publication for the taxon, but the name is given as sp. nov. The type is ex cult. F. A. Haage sub 98a.] [RPS -]

Frailea horstii Ritter, Succulenta 49: 124, 184-185, 1970. [In RPS 30: 6, 1979, erroneously reported has having been published in Kakt. Südamer. 1, 1979. Corrected in RPS 36: 4.] [RPS 21]

Frailea ignacionensis Buining & Moser, Succulenta 50(4): 63-64, (7): 135, 1971. Typus: *Friedrich* s.n. (ZSS). [See H. Krainz, Die Kakt. Lief. 48/49: C VIe, 1972, for details on the type collection.] [Paraguay] [RPS 22]

Frailea jajoiana Cervinka, Frailea 1971(2): 13-14, ill., 1971. [This may not be the original publication for the taxon, but the name is given as sp. nov. The type is ex cult. F. A. Haage sub 183.] [RPS -]

Frailea lepida Buining & Brederoo, in Krainz, Die Kakt. Lief. 54: C VIe, ills., 1973. Typus: *Horst et Uebelmann* HU 83 (U, ZSS). [Brazil: Rio Grande do Sul] [RPS 24]

Frailea mammifera Buining & Brederoo, in Krainz, Die Kakt. Lief. 50-51: C VIe, ills., 1972. Typus: *Horst et Uebelmann* HU 345 (ZSS). [Publication date variously also cited as 1973.] [Brazil: Rio Grande do Sul] [RPS 24]

Frailea matoana Buining & Brederoo, Cact. Succ. J. (US) 43: 139-142, 1971. [RPS 22]

Frailea melitae Buining & Brederoo, Kakt. and. Sukk. 25(6): 121-123, ills., 1974. Typus: *Horst et Uebelmann* HU 376 (1972) (U, ZSS). [Brazil: Mato Grosso] [RPS 25]

Frailea moseriana Buining & Brederoo, in Krainz, Die Kakt. Lief. 50-51: C VIe, ills., 1973. Typus: *Friedrich* s.n. (ZSS). [Paraguay] [RPS 24]

Frailea perbella Prestlé, Succulenta 59(5): 115-120, 1980. [RPS 31]

Frailea perumbilicata Ritter, Succulenta 49: 125, 184-185, 1970. [In RPS 30: 6, 1979, erroneously reported has having been published in Kakt. Südamer. 1, 1979. Corrected in RPS 36: 4.] [RPS 21]

Frailea perumbilicata var. **spinosior** Ritter, Succulenta 49: 125, 184-185, 1970. [In RPS 30: 6, 1979, erroneously reported has having been published in Kakt. Südamer. 1, 1979. Corrected in RPS 36: 4.] [RPS 21]

Frailea phaeodisca (Spegazzini) Y. Ito, Explan. Diagr. Austroechinocactinae, 237, 1957. Basionym: *Echinocactus phaeodiscus*. [RPS 8]

Frailea pseudogracillima Ritter, Kakt. Südamer. 1: 201, 1979. *Nom. inval.* (Art. 37.1). [= *Echinocactus gracillimus* Schumann *non* Lemaire. Erroneously included as valid in RPS 30:6, corrected in RPS 36: 4.] [RPS 30]

Frailea pseudopulcherrima Fric *ex* Y. Ito, Explan. Diag. Austroechinocactinae, 235, 294, 1957. [Sphalm. 'puseudopulcherrima'. Based on *Echinocactus pseudopulcherrimus* Fric *nom. nud.* 1934.] [RPS 8]

Frailea pullispina Backeberg, Descr. Cact. Nov. 3: 6, 1963. *Nom. inval.* (Art. 9.5). [Erroneously included as valid in RPS 14.] [RPS 14]

Frailea pullispina var. *atrispina* Backeberg, Descr. Cact. Nov. 3: 6, 1963. *Nom. inval.* (Art. 43.1). [Erroneously included as valid in RPS 14.] [RPS 14]

Frailea pullispina var. *centrispina* Backeberg, Descr. Cact. Nov. 3: 6, 1963. *Nom. inval.* (Art. 43.1). [Erroneously included as valid in RPS 14.] [RPS 14]

Frailea pumila var. **gracillima** (Lemaire) Y. Ito, Explan. Diagr. Austroechinocactinae, 233, 1957. Basionym: *Echinocactus gracillimus*. [RPS 8]

Frailea pumila var. **maior** Ritter, Kakt. Südamer. 1: 206, 1979. [RPS 30]

Frailea pygmaea var. **antigibbera** Ritter, Kakt. Südamer. 1: 203, 1979. [RPS 30]

Frailea pygmaea var. **aurea** (Backeberg) Backeberg, Die Cact. 3: 1661, 1959. Basionym: *Frailea aurea*. [RPS 10]

Frailea pygmaea var. **curvispina** Ritter, Kakt. Südamer. 1: 204, 1979. [RPS 30]

Frailea pygmaea var. **dadakii** (Fric) Backeberg, Die Cact. 3: 1660, 1959. Basionym: *Echinocactus dadakii*. [RPS 10]

Frailea pygmaea var. **lilalunula** Ritter, Kakt. Südamer. 1: 203, 1979. [RPS 30]

Frailea pygmaea var. **longispina** Ritter, Kakt. Südamer. 1: 204, 1979. [RPS 30]

Frailea pygmaea var. **maior** Ritter, Kakt. Südamer. 1: 204, 1979. [RPS 30]

Frailea pygmaea var. **phaeodisca** (Spegazzini) Y. Ito, Cacti, 70, 1952. Basionym: *Echinocactus phaeodiscus*. [RPS 4]

Frailea pygmaea var. **planicosta** Ritter, Kakt. Südamer. 1: 204, 1979. [RPS 30]

Frailea schilinzkyana var. **grandiflora** (Haage jr.) Y. Ito, Explan. Diagr. Austroechinocactinae, 233, 1957. Basionym: *Echinocactus schilinzkyanus* var. *grandiflorus*. [RPS 8]

Frailea uhligiana Backeberg, Descr. Cact. Nov.

3: 6, 1963. *Nom. inval.* (Art. 9.5). [Erroneously included as valid in RPS 14.] [RPS 14]

Frailea ybatensis Buining & Moser, Succulenta 50(3): 46-47, (7): 135, 1971. Typus: *Friedrich* s.n. (ZSS). [Sphalm. 'ybatense'. See H. Krainz in Die Kakt. Lief. 48/49: C VIe, 1972, for details on the typification.] [Paraguay] [RPS 22]

Furiolobivia Y. Ito, Explan. Diagr. Austroechinocactinae, 77, 286, 1957. Typus: *Echinopsis nigra.* [First invalidly described in Bull. Takarazuka Insectarium 71: 13-20, 1950 (cf. RPS 3).] [RPS 8]

Furiolobivia ducis-pauli (Fric) Y. Ito, Explan. Diagr. Austroechinocactinae, 80, 1957. *Nom. inval.* (Art. 43.1), based on *Echinopsis ducis-pauli.* [Sphalm. 'ducis pauli'; RPS 8 gives the epithet incorrectly as 'ducis-paulii'.] [RPS 8]

Furiolobivia ferox (Britton & Rose) Y. Ito, Explan. Diagr. Austroechinocactinae, 80, 1957. *Nom. inval.* (Art. 43.1), based on *Lobivia ferox.* [RPS 8]

Furiolobivia longispina (Britton & Rose) Y. Ito, Explan. Diagr. Austroechinocactinae, 79, 1957. *Nom. inval.* (Art. 43.1), based on *Lobivia longispina.* [RPS 8]

Furiolobivia nigra (Backeberg) Y. Ito, Explan. Diagr. Austroechinocactinae, 79, 1957. *Nom. inval.*, based on *Echinopsis nigra, nom. inval.* (Art. 36.1). [Erroneously included as valid in RPS 8.] [RPS 8]

Furiolobivia potosina (Werdermann) Y. Ito, Explan. Diagr. Austroechinocactinae, 80, 1957. *Nom. inval.* (Art. 43.1), based on *Echinopsis potosina.* [RPS 8]

Gerocephalus F. Ritter, Kakt. and. Sukk. 19: 156, 1968. *Nom. illeg.* (Art. 63). [The generic name *Espostoopsis* Buxbaum has priority by one month.] [RPS 19]

Glandulicactus uncinatus var. **wrightii** (Engelmann) Backeberg, Die Cact. 5: 2925, 1961. Basionym: *Echinocactus uncinatus* var. *wrightii.* [RPS 12]

Grusonia bulbispina (Engelmann) Robinson, Phytologia 26: 176, 1973. Basionym: *Opuntia bulbispina.* [RPS 24]

Grusonia clavata (Engelmann) Robinson, Phytologia 26: 176, 1973. Basionym: *Opuntia clavata.* [RPS 24]

Grusonia grahamii (Engelmann) Robinson, Phytologia 26: 176, 1973. Basionym: *Opuntia grahamii.* [RPS 24]

Grusonia hamiltonii Gates *ex* Backeberg, Die Cact. 1: 210, 1958. *Nom. inval.* (Art. 36.1, 37.1). [Based on *Opuntia hamiltonii* Gates (*nom. nud.*).] [RPS 9]

Grusonia pulchella (Engelmann) Robinson, Phytologia 26: 176, 1973. Basionym: *Opuntia pulchella.* [RPS 24]

Grusonia schottii (Engelmann) Robinson, Phytologia 26: 176, 1973. Basionym: *Opuntia schottii.* [RPS 24]

Grusonia stanlyi (Engelmann) Robinson, Phytologia 26: 176, 1973. Basionym: *Opuntia stanlyi.* [Validity of the name depends on the validity of the basionym, which might be invalid under Art. 34.] [RPS 24]

Grusonia vilis (Rose) Robinson, Phytologia 26: 176, 1973. Basionym: *Opuntia vilis.* [RPS 24]

Gymnantha Y. Ito, Explan. Diagr. Austroechinocactinae, 52, 284, 1957. Typus: *Lobivia cumingii.* [*Non Gymnanthus* Junghans 1840.] [RPS 8]

Gymnantha cumingii (Hopffer) Y. Ito, Explan. Diagr. Austroechinocactinae, 53, 1957. Basionym: *Echinocactus cumingii.* [The name given as basionym was already a combination, and is here corrected.] [RPS 8]

Gymnanthocereus altissimus Ritter, Cactus (Paris) 14(62): 119, 1959. Typus: *Ritter* 291 (ZSS). [Peru] [RPS 10]

Gymnanthocereus macranthus Ritter, Kakt. Südamer. 4: 1317, 1981. Typus: *Ritter* 1062 (U). [Peru: La Libertad] [RPS 32]

Gymnanthocereus pilleifer Ritter, Succulenta 46(2): 24, ill. (p. 17), 1967. [First mentioned (without type, but with Latin diagnosis) in l.c. 45: 118-119, 1966.] [RPS 17]

Gymnocactus aguirreanus Glass & Foster, Cact. Succ. J. (US) 44(2): 80-81, ill., 1972. Typus: *Glass et Foster* 2206 (POM). [Mexico: Coahuila] [RPS 23]

Gymnocactus beguinii (F. A. C. Weber) Backeberg, Die Cact. 5: 2851, 1961. *Nom. illeg.* (Art. 63.1), based on *Echinocactus beguinii, nom. inval.* (Art. 63.1). [First proposed as *comb. nud.* by Backeberg in Cact. Succ. J. (US) 23(5): 151 (RPS 3).] [RPS 12]

Gymnocactus beguinii var. *smithii* (Mühlenpfordt) Backeberg, Die Cact. 5: 2855, 1961. Incorrect name (Art. 11.3, 57), based on *Echinocactus smithii.* [RPS 12]

Gymnocactus conothelos (Regel & Klein) Backeberg, Die Cact. 5: 2859, 1961. Basionym: *Echinocactus conothelos.* [RPS 12]

Gymnocactus gielsdorfianus (Werdermann) Backeberg, Cact. Succ. J. (US) 23(5): 151, 1951. [Published as nom. comb. nud., validity not assessed.] [RPS 3]

Gymnocactus goldii (Bravo) Y. Ito, The Cactaceae, 475, 1981. *Nom. inval.* (Art. 33.2), based on *Thelocactus goldii.* [RPS -]

Gymnocactus horripilus (Lemaire) Backeberg, Cact. Succ. J. (US) 23(5): 151, 1951. Basionym: *Echinocactus horripilus.* [RPS 3]

Gymnocactus knuthianus (Bödeker) Backeberg, Cact. Succ. J. (US) 23(5): 151, 1951. Basionym: *Echinocactus knuthianus.* [RPS 3]

Gymnocactus mandragora (Fric) Backeberg, Die Cact. 5: 2862, 1961. Basionym: *Echinocactus mandragora.* [RPS 12]

Gymnocactus roseanus (Bödeker) Glass & Foster, Cact Succ. J. (US) 42: 234, 1970. Basionym: *Echinocactus roseanus.* [RPS 21]

Gymnocactus saussieri (F. A. C. Weber) Backeberg, Cact. Succ. J. (US) 23(5): 151, 1951. Basionym: *Echinocactus saussieri.* [Repeated in Die Cact. 5: 2860, 1960.] [RPS 3]

Gymnocactus subterraneus (Backeberg) Backeberg, Cact. Succ. J. (US) 23(5): 151,

1951. Basionym: *Thelocactus subterraneus*. [RPS 3]

Gymnocactus subterraneus var. **zaragosae** Glass & Foster, Cact. Succ. J. (US) 50(6): 283, ill., 1978. Typus: *Glass et Foster* 3919 (POM). [RPS 29]

Gymnocactus valdezianus var. **albiflorus** (Pazout) Backeberg, Kakt.-Lex. 163, 1966. Basionym: *Pelecyphora valdeziana* var. *albiflora*. [RPS 17]

Gymnocactus viereckii (Werdermann) Backeberg, Cact. Succ. J. (US) 23(5): 151, 1951. Basionym: *Echinocactus viereckii*. [RPS -]

Gymnocactus viereckii var. **major** Glass & Foster, Cact. Succ. J. (US) 50(6): 285, 1978. [RPS 29]

Gymnocactus ysabelae (Schlange) Backeberg, Die Cact. 5: 2856, 1961. Basionym: *Thelocactus ysabelae*. [RPS 12]

Gymnocactus ysabelae var. **brevispinus** (Schlange) Backeberg, Die Cact. 5: 2856, 1961. Basionym: *Thelocactus ysabelae* var. *brevispinus*. [RPS 12]

Gymnocactus John & Riha, Kaktusy 19(1): 22, 1981. *Nom. illeg.* (Art. 64.1). [*Non Gymnocactus* Backeberg 1938.] [RPS 34]

Gymnocalycium cv. **Jan Suba** Pazout, Friciana 1(7): 5, 11, 1962. [= *Gymnocalycium denudatum* × *G. baldianum*. Republished l.c. 15: 10, 1963; parentage given as *G. denudatum* var. *backebergii* × *G. baldianum*.] [RPS 13]

Gymnocalycium sect. *Abscisosemineae* Pazout, Friciana 5(29): 22, 1965. *Nom. illeg.* (Art. 63.1). [*Nom. nov. pro* sect. *Ovatisemineae*.] [RPS -]

Gymnocalycium sect. **Calochlora** Schütz, Friciana 46: 3-23, 1969. Typus: *G. calochlorum* Ito. Vol. for 1968, published Sept. 1969. [RPS 20]

Gymnocalycium sect. *Coactosemineae* Pazout, Friciana 5(29): 22, 1965. *Nom. illeg.* (Art. 63.1). [*Nom. nov. pro* sect. *Muscosemineae*.] [RPS -]

Gymnocalycium sect. *Conchatosemineae* Pazout, Friciana 5(29): 22, 1965. *Nom. illeg.* (Art. 63.1). [*Nom. nov. pro* sect. *Trichomosemineae*.] [RPS -]

Gymnocalycium sect. *Denudata* Schütz, Friciana 46: 8, 1969. *Nom. inval.* (Art. 22). [Equals sect. *Gymnocalycium*. Vol. for 1968, published Sept. 1969.] [RPS 20]

Gymnocalycium sect. **Gibbosa** Schütz, Friciana 46: 8, 12, 1969. Typus: *Gymnocalycium gibbosum*. Given as illegitimate name in RPS 20. [RPS 20]

Gymnocalycium sect. **Hybopleura** Schütz, Friciana 46: 9, 13, 1969. Typus: *G. hybopleurum* Backeberg. Vol. for 1968, published Sept. 1969. [RPS 20]

Gymnocalycium sect. **Lafaldensia** Schütz, Friciana 46: 3-23, 1969. Typus: *G. bruchii* Hosseus. Vol. for 1968, published Sept. 1969. [RPS 20]

Gymnocalycium sect. **Loricata** Schütz, Friciana 46: 3-23, 1969. Typus: *G. spegazzinii* Britton & Rose. Vol. for 1968, published Sept. 1969. [RPS 20]

Gymnocalycium sect. *Macrosemineae* Pazout, Friciana 5(29): 27, 1965. *Nom. inval.* (Art. 36.1, 37.1). [RPS -]

Gymnocalycium sect. **Mazanensia** Schütz, Friciana 46: 3-23, 1969. Typus: *G. hossei* A. Berger. Vol. for 1968, published Sept. 1969. [RPS 20]

Gymnocalycium sect. *Microsemineae* Pazout, Friciana 5(29): 27, 1965. *Nom. inval.* (Art. 36.1, 37.1). [RPS -]

Gymnocalycium sect. **Paraguayensia** Schütz, Friciana 46: 3-23, 1969. Typus: *G. fleischerianum* Backeberg. Vol. for 1968, published Sept. 1969. [RPS 20]

Gymnocalycium sect. **Periferalia** Schütz, Friciana 46: 3-23, 1969. Typus: *G. megatae* Ito. Vol. for 1968, published Sept. 1969. [RPS 20]

Gymnocalycium sect. *Saglionia* Schütz, Friciana 46: 3-23, 1969. *Nom. inval.* (Art. ?). [Equals sect. *Microsemineum*; validity of name not assessed (given as illeg. in RPS 20). Vol. for 1968, published Sept. 1969.] [RPS 20]

Gymnocalycium sect. *Terminalia* Schütz, Friciana 46: 3-23, 1969. *Nom. inval.* (Art. ?). [Equals sect. *Muscosemineum*; validity of name not assessed (given as illeg. in RPS 20). Vol. for 1968, published Sept. 1969.] [RPS 20]

Gymnocalycium ser. **Baldiana** Buxbaum, in Krainz, Die Kakt. C VIf (July 1968), 1968. Typus: *G. baldianum* Spegazzini. [RPS 19]

Gymnocalycium ser. **Castellanosiana** Buxbaum, in Krainz, Die Kakt. C VIf (July 1968), 1968. Typus: *G. castellanosii* Backeberg. [RPS 19]

Gymnocalycium ser. *Centralia* Pazout, Friciana 4(23): 17, 1964. *Nom. inval.* (Art. 35, 36, 37). [Rank of taxon not specified.] [RPS -]

Gymnocalycium ser. **Chiquitana** Buxbaum, in Krainz, Die Kakt. Lief. 38-39: C VIf (July 1968), 1968. Typus: *G. chiquitanum* Cardenas. [RPS 19]

Gymnocalycium ser. **Horridispina** Buxbaum, in Krainz, Die Kakt. Lief. 38-39: C VIf (July 1968), 1968. Typus: *G. horridispinum* Frank. [RPS 19]

Gymnocalycium ser. **Lafaldenses** Buxbaum, in Krainz, Die Kakt. Lief. 38-39: C VIf (July 1968), 1968. Typus: *G. lafaldense* Vaupel. [RPS 19]

Gymnocalycium ser. *Mihanovichia* Pazout, Friciana 4(23): 17, 1964. *Nom. inval.* (Art. 35, 36, 37). [Rank of taxon not specified.] [RPS -]

Gymnocalycium ser. **Mostiana** Buxbaum, in Krainz, Die Kakt. Lief. 38-39: C VIf (July 1968), 1968. Typus: *G. mostii* (Gürke) Britton & Rose. [RPS 19]

Gymnocalycium ser. *Occidentalia* Pazout, Friciana 4(23): 17, 1964. *Nom. inval.* (Art. 35, 36, 37). [Rank of taxon not specified. Several names are cited as synonyms at apparently series rank.] [RPS -]

Gymnocalycium ser. *Orientalia* Pazout, Friciana 4(23): 17, 1964. *Nom. inval.* (Art. 35, 36, 37). [Rank of taxon not specified.] [RPS -]

Gymnocalycium ser. **Pflanzianae** Buxbaum, in Krainz, Die Kakt. Lief. 38-39: C VIf (July

1968), 1968. Typus: *G. pflanzii* (Vaupel) Werdermann. [RPS 19]

Gymnocalycium ser. *Pileisperma* Buxbaum, in Krainz, Die Kakt. Lief. 38-39: C VIf (July 1968), 1968. *Nom. inval.* (Art. 37.1). [Published as *ser. prov.* and based on "G. sp. nov. × hort. Frank".] [RPS 19]

Gymnocalycium ser. **Quehliana** Buxbaum, in Krainz, Die Kakt. Lief. 38-39: C VIf (July 1968), 1968. Typus: *G. quehlianum* (Haage) A. Berger. [RPS 19]

Gymnocalycium ser. **Sagliones** Buxbaum, in Krainz, Die Kakt. Lief. 38-39: C VIf (July 1968), 1968. Typus: *G. saglione* (Cels) Britton & Rose. [RPS 19]

Gymnocalycium ser. **Schickendantzianae** Buxbaum, in Krainz, Die Kakt. Lief. 38-39: C VIf (July 1968), 1968. Typus: *G. schickendantzii* (F. A. C. Weber) Britton & Rose. [RPS 19]

Gymnocalycium ser. **Uruguayenses** Buxbaum, in Krainz, Die Kakt. Lief. 38-39: C VIf (July 1968), 1968. Typus: *G. uruguayense* (Arechavaleta) Britton & Rose. [RPS 19]

Gymnocalycium subgen. *Astrocephalum* Y. Ito, Bull. Takarazuka Insectarium 71: 13-20, 1950. *Nom. inval.* (Art. 36.1). [RPS 3]

Gymnocalycium subgen. *Cosmohybocephalum* Y. Ito, Bull. Takarazuka Insectarium 71: 13-20, 1950. *Nom. inval.* (Art. 36.1). [RPS 3]

Gymnocalycium subgen. *Discocephalum* Y. Ito, Bull. Takarazuka Insectarium 71: 13-20, 1950. *Nom. inval.* (Art. 36.1). [RPS 3]

Gymnocalycium subgen. *Erythrocephalum* Y. Ito, Bull. Takarazuka Insectarium 71: 13-20, 1950. *Nom. inval.* (Art. 36.1). [RPS 3]

Gymnocalycium subgen. *Goniocephalum* Y. Ito, Bull. Takarazuka Insectarium 71: 13-20, 1950. *Nom. inval.* (Art. 36.1). [RPS 3]

Gymnocalycium subgen. *Hybocephalum* Y. Ito, Bull. Takarazuka Insectarium 71: 13-20, 1950. *Nom. inval.* (Art. 36.1). [RPS 3]

Gymnocalycium subgen. *Macrohybocephalum* Y. Ito, Bull. Takarazuka Insectarium 71: 13-20, 1950. *Nom. inval.* (Art. 36.1). [RPS 3]

Gymnocalycium subgen. *Macrothelocephalum* Y. Ito, Bull. Takarazuka Insectarium 71: 13-20, 1950. *Nom. inval.* (Art. 36.1). [RPS 3]

Gymnocalycium subgen. *Microcephalum* Y. Ito, Bull. Takarazuka Insectarium 71: 13-20, 1950. *Nom. inval.* (Art. 36.1). [RPS 3]

Gymnocalycium subgen. **Microsemineum** Schütz, Friciana 46: 3-23, 1969. Typus: *G. saglione* (Cels) Britton & Rose. Vol. for 1968, published Sept. 1969. [RPS 20]

Gymnocalycium subgen. **Muscosemineum** Schütz, Friciana 46: 3-23, 1969. Typus: *G. mihanovichii* Britton & Rose. Vol. for 1968, published Sept. 1969. [RPS 20]

Gymnocalycium subgen. **Ovatisemineum** Schütz, Friciana 46: 3-23, 1969. Typus: *G. gibbosum* (Pfeiffer) Britton & Rose. Vol. for 1968, published Sept. 1969. [RPS 20]

Gymnocalycium subgen. **Pirisemineum** Till & Hesse, Pl. Syst. Evol. 149(1-2): 151, 1985. Typus: *G. zegarrae* Cardenas. [RPS 36]

Gymnocalycium subgen. *Pliohybocephalum* Y. Ito, Bull. Takarazuka Insectarium 71: 13-20, 1950. *Nom. inval.* (Art. 36.1). [RPS 3]

Gymnocalycium subgen. *Rhombohybocephalum* Y. Ito, Bull. Takarazuka Insectarium 71: 13-20, 1950. *Nom. inval.* (Art. 36.1). [RPS 3]

Gymnocalycium subgen. *Stephanocephalum* Y. Ito, Bull. Takarazuka Insectarium 71: 13-20, 1950. *Nom. inval.* (Art. 36.1). [RPS 3]

Gymnocalycium subgen. **Trichomosemineum** Schütz, Friciana 46: 3-23, 1969. Typus: *G. quehlianum* A. Berger. Vol. for 1968, published Sept. 1969. [RPS 20]

Gymnocalycium subgen. *Xanthocephalum* Y. Ito, Bull. Takarazuka Insectarium 71: 13-20, 1950. *Nom. inval.* (Art. 36.1). [RPS 3]

Gymnocalycium subser. **Castellanosiana** Buxbaum, in Krainz, Die Kakt. Lief. 38-39: C VIf (July 1968), 1968. Typus: *G. castellanosii* Backeberg. [RPS 19]

Gymnocalycium subser. *Denudata* Buxbaum, in Krainz, Die Kakt. Lief. 38-39: C VIf (July 1968), 1968. *Nom. inval.* (Art. 22.1). [Based on *G. denudatum* (Link & Otto) Pfeiffer (*typus generis*). Erroneously included as valid in RPS 19.] [RPS 19]

Gymnocalycium subser. **Mihanovichiana** Buxbaum, in Krainz, Die Kakt. Lief. 38-39: C VIf (July 1968), 1968. Typus: *G. mihanovichii* (Fric) Britton & Rose. [RPS 19]

Gymnocalycium subser. **Schickendantzianae** Buxbaum, in Krainz, Die Kakt. Lief. 38-39: C VIf (July 1968), 1968. Typus: *G. schickendantzii* (F. A. C. Weber) Britton & Rose. [RPS 19]

Gymnocalycium subser. **Uruguayenses** Buxbaum, in Krainz, Die Kakt. Lief. 38-39: C VIf (July 1968), 1968. Typus: *G. uruguayense* (Arechavaleta) Britton & Rose. [RPS 19]

Gymnocalycium achirasense Till & Schatzl, Kakt. and. Sukk. 38(8): 191, 1987. Typus: *Genser* B21 (WU). [= cult. BG Linz 2275] [First published invalidly (Art. 9.5) in l.c. 30(2): 25-28, ills., 1979.] [Argentina: San Luis] [RPS 38]

Gymnocalycium acorrugatum Lambert, Succulenta 67(1): 4-7, ills., SEM-ills., 1988. Typus: *Lambert* 69 (U). [German translation in Gymnos 5(9): 35-36, 1988.] [Argentina: San Juan] [RPS 39]

Gymnocalycium alboareolatum Rausch, Succulenta 64(10): 213-214, ill., 1985. Typus: *Rausch* 716 (ZSS). [Argentina: Catamarca] [RPS 36]

Gymnocalycium alboareolatum var. **ramosum** Rausch, AG Gymnocalycium 3(2): 31, ills., 1990. Typus: *Rausch* 716b (ZSS). [Argentina: La Rioja] [RPS 41]

Gymnocalycium ambatoense Piltz, Kakt. and. Sukk. 31(1): 10-13, 1980. Typus: *Piltz 22/5* (KOELN [Succulentarium]). [Argentina: Catamarca] [RPS 31]

Gymnocalycium andreae var. **grandiflorum** Krainz & Andreae, in Krainz, Die Kakt. Lief. 4: C VIe, 1957. Typus: *Hosseus* s.n. (ZSS). [2 specimens ex cult. ZSS.] [Argentina: Punilla] [RPS -]

Gymnocalycium andreae var. **svecianum** Pazout, in Pazout, Valnicek & Subik, Kaktusy, 132, 1960. [Sphalm. 'andrae'.]

[RPS 12]

Gymnocalycium antherostele Ritter, Kakt. Südamer. 2: 475-476, 1980. Typus: *Ritter* 963 (U, ZSS [seeds only]). [Type cited for U l.c. 1: iii, 1979.] [Argentina: Salta] [RPS 31]

Gymnocalycium armatum Ritter, Kakt. Südamer. 2: 662-663, 1980. [RPS 31]

Gymnocalycium armatum var. **albiflorum** (Pazout) Schütz, Friciana 51: 15, 17, 24, 1980. Basionym: *Gymnocalycium mihanovichii* var. *albiflorum*. [RPS 31]

Gymnocalycium artigas Herter, Rev. Sudam. Bot. 10(1): 1-2, 1951. [RPS 2]

Gymnocalycium asterium Y. Ito *ex* Castellanos, Revista Fac. Ci. Agron. 6(2): 2, 1957. *Nom. illeg.* (Art. 63). [*Nom. nov. pro Echinocactus stellatus* Spegazzini 1905 *non* Scheidweiler 1840; correct name is *Gymnocalycium stellatum* Spegazzini 1925. Mentioned first by Ito in Cacti, 89, 1952, as *G. asterum* (Spegazzini) Y. Ito (cf. RPS 4, given as *nom. nud.*); and again by Ito in Explan. Diagr. Austroechinocactinae, 191, 1957.] [RPS 9]

Gymnocalycium asterium var. **minimum** (Pazout) Pazout, Friciana 6: 7, 1962. Basionym: *Gymnocalycium stellatum* var. *minimum*. [Given as 'comb. subnud.' in RPS 13; validity not further assessed.] [RPS 13]

Gymnocalycium asterium var. *paucispinum* Backeberg, Das Kakt.-Lex. 164, 1966. *Nom. inval.* (Art. 9.5). [Erroneously included as valid in RPS 17.] [RPS 17]

Gymnocalycium asterum (Spegazzini) Y. Ito, Cacti, 89, 1952. *Nom. inval.* (Art. 36.1). [Probably only a spelling error for *Gymnocalycium asterium* ?] [RPS 4]

Gymnocalycium bayrianum Till, Kakt. and. Sukk. 38(8): 191, 1987. Typus: *Schickendantz s.n.* (WU). [Ex cult. H. Till acc. no. 473.] [First published invalidly (Art. 9.5) l.c. 18(12): 222-224, ills. 1967. Erroneously included as valid in RPS 18: 10.] [Argentina: Tucuman] [RPS 38]

Gymnocalycium bicolor Schütz, Friciana 1(7): 2-3, ill. (p. 8), 1962. Typus: *Ritter* 433 (Brno, ZSS). [The type is said to be in the "Succulent Collection" and may have been a living plant, which would invalidate the name.] [Argentina: Córdoba] [RPS 13]

Gymnocalycium borthii Koop *ex* Till, Kakt. and. Sukk. 38(8): 191, 1987. Typus: *Borth* BO 55 (WU). [ex cult. BG Linz 2270] [First published invalidly (Art. 9.5) in l.c. 27(2): 25-27, ills. 1976.] [Argentina: San Luis] [RPS 38]

Gymnocalycium bozsingianum Schütz, Kaktusy 13(6): 124-126, 1977. [RPS 28]

Gymnocalycium bruchii var. **brigittae** J. Piltz, Succulenta 66(10): 213-216, ills, SEM-ills., 1987. Typus: *Piltz* 214/3 (KOELN). [Reprinted in Gymnos 5(9): 30-34, 1988.] [Argentina: Córdoba] [RPS 38]

Gymnocalycium bruchii var. *hossei* Backeberg, Die Cact. 3: 1699-1700, 1959. *Nom. inval.* (Art. 37.1). [First published (without Latin diagnosis) in Backeberg & Knuth, Kaktus-ABC, 286, 1935. Erroneously included as valid in RPS 10.] [RPS 10]

Gymnocalycium bruchii var. **niveum** Rausch, Succulenta 68(9): 179-181, ills., 1989. Typus: *Rausch* 727 (ZSS). [Argentina: Córdoba] [RPS 40]

Gymnocalycium bruchii var. *spinosissimum* (Haage *fil.*) Y. Ito, Cacti, 91, 1952. *Nom. inval.* (Art. 33.2), based on *Gymnocalycium lafaldense* var. *spinosissimum*. [RPS 4]

Gymnocalycium buenekeri Swales, Cact. Succ. J. Gr. Brit. 40(4): 97-100, 1978. [RPS 29]

Gymnocalycium calochlorum (Bödeker) Y. Ito, Cacti, 90, 1952. Basionym: *Echinocactus calochlorus*. [RPS 4]

Gymnocalycium calochlorum var. **proliferum** (Backeberg) Backeberg, Die Cact. 3: 1718, 1959. Basionym: *Echinocactus prolifer*. [RPS 10]

Gymnocalycium cardenasianum Ritter, Taxon 13(4): 144, 1964. Typus: *Ritter* 88 (U, ZSS). [Type cited for U l.c. 12: 29, 1963.] [First mentioned as nude name in Backeberg, Die Cact. 3: 1747, 1959 (cf. RPS 10).] [Bolivia: Tarija] [RPS 15]

Gymnocalycium carminanthum Borth & Koop, Kakt. and. Sukk. 27(4): 73-76, ills., 1976. Typus: *Borth* 130 (LI, ZSS). [Argentina: Catamarca] [RPS 27]

Gymnocalycium chiquitanum Cardenas, Cactus (Paris) No. 78: 95-96, ills., 1963. Typus: *Hammerschmid* 5562 (). [Bolivia: Prov. Chiquitos] [RPS 14]

Gymnocalycium chuquisacanum Cardenas, Cact. Succ. J (US) 38: 146-147, 1966. [RPS 17]

Gymnocalycium cintiensis (Cardenas) P. C. Hutchison, Nation. Cact. Succ. J. 14: 38, 1959. Basionym: *Weingartia cintiensis*. [RPS 10]

Gymnocalycium cumingii (Britton & Rose) P. C. Hutchison, Nation. Cact. Succ. J. 14: 38, 1959. Basionym: *Lobivia cumingii*. [RPS 10]

Gymnocalycium damsii var. *centrispinum* Backeberg, Descr. Cact. Nov. 3: 6, 1963. *Nom. inval.* (Art. 9.5). [Erroneously included as valid in RPS 14.] [RPS 14]

Gymnocalycium damsii var. *rotundulum* Backeberg, Descr. Cact. Nov. 3: 6, 1963. *Nom. inval.* (Art. 9.5). [Erroneously included as valid in RPS 14.] [RPS 14]

Gymnocalycium damsii var. *torulosum* Backeberg, Descr. Cact. Nov. 3: 6, 1963. *Nom. inval.* (Art. 9.5). [Erroneously included as valid in RPS 14.] [RPS 14]

Gymnocalycium damsii var. *tucavocense* Backeberg, Descr. Cact. Nov. 3: 6, 1963. *Nom. inval.* (Art. 9.5). [Erroneously included as valid in RPS 14.] [RPS 14]

Gymnocalycium delaetii (Schumann) Y. Ito, Cacti, 86, 1952. Basionym: *Echinocactus delaetii*. [RPS 4]

Gymnocalycium denudatum var. **andersohnianum** (Haage jr.) Y. Ito, Explan. Diagr. Austroechinocactinae, 170, 1957. Basionym: *Echinocactus denudatus* var. *andersohnianus*. [RPS 8]

Gymnocalycium denudatum var. **backebergii** Pazout, Friciana 15: 6, 1963. [Sphalm. 'backebargii', 'backerbergii'.] [RPS 14]

Gymnocalycium denudatum var. **delaetianum**

(Haage jr.) Y. Ito, Explan. Diagr. Austroechinocactinae, 170, 1957. Basionym: *Echinocactus denudatus* var. *delaetianus*. [RPS 8]

Gymnocalycium denudatum var. **flavispinum** (Schelle) Y. Ito, Explan. Diagr. Austroechinocactinae, 170, 1957. Basionym: *Echinocactus denudatus* var. *flavispinus*. [RPS 8]

Gymnocalycium denudatum var. **golzianum** (Mundt) Y. Ito, Explan. Diagr. Austroechinocactinae, 170, 1957. Basionym: *Echinocactus denudatus* var. *golzianus*. [RPS 8]

Gymnocalycium denudatum var. **heuschkelianum** (Haage jr.) Y. Ito, Explan. Diagr. Austroechinocactinae, 170, 1957. Basionym: *Echinocactus denudatus* var. *heuschkelianus*. [RPS 8]

Gymnocalycium denudatum var. **meiklejohnianum** (Haage jr.) Y. Ito, Explan. Diagr. Austroechinocactinae, 170, 1957. Basionym: *Echinocactus denudatus* var. *meiklejohnianus*. [RPS 8]

Gymnocalycium denudatum var. **octogonum** (Schumann) Y. Ito, Explan. Diagr. Austroechinocactinae, 170, 1957. Basionym: *Echinocactus denudatus* var. *octogonus*. [RPS 8]

Gymnocalycium denudatum var. *paraguayense* Haage *fil. ex* Y. Ito, Cacti, 85, 1952. *Nom. inval.* (Art. 36.1). [Based on a nude name.] [RPS 4]

Gymnocalycium denudatum var. **roseiflorum** (Hildmann) Y. Ito, Explan. Diagr. Austroechinocactinae, 170, 1957. Basionym: *Echinocactus denudatus* var. *roseiflorus*. [RPS 8]

Gymnocalycium denudatum var. **scheidelianum** (Haage jr.) Y. Ito, Explan. Diagr. Austroechinocactinae, 170, 1957. Basionym: *Echinocactus denudatus* var. *scheidelianus*. [RPS 8]

Gymnocalycium denudatum var. **wagnerianum** (Haage jr.) Y. Ito, Explan. Diagr. Austroechinocactinae, 170, 1957. Basionym: *Echinocactus denudatus* var. *wagnerianus*. [RPS 8]

Gymnocalycium denudatum var. **wieditzianum** (Haage jr.) Y. Ito, Explan. Diagr. Austroechinocactinae, 170, 1957. Basionym: *Echinocactus denudatus* var. *wieditzianus*. [RPS 8]

Gymnocalycium erinaceum J. Lambert, Succulenta 64(3): 64-66, ill., 1985. Typus: *J. Lambert* 40 (U). [RPS 36]

Gymnocalycium eurypleurum Plesnik *ex* Ritter, Kakt. Südamer. 1: 268, 1979. Typus: *Ritter* 1178 (U, ZSS [seeds only]). [Type cited for U l.c. 1: iii, 1979.] [First published by F. Plesnik in Kaktusy 1972: 78, but the type is said to be a living plant in the collection of the author (cf. W. Weitzel in Gymnocalycium-Brief No. 2: 9-12, 1984). The name was first mentioned by Ritter in 1963.] [Paraguay] [RPS 30]

Gymnocalycium eytianum Cardenas, Kakt. and. Sukk. 9(2): 25, 1958. [RPS 9]

Gymnocalycium ferox (Backeberg) Slaba, Kaktusy 20(4): 80, 1984. *Nom. inval.*, based on *Gymnocalycium hybopleurum* var. *ferox*, *nom. inval.* (Art. 9.5, 37.1). [RPS 35]

Gymnocalycium ferox var. *ferocior* (Backeberg) Slaba, Kaktusy 20(4): 83, 1984. *Nom. inval.*, based on *Gymnocalycium hybopleurum* var. *ferocior*, *nom. inval.* (Art. 9.5, 37.1). [Specific epithet not given for this combination, deduced from the context.] [RPS 35]

Gymnocalycium ferrarii Rausch, Kakt. and. Sukk. 32(1): 6-7, ill., 1981. Typus: *Rausch* 718 (ZSS). [Argentina: Catamarca] [RPS 32]

Gymnocalycium fidaianum (Backeberg) P. C. Hutchison, Cact. Succ. J. (US) 29(1): 11, 1957. Basionym: *Echinocactus fidaianus*. [RPS 8]

Gymnocalycium fleischerianum Backeberg, Die Cact. 3: 1703, ill., 1959. *Nom. inval.* (Art. 37.1). [First published (without Latin diagnosis) in Backeberg & Knuth, Kaktus-ABC, 288, 1935, and subsequently used by several authors. The name was originally attributed to Jajo 1934. Erroneously included as valid in RPS 10.] [RPS 10]

Gymnocalycium fleischerianum var. *andersohnianum* (Haage) Schütz, Friciana 40: 11, 1967. *Nom. inval.* (Art. 43.1), based on *Echinocactus denudatus* var. *andersohnianus*. [Erroneously included as valid in RPS 18.] [RPS 18]

Gymnocalycium fleischerianum var. *heuschkelianum* (Haage) Schütz, Friciana 40: 11, 1967. *Nom. inval.* (Art. 43.1), based on *Echinocactus denudatus* var. *heuschkelianus*. [Erroneously included as valid in RPS 18.] [RPS 18]

Gymnocalycium fleischerianum var. *meiklejohnianum* (Haage) Schütz, Friciana 40: 11, 1967. *Nom. inval.* (Art. 43.1), based on *Echinocactus denudatus* var. *meiklejohnianus*. [Erroneously included as valid in RPS 18.] [RPS 18]

Gymnocalycium fricianum Plesnik, Kakt. and. Sukk. 15: 110, 1964. [RPS 15]

Gymnocalycium friedrichii (Werdermann) Pazout, Friciana 4(23): 18, 1964. *Nom. inval.* (Art. 33.2), based on *Gymnocalycium mihanovichii* var. *friedrichii*. [This citation refers to the German summary; the exact place in the text was not traceable. Erroneously included as valid in RPS 15.] [RPS 15]

Gymnocalycium friedrichii var. *angustostriatum* Pazout, Friciana 4(23): 18, 1964. *Nom. inval.* (Art. 33.2, 43.1). [This citation refers to the German summary; the exact place in the text was not traceable. No basionym was found.] [RPS -]

Gymnocalycium friedrichii var. *moserianum* Pazout, Succulenta 1966: 97, 1966. *Nom. inval.* (Art. 37.1, 43.1). [RPS 17]

Gymnocalycium friedrichii var. *pazoutianum* Moser & Valnicek, Kaktusy 67: 60, 1967. *Nom. inval.* (Art. 37.1, 43.1). [Given as *comb. nov.*, but with Latin diagnosis. Repeated in Friciana 44: 3-15, 1969 (vol. for 1967), but again without type.] [RPS 18]

Gymnocalycium friedrichii var. *piraretaense* (?) Pazout, Friciana 4(23): 18, 1964. *Nom. inval.* (Art. 43.1). [This citation refers to the German summary; the exact place in the text was not traceable. No basionym was found.] [RPS -]

Gymnocalycium gibbosum fa. *intermedium* Fleischer, Friciana 4(24): 4, 1964. *Nom. inval.* (Art. 36.1, 37.1). [Sphalm. 'intermedia'. Erroneously included as valid in RPS 15.] [RPS 15]

Gymnocalycium gibbosum fa. *minimum* Fleischer, Friciana 4(24): 4, 1964. *Nom. inval.* (Art. 36.1, 37.1). [Sphalm. 'minima'. Erroneously included as valid in RPS 15.] [RPS 15]

Gymnocalycium gibbosum var. *caespitosum* Fric *ex* Fleischer, Friciana 4(24): 3-6, 1964. *Nom. inval.* (Art. 9.5, 37.1). [Erroneously included as valid in RPS 15. Based on a nude name by Fric in 1926.] [RPS 15]

Gymnocalycium gibbosum var. **celsianum** (Labouret) Y. Ito, Explan. Diagr. Austroechinocactinae, 190, 1957. Basionym: *Echinocactus gibbosus* var. *celsianus*. [RPS 8]

Gymnocalycium gibbosum var. **cerebriforme** (Spegazzini) Y. Ito, Explan. Diagr. Austroechinocactinae, 190, 1957. Basionym: *Echinocactus gibbosus* var. *cerebriformis*. [RPS 8]

Gymnocalycium gibbosum var. **fennellii** (Haage jr.) Y. Ito, Explan. Diagr. Austroechinocactinae, 190, 1957. Basionym: *Echinocactus gibbosus* var. *fennellii*. [RPS 8]

Gymnocalycium gibbosum var. **ferox** (Labouret) Y. Ito, Explan. Diagr. Austroechinocactinae, 190, 1957. Basionym: *Echinocactus gibbosus* var. *ferox*. [RPS 8]

Gymnocalycium gibbosum var. *hyptiacanthum* Fric *ex* Y. Ito, Cacti, 89, 1952. *Nom. inval.* (Art. 36.1). [Basionym attributed to K. Schumann. Treated as *nom. nud.* in RPS 4, validity not assessed.] [RPS 4]

Gymnocalycium gibbosum var. **leonense** (Hildmann) Y. Ito, Explan. Diagr. Austroechinocactinae, 190, 1957. Basionym: *Echinocactus gibbosus* var. *leonensis*. [RPS 8]

Gymnocalycium gibbosum var. **leucacanthum** (Schumann) Y. Ito, Explan. Diagr. Austroechinocactinae, 190, 1957. Basionym: *Echinocactus gibbosus* var. *leucacanthus*. [RPS 8]

Gymnocalycium gibbosum var. **leucodictyon** (Schumann) Y. Ito, Explan. Diagr. Austroechinocactinae, 190, 1957. Basionym: *Echinocactus gibbosus* var. *leucodictyon*. [RPS 8]

Gymnocalycium gibbosum var. **pluricostatum** (Rümpler) Y. Ito, Explan. Diagr. Austroechinocactinae, 190, 1957. Basionym: *Echinocactus gibbosus* var. *pluricostatus*. [Original author of the variety erroneously given as 'Förster' in RPS 8.] [RPS 8]

Gymnocalycium gibbosum var. **polygonum** (Schumann) Y. Ito, Explan. Diagr. Austroechinocactinae, 190, 1957. Basionym: *Echinocactus gibbosus* var. *polygonus*. [RPS 8]

Gymnocalycium gibbosum var. *rostratum* Fleischer, Friciana 4(24): 3-6, 1964. *Nom. inval.* (Art. 9.5, 37.1). [Erroneously included as valid in RPS 15 (type cited to be a living plant).] [RPS 15]

Gymnocalycium gibbosum var. **ventanicola** (Spegazzini) Y. Ito, Explan. Diagr. Austroechinocactinae, 190, 1957. Basionym: *Echinocactus gibbosus* var. *ventanicola*. [Sphalm. 'ventanicolum'.] [RPS 8]

Gymnocalycium glaucum Ritter, Sukkulentenkunde 7/8: 37, 1963. [RPS 14]

Gymnocalycium griseopallidum Backeberg, Das Kakt.-Lex. 167, 1966. *Nom. inval.* (Art. 9.5). [Erroneously included as valid in RPS 17.] [RPS 17]

Gymnocalycium hamatum Ritter, Kakt. Südamer. 2: 663-664, fig. 650 (p. 823), 1980. Typus: *Ritter 819* (U, ZSS [seeds only]). [Type cited for U l.c. 1: iii, 1979.] [Bolivia: Gran Chaco] [RPS 31]

Gymnocalycium hammerschmidii Backeberg, Descr. Cact. Nov. 3: 7, 1963. *Nom. inval.* (Art. 9.5). [Erroneously included as valid in RPS 14.] [RPS 14]

Gymnocalycium horridispinum G. Frank *ex* Till, Kakt. and. Sukk. 38(8): 191, 1987. Typus: *Fechser* s.n. (WU). [Ex cult. H. Till acc. no. 119.] [First published invalidly (Art. 9.5) in l.c. 14(1): 8-10, ills. 1963. Included as valid in RPS 14: 7, as there was no indication in the publication that the type had not been conserved.] [Argentina: Córdoba] [RPS 38]

Gymnocalycium horstii Buining, Kakt. and. Sukk. 21(9): 162-165, ills., 1970. Typus: *Horst et Uebelmann HU 79* (U, ZSS). [Brazil: Rio Grande do Sul] [RPS 21]

Gymnocalycium horstii var. **buenekeri** Buining, Kakt. and. Sukk. 21: 162-165, 1970. Typus: *Bueneker* s.n. (not indicated). [Brazil: Rio Grande do Sul] [RPS 21]

Gymnocalycium hossei var. *crassispinum* Y. Ito, The Cactaceae, 394, ill., 1981. *Nom. inval.* (Art. 37.1). [RPS -]

Gymnocalycium hybopleurum var. *breviflorum* Backeberg, Das Kakt.-Lex. 168, 1966. *Nom. inval.* (Art. 9.5). [Erroneously included as valid in RPS 17.] [RPS 17]

Gymnocalycium hybopleurum var. *centrispinum* Backeberg, Das Kakt.-Lex. 168, 1966. *Nom. inval.* (Art. 9.5). [Erroneously included as valid in RPS 17.] [RPS 17]

Gymnocalycium hybopleurum var. *ferocior* Backeberg, Das Kakt.-Lex., 168, 1966. *Nom. inval.* (Art. 9.5, 37.1). [Erroneously included as valid in RPS 17.] [RPS 17]

Gymnocalycium hybopleurum var. *ferox* Backeberg, Das Kakt.-Lex. 168, 1966. *Nom. inval.* (Art. 9.5, 37.1). [RPS 17]

Gymnocalycium hyptiacanthum var. **eleutheracanthum** (Monville) Y. Ito, Explan. Diagr. Austroechinocactinae, 198, 1957. Basionym: *Echinocactus hyptiacanthus* var. *eleutheracanthus*. [RPS 8]

Gymnocalycium hyptiacanthum var.

megalothele (Monville) Y. Ito, Explan. Diagr. Austroechinocactinae, 198, 1957. Basionym: *Echinocactus hyptiacanthus* var. *megalothelos*. [Sphalm. 'megalothelum'.] [RPS 8]

Gymnocalycium hyptiacanthum var. **nitidum** (Monville) Y. Ito, Explan. Diagr. Austroechinocactinae, 198, 1957. Basionym: *Echinocactus hyptiacanthus* var. *nitidus*. [RPS 8]

Gymnocalycium intertextum Backeberg *ex* Till, Kakt. and. Sukk. 38(8): 191, 1987. Typus: *Fechser* s.n. (WU). [Ex cult. H. Till, ex coll. Uhlig & Backeberg sub U 2176] [First published invalidly (Art. 9.5) in Backeberg, Kakt.-Lex., 169, fig. 138, 1966. So far omitted from RPS.] [Argentina] [RPS 38]

Gymnocalycium izozogsii Cardenas, Cact. Succ. J. (US) 38: 145-146, 1966. [RPS 17]

Gymnocalycium kieslingii O. Ferrari, Cact. Succ. J. (US) 57(6): 244-246, ills., 1985. Typus: *O. Ferrari* 19/1980 (SI). [Argentina] [RPS 36]

Gymnocalycium kieslingii fa. **alboareolatum** O. Ferrari, Cact. Succ. J. (US) 57(6): 246, 1985. Typus: *O. Ferrari* s.n. (SI). [Argentina] [RPS 36]

Gymnocalycium kieslingii fa. **castaneum** O. Ferrari, Cact. Succ. J. (US) 57(6): 246, 1985. Typus: *O. Ferrari* s.n. (SI). [Argentina] [RPS 36]

Gymnocalycium lagunillasense Cardenas, Kakt. and. Sukk. 9(2): 22, 1958. [RPS 9]

Gymnocalycium leeanum var. **brevispinum** Backeberg, Cact. Succ. J. (US) 23(3): 87, 1951. [Republished in Die Cact. 3: 1736, 1959 (and reported as correct place of citation in RPS 10).] [RPS 3]

Gymnocalycium leeanum var. **roseiflorum** Y. Ito, Explan. Diagr. Austroechinocacti, 199, 293, 1957. [First invalidly published in Cacti, 90, 1952 (Art. 36.1, cf. RPS 4).] [RPS 8]

Gymnocalycium marquezii Cardenas, Kakt. and. Sukk. 9(2): 26, 1958. [RPS 9]

Gymnocalycium marquezii var. *argentinense* Backeberg, Das Kakt.-Lex., 169, 1966. *Nom. inval.* (Art. 9.5). [RPS 17]

Gymnocalycium marsoneri Fric *ex* Y. Ito, Explan. Diagr. Austroechinocacti, 175, 293, 1957. [= *Gymnocalycium marsoneri* Fric 1934 (*nom. nud.*).] [RPS 8]

Gymnocalycium matoense Buining & Brederoo, Kakt. and. Sukk. 26(12): 265-268, ills., 1975. Typus: *Horst et Uebelmann* HU 452 (1974) (U, ZSS). [Brazil: Mato Grosso] [RPS 26]

Gymnocalycium mazanense var. *ferox* Backeberg, Die Cact. 3: 1767, 1959. *Nom. inval.* (Art. 37.1). [First published (without Latin diagnosis and therefore invalid) in Backeberg & Knuth, Kaktus-ABC, 291, 1935. Erroneously included as valid in RPS 10.] [RPS 10]

Gymnocalycium megalothelos var. **delaetianum** (Haage) Schütz, Friciana 40: 11, 1966. Basionym: *Echinocactus denudatus* var. *delaetianus*. [RPS 18]

Gymnocalycium megatae Y. Ito, Explan. Diagr. Austroechinocacti, 172, 292, 1957. [RPS 8]

Gymnocalycium mesopotamicum Kiesling, Cact. Succ. J. Gr. Brit. 42(2): 39-42, ills., 1980. Typus: *Cutler et Lonsdale* 126-552 (K, SI, ZSS [seeds only]). [Argentina: Corrientes] [RPS 31]

Gymnocalycium michoga Fric *ex* Y. Ito, Explan. Diagr. Austroechinocacti, 175, 292, 1957. [= *Gymnocalycium michoga* Fric 1926 (*nom. nud.*).] [RPS 8]

Gymnocalycium mihanovichii var. **albiflorum** Pazout, Friciana 17: 5, 1963. [RPS 14]

Gymnocalycium mihanovichii var. **angustostriatum** Pazout, Friciana 1(7): 3-4, 1962. [RPS 13]

Gymnocalycium mihanovichii var. *filadelfiense* Backeberg, Kakt.-Lex., 170, 1966. *Nom. inval.* (Art. 37.1). [First mentioned by Pazout in Friciana 29: 10, 1965. Erroneously included as valid in RPS 1in] [RPS 17]

Gymnocalycium mihanovichii var. **fleischerianum** Pazout, Kakt. Listy. 16(10) 109-159, 1951. [RPS 3]

Gymnocalycium mihanovichii var. **melocactiforme** Pazout, Kakt. Listy. 16(10) 109-159, 1951. [Sphalm. 'melocactiformis'.] [RPS 3]

Gymnocalycium mihanovichii var. *roseiflorum* Y. Ito, Cacti, 91, 1952. *Nom. inval.* (Art. 36.1). [RPS 4]

Gymnocalycium mihanovichii var. **rysanekii** Pazout, Friciana 29: 11, 1965. [RPS 16]

Gymnocalycium mihanovichii var. **rysanekium** Pazout, Friciana 35: 19, 1965. [RPS 16]

Gymnocalycium mihanovichii var. **stenogonum** Fric & Pazout, Kakt. Listy 16(10): 109-159, 1951. [The typification of the taxon was not resolved. First mentioned as nude name by Fric in Price-List for 1928, [2].] [RPS 3]

Gymnocalycium mihanovichii var. **stenostriatum** Pazout, Friciana 1(7): 10, 1962. [RPS 13]

Gymnocalycium millaresii Cardenas, Cact. Succ. J. (US) 38: 144-145, 1966. [RPS 17]

Gymnocalycium multiflorum var. **albispinum** (Schumann) Y. Ito, Cacti, 86, 1952. Basionym: *Echinocactus multiflorus* var. *albispinus*. [RPS 4]

Gymnocalycium multiflorum var. **parisense** (Schumann) Y. Ito, Cacti, 86, 1952. Basionym: *Echinocactus multiflorus* var. *parisiensis*. [RPS 4]

Gymnocalycium neocumingii (Backeberg) P. C. Hutchison, Cact. Succ. J. (US) 29(1): 14, 1957. Basionym: *Weingartia neocumingii*. [RPS 8]

Gymnocalycium neumannianum (Backeberg) P. C. Hutchison, Cact. Succ. J. (US) 29(1): 11, 1957. Basionym: *Echinocactus neumannianus*. [RPS 8]

Gymnocalycium nigriareolatum var. *densispinum* Backeberg, Die Cact. 3: 1759, 1959. *Nom. inval.* (Art. 37.1). [First published (without Latin diagnosis and therefore invalid) in Backeberg, Kat. 10 Jahre Kakt.-Forsch., 16, 1937. Erroneously included as valid in RPS 10.] [RPS 10]

Gymnocalycium obductum Piltz, Succulenta 69(4): 73-78, ills., SEM-ills., 1990. Typus: *Piltz* 121/1 (U). [Argentina: Salta] [RPS 41]

Gymnocalycium occultum Fric *ex* Schütz, Friciana 1(7): 4-5, 1962. [RPS 13]

Gymnocalycium ochoterenae var. *polygonum* Backeberg, Das Kakt.-Lex. 171, 1966. *Nom. inval.* (Art. 9.5). [Erroneously included as valid in RPS 17.] [RPS 17]

Gymnocalycium ochoterenae var. *tenuispinum* Backeberg, Kakt.-Lex. 171, 1966. *Nom. inval.* (Art. 9.5). [Erroneously included as valid in RPS 17.] [RPS 17]

Gymnocalycium ochoterenae var. *varispinum* Backeberg, Das Kakt.-Lex. 172, 1966. *Nom. inval.* (Art. 9.5). [Sphalm. 'variispinum'. Erroneously included as valid in RPS 17.] [RPS 17]

Gymnocalycium onychacanthum Y. Ito, Explan. Diagr. Austroechinocacti, 173, 292, 1957. [RPS 8]

Gymnocalycium ourselianum (Cels *ex* Salm-Dyck) Y. Ito, Cacti, 87, 1952. Basionym: *Echinocactus ourselianus.* [RPS 4]

Gymnocalycium paediophilum Schütz, Kaktusy 13(5): 100-101, ill., 1977. *Ritter* 1177 (Plzen). [Sphalm. 'paediophylum'. Typification is unclear, the number *Ritter* 1179 is cited in the text (but not in the description), but Ritter reports his number 1177 as pertaining to this taxon. The taxon was redescribed by Ritter in Kakt. Südamer. 1: 269, fig. 215 (p. 340), 1979.] [Paraguay: Depart. Boquerón] [RPS 30]

Gymnocalycium paraguayense (Schumann) Schütz, Friciana 8(40): 11, 1967. Basionym: *Echinocactus paraguayensis.* [Combination repeated in l.c. 8(47): 3-15, but the basionym is ascribed to Mundt. However, no description is given at the place cited.] [RPS 18]

Gymnocalycium paraguayense var. **roseiflorum** (Hildmann) Schütz, Friciana 40: 11, 1967. Basionym: *Echinocactus denudatus* var. *roseiflorus.* [RPS 18]

Gymnocalycium paraguayense var. **scheidelianum** (Haage) Schütz, Friciana 40: 11, 1967. Basionym: *Echinocactus denudatus* var. *scheidelianus.* [RPS 18]

Gymnocalycium paraguayense var. **wagnerianum** (Haage) Schütz, Friciana 40: 11, 1967. Basionym: *Echinocactus denudatus* var. *wagnerianus.* [RPS 18]

Gymnocalycium paraguayense var. **wieditzianum** (Haage) Schütz, Friciana 40: 11, 1967. Basionym: *Echinocactus denudatus* var. *wieditzianus.* [RPS 18]

Gymnocalycium pflanzii fa. **chuquisacanum** (Cardenas) Donald, Nation. Cact. Succ. J. 26: 100, 1971. Basionym: *Gymnocalycium chuquisacanum.* [RPS 22]

Gymnocalycium pflanzii fa. **izozogsii** (Cardenas) Donald, Nation. Cact. Succ. J. 26: 100, 1971. Basionym: *Gymnocalycium izozogsii.* [RPS 22]

Gymnocalycium pflanzii ssp. **argentinense** Till & W. Till, Kakt. and. Sukk. 39(12): 273-277, ills., 1988. Typus: *Anonymus* in *H. Till* 684 (WU). [Argentina: Prov. Salta] [RPS 39]

Gymnocalycium pflanzii var. **albipulpa** Ritter, Kakt. Südamer. 2: 660-661, 1980. Typus: *Ritter* 397 (U, ZSS [seeds only]). [Type cited for U l.c. 1: iii, 1979.] [Bolivia] [RPS 31]

Gymnocalycium pflanzii var. **eytianum** (Cardenas) Donald, Nation. Cact. Succ. J. 26: 100, 1971. Basionym: *Gymnocalycium eytianum.* [RPS 22]

Gymnocalycium pflanzii var. **izozogsii** (Cardenas) Donald, Nation. Cact. Succ. J. 26: 100, 1971. Basionym: *Gymnocalycium izozogsii.* [RPS 22]

Gymnocalycium pflanzii var. **lagunillasense** (Cardenas) Donald, Nation. Cact. Succ. J. 26: 100, 1971. Basionym: *Gymnocalycium lagunillasense.* [RPS 22]

Gymnocalycium pflanzii var. **millaresii** (Cardenas) Donald, Nation. Cact. Succ. J. 26: 100, 1971. Basionym: *Gymnocalycium millaresii.* [RPS 22]

Gymnocalycium pflanzii var. **riograndense** (Cardenas) Donald, Nation. Cact. Succ. J. 26: 100, 1971. Basionym: *Gymnocalycium riograndense.* [RPS 22]

Gymnocalycium pflanzii var. **zegarrae** (Cardenas) Donald, Nation. Cact. Succ. J. 26: 100, 1971. Basionym: *Gymnocalycium zegarrae.* [RPS 22]

Gymnocalycium piltziorum B. Schütz, Kakt. and Sukk. 33(7): 144-145, ill., 1982. Typus: *Piltz* 38 (PZ, ZSS). [Argentina: La Rioja] [RPS 34]

Gymnocalycium platense var. **leptanthum** (Spegazzini) Y. Ito, Explan. Diagr. Austroechinocactinae, 194, 1957. Basionym: *Echinocactus platensis* var. *leptanthus.* [Sphalm. 'plantense'.] [RPS 8]

Gymnocalycium platense var. **parvulum** (Spegazzini) Y. Ito, Cacti, 89, 1952. Basionym: *Echinocactus platensis* var. *parvulus.* [RPS 4]

Gymnocalycium platense var. **quehlianum** (Haage jr.) Y. Ito, Explan. Diagr. Austroechinocactinae, 194, 1957. Basionym: *Echinocactus quehlianus.* [Sphalm. 'plantense'. The basionym given was *Echinocactus platensis* var. *quehlianus.*] [RPS 8]

Gymnocalycium platense var. **ventanicola** (Spegazzini) Kiesling, Darwiniana 24(1-4): 441, 1982. Basionym: *Echinocactus gibbosus* var. *ventanicola.* [RPS 33]

Gymnocalycium pseudo-malacocarpus Backeberg, Das Kakt.-Lex. 172, 1966. *Nom. inval.* (Art. 9.5). [Erroneously included as valid in RPS 17. The ending of the specific epithet is correctly -us, as the name alludes to the similarity of the taxon with the genus *Malacocarpus.*] [RPS 17]

Gymnocalycium pugionacanthum Backeberg *ex* Till, Kakt. and. Sukk. 38(8): 191, 1987. Typus: *Fechser* s.n. (WU, WU). [Ex cult. H. Till, ex coll. Uhlig & Backeberg sub U 2148] [First invalidly (Art. 9.5) published in Kakt.-Lex., 172, fig. 144, 1966. Erroneously included as valid in RPS 17: 10.] [Argentina: Cordoba (?)] [RPS 38]

Gymnocalycium pulquinense (Cardenas) P. C.

Hutchison, Cact. Succ. J. (US) 29(1): 13, 1957. Basionym: *Weingartia pulquinensis*. [RPS 8]

Gymnocalycium pulquinense var. **corroanum** (Cardenas) P. C. Hutchison, Cact. Succ. J. (US) 29(1): 13, 1957. Basionym: *Weingartia pulquinensis* var. *corroana*. [RPS 8]

Gymnocalycium pungens Fleischer, Friciana 1(7): 1, 8, 1962. [RPS 13]

Gymnocalycium quehlianum var. *albispinum* Bozsing, in Backeberg, Die Cact. 3: 1722, 1959. *Nom. inval.* (Art. 37.1). [Erroneously included as valid in RPS 10.] [RPS 10]

Gymnocalycium quehlianum var. *flavispinum* Bozsing, in Backeberg, Die Cact. 3: 1722, 1959. *Nom. inval.* (Art. 37.1). [Erroneously included as valid in RPS 10.] [RPS 10]

Gymnocalycium quehlianum var. **stellatum** (Spegazzini) Dölz, Sukkulentenkunde 6: 30, 1957. Basionym: *Echinocactus stellatus*. [RPS 8]

Gymnocalycium ragonesii Castellanos, Lilloa 23: 5-13, 1950. [Sphalm. 'ragonesi'.] [RPS 3]

Gymnocalycium rauschii H. Till & W. Till, Succulenta 69(2): 27-31, ills., SEM-ills., 1990. Typus: *Rausch* 350 p.p. (WU). [Ex cult. H. Till.] [German translation and emendation in AG Gymnocalycium 3(4): 41-42, ills., 1990.] [Uruguay: Dept. Tacuarembó] [RPS 41]

Gymnocalycium riograndense Cardenas, Kakt. and. Sukk. 9(2): 24, 1958. [RPS 9]

Gymnocalycium ritterianum Rausch, Kakt. and. Sukk. 23(7): 180-181, ill., 1972. Typus: *Rausch* 126 (ZSS). [Type cited for W but placed in ZSS.] [Argentina: La Rioja] [RPS 23]

Gymnocalycium schatzlianum Strigl & W. Till, Kakt. and. Sukk. 36(12): 250-253, ills., 1985. Typus: *Rausch* 541 (WU). [Ex cult. H. Till acc. no. 1326] [Argentina: La Pampa] [RPS 36]

Gymnocalycium schroederianum var. **bayense** Kiesling, Cact. Succ. J. (US) 58(2): 48-49, ills., 1987. Typus: *Kiesling et Lopez* 4323 (SI). [= *Gymnocalycium bayense* Rausch *nom. nud.*] [Argentina: Buenos Aires] [RPS 38]

Gymnocalycium schroederianum var. **paucicostatum** Kiesling, Cact. Succ. J. (US) 58(2): 49, ill. (p. 44), 1987. Typus: *Schinini et al.* 21678 (SI). [Argentina: Corrientes] [RPS 38]

Gymnocalycium schuetzianum Till & Schatzl, Kakt. and. Sukk. 32(10): 234-236, 1981. [RPS 32]

Gymnocalycium spegazzinii var. **major** Backeberg, Cact. Succ. J. (US) 23(3): 88, 1951. [RPS 3]

Gymnocalycium stellatum var. **minimum** Pazout, in Pazout, Valnicek & Subik, Kaktusy, 131, 1960. Basionym: *Gymnocalycium asterium*. [Republished (and given as combination) in Nation. Cact. Succ. J. 30: 49-50, 1975.] [RPS 12]

Gymnocalycium stellatum var. *paucispinum* (Backeberg) Strong, Nation. Cact. Succ. J. 30: 50, 1975. *Nom. inval.*, based on *Gymnocalycium asterium* var. *paucispinum*, *nom. inval.* (Art. 9.5). [Erroneously included as valid in RPS 26.] [RPS 26]

Gymnocalycium stenopleurum Ritter, Kakt. Südamer. 1: 265, 1979. [RPS 30]

Gymnocalycium striglianum Jeggle *ex* Till, Kakt. and. Sukk. 38(8): 191, 1987. Typus: *Rausch* s.n. (WU). [Ex cult. H. Till acc. no. 563.] [First published invalidly (Art. 9.5) in l.c. 24: 267, ills. 1973. Erroneously included as valid in RPS 14: 7.] [Argentina: Mendoza] [RPS 38]

Gymnocalycium taningaense Piltz, Kakt. and. Sukk. 41(2): 22-26, ills., SEM-ills., 1990. Typus: *Piltz* 212/6 (KOELN [Succulentarium]). [Argentina: Córdoba] [RPS 41]

Gymnocalycium tillianum Rausch, Kakt. and. Sukk. 21(4): 66, 1970. Typus: *Rausch* 227 (W, ZSS). [Argentina: Córdoba] [RPS 21]

Gymnocalycium tobuschianum Schick, Cact. Succ. J. Gr. Brit. 15: 44, 1953. [RPS 4]

Gymnocalycium tortuga Hort. *ex* Blossfeld, in Backeberg, Die Cact. 3: 1784, 1959. *Nom. inval.* (Art. 36.1, 37.1). [RPS 10]

Gymnocalycium triacanthum Backeberg, Die Cact. 3: 1730, 1959. *Nom. inval.* (Art. 9.5, 37.1). [Erroneously included as valid in RPS 10.] [RPS 10]

Gymnocalycium tudae Y. Ito, Explan. Diagr. Austroechinocacti, 172, 292, 1957. [RPS 8]

Gymnocalycium tudae var. **bolivianum** Ritter, Kakt. Südamer. 2: 663, 1980. Typus: *Ritter* 1133 (U, ZSS [seeds only]). [Type cited for U l.c. 1: iii, 1979.] [Bolivia: Cordillera] [RPS 31]

Gymnocalycium tudae var. *pseudomalacocarpus* (Backeberg) Donald, Nation. Cact. Succ. J. 26: 100, 1971. *Nom. inval.*, based on *Gymnocalycium pseudo-malacocarpus*, *nom. inval.* (Art. 9.5). [Erroneously included as valid in RPS 22.] [RPS 22]

Gymnocalycium uebelmannianum Rausch, Succulenta 51(4): 61-64, ill., 1972. Typus: *Rausch* 141 (ZSS). [Type cited for W but placed in ZSS.] [Argentina] [RPS 23]

Gymnocalycium uruguayense var. **roseiflorum** Y. Ito, Explan. Diagr. Austroechinocacti, 198, 293, 1957. [First invalidly published in Cacti, 90, 1952 (Art. 36.1, cf. RPS 4). = *Gymnocalycium uruguayense* var. *rosea* Fric 1928 (*nom. nud.*).] [RPS 8]

Gymnocalycium vatteri Buining, Succulenta 29(5): 65-67, ill., 1950. [No type cited in the publication.] [RPS 1]

Gymnocalycium vorwerkii (Fric) P. C. Hutchison, Nation. Cact. Succ. J. 14: 38, 1959. Basionym: *Neowerdermannia vorwerkii*. [RPS 10]

Gymnocalycium weissianum var. *atroroseum* Backeberg, Die Cact. 3: 1763, 1959. *Nom. inval.* (Art. 37.1). [First invalidly published (Art. 36.1) in Backeberg & Knuth, Kaktus-ABC, 297, 1935. Erroneously included as valid in RPS 10.] [RPS 10]

Gymnocalycium weissianum var. *cinerascens* Backeberg, Die Cact. 3: 1764, 1959. *Nom. inval.* (Art. 37.1). [First invalidly published (Art. 36.1) in Backeberg & Knuth, Kaktus-

ABC, 297, 1935. Erroneously included as valid in RPS 10.] [RPS 10]

Gymnocalycium westii P. C. Hutchison, Cact. Succ. J. (US) 29(1): 11-14, ills., 1957. Typus: *West* 6367 (UC, US, ZSS). [Bolivia: Potosi] [RPS 8]

Gymnocalycium zegarrae Cardenas, Kakt. and. Sukk. 9(2): 21, 1958. [RPS 9]

Gymnocereus Backeberg, Die Cact. 2: 920, 1959. Typus: *Gymnocereus microspermus*. [First treated as illegitimate new name for *Gymnanthocereus* Backeberg in RPS 9, but published as a replacement for *Gymnanthocereus* Backeberg 1938 *non* Backeberg 1937.] [First mentioned in l.c. 1: 53, 72, 1958.] [RPS 10]

Gymnocereus altissimus (Ritter) Backeberg, Die Cact. 6: 3667, 1962. Basionym: *Gymnanthocereus altissimus*. [RPS 13]

Gymnocereus amstutziae Rauh & Backeberg, in Backeberg, Die Cact. 2: 924, 1959. [First published in Backeberg, Descr. Cact. Nov. [1:] 14, 1957 (dated 1956) (cf. RPS 7), but there to be treated as invalid name (Art. 42) (cf. RPS 10).] [RPS 10]

Gymnocereus microspermus (Werdermann & Backeberg) Backeberg, Die Cact. 2: 923, 1959. Basionym: *Cereus microspermus*. [RPS 10]

Haageocereus acanthocladus Rauh & Backeberg, in Backeberg, Descr. Cact. Nov. [1:] 23, 1957. [Dated 1956, published 1957.] [RPS 7]

Haageocereus achaetus Rauh & Backeberg, in Backeberg, Descr. Cact. Nov. [1:] 25, 1957. [Dated 1956, published 1957.] [RPS 7]

Haageocereus acranthus var. **crassispinus** Rauh & Backeberg, in Backeberg, Descr. Cact. Nov. [1:] 22, 1957. [Dated 1956, published 1957.] [RPS 7]

Haageocereus acranthus var. **fortalezensis** Rauh & Backeberg, in Backeberg, Descr. Cact. Nov. [1:] 22, 1957. [Dated 1956, published 1957.] [RPS 7]

Haageocereus acranthus var. **fortalezensis** Backeberg, Descr. Cact. Nov. 1: 22, 1957. [RPS]

Haageocereus acranthus var. **metachrous** Rauh & Backeberg, in Backeberg, Descr. Cact. Nov. [1:] 22, 1957. [Dated 1956, published 1957.] [RPS 7]

Haageocereus akersii Backeberg, in Rauh, Beitr. Kenntn. peruan. Kakt.veg., 416, 1958. *Nom. inval.* (Art. 36.1, 37.1). ["*Comb. nov. pro Peruvocereus multangularis* (Willdenow) Akers 1950"; most probably invalid for lack of a Latin diagnosis, although given as valid in RPS 10.] [RPS 10]

Haageocereus albisetatus (Akers) Cullmann, Kakt. and. Sukk. 8(12): 180, 1957. *Nom. inval.* (Art. 33.2). [Given as comb. nov., but without any indication of a basionym, probably based on *Peruvocereus albisetus*.] [RPS -]

Haageocereus albispinus (Akers) Cullmann, Kakt. and. Sukk. 8(12): 180, 1957. *Nom. inval.* (Art. 33.2), based on *Peruvocereus albispinus*. [RPS -]

Haageocereus albispinus var. **floribundus** (Akers) Backeberg, Die Cact. 2: 1210, 1959. Basionym: *Peruvocereus albispinus* var. *floribundus*. [RPS 10]

Haageocereus albispinus var. **roseospinus** (Akers) Backeberg, Die Cact. 2: 1211, 1959. Basionym: *Peruvocereus albispinus* var. *roseospinus*. [RPS 10]

Haageocereus albus (Ritter) Rowley, Nation. Cact. Succ. J. 37(3): 76, 1982. Basionym: *Weberbauerocereus albus*. [RPS 33]

Haageocereus ambiguus Rauh & Backeberg, in Backeberg, Descr. Cact. Nov. [1:] 25, 1957. [Dated 1956, published 1957.] [RPS 7]

Haageocereus ambiguus var. **reductus** Rauh & Backeberg, in Backeberg, Descr. Cact. Nov. [1:] 25, 1957. [Dated 1956, published 1957.] [RPS 7]

Haageocereus andinus Ritter, Backeberg's Descr. & Erört. taxon. nomenklat. Fragen, 14, 1958. *Nom. inval.* (Art. 36.1, 37.1). [Published as provisional name.] [RPS -]

Haageocereus aticensis Ritter, Katalog H. Winter, [unpaged], 1958. *Nom. inval.* (Art. 36.1, 37.1). [Published as provisional name.] [RPS -]

Haageocereus aureispinus Rauh & Backeberg, in Rauh, Beitr. Kenntn. Peruan. Kakt.veg., 404, 1958. [RPS 10]

Haageocereus aureispinus var. **fuscispinus** Rauh & Backeberg, in Rauh, Beitr. Kenntn. Peruan. Kakt.veg., 407, 1958. [RPS 10]

Haageocereus aureispinus var. *rigidispinus* (Rauh & Backeberg) Rauh & Backeberg, in Rauh, Beitr. Kenntn. peruan. Kakt.veg., 407, 1958. *Nom. inval.* (Art. 33.2), based on *Haageocereus rigidispinus*. [RPS 10]

Haageocereus australis fa. **nanus** Ritter, Kakt. Südamer. 3: 1127, 1980. Typus: *Ritter* 126c (U, ZSS). [Type cited for U l.c. 1: iii, 1979.] [Peru: Moquegua] [RPS 31]

Haageocereus australis fa. **subtilispinus** Ritter, Kakt. Südamer. 3: 1127, 1980. Typus: *Ritter* 126a (U, ZSS). [Type cited for U l.c. 1: iii, 1979.] [Chile: Pisagua] [RPS 31]

Haageocereus australis var. **acinacispinus** Rauh & Backeberg, in Backeberg, Descr. Cact. Nov. [1:] 25, 1957. [Dated 1956, published 1957.] [RPS 7]

Haageocereus chalaensis Ritter, Kakt. Südamer. 4: 1389-1390, 1981. Typus: *Ritter* 187 (U, ZSS). [Type cited for U l.c. 1: iii, 1979.] [Peru: Arequipa] [RPS 32]

Haageocereus chosicensis fa. **rubrospinus** (Akers) Krainz, Die Kakt. C Va, 1964. Basionym: *Haageocereus chosicensis* var. *rubrospinus*. [RPS 15]

Haageocereus chosicensis var. **albispinus** (Akers) Backeberg, Cact. Succ. J. (US) 23(2): 47, 1951. Basionym: *Peruvocereus albispinus*. [Repeated by Krainz (and ascribed to Ritter) in Katalog ZSS ed. 2, 64, 1967 (cf. RPS 18).] [RPS 3]

Haageocereus chosicensis var. **dichromus** (Rauh & Backeberg) Ritter *ex* Krainz, Kat. ZSS ed. 2, 64, 1967. Basionym: *Haageocereus dichromus*. [RPS 18]

Haageocereus chosicensis var. **rubrospinus** (Akers) Backeberg, Cact. Succ. J (US) 23(2): 47, 1951. Basionym: *Peruvocereus* *rubrospinus*. [RPS 3]

Haageocereus chrysacanthus (Akers) Cullmann, Kakt. and. Sukk. 8(12): 180, 1957. *Nom. inval*. (Art. 33.2), based on *Peruvocereus chrysacanthus*. [Name attributed to Backeberg and used by Rauh in Beitr. Peruan. Kakt.-veg., 415, 1958.] [RPS -]

Haageocereus chrysacanthus Ritter, Katalog H. Winter, [unpaged], 1957. *Nom. inval*. (Art. 34, 36, 37). [Published as provisional name, *non Haageocereus chrysacanthus* (Akers) Backeberg.] [RPS -]

Haageocereus chryseus Ritter, Kakt. Südamer. 4: 1390-1391, 1981. *Nom. inval*. (Art. 37). [Based on two syntypes, *Ritter 585* and *Ritter 147a*.] [RPS 32]

Haageocereus clavatus (Akers) Cullmann, Kakt. and. Sukk. 8(12): 180, 1957. *Nom. inval*. (Art. 33.2). [Given as comb. nov., but without any indication of a basionym.] [RPS -]

Haageocereus clavispinus Rauh & Backeberg, in Backeberg, Descr. Cact. Nov. [1:] 21, 1957. Typus: *Rauh* K44 (1956) (?, ZSS). [Dated 1956, published 1957.] [Peru] [RPS 7]

Haageocereus comosus Rauh & Backeberg, in Backeberg, Descr. Cact. Nov. [1:] 21, 1957. [Dated 1956, published 1957.] [RPS 7]

Haageocereus convergens Ritter, Katalog H. Winter, [unpaged], 1956. *Nom. inval*. (Art. 36.1, 37.1). [Published as provisional name.] [RPS -]

Haageocereus crassiareolatus Rauh & Backeberg, in Backeberg, Descr. Cact. Nov. [1:] 24, 1957. [Dated 1956, published 1957.] [RPS 7]

Haageocereus crassiareolatus var. **smaragdisepalus** Rauh & Backeberg, in Backeberg, Descr. Cact. Nov. [1:] 24, 1957. [Dated 1956, published 1957.] [RPS 7]

Haageocereus decumbens fa. **spinosior** (Backeberg) Krainz, Die Kakt. C Va, 1965. Basionym: *Haageocereus decumbens* var. *spinosior*. [RPS 16]

Haageocereus decumbens var. **brevispinus** Ritter, Kakt. Südamer. 4: 1392, 1981. Typus: *Ritter* 1024 (U). [Peru: Arequipa] [RPS 32]

Haageocereus decumbens var. **multicolorispinus** (Buining) Krainz, Die Kakt. C Va, 1965. Basionym: *Haageocereus multicolorispinus*. [RPS 16]

Haageocereus decumbens var. **spinosior** Backeberg, Cact. Succ. J. (US) 23(2): 47, 1951. [First mentioned (*nom. inval*., Art. 36.1) by Backeberg in Backeberg & Knuth, Kaktus ABC, 208, 1935.] [RPS 3]

Haageocereus deflexispinus Rauh & Backeberg, in Backeberg, Descr. Cact. Nov. [1:] 22, 1957. [Dated 1956, published 1957.] [RPS 7]

Haageocereus dichromus Rauh & Backeberg, in Backeberg, Descr. Cact. Nov. [1:] 24, 1957. [Dated 1956, published 1957.] [RPS 7]

Haageocereus dichromus var. **pallidior** Rauh & Backeberg, in Backeberg, Descr. Cact. Nov. [1:] 24, 1957. [Dated 1956, published 1957.] [RPS 7]

Haageocereus divaricatispinus Rauh & Backeberg, in Backeberg, Descr. Cact. Nov. [1:] 26, 1957. Typus: *Rauh* K176 (1956) (ZSS). [Dated 1956, published 1957.] [Peru] [RPS 7]

Haageocereus fascicularis (Meyen) Ritter, Kakt. Südamer. 3: 1125-1126, 1980. Basionym: *Cereus fascicularis*. [RPS 31]

Haageocereus faustianus (Backeberg) Ritter, Backeberg's Descr. & Erört. taxon. Fragen, [unpaged], 1958. Basionym: *Borzicactus faustianus*. [RPS 15]

Haageocereus fulvus Ritter, Kakt. Südamer. 4: 1393, 1981. Basionym: *Haageocereus acranthus* var. *fortalezensis*. [RPS 32]

Haageocereus fulvus var. **yautanensis** Ritter, Kakt. Südamer. 4: 1393-1394, 1981. Typus: *Ritter* 1067 (U, ZSS [seeds only]). [Type cited for U l.c. 1: iii, 1979.] [Peru: Ancash] [RPS 32]

Haageocereus horrens Rauh & Backeberg, in Backeberg, Descr. Cact. Nov. [1:] 22, 1957. Typus: *Rauh* K68 (1956) (?, ZSS). [Dated 1956, published 1957. See W. Rauh in Beitr. Kenntn. peruan. Kakt.-veg., 411, 1958, for additional data on locality of type collection.] [Peru] [RPS 7]

Haageocereus horrens var. **sphaerocarpus** Rauh & Backeberg, in Backeberg, Descr. Cact. Nov. [1:] 22, 1957. Typus: *Rauh* K48 (1956) (ZSS). [Dated 1956, published 1957.] [Peru] [RPS 7]

Haageocereus hystrix Ritter, Katalog H. Winter, [unpaged], 1958. *Nom. inval*. (Art. 36.1, 37.1). [Published as provisional name.] [RPS -]

Haageocereus icensis Backeberg *ex* Ritter, Kakt. Südamer. 4: 1394, 1981. Typus: *Ritter* 146 p.p. (U, ZSS). [Type cited for U l.c. 1: iii, 1979.] [Peru: Ica] [RPS 32]

Haageocereus icosagonoides Rauh & Backeberg, in Backeberg, Descr. Cact. Nov. [1:] 23, 1957. [Dated 1956, published 1957.] [RPS 7]

Haageocereus icosagonoides fa. **heteracanthus** Ritter, Kakt. Südamer. 4: 1395, 1981. Typus: *Ritter* 169a (U). [Peru: Lambayeque] [RPS 32]

Haageocereus imperialensis Ritter, Katalog H. Winter, [unpaged], 1958. *Nom. inval*. (Art. 36.1, 37.1). [Published as provisional name.] [RPS -]

Haageocereus lackayensis Rauh & Backeberg, in Backeberg, Descr. Cact. Nov. [1:] 22, 1957. [Dated 1956, published 1957.] [RPS 7]

Haageocereus lanugispinus Ritter, Kakt. Südamer. 4: 1395-1396, 1981. Typus: *Ritter* 583 (U). [Peru: Lima/Ancash] [RPS 32]

Haageocereus laredensis var. **longispinus** Rauh & Backeberg, in Backeberg, Descr. Cact. Nov. [1:] 23, 1957. [Dated 1956, published 1957.] [RPS 7]

Haageocereus laredensis var. **pseudoversicolor** (Rauh & Backeberg) Ritter, Backeberg's Descr. & Erört. taxon. Fragen, [unpaged], 1958. Basionym: *Haageocereus pseudoversicolor*. [RPS 15]

Haageocereus limensis (Salm-Dyck) Ritter,

Backeberg's Descr. & Erört. taxon. Fragen, [unpaged], 1958. Basionym: *Cereus limensis*. [RPS 15]

Haageocereus limensis var. **andicola** Ritter, Kakt. Südamer. 4: 1397-1399, 1981. Typus: *Ritter* 145 (U, ZSS). [Type cited for U l.c. 1: iii, 1979.] [Peru: Lima] [RPS -]

Haageocereus limensis var. **brevispinus** Ritter, Kakt. Südamer. 4: 1399, 1981. Typus: *Ritter* 145c (U). [Peru: Lima] [RPS 32]

Haageocereus limensis var. **deflexispinus** (Backeberg) Ritter, Kakt. Südamer. 4: 1399-1400, 1981. Basionym: *Haageocereus deflexispinus*. [RPS 32]

Haageocereus limensis var. **metachrous** (Rauh & Backeberg) Ritter, Backeberg's Descr. & Erört. taxon. Fragen, [unpaged], 1958. Basionym: *Haageocereus acranthus* var. *metachrous*. [RPS 15]

Haageocereus limensis var. **zonatus** (Rauh & Backeberg) Ritter, Backeberg's Descr. & Erört. taxon. Fragen, [unpaged], 1958. Basionym: *Haageocereus zonatus*. [RPS 15]

Haageocereus litoralis Rauh & Backeberg, in Backeberg, Descr. Cact. Nov. [1:] 26, 1957. [Dated 1956, published 1957.] [RPS 7]

Haageocereus longiareolatus Rauh & Backeberg, in Backeberg, Descr. Cact. Nov. [1:] 21, 1957. [Dated 1956, published 1957.] [RPS 7]

Haageocereus mamillatus Rauh & Backeberg, in Backeberg, Descr. Cact. Nov. [1:] 25-26, 1957. Typus: *Rauh* K137 (1956) (ZSS). [Dated 1956, published 1957.] [Peru] [RPS 7]

Haageocereus mamillatus var. **brevior** Rauh & Backeberg, in Backeberg, Descr. Cact. Nov. [1:] 26, 1957. [Dated 1956, published 1957.] [RPS 7]

Haageocereus montanus Ritter, Katalog H. Winter, [unpaged], 1957. *Nom. inval.* (Art. 36.1, 37.1). [Published as provisional name.] [RPS -]

Haageocereus multangularis (Haworth) Ritter, Backeberg's Descr. & Erört. taxon. Fragen, [unpaged], 1958. Basionym: *Cereus multangularis*. [Probably invalid, if no basionym is cited. Validity not assessed (may affect subordinate taxa).] [RPS -]

Haageocereus multangularis subvar. **chrysacanthus** (Akers) Ritter, Backeberg's Descr. & Erört. taxon. Fragen, [unpaged], 1958. Basionym: *Peruvocereus chrysacanthus*. [RPS 15]

Haageocereus multangularis var. *aureus* Ritter, Kakt. Südamer. 4: 1405, 1981. Based on *Ritter* 147d. *Nom. inval.* (Art. 36.1). [Published as provisional name.] [Peru: Ancash] [RPS -]

Haageocereus multangularis var. **dichromus** (Rauh & Backeberg) Ritter, Backeberg's Descr. & Erört. taxon. Fragen, [unpaged], 1958. Basionym: *Haageocereus dichromus*. [RPS 15]

Haageocereus multangularis var. **pseudomelanostele** (Backeberg) Ritter, Backeberg's Descr. & Erört. taxon. Fragen, [unpaged], 1958. [RPS 15]

Haageocereus multangularis var. **turbidus** (Rauh & Backeberg) Ritter, Backeberg's Descr. & Erört. taxon. Fragen, [unpaged], 1958. Basionym: *Haageocereus turbidus*. [RPS 15]

Haageocereus multicolorispinus Buining, Sukkulentenkunde 7/8: 41-42, 1963. Typus: *Akers* s.n. (ZSS). [Ex cult. W. Cullmann.] [Peru] [RPS 14]

Haageocereus ocana-camanensis Rauh & Backeberg, in Backeberg, Descr. Cact. Nov. [1:] 26, 1957. [Dated 1956, published 1957.] [RPS 7]

Haageocereus olowinskianus subvar. **erythranthus** Rauh & Backeberg, in Backeberg, Descr. Cact. Nov. [1:] 24, 1957. [Dated 1956, published 1957.] [RPS 7]

Haageocereus olowinskianus subvar. *rubriflorior* (Rauh & Backeberg) Rauh & Backeberg, in Rauh, Beitr. Kenntn. peruan. Kakt.veg., 387, 1958. *Nom. inval.* (Art. 33.2), based on *Haageocereus olowinskianus* var. *rubriflorior*. [RPS 10]

Haageocereus olowinskianus var. **repandus** Rauh & Backeberg, in Backeberg, Descr. Cact. Nov. [1:] 24, 1957. [Dated 1956, published 1957.] [RPS 7]

Haageocereus olowinskianus var. **rubriflorior** Rauh & Backeberg, in Backeberg, Descr. Cact. Nov. [1:] 24, 1957. [Dated 1956, published 1957.] [RPS 7]

Haageocereus olowinskianus var. **subintertextus** Rauh & Backeberg, in Backeberg, Descr. Cact. Nov. [1:] 24, 1957. [Dated 1956, published 1957.] [RPS 7]

Haageocereus pacalaensis var. **laredensis** (Backeberg) Krainz, Die Kakt. C Va, 1962. Basionym: *Cereus pseudomelanostele* var. *laredensis*. [RPS 13]

Haageocereus pacalaensis var. **longispinus** (Rauh & Backeberg) Krainz, Die Kakt. C Va, 1962. Basionym: *Haageocereus laredensis* var. *longispinus*. [RPS 13]

Haageocereus pacalaensis var. *montanus* Ritter, Kakt. Südamer. 4: 1417, 1981. Based on *Ritter* 294a. *Nom. inval.* (Art. 36.1). [Published as provisional name.] [Peru] [RPS -]

Haageocereus pacalaensis var. **pseudoversicolor** (Rauh & Backeberg) Ritter *ex* Krainz, Kat. ZSS ed. 2, 64, 1967. Basionym: *Haageocereus pseudoversicolor*. [RPS 18]

Haageocereus pacalaensis var. **repens** (Rauh & Backeberg) Krainz, Die Kakt. C Va, 1962. Basionym: *Haageocereus repens*. [RPS 13]

Haageocereus pacaranensis Ritter, Katalog H. Winter, [unpaged], 1958. *Nom. inval.* (Art. 36.1, 37.1). [Published as provisional name.] [RPS -]

Haageocereus pachystele Rauh & Backeberg, in Backeberg, Descr. Cact. Nov. [1:] 24, 1957. Typus: *Rauh* K91 (1956) (?, ZSS [seeds only]). [Dated 1956, published 1957.] [Peru] [RPS 7]

Haageocereus paradoxus Rauh & Backeberg, in Backeberg, Descr. Cact. Nov. [1:] 21, 1957. [Dated 1956, published 1957.] [RPS 7]

Haageocereus peculiaris (Rauh & Backeberg) Ritter, Backeberg's Descr. & Erört. taxon.

Fragen, [unpaged], 1958. Basionym: *Loxanthocereus peculiaris*. [RPS 15]

Haageocereus peniculatus Rauh & Backeberg, in Backeberg, Descr. Cact. Nov. [1:] 21, 1957. [Dated 1956, published 1957.] [RPS 7]

Haageocereus piliger Rauh & Backeberg, in Backeberg, Descr. Cact. Nov. [1:] 26, 1957. [Dated 1956, published 1957.] [RPS 7]

Haageocereus platinospinus var. **pluriflorus** (Backeberg) Ritter, Kakt. Südamer. 4: 1418-1419, 1981. Basionym: *Haageocereus pluriflorus*. [RPS 32]

Haageocereus pluriflorus Rauh & Backeberg, in Backeberg, Descr. Cact. Nov. [1:] 23, 1957. [Dated 1956, published 1957.] [RPS 7]

Haageocereus pseudoacranthus Rauh & Backeberg, in Backeberg, Descr. Cact. Nov. [1:] 23, 1957. [Dated 1956, published 1957.] [RPS 7]

Haageocereus pseudomelanostele var. **carminiflorus** Rauh & Backeberg, in Backeberg, Descr. Cact. Nov. [1:] 21, 1957. [Dated 1956, published 1957.] [RPS 7]

Haageocereus pseudomelanostele var. **chrysacanthus** (Akers) Ritter *ex* Krainz, Kat. ZSS ed. 2, 65, 1967. Basionym: *Peruvocereus chrysacanthus*. [RPS 18]

Haageocereus pseudomelanostele var. **longicomus** (Akers) Backeberg, Cact. Succ. J. (US) 23(2): 47, 1951. Basionym: *Peruvocereus setosus* var. *longicomus*. [RPS 3]

Haageocereus pseudomelanostele var. **setosus** (Akers) Backeberg, Cact. Succ. J. (US) 23(2): 47, 1951. Basionym: *Peruvocereus setosus*. [Repeated by Krainz in Katalog ZSS ed. 2, 65, 1967 (cf. RPS 18).] [RPS 3]

Haageocereus pseudoversicolor Rauh & Backeberg, in Backeberg, Descr. Cact. Nov. [1:] 23, 1957. Typus: *Rauh* K85 (1956) (?, ZSS [seeds only]). [Dated 1956, published 1957.] [Peru] [RPS 7]

Haageocereus repens Rauh & Backeberg, in Backeberg, Descr. Cact. Nov. [1:] 26, 1957. [Dated 1956, published 1957.] [RPS 7]

Haageocereus rigidispinus Rauh & Backeberg, in Backeberg, Descr. Cact. Nov. [1:] 26, 1957. [Dated 1956, published 1957.] [RPS 7]

Haageocereus rubrospinus (Akers) Cullmann, Kakt. and. Sukk. 8(12): 180, 1957. *Nom. inval.* (Art. 33.2), based on *Peruvocereus rubrospinus*. [RPS -]

Haageocereus salmonoideus (Akers) Cullmann, Kakt. and. Sukk. 8(12): 180, 1957. *Nom. inval.* (Art. 33.2). [Given as comb. nov., but without any indication of a basionym.] [RPS -]

Haageocereus seticeps Rauh & Backeberg, in Backeberg, Descr. Cact. Nov. [1:] 21, 1957. [Dated 1956, published 1957.] [RPS 7]

Haageocereus seticeps var. **robustispinus** Rauh & Backeberg, in Backeberg, Descr. Cact. Nov. [1:] 21, 1957. [Dated 1956, published 1957.] [RPS 7]

Haageocereus setosus (Akers) Cullmann, Kakt. and. Sukk. 8(12): 180, 1957. *Nom. inval.* (Art. 33.2), based on *Peruvocereus setosus*. [RPS -]

Haageocereus smaragdiflorus Rauh & Backeberg, in Backeberg, Descr. Cact. Nov. [1:] 21, 1957. [Dated 1956, published 1957.] [RPS 7]

Haageocereus subtilispinus Ritter, Kakt. Südamer. 4: 1419-1420, 1981. Typus: *Ritter* 582 (U, ZSS). [Type cited for U l.c. 1: iii, 1979.] [Peru: Arequipa] [RPS 32]

Haageocereus symmetros Rauh & Backeberg, in Backeberg, Descr. Cact. Nov. [1:] 24, 1957. [Dated 1956, published 1957.] [RPS 7]

Haageocereus tenuis Ritter, Kakt. Südamer. 4: 1421-1422, 1981. Typus: *Ritter* 126e (U). [Peru: Lima] [RPS 32]

Haageocereus tenuispinus Rauh & Backeberg, in Backeberg, Descr. Cact. Nov. [1:] 22, 1957. [Dated 1956, published 1957.] [RPS 7]

Haageocereus turbidus Rauh & Backeberg, in Backeberg, Descr. Cact. Nov. [1:] 25, 1957. [Dated 1956, published 1957.] [RPS 7]

Haageocereus turbidus var. **maculatus** Rauh & Backeberg, in Backeberg, Descr. Cact. Nov. [1:] 25, 1957. [Dated 1956, published 1957.] [RPS 7]

Haageocereus versicolor fa. **aureispinus** (Backeberg) Krainz, Die Kakt. C Vb, 1963. Basionym: *Haageocereus versicolor* var. *aureispinus*. [RPS 14]

Haageocereus versicolor fa. **fuscus** (Backeberg) Krainz, Die Kakt. C Vb, 1963. Basionym: *Haageocereus versicolor* var. *fuscus*. [RPS 14]

Haageocereus versicolor fa. **lasiacanthus** (Werdermann & Backeberg) Krainz, Die Kakt. C Vb, 1963. Basionym: *Cereus versicolor* var. *lasiacanthus*. [RPS 14]

Haageocereus versicolor var. **aureispinus** Backeberg, Cact. & Succ. J. (US) 23(2): 47, 1951. [RPS 3]

Haageocereus versicolor var. **catacanthus** Rauh & Backeberg, in Backeberg, Descr. Cact. Nov. [1:] 23, 1957. [Dated 1956, published 1957.] [RPS 7]

Haageocereus viridiflorus (Akers) Cullmann, Kakt. and. Sukk. 8(12): 180, 1957. *Nom. inval.* (Art. 33.2). [Given as comb. nov., but without any indication of a basionym.] [RPS -]

Haageocereus vulpes Ritter, Kakt. Südamer. 4: 1423, 1981. Typus: *Ritter* 1059 (U, ZSS [seeds only]). [Type cited for U l.c. 1: iii, 1979.] [Peru: Chancay] [RPS 32]

Haageocereus weberbaueri (Schumann *ex* Vaupel) D. Hunt, Bradleya 5: 92, 1987. Basionym: *Cereus weberbaueri*. [RPS 38]

Haageocereus zangalensis Ritter, Kakt. Südamer. 4: 1224, 1981. Typus: *Ritter* 1074 (U). [Peru: Cajamarca] [RPS 32]

Haageocereus zehnderi Rauh & Backeberg, in Backeberg, Descr. Cact. Nov. [1:] 23, 1957. Typus: *Rauh* K67 (1956) (ZSS). [Dated 1956, published 1957.] [Peru] [RPS 7]

Haageocereus zonatus Rauh & Backeberg, in Backeberg, Descr. Cact. Nov. [1:] 22, 1957. [Dated 1956, published 1957.] [RPS 7]

× **Haagespostoa** Rowley, Nation. Cact. Succ. J. 37(3): 76, 1982. [= *Haageocereus* × *Espostoa*.] [RPS 33]

× **Haagespostoa albiseta** (Akers) Rowley, Nation. Cact. Succ. J. 37(3): 76, 1982. Basionym: *Peruvocereus albisetus*. [= *Espostoa haagei* × *Haageocereus multangularis*.] [RPS 33]

× **Haagespostoa climaxantha** (Werdermann) Rowley, Nation. Cact. Succ. J. 37(3): 76, 1982. Basionym: *Binghamia climaxantha*. [= *Espostoa haagei* × *Haageocereus albispinus*.] [RPS 33]

Hamatocactus subgen. **Glandulicactus** (Backeberg) Buxbaum, in Krainz, Die Kakt. C VIIIb, 1963. Basionym: *Glandulicactus*. [RPS 14]

Hamatocactus crassihamatus (F. A. C. Weber) Buxbaum, Österr. Bot. Zeitschr. 98(1-2): 60, 1951. *Nom. inval.*, based on *Echinocactus crassihamatus*, *nom. inval.* (Art. 35). [RPS 2]

Hamatocactus davisii (Houghton) Y. Ito, Cacti, 100, 1952. [Described as *nomen nudum*. Basionym information not avilable and validity not assessed.] [RPS 4]

Hamatocactus hamatacanthus var. **crassispinus** (Engelmann) Y. Ito, Cacti, 99, 1952. Basionym: *Echinocactus hamatacanthus* var. *crassispinus*. [RPS 4]

Hamatocactus hamatacanthus var. **gracilispinus** (Engelmann) Y. Ito, Cacti, 99, 1952. [Basionym information not available; validity not assessed.] [RPS 4]

Hamatocactus hamatacanthus var. **insignis** (Haage *fil.*) Y. Ito, Cacti, 99, 1952. [Basionym information not available; validity not assessed.] [RPS 4]

Hamatocactus hamatacanthus var. **sinuatus** (Dietrich) Y. Ito, Cacti, 99, 1952. Basionym: *Echinocactus sinuatus*. [Bracket author given as 'F. A. C. Weber'.] [RPS 4]

Hamatocactus setispinus fa. **cachetianus** (Labouret) Krainz, Die Kakt. C VIIIb, 1965. Basionym: *Echinocactus setispinus* var. *cachetianus*. [RPS 16]

Hamatocactus setispinus fa. **flavibaccatus** Unger, Kakt. and. Sukk. 32(12): 290-291, 1981. [Based on cultivated material.] [RPS 32]

Hamatocactus setispinus fa. **orcuttii** (Schumann) Krainz, Die Kakt. C VIIIb, 1965. Basionym: *Echinocactus setispinus* var. *orcuttii*. [RPS 16]

Hamatocactus setispinus var. **cachetianus** (Labouret) Y. Ito, Cacti, 99, 1952. Basionym: *Echinocactus setispinus* var. *cachetianus*. [RPS 4]

Hamatocactus setispinus var. **orcuttii** (Schumann) Y. Ito, Cacti, 99, 1952. Basionym: *Echinocactus setispinus* var. *orcuttii*. [RPS 4]

Hamatocactus uncinatus var. **wrightii** (Engelmann) Bravo, Cact. Suc. Mex. 25(3): 65, 1980. Basionym: *Echinocactus uncinatus* var. *wrightii*. [RPS 31]

× **Harricereus** Rowley, Nation. Cact. Succ. J. 37(3): 76, 1982. [= *Harrisia* × *Cereus*.] [RPS 33]

Harrisia divaricata (Lamarck) Backeberg, Die Cact. 4: 2101, 1960. Basionym: *Cactus divaricatus*. [RPS 11]

Harrisia hahniana (Backeberg) Kimnach & P. C. Hutchison, Cact. Succ. J. (US) 59(2): 59, ill. (p. 60), 1987. Basionym: *Mediocactus hahnianus*. [RPS 38]

Harrisia nashii var. *straminea* Backeberg, Die Cact. 4: 2101, 1960. *Nom. inval.* (Art. 37.1). [Erroneously included as valid in RPS 11. Based on *Harrisia fimbriata* var. *straminea* ('straminia') Marshall, *nom. nud.*).] [RPS 11]

Harrisia taetra Areces, Rev. Jard. Bot. Nac. Univ. Habana 1(1): 13-29, 1981. [Volume for 1980, published 1981.] [RPS 32]

Harrisia tetracantha (Labouret) D. Hunt, Bradleya 5: 92, 1987. Basionym: *Cereus tetracanthus*. [RPS 38]

Harrisiae Buxbaum, Madroño 14(6): 182, 1958. Typus: *Harrisia*. [Published as 'linea'.] [RPS 9]

× **Harrisinopsis** Rowley, Nation. Cact. Succ. J. 37(3): 77, 1982. [= *Harrisia* × *Echinopsis*.] [RPS 33]

Haseltonia columna-trajani (Karwinski) Backeberg, Die Cact. 4: 2263, 1960. Basionym: *Cereus columna-trajani*. [RPS 11]

Hatiora subgen. **Rhipsalidopsis** (Britton & Rose) Barthlott, Bradleya 5: 100, 1987. Basionym: *Rhipsalidopsis*. [RPS 38]

Hatiora clavata (F. A. C. Weber) Moran, Gent. Herb. 8(4): 343, 1953. Basionym: *Rhipsalis clavata*. [RPS 4]

Hatiora epiphylloides (Campos-Porto & Werdermann) Buxbaum, Kakt. and. Sukk. 8(8): 116, 1957. Basionym: *Rhipsalis epiphylloides*. [Repeated by P. V. Heath in Epiphytes 7(28): 89, 1983 (cf. RPS 34).] [RPS 34]

Hatiora epiphylloides fa. bradei (Campos-Porto & Castellanos) P. V. Heath, Epiphytes 7(28): 90, 1983. Basionym: *Hariota epiphylloides* var. *bradei*. [RPS 34]

Hatiora gaertneri (Regel) Barthlott, Bradleya 5: 100, 1987. Basionym: *Epiphyllum russellianum* var. *gaertneri*. [RPS 38]

Hatiora gaertneri fa. *serrata* (Lindinger) Süpplie, Rhipsalidinae, [63], 1990. *Nom. inval.*, based on *Rhipsalidopsis serrata*, *nom. inval.* (Art. 36.1). [Basionym not actually given in the publication.] [RPS 41]

Hatiora × **graeseri** (Werdermann) Barthlott, Bradleya 5: 100, 1987. Basionym: × *Rhipsaphyllopsis graeseri*. [= *Hatiora gaertneri* × *H. rosea*.] [RPS 38]

Hatiora herminiae (Campos-Porto & Castellanos) Barthlott, Bradleya 5: 100, 1987. Basionym: *Hariota herminiae*. [First published invalidly (Art. 33.2) by Backeberg, Die Cact. 2: 710. 1959.] [RPS 38]

Hatiora rosea (Lagerheim) Barthlott, Bradleya 5: 100, 1987. Basionym: *Rhipsalis rosea*. [RPS 38]

Hatiora rosea fa. *remanens* (Backeberg) Süpplie, Rhipsalidinae, [64], 1990. *Nom. inval.*, based on *Rhipsalidopsis rosea* var. *remanens*, *nom. inval.* (Art. 9.5, 37.1). [RPS 41]

Hatiora salicornioides fa. **bambusoides** (F. A. C. Weber) Süpplie, Rhipsalidinae, [65], 1990.

Basionym: *Rhipsalis salicornioides* var. *bambusoides*. [RPS 41]

Hatiora salicornioides fa. **cylindrica** (Britton & Rose) Süpplie, Rhipsalidinae, [65], 1990. Basionym: *Hatiora cylindrica*. [RPS 41]

Hatiora salicornioides fa. *gracilis* (F. A. C. Weber) Süpplie, Rhipsalidinae, [66], 1990. *Nom. inval.* (Art. 33.2), based on *Rhipsalis salicornioides* var. *gracilis*. [Sphalm. 'gracilles'. The basionym citation is completely wrong, and the name may not be valid at all.] [RPS 41]

Hatiora salicornioides fa. *stricta* (F. A. C. Weber) Süpplie, Rhipsalidinae, [66], 1990. *Nom. inval.* (Art. 33.2), based on *Rhipsalis stricta*. [The place given as basionym citation is completely wrong.] [RPS 41]

Hatiora salicornioides fa. **villigera** (Schumann) Süpplie, Rhipsalidinae, [66], 1990. Basionym: *Hariota villigera*. [RPS 41]

Hatiora salicornioides var. **gracilis** (F. A. C. Weber) Backeberg, Die Cact. 2: 708, 1959. Basionym: *Rhipsalis salicornioides* var. *gracilis*. [RPS 10]

Hatiora salicornioides var. **stricta** (F. A. C. Weber) Backeberg, Die Cact. 2: 708, 1959. Basionym: *Rhipsalis salicornioides* var. *stricta*. [RPS 10]

Hatiora salicornioides var. **villigera** (Schumann) Backeberg, Die Cact. 2: 708, 1959. Basionym: *Hariota villigera*. [RPS 10]

Heliabravoa Backeberg, Cact. Succ. J. (US) 28(1): 12, 23, 1956. Typus: *Heliabravoa chende*. [RPS 7]

Heliabravoa chende (Roland Gosselin) Backeberg, Cact. Succ. J. Gr. Brit. 18(1): 12, 23, 1956. Basionym: *Cereus chende*. [RPS 7]

Helianthocereus cv. **Bodensee** E. Kleiner, Kakt. and. Sukk. 31(4): 106-112, 1980. [RPS 31]

Helianthocereus cv. **Hegau** E. Kleiner, Kakt. and. Sukk. 31(4): 106-112, 1980. [RPS 31]

Helianthocereus cv. **Schwarzwald** E. Kleiner, Kakt. and. Sukk. 31(4): 106-112, 1980. [RPS 31]

Helianthocereus subgen. *Euhelianthocereus* Backeberg, Cact. Succ. J. (US) 22(5): 153, 1950. *Nom. inval.* (Art. 21.3). [RPS 4]

Helianthocereus subgen. **Neohelianthocereus** Backeberg, Cact. Succ. J. (US) 22(5): 153, 1950. Typus: *Cereus huascha* F. A. C. Weber. [RPS 1]

Helianthocereus andalgalensis (F. A. C. Weber) Backeberg, Cact. Succ. J. (US) 23(2): 48, 1951. Basionym: *Cereus andalgalensis*. [RPS 3]

Helianthocereus antezanae (Cardenas) Backeberg, Cactus (Paris) No. 45: 208, 1955. Basionym: *Trichocereus antezanae*. [RPS 6]

Helianthocereus atacamensis (Philippi) Backeberg, Die Cact. 2: 1315, 1959. Basionym: *Cereus atacamensis*. [RPS 10]

Helianthocereus conaconensis (Cardenas) Backeberg, Cactus (Paris) No. 45: 208, 1955. Basionym: *Trichocereus conaconensis*. [RPS 6]

Helianthocereus crassicaulis Backeberg, Kakt.-Lex., 185-186, 1966. *Nom. inval.* (Art. 9.5). [Later validated by Kiesling as *Lobivia crassicaulis*.] [RPS 17]

Helianthocereus escayachensis (Cardenas) Backeberg, Kakt.-Lex. 186, 1966. Basionym: *Trichocereus escayachensis*. [RPS 17]

Helianthocereus grandiflorus (Britton & Rose) Backeberg, Cactus (Paris) No. 46-47: 274, 1955. Basionym: *Lobivia grandiflora*. [RPS 6]

Helianthocereus herzogianus (Cardenas) Backeberg, Cactus (Paris) No. 45: 208, 1955. Basionym: *Trichocereus herzogianus*. [RPS 6]

Helianthocereus herzogianus var. **totorensis** (Cardenas) Backeberg, Cactus (Paris) No. 45: 208, 1955. Basionym: *Trichocereus herzogianus* var. *totorensis*. [RPS 6]

Helianthocereus huascha (F. A. C. Weber) Backeberg, Cact. Succ. J. (US) 23(2): 48, 1951. Basionym: *Cereus huascha*. [RPS 3]

Helianthocereus huascha var. **auricolor** (Backeberg) Backeberg, Cactus (Paris) No. 46-47: 274, 1955. Basionym: *Trichocereus auricolor*. [RPS 6]

Helianthocereus huascha var. *macranthus* Backeberg, Kakt.-Lex. 186, 1966. *Nom. inval.* (Art. 9.5, 37.1). [Erroneously included as valid in RPS 17.] [RPS 17]

Helianthocereus huascha var. *rosiflorus* (Y. Ito) Backeberg, Kakt.-Lex. 186, 1966. *Nom. inval.* (Art. 33.2), based on *Soehrensia rosiflora*. [RPS 17]

Helianthocereus huascha var. **rubriflorus** (F. A. C. Weber) Backeberg, Cactus (Paris) No. 46-47: 273, 1955. Basionym: *Cereus huascha* var. *rubriflorus*. [RPS 6]

Helianthocereus hyalacanthus (Spegazzini) Backeberg, Die Cact. 2: 1333, 1959. Basionym: *Lobivia hyalacantha*. [RPS 10]

Helianthocereus narvaecensis (Cardenas) Backeberg, Cactus (Paris) No. 45: 208, 1955. Basionym: *Trichocereus narvaecensis*. [RPS 6]

Helianthocereus orurensis (Cardenas) Backeberg, Cactus (Paris) No. 45: 208, 1955. Basionym: *Trichocereus orurensis*. [RPS 6]

Helianthocereus orurensis var. **albiflorus** (Cardenas) Backeberg, Cactus (Paris) No. 45: 208, 1955. Basionym: *Trichocereus orurensis* var. *albiflorus*. [RPS 6]

Helianthocereus pasacana (F. A. C. Weber) Backeberg, Die Cact. 2: 1314, 1959. Basionym: *Pilocereus pasacanus*. [RPS 10]

Helianthocereus pecheretianus Backeberg, Cactus (Paris) No. 45: 210; No. 46-47: 278, 1955. [RPS 6]

Helianthocereus pecheretianus var. *viridior* Backeberg, Kakt.-Lex., 187, 1966. *Nom. inval.* (Art. 37.1). [Erroneously included as valid in RPS 17.] [RPS 17]

Helianthocereus poco var. **albiflorus** (Cardenas) Backeberg, Cactus (Paris) No. 45: 208, 1955. Basionym: *Trichocereus poco* var. *albiflorus*. [RPS 6]

Helianthocereus poco var. **fricianus** (Cardenas) Backeberg, Cactus (Paris) No. 45: 208, 1955. Basionym: *Trichocereus poco* var. *fricianus*. [RPS 6]

Helianthocereus poco var. *sanguiniflorus* Backeberg, Kakt.-Lex., 187, 1966. *Nom. inval.* (Art. 37.1). [Erroneously included as valid in RPS 17.] [RPS 17]
Helianthocereus pseudocandicans (Backeberg) Backeberg, Kakt.-Lex., 187, 1966. *Nom. inval.* (Art. 37.1). [Erroneously included as valid in RPS 17.] [RPS 17]
Helianthocereus pseudocandicans var. *flaviflorus* Backeberg, Kakt.-Lex., 187, 1966. *Nom. inval.* (Art. 37.1, 43.1). [Erroneously included as valid in RPS 17.] [RPS 17]
Helianthocereus pseudocandicans var. *roseoflorus* Backeberg, Kakt.-Lex., 187, 1966. *Nom. inval.*, based on *Trichocereus candicans* var. *roseoflorus*, *nom. inval.* (Art. 9.5, 37.1). [RPS 17]
Helianthocereus randallii (Cardenas) Backeberg, Kakt.-Lex., 187, 1966. Basionym: *Trichocereus randallii*. [RPS 17]
Helianthocereus tarijensis (Vaupel) Backeberg, Cactus (Paris) 10(45): 208, 1955. *Nom. inval.* (Art. 33.2), based on *Cereus tarijensis*. [The name given as basionym was itself a combination, and is here corrected.] [RPS 6]
×*Heliapocryptus* Meier, Kakt. and. Sukk. 32(7): 150-152, 1981. *Nom. inval* (Art. H.6.4). [= *Heliocereus* × *Aporocactus* × *Cryptocereus*.] [RPS 34]
×*Heliaporus* Rowley, Cact. Succ. J. Gr. Brit. 13: 54, 1951. *Nom. inval.* (Art. H6.4). [= *Aporocactus* × *Heliocereus*.] [RPS 2]
×*Heliaporus smithii* (Pfeiffer) Rowley, Cact. Succ. J. Gr. Brit. 13: 54, 1951. *Nom. inval.* (Art. 43.1), based on *Cereus smithii*. [RPS 2]
Heliocerei Buxbaum, Madroño 14(6): 182, 1958. Typus: *Heliocereus*. [Published as 'linea'.] [RPS 9]
Heliocereinae Bravo H., Cact. Suc. Mex. 7(2): 35, 1962. *Nom. inval.* (Art. 36.1, 37.1). [Given as "comb. nov."] [RPS -]
Heliocereus aurantiacus Kimnach, Cact. Succ. J. (US) 46: 66-69, 1974. [RPS 25]
Heliocereus aurantiacus var. **blomianus** Kimnach, Cact. Succ. J. (US) 62(6): 268-271, ills., 1990. Typus: *MacDougall* A202 (HNT, CAS, US). [Mexico: Chiapas] [RPS 41]
Heliocereus elegantissimus var. **helenae** Scheinvar, Phytologia 49(4): 317-321, 1981. [RPS 32]
Heliocereus elegantissimus var. *stenopetalus* Bravo, Cact Succ. J. (US) 38: 3-4, 1966. *Nom. inval.* (Art. 37.1). [RPS 17]
Heliocereus luzmariae Scheinvar, Cact. Succ. J. (US) 57(6): 268-275, ills., 1985. Typus: *Arreola et Guzman S.* 33 (MEXU). [Mexico] [RPS 36]
Heliocereus speciosissimus (De Candolle) Y. Ito, Cacti, 146, 1952. Basionym: *Cereus speciosissimus*. [RPS 4]
Heliocereus speciosus var. **amecamensis** (Heese) Bravo, Cact. Suc. Mex. 19: 47, 1974. Basionym: *Cereus amecamensis*. [First published invalidly (Art. 34.1) by Weingart in A. Berger, Kakteen, 131, 1929. Erroneously included as illegitimate in RPS 25.] [RPS 25]
Heliocereus speciosus var. **elegantissimus** (Britton & Rose) Backeberg, Die Cact. 4: 2121, 1960. Basionym: *Heliocereus elegantissimus*. [Given as "comb. illeg." in RPS 11 without apparent reason.] [RPS 11]
Heliocereus speciosus var. **serratus** (Weingart) Backeberg, Die Cact. 4: 2121, 1960. Basionym: *Cereus serratus*. [RPS 11]
Heliocereus speciosus var. **superbus** (Ehrenberg) Backeberg, Die Cact. 4: 2120, 1960. Basionym: *Cereus superbus*. [RPS 11]
×**Heliochia** Rowley, in Backeberg, Die Cact. 6: 3551, 1962. [= *Heliocereus* × *Nopalxochia*.] [RPS 13]
×**Heliochia** cv. **Holly Gate** Innes, Ashingtonia 1: 134, ill. (front cover), 1975. [Published as *Heliophyllum*; = ×*Heliochia* 'Ackermannii' × *Heliocereus speciosus*.] [RPS 26]
×**Heliochia** cv. **Maiden Erlegh** Rowley, Nation. Cact. Succ. J. 24: 77, ill. (front cover), 1969. [= *Heliocereus speciosus* × 'Deutsche Kaiserin'.] [RPS 20]
×**Heliochia hybrida** (Geel) Heath, Taxon 38(1): 126, 1989. Basionym: *Cactus hybridus*. [= *Heliocereus speciosus* × *Nopalxochia phyllanthoides*.] [RPS 40]
×**Heliochia vandesii** (G. Don *pro sp.*) Rowley, in Backeberg, Die Cact. 6: 3551, 1962. Basionym: *Epiphyllum vandesii*. [= *Heliocereus speciosus* × *Nopalxochia phyllanthoides*.] [RPS 13]
×**Heliochia violacea** (F. & T. Smith *ex* Anonymus) Rowley, in Backeberg, Die Cact. 6: 3552, 1962. Basionym: *Phyllocactus violaceus*. [Basionym reference here corrected; = *Heliocereus speciosus* × *Nopalxochia ackermannii*.] [RPS 13]
×*Heliocryptus* Meier, Kakt. and. Sukk. 32(7): 150-152, 1981. *Nom. inval.* (Art. H.6.2). [= *Heliocereus* × *Cryptocereus*.] [RPS 34]
×*Heliophyllum pseudoackermannii* Simms, Gard. Chron. 1965 (1. May): 430, 1965. *Nom. inval.* (Art. 36.1). [= × *Heliochia* vandesii 'Ackermannii'.] [RPS 16]
×*Helioselenius* Rowley, Cact. Succ. J. Gr. Brit. 13: 54, 1951. *Nom. inval.* (Art. H6.2). [= *Heliocereus* × *Selenicereus*.] [RPS 2]
×*Helioselenius fulgidus* (Hooker *pro sp.*) Hoevel, Kakt. and. Sukk. 18: 51-53, 1967. *Nom. inval.* (Art. 43.1), based on *Cereus fulgidus*. [RPS 18]
×*Helioselenius maynardii* (Paxton) Rowley, Cact. Succ. J. Gr. Brit. 13: 54, 1951. *Nom. inval.* (Art. 43.1), based on *Cereus maynardii*. [RPS 2]
×**Heliphyllum** Rowley, in Backeberg, Die Cact. 6: 3555, 1962. [= *Epiphyllum* × *Heliocereus*; = *Phyllocereus* Worsley & Knebel et al. (*non* Miquel).] [RPS 13]
×**Heliphyllum charltonii** Hort. *ex* Rowley, in Backeberg, Die Cact. 6: 3555, 1962. Basionym: *Phyllocactus charltonii*. [= *Heliocereus speciosus* × *Epiphyllum crenatum*.] [RPS 13]
×**Heptocereus** P. V. Heath, Epiphytes 7(28): 91, 1983. [= *Cryptocereus* × *Heliocereus*.] [RPS 34]
Hertrichocereus Backeberg, Cact. Succ. J. (US) 22(5): 153, 1950. Typus: *Cereus beneckei*. [RPS 1]

Heterolobivia Y. Ito, Bull. Takarazuka Insectarium 71: 13-20, 1950. *Nom. inval.* (Art. 36.1). [RPS 3]

Hildewintera Ritter, Kakt. and. Sukk. 17: 4 (January), 1966. Typus: *Hildewintera aureispina.* [This name has priority over *Winterocereus* Backeberg (March 1966).] [*Nom. nov. pro Winteria* Ritter 1962 *non* Murr. 1784 *nec* van Tiegh. 1900.] [RPS 17]

Hildewintera aureispina (Ritter) Ritter, Kakt. and. Sukk. 17: 11, 1966. Basionym: *Winteria aureispina.* [RPS 17]

Homalocephala texensis var. **gourgensii** (Cels) Y. Ito, Cacti, 108, 1952. [Basionym information not available; validity of name not assessed.] [RPS 4]

Horridocactus aconcaguensis (Ritter) Backeberg, Die Cact. 6: 3791, 1962. Basionym: *Pyrrhocactus aconcaguensis.* [RPS 13]

Horridocactus aconcaguensis var. **orientalis** (Ritter) Backeberg, Die Cact. 6: 3791, 1962. Basionym: *Pyrrhocactus aconcaguensis* var. *orientalis.* [RPS 13]

Horridocactus andicola Ritter, Succulenta 1959(7): 97-100, 1959. [RPS 10]

Horridocactus andicola var. **descendens** Ritter, Succulenta 1959(7): 97, 99, 1959. [Sphalm. 'andicolus'.] [RPS 10]

Horridocactus andicola var. **mollensis** Ritter, Succulenta 1959(7): 97, 99, 1959. [Sphalm. 'andicolus'.] [RPS 10]

Horridocactus andicola var. **robustus** Ritter, Succulenta 1959(7): 97, 99, 1959. [Sphalm. 'andicolus'.] [RPS 10]

Horridocactus armatus (Ritter) Backeberg, Die Cact. 6: 3792, 1962. Basionym: *Pyrrhocactus armatus.* [RPS 13]

Horridocactus atroviridis (Ritter) Backeberg, Die Cact. 6: 3793, 1962. Basionym: *Pyrrhocactus atroviridis.* [RPS 13]

Horridocactus carrizalensis (Ritter) Backeberg, Descr. Cact. Nov. 3: 7, 1963. Basionym: *Pyrrhocactus carrizalensis.* [RPS 14]

Horridocactus choapensis (Ritter) Backeberg, Die Cact. 6: 3793, 1962. Basionym: *Pyrrhocactus choapensis.* [RPS 13]

Horridocactus crispus (Ritter) Backeberg, Die Cact. 6: 3795, 1962. Basionym: *Pyrrhocactus crispus.* [RPS 13]

Horridocactus echinus (Ritter) Backeberg, Descr. Cact. Nov. 3: 5, 1963. Basionym: *Pyrrhocactus echinus.* [RPS 14]

Horridocactus echinus var. **minor** (Ritter) Backeberg, Kakt.-Lex., 190, 1966. Basionym: *Pyrrhocactus echinus* var. *minor.* [RPS 17]

Horridocactus engleri Ritter, Succulenta 38(6): 76-77, 1959. Typus: *Ritter* 235 (ZSS). [Chile] [RPS 10]

Horridocactus engleri var. *krausii* Backeberg, Die Cact. 6: 3796, 1962. *Nom. inval.* (Art. 37.1). [Erroneously included as valid in RPS 13.] [RPS 13]

Horridocactus eriosyzoides Ritter, Succulenta 1959(5): 49, 1959. [RPS 10]

Horridocactus garaventae Ritter, Succulenta 1959(4): 41, 1959. [Sphalm. 'garaventai'.] [RPS 10]

Horridocactus grandiflorus (Ritter) Backeberg, Die Cact. 6: 3796, 1962. Basionym: *Pyrrhocactus grandiflorus.* [RPS 13]

Horridocactus kesselringianus var. *subaequalis* Backeberg, Die Cact. 3: 1841, 1959. *Nom. inval.* (Art. 37.1). [Erroneously included as valid in RPS 10.] [RPS 10]

Horridocactus lissocarpus (Rittter) Backeberg, Die Cact. 6: 3796, 1962. Basionym: *Pyrrhocactus lissocarpus.* [RPS 13]

Horridocactus lissocarpus var. **gracilis** (Ritter) Backeberg, Die Cact. 6: 3797, 1962. Basionym: *Pyrrhocactus lissocarpus* var. *gracilis.* [RPS 13]

Horridocactus marksianus (Ritter) Backeberg, Die Cact. 6: 3797, 1962. Basionym: *Pyrrhocactus marksianus.* [RPS 13]

Horridocactus marksianus var. **tunensis** (Ritter) Backeberg, Die Cact. 6: 3798, 1962. Basionym: *Pyrrhocactus marksianus* var. *tunensis.* [RPS 13]

Horridocactus paucicostatus Ritter, Succulenta 1959(9): 113, 1959. [RPS 10]

Horridocactus paucicostatus var. **viridis** Ritter, Succulenta 1959(9): 113, 1959. [RPS 10]

Horridocactus pygmaeus Ritter, Katalog H. Winter, [unpaged], 1957. *Nom. inval.* (Art. 36.1, 37.1). [Published as provisional name. Year of first usage not established.] [RPS -]

Horridocactus robustus Ritter, Katalog H. Winter, [unpaged], 1962. *Nom. inval.* (Art. 36.1, 37.1). [Published as provisional name.] [RPS -]

Horridocactus taltalensis Ritter, Katalog H. Winter, 15, 1959. *Nom. inval.* (Art. 36.1, 37.1). [Published as provisional name.] [RPS -]

Horridocactus trapichensis Ritter, Winter Kat., [], 1959. *Nom. inval.* (Art. 36.1, 37.1). [A catalogue name.] [RPS -]

Horridocactus tuberisulcatus (Jacobi) Y. Ito, Cacti, 80, 1952. Basionym: *Echinocactus tuberisulcatus.* [RPS 4]

Horridocactus vallenarensis (Ritter) Backeberg, Die Cact. 6: 3798, 1962. Basionym: *Pyrrhocactus vallenarensis.* [RPS 13]

+ **Hylocalycium singulare** P. V. Heath, Sussex Cact. Succ. Yearb. 1987: 46, 1987. [= graft chimaera between *Gymnocalycium mihanovichii* var. *friedrichii* 'Hibotan' and *Hylocereus undatus.*] [RPS 38]

Hylocereeae Buxbaum, Madroño 14(6): 179, 1958. Typus: *Hylocereus.* [Sphalm. 'Hylocereae'.] [Sphalm. 'Hylocereae'.] [RPS 9]

Hylocereus escuintlensis Kimnach, Cact. Succ. J. (US) 56(4): 177-180, ills., 1984. Typus: *Birdsey* 313 (HNT, F, K, MEXU, UC, US). [= Univ. California BG 53.511 = Huntington BG 15902] [RPS 35]

Hylocereus estebanensis Backeberg, Descr. Cact. Nov. [1:] 11, 1957. [Dated 1956, published 1957.] [RPS 7]

Hylocereus scandens (Salm-Dyck) Backeberg, Die Cact. 2: 817, 1959. Basionym: *Cereus scandens.* [RPS 10]

Hylocereus schomburgkii (Otto) Backeberg, Die Cact. 2: 816, 1959. Basionym: *Cereus* *schomburgkii*. [RPS 10]

Hymenorebutia aurea (Britton & Rose) Ritter, Kakt. Südamer. 2: 467, 1980. Basionym: *Echinopsis aurea*. [RPS 31]

Hymenorebutia aurea var. **callochrysea** Ritter, Kakt. Südamer. 2: 468, 1980. [RPS 31]

Hymenorebutia aurea var. **catamarcensis** Ritter, Kakt. Südamer. 2: 468, 1980. [RPS 31]

Hymenorebutia aurea var. **cylindrica** (Backeberg) Ritter, Kakt. Südamer. 2: 467, 1980. Basionym: *Lobivia cylindrica*. [RPS 31]

Hymenorebutia aurea var. **depressicostata** Ritter, Kakt. Südamer. 2: 468, 1980. [RPS 31]

Hymenorebutia aurea var. **lariojensis** Ritter, Kakt. Südamer. 2: 467-468, 1980. [RPS 31]

Hymenorebutia chlorogona (Wessner) Ritter, Kakt. Südamer. 2: 472, 1980. Basionym: *Lobivia chlorogona*. [RPS 31]

Hymenorebutia chrysantha (Werdermann) Ritter, Kakt. Südamer. 2: 473, 1980. Basionym: *Echinopsis chrysantha*. [RPS 31]

Hymenorebutia cintiensis (Cardenas) Ritter, Kakt. Südamer. 2: 583-586, 1980. Basionym: *Lobivia cintiensis*. [RPS 31]

Hymenorebutia drijveriana (Backeberg) Ritter, Kakt. Südamer. 4: 473, 1980. Basionym: *Lobivia drijveriana*. [RPS 31]

Hymenorebutia kuehnrichii (Fric) Ritter, Kakt. Südamer. 2: 473, 1980. Basionym: *Lobivia kuehnrichii*. [RPS 31]

Hymenorebutia napina (Pazout) Pazout, Kakt. and. Sukk. 15: 125, 1964. Basionym: *Lobivia napina*. [RPS 15]

Hymenorebutia pusilla (Ritter) Ritter, Kakt. Südamer. 2: 588, 1980. Basionym: *Lobivia pusilla*. [RPS 31]

Hymenorebutia pusilla fa. **flaviflora** (Ritter) Ritter, Kakt. Südamer. 2: 589, 1980. Basionym: *Lobivia pusilla* fa. *flaviflora*. [RPS 31]

Hymenorebutia quinesensis (Rausch) Ritter, Kakt. Südamer. 2: 469-472, 1980. Basionym: *Echinopsis aurea* var. *quinesensis*. [RPS 31]

Hymenorebutia tiegeliana (Wessner) Ritter, Kakt. Südamer. 2: 586-587, 1980. Basionym: *Lobivia tiegeliana*. [RPS 31]

Hymenorebutia tiegeliana var. **dimorphipetala** Ritter, Kakt. Südamer. 2: 588, 1980. [RPS 31]

Hymenorebutia tiegeliana var. **distefanoiana** (Cullmann & Ritter) Ritter, Kakt. Südamer. 2: 587-588, 1980. Basionym: *Lobivia tiegeliana* var. *distefanoiana*. [RPS 31]

Hymenorebutia tiegeliana var. **ruberrima** (Rausch) Ritter, Kakt. Südamer. 2: 588, 1980. Basionym: *Lobivia tiegeliana* var. *ruberrima*. [RPS 31]

Hymenorebutia torataensis Ritter, Kakt. Südamer. 2: 589, 1980. [RPS 31]

Hymenorebutia torreana Ritter, Kakt. Südamer. 2: 589-590, 1980. Typus: *Ritter* 383 (U, ZSS). [Type cited for U l.c. 1: iii, 1979.] [Bolivia: Sud-Cinti] [RPS 31]

Irechocereus Krainz, Katalog ZSS ed. 2, 67, 1967. *Nom. inval.* (Art. 36.1, 37.1). [Published as provisional name.] [RPS -]

Irechocereus akersii Krainz, Katalog ZSS ed. 2, 67, 1967. *Nom. inval.* (Art. 36.1, 37.1). [Published as provisional name.] [RPS -]

Islaya bicolor Akers & Buining, Succulenta 1951(3): 38-41, 1951. [RPS 2]

Islaya brevicylindrica Rauh & Backeberg, in Backeberg, Descr. Cact. Nov. [1:] 33, 1957. [Dated 1956, published 1957.] [RPS 7]

Islaya chalaensis Ritter, Katalog H. Winter, [unpaged], 1957. *Nom. inval.* (Art. 36.1, 37.1). [Published as provisional name.] [RPS -]

Islaya copiapoides Rauh & Backeberg, in Backeberg, Descr. Cact. Nov. [1:] 33, 1957. [Dated 1956, published 1957.] [RPS 7]

Islaya copiapoides var. **chalaensis** Ritter, Kakt. Südamer. 4: 1295-1297, 1981. Typus: *Ritter* 128 (U, ZSS). [Peru: Arequipa] [RPS 32]

Islaya copiapoides var. **pseudomollendensis** Ritter, Kakt. Südamer. 4: 1297, 1981. Typus: *Ritter* 128a (U). [Peru: Arequipa] [RPS 32]

Islaya divaricatiflora Ritter, Taxon 13(4): 144, 1964. Typus: *Ritter* 588 (U, ZSS [seeds only]). [Type cited for U l.c. 12: 29, 1963.] [Peru: Arequipa] [RPS 15]

Islaya flavida Ritter, Kakt. Südamer. 4: 1298-1299, 1981. Typus: *Ritter* 186 (U, ZSS). [Type cited for U l.c. 1: iii, 1979.] [Peru: Caraveli] [RPS 32]

Islaya grandiflorens Rauh & Backeberg, in Backeberg, Descr. Cact. Nov. [1:] 33, 1957. [Dated 1956, published 1957.] [RPS 7]

Islaya grandiflorens var. **spinosior** Rauh & Backeberg, in Backeberg, Die Cact. 3: 1894, fig., 1959. Typus: *Ritter* 128 (U ?). [Peru: Arequipa] [RPS -]

Islaya grandiflorens var. **tenuispina** Rauh & Backeberg, in Rauh, Beitr. Kenntn. Peruan. Kakt.veg., 500, 1958. [RPS 10]

Islaya grandis Rauh & Backeberg, in Backeberg, Descr. Cact. Nov. [1:] 33, 1957. Typus: *Rauh* K150 (1956) (ZSS). [Dated 1956, published 1957.] [Peru] [RPS 7]

Islaya grandis var. **brevispina** Rauh & Backeberg, in Rauh, Beitr. Kenntn. Peruan. Kakt.veg., 494, 1958. Typus: *Rauh* K150a (1956) (ZSS). [Peru] [RPS 10]

Islaya grandis var. *neglecta* Simo, Kakt. and. Sukk. 9(7): 103, 1958. *Nom. inval.* (Art. 9.5). [RPS 9]

Islaya islayensis var. **copiapoides** (Rauh & Backeberg) Ritter, Backeberg's Descr. & Erört. taxon. Fragen, [unpaged], 1958. Basionym: *Islaya copiapoides*. [RPS 15]

Islaya islayensis var. **minor** (Backeberg) Ritter, Kakt. Südamer. 4: 1300, 1981. Basionym: *Islaya minor*. [RPS 32]

Islaya krainziana Ritter, Sukkulentenkunde 7/8: 31-33, ill., 1963. Typus: *Ritter* 200 p.p. (ZSS). [Chile] [RPS 14]

Islaya maritima Ritter, Kakt. Südamer. 4: 1301-1303, 1981. Typus: *Ritter* 590 (U). [*Sp. nov. pro Islaya grandiflorens* Rauh 1958 (*non* Rauh & Backeberg 1956).] [Peru] [RPS 32]

Islaya minuscula Ritter, Kakt. Südamer. 4: 1303, 1981. Typus: *Ritter* 1462 (U). [Peru: Arequipa] [RPS 32]

Islaya molendensis (Vaupel) Backeberg, Cact. Succ. J. (US) 23(4): 119, 1951. [RPS 3]

Islaya omasensis C. Ostolaza & T. Mischler, Kakt. and Sukk. 34(3): 54-57, 1983. Typus: *Anonymus* Cactaceae 5407 (M). [RPS 34]

Islaya paucispina var. **curvispina** Rauh & Backeberg, in Rauh, Beitr. Kenntn. Peruan. Kakt.veg., 502, 1958. [RPS 10]

Islaya paucispinosa Rauh & Backeberg, in Backeberg, Descr. Cact. Nov. [1:] 33, 1957. Typus: *Rauh* K42 (1954) (ZSS). [Dated 1956, published 1957.] [Peru] [RPS 7]

Islaya raphidophora Y. Ito, The Cactaceae, 463, ill., 1981. Typus: *Ritter* 128 (). [Ascribed to "Ritter & Ito". Based on the material cited of which a specimen is at U but this was most probably not seen by Ito. Validity of name not assessed.] [Peru: Arequipa] [RPS -]

Islaya unguispina Ritter, Kakt. Südamer. 4: 1304, 1981. Typus: *Ritter* 591 (U). [Peru] [RPS 32]

Jaenocereus Krainz, Katalog ZSS ed. 2, 67, 1967. *Nom. inval.* (Art. 36.1, 37.1). [Published as provisional name.] [RPS -]

Jaenocereus nigripilis Krainz, Katalog ZSS ed. 2, 67, 1967. *Nom. inval.* (Art. 36.1, 37.1). [Published as provisional name.] [RPS -]

Jasminocereus howellii Dawson, Cact. Succ. J. (US) 34: 71, 1962. [RPS 13]

Jasminocereus howellii var. **delicatus** Dawson, Cact. Succ. J. (US) 34: 72, 1962. [RPS 13]

Jasminocereus thouarsii (F. A. C. Weber) Backeberg, Die Cact. 2: 912, 1959. Basionym: *Cereus thouarsii*. [Repeated by Buxbaum in Krainz, Die Kakt. Lief. 16: C IV, 1961.] [RPS 10]

Jasminocereus thouarsii var. **chathamensis** Dawson, Cact. Succ. J. (US) 34: 73, 1962. [RPS 13]

Jasminocereus thouarsii var. **delicatus** (Dawson) Anderson & Walkington, Madroño 20: 256, 1970. Basionym: *Jasminocereus howellii* var. *delicatus*. [RPS 21]

Jasminocereus thouarsii var. **sclerocarpus** (Schumann) Anderson & Walkington, Madroño 20: 256, 1970. Basionym: *Cereus sclerocarpus*. [RPS 21]

Krainzia guelzowiana (Werdermann) Backeberg, Cact. Succ. J. (US) 23(5): 152, 1951. Basionym: *Mammillaria guelzowiana*. [Published as *comb. nud.*, validity not assessed.] [RPS 3]

Krainzia guelzowiana var. *comocephala* Y. Ito, The Cactaceae, 564, ill., 1981. *Nom. inval.* (Art. 37.1). [Spelling of the varietal epithet exactly as given in the protologue.] [RPS -]

Lasiocereus Ritter, Kakt. Südamer. 4: 1477-1478, 1981. Typus: *L. rupicola*. [First mentioned in Backeberg, Die Cact. 2: 1359, 1959 (cf. RPS 10), and in l.c. 6: 3729, 1962, and then first published in Succulenta 45(8): 119, 1966, but the type given, *L. rupicola* (sphalm. *rupicolus*) was not validly described at the time.] [RPS -]

Lasiocereus fulvus Ritter, Kakt. Südamer. 4: 1479, 1981. Typus: *Ritter* 1303 (U, ZSS [seeds only]). [Type cited for U l.c. 1: iii, 1979.] [First mentioned (without type, but with Latin diagnosis) in Succulenta 45(8): 119, 1966.] [Peru: Amazonas] [RPS -]

Lasiocereus rupicola Ritter, Kakt. Südamer. 4: 1478, 1981. Typus: *Ritter* 661 (U, ZSS [seeds only]). [Type cited for U l.c. 1: iii, 1979.] [First mentioned in Backeberg, Die Cact. 6: 3729; then published without type in Succulenta 45(8): 119, 1966.] [Peru: Cajamarca] [RPS -]

Lemaireocereus chlorocarpus (Humboldt, Bonpland & Knuth) Borg, Cacti, ed. 2, 167, 1951. Basionym: *Cactus chlorocarpus*. [RPS 9]

Lemaireocereus laevigatus (Salm-Dyck) Borg, Cacti, ed. 2, 163, 1951. Basionym: *Cereus laevigatus*. [RPS 9]

Lemaireocereus littoralis (K. Brandegee) Gates, Cact. Succ. J. (US) 30(4): 114, 1958. Basionym: *Cereus thurberi* var. *littoralis*. [RPS 9]

Leocereus bahiensis var. **barreirensis** P. J. Braun & E. Esteves Pereira, Kakt. and. Sukk. 41(9): 205, 1990. Typus: *Esteves Pereira* 118 (UFG 12.360, ZSS). [Brazil: Bahia] [RPS 41]

Leocereus bahiensis var. **exiguospinus** P. J. Braun & E. Esteves Pereira, Kakt. and. Sukk. 41(9): 205, 1990. Typus: *Esteves Pereira* 135 (UFG [not found Jan. 1991], ZSS). [Brazil: Bahia] [RPS 41]

Leocereus bahiensis var. **robustispinus** P. J. Braun & E. Esteves Pereira, Kakt. and. Sukk. 41(9): 205, 1990. Typus: *Braun* 702 (ZSS). [Brazil: Bahia] [RPS 41]

Leocereus bahiensis var. **urandianus** (Ritter) P. J. Braun & E. Esteves Pereira, Kakt. and. Sukk. 41(9): 205, 1990. Basionym: *Leocereus urandianus*. [RPS 41]

Leocereus estevesii P. J. Braun, Kakt. and. Sukk. 41(9): 204-205, 1990. Typus: *Esteves Pereira* 207 (UFG 12.380, B, ZSS). [Brazil: Piauí] [RPS 41]

Leocereus urandianus Ritter, Kakt. Südamer. 1: 222, 1979. Typus: *Ritter* 1231 (U). [Erroneously given as 'urandiensis' in RPS 30.] [Brazil: Bahia] [RPS 30]

Leocereus urandiensis hort., Rep. Pl. Succ. 30: 6, 1981. *Nom. inval.* (Art. 36.1, 37.1). [A spelling variant for *Leocereus urandianus*.] [RPS -]

Lepidocoryphantha runyonii (Britton & Rose) Backeberg, Die Cact. 5: 2957, 1961. Basionym: *Coryphantha runyonii*. [RPS 12]

Lepismium subgen. **Acanthorhipsalis** (Schumann) Barthlott, Bradleya 5: 99, 1987. Basionym: *Rhipsalis* subgen. *Acanthorhipsalis*. [RPS 38]

Lepismium subgen. *Heteropodium* Backeberg, Descr. Cact. Nov. 3: 7, 1963. *Nom. inval.* (Art. 10.1, 37.1). [Based on *Lepismium marnierianum* Backeberg (*nom. inval.*).] [RPS 14]

Lepismium subgen. **Lymanbensonia** (Kimnach) Barthlott, Bradleya 5: 99, 1987. Basionym: *Lymanbensonia*. [RPS 38]

Lepismium subgen. **Ophiorhipsalis** (Schumann) Barthlott, Bradleya 5: 99, 1987. Basionym: *Rhipsalis* subgen. *Ophiorhipsalis*. [RPS 38]

Lepismium subgen. **Pfeiffera** (Salm-Dyck)

Barthlott, Bradleya 5: 99, 1987. Basionym: *Pfeiffera*. [RPS 38]

Lepismium aculeatum (F. A. C. Weber) Barthlott, Bradleya 5: 99, 1987. Basionym: *Rhipsalis aculeata*. [RPS 38]

Lepismium anceps (F. A. C. Weber) Y. Ito, Cacti, 163, 1952. Basionym: *Rhipsalis anceps*. [RPS 4]

Lepismium bolivianum (Britton) Ewald, Epiphytes 7(28): 79, 93, 1983. *Nom. inval.* (Art. 33.2), based on *Hariota boliviana*. [RPS 34]

Lepismium brevispinum Barthlott, Bradleya 5: 99, 1987. Typus: *Ritter* s.n. (fig. 1114 in F. Ritter, Kakt. Südamer. 4: 1529). [= *Acanthorhipsalis brevispina* F. Ritter 1981, *nom. inval.* (Art. 37.1).] [Peru: Amazonas] [RPS 38]

Lepismium chrysanthum (Löfgren) Backeberg, Die Cact. 2: 694, 1959. Basionym: *Rhipsalis chrysantha*. [RPS 10]

Lepismium crenatum (Britton) Barthlott, Bradleya 5: 99, 1987. Basionym: *Hariota crenata*. [RPS 38]

Lepismium cruciforme fa. **anceps** (F. A. C. Weber) Süpplie, Rhipsalidinae, [11], 1990. Basionym: *Rhipsalis anceps*. [RPS 41]

Lepismium cruciforme fa. **myosurus** (Salm-Dyck) Süpplie, Rhipsalidinae, [11], 1990. Basionym: *Cereus myosurus*. [RPS 41]

Lepismium cruciforme subvar. **vollii** (Backeberg) Backeberg, Die Cact. 2: 688, 1959. Basionym: *Lepismium vollii*. [RPS 10]

Lepismium cruciforme var. **anceps** (F. A. C. Weber) Backeberg, Die Cact. 2: 689, 1959. Basionym: *Rhipsalis anceps*. [RPS 10]

Lepismium cruciforme var. **cavernosum** (Lindinger) Backeberg, Die Cact. 2: 689, 1959. Basionym: *Lepismium cavernosum*. [RPS 10]

Lepismium cruciforme var. **knightii** (Salm-Dyck) Boom, Succulenta 1959(11): 142, 1959. Basionym: *Lepismium myosurus* var. *knightii*. [RPS 10]

Lepismium cruciforme var. **myosurus** (Salm-Dyck) Backeberg, Die Cact. 2: 688, 1959. Basionym: *Cereus myosurus*. [RPS 10]

Lepismium epiphyllanthoides Backeberg, Cact. Succ. J. (US) 23(1): 16, 1951. [RPS 3]

Lepismium erectum (Ritter) Süpplie, Rhipsalidinae, [15], 1990. Basionym: *Pfeiffera erecta*. [Sphalm. 'erecta'.] [RPS 41]

Lepismium grandiflorum (Haworth) Backeberg, Die Cact. 2: 691, 1959. Basionym: *Rhipsalis grandiflora*. [RPS 10]

Lepismium houlletianum (Lemaire) Barthlott, Bradleya 5: 99, 1987. Basionym: *Rhipsalis houlletiana*. [RPS 38]

Lepismium houlletianum var. **regnellii** (Lindberg) Barthlott, Bradleya 5: 99, 1987. Basionym: *Rhipsalis regnellii*. [RPS 38]

Lepismium ianthothele (Monville) Barthlott, Bradleya 5: 99, 1987. Basionym: *Cereus ianthothele*. [RPS 38]

Lepismium incachacanum (Cardenas) Barthlott, Bradleya 5: 99, 1987. Basionym: *Rhipsalis incachacana*. [RPS 38]

Lepismium lorentzianum (Grisebach) Barthlott, Bradleya 5: 99, 1987. Basionym: *Rhipsalis lorentziana*. [RPS 38]

Lepismium lumbricoides (Lemaire) Barthlott, Bradleya 5: 99, 1987. Basionym: *Cereus lumbricoides*. [RPS 38]

Lepismium marnierianum Backeberg, Descr. Cact. Nov. 3: 7, 1963. *Nom. inval.* (Art. 9.5). [Erroneously included as valid in RPS 14.] [RPS 14]

Lepismium mataralense (Ritter) Süpplie, Rhipsalidinae, [13], 1990. Basionym: *Pfeiffera mataralensis*. [Sphalm. 'mataralensis'.] [RPS 41]

Lepismium mataralense var. **floccosum** (Ritter) Süpplie, Rhipsalidinae, [13], 1990. Basionym: *Pfeiffera mataralensis* var. *floccosa*. [Sphalm. 'mataralensis', 'floccosa'. Basionym erroneously given as *Pfeiffera mataralensis*, treated as bibliographic error.] [RPS 41]

Lepismium micranthum (Vaupel) Barthlott, Bradleya 5: 99, 1987. Basionym: *Cereus micranthus*. [RPS 38]

Lepismium miyagawae (Barthlott & Rauh) Barthlott, Bradleya 5: 99, 1987. Basionym: *Pfeiffera miyagawae*. [RPS 38]

Lepismium monacanthum (Grisebach) Barthlott, Bradleya 5: 99, 1987. Basionym: *Rhipsalis monacantha*. [RPS 38]

Lepismium paranganiense (Cardenas) Barthlott, Bradleya 5: 99, 1987. Basionym: *Acanthorhipsalis paranganiensis*. [RPS 38]

Lepismium pittieri (Britton & Rose) Backeberg, Die Cact. 2: 692, 1959. Basionym: *Rhipsalis pittieri*. [RPS 10]

Lepismium puniceo-discus var. **chrysocarpum** (Löfgren) Backeberg, Die Cact. 2: 691, 1959. Basionym: *Rhipsalis chrysocarpa*. [RPS 10]

Lepismium rigidum (Löfgren) Backeberg, Die Cact. 2: 691, 1959. Basionym: *Rhipsalis rigida*. [RPS 10]

Lepismium saxatile Friedrich & Redecker *ex* Backeberg, Nation. Cact. Succ. J. 20(4): 57, 1965. Typus: *Friedrich* s.n. (K). [See also Kakt. and. Sukk. 17(5): 91-92, 1966.] [?] [RPS 16]

Lepismium warmingianum (Schumann) Barthlott, Bradleya 5: 99, 1987. Basionym: *Rhipsalis warmingiana*. [RPS 38]

Leptocereeae Buxbaum, Madroño 14(6): 178, 1958. Typus: *Leptocereus*. [Sphalm. 'Leptocereae'.] [RPS 9]

Leptocladia Buxbaum *non* Agardh, Österr. Bot. Zeitschr. 98(1-2): 81-82, 1951. *Nom. illeg.* (Art. 64.1). [The following combinations under this illegitimate generic name are all incorrect (*nomina mendosa*)..] [*Non Leptocladia* Agardh 1892.] [RPS 2]

Leptocladia densispina (Coulter) Buxbaum, Österr. Bot. Zeitschr. 98(1-2): 82, 1951. Incorrect name (Art. 11.3), based on *Mammillaria densispina*. [RPS 2]

Leptocladia echinaria (De Candolle) Buxbaum, Österr. Bot. Zeitschr. 98(1-2): 82, 1951. Incorrect name (Art. 11.3), based on *Mammillaria echinaria*. [RPS 2]

Leptocladia elongata (De Candolle) Buxbaum, Österr. Bot. Zeitschr. 98(1-2): 82, 1951.

Incorrect name (Art. 11.3), based on *Mammillaria elongata*. [RPS 2]

Leptocladia leona (Poselger) Buxbaum, Österr. Bot. Zeitschr. 98(1-2): 82, 1951. Incorrect name (Art. 11.3), based on *Mammillaria leona*. [RPS 2]

Leptocladia microhelia (Werdermann) Buxbaum, Österr. Bot. Zeitschr. 98(1-2): 82, 1951. Incorrect name (Art. 11.3), based on *Mammillaria microhelia*. [RPS 2]

Leptocladia microheliopsis (Werdermann) Buxbaum, Österr. Bot. Zeitschr. 98(1-2): 82, 1951. Incorrect name (Art. 11.3), based on *Mammillaria microheliopsis*. [RPS 2]

Leptocladia mieheana (Tiegel) Buxbaum, Österr. Bot. Zeitschr. 98(1-2): 82, 1951. Incorrect name (Art. 11.3), based on *Mammillaria mieheana*. [RPS 2]

Leptocladia viperina (Purpus) Buxbaum, Österr. Bot. Zeitschr. 98(1-2): 82, 1951. Incorrect name (Art. 11.3), based on *Mammillaria viperina*. [RPS 2]

Leptocladodia F. Buxbaum, Österr. Bot. Zeitschr. 101: 601, 1954. Typus: *Mammillaria elongata* De Candolle. [*Nom. nov. pro Leptocladia* Buxbaum 1951 *non* Agardh 1892. In subsequent publications, Buxbaum very often cited *Leptocladodia*-names with the references for his combinations under *Leptocladia*, which has resulted in many later errors and mistakes.] [RPS 6]

Leptocladodia elongata fa. *densa* (Link & Otto) Krainz, Kat. ZSS ed. 2, 68, 1967. *Nom. inval.* (Art. 43.1), based on *Mammillaria densa*. [Erroneously included as valid in RPS 18.] [RPS 18]

Leptocladodia elongata fa. *intertexta* (De Candolle) Krainz, Kat. ZSS ed. 2, 68, 1967. *Nom. inval.* (Art. 43.1), based on *Mammillaria intertexta*. [Erroneously included as valid in RPS 18.] [RPS 18]

Leptocladodia elongata fa. *rufo-crocea* (Salm-Dyck) Krainz, Kat. ZSS ed. 2, 68, 1967. *Nom. inval.* (Art. 43.1), based on *Mammillaria rufo-crocea*. [Erroneously included as valid in RPS 18.] [RPS 18]

Leptocladodia elongata fa. *stellata* (Martius) Krainz, Kat. ZSS ed. 2, 68, 1967. *Nom. inval.* (Art. 43.1), based on *Mammillaria stella-aurata*. [Erroneously included as valid in RPS 18.] [RPS 18]

Leptocladodia elongata fa. *tenuis* (De Candolle) Krainz, Kat. ZSS ed. 2, 68, 1967. *Nom. inval.* (Art. 43.1), based on *Mammillaria tenuis*. [Erroneously included as valid in RPS 18.] [RPS 18]

Leptocladodia elongata var. *echinaria* (De Candolle) Krainz, Kat. ZSS ed. 2, 68, 1967. *Nom. inval.* (Art. 43.1), based on *Mammillaria echinaria*. [Erroneously included as valid in RPS 18.] [RPS 18]

Leptocladodia leona (Poselger) Buxbaum, in Krainz, Die Kakt. Lief. 20: C VIIIc (1. 4. 1962), 1962. Basionym: *Mammillaria leona*. [Combination by inference; the name is accepted, full basionym information is given and a reference to the earlier incorrect name *Leptocladia leona* is included.] [RPS -]

Leptocladodia microhelia (Werdermann) Buxbaum, in Krainz, Die Kakt. Lief. 63: C IIId (1. 10. 1975), 1975. Basionym: *Mammillaria microhelia*. [Combination by inference; the name is accepted and full basionym information is given.] [RPS -]

Leptocladodia microhelia fa. **microheliopsis** (Werdermann) Krainz, Die Kakt. Lief. 63: C IIId (1. 10. 1975), 1975. Basionym: *Mammillaria microheliopsis*. [First invalidly (Art. 43.1) published in Kat. ZSS ed. 2, 68, 1967 (and included as valid in RPS 18).] [RPS -]

Leptocladodia sphacelata (Martius) Buxbaum, in Krainz, Die Kakt. Lief. 20: C VIIIc, 1962. Basionym: *Mammillaria sphacelata*. [RPS 13]

Leptocladodia viperina (Purpus) Buxbaum, in Krainz, Die Kakt. Lief. 20: C VIIIc, 1962. Basionym: *Mammillaria viperina*. [Combination by inference; the name is accepted and full basionym information is given, but the author was of the opinion that he already published the combination in 1951 (this citation relates to *Leptocladia viperina, nom. mendosum*).] [RPS -]

×*Leuchtenfera* Arakawa & Ito, The Cactaceae, 653, 1981. *Nom. inval.* (Art. H6.2). [The correct name for this cross is *Ferobergia*..] [= *Leuchtenbergia* × *Ferocactus*.] [RPS -]

Leucostele Backeberg, Kakt. and. Sukk. 4(3): 36-41, 1953. Typus: *Leucostele rivierei*. [RPS 4]

Leucostele rivierei Backeberg, Kakt. and. Sukk. 4(3): 36-41, 1953. [RPS 4]

Lobivia cv. **Golden Flower** Timmermans, Succulenta 1954: 28, 1954. [= *Chamaecereus silvestrii* × *Lobivia aurea*.] [RPS 5]

Lobivia cv. **Grullemans** Timmermans, Succulenta 1954: 27, 1954. [(*Chamaecereus silvestrii* × *Lobivia famatimensis*) × *Lobivia sp.*] [RPS 5]

Lobivia cv. **Jupiter** Timmermans, Succulenta 1954: 28, 1954. [= *Lobivia allegraiana* × *Pseudolobivia polyancistra*.] [RPS 5]

Lobivia cv. **Mars** Timmermans, Succulenta 1954: 28, 1954. [= *Lobivia allegraiana* × *Pseudolobivia polyancistra*.] [RPS 5]

Lobivia cv. **Persus** Timmermans, Succulenta 1954: 28, 1954. [= *Lobivia allegraiana* × *Pseudolobivia polyancistra*.] [RPS 5]

Lobivia cv. **Vandenthoorn** Timmermans, Succulenta 1954: 27, 1954. [= (*Pseudoechinopsis aurea* × *Chamaecereus silvestrii*) × (*Lobivia pseudocachensis* var. *sanguinea* × *Mediolobivia sarothroides*).] [RPS 5]

Lobivia subgen. *Chrysolobivia* Y. Ito, Bull. Takarazuka Insectarium 71: 13-20, 1950. *Nom. inval.* (Art. 36.1). [RPS 3]

Lobivia subgen. *Cirrhilobivia* Y. Ito, Bull. Takarazuka Insectarium 71: 13-20, 1950. *Nom. inval.* (Art. 36.1). [RPS 3]

Lobivia subgen. *Cyclolobivia* Y. Ito, Bull. Takarazuka Insectarium 71: 13-20, 1950. *Nom. inval.* (Art. 36.1). [RPS 3]

Lobivia subgen. *Dolabrilobivia* Y. Ito, Bull. Takarazuka Insectarium 71: 13-20, 1950.

Nom. inval. (Art. 36.1). [RPS 3]

Lobivia subgen. **Hymenolobivia** Buining & Kreuzinger *ex* Ritter, Kakt. Südamer. 2: 460, 1980. Typus: *L. buiningiana* Ritter. [RPS 31]

Lobivia subgen. *Ionolobivia* Y. Ito, Bull. Takarazuka Insectarium 71: 13-20, 1950. *Nom. inval.* (Art. 36.1). [RPS 3]

Lobivia subgen. *Leucolobivia* Y. Ito, Bull. Takarazuka Insectarim 71: 13-20, 1950. *Nom. inval.* (Art. 36.1). [RPS 3]

Lobivia subgen. *Reblobivia* Y. Ito, Bull. Takarazuka Insectarium 71: 13-20, 1950. *Nom. inval.* (Art. 36.1). [Sphalm. for 'Rebulobivia' ?.] [RPS 3]

Lobivia subgen. *Stenolobivia* Y. Ito, Bull. Takarazuka Insectarium 71: 13-20, 1950. *Nom. inval.* (Art. 36.1). [RPS 3]

Lobivia subgen. *Thiolobivia* Y. Ito, Bull. Takarazuka Insectarium 71: 13-20, 1950. *Nom. inval.* (Art. 36.1). [RPS 3]

Lobivia abrantha Y. Ito, The Full Bloom of Cactus Flowers, 44, 47, ill., 1962. *Nom. inval.* (Art. 37.1). [Included as valid in RPS 13.] [RPS 13]

Lobivia acanthoplegma (Backeberg) Backeberg, Descr. Cact. Nov. 3: 7, 1963. *Nom. inval.*, based on *Pseudolobivia acanthoplegma, nom. inval.* (Art. 37.1). [Erroneously included as valid in RPS 14.] [RPS 14]

Lobivia acanthoplegma var. *patula* Rausch, Succulenta 53: 150, 1974. Based on *Rausch* 54. *Nom. inval.* (Art. 43.1). [Erroneously included as valid in RPS 25.] [Bolivia: Cochabamba] [RPS 25]

Lobivia acanthoplegma var. *pilosa* Rausch, Kakt. and. Sukk. 31(7): 242-243, ill., 1980. Based on *Rausch* 667. *Nom. inval.* (Art. 43.1). [Erroneously included as valid in RPS 31.] [Bolivia: Cochabamba] [RPS 31]

Lobivia acanthoplegma var. *roseiflora* Rausch, Kakt. and. Sukk. 28(4): 75, 1977. Based on *Rausch* 457. *Nom. inval.* (Art. 43.1). [Erroneously included as valid in RPS 28.] [Bolivia: Cochabamba] [RPS 27]

Lobivia aculeata var. **walterspielii** (Bödeker) Backeberg, Die Cact. 3: 1406, 1959. Basionym: *Lobivia walterspielii.* [RPS 10]

Lobivia adpressispina Ritter, Kakt. Südamer. 2: 576-577, 1980. [RPS 31]

Lobivia aguilarii Vasquez, Kakt. and. Sukk. 25(1): 1-2, ill., 1974. Typus: *Vasquez* 558 (W [not found], ZSS). [Bolivia: Cochabamba] [RPS 25]

Lobivia akersii Rausch, Kakt. and. Sukk. 24: 25, 28, 1973. [RPS 24]

Lobivia albicentra (Fric) Y. Ito, Explan. Diagr. Austroechinocactinae, 102, 1957. Basionym: *Hymenorebulobivia albicentra.* [RPS 8]

Lobivia amblayensis Rausch, Kakt. and. Sukk. 23(3): 67-88, ill., 1972. Typus: *Rausch* 19 (W, ZSS). [Argentina: Salta] [RPS 23]

Lobivia amblayensis var. **albispina** Rausch, Kakt. and. Sukk. 23(3): 68, 1972. Typus: *Rausch* 239 (ZSS). [Type cited for W but placed in ZSS.] [Argentina: Salta] [RPS 23]

Lobivia arachnacantha Buining & Ritter, Succulenta 35(3): 37-38, 1956. Typus: *Ritter* 360 (U, ZSS). [Bolivia: Santa Cruz] [RPS 7]

Lobivia arachnacantha var. **densiseta** Rausch, Kakt. and. Sukk. 21(3): 45, 1970. Typus: *Rausch* 186 (W, ZSS). [First invalidly published (Art. 37.1) in l.c. 19(3): 49, ill., 1968.] [Bolivia: Santa Cruz] [RPS 21]

Lobivia arachnacantha var. **sulphurea** Vasquez, Succulenta 53: 108-109, 1974. [RPS 25]

Lobivia atrovirens var. **haefneriana** (Cullmann) Rausch, Lobivia 85, 12, ill. (p. 14), 1987. Basionym: *Mediolobivia haefneriana.* [RPS 38]

Lobivia atrovirens var. **huasiensis** (Rausch) Rausch, Lobivia 85, 13, 1987. Basionym: *Rebutia huasiensis.* [Basionym erroneously cited as *Aylostera huasiensis*, treated as bibliographical error (Art. 33.2, Ex. 4).] [RPS 38]

Lobivia atrovirens var. **pseudoritteri** Rausch, Lobivia 85, 13, 137, 1987. Typus: *Rausch* 506 (ZSS). [Syn. *Rebutia ritteri* sensu Ritter (*non* Wessner), Kakt. Südamer. 2: 603. 1980.] [Bolivia: Tarija] [RPS 38]

Lobivia atrovirens var. **raulii** (Rausch) Rausch, Lobivia 85, 13, 1987. Basionym: *Rebutia raulii.* [Basionym erroneously cited as *Digitorebutia raulii*, treated as bibliographical error (Art. 33.2, Ex. 4).] [RPS 38]

Lobivia atrovirens var. **ritteri** (Wessner) Rausch, Lobivia 85, 13, 1987. Basionym: *Lobivia ritteri.* [RPS 38]

Lobivia atrovirens var. **yuncharasensis** Rausch, Lobivia 85, 13, 137, 1987. Typus: *Rausch* 91 (ZSS). [Bolivia: Tarija] [RPS 38]

Lobivia atrovirens var. **yuquinensis** (Rausch) Rausch, Lobivia 85, 13, 1987. Basionym: *Rebutia yuquinensis.* [Basionym erroneously cited as *Digitorebutia yuquinensis*, treated as bibliographical error (Art. 33.2, Ex. 4).] [RPS 38]

Lobivia atrovirens var. **zecheri** (Rausch) Rausch, Lobivia 85, 13, 1987. Basionym: *Rebutia zecheri.* [Basionym erroneously cited as *Aylostera zecheri*, treated as bibliographical error (Art. 33.2, Ex. 4).] [RPS 38]

Lobivia aurantiaca Backeberg, Die Cact. 3: 1400, 1959. *Nom. inval.* (Art. 36.1, 37.1). [*Sp. nov. pro Lobivia boliviensis sensu* Werdermann in Blüh. Kakt. and. Sukk., t. 149, 1938. Erroneously included as valid in RPS 10.] [RPS 10]

Lobivia aurea var. **albiflora** Rausch, Succulenta 58(7): 161-162, ill., 1979. Typus: *Rausch* 710 (ZSS). [Argentina: Córdoba] [RPS 30]

Lobivia aurea var. **callochrysea** (Ritter) Rausch, Lobivia 85, 16, 1987. Basionym: *Hymenorebutia aurea* var. *callochrysea.* [RPS 38]

Lobivia aurea var. **catamarcensis** (Ritter) Rausch, Lobivia 85, 16, ill. (p. 15), 1987. Basionym: *Hymenorebutia aurea* var. *catamarcensis.* [RPS 38]

Lobivia aurea var. **dobeana** (Dölz) Rausch, Lobivia 75, 3: 148, 1975. Basionym: *Lobivia dobeana.* [RPS -]

Lobivia aurea var. **quinesensis** (Rausch) Rausch *ex* Rowley, Cact. Succ. J. Gr. Brit. 44(4): 80,

1982. Basionym: *Echinopsis aurea* var. *quinesensis*. [RPS 33]
Lobivia aurea var. **sierragrandensis** Rausch, Lobivia 85, 17, 137, ill. (p. 138), 1987. Typus: *Rausch* 711b (ZSS). [Argentina: Córdoba] [RPS 38]
Lobivia aurea var. *spinosissima* Y. Ito, Cacti, 67, 1952. *Nom. inval.* (Art. 36.1). [RPS 4]
Lobivia aurea var. **tortuosa** Rausch, Lobivia 85, 17, 137-138, ill., 1987. Typus: *Rausch* 711e (ZSS). [Argentina: Santiago del Estero] [RPS 38]
Lobivia aureolilacina Cardenas, Cact. Succ. J. (US) 33: 110, 1961. [RPS 12]
Lobivia aureosenilis Knize, Biota 7(57): 253, 1968. Typus: *Knize* 437 (Z [not found]). [Publication date given as 1969 in RPS 25.] [Peru: Arequipa] [RPS 20]
Lobivia backebergiana Y. Ito, The Full Bloom of Cactus Flowers, 48, 50, ill., 1962. *Nom. inval.* (Art. 37.1). [Included as valid in RPS 13.] [RPS 13]
Lobivia backebergii ssp. **hertrichiana** (Backeberg) Rausch *ex* Rowley, Cact. Succ. J. Gr. Brit. 44(4): 80, 1982. Basionym: *Lobivia hertrichiana*. [First mentioned by Rausch in Lobivia 3: 178, 1977.] [RPS 33]
Lobivia backebergii ssp. **schieliana** (Backeberg) Rausch *ex* Rowley, Cact. Succ. J. Gr. Brit. 44(4): 80, 1982. Basionym: *Lobivia schieliana*. [First mentioned by Rausch in Lobivia 3: 178, 1977.] [RPS 33]
Lobivia backebergii ssp. **wrightiana** (Backeberg) Rausch *ex* Rowley, Cact. Succ. J. Gr. Brit. 44(4): 80, 1982. Basionym: *Lobivia wrightiana*. [First mentioned by Rausch in Lobivia 3: 178, 1977.] [RPS 33]
Lobivia backebergii ssp. **zecheri** (Rausch) Rausch *ex* Rowley, Cact. Succ. J. Gr. Brit. 44(4): 80, 1982. Basionym: *Lobivia zecheri*. [First mentioned by Rausch in Lobivia 3: 178, 1977.] [RPS 33]
Lobivia backebergii var. **hertrichiana** (Backeberg) Rausch, Lobivia 1: 20, 1975. Basionym: *Lobivia hertrichiana*. [RPS 26]
Lobivia backebergii var. **larae** (Cardenas) Rausch, Lobivia 85, 20, ill. (p. 19), 1987. Basionym: *Lobivia larae*. [RPS 38]
Lobivia backebergii var. **laui** (Donald) Rausch, Lobivia 1: 20, 1975. Basionym: *Lobivia laui*. [RPS 26]
Lobivia backebergii var. **oxyalabastra** (Cardenas & Rausch) Rausch, Lobivia 1: 20, 1975. Basionym: *Lobivia oxyalabastra*. [RPS 26]
Lobivia backebergii var. **schieliana** (Backeberg) Rausch, Lobivia 1: 20, 1975. Basionym: *Lobivia schieliana*. [Erroneously included as invalid in RPS 26. Often misspelled as 'schieleana'.] [RPS 26]
Lobivia backebergii var. **simplex** Rausch, Lobivia 1: 20, 1975. Basionym: *Lobivia simplex*. [RPS 26]
Lobivia backebergii var. **winteriana** (Ritter) Rausch, Lobivia 1: 20, 1975. Basionym: *Lobivia winteriana*. [RPS 26]
Lobivia backebergii var. **wrightiana** (Backeberg) Rausch, Lobivia 1: 20, 1975. Basionym: *Lobivia wrightiana*. [Erroneously included as invalid in RPS 26.] [RPS 26]
Lobivia backebergii var. **zecheri** (Rausch) Rausch, Lobivia 1: 20, 1975. Basionym: *Lobivia zecheri*. [Erroneously given as 'comb. illeg.' in RPS 26 because *L. zecheri* var. *fungiflora* Braun (*nom. inval.*) is cited in the synonymy.] [RPS 26]
Lobivia boliviensis var. **croceantha** Y. Ito, Explan. Diagr. Austroechinocacti, 94, 287, 1957. [RPS 8]
Lobivia boliviensis var. **rubriflora** Y. Ito, Explan. Diagr. Austroechinocacti, 94, 287, 1957. [RPS 8]
Lobivia boliviensis var. **violaciflora** Y. Ito, Explan. Diagr. Austroechinocacti, 94, 287, 1957. [RPS 8]
Lobivia brunneo-rosea Backeberg, Descr. Cact. Nov. [1:] 28, 1957. [Dated 1956, published 1957.] [RPS 7]
Lobivia buiningiana Ritter, Kakt. Südamer. 2: 461-462, 1980. Typus: *Ritter* 55 (U, ZSS [seeds only]). [Type cited for U l.c. 1: iii, 1979.] [Argentina: Jujuy] [RPS 31]
Lobivia caespitosa var. **altiplani** Ritter, Kakt. Südamer. 2: 569-570, 1980. Typus: *Ritter* 99a (U, ZSS). [Type cited for U l.c. 1: iii, 1979.] [Bolivia: Aroma] [RPS 31]
Lobivia caespitosa var. *columnaris* Y. Ito, Shaboten 94: 24-25, 1977. *Nom. inval.* (Art. 37.1). [Repeated invalidly (Art. 37.1) in The Cactaceae, 642, 1981.] [RPS 17]
Lobivia caespitosa var. **rinconadensis** Ritter, Kakt. Südamer. 2: 569, 1980. [RPS 31]
Lobivia caespitosa var. **violacea** Rausch, Kakt. and. Sukk. 30(7): 161-162, 1979. Typus: *Rausch* 735 (ZSS). [Bolivia: Ayopaya] [RPS 30]
Lobivia caineana Cardenas, Cact. Succ. J. (US) 24(6): 184-185, 1952. [RPS 3]
Lobivia calorubra (Cardenas) Rausch, Lobivia 85, 21, 1987. Basionym: *Echinopsis calorubra*. [RPS 38]
Lobivia calorubra var. **grandiflora** (Rausch) Rausch, Lobivia 85, 21, 1987. Basionym: *Lobivia pojoensis* var. *grandiflora*. [RPS 38]
Lobivia calorubra var. **megalocephala** (Rausch) Rausch, Lobivia 85, 21, 1987. Basionym: *Echinopsis rauschii* var. *megalocephala*. [RPS 38]
Lobivia calorubra var. **mizquensis** (Rausch) Rausch, Lobivia 85, 21, 1987. Basionym: *Lobivia mizquensis*. [RPS 38]
Lobivia calorubra var. **pojoensis** (Rausch) Rausch, Lobivia 85, 21, ill. (p. 22), 1987. Basionym: *Lobivia pojoensis*. [RPS 38]
Lobivia camataquiensis Cardenas, Cactus 78: 90, 1963. [RPS 14]
Lobivia campicola Ritter, Kakt. Südamer. 2: 576, 1980. [RPS 31]
Lobivia cardenasiana Rausch, Kakt. and. Sukk. 23(2): 32-33, ill., 1972. Typus: *Rausch* 498 (W, ZSS). [Bolivia: Tarija] [RPS 23]
Lobivia cariquinensis Cardenas, Cactus (Paris) No. 65: 181, 1959. [RPS 10]
Lobivia charazanensis Cardenas, Cactus (Paris) No. 57: 256-257, 1957. [RPS 8]
Lobivia charcasina Cardenas, Cactus (Paris) No. 82: 42, 1964. [RPS 15]
Lobivia chorrillosensis Rausch, Kakt. and.

Sukk. 25(7): 145-146, ill., 1974. Typus: *Rausch* 157 (ZSS). [Type cited for W but placed in ZSS.] [Argentina: Salta] [RPS 25]

Lobivia chrysantha ssp. **jajoiana** (Backeberg) Rausch *ex* Rowley, Cact. Succ. J. Gr. Brit. 44(4): 80, 1982. Basionym: *Lobivia jajoiana*. [First mentioned by Rausch in Lobivia 3: 179, 1976.] [RPS 33]

Lobivia chrysantha ssp. **marsoneri** (Werdermann) Rausch *ex* Rowley, Cact. Succ. J. Gr. Brit. 44(4): 80, 1982. Basionym: *Echinopsis marsoneri*. [First mentioned by Rausch in Lobivia 3: 179, 1976.] [RPS 33]

Lobivia chrysantha subvar. **fleischeriana** (Backeberg) Rausch, Lobivia 2: 112, 1976. Basionym: *Lobivia jajoiana* var. *fleischeriana*. [RPS 27]

Lobivia chrysantha subvar. **hypocyrta** (Rausch) Rausch, Lobivia 2: 112, 1976. Basionym: *Lobivia chrysantha* var. *hypocyrta*. [RPS 27]

Lobivia chrysantha subvar. **klusacekii** (Fric) Rausch, Lobivia 2: 112, 1976. Basionym: *Lobivia klusacekii*. [RPS 27]

Lobivia chrysantha subvar. **paucicostata** (Rausch) Rausch, Lobivia 2: 112, 1976. Basionym: *Lobivia glauca* var. *paucicostata*. [RPS 27]

Lobivia chrysantha subvar. **rubescens** (Backeberg) Rausch, Lobivia 2: 112, 1976. Basionym: *Lobivia rubescens*. [RPS 27]

Lobivia chrysantha subvar. **vatteri** (Krainz) Rausch, Lobivia 2: 112, 1976. Basionym: *Lobivia vatteri*. [RPS 27]

Lobivia chrysantha var. **caspalasensis** (Rausch) Rowley, Cact. Succ. J. Gr. Brit. 44(4): 80, 1982. Basionym: *Lobivia jajoiana* var. *caspalasensis*. [First mentioned by Rausch in Lobivia 3: 179, 1976.] [RPS 33]

Lobivia chrysantha var. **glauca** (Rausch) Rowley, Cact. Succ. J. Gr. Brit. 44(4): 80, 1982. Basionym: *Lobivia glauca*. [RPS 33]

Lobivia chrysantha var. **hypocyrta** Rausch, Kakt. and. Sukk. 23(11): 292-293, ill., 1972. Typus: *Rausch* 161 (ZSS). [Type cited for W but placed in ZSS.] [Argentina: Salta] [RPS 23]

Lobivia chrysantha var. **jajoiana** (Backeberg) Rausch, Lobivia 2: 112, 1976. Basionym: *Lobivia jajoiana*. [RPS 27]

Lobivia chrysantha var. **janseniana** (Backeberg) Backeberg, Cact. Succ. J. (US) 23(2): 51, 1951. Basionym: *Lobivia janseniana*. [RPS 3]

Lobivia chrysantha var. **klusacekii** (Fric) Rausch *ex* Rowley, Cact. Succ. J. Gr. Brit. 44(49: 80, 1982. Basionym: *Lobivia klusacekii*. [First mentioned by Rausch in Lobivia 3: 179, 1976.] [RPS 33]

Lobivia chrysantha var. **leucacantha** (Backeberg) Backeberg, Cact. Succ. J. (US) 23(2): 51, 1951. Basionym: *Lobivia janseniana* var. *leucacantha*. [RPS 3]

Lobivia chrysantha var. **marsoneri** (Werdermann) Rausch, Lobivia 2: 112, 1976. Basionym: *Echinopsis marsoneri*. [RPS 27]

Lobivia chrysantha var. *muhriae* Rowley, Cact. Succ. J. Gr. Brit. 44(4): 80, 1982. *Nom. inval.*, based on *Lobivia muhriae, nom. inval.* (Art. 9.5). [RPS 33]

Lobivia chrysantha var. *paucicostata* (Rausch) Rausch, Lobivia 3: 179, 1976. *Nom. inval.* (Art. 33.2), based on *Lobivia glauca* var. *paucicostata*. [Erroneously included as valid in RPS 28.] [RPS 28]

Lobivia chrysantha var. **rubescens** (Backeberg) Rausch *ex* Rowley, Cact. Succ. J. Gr. Brit. 44(4): 80, 1982. Basionym: *Lobivia rubescens*. [First mentioned by Rausch in Lobivia 3: 179, 1976.] [RPS 33]

Lobivia chrysantha var. **vatteri** (Krainz) Rausch *ex* Rowley, Cact. Succ. J. Gr. Brit. 44(4): 80, 1982. Basionym: *Lobivia vatteri*. [First mentioned by Rausch in Lobivia 3: 179, 1976.] [RPS 33]

Lobivia chrysochete var. *markusii* (Rausch) Rausch, Lobivia 2: 96, 1976. *Nom. inval.* (Art. 33), based on *Lobivia markusii*. [Given as invalid combination, because the basionym is cited where it was first published invalidly, instead of where it was later validated.] [RPS 27]

Lobivia chrysochete var. **minutiflora** Rausch, Kakt. and. Sukk. 28(4): 74, 1977. Typus: *Rausch* 512 (ZSS). [Sphalm. 'minutiflorara'.] [Argentina: Salta] [RPS 28]

Lobivia chrysochete var. **subtilis** Rausch, Succulenta 59(3): 53-54, ill., 1980. Typus: *Rausch* 691 (ZSS). [Argentina: Salta] [RPS 31]

Lobivia chrysochete var. **tenuispina** (Ritter) Rausch, Lobivia 2: 96, 1976. Basionym: *Lobivia tenuispina*. [RPS 27]

Lobivia cinnabarina ssp. *acanthoplegma* (Backeberg) Rausch, Lobivia 3: 178, 1976. *Nom. inval.*, based on *Pseudolobivia acanthoplegma, nom. inval.* (Art. 37.1). [RPS 28]

Lobivia cinnabarina ssp. **prestoana** (Cardenas) Rausch *ex* Rowley, Cact. Succ. J. Gr. Brit. 44(4): 80, 1982. Basionym: *Lobivia prestoana*. [First invalidly published (Art. 33.2) by Rausch in Lobivia 3: 178, 1976.] [RPS 33]

Lobivia cinnabarina ssp. **taratensis** (Cardenas) Rowley, Cact. Succ. J. Gr. Brit. 44(4): 80, 1982. Basionym: *Lobivia taratensis*. [RPS 33]

Lobivia cinnabarina subvar. **draxleriana** (Rausch) Rausch, Lobivia 1: 64, 1975. Basionym: *Lobivia draxleriana*. [Erroneously included as invalid in RPS 26.] [RPS 26]

Lobivia cinnabarina subvar. *neocinnabarina* (Backeberg) Rausch, Lobivia 1: 64, 1975. *Nom. inval.*, based on *Lobivia neocinnabarina, nom. inval.* (Art. 9.5). [RPS 26]

Lobivia cinnabarina subvar. **oligotricha** (Cardenas) Rausch, Lobivia 1: 64, 1975. Basionym: *Lobivia oligotricha*. [Erroneously included as invalid in RPS 26.] [RPS 26]

Lobivia cinnabarina subvar. **patula** (Rausch) Rausch, Lobivia 1: 64, 1975. Basionym: *Lobivia patula*. [Erroneously included as invalid in RPS 26.] [RPS 26]

Lobivia cinnabarina subvar. **walterspielii** (Bödeker) Rausch, Lobivia 1: 64, 1975.

Basionym: *Lobivia walterspielii.* [RPS 26]

Lobivia cinnabarina subvar. **zudanensis** (Cardenas) Rausch, Lobivia 1: 64, 1975. Basionym: *Lobivia zudanensis.* [RPS 26]

Lobivia cinnabarina var. *acanthoplegma* (Backeberg) Rausch, Lobivia 1: 64, 1975. *Nom. inval.*, based on *Pseudolobivia acanthoplegma, nom. inval.* (Art. 37.1). [RPS 26]

Lobivia cinnabarina var. **draxleriana** (Rausch) Rausch *ex* Rowley, Cact. Succ. J. Gr. Brit. 44(4): 80, 1982. Basionym: *Lobivia draxleriana.* [Republished by Rausch in Lobivia 85, 28, 1987 (reported as correct citation in RPS 38). First published invalidly (Art. 33.2) by Rausch in Lobivia 3: 178, 1976 (RPS 28).] [RPS 33]

Lobivia cinnabarina var. **grandiflora** Rausch, Kakt. and. Sukk. 23(11): 291-292, ill., 1972. Typus: *Rausch* 265 (ZSS). [Type cited for W but placed in ZSS.] [Bolivia] [RPS 23]

Lobivia cinnabarina var. **oligotricha** (Cardenas) Rausch *ex* Rowley, Cact. Succ. J. Gr. Brit. 44(4): 80, 1982. Basionym: *Lobivia oligotricha.* [First invalidly published (Art. 33.2) by Rausch in Lobivia 3: 178, 1976.] [RPS 33]

Lobivia cinnabarina var. *prestoana* (Cardenas) Rausch, Lobivia 1: 64, 1975. *Nom. inval.* (Art. 57), based on *Lobivia prestoana.* [RPS 26]

Lobivia cinnabarina var. *roseiflora* Rausch, Lobivia 3: 178, 1976. *Nom. inval.* (Art. 33.2). [RPS 28]

Lobivia cinnabarina var. **spinosior** (Salm-Dyck) Y. Ito, Explan. Diagr. Austroechinocactinae, 91, 1957. Basionym: *Echinocactus cinnabarinus* var. *spinosior.* [RPS 8]

Lobivia cinnabarina var. **taratensis** (Cardenas) Rowley, Cact. Succ. J. Gr. Brit. 44(4): 80, 1982. Basionym: *Lobivia taratensis.* [RPS 33]

Lobivia cinnabarina var. **walterspielii** (Bödeker) Rausch *ex* Rowley, Cact. Succ. J. Gr. Brit. 44(4): 80, 1982. Basionym: *Lobivia walterspielii.* [First invalidly published (Art. 33.2) by Rausch in Lobivia 3: 178, 1976.] [RPS 33]

Lobivia cinnabarina var. **zudanensis** (Cardenas) Rausch *ex* Rowley, Cact. Succ. J. Gr. Brit. 44(4): 80, 1982. Basionym: *Lobivia zudanensis.* [First invalidly published (Art. 33.2) by Rausch in Lobivia 3: 178, 1976.] [RPS 33]

Lobivia cintiensis Cardenas, Cactus (Paris) 14(65): 179, 1959. [RPS 10]

Lobivia cintiensis var. **elongata** Ritter, Taxon 12: 124, 1963. [RPS 14]

Lobivia cornuta Rausch, Succulenta 51(3): 41-43, ill., 1972. Typus: *Rausch* 500 (ZSS). [Type cited for W but placed in ZSS.] [Bolivia: Tarija] [RPS 23]

Lobivia cotagaitensis Rausch, Lobivia 2: 78, ill., 1976. *Nom. inval.* (Art. 36.1, 37.1). [RPS -]

Lobivia crassicaulis Kiesling, Darwiniana 21: 324-325, 1978. [= *Helianthocereus crassicaulis* Backeberg (*nom. inval.*, Art. 9.5).] [RPS 32]

Lobivia cruciaureispina Knize, Biota 7(57): 253-254, 1968. [Publication date given as 1969 in RPS 25.] [RPS 20]

Lobivia culpinensis Ritter, Succulenta 45: 83-84, 1966. [RPS 17]

Lobivia cylindracea Backeberg, Descr. Cact. Nov. [1:] 29, 1957. [Dated 1956, published 1957.] [RPS 7]

Lobivia densispina fa. **albiflora** (Wessner) Buining, Sukkulentenkunde 7/8: 95, 1963. Basionym: *Lobivia pectinifera* var. *albiflora.* [RPS 14]

Lobivia densispina fa. **albolanata** (Buining) Buining, Sukkulentenkunde 7/8: 96, 1963. Basionym: *Hymenorebutia albolanata.* [RPS 14]

Lobivia densispina fa. **aurantiaca** (Wessner) Buining, Sukkulentenkunde 7/8: 96, 1963. Basionym: *Lobivia pectinifera* var. *aurantiaca.* [RPS 14]

Lobivia densispina fa. **blossfeldii** (Wessner) Buining, Sukkulentenkunde 7/8: 95, 1963. Basionym: *Lobivia densispina* var. *blossfeldii.* [RPS 14]

Lobivia densispina fa. **chlorogona** (Wessner) Buining, Sukkulentenkunde 7/8: 95, 1963. Basionym: *Lobivia chlorogona.* [RPS 14]

Lobivia densispina fa. **cinnabarina** (Wessner) Buining, Sukkulentenkunde 7/8: 96, 1963. Basionym: *Lobivia pectinifera* var. *cinnabarina.* [RPS 14]

Lobivia densispina fa. **citriflora** (Wessner) Buining, Sukkulentenkunde 7/8: 96, 1963. Basionym: *Lobivia pectinifera* var. *citriflora.* [RPS 14]

Lobivia densispina fa. **citriniflora** (Backeberg) Buining, Sukkulentenkunde 7/8: 95, 1963. Basionym: *Lobivia rebutioides* var. *citriniflora.* [RPS 14]

Lobivia densispina fa. **cupreoviridis** (Wessner) Buining, Sukkulentenkunde 7/8: 95, 1963. Basionym: *Lobivia chlorogona* var. *cupreoviridis.* [RPS 14]

Lobivia densispina fa. **eburnea** (Wessner) Buining, Sukkulentenkunde 7/8: 96, 1963. Basionym: *Lobivia pectinifera* var. *eburnea.* [RPS 14]

Lobivia densispina fa. **haematantha** (Backeberg) Buining, Sukkulentenkunde 7/8: 96, 1963. Basionym: *Lobivia pectinifera* var. *haematantha.* [RPS 14]

Lobivia densispina fa. **kraussiana** (Backeberg) Buining, Sukkulentenkunde 7/8: 95, 1963. Basionym: *Lobivia rebutioides* var. *kraussiana.* [RPS 14]

Lobivia densispina fa. **leucomalla** (Wessner) Buining, Sukkulentenkunde 7/8: 95, 1963. Basionym: *Lobivia leucomalla.* [RPS 14]

Lobivia densispina fa. **pectinifera** (Wessner) Buining, Sukkulentenkunde 7/8: 95, 1963. Basionym: *Lobivia pectinifera.* [RPS 14]

Lobivia densispina fa. **purpureostoma** (Wessner) Buining, Sukkulentenkunde 7/8: 95, 1963. Basionym: *Lobivia chlorogona* var. *purpureostoma.* [RPS 14]

Lobivia densispina fa. **rubroviridis** (Wessner) Buining, Sukkulentenkunde 7/8: 95, 1963. Basionym: *Lobivia chlorogona* var. *rubroviridis.* [RPS 14]

Lobivia densispina fa. **sanguinea** (Wessner) Buining, Sukkulentenkunde 7/8: 95, 1963. Basionym: *Lobivia densispina* var. *sanguinea*. [RPS 14]

Lobivia densispina fa. **setosa** (Backeberg) Buining, Sukkulentenkunde 7/8: 95, 1963. Basionym: *Lobivia famatimensis* var. *setosa*. [RPS 14]

Lobivia densispina fa. **subcarnea** (Wessner) Buining, Sukkulentenkunde 7/8: 96, 1963. Basionym: *Lobivia pectinifera* var. *subcarnea*. [RPS 14]

Lobivia densispina fa. **sufflava** (Wessner) Buining, Sukkulentenkunde 7/8: 96, 1963. Basionym: *Lobivia pectinifera* var. *sufflava*. [RPS 14]

Lobivia densispina fa. **versicolor** (Wessner) Buining, Sukkulentenkunde 7/8: 95, 1963. Basionym: *Lobivia chlorogona* var. *versicolor*. [RPS 14]

Lobivia densispina fa. **wessneriana** (Fritzen) Buining, Sukkulentenkunde 7/8: 95, 1963. Basionym: *Lobivia wessneriana*. [RPS 14]

Lobivia densispina var. **kreuzingeri** (Fric) Buining, Sukkulentenkunde 7/8: 95, 1963. Basionym: *Hymenorebutia kreuzingeri*. [RPS 14]

Lobivia densispina var. **rebutioides** (Backeberg) Buining, Sukkulentenkunde 7/8: 95, 1963. Basionym: *Lobivia rebutioides*. [RPS 14]

Lobivia draxleriana Rausch, Succulenta 50(10): 193-194, ill., 1971. Typus: *Rausch* 279 (W, ZSS). [Bolivia] [RPS 22]

Lobivia echinata Rausch, Kakt. and. Sukk. 24(8): 169-170, ill., 1973. Typus: *Rausch* 416 (ZSS). [Type cited for W but placed in ZSS.] [Peru] [RPS 24]

Lobivia einsteinii (Fric) Rausch, Lobivia 85, 28, ills. (p. 30-31), 1987. Basionym: *Rebutia einsteinii*. [RPS 38]

Lobivia einsteinii var. **atrospinosa** Rausch, Lobivia 85, 32, 138, 1987. Typus: *Rausch* 163 (ZSS). [Argentina: Salta] [RPS 38]

Lobivia einsteinii var. **aureiflora** (Backeberg) Rausch, Lobivia 85, 33, ills. (p. 34), 1987. Basionym: *Rebutia aureiflora*. [RPS 38]

Lobivia einsteinii var. **elegans** (Backeberg) Rausch, Lobivia 85, 33, 1987. Basionym: *Mediolobivia elegans*. [RPS 38]

Lobivia einsteinii var. **gonjianii** (Kiesling) Rausch, Lobivia 85, 32, ill. (p. 31), 1987. Basionym: *Rebutia gonjianii*. [Combination erroneously attributed to Donald, who published the combination *Rebutia einsteinii* var. *gonjianii* in 1974.] [RPS 38]

Lobivia elongata Backeberg, Descr. Cact. Nov. [1:] 29, 1957. [Dated 1956, published 1957.] [RPS 7]

Lobivia euanthema var. **tilcarensis** Rausch, Lobivia 85, 36, 138-139, ill. (p. 35), 1987. Typus: *Rausch* 700 (ZSS). [Argentina: Jujuy] [RPS 38]

Lobivia famatimensis var. **albiflora** (Wessner) Backeberg, Cact. Succ. J. (US) 23(2): 51, 1951. Basionym: *Lobivia pectinifera* var. *albiflora*. [RPS 3]

Lobivia famatimensis var. **aurantiaca** (Backeberg) Backeberg, Cact. Succ. J. (US) 23(2): 51, 1951. Basionym: *Lobivia pectinifera* var. *aurantiaca*. [RPS 3]

Lobivia famatimensis var. **cinnabarina** (Backeberg) Backeberg, Cact. Succ. J. (US) 23(2): 51, 1951. Basionym: *Lobivia pectinifera* var. *cinnabarina*. [RPS 3]

Lobivia famatimensis var. **densispina** (Werdermann) Backeberg, Die Cact. 3: 1453, 1959. Basionym: *Echinopsis densispina*. [RPS 10]

Lobivia famatimensis var. **haematantha** (Backeberg) Backeberg, Cact. Succ. J. (US) 23(2): 51, 1951. Basionym: *Lobivia pectinifera* var. *haematantha*. [RPS 3]

Lobivia famatimensis var. **jachalensis** Rausch, Kakt. and. Sukk. 28(12): 289, 1977. Typus: *Rausch* 557a (ZSS). [Argentina: San Juan] [RPS 28]

Lobivia famatimensis var. **leucomalla** (Wessner) Backeberg, Die Cact. 3: 1454, 1959. Basionym: *Lobivia leucomalla*. [RPS 10]

Lobivia famatimensis var. **sanjuanensis** Rausch, Kakt. and. Sukk. 28(4): 75, 1977. Typus: *Rausch* 557 (ZSS). [Argentina: San Juan] [RPS 28]

Lobivia famatimensis var. **setosa** Backeberg, Descr. Cact. Nov. [1:] 29, 1957. [Dated 1956, published 1957.] [RPS 7]

Lobivia ferox var. **camargensis** Ritter, Kakt. Südamer. 2: 571, 1980. [RPS 31]

Lobivia ferox var. **longispina** (Britton & Rose) Rausch, Lobivia 2: 72, 1976. Basionym: *Lobivia longispina*. [RPS 27]

Lobivia ferox var. *nigra* (Backeberg) Friedrich, Kakt. and. Sukk. 25: 83, 1974. *Nom. inval.*, based on *Echinopsis nigra, nom. inval.* (Art. 36.1). [Erroneously included as valid in RPS 25] [RPS 25]

Lobivia ferox var. **potosina** (Werdermann) Rausch, Lobivia 2: 72, 1976. Basionym: *Echinopsis potosina*. [RPS 27]

Lobivia formosa ssp. **bruchii** (Britton & Rose) Rausch, Lobivia 3: 164, 1976. Basionym: *Lobivia bruchii*. [RPS 27]

Lobivia formosa ssp. **grandis** (Britton & Rose) Rausch, Lobivia 3: 164, 1976. Basionym: *Lobivia grandis*. [RPS 27]

Lobivia formosa ssp. **tarijensis** (Vaupel) Rausch, Lobivia 3: 164, 1976. Basionym: *Cereus tarijensis*. [RPS 27]

Lobivia formosa var. **bertramiana** (Backeberg) Rausch, Lobivia 85, 45, ill. (p. 51), 1987. Basionym: *Trichocereus bertramianus*. [First published by Rausch in Lobivia 3: 164, 1976, but treated as *nom. illeg.* (Art. 60) by RPS 27.] [RPS 38]

Lobivia formosa var. **grandis** (Britton & Rose) Rausch, Lobivia 85, 49, 1987. Basionym: *Lobivia grandis*. [RPS 38]

Lobivia formosa var. **hyalacantha** (Spegazzini) Rausch, Lobivia 85, 45, 1987. Basionym: *Lobivia hyalacantha*. [RPS 38]

Lobivia formosa var. **kieslingii** (Rausch) Rausch, Lobivia 85, 48, 1987. Basionym: *Lobivia kieslingii*. [Name first used (*nom. nud.*) in Lobivia 3: 164, 1976.] [RPS 38]

Lobivia formosa var. **nivalis** (Fric) Rausch, Lobivia 85, 45, 1987. Basionym: *Lobivia* *bruchii* var. *nivalis*. [Probably based on a *nomen nudum* (fide RPS 27); validity not assessed. First published by Rausch in Lobivia 3: 164, 1976 (*nom. inval.* ?, no comment in RPS 27).] [RPS 38]

Lobivia formosa var. **orurensis** (Cardenas) Rowley, Cact. Succ. J. Gr. Brit. 44(4): 80, 1982. Basionym: *Trichocereus orurensis*. [RPS 33]

Lobivia formosa var. *pinchasensis* Rausch, Lobivia 3: 164, 1976. *Nom. inval.* (Art. 36.1, 37.1). [RPS -]

Lobivia formosa var. **poco** (Backeberg) Rowley, Cact. Succ. J. Gr. Brit. 44(4): 80, 1982. Basionym: *Trichocereus poco*. [RPS 33]

Lobivia formosa var. **randallii** (Cardenas) Rausch, Lobivia 85, 45, 1987. Basionym: *Trichocereus randallii*. [RPS 38]

Lobivia formosa var. **rosarioana** (Rausch) Rausch, Lobivia 85, 48, ill. (p. 49), 1987. Basionym: *Lobivia rosarioana*. [RPS 38]

Lobivia formosa var. **tarijensis** (Vaupel) Rausch, Lobivia 85, 44, ill. (p. 50), 1987. Basionym: *Cereus tarijensis*. [RPS 38]

Lobivia formosa var. **totorensis** (Cardenas) Rausch, Lobivia 3: 164, 1976. Basionym: *Trichocereus herzogianus* var. *totorensis*. [RPS 27]

Lobivia formosa var. *uebelmanniana* (Lembcke & Backeberg) Rausch, Lobivia 85, 48, 1987. *Nom. inval.*, based on *Soehrensia uebelmanniana*, *nom. inval.* (Art. 37.1). [RPS 38]

Lobivia fricii Rausch, Kakt. and. Sukk. 24: 220-222, 1973. [RPS 24]

Lobivia fungiflora Braun, Stachelpost 9: 101-102, 120, 1973. *Nom. inval.* (Art. 34.1, 36.1). [Described as provisional name.] [RPS 24]

Lobivia glauca Rausch, Succulenta 50: 168, 1971. Typus: *Rausch* 218 (W, ZSS). [Argentina: Jujuy] [RPS 22]

Lobivia glauca var. **paucicostata** Rausch, Succulenta 50: 169, 1971. Typus: *Rausch* 217 (W, ZSS). [Argentina: Jujuy] [RPS 22]

Lobivia glaucescens Ritter, Kakt. Südamer. 4: 1331-1332, 1981. Typus: *Ritter* 1460 (U). [Peru: Moqugua] [RPS 32]

Lobivia grandiflora var. **crassicaulis** (Kiesling) Rausch, Lobivia 85, 53, ill. (p. 54), 1987. Basionym: *Lobivia crassicaulis*. [Erroneously based on *Helianthocereus crassicaulis* Backeberg (*nom. inval.*, Art. 9.5), later validly described by Kiesling, and treated as bibliographic error.] [RPS 38]

Lobivia grandiflora var. **herzogii** Rausch, Lobivia 85, 159, ill., 1987. Typus: *Rausch* 795 (ZSS). [Argentina: Tucuman] [RPS 38]

Lobivia grandiflora var. **lobivioides** (Gräser & Ritter) Rausch, Lobivia 85, 53, ill. (p. 55), 1987. Basionym: *Trichocereus lobivioides*. [RPS 38]

Lobivia grandiflora var. **longispina** Rausch, Lobivia 85, 53, 139, ill. (p. 138), 1987. Typus: *Rausch* 151a (ZSS). [Argentina: Catamarca] [RPS 38]

Lobivia grandiflora var. **pumila** Rausch, Lobivia 85, 53, 139-140, ill. (p. 138), 1987. Typus: *Rausch* 758 (ZSS). [Argentina: Catamarca] [RPS 38]

Lobivia grandis var. *flaviflora* Rausch, Lobivia 3: 185, 1976. *Nom. inval.* (Art. 36.1, 37.1). [RPS -]

Lobivia graulichii (Fric) Backeberg, Die Cact. 3: 1469, 1959. Basionym: *Cinnabarinea graulichii*. [The name may be invalid, as the basionym is most probably a nude name (cf. Backeberg, l.c.).] [RPS 10]

Lobivia haageana fa. **albihepatica** (Backeberg) Krainz, Die Kakt. C Vc, 1960. Basionym: *Lobivia haageana* var. *albihepatica*. [RPS 11]

Lobivia haageana fa. **bicolor** (Backeberg) Krainz, Die Kakteen C Vc, 1960. Basionym: *Lobivia haageana* var. *bicolor*. [RPS 11]

Lobivia haageana fa. **chrysantha** (Backeberg) Krainz, Die Kakteen C Vc, 1960. Basionym: *Lobivia haageana* var. *chrysantha*. [RPS 11]

Lobivia haageana fa. **cinnabarina** (Backeberg) Krainz, Die Kakteen C Vc, 1960. Basionym: *Lobivia haageana* var. *cinnabarina*. [RPS 11]

Lobivia haageana fa. **croceantha** (Backeberg) Krainz, Die Kakteen C Vc, 1960. Basionym: *Lobivia haageana* var. *croceantha*. [RPS 11]

Lobivia haageana fa. **durispina** (Backeberg) Krainz, Die Kakteen C Vc, 1960. Basionym: *Lobivia haageana* var. *durispina*. [RPS 11]

Lobivia haageana fa. **grandiflora-stellata** (Backeberg) Krainz, Die Kakteen C Vc, 1960. Basionym: *Lobivia haageana* var. *grandiflora-stellata*. [RPS 11]

Lobivia haageana fa. **leucoerythrantha** (Backeberg) Krainz, Die Kakteen C Vc, 1960. Basionym: *Lobivia haageana* var. *leucoerythrantha*. [RPS 11]

Lobivia haageana var. **albihepatica** Backeberg, Cactus (Paris) No. 41: 84, 1954. [Repeated in Descr. Cact. Nov. [1:] 29, 1957 (dated 1956) (cf. RPS 7).] [RPS 5]

Lobivia haageana var. **bicolor** Backeberg, Descr. Cact. Nov. [1:] 29, 1957. [Dated 1956, published 1957.] [RPS 7]

Lobivia haageana var. **chrysantha** Backeberg, Cactus (Paris) No. 41: 84, 1954. [Repeated in Descr. Cact. Nov. [1:] 29, 1957 (dated 1956) (cf. RPS 7).] [RPS 5]

Lobivia haageana var. **cinnabarina** Backeberg, Cactus (Paris) No. 41: 84, 1954. [Repeated in Descr. Cact. Nov. [1:] 29, 1957 (dated 1956) (cf. RPS 7).] [RPS 5]

Lobivia haageana var. **croceantha** Backeberg, Descr. Cact. Nov. [1:] 29, 1957. [Dated 1956, published 1957.] [RPS 7]

Lobivia haageana var. **durispina** Backeberg, Descr. Cact. Nov. [1:] 29, 1957. [Dated 1956, published 1957.] [RPS 7]

Lobivia haageana var. **grandiflora-stellata** Backeberg, Descr. Cact. Nov. [1:] 29, 1957. [Dated 1956, published 1957.] [RPS 7]

Lobivia haageana var. **leucoerythrantha** Backeberg, Descr. Cact. Nov. [1:] 29, 1957. [Dated 1956, published 1957.] [RPS 7]

Lobivia haagei var. **canacruzensis** (Rausch) Rausch, Lobivia 85, 56, 1987. Basionym: *Rebutia canacruzensis*. [Basionym erroneously cited as *Digitorebutia canacruzensis*, treated as bibliographical error (Art. 33.2, Ex. 4).] [RPS 38]

Lobivia haagei var. **crassa** Rausch, Lobivia 85, 56, 140, ill. (p. 139), 1987. Typus: *Rausch* 501 (ZSS). [Bolivia: Tarija] [RPS 38]

Lobivia haagei var. **elegantula** Rausch, Lobivia 85, 57, 140, 1987. Typus: *Rausch* 502 (ZSS). [Argentina: Jujuy] [RPS 38]

Lobivia haagei var. *eos* (Rausch) Rausch, Lobivia 85, 57, 1987. *Nom. inval*. (Art. 33.2), based on *Rebutia eos*. [Basionym erroneously cited as *Digitorebutia eos*, treated as bibliographical error (Art. 33.2, Ex. 4). Incorrect bibliographic citation invalidates the combination, however.] [RPS 38]

Lobivia haagei var. **mudanensis** (Rausch) Rausch, Lobivia 85, 57, ill. (p. 59), 1987. Basionym: *Rebutia mudanensis*. [Basionym erroneously cited as *Digitorebutia mudanensis*, treated as bibliographical error (Art. 33.2, Ex. 4).] [RPS 38]

Lobivia haagei var. **nazarenoensis** (Rausch) Rausch, Lobivia 85, 57, ill. (p. 59), 1987. Basionym: *Digitorebutia nazarenoensis*. [RPS 38]

Lobivia haagei var. **orurensis** (Backeberg) Rausch, Lobivia 85, 57, ill. (p. 58), 1987. Basionym: *Lobivia orurensis*. [RPS 38]

Lobivia haagei var. **pallida** (Rausch) Rausch, Lobivia 85, 57, 1987. Basionym: *Rebutia pallida*. [Basionym erroneously cited as *Digitorebutia pallida*, treated as bibliographical error (Art. 33.2, Ex. 4).] [RPS 38]

Lobivia haagei var. **pelzliana** Rausch, Lobivia 85, 57, 140, ill. (p. 139), 1987. Typus: *Rausch* 333a (ZSS). [Argentina: Jujuy] [RPS 38]

Lobivia haagei var. **violascens** (Ritter) Rausch, Lobivia 85, 57, 1987. Basionym: *Rebutia violascens*. [RPS 38]

Lobivia haematantha fa. **sublimiflora** (Backeberg) Ullmann, Kaktusy 26(1): 19, 1990. Basionym: *Lobivia sublimiflora*. [RPS 41]

Lobivia haematantha ssp. **chorrillosensis** (Rausch) Rausch *ex* Rowley, Cact. Succ. J. Gr. Brit. 44(4): 81, 1982. Basionym: *Lobivia chorrillosensis*. [First invalidly published by Rausch in Lobivia 3: 179, 1976.] [RPS 33]

Lobivia haematantha ssp. **densispina** (Werdermann) Rausch *ex* Rowley, Cact. Succ. J. Gr. Brit. 44(4): 81, 1982. Basionym: *Lobivia densispina*. [First invalidly published by Rausch in Lobivia 3: 179, 1976.] [RPS 33]

Lobivia haematantha ssp. **kuehnrichii** (Fric) Rausch *ex* Rowley, Cact. Succ. J. Gr. Brit. 44(4): 81, 1982. Basionym: *Lobivia kuehnrichii*. [First invalidly published by Rausch in Lobivia 3: 179, 1976.] [RPS 33]

Lobivia haematantha subvar. **amblayensis** (Rausch) Rausch, Lobivia 2: 124, 1976. Basionym: *Lobivia amblayensis*. [RPS 27]

Lobivia haematantha subvar. **chorrillosensis** (Rausch) Rausch, Lobivia 2: 124, 1976. Basionym: *Lobivia chorrillosensis*. [RPS 27]

Lobivia haematantha subvar. **elongata** (Backeberg) Rausch, Lobivia 2: 124, 1976. Basionym: *Lobivia elongata*. [RPS 27]

Lobivia haematantha subvar. **fechseri** (Rausch) Rausch, Lobivia 2: 124, 1976. Basionym: *Lobivia hualfinensis* var. *fechseri*. [RPS 27]

Lobivia haematantha subvar. **hualfinensis** (Rausch) Rausch, Lobivia 2: 124, 1976. Basionym: *Lobivia hualfinensis*. [RPS 27]

Lobivia haematantha subvar. **pectinifera** (Wessner) Rausch, Lobivia 2: 125, 1976. Basionym: *Lobivia pectinifera*. [RPS 27]

Lobivia haematantha subvar. **rebutioides** (Backeberg) Rausch, Lobivia 2: 125, 1976. Basionym: *Lobivia rebutioides*. [RPS 27]

Lobivia haematantha subvar. **sublimiflora** (Backeberg) Rausch, Lobivia 2: 125, 1976. Basionym: *Lobivia sublimiflora*. [RPS 27]

Lobivia haematantha var. **amblayensis** (Rausch) Rausch *ex* Rowley, Cact. Succ. J. Gr. Brit. 44(4): 81, 1982. Basionym: *Lobivia amblayensis*. [First invalidly published by Rausch in Lobivia 3: 179, 1976.] [RPS 33]

Lobivia haematantha var. **chorrillosensis** (Rausch) Rausch, Lobivia 85, 61, 1987. Basionym: *Lobivia chorrillosensis*. [RPS 38]

Lobivia haematantha var. **densispina** (Werdermann) Rausch, Lobivia 2: 124, 1976. Basionym: *Lobivia densispina*. [Given as illegitimate in RPS 27: 6, validity not assessed.] [RPS 27]

Lobivia haematantha var. **drijveriana** (Backeberg) Rowley, Cact. Succ. J. Gr. Brit. 44(4): 81, 1982. Basionym: *Lobivia drijveriana*. [RPS 33]

Lobivia haematantha var. **elongata** (Backeberg) Rausch *ex* Rowley, Cact. Succ. J. Gr. Brit. 44(4): 80, 1982. Basionym: *Lobivia elongata*. [First invalidly published by Rausch in Lobivia 3: 179, 1976.] [RPS 33]

Lobivia haematantha var. **fechseri** (Rausch) Rausch *ex* Rowley, Cact. Succ. J. Gr. Brit. 44(4): 80, 1982. Basionym: *Lobivia hualfinensis* var. *fechseri*. [First invalidly published by Rausch in Lobivia 3: 179, 1976.] [RPS 33]

Lobivia haematantha var. **hualfinensis** (Rausch) Rausch *ex* Rowley, Cact. Succ. J. Gr. Brit. 44(4): 81, 1982. Basionym: *Lobivia hualfinensis*. [First invalidly published by Rausch in Lobivia 3: 179, 1976.] [RPS 33]

Lobivia haematantha var. **jasimanensis** Rausch, Lobivia 85, 60, 140, 1987. Typus: *Rausch* 792 (ZSS). [Argentina: Salta] [RPS 38]

Lobivia haematantha var. **kuehnrichii** (Fric) Rausch, Lobivia 2: 124, 1976. Basionym: *Lobivia kuehnrichii*. [Given as invalid in RPS 27: 6, validity not assessed.] [RPS 27]

Lobivia haematantha var. **pectinifera** (Wessner) Rausch *ex* Rowley, Cact. Succ. J. Gr. Brit. 44(4): 81, 1982. Basionym: *Lobivia pectinifera*. [First invalidly published by Rausch in Lobivia 3: 179, 1976.] [RPS 33]

Lobivia haematantha var. **rebutioides** (Backeberg) Rausch *ex* Rowley, Cact. Succ. J. Gr. Brit. 44(4): 81, 1982. Basionym: *Lobivia rebutioides*. [Republished by Rausch in Lobivia 85, 61, 1987 (reported as correct citation in RPS 38). First invalidly published (Art. 33.2) by Rausch in Lobivia 3:

179, 1976 (RPS 28).] [RPS 33]

Lobivia haematantha var. **sublimiflora** (Backeberg) Rausch *ex* Rowley, Cact. Succ. J. Gr. Brit. 44(4): 81, 1982. Basionym: *Lobivia sublimiflora*. [First invalidly published by Rausch in Lobivia 3: 179, 1976.] [RPS 33]

Lobivia haematantha var. **viridis** Rausch, Lobivia 85, 60, 140-141, ill. (p. 139), 1987. Typus: *Rausch* 709a (ZSS). [Argentina: Salta] [RPS 38]

Lobivia hermanniana var. *breviflorior* Backeberg, Kakt.-Lex., 208, 1966. *Nom. inval.* (Art. 37.1). [Erroneously included as valid in RPS 17.] [RPS 17]

Lobivia hertrichiana fa. **binghamiana** (Backeberg) Ullmann, Kaktusy 26(1): 19, 1990. Basionym: *Lobivia binghamiana*. [RPS 41]

Lobivia hertrichiana fa. **incaica** (Backeberg) Ullmann, Kaktusy 26(1): 19, 1990. Basionym: *Lobivia incaica*. [RPS 41]

Lobivia hertrichiana fa. **minuta** (Ritter) Ullmann, Kaktusy 26(1): 19, 1990. Basionym: *Lobivia minuta*. [RPS 41]

Lobivia hertrichiana var. **laui** (Donald) Rausch, Lobivia 85, 65, 1987. Basionym: *Lobivia laui*. [sphalm. 'lauii'.] [RPS 38]

Lobivia hertrichiana var. **simplex** (Rausch) Rausch, Lobivia 85, 65, ill. (p. 66), 1987. Basionym: *Lobivia simplex*. [RPS 38]

Lobivia higginsiana var. **carnea** Y. Ito, Explan. Diagr. Austroechinocacti, 84, 286, 1957. [RPS 8]

Lobivia hoffmanniana Backeberg, Cact. 3: 1434, 1959. *Nom. inval.* (Art. 9.5). [Erroneously included as valid in RPS 10. Validated as *Weingartia hoffmanniana* F. Brandt in 1984.] [RPS 10]

Lobivia horrida Ritter, Taxon 12: 124, 1963. [RPS 14]

Lobivia hualfinensis Rausch, Kakt. and. Sukk. 21: 45, 1970. Typus: *Rausch* 146 (W). [First invalidly published (Art. 37.1) in l.c. 19: 67, 1968.] [Argentina: Catamarca] [RPS 21]

Lobivia hualfinensis var. **fechseri** Rausch, Succulenta 51(7): 123-124, ills., 1972. Typus: *Rausch* 230 (W, ZSS). [Argentina: Catamarca] [RPS 23]

Lobivia huascha ssp. **narvaecensis** (Cardenas) Rausch, Lobivia 3: 172, 1976. Basionym: *Trichocereus narvaecensis*. [RPS 27]

Lobivia huascha var. **andalgalensis** (F. A. C. Weber) Rausch, Lobivia 3: 172, 1976. Basionym: *Cereus andalgalensis*. [Given as invalid in RPS 27: 6, validity not assessed.] [RPS 27]

Lobivia huascha var. **calliantha** (Ritter) Rausch, Lobivia 85, 69, 1987. Basionym: *Trichocereus callianthus*. [RPS 38]

Lobivia huascha var. *crassicaulis* (Backeberg) Rausch, Lobivia 3: 172, 1976. *Nom. inval.*, based on *Helianthocereus crassicaulis, nom. inval.* (Art. 9.5). [Erroneously included as valid in RPS 27: 6.] [RPS 27]

Lobivia huascha var. **grandiflora** (Britton & Rose) Rausch, Lobivia 3: 172, 1976. Basionym: *Lobivia grandiflora*. [RPS 27]

Lobivia huascha var. **robusta** Rausch, Lobivia 85, 72, 141, 1987. Typus: *Rausch* 229 (ZSS). [Argentina: Catamarca] [RPS 38]

Lobivia huascha var. **rubriflora** (F. A. C. Weber) Rowley, Cact. Succ. J. Gr. Brit. 44(4): 81, 1982. Basionym: *Cereus huascha* var. *rubriflorus*. [Republished by Rausch in Lobivia 85, 72, 1987 (reported as correct citation in RPS 38).] [RPS 33]

Lobivia huascha var. **walteri** (Kiesling) Rausch, Lobivia 85, 69, ills. (p. 67), 1987. Basionym: *Lobivia walteri*. [RPS 38]

Lobivia huilcanota Rauh & Backeberg, in Backeberg, Descr. Cact. Nov. [1:] 28, 1957. Typus: *Rauh* K60 (1954) (ZSS). [Dated 1956, published 1957.] [Peru] [RPS 7]

Lobivia hystrichacantha Y. Ito, The Full Bloom of Cactus Flowers, 45, 47, ill., 1962. *Nom. inval.* (Art. 37.1). [Included as valid in RPS 13.] [RPS 13]

Lobivia hystrix Ritter, Succulenta 1966: 84, 1966. [RPS 17]

Lobivia ikedae Y. Ito, Explan. Diagr. Austroechinocacti, 107, 288, 1957. [RPS 8]

Lobivia ikedae var. **cinnabarina** Y. Ito, Explan. Diagr. Austroechinocacti, 108, 288, 1957. [RPS 8]

Lobivia ikedae var. **erythrantha** Y. Ito, Explan. Diagr. Austroechinocacti, 108, 288, 1957. [RPS 8]

Lobivia imporana var. *elongata* Köhler, Kakt. and. Sukk. 17: 196-197, 1966. *Nom. inval.* (Art. 36.1). [RPS 17]

Lobivia incuiensis Rauh & Backeberg, in Backeberg, Descr. Cact. Nov. [1:] 28, 1957. [Dated 1956, published 1957.] [RPS 7]

Lobivia intermedia Rausch, Kakt. and. Sukk. 23(10): 263-264, 1972. Typus: *Rausch* 409 (ZSS). [Type cited for W but placed in ZSS.] [Peru] [RPS 23]

Lobivia jajoiana var. **aurata** Rausch, Lobivia 85, 73, 141, 1987. Typus: *Rausch* 701 (ZSS). [Argentina: Jujuy] [RPS 38]

Lobivia jajoiana var. **caspalasensis** Rausch, Kakt. and. Sukk. 28(4): 75, 1977. Typus: *Rausch* 693 (ZSS). [Argentina: Jujuy] [RPS 28]

Lobivia jajoiana var. **elegans** Rausch, Lobivia 85, 73, 141, ill. (p. 142), 1987. Typus: *Rausch* 36 (ZSS). [Argentina: Jujuy] [RPS 38]

Lobivia jajoiana var. **nidularis** Rausch, Lobivia 85, 76, 141, ill. (p. 142), 1987. Typus: *Rausch* 702 (ZSS). [Argentina: Salta] [RPS 38]

Lobivia jajoiana var. **nigristoma** (Kreuzinger & Buining) Backeberg, Die Cact. 3: 1464, 1959. Basionym: *Lobivia nigrostoma*. [RPS 10]

Lobivia jajoiana var. *paucicostata* (Rausch) Rausch, Lobivia 85, 73, ill. (p. 75), 1987. Incorrect name (Art. 57.1, 57.3), based on *Lobivia glauca* var. *paucicostata*. [RPS 38]

Lobivia jajoiana var. **pungens** Rausch, Lobivia 85, 76, 141, ill. (p. 74), 1987. Typus: *Rausch* 516 (ZSS). [Argentina: Salta] [RPS 38]

Lobivia jajoiana var. *striatipetala* Y. Ito, The Cactaceae, 303, 1981. *Nom. inval.* (Art. 37.1). [RPS -]

Lobivia jajoiana var. **vatteri** (Krainz) Ullmann, Kaktusy 26(1): 19, 1990. Basionym: *Lobivia vatteri*. [RPS 41]
Lobivia katagirii Y. Ito, Explan. Diagr. Austroechinocacti, 87, 286, 1957. [RPS 8]
Lobivia katagirii var. **aureorubriflora** Y. Ito, Explan. Diagr. Austroechinocacti, 87, 286, 1957. [RPS 8]
Lobivia katagirii var. **chrysantha** Y. Ito, Explan. Diagr. Austroechinocacti, 87, 286, 1957. [RPS 8]
Lobivia katagirii var. **croceantha** Y. Ito, Explan. Diagr. Austroechinocacti, 87, 286, 1957. [RPS 8]
Lobivia katagirii var. **salmonea** Y. Ito, Explan. Diagr. Austroechinocacti, 87, 287, 1957. [RPS 8]
Lobivia kieslingii Rausch, Kakt. and. Sukk. 28(11): 249-250, ill., 1977. Typus: *Rausch* 573 (ZSS). [Argentina: Tucuman] [RPS 28]
Lobivia klusacekii var. **roseiflora** Subik, in Pazout, Valnicek & Subik, Kaktusy, 13, 1960. [RPS 12]
Lobivia korethroides (Werdermann) Y. Ito, Cacti, 67, 1952. Basionym: *Echinopsis korethroides*. [RPS 4]
Lobivia kupperiana var. *rubriflora* Backeberg, Die Cact. 3: 1441, 1959. *Nom. inval.* (Art. 9.5). [Erroneously included as valid in RPS 10.] [RPS 10]
Lobivia larae Cardenas, Cact. Succ. J. (US): 36: 24-25, 1964. [RPS 15]
Lobivia lateritia var. **cintiensis** (Cardenas) Ullmann, Kaktusy 26(1): 19, 1990. Basionym: *Lobivia cintiensis*. [RPS 41]
Lobivia lateritia var. **citriflora** Rausch, Succulenta 59(2): 29-30, ill., 1980. Typus: *Rausch* 748 (ZSS). [Bolivia] [RPS 31]
Lobivia lateritia var. **cotagaitensis** Rausch, Kakt. and. Sukk. 28(10): 235, 1977. Typus: *Rausch* 674 (ZSS). [Bolivia: Potosi] [RPS 28]
Lobivia lateritia var. **kupperiana** (Backeberg) Rausch, Lobivia 2: 76, 1976. Basionym: *Lobivia kupperiana*. [Given as invalid in RPS 27: 6, validity not assessed.] [RPS 27]
Lobivia lateritia var. *rubriflora* (Backeberg) Rausch, Lobivia 85, 81, 1987. *Nom. inval.*, based on *Lobivia kupperiana* var. *rubriflora*, *nom. inval.* (Art. 9.5). [RPS 38]
Lobivia laui Donald, Ashingtonia 1: 40, 1974. [sphalm. 'lauii'.] [RPS 25]
Lobivia lauramarca Rauh & Backeberg, in Backeberg, Descr. Cact. Nov. [1:] 28, 1957. [Dated 1956, published 1957.] [RPS 7]
Lobivia leptacantha Rausch, Kakt. and. Sukk. 23(8): 207-208, ill., 1972. Typus: *Rausch* 422 (W, ZSS). [Peru: Cuzco] [RPS 23]
Lobivia leucosiphus Cardenas, Cact. Succ. J. (US) 38: 142, 1966. *Nom. inval.* (Art. 34.1, 36.1). [Name erroneously used in the caption of an illustration instead of *Lobivia taratensis* var. *leucosiphus*.] [RPS 17]
Lobivia markusii Rausch, Kakt. and. Sukk. 21: 45, 1970. Typus: *Rausch* 215 (W). [First invalidly published (Art. 37.1) in l.c. 17: 121, 1966.] [Bolivia: Jujuy] [RPS 21]
Lobivia marsoneri var. **iridescens** (Backeberg) Rausch, Lobivia 85, 84, ills. (p. 83), 1987. Basionym: *Lobivia iridescens*. [RPS 38]
Lobivia matuzawae Y. Ito, Explan. Diagr. Austroechinocactinae, 109, 289, 1957. [RPS 8]
Lobivia maximiliana ssp. **caespitosa** (J. Purpus) Rausch *ex* Rowley, Cact. Succ. J. Gr. Brit. 44(49: 81, 1982. Basionym: *Echinopsis caespitosa*. [RPS 33]
Lobivia maximiliana ssp. **quiabayensis** (Rausch) Rausch *ex* Rowley, Cact. Succ. J. Gr. Brit. 44(4): 81, 1982. Basionym: *Lobivia quiabayensis*. [RPS 33]
Lobivia maximiliana ssp. **westii** (P. C. Hutchison) Rausch *ex* Rowley, Cact. Succ. J. Gr. Brit. 44(4): 81, 1982. Basionym: *Lobivia westii*. [RPS 33]
Lobivia maximiliana var. **caespitosa** (J. Purpus) Rausch, Lobivia 1: 64, 1975. Basionym: *Echinopsis caespitosa*. [RPS 26]
Lobivia maximiliana var. **charazanensis** (Cardenas) Rausch, Lobivia 1: 36, 1975. Basionym: *Lobivia charazanensis*. [RPS 26]
Lobivia maximiliana var. **corbula** (Herrera) Rausch, Lobivia 1: 36, 1975. Basionym: *Mammillaria corbula*. [RPS 26]
Lobivia maximiliana var. **durispina** Rausch, Lobivia 85, 88, 141-142, ill., 1987. Typus: *Rausch* 204 (ZSS). [Bolivia: Cochabamba] [RPS 38]
Lobivia maximiliana var. **hermanniana** (Backeberg) Rausch, Lobivia 1: 36, 1975. Basionym: *Lobivia hermanniana*. [RPS 26]
Lobivia maximiliana var. **intermedia** (Rausch) Rausch, Lobivia 1: 36, 1975. Basionym: *Lobivia intermedia*. [RPS 26]
Lobivia maximiliana var. **leptacantha** (Rausch) Rausch, Lobivia 1: 36, 1975. Basionym: *Lobivia leptacantha*. [RPS 26]
Lobivia maximiliana var. **miniatiflora** (Ritter) Rausch, Lobivia 1: 36, 1975. Basionym: *Lobivia miniatiflora*. [RPS 26]
Lobivia maximiliana var. **quiabayensis** (Rausch) Rausch, Lobivia 1: 36, 1975. Basionym: *Lobivia quiabayensis*. [RPS 26]
Lobivia maximiliana var. **sicuaniensis** (Rausch) Rausch, Lobivia 85, 89, 1987. Basionym: *Lobivia sicuaniensis*. [RPS 38]
Lobivia maximiliana var. **violacea** (Rausch) Rowley, Cact. Succ. J. Gr. Brit. 44(4): 81, 1982. Basionym: *Lobivia caespitosa* var. *violacea*. [Republished by Rausch in Lobivia 85, 88, 1987 (reported as correct citation in RPS 38).] [RPS 33]
Lobivia maximiliana var. **westii** (P. C. Hutchison) Rausch, Lobivia 1: 36, 1975. Basionym: *Lobivia westii*. [RPS 26]
Lobivia megatae Y. Ito, Explan. Diagr. Austroechinocactinae, 103, 287, 1957. [RPS 8]
Lobivia miniatiflora Ritter, Taxon 12(3): 124, 1963. Typus: *Ritter* 330 (U, ZSS). [Type cited for U l.c. 12: 28.] [Bolivia: La Paz] [RPS 14]
Lobivia miniatinigra Ritter, Kakt. Südamer. 2: 462-463, 1980. [RPS 31]
Lobivia minuta Ritter, Taxon 12(3): 124, 1963. [RPS 14]
Lobivia mirabunda Backeberg, Descr. Cact. Nov. [1:] 29, 1957. [Dated 1956, published

1957.] [RPS 7]

Lobivia mizquensis Rausch, Kakt. and. Sukk. 23(6): 151-152, ill., 1972. Typus: *Rausch* 463 (W, ZSS). [Bolivia] [RPS 23]

Lobivia muhriae Backeberg, Descr. Cact. Nov. 3: 7, 1963. *Nom. inval.* (Art. 9.5). [RPS 14]

Lobivia muhriae var. *flaviflora* Backeberg, Descr. Cact. Nov. 3: 7, 1963. *Nom. inval.* (Art. 43.1). [Erroneously included as valid in RPS 14.] [RPS 14]

Lobivia multicostata Backeberg, Descr. Cact. Nov. 3: 7, 1963. *Nom. inval.* (Art. 37.1). [RPS 14]

Lobivia napina Pazout, in Pazout, Valnicek & Subik, Kaktusy, 130-131, 1960. [RPS 12]

Lobivia nealeana var. **grandiflora** Y. Ito, Explan. Diagr. Austroechinocactinae, 103, 287, 1957. [RPS 8]

Lobivia nealeana var. **purpureiflora** Y. Ito, Explan. Diagr. Austroechinocactinae, 103, 287, 1957. [RPS 8]

Lobivia neocinnabarina Backeberg, Descr. Cact. Nov. 3: 7, 1963. *Nom. inval.* (Art. 9.5). [Erroneously included as valid in RPS 14.] [RPS 14]

Lobivia nigricans var. **albispina** Rausch, Lobivia 85, 100, 142, ill. (p. 98), 1987. Typus: *Rausch* 771 (ZSS). [Argentina: Salta] [RPS 38]

Lobivia nigricans var. **carmeniana** (Rausch) Rausch, Lobivia 85, 100, ill. (p. 98), 1987. Basionym: *Rebutia carmeniana.* [Basionym erroneously cited as *Digitorebutia carmeniana*, treated as bibliographical error (Art. 33.2, Ex. 4).] [RPS 38]

Lobivia nigricans var. **peterseimii** (Fric) Rausch, Lobivia 85, 100, 1987. Basionym: *Rebulobivia peterseimii.* [Basionym not checked for validity.] [RPS 38]

Lobivia nigrispina var. **rubriflora** Backeberg, Descr. Cact. Nov. [1:] 30, 1957. [Dated 1956, published 1957.] [RPS 7]

Lobivia nigrostoma Kreuzinger & Buining, Succulenta 1950(1): 2-4, 1950. [RPS 1]

Lobivia oligotricha Cardenas, Cactus 78: 91-92, ill., 1963. [RPS 14]

Lobivia omasuyana Cardenas, Kakt. and. Sukk. 16: 21-22, ill., 1965. [Intermittently spelled 'omasuyana' and 'omasuyensis'; here Index Kewensis is followed.] [RPS 16]

Lobivia omasuyensis Cardenas, Kakt. and. Sukk. 16: 21-22, 1965. *Nom. inval.* (Art. 34). [An orthographic variant for *L. omasuyana*.] [RPS 16]

Lobivia otukae Y. Ito, Explan. Diagr. Austroechinocactinae, 100, 287, 1957. [RPS 8]

Lobivia otukae var. **cinnabarina** Y. Ito, Explan. Diagr. Austroechinocactinae, 100, 287, 1957. [RPS 8]

Lobivia otukae var. **croceantha** Y. Ito, Explan. Diagr. Austroechinocactinae, 100, 287, 1957. [RPS 8]

Lobivia oxyalabastra Cardenas & Rausch, Kakt. and. Sukk. 17(4): 76-77, ills., 1966. Typus: *Rausch et Markus* s.n. (BOLV, ZSS [authentic material ?]). [Material at ZSS is *Rausch 200*.] [Bolivia: Cochabamba] [RPS 17]

Lobivia pachyacantha Y. Ito, The Full Bloom of Cactus Flowers, 49, 50, ill., 1962. *Nom. inval.* (Art. 37.1). [Included as valid in RPS 13.] [RPS 13]

Lobivia pampana var. **borealis** Rausch, Lobivia 85, 101, 142-143, ill. (p. 102), 1987. Typus: *Rausch* 387a (ZSS). [Peru] [RPS 38]

Lobivia pentlandii fa. **brunneo-rosea** (Backeberg) Ullmann, Kaktusy 26(1): 19, 1990. Basionym: *Lobivia brunneo-rosea.* [Sphalm. 'bruneo-rosea'.] [RPS 41]

Lobivia pentlandii fa. **schneideriana** (Backeberg) Ullmann, Kaktusy 26(1): 19, 1990. Basionym: *Lobivia schneideriana.* [RPS 41]

Lobivia pentlandii var. **albiflora** (Werdermann) Y. Ito, Explan. Diagr. Austroechinocactinae, 83, 1957. Basionym: *Echinopsis pentlandii* var. *albiflora.* [RPS 8]

Lobivia pentlandii var. **cavendishii** (Rümpler) Y. Ito, Explan. Diagr. Austroechinocactinae, 82, 1957. Basionym: *Echinopsis pentlandii* var. *cavendishii.* [RPS 8]

Lobivia pentlandii var. **coccinea** (Salm-Dyck) Y. Ito, Explan. Diagr. Austroechinocactinae, 83, 1957. Basionym: *Echinopsis coccinea.* [RPS 8]

Lobivia pentlandii var. **corbula** (Herrera) Krainz, Kat. ZSS ed. 2, 71, 1967. Basionym: *Mammillaria corbula.* [RPS 18]

Lobivia pentlandii var. **elegans** (Hildmann) Y. Ito, Explan. Diagr. Austroechinocactinae, 82, 1957. Basionym: *Echinopsis elegans.* [RPS 8]

Lobivia pentlandii var. **forbesii** (R. Meyer) Y. Ito, Explan. Diagr. Austroechinocactinae, 83, 1957. Basionym: *Echinopsis forbesii.* [RPS 8]

Lobivia pentlandii var. **gracilispina** (Lemaire) Y. Ito, Explan. Diagr. Austroechinocactinae, 82, 1957. Basionym: *Echinopsis pentlandii* var. *gracilispina.* [Sphalm. 'gracilipina' in RPS 8.] [RPS 8]

Lobivia pentlandii var. *grandiflora* Y. Ito, Full Bloom of Cactus Flowers, 31, ill., 1962. *Nom. inval.* (Art. 37.1). [RPS -]

Lobivia pentlandii var. **hardeniana** (Bödeker) Rausch, Lobivia 1: 44, 1975. Basionym: *Echinopsis hardeniana.* [RPS 26]

Lobivia pentlandii var. **larae** (Cardenas) Rausch, Lobivia 1: 44, 1975. Basionym: *Lobivia larae.* [RPS 26]

Lobivia pentlandii var. **longispina** (Rümpler) Y. Ito, Explan. Diagr. Austroechinocactinae, 82, 1957. Basionym: *Echinopsis pentlandii* var. *longispina.* [RPS 8]

Lobivia pentlandii var. **maximiliana** (Heyder) Backeberg, Cact. Succ. J. (US) 23(2): 50, 1951. [The basionym given is already a combination.] [RPS 3]

Lobivia pentlandii var. **neubertii** (Rümpler) Y. Ito, Explan. Diagr. Austroechinocactinae, 83, 1957. Basionym: *Echinopsis neubertii.* [RPS 8]

Lobivia pentlandii var. **ochroleuca** (R. Meyer) Y. Ito, Explan. Diagr. Austroechinocactinae, 82, 1957. Basionym: *Echinopsis pentlandii* var. *ochroleuca.* [RPS 8]

Lobivia pentlandii var. **pfersdorffii** (Rümpler)

Y. Ito, Explan. Diagr. Austroechinocactinae, 82, 1957. Basionym: *Echinopsis pfersdorffii*. [RPS 8]

Lobivia pentlandii var. **pyracantha** (Lemaire) Y. Ito, Explan. Diagr. Austroechinocactinae, 82, 1957. Basionym: *Echinopsis pentlandii* var. *pyracantha*. [RPS 8]

Lobivia pentlandii var. **radians** (Lemaire) Y. Ito, Explan. Diagr. Austroechinocactinae, 83, 1957. Basionym: *Echinopsis pentlandii* var. *radians*. [RPS 8]

Lobivia pentlandii var. **scheeri** (Lemaire) Y. Ito, Explan. Diagr. Austroechinocactinae, 83, 1957. Basionym: *Echinopsis pentlandii* var. *scheeri*. [Sphalm. 'scheerii'.] [RPS 8]

Lobivia pentlandii var. **tricolor** (Rümpler) Y. Ito, Explan. Diagr. Austroechinocactinae, 83, 1957. Basionym: *Echinopsis tricolor*. [RPS 8]

Lobivia pentlandii var. **vitellina** (Hildmann) Y. Ito, Explan. Diagr. Austroechinocactinae, 83, 1957. Basionym: *Echinopsis pentlandii* var. *vitellina*. [RPS 8]

Lobivia pictiflora Ritter, Succulenta 1966: 84-85, 1966. [Sphalm. 'pictiflorea'.] [RPS 17]

Lobivia pojoensis Rausch, Kakt. and. Sukk. 21(3): 45, 1970. Typus: *Rausch* 188 (W, ZSS). [First invalidly published (Art. 37.1) in l.c. 19(1): 8, ill., 1968.] [Bolivia: Cochabamba] [RPS 21]

Lobivia pojoensis var. **grandiflora** Rausch, Kakt. and. Sukk. 19(1): 8-9, 1968. Typus: *Rausch* 193 (ZSS [type material]). [No herbarium cited for type.] [Validity of name questionable as *L. pojoensis* was first validly described in 1970 according to Rausch 1970.] [Bolivia: Cochabamba] [RPS -]

Lobivia polyantha Y. Ito, Explan. Diagr. Austroechinocactinae, 110, 289, 1957. [RPS 8]

Lobivia potosina (Werdermann) Friedrich, Kakt. and. Sukk. 25: 83, 1974. Basionym: *Echinopsis potosina*. [RPS 25]

Lobivia prestoana Cardenas, Cact. Succ. J. (US) 42: 185, 1970. [RPS 21]

Lobivia pseudocachensis var. **cinnabarina** Backeberg, Descr. Cact. Nov. [1:] 29, 1957. [Dated 1956, published 1957.] [RPS 7]

Lobivia pseudocachensis var. **sanguinea** Backeberg, Descr. Cact. Nov. [1:] 29, 1957. [Dated 1956, published 1957.] [RPS 7]

Lobivia pseudocariquinensis Cardenas, Cact. Succ. J. (US) 33: 111, 1961. [RPS 12]

Lobivia pseudocinnabarina Backeberg, Descr. Cact. Nov. 3: 7, 1963. *Nom. inval.* (Art. 9.5). [Erroneously included as valid in RPS 14.] [RPS 14]

Lobivia pugionacantha var. **cornuta** (Rausch) Rausch, Lobivia 2: 84, 1976. Basionym: *Lobivia cornuta*. [RPS 27]

Lobivia pugionacantha var. **corrugata** Rausch, Lobivia 85, 105, 143, 1987. Typus: *Rausch* 176 (ZSS). [Argentina: Jujuy] [RPS 38]

Lobivia pugionacantha var. **culpinensis** (Ritter) Rausch, Lobivia 2: 84, 1976. Basionym: *Lobivia culpinensis*. [RPS 27]

Lobivia pugionacantha var. **flaviflora** Backeberg, Descr. Cact. Nov. [1:] 29, 1957. [Dated 1956, published 1957.] [RPS 7]

Lobivia pugionacantha var. **haemantha** Rausch, Lobivia 85, 108, 143-144, ill. (p. 111), 1987. Typus: *Rausch* 639 (ZSS). [Bolivia: Sud Cinti] [RPS 38]

Lobivia pugionacantha var. **rossii** (Bödeker) Rausch, Lobivia 2: 84, 1976. Basionym: *Lobivia rossii*. [Erroneously treated as invalid in RPS 27: 6.] [RPS 27]

Lobivia pugionacantha var. **salitrensis** (Rausch) Rausch, Lobivia 2: 84, 1976. Basionym: *Lobivia salitrensis*. [Erroneously treated as invalid in RPS 27: 6.] [RPS 27]

Lobivia pugionacantha var. **versicolor** (Rausch) Rausch, Lobivia 2: 84, 1976. Basionym: *Lobivia versicolor*. [RPS 27]

Lobivia purpurea (Fric) Y. Ito, Explan. Diagr. Austroechinocactinae, 102, 1957. Basionym: *Hymenorebulobivia purpurea*. [RPS 8]

Lobivia purpureominiata Ritter, Kakt. Südamer. 2: 460-461, 1980. [RPS 31]

Lobivia pusilla Ritter, Succulenta 1966: 85-86, 1966. [RPS 17]

Lobivia pusilla fa. **flaviflora** Ritter, Succulenta 1966: 86, 1966. [RPS 17]

Lobivia pygmaea var. **colorea** (Ritter) Rausch, Lobivia 85, 116, 1987. Basionym: *Rebutia colorea*. [RPS 38]

Lobivia pygmaea var. **diersiana** (Rausch) Rausch, Lobivia 85, 116, ill. (p. 115), 1987. Basionym: *Rebutia diersiana*. [Basionym erroneously cited as *Digitorebutia diersiana*, treated as bibliographical error (Art. 33.2, Ex. 4).] [RPS 38]

Lobivia pygmaea var. **friedrichiana** (Rausch) Rausch, Lobivia 85, 116, ill. (p. 115), 1987. Basionym: *Rebutia friedrichiana*. [Basionym erroneously cited as *Digitorebutia friedrichiana*, treated as bibliographical error (Art. 33.2, Ex. 4).] [RPS 38]

Lobivia pygmaea var. **iscayachensis** (Rausch) Rausch, Lobivia 85, 116, 1987. Basionym: *Rebutia iscayachensis*. [Basionym erroneously cited as *Digitorebutia iscayachensis*, treated as bibliographical error (Art. 33.2, Ex. 4).] [RPS 38]

Lobivia pygmaea var. **knizei** Rausch, Lobivia 85, 116, 144, ill. (p. 143), 1987. Typus: *Rausch* 676a (ZSS). [sphalm. 'knizeii'.] [Bolivia] [RPS 38]

Lobivia pygmaea var. **minor** (Rausch) Rausch, Lobivia 85, 116, 1987. Basionym: *Rebutia diersiana* var. *minor*. [Basionym erroneously cited as *Digitorebutia diersiana* var. *minor*, treated as bibliographical error (Art. 33.2, Ex. 4).] [RPS 38]

Lobivia pygmaea var. *nigrescens* Rausch, Lobivia 85, 116, ill. (p. 115), 1987. *Nom. illeg.* (Art. 56), based on *Digitorebutia diersiana* var. *atrovirens*. [RPS 38]

Lobivia pygmaea var. **polypetala** Rausch, Lobivia 85, 116, 144, ill. (p. 143), 1987. Typus: *Rausch* 301 (ZSS). [Bolivia: Potosi] [RPS 38]

Lobivia pygmaea var. **tafnaensis** Rausch, Lobivia 85, 116, 144, ill. (p. 114), 1987. Typus: *Rausch* 508a (ZSS). [Argentina: Jujuy] [RPS 38]

Lobivia pygmaea var. **violacistaminea** Rausch,

Lobivia 85, 116, 144, ill. (p. 143), 1987. Typus: *Rausch* 742 (ZSS). [Bolivia: Sud Cinti] [RPS 38]

Lobivia quiabayensis Rausch, Kakt. and. Sukk. 21: 45, 1970. Typus: *Rausch* 205 (). [First invalidly published (Art. 37.1) in l.c. 19: 22, 1968.] [Bolivia: La Paz] [RPS 21]

Lobivia rauschii Zecher, Succulenta 53(3): 42-43, ill., 1974. Typus: *Rausch* 635 (ZSS). [Bolivia: Sud-Cinti] [RPS 25]

Lobivia rebutioides var. **chlorogona** (Wessner) Backeberg, Die Cact. 3: 1456, 1959. Basionym: *Lobivia chlorogona*. [RPS 10]

Lobivia rebutioides var. **sublimiflora** (Backeberg) Backeberg, Die Cact. 3: 1456, 1959. Basionym: *Lobivia sublimiflora*. [RPS 10]

Lobivia rigidispina Backeberg, Descr. Cact. Nov. 3: 7, 1963. *Nom. inval.* (Art. 37.1). [RPS 14]

Lobivia rosarioana Rausch, Kakt. and. Sukk. 30(12): 284-286, ills., 1979. Typus: *Rausch* 129 (ZSS). [Incorrectly reported as being based on 2 syntypes by Eggli in Trop. subtrop. Pfl.-welt 59: 72, 1987.] [Argentina: La Rioja] [RPS 30]

Lobivia rosarioana var. **rubriflora** Rausch, Kakt. and. Sukk. 30(12): 286, ill. (p. 285), 1979. Typus: *Rausch* 528 (ZSS). [Argentina: Catamarca] [RPS 30]

Lobivia rossei var. **boedekeriana** (Harden) Backeberg, Cact. Succ. J. (US) 23(2): 52, 1951. Basionym: *Lobivia boedekeriana*. [RPS 3]

Lobivia rossei var. **hardeniana** (Bödeker) Backeberg, Cact. Succ. J. (US) 23(2): 52, 1951. Basionym: *Lobivia hardeniana*. [RPS 3]

Lobivia rossei var. **stollenwerkiana** (Bödeker) Backeberg, Cact. Succ. J. (US) 23(2): 52, 1951. Basionym: *Lobivia stollenwerkiana*. [RPS 3]

Lobivia rossii var. **bustilloensis** Ritter, Kakt. Südamer. 2: 579-580, 1980. Typus: *Ritter* 348 (U, ZSS). [Type cited for U l.c. 1: iii, 1979.] [Bolivia: Bustillo] [RPS 31]

Lobivia rossii var. **carminata** Backeberg, Descr. Cact. Nov. [1:] 28, 1957. [Dated 1956, published 1957.] [RPS 7]

Lobivia rossii var. **salmonea** Backeberg, Descr. Cact. Nov. [1:] 28, 1957. [Dated 1956, published 1957.] [RPS 7]

Lobivia rossii var. **sanguinea** Backeberg, Descr. Cact. Nov. [1:] 28, 1957. [Dated 1956, published 1957.] [RPS 7]

Lobivia rossii var. **sayariensis** Ritter, Kakt. Südamer. 2: 580, 1980. [RPS 31]

Lobivia rowleyi Y. Ito, Explan. Diagr. Austroechinocactinae, 103, 288, 1957. [RPS 8]

Lobivia rowleyi var. **longispina** Y. Ito, Explan. Diagr. Austroechinocactinae, 104, 288, 1957. [RPS 8]

Lobivia rowleyi var. **rubraurantiaca** Y. Ito, Explan. Diagr. Austroechinocactinae, 104, 288, 1957. [RPS 8]

Lobivia ruberrima (Fric) Y. Ito, Explan. Diagr. Austroechinocactinae, 102, 1957. Basionym: *Hymenorebulobivia ruberrima*. [RPS 8]

Lobivia salitrensis Rausch, Succulenta 53(4): 61-62, ill., 1974. Typus: *Rausch* 636 (ZSS). [Type cited for W but placed in ZSS.] [Bolivia: Sud-Cinti] [RPS 25]

Lobivia salitrensis var. **flexuosa** Rausch, Succulenta 53(4): 63, 1974. Typus: *Rausch* 647 (ZSS). [Type cited for W but placed in ZSS.] [Bolivia: Sud-Cinti] [RPS 25]

Lobivia saltensis var. **emmae** (Backeberg) Rowley, Cact. Succ. J. Gr. Brit. 44(4): 81, 1982. Basionym: *Lobivia emmae*. [RPS 33]

Lobivia saltensis var. **multicostata** Rausch, Kakt. and. Sukk. 28(4): 75, 1977. Typus: *Rausch* 662 (ZSS). [Argentina: Salta] [RPS 28]

Lobivia saltensis var. *nealeana* (Backeberg) Rausch, Lobivia 3: 132, 1976. Incorrect name (Art. 57.1), based on *Lobivia nealeana*. [RPS 27]

Lobivia saltensis var. **pseudocachensis** (Backeberg) Rausch, Lobivia 85, 117, 1987. Basionym: *Lobivia pseudocachensis*. [RPS 38]

Lobivia saltensis var. **schreiteri** (Castellanos) Rausch, Lobivia 3: 132, 1976. Basionym: *Lobivia schreiteri*. [RPS 27]

Lobivia saltensis var. **stilowiana** (Backeberg) Rausch, Lobivia 3: 132, 1976. Basionym: *Lobivia stilowiana*. [RPS 27]

Lobivia saltensis var. **zapallarensis** Rausch, Lobivia 85, 120, 144-145, 1987. Typus: *Rausch* 16 (ZSS). [Argentina: Salta] [RPS 38]

Lobivia sanguiniflora var. **breviflora** (Backeberg) Rausch, Lobivia 2: 104, 1976. Basionym: *Lobivia breviflora*. [RPS 27]

Lobivia sanguiniflora var. **duursmaiana** (Backeberg) Rausch, Lobivia 2: 104, 1976. Basionym: *Lobivia duursmaiana*. [RPS 27]

Lobivia sanguiniflora var. **pseudolateritia** Backeberg, Descr. Cact. Nov. [1], 30, 1957. [Dated 1956, published 1957.] [RPS 7]

Lobivia schieliana Backeberg, Descr. Cact. Nov. [1:] 30, 1957. [Dated 1956, published early in 1957 according to correspondence at Kew. Often mis-spelled 'schieleana'.] [RPS 7]

Lobivia schieliana var. *albescens* Backeberg, Die Cact. 3: 1478, 1959. *Nom. inval.* (Art. 9.5). [Erroneously included as valid in RPS 10.] [RPS 10]

Lobivia schieliana var. **leptacantha** (Rausch) Rausch, Lobivia 85, 121, ill. (p. 122), 1987. Basionym: *Lobivia leptacantha*. [RPS 38]

Lobivia schieliana var. **quiabayensis** (Rausch) Rausch, Lobivia 85, 121, 1987. Basionym: *Lobivia quiabayensis*. [RPS 38]

Lobivia schneideriana var. **carnea** Backeberg, Descr. Cact. Nov. [1], 29, 1957. [Dated 1956, published 1957.] [RPS 7]

Lobivia schneideriana var. **cuprea** Backeberg, Descr. Cact. Nov. [1], 28, 1957. [Dated 1956, published 1957.] [RPS 7]

Lobivia schreiteri var. **riolarensis** Rausch, Lobivia 85, 124, 145, 1987. Typus: *Rausch* 793 (ZSS). [Argentina: Catamarca] [RPS 38]

Lobivia schreiteri var. **stilowiana** (Backeberg) Rausch, Lobivia 85, 121, ill. (p. 123), 1987.

Basionym: *Lobivia stilowiana.* [RPS 38]
Lobivia scopulina Backeberg, Die Cact. 6: 3735, 1962. *Nom. inval.* (Art. 37.1). [Erroneously included as valid in RPS 13.] [RPS 13]
Lobivia shaferi fa. **cylindracea** (Backeberg) E. Herzog, Kakt. / Sukk. 19(1-2): 3, 1985. Basionym: *Lobivia cylindracea.* [Vol. for 1984, published 1985.] [RPS 36]
Lobivia shaferi fa. *luteiflora* (Backeberg) E. Herzog, Kakt. / Sukk. 19(1-2): 7, 1985. *Nom. inval.*, based on *Pseudolobivia luteiflora, nom. inval.* (Art. 9.5). [Vol. for 1984, published 1985.] [RPS 36]
Lobivia shaferi ssp. **aurea** (Britton & Rose) E. Herzog, Kakt. / Sukk. 19(1-2): 6, 1985. Basionym: *Echinopsis aurea.* [Vol. for 1984, published 1985.] [RPS 36]
Lobivia shaferi ssp. *dobeana* (Dölz) E. Herzog, Kakt. / Sukk. 19(1-2): 13, 1985. *Nom. inval.* (Art. 33.2), based on *Lobivia dobeana.* [Vol. for 1984, published 1985.] [RPS 36]
Lobivia shaferi ssp. **fallax** (Oehme) E. Herzog, Kakt. / Sukk. 19(1-2): 14, 1985. Basionym: *Lobivia fallax.* [Vol. for 1984, published 1985.] [RPS 36]
Lobivia shaferi ssp. **leucomalla** (Wessner) E. Herzog, Kakt. / Sukk. 19(1-2): 10, 1985. Basionym: *Lobivia leucomalla.* [Vol. for 1984, published 1985.] [RPS 36]
Lobivia shaferi ssp. *rubriflora* Herzog, Kakt. / Sukk. 19(1-2): 13, 1985. *Nom. inval.* (Art. 34.1, 36.1). [Vol. for 1984, published 1985.] [RPS 36]
Lobivia shaferi var. **albiflora** (Rausch) E. Herzog, Kakt. / Sukk. 19(1-2): 9, 1985. Basionym: *Lobivia aurea* var. *albiflora.* [Vol. for 1984, published 1985.] [RPS 36]
Lobivia shaferi var. *flaviflora* (Spegazzini) E. Herzog, Kakt. / Sukk. 19(1-2): 3, 1985. *Nom. inval.*, based on *Cereus huascha* var. *flaviflorus, nom. inval.* (Art. ?). [Vol. for 1984, published 1985.] [RPS 36]
Lobivia shaferi var. **quinesensis** (Rausch) E. Herzog, Kakt. / Sukk. 19(1-2): 11, 1985. Basionym: *Echinopsis aurea* var. *quinesensis.* [Vol. for 1984, published 1985.] [RPS 36]
Lobivia sicuaniensis Rausch, Succulenta 50(12): 221, 229, ill., 1971. Typus: *Rausch* 426 (W, ZSS). [Peru] [RPS 22]
Lobivia silvestrii (Spegazzini) Rowley, Nation. Cact. Succ. J. 22: 68, 1967. Basionym: *Echinopsis silvestrii.* [RPS 18]
Lobivia simplex Rausch, Succulenta 51(7): 134-135, ill., 1972. Typus: *Rausch* 423 (W [not found], ZSS). [Peru] [RPS 23]
Lobivia spiniflora (Schumann) Rausch, Lobivia 85, 157, 1987. Basionym: *Echinocactus spiniflorus.* [RPS 38]
Lobivia spiniflora var. **klimpeliana** (Weidlich & Werdermann) Rausch, Lobivia 85, 157, 1987. Basionym: *Echinopsis klimpeliana.* [RPS 38]
Lobivia spiniflora var. **macrantha** Rausch, Lobivia 85, 158, ill., 1987. Typus: *Rausch* 761 (ZSS). [Argentina: Catamarca] [RPS 38]
Lobivia spiniflora var. **peitscheriana** (Backeberg) Rausch, Lobivia 85, 158, 1987. Basionym: *Acanthocalycium peitscherianum.* [RPS 38]
Lobivia spiniflora var. **violacea** (Werdermann) Rausch, Lobivia 85, 157, 1987. Basionym: *Echinopsis violacea.* [RPS 38]
Lobivia steinmannii var. **applanata** Rausch, Lobivia 85, 129, 145, 1987. Typus: *Rausch* 486 (ZSS). [Bolivia: Chuquisaca] [RPS 38]
Lobivia steinmannii var. **brachyantha** (Wessner) Rausch, Lobivia 85, 129, ill. (p. 127), 1987. Basionym: *Lobivia brachyantha.* [RPS 38]
Lobivia steinmannii var. **camargoensis** (Rausch) Rausch, Lobivia 85, 129, 1987. Basionym: *Rebutia camargoensis.* [Basionym erroneously cited as *Aylostera camargoensis*, treated as bibliographical error (Art. 33.2, Ex. 4).] [RPS 38]
Lobivia steinmannii var. **christinae** (Rausch) Rausch, Lobivia 85, 129, 1987. Basionym: *Rebutia christinae.* [Basionym erroneously cited as *Digitorebutia christinae*, treated as bibliographical error (Art. 33.2, Ex. 4).] [RPS 38]
Lobivia steinmannii var. **cincinnata** (Rausch) Rausch, Lobivia 85, 129, ill. (p. 126), 1987. Basionym: *Rebutia cincinnata.* [Basionym erroneously cited as *Digitorebutia cincinnata*, treated as bibliographical error (Art. 33.2, Ex. 4).] [RPS 38]
Lobivia steinmannii var. **costata** (Werdermann) Rausch, Lobivia 85, 129, ill. (p. 127), 1987. Basionym: *Rebutia costata.* [RPS 38]
Lobivia steinmannii var. **leucacantha** Rausch, Lobivia 85, 129, 145-146, 1987. Typus: *Rausch* 644 (ZSS). [Bolivia: Sud Cinti] [RPS 38]
Lobivia steinmannii var. **major** Rausch, Lobivia 85, 129, 146, ill., 1987. Typus: *Rausch* 334 (ZSS). [Argentina: Jujuy] [RPS 38]
Lobivia steinmannii var. **melanocentra** Rausch, Lobivia 85, 129, 146, 1987. Typus: *Rausch* 744 (ZSS). [Bolivia: Tarija] [RPS 38]
Lobivia steinmannii var. **parvula** Rausch, Lobivia 85, 129, 146, ill., 1987. Typus: *Rausch* 296 (ZSS). [Bolivia: Potosi] [RPS 38]
Lobivia steinmannii var. **rauschii** (Zecher) Rausch, Lobivia 85, 129, 1987. Basionym: *Rebutia rauschii.* [Basionym erroneously cited as *Digitorebutia rauschii*, treated as bibliographical error (Art. 33.2, Ex. 4).] [RPS 38]
Lobivia steinmannii var. **tuberculata** Rausch, Lobivia 85, 129, 145, 1987. Typus: *Rausch* 743 (ZSS). [Bolivia: Sud Cinti] [RPS 38]
Lobivia taratensis Cardenas, Cact. Succ. J. (US) 38: 141, 1966. [RPS 17]
Lobivia taratensis var. **leucosiphus** Cardenas, Cact. & Succ. J. (US) 38: 142, 1966. [RPS 17]
Lobivia tegeleriana var. **akersii** (Rausch) Rausch, Lobivia 1: 8, 1975. Basionym: *Lobivia akersii.* [RPS 26]
Lobivia tegeleriana var. **incuiensis** (Rauh & Backeberg) Rausch, Lobivia 1: 8, 1975. Basionym: *Lobivia incuiensis.* [RPS 26]
Lobivia tegeleriana var. **puquiensis** Ritter,

Kakt. Südamer. 4: 1329-1330, 1981. Typus: *Ritter* 175 (U, ZSS). [Type cited for U l.c. 1: iii, 1979.] [Peru: Ayacucho] [RPS 32]

Lobivia tenuispina Ritter, Succulenta 1966: 85, 1966. [RPS 17]

Lobivia thionantha var. **aurantiaca** (Rausch) Rausch, Lobivia 85, 156, 1987. Basionym: *Acanthocalycium aurantiacum.* [RPS 38]

Lobivia thionantha var. **brevispina** (Ritter) Rausch, Lobivia 85, 153, 1987. Basionym: *Acanthocalycium brevispinum.* [RPS 38]

Lobivia thionantha var. **catamarcensis** (Ritter) Rausch, Lobivia 85, 156, 1987. Basionym: *Acanthocalycium catamarcense.* [RPS 38]

Lobivia thionantha var. **chionantha** (Spegazzini) Rausch, Lobivia 85, 156, 1987. Basionym: *Echinocactus chionanthus.* [RPS 38]

Lobivia thionantha var. **erythrantha** Rausch, Lobivia 85, 155, ill., 1987. Typus: *Rausch* 763 (ZSS). [Argentina: Salta] [RPS 38]

Lobivia thionantha var. **ferrarii** (Rausch) Rausch, Lobivia 85, 156, 1987. Basionym: *Acanthocalycium ferrarii.* [RPS 38]

Lobivia thionantha var. **glauca** (Ritter) Rausch, Lobivia 85, 156, 1987. Basionym: *Acanthocalycium glaucum.* [RPS 38]

Lobivia thionantha var. **munita** Rausch, Lobivia 85, 154, ill., 1987. Typus: *Rausch* 772 (ZSS). [Argentina: Salta] [RPS 38]

Lobivia thionantha var. *variflora* (Backeberg) Rausch, Lobivia 85, 156, 1987. *Nom. inval.*, based on *Acanthocalycium variflorum, nom. inval.* (Art. 9.5). [Sphalm. 'variiflora'.] [RPS 38]

Lobivia tiegeliana cv. **Peclardiana** (Krainz *pro sp.*) Herzog, Kakt. / Sukk. 22(1-2): 40, 1987. Basionym: *Lobivia peclardiana.* [RPS 38]

Lobivia tiegeliana fa. **albiflora** (Krainz) Krainz, Die Kakt. C Vc, 1964. Basionym: *Lobivia peclardiana* var. *albiflora.* [RPS 15]

Lobivia tiegeliana fa. **winteriae** (Krainz) Krainz, Die Kakt. C Vc, 1964. Basionym: *Lobivia peclardiana* var. *winteriae.* [RPS 15]

Lobivia tiegeliana var. **cinnabarina** (Fric) Rowley, Cact. Succ. J. Gr. Brit. 44(4): 81, 1982. Basionym: *Lobivia graulichii* var. *cinnabarina.* [RPS 33]

Lobivia tiegeliana var. **distefanoiana** Cullmann & Ritter, Kakt. and. Sukk. 12(1): 7-8, ill., 1961. Typus: *Ritter* 620 (ZSS [type number]). [No herbarium cited for type.] [Bolivia: Tarija] [RPS 12]

Lobivia tiegeliana var. **flaviflora** (Ritter) Rausch, Lobivia 2: 92, 1976. Basionym: *Lobivia pusilla* fa. *flaviflora.* [RPS 27]

Lobivia tiegeliana var. *fricii* (Rausch) Rausch, Lobivia 2: 92, 1976. Incorrect name (Art. 57.1), based on *Lobivia fricii.* [RPS 27]

Lobivia tiegeliana var. **peclardiana** (Krainz) Krainz, Die Kakt. C Vc, 1964. Basionym: *Lobivia peclardiana.* [RPS 15]

Lobivia tiegeliana var. **pusilla** (Ritter) Rausch, Lobivia 2: 92, 1976. Basionym: *Lobivia pusilla.* [RPS 27]

Lobivia tiegeliana var. **ruberrima** Rausch, Succulenta 51(6): 101-102, ill., 1972. Typus: *Rausch* 84b (W, ZSS). [Material at ZSS is *Rauschfl 84a.] [Bolivia] [RPS 23]*

Lobivia tiegeliana var. **uriondoensis** Rausch, Succulenta 58(10): 233-234, ill., 1979. Typus: *Rausch* 90 (ZSS). [Bolivia] [RPS 30]

Lobivia titicacensis Cardenas, Cactus (Paris) 14(65): 183, 1959. [RPS 10]

Lobivia tuberculosa Ritter, Kakt. Südamer. 2: 582-583, 1980. [RPS 31]

Lobivia uitewaaliana Buining, Nation. Cact. Succ. J. 6(1): 15, 1951. [Sphalm. 'uitewaaleana'.] [RPS 2]

Lobivia vanurkiana Backeberg, Descr. Cact. Nov. 3: 7, 1963. *Nom. inval.* (Art. 37.1). [RPS 14]

Lobivia varians var. **rubro-alba** Backeberg, Descr. Cact. Nov. [1], 28, 1957. [Dated 1956, published 1957.] [RPS 7]

Lobivia varispina Ritter, Kakt. Südamer. 2: 577-578, 1980. [Sphalm. 'variispina'.] [RPS 31]

Lobivia vatteri var. **robusta** Backeberg, Descr. Cact. Nov. [1], 30, 1957. [Dated 1956, published 1957.] [RPS 7]

Lobivia versicolor Rausch, Kakt. and. Sukk. 25(8): 169-170, ill., 1974. Typus: *Rausch* 299 (ZSS). [Type cited for W but placed in ZSS.] [Bolivia: Potosi] [RPS 25]

Lobivia vilcabambae Ritter, Taxon 12(3): 124, 1963. [RPS 14]

Lobivia walteri Kiesling, Hickenia 1: 35-36, 1976. [Sphalm. 'walterii'.] [RPS 27]

Lobivia watadae Y. Ito, Explan. Diagr. Austroechinocactinae, 109, 288, 1957. [RPS 8]

Lobivia watadae var. **salmonea** Y. Ito, Explan. Diagr. Austroechinocactinae, 109, 289, 1957. [RPS 8]

Lobivia wegneriana Gruner & Klügling, Kakt. / Sukk. 1970(1): 10-12, ill., 1970. *Nom. inval.* (Art. 34.1, 37.1). [Published as provisional name.] [RPS 24]

Lobivia westii P. C. Hutchison, Cact. Succ. J. (US) 26(3): 81-83, ills., 1954. Typus: *West* 3741 (UC, ZSS [authentic material ?]). [Peru: Apurimac] [RPS 5]

Lobivia westii var. **intermedia** (Rausch) Ritter, Kakt. Südamer. 4: 1330, 1981. Basionym: *Lobivia intermedia.* [RPS 32]

Lobivia wilkeae (Backeberg) Friedrich, Kakt. and. Sukk. 25: 83, 1974. *Nom. inval.*, based on *Pseudolobivia wilkeae, nom. inval.* (Art. 37.1). [Erroneously included as valid in RPS 25.] [RPS 25]

Lobivia winteriana Ritter, Kakt. and. Sukk. 21(8): 146-147, ill., 1970. Typus: *Ritter* 1312 (U, ZSS). [Peru: Huancavelica] [RPS 21]

Lobivia wrightiana fa. *chilensis* hort. *ex* Herzog, Kakt. / Sukk. 24: 102, 1990. *Nom. inval.* (Art. 36.1, 37.1). [Dated 1989, published 1990. Published as nude name.] [RPS 41]

Lobivia wrightiana var. **brevispina** Backeberg, Descr. Cact. Nov. [1], 30, 1957. [Dated 1956, published 1957.] [RPS 7]

Lobivia wrightiana var. **winteriana** (Ritter) Rausch, Lobivia 85, 136, 1987. Basionym: *Lobivia winteriana.* [RPS 38]

Lobivia zecheri Rausch, Succulenta 50(8): 141, 146, ill., 1971. Typus: *Rausch* 407 (W,

ZSS). [Peru: Ayacucho] [RPS 22]

Lobivia zecheri fa. *ferruginea* hort. *ex* Herzog, Kakt. / Sukk. 24(3/4): 108, 1990. *Nom. inval.* (Art. 36.1, 37.1). [Dated 1989, published 1990. Published as nude name.] [RPS 41]

Lobivia zecheri fa. *fungiflora* (Braun) Herzog, Kakt. / Sukk. 24: 107, 1990. *Nom. inval.*, based on *Lobivia fungiflora, nom. inval.* (Art. 34.1, 36.1). [Dated 1989, published 1990. Published as provisional invalid name.] [RPS 41]

Lobivia zecheri var. *fungiflora* Braun, Stachelpost 9: 138-139, 1973. *Nom. inval.* (Art. 37.1). [Described as new variety, but based on *Lobivia fungiflora.*] [RPS 24]

Lobivia zudanensis Cardenas, Cact. Succ. J. (US) 42: 34, 1970. [RPS 21]

×*Lobivopsis* Johnson & Ito *ex* Ito, The Cactaceae, 265, 1981. *Nom. inval.* (Art. H9). [Based on a catalogue name by Johnson hused first in 1958.] [RPS -]

Lophocereus mieckleyanus (Weingart) Backeberg, Die Cact. 4: 2285, 1960. Basionym: *Cereus mieckleyanus*. [RPS 11]

Lophocereus schottii fa. **mieckleyanus** Lindsay, Cact. Succ. J. (US) 35: 184, 1963. [May or may not be the same taxon as *Cereus mieckleyanus* Weingart; described as new taxon to avoid confusion without loosing the epithet.] [RPS 14]

Lophocereus schottii fa. **spiralis** Carter, Cact. Suc. Mex. 11: 13-17, 27-28, 1966. [RPS 17]

Lophocereus schottii var. **tenuis** Lindsay, Cact. Succ. J. (US) 35: 187, 1963. [RPS 14]

Lophophora diffusa (Croizat) Bravo, Cact. Suc. Mex. 12: 13, 1967. Basionym: *Lophophora echinata* var. *diffusa*. [RPS 18]

Lophophora fricii Habermann, Kaktusy 10: 123-127, 1974. [RPS 25]

Lophophora jourdaniana Habermann, Kaktusy 11: 3-6, 1975. [RPS 26]

Lophophora lutea (Rouhier) Backeberg, Die Cact. 5: 2901, 1961. Basionym: *Echinocactus williamsii* var. *lutea*. [RPS 12]

Lophophora lutea var. *texana* (Kreuzinger) Backeberg, Die Cact. 5: 2903, 1961. *Nom. inval.* (Art. 37.1), based on *Lophophora williamsii* var. *texana*. [Erroneously included as valid in RPS 12. The basionym author of the variety is variously also cited as 'Fric'.] [RPS 12]

Lophophora williamsii var. *caespitosa* Y. Ito, Cacti, 96, 1952. *Nom. inval.* (Art. 36.1, 37.1). [RPS 4]

Lophophora williamsii var. **diffusa** (Croizat) Rowley, Rep. Pl. Succ. 27: 7, 1979. Basionym: *Lophophora echinata* var. *diffusa*. [RPS 27]

Lophophora williamsii var. **echinata** (Croizat) Bravo, Cact. Suc. Mex. 12: 12, 1967. Basionym: *Lophophora echinata*. [RPS 18]

Lophophora williamsii var. *heptagona* Y. Ito, The Cactaceae, 467-468, ill., 1981. *Nom. inval.* (Art. 37.1). [RPS -]

Loxanthocereus subgen. **Anhaloniopsis** Buxbaum, in Krainz, Die Kakt. part. 58 C Vc, 1974. Typus: *L. madisoniorum* (P. C. Hutchison) Buxbaum. [RPS 27]

Loxanthocereus subgen. **Hildewintera** (Ritter) Buxbaum, in Krainz, Die Kakt. part. 58 C Vc, 1974. Basionym: *Hildewintera*. [RPS 27]

Loxanthocereus acanthurus var. **ferox** Backeberg, Cact. Succ. J. (US) 23(1): 19, 1951. [RPS 3]

Loxanthocereus aticensis Rauh & Backeberg, in Backeberg, Descr. Cact. Nov. [1:] 15-16, 1957. [Dated 1956, published 1957.] [RPS 7]

Loxanthocereus aureispinus (Ritter) Buxbaum, in Krainz, Die Kakt. Lief. 57: C Vb, 1974. Basionym: *Hildewintera aureispina*. [Basionym reference here corrected.] [RPS 25]

Loxanthocereus bicolor Ritter, Kakt. Südamer. 4: 1455-1456, 1981. Typus: *Ritter* 173 p.p. (U, ZSS). [Type cited for U l.c. 1: iii, 1979.] [Peru: Lima] [RPS 32]

Loxanthocereus brevispinus Rauh & Backeberg, in Rauh, Beitr. Kenntn. Peruan. Kakt.veg., 317, 1958. [RPS 10]

Loxanthocereus camanaensis Rauh & Backeberg, in Backeberg, Descr. Cact. Nov. [1:] 15, 1957. [Dated 1956, published 1957.] [RPS 7]

Loxanthocereus canetensis Rauh & Backeberg, in Backeberg, Descr. Cact. Nov. [1:] 16, 1957. Typus: *Rauh* K166 (1956) (ZSS). [Dated 1956, published 1957.] [Peru] [RPS 7]

Loxanthocereus cantaensis Rauh & Backeberg, in Rauh, Beitr. Kenntn. Peruan. Kakt.veg., 314, 1958. [RPS 10]

Loxanthocereus casmaensis Backeberg, Cactus (Paris) No. 53: 139 et seq, 1957. *Nom. inval.* (Art. 36.1). [RPS 8]

Loxanthocereus clavispinus Rauh & Backeberg, in Backeberg, Descr. Cact. Nov. [1:] 15, 1957. Typus: *Rauh* K106 (1956) (ZSS). [Dated 1956, published 1957.] [Peru] [RPS 7]

Loxanthocereus convergens Ritter, Kakt. Südamer. 4: 1456-1457, 1981. Typus: *Ritter* 671 (ZSS). [Peru: Lima] [RPS 32]

Loxanthocereus crassiserpens (Rauh & Backeberg) Backeberg, Die Cact. 6: 3674, 1962. Basionym: *Cleistocactus crassiserpens*. [RPS 13]

Loxanthocereus cullmannianus Backeberg, Die Cact. 6: 3674, 1962. *Nom. inval.* (Art. 37.1). [Erroneously included as valid in RPS 13.] [RPS 13]

Loxanthocereus deserticola Ritter, Kakt. Südamer. 4: 1458-1459, 1981. Typus: *Ritter* 185 p.p. (U, ZSS). [Type cited for U l.c. 1: iii, 1979.] [Peru: Ica] [RPS 32]

Loxanthocereus erectispinus Rauh & Backeberg, in Backeberg, Descr. Cact. Nov. [1:] 15, 1957. [Dated 1956, published 1957.] [RPS 7]

Loxanthocereus eremiticus Ritter, Kakt. Südamer. 4: 1459, 1981. Typus: *Ritter* 174 (U, ZSS). [Type cited for U l.c. 1: iii, 1979.] [Peru: Lima] [RPS 32]

Loxanthocereus erigens Rauh & Backeberg, in Backeberg, Descr. Cact. Nov. [1:] 16, 1957. [Dated 1956, published 1957.] [RPS 7]

Loxanthocereus eulalianus Rauh & Backeberg,

in Backeberg, Descr. Cact. Nov. [1:] 14, 1957. [Dated 1956, published 1957.] [RPS 7]

Loxanthocereus ferrugineus Rauh & Backeberg, in Backeberg, Descr. Cact. Nov. [1:] 15, 1957. [Dated 1956, published 1957.] [RPS 7]

Loxanthocereus formosus (Ritter) Buxbaum, in Krainz, Die Kakt. Lief. 58: C Vc, 1974. Basionym: *Matucana formosa.* [RPS 27]

Loxanthocereus gracilis (Akers & Buining) Ritter, Backeberg's Descr. & Erört. taxon. nomenkl. Fragen, [unpaged], 1958. Basionym: *Maritimocereus gracilis.* [Concurrently published by Rauh & Backeberg in Rauh, Beitr. Kenntn. peruan. Kakt.-veg., 300, 1958 (cf. RPS 10).] [RPS 15]

Loxanthocereus gracilispinus Rauh & Backeberg, in Backeberg, Descr. Cact. Nov. [1:] 16, 1957. [Dated 1956, published 1957.] [RPS 7]

Loxanthocereus granditessellatus Rauh & Backeberg, in Backeberg, Descr. Cact. Nov. [1:] 15, 1957. [Dated 1956, published 1957.] [RPS 7]

Loxanthocereus hoffmannii Ritter, Kakt. Südamer. 4: 1461-1462, 1981. Based on *Hoffmann* s.n.. *Nom. inval.* (Art. 37.1). [Erroneously included as valid in RPS 32.] [Peru: Lima] [RPS 32]

Loxanthocereus hystrix Rauh & Backeberg, in Backeberg, Descr. Cact. Nov. [1:] 15, 1957. [Dated 1956, published 1957.] [RPS 7]

Loxanthocereus hystrix var. **brunnescens** Rauh, Beitr. Kenntn. Peruan. Kakt.veg., 312, 1958. [RPS 10]

Loxanthocereus madisoniorum (P. C. Hutchison) Buxbaum, in H. Krainz, Die Kakt. Lief. 58: C Vc, 1974. Basionym: *Borzicactus madisoniorum.* [RPS 27]

Loxanthocereus montanus Ritter, Kakt. Südamer. 4: 1463-1464, 1981. Typus: *Ritter* 581 (U, ZSS [seeds only]). [Type cited for U l.c. 1: iii, 1979.] [Peru: Ayacucho] [RPS 32]

Loxanthocereus multifloccosus Rauh & Backeberg, in Backeberg, Descr. Cact. Nov. [1:] 16, 1957. [Dated 1956, published 1957.] [RPS 7]

Loxanthocereus nanus Akers *ex* Backeberg, Descr. Cact. Nov. [1:] 17, 1957. [Dated 1956, published 1957.] [RPS 7]

Loxanthocereus neglectus Ritter, Taxon 13(3): 116, 1964. Typus: *Ritter* 135 (U, ZSS). [Type cited for U in original description, for ZSS in Kakt. Südamer. 4: 1466, 1981.] [Peru: Lima] [RPS 15]

Loxanthocereus neglectus var. **chimbotensis** Ritter, Taxon 13(3): 116, 1964. Typus: *Ritter* 277a (U, ZSS). [Type cited for U in original description, for ZSS in Kakt. Südamer. 4: 1466, 1981.] [Peru] [RPS 15]

Loxanthocereus otuscensis Ritter, Kakt. Südamer. 4: 1467, 1981. Typus: *Ritter* 579 (U, ZSS [seeds only]). [Type cited for U l.c. 1: iii, 1979.] [Erroneously listed as *L. obtuscensis* in RPS 32.] [Peru: La Libertad] [RPS 32]

Loxanthocereus pacaranensis Ritter, Kakt. Südamer. 4: 1468-1469, 1981. Typus: *Ritter* 277 (ZSS). [Peru: Lima] [RPS 32]

Loxanthocereus pachycladus Rauh & Backeberg, in Backeberg, Descr. Cact. Nov. [1:] 16, 1957. [Dated 1956, published 1957.] [RPS 7]

Loxanthocereus parvitessellatus Ritter, Kakt. Südamer. 4: 1469, 1981. Typus: *Ritter* 578 (U, ZSS). [Type cited for U l.c. 1: iii, 1979.] [Sphalm. 'parvitesselatus'.] [Peru: La Libertad] [RPS 32]

Loxanthocereus peculiaris Rauh & Backeberg, in Backeberg, Descr. Cact. Nov. [1:] 16, 1957. [Dated 1956, published 1957.] [RPS 7]

Loxanthocereus piscoensis Rauh & Backeberg, in Backeberg, Descr. Cact. Nov. [1:] 16, 1957. Typus: *Rauh* K161 (1956) (ZSS). [Dated 1956, published 1957.] [Peru] [RPS 7]

Loxanthocereus pullatus Rauh & Backeberg, in Backeberg, Descr. Cact. Nov. [1:] 14, 1957. Typus: *Rauh* K46 (1956) (ZSS). [Dated 1956, published 1957.] [Peru] [RPS 7]

Loxanthocereus pullatus var. **brevispinus** Rauh & Backeberg, in Rauh, Beitr. Kenntn. Peruan. Kakt.veg., 316, 1958. [RPS 10]

Loxanthocereus pullatus var. **fulviceps** Rauh & Backeberg, in Backeberg, Descr. Cact. Nov. [1:] 15, 1957. [Dated 1956, published 1957.] [RPS 7]

Loxanthocereus puquiensis Ritter, Kakt. Südamer. 4: 1470-1471, 1981. Typus: *Ritter* 181 (U, ZSS). [Type cited for U l.c. 1: iii, 1979.] [Peru: Ayacucho] [RPS 32]

Loxanthocereus riomajensis Rauh & Backeberg, in Backeberg, Descr. Cact. Nov. [1:] 15, 1957. [Dated 1956, published 1957.] [RPS 7]

Loxanthocereus splendens Akers *ex* Backeberg, Descr. Cact. Nov. [1:] 14, 1957. [Dated 1956, published 1957.] [RPS 7]

Loxanthocereus sulcifer Rauh & Backeberg, in Backeberg, Descr. Cact. Nov. [1:] 14, 1957. [Dated 1956, published 1957.] [RPS 7]

Loxanthocereus sulcifer var. **longispinus** Rauh & Backeberg, in Backeberg, Descr. Cact. Nov. [1:] 14, 1957. [Dated 1956, published 1957.] [RPS 7]

Loxanthocereus trujilloensis Ritter, Kakt. Südamer. 4: 1472-1473, 1981. Typus: *Ritter* 1469 (U, ZSS [type status ?]). [Type cited for U l.c. 1: iii, 1979.] [Peru: La Libertad] [RPS 32]

Loxanthocereus variabilis Ritter, Kakt. Südamer. 4: 1473-1475, 1981. Typus: *Ritter* 148 (U, ZSS). [Type cited for U l.c. 1: iii, 1979.] [Peru: Ica] [RPS 32]

Loxanthocereus xylorhizus Ritter, Kakt. Südamer. 4: 1475-1476, 1981. Typus: *Ritter* 321 (U, ZSS). [Type cited for U l.c. 1: iii, 1979.] [Peru: Lima] [RPS 32]

Loxanthocereus yauyosensis Ritter, Kakt. Südamer. 4: 1476-1477, 1981. Typus: *Ritter* 636 (U). [Peru: Lima] [RPS 32]

Lymanbensonia Kimnach, Cact. Succ. J. (US) 56(3): 101, 1984. Typus: *Cereus micranthus* Vaupel. [= *Acanthorhipsalis* Kimnach *non* (Schumann) Britton & Rose.] [RPS 35]

Lymanbensonia micrantha (Vaupel) Kimnach, Cact. Succ. J. (US) 56(3): 101, 1984. Basionym: *Cereus micranthus*. [RPS 35]
Maihuenia albolanata Ritter, Kakt. Südamer. 2: 377-378, 1980. Typus: *Ritter* 414 (U, ZSS). [Type cited for U l.c. 1: iii, 1979.] [Argentina: Neuquen] [RPS 31]
Maihuenia albolanata fa. **viridulispina** Ritter, Kakt. Südamer. 2: 378, 1980. Typus: *Ritter* 414a (U, ZSS). [Type cited for U l.c. 1: iii, 1979.] [Argentina: Neuquen] [RPS 31]
Maihuenia cumulata Ritter, Kakt. Südamer. 2: 378-379, 1980. Typus: *Ritter* 415 (U, ZSS). [Type cited for U l.c. 1: iii, 1979.] [Argentina: Neuquen] [RPS 31]
Maihuenia latispina Ritter, Kakt. Südamer. 2: 378, 1980. Typus: *Ritter* 416 (U, ZSS). [Type cited for U l.c. 1: iii, 1979.] [Argentina: Mendoza] [RPS 31]
Maihueniopsis albomarginata Ritter, Kakt. Südamer. 2: 389, 1980. [RPS 31]
Maihueniopsis archiconoidea Ritter, Kakt. Südamer. 3: 877, 1980. [RPS 31]
Maihueniopsis atacamensis (Philippi) Ritter, Kakt. Südamer. 3: 872-873, 1980. Basionym: *Opuntia atacamensis*. [RPS 31]
Maihueniopsis boliviana (Salm-Dyck) Kiesling, Darwiniana 25(1-4): 207, 1984. Basionym: *Opuntia boliviana*. [RPS 35]
Maihueniopsis camachoi (Espinosa) Ritter, Kakt. Südamer. 3: 873, 1980. Basionym: *Opuntia camachoi*. [RPS 31]
Maihueniopsis colorea (Ritter) Ritter, Kakt. Südamer. 3: 875, 1980. Basionym: *Tephrocactus coloreus*. [RPS 31]
Maihueniopsis conoidea (Backeberg) Ritter, Kakt. Südamer. 3: 873-874, 1980. *Nom. inval.*, based on *Tephrocactus conoideus*, *nom. inval.* (Art. 37.1). [Erroneously included as valid in RPS 31.] [RPS 31]
Maihueniopsis crassispina Ritter, Kakt. Südamer. 3: 879-880, 1980. [RPS 31]
Maihueniopsis darwinii (Henslow) Ritter, Kakt. Südamer. 2: 389, 1980. Basionym: *Opuntia darwinii*. [RPS 31]
Maihueniopsis darwinii var. **hickenii** (Britton & Rose) Kiesling, Darwiniana 25(1-4): 201, 1984. Basionym: *Opuntia hickenii*. [RPS 35]
Maihueniopsis domeykoensis Ritter, Kakt. Südamer. 3: 878-879, 1980. [RPS 31]
Maihueniopsis glomerata (Haworth) Kiesling, Darwiniana 25(1-4): 202, 1984. Basionym: *Opuntia glomerata*. [RPS 35]
Maihueniopsis grandiflora Ritter, Kakt. Südamer. 3: 877-878, 1980. [RPS 31]
Maihueniopsis hypogaea (Werdermann) Ritter, Kakt. Südamer. 2: 386-388, 1980. Basionym: *Opuntia hypogaea*. [RPS 31]
Maihueniopsis leoncito (Werdermann) Ritter, Kakt. Südamer. 3: 875-876, 1980. Basionym: *Opuntia leoncito*. [RPS 31]
Maihueniopsis leptoclada Ritter, Kakt. Südamer. 2: 388, 1980. [RPS 31]
Maihueniopsis mandragora (Backeberg) Ritter, Kakt. Südamer. 2: 389-390, 1980. Basionym: *Tephrocactus mandragora*. [RPS 31]
Maihueniopsis minuta (Backeberg) Kiesling, Darwiniana 25(1-4): 204, 1984. Basionym: *Tephrocactus minutus*. [RPS 35]
Maihueniopsis molinensis (Spegazzini) Ritter, Kakt. Südamer. 2: 390, 1980. Basionym: *Opuntia molinensis*. [RPS 31]
Maihueniopsis neuquensis (Borg) Ritter, Kakt. Südamer. 2: 389, 1980. Basionym: *Opuntia neuquensis*. [RPS 31]
Maihueniopsis nigrispina (Schumann) Kiesling, Darwiniana 25(1-4): 209, 1984. Basionym: *Opuntia nigrispina*. [RPS 35]
Maihueniopsis ovallei (Remy) Ritter, Kakt. Südamer. 3: 876, 1980. Basionym: *Opuntia ovallei*. [RPS 31]
Maihueniopsis ovata (Pfeiffer) Ritter, Kakt. Südamer. 2: 389, 1980. Basionym: *Opuntia ovata*. [RPS 31]
Maihueniopsis ovata fa. **calva** Ritter, Kakt. Südamer. 2: 389, 1980. [RPS 31]
Maihueniopsis ovata fa. **sterilis** Ritter, Kakt. Südamer. 3: 871-872, 1980. [RPS 31]
Maihueniopsis pentlandii (Salm-Dyck) Kiesling, Darwiniana 25(1-4): 211, 1984. Basionym: *Opuntia pentlandii*. [RPS 35]
Maihueniopsis rahmeri (Philippi) Ritter, Kakt. Südamer. 2: 874-875, 1980. Basionym: *Opuntia rahmeri*. [RPS 31]
Maihueniopsis tarapacana (Philippi) Ritter, Kakt. Südamer. 3: 874, 1980. Basionym: *Opuntia tarapacana*. [RPS 31]
Maihueniopsis wagenknechtii Ritter, Kakt. Südamer. 3: 878, 1980. [RPS 31]
Malacocarpus erinaceus var. **elatior** (Monville) Y. Ito, Explan. Diagr. Austroechinocactinae, 257, 1957. Basionym: *Echinocactus erinaceus* var. *elatior*. [RPS 8]
Malacocarpus orthacanthus (Link & Otto) Herter, Cactus (Paris) No. 41: 92, 1954. Basionym: *Echinocactus orthacanthus*. [RPS 5]
Malacocarpus sellowii var. **acutatus** (Arechavaleta) Y. Ito, Explan. Diagr. Austroechinocactinae, 257, 1957. Basionym: *Echinocactus sellowii* var. *acutatus*. [RPS 8]
Malacocarpus sellowii var. **courantii** (Gürke) Y. Ito, Explan. Diagr. Austroechinocactinae, 257, 1957. Basionym: *Echinocactus sellowii* var. *courantii*. [RPS 8]
Malacocarpus sellowii var. **macracanthus** (Arechavaleta) Y. Ito, Explan. Diagr. Austroechinocactinae, 256, 1957. Basionym: *Echinocactus sellowii* var. *macracanthus*. [Sphalm. 'macrocanthus'.] [RPS 8]
Malacocarpus sellowii var. **macrogonus** (Arechavaleta) Y. Ito, Cacti, 74, 1952. Basionym: *Echinocactus sellowii* var. *macrogonus*. [RPS 4]
Malacocarpus sellowii var. **martinii** (Schumann) Y. Ito, Explan. Diagr. Austroechinocactinae, 256, 1957. Basionym: *Echinocactus sellowii* var. *martinii*. [RPS 8]
Malacocarpus sellowii var. **turbinatus** (Arechavaleta) Y. Ito, Cacti, 74, 1952. Basionym: *Echinocactus sellowii* var. *turbinatus*. [RPS 4]
Malacocarpus sessiliflorus (Mackie in Hooker) Backeberg, Die Cact. 3: 1621, 1959. Basionym: *Echinocactus sessiliflorus*. [RPS 10]

Malacocarpus sessiliflorus var. **martinii** (Rümpler) Backeberg, Die Cact. 3: 1622, 1959. Basionym: *Echinocactus martinii*. [RPS 10]

Malacocarpus stegmannii Backeberg, Die Cact. 3: 1623, 1959. *Nom. inval.* (Art. 37.1). [Sphalm. 'stegmanni'.] [RPS -]

Malacocarpus tephracanthus var. **courantii** (Salm-Dyck) Backeberg, Die Cact. 3: 1619, 1959. [RPS -]

Malacocarpus tephracanthus var. **depressus** (Spegazzini) Backeberg, Die Cact. 3: 1620, 1959. Basionym: *Echinocactus acuatus* var. *depressus*. [RPS 10]

Mamillopsis senilis var. **diguetii** (F. A. C. Weber) Krainz, Kat. ZSS ed. 2, 83, 1967. Basionym: *Mammillaria senilis* var. *diguetii*. [RPS 18]

Mammillaria sect. **Phellosperma** (Britton & Rose) Moran, Gentes Herbar. 8(4): 324, 1953. Basionym: *Phellosperma*. [RPS 4]

Mammillaria sect. **Pseudomammillaria** (Buxbaum) Moran, Gentes Herbar. 8(4): 324, 1953. Basionym: *Pseudomammillaria*. [RPS 4]

Mammillaria sect. **Setosae** Bravo, Anales Inst. Biol. UNAM 28: 38, 1957. Typus: *Mammillaria pitcayensis* (only taxon assigned to the series). [RPS -]

Mammillaria sect. **Subhydrochilus** Backeberg *ex* D. Hunt, Cact. Succ. J. Gr. Brit. 39(3): 74, 1977. Typus: *M. guerreronis* Bödeker. [RPS 28]

Mammillaria ser. **Decipientes** D. Hunt, Cact. Succ. J. Gr. Brit. 41(4): 95, 1979. Typus: *Mammillaria decipiens* Scheidweiler. [RPS 30]

Mammillaria ser. **Lasiacanthae** D. Hunt, Cact. Succ. J. Gr. Brit. 33(3): 63, 1971. Typus: *Mammillaria lasiacantha* Engelmann. [RPS 22]

Mammillaria ser. **Longiflorae** D. Hunt, Cact. Succ. J. Gr. Brit. 33: 59, 1971. Typus: *Mammillaria longiflora* (Britton & Rose) A. Berger. [RPS 22]

Mammillaria ser. **Megastigmatae** Neutelings, Succulenta 65(1): 3, 1986. Typus: *M. microcarpa* Engelmann *ex* Britton & Rose, *nom. illeg.* (Art. 63) (legitimate name seems to be *M. milleri* (Britton & Rose) Bödeker). [RPS 37]

Mammillaria ser. **Pectiniferae** E. Kuhn & B. Hofmann, Inform.-Brief ZAG Mammillaria 5(4): 59-61, 1979. Typus: *Mammillaria pectinifera* F. A. C. Weber. [RPS 32]

Mammillaria ser. **Proliferae** D. Hunt, Cact. Succ. J. Gr. Brit. 39(3): 73, 1977. Typus: *M. prolifera* (Miller) Haworth. [RPS 28]

Mammillaria ser. **Sphacelatae** D. Hunt, Cact. Succ. J. Gr. Brit. 39(3): 73, 1977. Typus: *Mammillaria sphacelata* Martius. [RPS 28]

Mammillaria ser. **Supertextae** D. Hunt, Cact. Succ. J. Gr. Brit. 39(4): 98, 1977. Typus: *M. supertexta* Pfeiffer. [RPS 28]

Mammillaria ser. **Zephyranthoides** E. Kuhn & B. Hofmann, Inform.-Brief ZAG Mammillaria 5(3): 41-42, 1979. Typus: *Mammillaria zephyranthoides* Scheidweiler. [RPS 32]

Mammillaria subgen. **Austroebnerella** Buxbaum, Sukkulentenkunde 5: 28, 1954. Typus: *Mammillaria solisii* (Britton & Rose) Bödeker. [RPS 5]

Mammillaria subgen. **Bartschella** (Britton & Rose) Moran, Gentes Herbar. 8(4): 324, 1953. Basionym: *Bartschella*. [Repeated by Soulaire in Cactus (Paris) No. 43: 159, 1955 (cf. RPS 6).] [RPS 4]

Mammillaria subgen. *Beneckia* Bravo, Cact. Suc. Mex. 7(4): 84, 1962. *Nom. inval.* (Art. 36.1, 37.1). [Given as "comb. nov.".] [RPS -]

Mammillaria subgen. **Chilita** (Orcutt) Moran, Gentes Herbar. 8(4): 324, 1953. Basionym: *Chilita*. [RPS 4]

Mammillaria subgen. *Cryptocarpa* Keizer, Succulenta 69(5): 102, 1990. *Nom. illeg.* (Art. 63.1). [Typus: *Mammillaria longiflora*.] [RPS 41]

Mammillaria subgen. *Galactochilia* Bravo, Cact. Suc. Mex. 7(4): 84, 1962. *Nom. inval.* (Art. 36.1, 37.1). [Given as "comb. nov.".] [RPS -]

Mammillaria subgen. **Krainzia** (Backeberg) Bravo, Cact. Suc. Mex. 21: 66, 1976. Basionym: *Krainzia*. [RPS 27]

Mammillaria subgen. **Leptocladodia** (Buxbaum) Bravo, Cact. Suc. Mex. 17: 120, 1972. Basionym: *Leptocladodia*. [RPS 23]

Mammillaria subgen. **Longiflorae** (D. Hunt) Bravo, Cact. Suc. Mex. 27(1): 17, 1982. Basionym: *Mammillaria* ser. *Longiflorae*. [Sphalm. 'Longiflora'.] [RPS 33]

Mammillaria subgen. **Mamillopsis** (Morren *ex* Britton & Rose) D. Hunt, Cact. Succ. J. Gr. Brit. 39(2): 39, 1977. Basionym: *Mamillopsis*. [Mammillaria senilis Loddiges] [RPS 28]

Mammillaria subgen. **Mammilloydia** (Buxbaum) Moran, Gentes Herbar. 8(4): 324, 1953. Basionym: *Mammilloydia*. [Mammillaria candida Scheidweiler] [RPS 4]

Mammillaria subgen. **Oehmea** (Buxbaum) D. Hunt, Cact. Succ. J. Gr. Brit. 39(2): 38, 1977. Basionym: *Oehmea*. [Mammillaria beneckei Ehrenberg] [RPS 28]

Mammillaria subgen. **Porfiria** (Bödeker) Moran, Gentes Herbar. 8(4): 324, 1953. Basionym: *Porfiria*. [RPS 4]

Mammillaria subgen. **Solisia** (Britton & Rose) Moran, Gent. Herbar. 8(4): 324, 1953. Basionym: *Solisia*. [RPS 4]

Mammillaria albata Reppenhagen, Gattung Mammillaria, 130-131, fig. 40 (p. 192), SEM-ills. 109-110 (p. 214), 1987. Typus: *Reppenhagen* 1129 (ZSS). [Mexico: San Luis Potosi] [RPS 38]

Mammillaria albata var. **longispina** Reppenhagen, Gattung Mammillaria, 132, SEM-ills. 111-112 (p. 214), 1987. Typus: *Reppenhagen* 1149 (ZSS). [Published type is *Reppenhagen* 1449, but this is an error according to a letter by W. Reppenhagen to ZSS, Feb. 22, 1988.] [Mexico: San Luis Potosi] [RPS 38]

Mammillaria albata var. **sanciro** Reppenhagen, Gattung Mammillaria, 132-133, fig. 41 (p. 193), SEM-ills. 113-114 (p. 215), 1987. Typus: *Reppenhagen* 1132 (ZSS). [Mexico: San Luis Potosi] [RPS 38]

Mammillaria albicans fa. **slevinii** (Britton & Rose) Neutelings, Succulenta 65(5): 119, 1986. Basionym: *Neomammillaria slevinii.* [First invalidly published l.c. (1): 3-5. Basionym erroneously cited as *Mammillaria slevinii.*] [RPS 37]

Mammillaria albicolumnaria (Hester) Weniger, Cacti of the Southwest, 137, 1970. *Nom. inval.* (Art. 33.2), based on *Escobaria albicolumnaria.* [Not dated.] [RPS -]

Mammillaria albicolumnaria E. Kuhn, Info.-brief ZAG Mammillaria 7(4): 53-54, ills., 1981. *Nom. inval.* (Art. 36.1, 37.1). [Published as provisional name. *Non Mammillaria albicolumnaria* (Hester) Weniger.] [RPS -]

Mammillaria albidula Backeberg, Die Cact. 5: 3429-3430, 1961. *Nom. inval.* (Art. 9.5, 37.1). [Erroneously included as valid in RPS 12.] [RPS 12]

Mammillaria albilanata var. *fuauxiana* (Backeberg) E. Kuhn, Info.-brief ZAG Mammillaria 1984: 60, 1984. *Nom. inval.* (Art. 33.2), based on *Mammillaria fuauxiana.* [RPS -]

Mammillaria albilanata var. *gracilis* B. Hofmann, Info.-brief ZAG Mammillaria 1985: 25, 1985. *Nom. inval.* (Art. 36.1, 37.1). [Published as provisional name.] [RPS -]

Mammillaria albilanata var. *tegelbergiana* (Lindsay) E. Kuhn, Info.-brief ZAG Mammillaria 1984: 60, 1984. *Nom. inval.* (Art. 33.2), based on *Mammillaria tegelbergiana.* [RPS -]

Mammillaria albrechtiana Wohlschlager, Kakt. and. Sukk. 40(9): 218-219, ill., 1989. Typus: *Wohlschlager* 886 (WU). [Mexico: Oaxaca] [RPS 40]

Mammillaria angelensis var. **estebanensis** (Lindsay) Reppenhagen, Gattung Mammillaria, 34, 1987. Basionym: *Mammillaria estebanensis.* [RPS 38]

Mammillaria anniana Glass & Foster, Cact. Succ. J. (US) 53(2): 79-80, 1981. Typus: *Lau* 1332 (POM). [Mexico: Tamaulipas] [RPS 32]

Mammillaria apamensis Reppenhagen, Gattung Mammillaria, 71-72, fig.15 (p. 186), SEM-ills. 51-52 (p. 207), 1987. Typus: *Reppenhagen* 910 (ZSS). [Mexico: Hidalgo] [RPS 38]

Mammillaria apamensis var. **pratensis** Reppenhagen, Gattung Mammillaria, 73-74, fig.15 (p. 186), SEM-ills. 53-54 (p. 207), 1987. Typus: *Reppenhagen* 1099 (ZSS). [Mexico: Hidalgo] [RPS 38]

Mammillaria apozolensis Reppenhagen, Gattung Mammillaria, 143-144, fig. 47 (p. 194), SEM-ills. 125-126 (p. 216), 1987. Typus: *Reppenhagen* 978 (ZSS). [Mexico: Zacatecas] [RPS 38]

Mammillaria apozolensis var. **saltensis** Reppenhagen, Gattung Mammillaria, 145-146, fig. 51 (p. 195), SEM-ills. 127-128 (p. 216), 1987. Typus: *Reppenhagen* 1510 (ZSS). [Mexico: Zacatecas] [RPS 38]

Mammillaria arroyensis Reppenhagen, Kakt. and. Sukk. 40(12): 290-292, ills., 1989. Typus: *Reppenhagen* 1054 (KL). [Mexico: Nuevo Leon] [RPS 40]

Mammillaria ascensionis Reppenhagen, Kakt. and. Sukk. 30(3): 61-62, ill., 1979. Typus: *Reppenhagen* 1062 p.p. (ZSS). [Mexico: Nuevo Leon] [RPS -]

Mammillaria ascensionis var. **nominis-dulcis** (Lau) Reppenhagen, Gattung Mammillaria, 58, 1987. Basionym: *Mammillaria glassii* var. *nominis-dulcis.* [RPS 38]

Mammillaria atroflorens Backeberg, Die Cact. 6: 3892, 1962. *Nom. inval.* (Art. 9.5). [Erroneously included as valid in RPS 13.] [RPS 13]

Mammillaria aureilanata fa. **alba** (Backeberg) Krainz, Die Kakt. C VIIIc, 1964. Basionym: *Mammillaria aureilanata* var. *alba.* [RPS 15]

Mammillaria aureispina (Lau) Reppenhagen, Gattung Mammillaria, 101, SEM-ills. 79-80 (p. 210), 1987. Basionym: *Mammillaria rekoi* var. *aureispina.* [Erroneously included as *M. aurispina* in RPS 38.] [RPS 38]

Mammillaria aurisaeta Backeberg, Die Cact. 6: 3892, 1962. *Nom. inval.* (Art. 9.5). [Erroneously included as valid in RPS 13.] [RPS 13]

Mammillaria avila-camachoi Shurly, in Backeberg, Die Cact. 5: 3464, 1961. *Nom. inval.* (Art. 37.1). [Erroneously included as valid in RPS 12.] [RPS 12]

Mammillaria backebergiana Buchenau, Nation. Cact. Succ. J. 21: 47, 90, 1966. [Concurrently published in Cact. Suc. Mex. 11: 63-64, 74, 76] [RPS 17]

Mammillaria backebergiana var. **ernestii** (Fittkau) Glass & Foster, Cact. Succ. J. (US) 51(3): 126, 1979. Basionym: *Mammillaria ernestii.* [RPS 30]

Mammillaria bambusiphila Reppenhagen, Mitteilungsbl. AfM 10(5): 161-166, ills., 1986. Typus: *Reppenhagen* 748 (K). [Mexico] [RPS 37]

Mammillaria bambusiphila var. **parva** Reppenhagen, Mitteilungsbl. AfM 10(5): 167-171, ills., 1986. Typus: *Reppenhagen* 663 (K). [Mexico: Michoacan] [RPS 37]

Mammillaria barbata var. *viridiflora* (W. T. Marshall) E. Kuhn, Inform.-brief ZAG Mammillarien 5(3): 40, 1979. *Nom. inval.* (Art. 33.2), based on *Mammillaria wilcoxii* var. *viridiflora.* [RPS -]

Mammillaria barkeri Shurly, in Backeberg Die Cact. 5: 3464, 1961. *Nom. inval.* (Art. 37.1). [Erroneously included as valid in RPS 12.] [RPS 12]

Mammillaria baumii var. *radiaissima* (Lindsay) Neutelings, Succulenta 64(11): 232, ill., 1985. *Nom. inval.* (Art. 33.2), based on *Mammillaria radiaissima.* [RPS 36]

Mammillaria beiselii Diers, Kakt. and. Sukk. 30(3): 57-60, ills., 1979. Typus: *Beisel et Krasucka* 100 (KOELN [Succulentarium], ZSS). [Mexico: Colima / Michoacan] [RPS 30]

Mammillaria bella (Britton & Rose) Weniger, Cacti of the Southwest, 143, 1970. *Nom. inval.* (Art. 33.2, 64.1), based on *Escobaria bella.* [Not dated. *Non Mammillaria bella* Backeberg.] [RPS -]

Mammillaria beneckei var. **multiceps** Reppenhagen, Gattung Mammillaria, 29-30, 1987. Typus: *Reppenhagen* 671 (ZSS). [Mexico: Colima] [RPS 38]

Mammillaria berkiana Lau, Kakt. and. Sukk. 37(2): 30-33, ills., 1986. Typus: *Lau* 1245 (ZSS). [Mexico: Jalisco] [RPS 37]

Mammillaria bernalensis Reppenhagen, Gattung Mammillaria, 151-152, fig. 53 (p. 196), SEM-ills. 137-138 (p. 218), 1987. Typus: *Reppenhagen* 790 (ZSS). [Mexico: Querétaro] [RPS 38]

Mammillaria × birmandreis Bertrand, Cactus (Paris) No. 33: 93-93, 1952. [RPS 3]

Mammillaria bisbeeana (Orcutt) Backeberg, Die Cact. 5: 2971, in synon, 1961. *Nom. inval.* (Art. 33.2, 34.1), based on *Coryphantha bisbeeana*. [RPS -]

Mammillaria blossfeldiana var. **shurliana** (Gates) Wiggins, in Shreve & Wiggins, Veg. Fl. Sonoran Desert 2: 1030, 1964. Basionym: *Mammillaria shurliana*. [First mentioned as *nomen nudum* by Gates, Cact. Succ. J. (US) 13: 78, 1941.] [RPS 18]

Mammillaria bocasana cv. **Edward Hummel** P. C. Hutchison, Cact. Succ. J. (US) 27(3): 85, 1955. [RPS 6]

Mammillaria bocasana cv. **Michael Peachey** G. Barker, Cact. Succ. J. (Australia) 17(12): [page unknown], 1989. [Cited from the German translation published in Mitteilungsbl. AfM 15(2): 78-79, 1991. A form with rose-red coloured epidermis. CAN YOU PLEASE check the validity ????] [RPS 42]

Mammillaria bocasana cv. **Morgan's Cristate** P. C. Hutchison, Cact. Succ. J. (US) 27(3): 84-85, 1955. [RPS 6]

Mammillaria bocasana var. **murivora** Broogh, Kakt. and. Sukk. 28(4): 79, 1977. [*Nomen nugax.*] [RPS 28]

Mammillaria bocensis var. **movasana** Reppenhagen, Gattung Mammillaria, 148, fig. 52 (p. 195), SEM-ills. 131-132 (p. 217), 1987. Typus: *Reppenhagen* 829 (ZSS). [Mexico: Sonora] [RPS 38]

Mammillaria bocensis var. **rubida** (Schwarz *ex* Backeberg) Reppenhagen, Gattung Mammillaria, 146, SEM-ills. 129-130 (p. 217), 1987. Basionym: *Mammillaria rubida*. [RPS 38]

Mammillaria boelderliana Wohlschlager, Kakt. and. Sukk. 39(4): 78-79, ills., 1988. Typus: *Wohlschlager* 369 (WU). [Mexico: Zacatecas] [RPS 39]

Mammillaria bonavitii Reppenhagen, Gattung Mammillaria, 66-68, fig. 13 (p. 186), SEM-ills. 45-46 (p. 206), 1987. Typus: *Reppenhagen* 907 (ZSS). [A name first mentioned in a catalogue by Schmoll and included by Backeberg as provisional in Die Cact. 5: 3466. 1961.] [Mexico: Mexico] [RPS 38]

Mammillaria boolii Lindsay, Cact. Succ. J. (US) 25: 48-49, 1953. Typus: *Lindsay* 2220 (US). [Mexico: Sonora] [RPS 4]

Mammillaria × bosshardii Bosshard, Cactus (Paris) No. 24: 53-54, 1950. [= *Mammillaria crucigera* × *M. sempervivi*.] [RPS 3]

Mammillaria brachytrichion J. Lüthy, Kakt. and. Sukk. 38(12): 294-297, ills., SEM-ills., 1987. Typus: *Lau* 1337 (ZSS). [Mexico: Durango] [RPS 38]

Mammillaria brevicrinita Reppenhagen, Gattung Mammillaria, 42-43, fig. 4 (p. 183), SEM-ills. 25-26 (p. 204), 1987. Typus: *Reppenhagen* 1127 (ZSS). [Mexico: San Luis Potosi] [RPS 38]

Mammillaria bucareliensis var. **multiflora** Reppenhagen, Gattung Mammillaria, 164-165, fig. 61 (p. 198), SEM-ills. 153-154 (p. 220), 1987. Typus: *Reppenhagen* 355a (ZSS). [Mexico: San Luis Potosi] [RPS 38]

Mammillaria bucareliensis var. **tamaulipa** Reppenhagen, Gattung Mammillaria, 166-167, fig. 62 (p. 198), SEM-ills. 155-156 (p. 220), 1987. Typus: *Reppenhagen* 1864 (ZSS). [The collection number of the type is said to be *Reppenhagen* 1866 in the erratum sheet published separately as supplement to Mitteilungsbl. AfM 12(4), 1988, but the type material received at ZSS has the number 1864 as given in the book.] [Mexico: Tamaulipas] [RPS 38]

Mammillaria buchenaui Backeberg *ex* Mottram, Mammillaria Index, 11, 1980. [First published by Backeberg in Descr. Cact. Nov. 3: 8, 1963.] [RPS 14]

Mammillaria busonii Bachel, Kakt. and. Sukk. 22: 78-79, ill., 1971. [*Nomen nugax.*] [RPS 22]

Mammillaria buxbaumiana Reppenhagen, Gattung Mammillaria, 63-64, fig. 12 (p. 185), SEM-ills. 43-44 (p. 206), 1987. Typus: *Reppenhagen* 2037 (ZSS). [Sphalm. 'buxbaumeriana' in error according to a letter from W. Reppenhagen to ZSS, Feb. 22, 1988.] [Mexico: Guanajuato] [RPS 38]

Mammillaria calleana Backeberg, Cactus (Paris) No. 31, Suppl.: 2-3, ill., 1952. [The supplement is dated Dec. 1951 and numbered "Cactus No. 30" but was actually distributed only with No. 31 in March 1952.] [RPS 3]

Mammillaria camptotricha cv. **Mme. Marnier** Glass & Foster, Cact. Succ. J. (US) 53(2): 77, 1981. [RPS 32]

Mammillaria camptotricha var. *albescens* (Tiegel) E. Kuhn, Inform.-brief ZAG Mammillarien 8(4): 47, 1982. *Nom. inval.* (Art. 33.2), based on *Mammillaria albescens*. [RPS -]

Mammillaria camptotricha var. *subinermis* hort. *ex* E. Kuhn, Inform.-brief ZAG Mammillarien 8(4): 47, 1982. *Nom. inval.* (Art. 36.1, 37.1). [RPS -]

Mammillaria candida ssp. **ortizrubiana** (Bravo) Krainz, Die Kakt. Lief.: 55/56: C VIIIc, 1973. Basionym: *Neomammillaria ortizrubiana*. [RPS -]

Mammillaria candida var. **caespitosa** Voss, Cact. Succ. J. (US) 42(6): 280, 1970. [RPS 21]

Mammillaria candida var. **estanzuelensis** Reppenhagen, Gattung Mammillaria, 27, SEM-ills. 3-4 (p. 201), 1987. Typus: *Reppenhagen* 1206 (ZSS). [Mexico: Coahuila] [RPS 38]

Mammillaria candida var. **ortizrubiana**

(Bravo) B. Hofmann, Inform.-brief ZAG Mammillarien 1986: 21, 1986. Basionym: *Neomammillaria ortizrubiana*. [Sphalm. 'ortiz-rubiona' (the spelling of the original publication is re-instated here). Erroneously ommitted from RPS.] [RPS -]

Mammillaria capensis var. *pallida* Remski, Bot. Gaz. 116(2): 165, 1954. *Nom. inval.* (Art. 36.1). [Name attributed to Gates as comb. nov.] [RPS -]

Mammillaria carmenae Castañeda, Anales Inst. Biol. UNAM 24(2): 233-235, 1953. Neotypus: *Lau* 1192 (MEXU [neo]). [Neotype designated by Reppenhagen, Gattung Mammillaria 1: 296, 1991.] [Mexico: Tamaulipas] [RPS 5]

Mammillaria carnea var. **subtetragona** (Dietrich) Backeberg, Die Cact. 5: 3145, 1961. Basionym: *Mammillaria subtetragona*. [RPS 12]

Mammillaria casoi Bravo, Anales Inst. Biol. UNAM 25(1-2): 540, ills. (pp. 537-538), 1954. Typus: *Anonymus* s.n. (MEXU). [Mexico: Oaxaca] [RPS 6]

Mammillaria casoi fa. *longispina* Bravo, Anales Inst. Biol. UNAM 25(1-2): 539, ill., 1954. *Nom. inval.* (Art. 36.1). [Illustration and name in caption only.] [RPS 6]

Mammillaria centralifera Reppenhagen, Gattung Mammillaria, 153-154, fig. 54 (p. 196), SEM-ills. 139-140 (p. 218), 1987. Typus: *Reppenhagen* 1140 (ZSS). [Mexico: Querétaro] [RPS 38]

Mammillaria centraliplumosa Fittkau, Cact. Suc. Mex. 16: 39-41, 47-48, 1971. [RPS 22]

Mammillaria centricirrha var. *inermis* Y. Ito, Cacti, 128, 1952. *Nom. inval.* (Art. 36.1). [RPS 4]

Mammillaria chavezii Cowper, Nation. Cact. Succ. J. 18: 8, 1963. *Nom. inval.* (Art. 36.1, 37.1). [RPS 14]

Mammillaria chica Reppenhagen, Gattung Mammillaria, 52-53, fig. 8 (p. 184), SEM-ills. 35-36 (p. 205), 1987. Typus: *Reppenhagen* 1093 (ZSS). [Mexico: Coahuila] [RPS 38]

Mammillaria claviformis Reppenhagen, Gattung Mammillaria, 91-92, fig. 22 (p. 188), SEM-ills. 73-74 (p. 210), 1987. Typus: *Reppenhagen* 1462 (ZSS). [Mexico: Puebla] [RPS 38]

Mammillaria coahuilensis (Bödeker) Moran, Gentes Herbar. 8(4): 324, 1953. Basionym: *Porfiria coahuilensis*. [RPS 4]

Mammillaria coahuilensis var. **albiarmata** (Bödeker) Hofmann, Lieferungswerk Mammillaria, [part 4], unnumbered page, 1989. Basionym: *Mammillaria albiarmata*. [Dated 1. 12. 1988, effectively published in late summer 1989.] [RPS 40]

Mammillaria coahuilensis var. **albiflora** (Bödeker) Boom & Wouters, Succulenta 1963: 163, 1963. Basionym: *Porfiria schwartzii* var. *albiflora*. [RPS 14]

Mammillaria columbiana var. *albescens* Haage & Backeberg, Die Cact. 6: 3887, 1962. *Nom. inval.* (Art. 37.1). [Erroneously included as valid in RPS 12 (and again in RPS 13). First mentioned in l. c. 5: 3390, 1961; and republished in Kakt.-Lex, 233, 1966 (cf. RPS 17).] [RPS 12]

Mammillaria columbiana var. **bogotensis** (Werdermann) Dugand, Mutisia (Acta Bot. Colomb.) 20: 9, 1954. Basionym: *Mammillaria bogotensis*. [Combination repeated by Backeberg, Die Cact. 5: 3389, 1961 (and reported in RPS 13 as correct).] [RPS 9]

Mammillaria compacticaulis Reppenhagen, Gattung Mammillaria, 80-81, fig. 18 (p. 187), SEM-ills. 63-64 (p. 208), 1987. Typus: *Reppenhagen* 1047 (ZSS). [Mexico: Michoacan] [RPS 38]

Mammillaria compressa var. *tolimensis* (Craig) E. Kuhn, Inform.-brief ZAG Mammillarien 5(2): 20, 1979. *Nom. inval.* (Art. 33.2), based on *Mammillaria tolimensis*. [RPS -]

Mammillaria confusa fa. **strobilina** (Tiegel) Fittkau, Cact. Succ. J. (US) 42: 182-183, 1970. Basionym: *Mammillaria strobilina*. [RPS 21]

Mammillaria conspicua var. **vaupelii** (Tiegel) Reppenhagen, Gattung Mammillaria, 113, SEM-ills. 91-92 (p. 212), 1987. Basionym: *Mammillaria vaupelii*. [RPS 38]

Mammillaria cowperae Shurly, Cact. Succ. J. Gr. Brit. 21: 58-59, 67, 1959. [RPS 10]

Mammillaria crassa Reppenhagen, Gattung Mammillaria, 139-140, fig. 45 (p. 194), SEM-ills. 121-122 (p. 216), 1987. Typus: *Reppenhagen* 980 (ZSS). [Mexico: Aguascalientes] [RPS 38]

Mammillaria crassimammillis Reppenhagen, Gattung Mammillaria, 169-171, fig. 67 (p. 199), SEM-ills. 159-160 (p. 220), 1987. Typus: *Reppenhagen* 1790 (ZSS). [*Reppenhagen* 1620 originally given as collection number of the type, but corrected in the erratum published in Mitteilungsbl. AfM 11(3): 92, 1987.] [Mexico: Nuevo Leon] [RPS 38]

Mammillaria crassior Reppenhagen, Gattung Mammillaria, 85-86, fig. 19 (p. 187), SEM-ills. 67-68 (p. 209), 1987. Typus: *Reppenhagen* 761 (ZSS). [Mexico: Morelos] [RPS 38]

Mammillaria cylindrica var. *coronaria* (Willdenow) B. Hofmann, Inform.-brief ZAG Mammillarien 1984: 55, 1984. *Nom. inval.* (Art. 33.2), based on *Mammillaria coronaria*. [Name used in the transcript of a previously given lecture and probably not accepted by the author.] [RPS -]

Mammillaria dasyacantha (D. Hunt) Reppenhagen, Kakt. and. Sukk. 41(2): 39, 1990. *Nom. inval.* (Art. 33.2, 64.1), based on *Mammillaria laui* fa. *dasyacantha*. [*Non Mammillaria dasyacantha* Engelmann, cf. Art. 64.1.] [RPS 41]

Mammillaria decipiens var. **albescens** (Tiegel) Reppenhagen, Gattung Mammillaria, 66, 1987. Basionym: *Mammillaria albescens*. [Erroneously ommitted from previous issues.] [RPS 42]

Mammillaria deherdtiana Farwig, Cact. Succ. J. (US) 41(1): 27-29, 1969. [RPS 20]

Mammillaria deherdtiana var. **dodsonii** (Bravo) Glass & Foster, Cact. Succ. J. (US) 51(3): 125-126, 1979. Basionym:

Mammillaria dodsonii. [RPS 30]

Mammillaria diacentra Bravo, Anales Inst. Biol. UNAM 24: 47-50, 1953. [RPS 4]

Mammillaria diguetii (F. A. C. Weber) D. Hunt, J. Mamm. Soc. 11(5): 59, 1971. Basionym: *Mammillaria senilis* var. *diguetii.* [RPS 22]

Mammillaria dioica fa. **angelensis** (Craig) Neutelings, Succulenta 65(5): 119, 1986. Basionym: *Mammillaria angelensis.* [First invalidly published l.c. 65(1): 3-5.] [RPS 37]

Mammillaria dioica fa. **estebanensis** (Lindsay) Neutelings, Succulenta 65(5): 119, 1986. Basionym: *Mammillaria estebanensis.* [First invalidly published l.c. 65(1): 3-5.] [RPS 37]

Mammillaria dioica fa. **incerta** (Parish) Neutelings, Succulenta 65(5): 119, 1986. Basionym: *Mammillaria incerta.* [First invalidly published l.c. 65(1): 3-5.] [RPS 37]

Mammillaria dioica fa. **phitauiana** (Baxter) Neutelings, Succulenta 65(5): 119, 1986. Basionym: *Neomammillaria phitauiana.* [First invalidly published l.c. 65(1): 3-5.] [RPS 37]

Mammillaria dioica fa. **verhaertiana** (Bödeker) Neutelings, Succulenta 65(5): 119, 1986. Basionym: *Mammillaria verhaertiana.* [First invalidly published l.c. 65(1): 3-5.] [RPS 37]

Mammillaria dioica var. **armillata** (K. Brandegee) Neutelings, Succulenta 65(5): 119, 1986. Basionym: *Mammillaria armillata.* [First invalidly published l.c. 65(1): 3-5.] [RPS 37]

Mammillaria dioica var. **capensis** (Gates) Neutelings, Succulenta 65(5): 119, 1986. Basionym: *Neomammillaria capensis.* [First invalidly published l.c. 65(1): 3-5.] [RPS 37]

Mammillaria dioica var. **cerralboa** (Britton & Rose) Neutelings, Succulenta 65(5): 119, 1986. Basionym: *Neomammillaria cerralboa.* [First invalidly published in l.c. 65(1): 3-5.] [RPS 37]

Mammillaria dioica var. **incerta** (Parish) Munz, Aliso 4(1): 94, 1957. Basionym: *Mammillaria incerta.* [RPS 8]

Mammillaria dioica var. **multidigitata** (Radley *ex* Lindsay) Neutelings, Succulenta 65(5): 119, 1986. Basionym: *Mammillaria multidigitata.* [First invalidly published in l.c. 65(1): 3-5.] [RPS 37]

Mammillaria dioica var. **neopalmeri** (Craig) Neutelings, Succulenta 65(5): 119, 1986. Basionym: *Mammillaria neopalmeri.* [First invalidly published in l.c. 65(1): 3-5.] [RPS 37]

Mammillaria discolor var. **longispina** Reppenhagen, Gattung Mammillaria, 77, fig. 16 (p. 186), SEM-ills. 57-58 (p. 208), 1987. Typus: *Reppenhagen* 1100 (ZSS). [Mexico: Hidalgo] [RPS 38]

Mammillaria discolor var. **multispina** Reppenhagen, Gattung Mammillaria, 77, fig. 17 (p. 187), 1987. Typus: *Reppenhagen* 1725 (ZSS). [Mexico: Hidalgo] [RPS 38]

Mammillaria discolor var. **ochoterenae** (Bravo) Reppenhagen, Gattung Mammillaria, 77, 1987. Basionym: *Neomammillaria ochoterenae.* [RPS 38]

Mammillaria discolor var. **schmollii** (Bravo) Reppenhagen, Gattung Mammillaria, 78, 1987. Basionym: *Neomammillaria schmollii.* [RPS 38]

Mammillaria dixanthocentron Backeberg *ex* Mottram, Mammillaria-Index, 24, 1980. Typus: *Anonymus* s.n. (fig. 212 p.p. in C. Backeberg, Kakt.-Lex., p. 656.). [Designated by R. Mottram in Mammillaria-Index, 24, 1980.] [Based on *Mammillaria dixanthocentron* Backeberg in Descr. Cact. III: 8 (*nom. inval.*, Art. 9.5) which is based on the collection Buchenau 4 (not preserved).] [Mexico] [RPS -]

Mammillaria dixanthocentron var. **flavicentra** (Backeberg *ex* Mottram) Reppenhagen, Gattung Mammillaria, 122-123, fig. 35 (p. 191), 1987. Basionym: *Mammillaria flavicentra.* [The neotype designation (*Reppenhagen* 1636) is unnecessary, as *M. flavicentra* has already been typified by R. Mottram in 1980.] [RPS 38]

Mammillaria dixanthocentron var. **rubrispina** R. Wolf, Kakt. and. Sukk. 38(2): 46-47, ills., 1987. Typus: *R. & F. Wolf* 19/84 (WU). [Mexico: Oaxaca] [RPS 38]

Mammillaria dodsonii Bravo, Cact. Suc. Mex. 15: 3-6, 1970. [RPS 21]

Mammillaria droegeana Hildmann *ex* Reppenhagen, Gattung Mammillaria, 62, 1987. Based on *Reppenhagen* 1819. *Nom. inval.* (Art. 36.1). [Erroneously ommitted so far. Based on a catalogue name by Hildmann.] [Mexico: Querétaro] [RPS 42]

Mammillaria duncanii (Hester) Weniger, Cacti of the Southwest, 136, 1970. *Nom. inval.* (Art. 33.2), based on *Escobesseya duncanii.* [Not dated.] [RPS -]

Mammillaria duoformis var. **xuchiapensis** Reppenhagen, Gattung Mammillaria, 104-105, fig. 26 (p. 189), SEM-ills. 83-84 (p. 211), 1987. Typus: *Reppenhagen* 1677 (ZSS). [Mexico: Puebla] [RPS 38]

Mammillaria durangicola Reppenhagen, Gattung Mammillaria, 135-136, fig. 43 (p. 193), SEM-ills. 117-118 (p. 215), 1987. Typus: *Reppenhagen* 545 (ZSS). [Mexico: Durango] [RPS 38]

Mammillaria duwei Rogozinski & P. J. Braun, Kakt. and. Sukk. 36(8): 158-164, ills., 1985. Typus: *Duwe et Rogozinski* 1 (KOELN [Succulentarium], ZSS). [Mexico: Guanajuato] [RPS 36]

Mammillaria ebenacantha Shurly *ex* Backeberg, Die Cact. 5: 3469, 1961. *Nom. inval.* (Art. 36.1, 37.1). [Erroneously included as valid in RPS 12.] [RPS 12]

Mammillaria egregia Backeberg *ex* Rogozinski & Appenzeller, Mitteilungsbl. AfM 13(3): 108-116, (5): 214, ills., SEM-ills., 1989. Typus: *Rogozinski* 14/86 (ZSS). [First invalidly (Art. 37.1) published by Backeberg, Die Cact. 5: 3261, 1961; and previously used as a catalogue name by Schwarz.] [Mexico: Chihuahua] [RPS 40]

Mammillaria elegans var. **haageana** (Pfeiffer) Krainz, Kat. ZSS ed. 2, 76, 1967. Basionym: *Mammillaria haageana.* [RPS 18]

Mammillaria elegans var. **lanata** (Britton & Rose) Hofmann, Inform.-Brief ZAG

Mammillaria 1985: 25, 1986. Basionym: *Neomammillaria lanata*. [Issue for 1985, published 1986.] [RPS 37]

Mammillaria elegans var. **longicaudata** Reppenhagen, Gattung Mammillaria, 109-110, fig. 30 (p. 190), SEM-ills. 87-88 (p. 211), 1987. Typus: *Reppenhagen* 918 (ZSS). [Mexico: Morelos] [RPS 38]

Mammillaria elegans var. **lupina** Reppenhagen, Gattung Mammillaria, 108-109, fig. 29 (p. 190), 1987. Typus: *Reppenhagen* 1670 (ZSS). [Mexico: Morelos] [RPS 38]

Mammillaria elegans var. **meissneri** (Ehrenberg) Hofmann, Lieferungswerk Mammillaria, [part 4], unnumbered page, 1989. Basionym: *Mammillaria meissneri*. [Dated 1. 12. 1988, but effectively published in late summer 1989.] [RPS 40]

Mammillaria elegans var. **teyuca** Reppenhagen, Gattung Mammillaria, 111-112, fig. 50 (p. 195), SEM-ills. 89-90 (p. 212), 1987. Typus: *Reppenhagen* 956 (ZSS). [Mexico: Puebla] [RPS 38]

Mammillaria elongata var. **densa** (Link & Otto) Backeberg, Die Cact. 5: 3258, 1961. Basionym: *Mammillaria densa*. [RPS 12]

Mammillaria elongata var. **echinaria** (De Candolle) Backeberg, Die Cact. 5: 3257, 1961. Basionym: *Mammillaria echinaria*. [RPS 12]

Mammillaria elongata var. *obscurior* Heinrich, in Backeberg, Kakt.-Lex., 236, 1966. *Nom. inval.* (Art. 37.1). [RPS 17]

Mammillaria erectacantha var. *mundtii* (Schumann) E. Kuhn, Inform.-brief ZAG Mammillarien 1984: 59, 1984. *Nom. inval.* (Art. 33.2), based on *Mammillaria mundtii*. [Name used in the transcript of a previously given lecture and probably not accepted by the author.] [RPS -]

Mammillaria ernestii Fittkau, Cact. Suc. Mex. 16: 36-38, 1971. [RPS 22]

Mammillaria erythra Reppenhagen, Gattung Mammillaria, 180-182, fig. 68 (p. 199), SEM-ills. 171-172 (p. 222), 1987. Typus: *Reppenhagen* 1350 (ZSS). [Mexico: Puebla / Veracruz] [RPS 38]

Mammillaria erythrocalix Buchenau, Cact. Suc. Mex. 11: 1, 17-21, 27, 42, 1966. [RPS 17]

Mammillaria estebanensis Lindsay, Cact. Succ. J. (US) 39(1): 31, 1967. Typus: *Lindsay* 3002 (SD). [Mexico: Baja California] [RPS 18]

Mammillaria falsicrucigera Backeberg, Die Cact. 6: 3895, 1962. *Nom. inval.* (Art. 36.1, 37.1). [RPS 13]

Mammillaria felipensis Reppenhagen, Gattung Mammillaria, 38-39, fig. 2 (p. 183), SEM-ills. 19-20 (p. 203), 1987. Typus: *Reppenhagen* 636 (ZSS). [Mexico: Guanajuato] [RPS 38]

Mammillaria fittkaui Glass & Foster, Cact. Succ. J. (US) 43(3): 115-117, 1971. Typus: *Glass & Foster in Abbey Garden* 69-1169 (POM). [Mexico: Jalisco] [RPS 22]

Mammillaria flavescens var. **nivosa** (Link) Backeberg, Die Cact. 5: 3168-3169, 1961. Basionym: *Mammillaria nivosa*. [RPS 12]

Mammillaria flavicentra Backeberg *ex* Mottram, Mammillaria-Index, 31, 1980. Typus: *Anonymus* s.n. (fig. 212 p.p. in C. Backeberg, Kakt.-Lex., p. 656). [Said to be a 'lectotype' by Mottram, l.c.] [Based on *Mammillaria flavicentra* Backeberg, Descr. Cact. III: 8 (*Nom. inval.*, Art. 9.5), based on the collection *Buchenau* 3 (not preserved).] [Mexico] [RPS -]

Mammillaria flavihamata Backeberg, Die Cact. 6: 3895, 1962. *Nom. inval.* (Art. 9.5). [Erroneously included as valid in RPS 13.] [RPS 13]

Mammillaria floresii var. *hexacentra* Backeberg, Kakt.-Lex., 238, 1966. *Nom. inval.* (Art. 37.1). [Erroneously included as valid in RPS 17.] [RPS 17]

Mammillaria formosa var. *brauneana* (Bödeker) E. Kuhn, Inform.-brief ZAG Mammillarien 7(3): 33, 35, (4): 55, 1981. *Nom. inval.* (Art. 33.2), based on *Mammillaria brauneana*. [RPS -]

Mammillaria fragrans (Hester) Weniger, Cacti of the Southwest, 132, 1970. *Nom. inval.* (Art. 33.2), based on *Coryphantha fragrans*. [Not dated.] [RPS -]

Mammillaria freudenbergeri Reppenhagen, Kakt. and. Sukk. 38(10): 254-255, ills., 1987. Typus: *Freudenberger* s.n. (K). [Reprinted in Mitteilungsbl. AfM 11(6): 198-201, ills., with additional SEM-ills. of seed.] [Mexico: Coahuila] [RPS 38]

Mammillaria fuauxiana Backeberg, Fuaux. Herb. Bull. 1(4): 53, 1950. [RPS 1]

Mammillaria fuscata var. **russea** (Dietrich) Backeberg, Die Cact. 5: 3388, 1961. Basionym: *Mammillaria russea*. [RPS 12]

Mammillaria fuscata var. **sulphurea** (Sencke) Backeberg, Die Cact. 5: 3388, 1961. Basionym: *Mammillaria sulphurea*. [RPS 12]

Mammillaria garessii Cowper, Cact. Succ. J. (US) 42(1): 14-15, 1970. [RPS 21]

Mammillaria gasterantha Reppenhagen, Kakt. and. Sukk. 31(5): 138-139, ill., 1980. Typus: *Reppenhagen* 934 (ZSS). [Mexico: Guerrero] [RPS 31]

Mammillaria geminispina var. **brevispina** (Hildmann) Backeberg, Die Cact. 5: 3186, 1961. Basionym: *Mammillaria nivea* var. *brevispina*. [RPS 12]

Mammillaria geminispina var. *infernillensis* (Craig) E. Kuhn, Inform.-brief ZAG Mammillarien 7(4): 50, 56, 1981. *Nom. inval.* (Art. 33.2), based on *Mammillaria infernillensis*. [RPS -]

Mammillaria geminispina var. **nivea** (Schumann) Y. Ito, Cacti, 122, 1952. [No basionym traced; validity of name not assessed.] [RPS 4]

Mammillaria geminispina var. **nobilis** (Pfeiffer) Backeberg, Die Cact. 5: 3185, 1961. Basionym: *Mammillaria nobilis*. [RPS 12]

Mammillaria glassii Foster, Cact. Succ. J. (US) 40(4): 129, 132-134, 1968. Typus: *Glass et Foster* 631 (POM). [Mexico: Nuevo Leon] [RPS 19]

Mammillaria glassii var. **ascensionis** (Reppenhagen) Glass & Foster, Cact. Succ. J. (US) 51(3): 126, 1979. Basionym: *Mammillaria ascensionis*. [RPS 30]

Mammillaria glassii var. **nominis-dulcis** Lau, Cact. Succ. J. (US) 57(5): 198, ills., 1985. Typus: *Lau* 1186a (UNAM). [Mexico: Nuevo Leon] [RPS 36]

Mammillaria glassii var. **siberiensis** Lau, Cact. Succ. J. (US) 57(5): 198, ills., 1985. Typus: *Lau* 1322 (MEXU). [Mexico: Nuevo Leon] [RPS 36]

Mammillaria glochidiata var. **xiloensis** Reppenhagen, Gattung Mammillaria, 40-41, fig. 3 (p. 183), SEM-ills. 21-22 (p. 203), 1987. Typus: *Reppenhagen* 1924 (ZSS). [Mexico: Hidalgo] [RPS 38]

Mammillaria goldii Glass & Foster, Cact. Succ. J. (US) 40: 149-151, 1968. [RPS 19]

Mammillaria goodridgei var. **rectispina** Dawson, Cact. Succ. J. (US) 24(3): 80, 1952. [RPS 3]

Mammillaria goodridgii fa. **shurliana** (Gates) Neutelings, Succulenta 65(5): 120, 1986. Basionym: *Mammillaria shurliana*. [First invalidly published in l.c. 65(1): 3-5.] [RPS 37]

Mammillaria goodridgii var. **blossfeldiana** (Bödeker) Neutelings, Succulenta 65(5): 120, 1986. Basionym: *Mammillaria blossfeldiana*. [First invalidly published in l.c. 65(1): 3-5.] [RPS 37]

Mammillaria goodridgii var. **bullardiana** (Gates) Neutelings, Succulenta 65(5): 120, 1986. Basionym: *Neomammillaria bullardiana*. [First invalidly published in l.c. 65(1): 3-5.] [RPS 37]

Mammillaria goodridgii var. **hutchisoniana** (Gates) Neutelings, Succulenta 65(5): 119, 1986. Basionym: *Neomammillaria hutchisoniana*. [First invalidly published in l.c. 65(1): 3-5.] [RPS 37]

Mammillaria goodridgii var. **louisae** (Lindsay) Neutelings, Succulenta 65(5): 119, 1986. Basionym: *Mammillaria louisae*. [First invalidly published in l.c. 65(1): 3-5.] [RPS 37]

Mammillaria goodridgii var. **rectispina** Dawson, Cact. Succ. J. (US) 24(3): 80, 1952. Typus: *Dawson* 10631 (herb. Allan Hancock Foundation). [Mexico: Baja California] [RPS -]

Mammillaria grahamii var. **oliviae** (Orcutt) L. Benson, Cacti of Arizona, ed. 3, 22, 161, 1969. Basionym: *Mammillaria oliviae*. [RPS 20]

Mammillaria grusonii var. *pachycylindrica* (Backeberg) E. Kuhn, Inform.-brief ZAG Mammillarien 6(4): 40, 1980. *Nom. inval.* (Art. 33.2), based on *Mammillaria pachycylindrica*. [RPS -]

Mammillaria grusonii var. *zeyeriana* (Haage *ex* Schumann) E. Kuhn, Inform.-brief ZAG Mammillarien 6(4): 40, 1980. *Nom. inval.* (Art. 33.2), based on *Mammillaria zeyeriana*. [RPS -]

Mammillaria guelzowiana var. **robustior** R. Wolf, Kakt. and. Sukk. 37(12): 256-257, ills., 1986. Typus: *R. & F. Wolf* 60/86 (WU). [Mexico] [RPS 37]

Mammillaria guerreronis var. **zopilotensis** (Craig) Backeberg, Cact. Succ. J. (US) 23(5): 152, 1951. [Published as *comb. nud.*, validity not assessed.] [RPS 3]

Mammillaria guillauminiana Backeberg, Cactus (Paris) No. 33: 81-82 (= Suppl. Cactus No. 33: 9-10) ill., 1952. [RPS 3]

Mammillaria gummifera var. **applanata** (Engelmann) L. Benson, Cacti of Arizona, ed. 3, 22, 152, 1969. Basionym: *Mammillaria applanata*. [RPS 20]

Mammillaria gummifera var. *brandegeei* (Coulter) E. Kuhn, Inform.-brief ZAG Mammillarien 7(1): 8, 1981. *Nom. inval.* (Art. 33.2), based on *Mammillaria brandegeei*. [RPS -]

Mammillaria gummifera var. *gaumeri* (Britton & Rose) E. Kuhn, Inform.-brief ZAG Mammillarien 7(1): 8, 1981. *Nom. inval.* (Art. 33.2), based on *Mammillaria gaumeri*. [RPS -]

Mammillaria gummifera var. **hemisphaerica** (Engelmann) L. Benson, Cact. Succ. J. (US) 41: 128, 1969. Basionym: *Mammillaria hemisphaerica*. [Also proposed (invalidly, Art. 33.2) by B. Hofmann in Inform.-brief ZAG Mammillarien 8(1): [titlepage], 1982.] [RPS 20]

Mammillaria gummifera var. *heyderi* (Engelmann) E. Kuhn, Inform.-brief ZAG Mammillarien 7(1): 8, 1981. *Nom. inval.* (Art. 33.2), based on *Mammillaria heyderi*. [First used in caption to an illustration in l.c. 6(3): 31, 1980.] [RPS -]

Mammillaria gummifera var. *lewisiana* (Lindsay) E. Kuhn, Inform.-brief ZAG Mammillarien 7(1): 8, 1981. *Nom. inval.* (Art. 33.2), based on *Mammillaria lewisiana*. [RPS -]

Mammillaria gummifera var. **macdougalii** (Rose) L. Benson, Cacti of Arizona, ed. 3, 22, 151, 1969. Basionym: *Mammillaria macdougalii*. [RPS 20]

Mammillaria gummifera var. **meiacantha** (Engelmann) L. Benson, Cacti of Arizona, ed. 3, 22, 151, 1969. Basionym: *Mammillaria meiacantha*. [RPS 20]

Mammillaria gummifera var. *peninsularis* (Britton & Rose) E. Kuhn, Inform.-brief ZAG Mammillarien 7(1): 8, 1981. *Nom. inval.* (Art. 33.2), based on *Neomammillaria peninsularis*. [RPS -]

Mammillaria gummifera var. *petrophila* (K. Brandegee) E. Kuhn, Inform.-brief ZAG Mammillarien 7(1): 8, 1981. *Nom. inval.* (Art. 33.2), based on *Mammillaria petrophila*. [RPS -]

Mammillaria haageana var. **schmollii** (Craig) D. Hunt, Cact. Succ. J. Gr. Brit. 41(3): 63, 1979. Basionym: *Mammillaria elegans* var. *schmollii*. [RPS 30]

Mammillaria hahniana var. *chionocephala* (Purpus) E. Kuhn, Inform.-brief ZAG Mammillarien 7(3): 36, (4): 56, 1981. *Nom. inval.* (Art. 33.2), based on *Mammillaria chionocephala*. [RPS -]

Mammillaria hahniana var. *klissingiana* (Bödeker) E. Kuhn, Inform.-brief ZAG Mammillarien 7(3): 36, (4): 56, 1981. *Nom. inval.* (Art. 33.2), based on *Mammillaria klissingiana*. [RPS -]

Mammillaria hahniana var. *mendeliana* (Bravo)

E. Kuhn, Inform.-brief ZAG Mammillarien 7(3): 36, (4): 55, 1981. *Nom. inval.* (Art. 33.2), based on *Neomammillaria mendeliana*. [RPS -]

Mammillaria hamilton-hoytea var. **pilensis** (Shurly *ex* Eggli) Reppenhagen, Gattung Mammillaria, 138-139, 1987. Basionym: *Mammillaria pilensis*. [The 'type locality' cited by Reppenhagen is not based on previously published information from either Backeberg or Shurly.] [RPS 38]

Mammillaria haudeana Lau & Wagner, Kakt. and. Sukk. 29(11): 250-253, ill., 1978. Typus: *Lau* 777 (HEID, ZSS). [Mexico: Sonora] [RPS 29]

Mammillaria heidiae Krainz, Kakt. and. Sukk. 26(10): 217-218, ill., 1975. Typus: *Krähenbühl* 280174 (ZSS). [Mexico: Puebla] [RPS 26]

Mammillaria hemisphaerica var. *waltheri* Remski, Bot. Gaz. 116(2): 165, 1954. *Nom. inval.* (Art. 36.1). [Given as "comb. nov.".] [RPS -]

Mammillaria hernandezii Glass & Foster, Cact. Succ. J. (US) 55(1): 22, 36, ills., 1983. Typus: *Hernandez in Otero* 23 (POM). [Mexico: Oaxaca] [RPS 34]

Mammillaria hesteri (Wright) Weniger, Cacti of the Southwest, 139, 1970. *Nom. inval.* (Art. 33.2), based on *Coryphantha hesteri*. [Not dated.] [RPS -]

Mammillaria heyderi var. **bullingtoniana** Castetter, Pierce & Schwerin, Cact. Succ. J. (US) 48(3): 138-139, 1976. [RPS 27]

Mammillaria heyderi var. **gummifera** (Engelmann) L. Benson, Cact. Succ. J. (US) 47(1): 40, 1975. Basionym: *Mammillaria gummifera*. [RPS 27]

Mammillaria heyderi var. **meiacantha** (Engelmann) L. Benson, Cact. Succ. J. (US) 47(1): 40, 1975. Basionym: *Mammillaria meiacantha*. [RPS 27]

Mammillaria heyderi var. **waltheri** (Bödeker) Mottram, Mammillaria Index, 41, 1980. Basionym: *Mammillaria waltheri*. [RPS 31]

Mammillaria hirsuta var. **grandis** Reppenhagen, Gattung Mammillaria, 36-37, fig. 1 (p. 183), SEM-ills. 17-18 (p. 203), 1987. Typus: *Reppenhagen* 1954 (ZSS). [Mexico: Zacatecas] [RPS 38]

Mammillaria huajuapensis Bravo, Anales Inst. Biol. UNAM 25(1-2): 535-536, ill., 1954. Typus: *Anonymus* s.n. (MEXU). [Mexico: Oaxaca] [RPS 6]

Mammillaria hubertmulleri Reppenhagen, Kakt. and. Sukk. 35(8): 182-184, ills., 1984. Typus: *Reppenhagen* 1460 (K). [Mexico: Morelos] [RPS 35]

Mammillaria huiguerensis Reppenhagen, Gattung Mammillaria, 141-142, fig. 46 (p. 194), SEM-ills. 123-124 (p. 216), 1987. Typus: *Reppenhagen* 975 (ZSS). [Mexico: Zacatecas] [RPS 38]

Mammillaria huitzilopochtli D. Hunt, Cact. Succ. J. Gr. Brit. 41(4): 106-107, 1979. [RPS 30]

Mammillaria ignota Reppenhagen, Gattung Mammillaria, 126-127, fig. 38 (p. 192), SEM-ills. 105-106 (p. 214), 1987. Typus: *Reppenhagen* 1644 (ZSS). [Mexico: Oaxaca] [RPS 38]

Mammillaria igualensis Reppenhagen, Gattung Mammillaria, 123-124, fig. 36 (p. 191), SEM-ills. 101-102 (p. 213), 1987. Typus: *Reppenhagen* 932 (ZSS). [Mexico: Guerrero] [RPS 38]

Mammillaria igualensis var. **palmillensis** Reppenhagen, Gattung Mammillaria, 125-126, fig. 37 (p. 192), SEM-ills. 103-104 (p. 213), 1987. Typus: *Reppenhagen* 1668 (ZSS). [Mexico: Guerrero] [RPS 38]

Mammillaria isotensis Reppenhagen, Gattung Mammillaria, 78-79, fig. 49 (p. 195), SEM-ills. 61-62 (p. 208), 1987. Typus: *Reppenhagen* 771 (ZSS). [Mexico: Mexico] [RPS 38]

Mammillaria jozef-bergeri Wojnański & Trajer, Polsk. Towarz. Milosn. Kakt. 1969: 84, 1969. *Nom. inval.* (Art. 36.1, 37.1). [Validity of name not assessed.] [RPS -]

Mammillaria kleiniorum Appenzeller, Mitteilungsbl. AfM 10(6): 197-203, ills., 1986. Typus: *Klein* 190a (ZSS). [Mexico] [RPS 37]

Mammillaria kraehenbuehlii (Krainz) Krainz, Die Kakt. part 46/47, C VIIIc (June 1971), 1971. Basionym: *Pseudomammillaria kraehenbuehlii*. [First published as nude combination in Kakt. and. Sukk. 22(5): 93-95, 1971. Repeated by D. Hunt in J. Mammillaria Soc. 11: 61 (Oct. 1971).] [RPS 22]

Mammillaria krasuckae Reppenhagen, Gattung Mammillaria, 99-100, fig. 25 (p. 189), 1987. Typus: *Reppenhagen* 1647 (ZSS). [Mexico: Oaxaca] [RPS 38]

Mammillaria kuentziana Fearn, Nation. Cact. Succ. J. 18: 33, 1963. *Nom. inval.* (Art. 37.1). [RPS 14]

Mammillaria lanigera Reppenhagen, Gattung Mammillaria, 116-117, fig. 32 (p. 190), SEM-ills. 95-96 (p. 212), 1987. Typus: *Reppenhagen* 944 (ZSS). [Mexico: Oaxaca] [RPS 38]

Mammillaria lanigera var. **juxtlahuacensis** Reppenhagen, Gattung Mammillaria, 118-119, fig. 33 (p. 191), SEM-ills. 97-98 (p. 213), 1987. Typus: *Reppenhagen* 877 (ZSS). [Mexico: Oaxaca] [RPS 38]

Mammillaria lasiacantha var. *egregia* (Backeberg) Linzen, Inform.-brief ZAG Mammillarien 8(3): 39, (4): 49, 1981. *Nom. inval.* (Art. 33.2), based on *Mammillaria egregia*. [RPS -]

Mammillaria lasiacantha var. *hamatispina* E. Kuhn *ex* Linzen, Inform.-brief ZAG Mammillarien 7(2): 25, 1981. *Nom. inval.* (Art. 36.1, 37.1). [RPS -]

Mammillaria lasiacantha var. *stella-de-tacubaya* (Heese) Linzen, Inform.-brief ZAG Mammillarien 7(2): 25, 1981. *Nom. inval.* (Art. 33.2), based on *Mammillaria stella-de-tacubaya*. [Repeatedly mentioned again in l.c. 8(3): 36, (4): 49, 1982.] [RPS -]

Mammillaria laui D. Hunt, Cact. Succ. J. Gr. Brit. 41(4): 100, SEM-ill, col. ill. (pp. 101, 103), 1979. Typus: *Lau* 1171 (K). [Mexico: Tamaulipas] [RPS 30]

Mammillaria laui fa. **dasyacantha** D. Hunt, Cact. Succ. J. Gr. 41(4): 100, 1979. Typus: *Lau* 1219 p.p. (K). [Mexico: Tamaulipas] [RPS 30]

Mammillaria laui fa. **subducta** D. Hunt, Cact. Succ. J. Gr. Brit. 41(4): 100, 1979. Typus: *Lau* 1222 (K). [Mexico: Tamaulipas] [RPS 30]

Mammillaria laui var. *subducta* (D. Hunt) Reppenhagen, Kakt. and. Sukk. 41(2): 39, 1990. *Nom. inval.* (Art. 33.2), based on *Mammillaria laui* fa. *subducta*. [RPS 41]

Mammillaria leei (Rose *ex* Bödeker) Weniger, Cacti of the Southwest, 142, 1970. *Nom. inval.* (Art. 33.2), based on *Escobaria leei*. [Not dated.] [RPS -]

Mammillaria leptacantha (Lau) Reppenhagen, Gattung Mammillaria, 103, SEM-ills. 81-82 (p. 211), 1987. Basionym: *Mammillaria rekoi* var. *leptacantha*. [RPS 38]

Mammillaria lewisiana Gates, Cact. Succ. J. (US) 27(6): 185-186, 1955. [First mentioned as nude name in a list by Remski in Bot. Gaz. 116(2): 165, 1954.] [RPS 6]

Mammillaria limonensis Reppenhagen, Kakt. and. Sukk. 36(3): 44-46, ills., 1985. Typus: *Reppenhagen* 1620 (K). [Mexico: Jalisco] [RPS 36]

Mammillaria linaresensis R. & F. Wolf, Kakt. and. Sukk. 41(11): 262-264, ills., 1990. Typus: *Wolf* 84/84 (WU). [Mexico: Nuevo Leon] [RPS 41]

Mammillaria longiflora fa. **stampferi** (Reppenhagen) D. Hunt, Cact. Succ. J. Gr. Brit. 41(4): 97-98, 1979. Basionym: *Mammillaria stampferi*. [RPS 30]

Mammillaria longimamma var. *melaleuca* (Karwinski) E. Kuhn, Inform.-brief ZAG Mammillarien 8(4): 47, 1982. *Nom. inval.* (Art. 33.2), based on *Mammillaria melaleuca*. [RPS -]

Mammillaria louisae Lindsay, Cact. Succ. J. (US) 32(6): 169, 1960. Typus: *Lindsay* 2022 (SD). [Mexico: Baja California] [RPS 11]

Mammillaria magallanii fa. *hamatispina* (Backeberg) Linzen, Inform.-brief ZAG Mammillarien 8(3): 38, (4): 49, 1982. *Nom. inval.*, based on *Mammillaria magallanii* var. *hamatispina*, *nom. inval.* (Art. 9.5). [RPS -]

Mammillaria magallanii var. *hamatispina* Backeberg, Die Cact. 6: 3882, 1962. *Nom. inval.* (Art. 9.5). [Erroneously included as valid in RPS 13.] [RPS 13]

Mammillaria magallanii var. *roseoalba* (Bödeker & Ritter) Linzen, Inform.-brief ZAG Mammillarien 8(3): 39, 1982. *Nom. inval.* (Art. 33.2), based on *Mammillaria roseo-alba*. [RPS -]

Mammillaria magneticola Meyran, Cact. Suc. Mex. 6: 17, 1961. [RPS 12]

Mammillaria magnifica Buchenau, Cact. Suc. Mex. 12: 3-7, 20-22, 24, 1967. [RPS 18]

Mammillaria magnifica var. **minor** Buchenau, Cact. Suc. Mex. 12: 3-7, 20-22, 1967. [RPS 18]

Mammillaria magnimamma var. *rubriflora* E. Kuhn, Inform.-brief ZAG Mammillarien 6(3): 30, 1980. *Nom. inval.* (Art. 36, 37, 43). [RPS -]

Mammillaria maritima (Lindsay) D. Hunt, Cact. Succ. J. Gr. Brit. 33: 58, 1971. Basionym: *Cochemiea maritima*. [RPS 22]

Mammillaria marnieriana Backeberg, Cactus (Paris) 31, Suppl.: 2, ill., 1952. [The supplement is dated Dec. 1951 and numbered "Cactus No. 30" but was actually distributed only with No. 31 in March 1952.] [RPS 3]

Mammillaria martinezii Backeberg, Cactus (Paris) 31, Suppl.: 6, ill., 1952. [The supplement is dated Dec. 1951 and numbered "Cactus No. 30" but was actually distributed only with No. 31 in March 1952.] [RPS 3]

Mammillaria mathildae Krähenbühl & Krainz, Kakt. and. Sukk. 24(12): 265-266, ill., 1973. Typus: *Wagner* s.n. (ZSS). [Description repeated in Krainz, Die Kakt. Lief. 55-56: C VIIIc, 1973.] [Mexico: Querétaro] [RPS 24]

Mammillaria matudae Bravo, Cact. Succ. J. Gr. Brit. 18(4): 83-84, 98, 1956. [RPS 7]

Mammillaria matudae fa. *duocentralis* Fittkau, Cact. Suc. Mex. 16: 43-44, 1971. *Nom. inval.* (Art. 37.1). [RPS 22]

Mammillaria matudae var. **serpentiformis** Fittkau, Cact. Suc. Mex. 16: 42-44, 1971. [RPS 22]

Mammillaria mazatlanensis fa. **patonii** (Bravo) Neutelings, Succulenta 65(5): 119, 1986. Basionym: *Neomammillaria patonii*. [First invalidly published l.c. 65(1): 3-5.] [RPS 37]

Mammillaria mazatlanensis fa. **sinalensis** (Craig) Neutelings, Succulenta 65(5): 119, 1986. Basionym: *Mammillaria occidentalis* var. *sinalensis*. [First invalidly published l.c. 65(1): 3-5.] [RPS 37]

Mammillaria mazatlanensis var. **occidentalis** (Britton & Rose) Neutelings, Succulenta 65(5): 119, 1986. Basionym: *Neomammillaria occidentalis*. [First invalidly published l.c. 65(1): 3-5.] [RPS 37]

Mammillaria mercadensis var. **guillauminiana** (Backeberg) Reppenhagen, Gattung Mammillaria, 35, SEM-ills. 11-12 (p. 202), 1987. Basionym: *Mammillaria guillauminiana*. [RPS 38]

Mammillaria meridiorosei Castetter, Pierce & Schwerin , Cact. Succ. J. (US) 50(4): 176-178, 1978. Typus: *Brack* 3 (UNM 63436). [USA: New Mexico] [RPS 29]

Mammillaria meyranii Bravo, Cact. Succ. J. Gr. Brit. 18(4): 84, 98, 1956. [RPS 7]

Mammillaria meyranii var. *michoacana* Buchenau, Cact. Suc. Mex. 14: 75-76, 93, 1969. *Nom. inval.* (Art. 37.1). [RPS 20]

Mammillaria microcarpa var. **auricarpa** W. T. Marshall, Arizona's Cactuses, 109, 1950. Typus: *Earle* s.n. (DES 101). [Type not found by Benson in 1982.] [Validity of this combination depends on the validty of the basionym which bight be invalid under Art. 34.] [USA: Arizona] [RPS 2]

Mammillaria microcarpa var. *grahamii* (Engelmann) Backeberg, Kakteenlexikon, 247, 1966. *Nom. inval.* (Art. 33.2), based on *Mammillaria grahamii*. [Validity of this combination depends on the validty of the basionym which bight be invalid under Art. 34.] [RPS -]

Mammillaria microcarpa var. **milleri** (Britton

& Rose) W. T. Marshall, Arizona's Cactuses, 108, 1950. Basionym: *Neomammillaria milleri*. [Validity of this combination depends on the validty of the basionym which bight be invalid under Art. 34.] [RPS 4]

Mammillaria miegiana Earle, Sag. Bull. 26: 77-79, 1972. [RPS 23]

Mammillaria milleri fa. **auricarpa** (W. T. Marshall) Neutelings, Succulenta 65(5): 119, 1986. Basionym: *Mammillaria microcarpa* var. *auricarpa*. [Sphalm. 'aureicarpa'. First invalidly published in l.c. 65(1): 3-5.] [RPS 37]

Mammillaria milleri fa. **gueldemanniana** (Backeberg) Neutelings, Succulenta 65(5): 119, 1986. Basionym: *Mammillaria gueldemanniana*. [First invalidly published in l.c. 65(1): 3-5.] [RPS 37]

Mammillaria milleri fa. **multancistra** de Morree, Succulenta 69(3): 69-72, ills., 1990. Typus: *de Morree* 801 (L). [USA: Arizona] [RPS f1]

Mammillaria milleri fa. **oliviae** (Orcutt) Neutelings, Succulenta 65(5): 119, 1986. Basionym: *Mammillaria oliviae*. [First invalidly published in l.c. 65(1): 3-5.] [RPS 37]

Mammillaria milleri fa. **swinglei** (Britton & Rose) Neutelings, Succulenta 65(5): 119, 1986. Basionym: *Neomammillaria swinglei*. [First invalidly published in l.c. 65(1): 3-5.] [RPS 37]

Mammillaria milleri var. *grahamii* (Engelmann) Neutelings, Succulenta 65(5): 119, 1986. Incorrect name (Art. 57.1), based on *Mammillaria grahamii*. [First invalidly published in l.c. 65(1): 3-5.] [RPS 37]

Mammillaria milleri var. **sheldonii** (Britton & Rose) Neutelings, Succulenta 65(5): 119, 1986. Basionym: *Neomammillaria sheldonii*. [First invalidly published in l.c. 65(1): 3-5.] [RPS 37]

Mammillaria mitlensis Bravo, Cact. Suc. Mex. 1(5): 85-87, 1956. [RPS 7]

Mammillaria mixtecensis Bravo, Anales Inst. Biol. UNAM 25(1-2): 533-534, ill., 1954. Typus: *Bravo Hollis* s.n. (MEXU). [Mexico: Oaxaca] [RPS 6]

Mammillaria moeller-valdeziana Appenzeller, Mitteilungsbl. AfM 12(5): 161-167, ills., SEM-ills., 1988. Typus: *Moeller-Valdez et H. J. Klein in Klein* 217 (ZSS). [Additional information in l.c. (6): 212-213, SEM-ill., 1988.] [Mexico: San Luis Potosi] [RPS 39]

Mammillaria mollihamata Shurly, Cact. Succ. Gr. Brit. 22(3): 52, 57, 1960. [RPS 11]

Mammillaria monticola Reppenhagen, Gattung Mammillaria, 115-116, fig. 31 (p. 190), SEM-ills. 93-94 (p. 212), 1987. Typus: *Reppenhagen* 844 (ZSS). [Mexico: Puebla] [RPS 38]

Mammillaria morricalii Cowper, Cact. Succ. J. (US) 41(5): 208-209, 1969. [RPS 20]

Mammillaria multihamata var. *fittkaui* (Glass & Foster) E. Kuhn, Kakt. / Sukk. 10(3): 44-46, ill., 1975. *Nom. inval.* (Art. 43.1), based on *Mammillaria fittkaui*. [RPS -]

Mammillaria mystax var. *mixtecensis* (Bravo) E. Kuhn, Inform.-brief ZAG Mammillarien 5(2): 19, 21, 1979. *Nom. inval.* (Art. 33.2), based on *Mammillaria mixtecensis*. [RPS -]

Mammillaria mystax var. *mutabilis* (Scheidweiler) E. Kuhn, Inform.-brief ZAG Mammillarien 5(2): 20-21, 1979. *Nom. inval.* (Art. 33.2), based on *Mammillaria mutabilis*. [RPS -]

Mammillaria mystax var. *neomystax* (Backeberg) E. Kuhn, Inform.-brief ZAG Mammillarien 5(2): 21, 1979. *Nom. inval.* (Art. 33.2), based on *Mammillaria neomystax*. [RPS -]

Mammillaria mystax var. *varieaculeata* (Buchenau) E. Kuhn, Inform.-brief ZAG Mammillarien 5(2): 21, 1979. *Nom. inval.* (Art. 33.2), based on *Mammillaria varieaculeata*. [RPS -]

Mammillaria nagliana Reppenhagen, Gattung Mammillaria, 173-174, fig. 65 (p. 199), SEM-ills. 163-164 (p. 221), 1987. Typus: *Reppenhagen* 1045 (ZSS). [Mexico: Michoacan] [RPS 38]

Mammillaria nana Backeberg *ex* Mottram, Mammillaria-Index, 59, 1980. Typus: *Anonymus* s.n. (icono). [Backeberg, Kakt.-Lex., 658, fig. 215, 1966. Originally based on Buchenau 7 (not preserved).] [First published by C. Backeberg in Descr. Cact. Nov. 3: 8, 1963, *nom. inval.* (Art. 9.5).] [RPS -]

Mammillaria nazasensis (Glass & Foster) Reppenhagen, Gattung Mammillaria, 36, SEM-ills. 13-14 (p. 202), 1987. Basionym: *Mammillaria pennispinosa* var. *nazasensis*. [RPS 38]

Mammillaria nejapensis fa. **brevispina** (Craig & Dawson) Krainz, Die Kakt. C VIIIc, 1961. Basionym: *Mammillaria nejapensis* var. *brevispina*. [RPS 12]

Mammillaria nejapensis fa. **longispina** (Craig & Dawson) Krainz, Die Kakt. C VIIIc, 1961. Basionym: *Mammillaria nejapensis* var. *longispina*. [RPS 12]

Mammillaria nellieae (Croizat) Weniger, Cacti of the Southwest, 139, 1970. *Nom. inval.* (Art. 33.2), based on *Coryphantha nellieae*. [Not dated, and attributed to Croizat.] [RPS -]

Mammillaria neobertrandiana Backeberg, Cactus (Paris) No. 33: 82 (= Suppl. Cactus No. 33: 10), 1952. Neotypus: *Reppenhagen* 686 (ZSS [neo]). [Neotype designated by Reppenhagen, Gattung Mammillaria 1: 280, 1991.] [Mexico: Durango] [RPS 3]

Mammillaria neocrucigera Backeberg, Die Cact. 5: 3426, 1961. *Nom. inval.* (Art. 36.1, 37.1), based on *Mammillaria crucigera*. [An invalid *nomen novum* for *Mammillaria crucigera* Bödeker *non* Martius.] [RPS 12]

Mammillaria neomystax Backeberg, Cactus (Paris) 31, Suppl.: 5-6, ill., 1952. [The supplement is dated Dec. 1951 and numbered "Cactus No. 30" but was actually distributed only with No. 31 in March 1952.] [RPS 3]

Mammillaria neopotosina Remski, Bot. Gaz. 116(2): 165, 1954. *Nom. inval.* (Art. 36.1). [Given as "nom. nov.".] [RPS -]

Mammillaria noureddineana Reppenhagen, Gattung Mammillaria, 128-129, fig. 39 (p.

192), SEM-ills. 107-108 (p. 214), 1987. Typus: *Reppenhagen* 1646 (ZSS). [Mexico: Oaxaca] [RPS 38]

Mammillaria nunezii var. **solisii** (Britton & Rose) Backeberg, Die Cact. 5: 3374, 1961. Basionym: *Neomammillaria solisii.* [RPS 12]

Mammillaria obconella var. *durispina* (Bödeker) E. Kuhn, Inform.-brief ZAG Mammillarien 1984: 59, 1984. *Nom. inval.* (Art. 33.2), based on *Mammillaria durispina.* [RPS -]

Mammillaria obconella var. **galeottii** (Scheidweiler) Backeberg, Die Cact. 5: 3343, 1961. Basionym: *Mammillaria galeottii.* [RPS 12]

Mammillaria obconella var. *ingens* (Backeberg) E. Kuhn, Inform.-brief ZAG Mammillarien 1984: 59, 1984. *Nom. inval.* (Art. 33.2), based on *Mammillaria ingens.* [RPS -]

Mammillaria obconella var. *kelleriana* (Schmoll) E. Kuhn, Inform.-brief ZAG Mammillarien 1984: 59, 1984. *Nom. inval.* (Art. 33.2), based on *Mammillaria kelleriana.* [RPS -]

Mammillaria obscura var. *heeseana* (MacDowell) E. Kuhn, Inform.-brief ZAG Mammillarien 7(1): 6, 1981. *Nom. inval.* (Art. 33.2), based on *Mammillaria heeseana.* [RPS -]

Mammillaria occidentalis var. **monocentra** (Craig) Backeberg, Die Cact. 5: 3289, 1961. Basionym: *Mammillaria mazatlanensis* var. *monocentra.* [RPS 12]

Mammillaria orestera L. Benson, The Cacti of Arizona ed. 3, 22, 155, 1969. [RPS 20]

Mammillaria oteroi Glass & Foster, Cact. Succ. J. (US) 47(2): 94, 1975. Typus: *Otero s.n. in Abbey Garden* 72-003 (POM). [Mexico: Oaxaca] [RPS 27]

Mammillaria pachycylindrica Backeberg, Cact. Succ. J. Gr. Brit. 21: 82-83, 1959. [Sphalm. 'pachypcylindrica'.] [RPS 10]

Mammillaria papasquiarensis (Bravo) Reppenhagen, Gattung Mammillaria, 137-138, fig. 44 (p. 193), SEM-ills. 119-120 (p. 215), 1987. Basionym: *Mammillaria tesopacensis* var. *papasquiarensis.* [RPS 38]

Mammillaria parkinsonii var. *brevispina* Remski, Bot. Gaz. 116(2): 165, 1954. *Nom. inval.* (Art. 36.1). [RPS -]

Mammillaria parkinsonii var. **dietrichiae** (Tiegel) Backeberg, Die Cact. 5: 3217, 1961. Basionym: *Mammillaria dietrichiae.* [RPS 12]

Mammillaria parrasensis Reppenhagen, Gattung Mammillaria, 133-134, fig. 42 (p. 193), SEM-ills. 115-116 (p. 215), 1987. Typus: *Reppenhagen* 1083 (ZSS). [Mexico: Coahuila] [RPS 38]

Mammillaria patonii var. **sinalensis** (Craig) Backeberg, Die Cact. 5: 3291, 1961. Basionym: *Mammillaria occidentalis* var. *sinalensis.* [RPS 12]

Mammillaria pectinifera fa. **solisioides** (Backeberg) Sanchez-Mejorada, Cact. Suc. Mex. 25(3): 65, 1980. Basionym: *Mammillaria solisioides.* [RPS 31]

Mammillaria pectinifera var. *solisioides* (Backeberg) Linzen, Inform.-brief ZAG Mammillarien 8(3): 36, 1982. *Nom. inval.* (Art. 33.2), based on *Mammillaria solisioides.* [RPS -]

Mammillaria pennispinosa var. **nazasensis** Glass & Foster, Cact. Succ. J. (US) 47(2): 96, ill., 1975. Typus: *Glass et Foster* 802 (ZSS). [Mexico: Durango] [RPS 27]

Mammillaria perezdelarosae Bravo & Scheinvar, Cact. Suc. Mex. 30(4): 76-80, ills., SEM-ills., 1985. Typus: *Bravo-H. et Scheinvar* 3938 (MEXU). [Mexico: Jalisco] [RPS 36]

Mammillaria picta var. *dumetorum* (Purpus) E. Kuhn, Inform.-brief ZAG Mammillarien 1975(3): 30, 1975. *Nom. inval.* (Art. 33.2), based on *Mammillaria dumetorum.* [RPS -]

Mammillaria picta var. *viereckii* (Bödeker) E. Kuhn, Inform.-brief ZAG Mammillarien 1975(3): 30, 1975. *Nom. inval.* (Art. 33.2), based on *Mammillaria viereckii.* [RPS -]

Mammillaria pilensis Shurly *ex* Eggli, Bradleya 3: 102, 1985. Typus: *Schwarz* s.n. (K). [Type ex hort. Shurly.] [First published invalidly by Backeberg, Die Cact. 5: 3408-3410, fig. 3149, 1961.] [Mexico] [RPS 36]

Mammillaria pilispina var. *albicoma* (Bödeker) E. Kuhn, Kakt. / Sukk. 11(1): 10, 1976. *Nom. inval.* (Art. 33.2, 43.1), based on *Mammillaria albicoma.* [RPS -]

Mammillaria pitcayensis Bravo, Anales Inst. Biol. UNAM 28: 37-38, ill., 1957. Typus: *Matuda* s.n. (?). [Description reprinted in Cact. Suc. Mex. 3(4): 75-76, 1957. The epithet is spelled 'pitcayensis' throughout both papers, although it is presumably named for the village 'Pilcaya', which is similarly mis-spelled.] [Mexico: Guerrero] [RPS 9]

Mammillaria polyedra var. *multiseta* (Ehrenberg) E. Kuhn, Inform.-brief ZAG Mammillarien 5(2): 20, 1979. *Nom. inval.* (Art. 33.2), based on *Mammillaria multiseta.* [RPS -]

Mammillaria polyedra var. *polygona* (Salm-Dyck) E. Kuhn, Inform.-brief ZAG Mammillarien 5(2): 20, 1979. *Nom. inval.* (Art. 33.2), based on *Mammillaria polygona.* [RPS -]

Mammillaria pottsii var. **gigas** Reppenhagen, Gattung Mammillaria, 61, fig. 10 (p. 185), SEM-ills. 39-40 (p. 205), 1987. Typus: *Reppenhagen* 2135 (ZSS). [Mexico: Zacatecas] [RPS 38]

Mammillaria pottsii var. **multicaulis** Reppenhagen, Gattung Mammillaria, 60-61, fig. 9 (p. 185), SEM-ills. 37-38 (p. 205), 1987. Typus: *Reppenhagen* 687 (ZSS). [Mexico: Durango] [RPS 38]

Mammillaria priessnitzii Reppenhagen, Gattung Mammillaria, 158-160, fig. 58 (p. 197), SEM-ills. 147-148 (p. 219), 1987. Typus: *Reppenhagen* 1142 (ZSS). [Mexico: Querétaro] [RPS 38]

Mammillaria pringlei var. *longicentra* Backeberg, Descr. Cact. Nov. 3: 8, 1963. *Nom. inval.* (Art. 9.5). [Erroneously included as valid in RPS 14.] [RPS 14]

Mammillaria prolifera fa. **grisea** (Meinshausen) B. Hofmann, Inform.-brief ZAG Mammillarien 1986: 21, 1986.

Basionym: *Mammillaria multiceps* var. *grisea*. [RPS -]

Mammillaria prolifera fa. **haitiensis** (Schumann) Krainz, Die Kakt. C VIIIc, 1964. Basionym: *Mammillaria pusilla* var. *haitiensis*. [RPS 15]

Mammillaria prolifera var. **arachnoidea** D. Hunt, Cact. Succ. J. Gr. Brit. 40(1): 11, 1978. [RPS 29]

Mammillaria prolifera var. **haitiensis** (Schumann) Y. Ito, Cacti, 127, 1952. [Sphalm. 'haitensis'. Given as *nom. nud.* in RPS 4, validity not assessed. Name first used by Borg, Cacti, 316, 1937.] [RPS -]

Mammillaria prolifera var. *nivosa* (Link) B. Hofmann, Inform.-brief ZAG Mammillarien 5(2): 26, 1979. *Nom. inval.* (Art. 33.2), based on *Mammillaria nivosa*. [RPS -]

Mammillaria prolifera var. **perpusilla** (Meinshausen) B. Hofmann, Lieferungswerk Mammillaria, [part 2], unnumbered page, 1987. Basionym: *Mammillaria multiceps* var. *perpusilla*. [Dated 1. 7. 1986, but effectively distributed only in 1987.] [RPS 38]

Mammillaria pseudoalamensis Backeberg, Cactus (Paris) No. 37: 210, 1953. [RPS 4]

Mammillaria pseudogoldii hort., J. Mammillaria Soc. 19(3): 31, 1979. *Nom. inval.* (Art. 36.1, 37.1). [Provisional name.] [RPS -]

Mammillaria pseudorekoi var. *duoformis* (Craig & Dawson) E. Kuhn, Inform.-brief ZAG Mammillarien 1984: 60, 1984. *Nom. inval.* (Art. 33.2), based on *Mammillaria duoformis*. [Used in transcript of a lecture.] [RPS -]

Mammillaria pseudoscrippsiana Backeberg, Cactus (Paris) 31, Suppl.: 4, ill., 1952. [The supplement is dated Dec. 1951 and numbered "Cactus No. 30" but was actually distributed only with No. 31 in March 1952.] [RPS 3]

Mammillaria pseudoscrippsiana var. **rooksbyana** Backeberg, Cactus (Paris) 31, Suppl.: 4-5, ill., 1952. [The supplement is dated Dec. 1951 and numbered "Cactus No. 30" but was actually distributed only with No. 31 in March 1952.] [RPS 3]

Mammillaria pseudosimplex Haage & Backeberg, Die Cact. 6: 3888, 1962. *Nom. inval.* (Art. 37.1). [Erroneously included as valid in RPS 12 (and again in RPS 13). First mentioned in l.c. 5: 3390, 1961.] [RPS 12]

Mammillaria puberula Reppenhagen, Gattung Mammillaria, 44-45, fig. 5 (p. 184), SEM-ills. 23-24 (p. 203), 1987. Typus: *Reppenhagen* 1059 (ZSS). [Mexico: San Luis Potosi] [RPS 38]

Mammillaria pullihamata Backeberg *ex* Reppenhagen, Gattung Mammillaria, 96-97, fig. 23 (p. 188), 1987. Typus: *Reppenhagen* 195 (ZSS). [Based on a provisional name by C. Backeberg, Die Cact. 6: 3898, 1962 (included as 'n. sp. subnud.' in RPS 13).] [Mexico: Oaxaca] [RPS 38]

Mammillaria ramifera Remski, Bot. Gaz. 116(2): 165, 1954. *Nom. inval.* (Art. 36.1). [Attributed to Clover.] [RPS -]

Mammillaria ramillosa (Cutak) Weniger, Cacti of the Southwest, 119, 1970. *Nom. inval.* (Art. 33.2), based on *Coryphantha ramillosa*. [RPS -]

Mammillaria rectispina (Dawson) Reppenhagen, Gattung Mammillaria, 33, SEM-ills. 9-10 (p. 202), 1987. Basionym: *Mammillaria goodridgii* var. *rectispina*. [RPS 38]

Mammillaria rekoi var. **aureispina** Lau, Cact. Suc. Mex. 28(1): 19-20, ill., 1983. Typus: *Lau* 1055 (MEXU, ZSS). [Mexico: Oaxaca] [RPS 34]

Mammillaria rekoi var. **leptacantha** Lau, Cact. Succ. J. (US) 55(2): 69-70, ills., 1983. Typus: *Lau* 1314 (ZSS). [Mexico: Oaxaca] [RPS 34]

Mammillaria reppenhagenii D. Hunt, Kakt. and. Sukk. 28(7): 129-130, ills., 1977. Typus: *Reppenhagen* 382 (ZSS). [Mexico: Colima] [RPS 28]

Mammillaria rhodantha var. *rutila* (Zuccarini) E. Kuhn, Inform.-brief ZAG Mammillarien 1984: 59, 1984. *Nom. inval.* (Art. 33.2), based on *Mammillaria rutila*. [Used in transcript of a lecture.] [RPS -]

Mammillaria rioverdense Reppenhagen, Gattung Mammillaria, 162-163, fig. 60 (p. 197), SEM-ills. 151-152 (p. 219), 1987. Typus: *Reppenhagen* 1123 (ZSS). [*Reppenhagen* 1133 erroneously given as type, according to the erratum sheet published as supplement to Mitteilungsbl. AfM 12(4), 1988.] [Mexico: San Luis Potosi] [RPS 38]

Mammillaria robertii (A. Berger) Weniger, Cacti of the Southwest, 140, 1970. *Nom. inval.* (Art. 33.2), based on *Coryphantha robertii*. [Not dated. Sphalm. 'roberti'.] [RPS -]

Mammillaria roseoalba var. *winterae* (Bödeker) E. Kuhn, Inform.-brief ZAG Mammillarien 6(4): 41, 1980. *Nom. inval.* (Art. 33.2), based on *Mammillaria winterae*. [RPS -]

Mammillaria rosiflora (Lahman) Weniger, Cacti of the Southwest, 124, 1970. *Nom. inval.*, based on *Neobesseya rosiflora, nom. inval.* (Art. ?). [RPS -]

Mammillaria rossiana W. Heinrich, Kakt. and. Sukk. 9(8): 119-121, 1958. *Nom. inval.* (Art. 9.5). [Material at ZSS was prepared some years after the publication and can therefore not be regarded as holotype.] [RPS 9]

Mammillaria rubrograndis Reppenhagen & Lau *ex* Reppenhagen, Kakt. and. Sukk. 30(4): 82-83, ill., 1979. Typus: *Lau s.n.* (ZSS). [Mexico: Tamaulipas] [RPS 30]

Mammillaria runyonii (Britton & Rose) Weniger, Cacti of the Southwest, 121, 1970. *Nom. inval.* (Art. 33.2), based on *Coryphantha runyonii*. [Not dated.] [RPS -]

Mammillaria saboae Glass, Cact. Suc. Mex. 11: 55-60, 73, 1966. [RPS 17]

Mammillaria saboae fa. **haudeana** (Lau & Wagner) D. Hunt, Cact. Succ. Gr. Brit. 41(4): 97, 1979. Basionym: *Mammillaria haudeana*. [RPS 30]

Mammillaria saboae var. **goldii** (Glass & Foster) Glass & Foster, Cact. Succ. J. (US) 51(3): 123, 1979. Basionym: *Mammillaria goldii*. [RPS 30]

Mammillaria saboae var. **haudeana** (Lau &

Wagner *ex* Wagner) Glass & Foster, Cact. Succ. J. (US) 51(3): 124, 1979. Basionym: *Mammillaria haudeana*. [RPS 30]

Mammillaria saboae var. **theresae** (Cutak) Rowley, J. Mamm. Soc. 19(3): 30, 1979. Basionym: *Mammillaria theresae*. [RPS 30]

Mammillaria saint-pieana Backeberg *ex* Mottram, Mammillaria Index, 80, 1980. [First invalidly published (*nom. inval.* Art. 9.5, 37.1) by Backeberg in Descr. Cact. Nov. 3: 8, 1963, and reported as valid in RPS 14.] [RPS 14]

Mammillaria san-angelensis Sanchez-Mejorada, Cact. Suc. Mex. 26(1): 8, 1981. [RPS 32]

Mammillaria sanjuanensis Reppenhagen, Gattung Mammillaria, 98-99, fig. 24 (p. 188), SEM-ills. 77-78 (p. 210), 1987. Typus: *Reppenhagen* 946 (ZSS). [Mexico: Oaxaca] [RPS 38]

Mammillaria santaclarensis Cowper, Cact. Succ. J. (US) 41(6): 248-250, 1969. [RPS 20]

Mammillaria saxicola Reppenhagen, Gattung Mammillaria, 160-162, fig. 59 (p. 197), SEM-ills. 149-150 (p. 219), 1987. Typus: *Reppenhagen* 1139 (ZSS). [Mexico: Querétaro] [RPS 38]

Mammillaria schiedeana var. **dumetorum** (J. Purpus) Glass & Foster, Cact. Succ. J. (US) 53(3): 143-144, 1981. Basionym: *Mammillaria dumetorum*. [Combination also proposed by T. Linzen in Inform.-brief ZAG Mammillarien 8(4): 49, 1982.] [RPS 32]

Mammillaria schumannii var. **globosa** R. Wolf, Kakt. and. Sukk. 38(6): 147, ills., 1987. Typus: *R. & F. Wolf* 47/83 (WU). [Mexico: Baja California Sur] [RPS 38]

Mammillaria schwartzii (Fric) Backeberg, Kakt.-Lex., 258, 1966. *Nom. illeg.* (Art. 64.1), based on *Haagea schwartzii*. [*Non M. schwarzii* Shurly 1960.] [RPS 17]

Mammillaria schwarzii Buxbaum, in Krainz, Die Kakt. C VIIIc (23), 1961. *Nom. inval.* (Art. 64). [*Non Mammillaria schwarzii* Shurly, based on *Haagea schwarzii* Fric, attributed to Moran and given in error instead of *M. coahuilensis*.] [RPS 12]

Mammillaria scrippsiana var. **armeria** Reppenhagen, Gattung Mammillaria, 149-150, fig. 48 (p. 194), SEM-ills. 135-136 (p. 217), 1987. Typus: *Reppenhagen* 1621 (ZSS). [Mexico: Jalisco] [RPS 38]

Mammillaria scrippsiana var. *pseudoscrippsiana* (Backeberg) E. Kuhn, Inform.-brief ZAG Mammillarien 6(4): 44, 1980. *Nom. inval.* (Art. 33.2), based on *Mammillaria pseudoscrippsiana*. [RPS -]

Mammillaria seitziana var. **tolantongensis** Reppenhagen, Gattung Mammillaria, 154-156, fig. 55 (p. 196), SEM-ills. 141-142 (p. 218), 1987. Typus: *Reppenhagen* 1689 (ZSS). [Mexico: Hidalgo] [RPS 38]

Mammillaria sempervivi var. **caput-medusae** (Otto) Backeberg, Die Cact. 5: 3160, 1961. Basionym: *Mammillaria caput-medusae*. [RPS 12]

Mammillaria setispina var. *maritima* (Lindsay) E. Kuhn, Inform.-brief ZAG Mammillarien 8(4): 47, 1982. *Nom. inval.* (Art. 33.2), based on *Cochemiea maritima*. [RPS -]

Mammillaria shurliana Gates, Cact. Succ. J. Gr. Brit. 18: 30, 1956. Typus: *Gates* 518 (DS). [First mentioned as *nom. nud.* at varietal level in Cact. Succ. J. (US) 13: 78, 1941. RPS 7 incorrectly gives the 1956-publication as publication place for a new combination.] [Mexico: Baja California] [RPS 7]

Mammillaria shurlyi Buxbaum, In Krainz, Die Kakt. C VIIc, 1961. [*Nom. nov. pro M. schwarzii* Shurly; incorrectly considered invalid / illegitimate in RPS 12 because of assumed homonymy with *Mammillaria shurliana* Gates.] [RPS 12]

Mammillaria silvatica Reppenhagen, Gattung Mammillaria, 89-90, fig. 21 (p. 188), SEM-ills. 71-72 (p. 209), 1987. Typus: *Reppenhagen* 1603 (ZSS). [Mexico: Jalisco] [RPS 38]

Mammillaria simplex var. *albidispina* Backeberg, Kakt.-Lex., 259, 1966. *Nom. inval.* (Art. 37.1). [Erroneously included as valid in RPS 17.] [RPS 17]

Mammillaria sneedii (Britton & Rose) Weniger, Cacti of the Southwest, 141, 1970. *Nom. inval.* (Art. 33.2), based on *Escobaria sneedii*. [Not dated.] [RPS -]

Mammillaria solisioides Backeberg, Cactus (Paris) No. 31, Suppl.: 3, ill., 1952. [The supplement is dated Dec. 1951 and numbered "Cactus No. 30" but was actually distributed only with No. 31 in March 1952.] [RPS 3]

Mammillaria sonorensis var. *bocensis* (Craig) E. Kuhn, Inform.-brief ZAG Mammillarien 7(1): 7, 1981. *Nom. inval.* (Art. 33.2), based on *Mammillaria bocensis*. [RPS -]

Mammillaria sonorensis var. *canelensis* (Craig) E. Kuhn, Inform.-brief ZAG Mammillarien 6(4): 43, 1980. *Nom. inval.* (Art. 33.2), based on *Mammillaria canelensis*. [RPS -]

Mammillaria sonorensis var. *evermanniana* (Britton & Rose) E. Kuhn, Inform.-brief ZAG Mammillarien 6(4): 43, 1980. *Nom. inval.* (Art. 33.2), based on *Neomammillaria evermanniana*. [RPS -]

Mammillaria sonorensis var. *marksiana* (Krainz) E. Kuhn, Inform.-brief ZAG Mammillarien 6(4): 43, 1980. *Nom. inval.* (Art. 33.2), based on *Mammillaria marksiana*. [RPS -]

Mammillaria sonorensis var. *tayloriorum* (Glass & Foster) E. Kuhn, Inform.-brief ZAG Mammillarien 6(4): 43, 1980. *Nom. inval.* (Art. 33.2), based on *Mammillaria tayloriorum*. [RPS -]

Mammillaria sphacelata var. *viperina* (J. A. Purpus) E. Kuhn, Inform.-brief ZAG Mammillarien 8(1): 7, 1982. *Nom. inval.* (Art. 33.2), based on *Mammillaria viperina*. [RPS -]

Mammillaria spinosissima var. *bella* (Backeberg) B. Hofmann, Inform.-brief ZAG Mammillarien 1984: 21, 1984. *Nom. inval.* (Art. 33.2), based on *Mammillaria bella*. [Only used in caption to illustration.] [RPS -]

Mammillaria spinosissima var. *centraliplumosa* (Fittkau) E. Kuhn, Inform.-brief ZAG

Mammillarien 1984: 59, 1984. *Nom. inval.* (Art. 33.2), based on *Mammillaria centraliplumosa.* [RPS -]

Mammillaria spinosissima var. *meyranii* (Bravo) E. Kuhn, Inform.-brief ZAG Mammillarien 1984: 59, 1984. *Nom. inval.* (Art. 33.2), based on *Mammillaria meyranii.* [RPS -]

Mammillaria spinosissima var. *pilcayensis* (Bravo) E. Kuhn, Inform.-brief ZAG Mammillarien 1984: 59, 1984. *Nom. inval.* (Art. 33.2), based on *Mammillaria pilcayensis.* [RPS -]

Mammillaria spinosissima var. *virginis* (Fittkau & Kladiwa) E. Kuhn, Inform.-brief ZAG Mammillarien 1984: 59, 1984. *Nom. inval.* (Art. 33.2), based on *Mammillaria virginis.* [RPS -]

Mammillaria stampferi Reppenhagen, Kakt. and. Sukk. 30(8): 185-187, ills., 1979. Typus: *Reppenhagen* 1358 (ZSS). [Mexico: Durango] [RPS 30]

Mammillaria standleyi var. *lindsayi* (Craig) E. Kuhn, Inform.-brief ZAG Mammillarien 7(1): 7, 1981. *Nom. inval.* (Art. 33.2), based on *Mammillaria lindsayi.* [RPS -]

Mammillaria subcrocea var. *echinata* (Schumann) Y. Ito, Cacti, 127, 1952. *Nom. inval.* (Art.). [Described as *nomen nudum*, no basionym traceable.] [RPS 4]

Mammillaria subducta (D. Hunt) Reppenhagen, Gattung Mammillaria, 55, 1987. Basionym: *Mammillaria laui* fa. *subducta.* [Basionym is cited via the literature references given.] [RPS 38]

Mammillaria subducta var. **dasyacantha** (D. Hunt) Reppenhagen, Gattung Mammillaria, 55, 1987. Basionym: *Mammillaria laui* fa. *dasyacantha.* [Basionym is cited indirectly and via the literature references given.] [RPS 38]

Mammillaria subtilis Backeberg, Cact. Succ. J. Gr. Brit. 12(4): 81, 1950. [RPS 1]

Mammillaria supertexta var. *huitzilopochtli* (D. Hunt) E. Kuhn, Inform.-brief ZAG Mammillarien 11: 60, 1984. *Nom. inval.* (Art. 33.2), based on *Mammillaria huitzilopochtli.* [Name used in transcript of lecture.] [RPS -]

Mammillaria supertexta var. *lanata* (Britton & Rose) E. Kuhn, Inform.-brief ZAG Mammillarien 11: 60, 1984. *Nom. inval.* (Art. 33.2), based on *Neomammillaria lanata.* [Name used in transcript of lecture.] [RPS -]

Mammillaria supertexta var. *leucostoma* Backeberg, Die Cact. 5: 3230, 1961. *Nom. inval.* (Art. 9.5). [Erroneously included as valid in RPS 12.] [RPS 12]

Mammillaria supraflumen Reppenhagen, Gattung Mammillaria, 87-88, fig. 20 (p. 187), SEM-ills. 69-70 (p. 209), 1987. Typus: *Reppenhagen* 1609 (ZSS). [Mexico: Jalisco] [RPS 38]

Mammillaria tayloriorum Glass & Foster, Cact. Succ. J. (US) 47(4): 175, 1975. [RPS 27]

Mammillaria tegelbergiana Lindsay, Cact. Succ. J. (US) 38: 196-197, 1966. [RPS 17]

Mammillaria tesopacensis var. **papasquiarensis** Bravo, Cact. Suc. Mex. 11: 84-87, 103, 1966. [RPS 17]

Mammillaria theresae Cutak, Cact. Succ. J. (US) 39: 237-241, ills., 1967. [RPS 18]

Mammillaria thornberi var. **yaquensis** (Craig) Neutelings, Succulenta 65(5): 119, 1986. Basionym: *Mammillaria yaquensis.* [First invalidly published in l.c. 65(1): 3-5.] [RPS 37]

Mammillaria tiegeliana Backeberg, Die Cact. 5: 3402, 1961. *Nom. inval.* (Art. 37.1). [Erroneously included as valid in RPS 12.] [RPS 12]

Mammillaria tlalocii Reppenhagen, Gattung Mammillaria, 120-121, fig. 34 (p. 191), SEM-ills. 99-100 (p. 213), 1987. Typus: *Reppenhagen* 901 (ZSS). [Mexico: Oaxaca] [RPS 38]

Mammillaria tobuschii W. T. Marshall, Saguaroland Bull. 5: 78-81, 1952. [RPS 3]

Mammillaria tonalensis D. Hunt, Cact. Succ. J. Gr. Brit. 41(4): 103-104, 1979. Typus: *Lau* 1114 (K). [Mexico: Oaxaca] [RPS 30]

Mammillaria tropica Reppenhagen, Gattung Mammillaria, 171-173, fig. 64 (p. 198), SEM-ills. 161-162 (p. 221), 1987. Typus: *Reppenhagen* 677 (ZSS). [Mexico: Jalisco] [RPS 38]

Mammillaria uncinata var. **bihamata** (Pfeiffer) Backeberg, Die Cact. 5: 3143, 1961. Basionym: *Mammillaria bihamata.* [RPS 12]

Mammillaria vallensis Reppenhagen, Gattung Mammillaria, 156-158, fig. 56 (p. 196), SEM-ills. 143-144 (p. 218), 1987. Typus: *Reppenhagen* 1117 (ZSS). [Mexico: San Luis Potosi] [RPS 38]

Mammillaria vallensis var. **brevispina** Reppenhagen, Gattung Mammillaria, 158, fig. 57 (p. 197), SEM-ills. 145-146 (p. 219), 1987. Typus: *Reppenhagen* 1119 (ZSS). [Mexico: Tamaulipas] [RPS 38]

Mammillaria variabilis Reppenhagen, Kakt. and. Sukk. 36(10): 206-207, ill., 1985. Typus: *Reppenhagen* 1393 (K). [Mexico: Guanajuato] [RPS 36]

Mammillaria varicolor (Tiegel) Weniger, Cacti of the Southwest, 138, 1970. *Nom. inval.* (Art. 33.2), based on *Coryphantha varicolor.* [Not dated. The name given as basionym is itself a combination (here corrected).] [RPS -]

Mammillaria varieaculeata Buchenau, Cact. Suc. Mex. 11: 77, 79-81, 102, 1966. [RPS 17]

Mammillaria verticealba Reppenhagen, Gattung Mammillaria, 68-69, fig. 14 (p. 186), SEM-ills. 47-48 (p. 206), 1987. Typus: *Reppenhagen* 719 (ZSS). [Mexico: Michoacan] [RPS 38]

Mammillaria verticealba var. **zacatecasensis** Reppenhagen, Gattung Mammillaria, 70-71, SEM-ills. 49-50 (p. 207), 1987. Typus: *Reppenhagen* 973 (ZSS). [No type is actually indicated, but the reference to *Reppenhagen* 973 in the caption to the SEM-ills. is here regarded as being sufficient in the context of ICBN Arts. 37.3 and 37.4.] [Mexico: Zacatecas] [RPS 38]

Mammillaria viescensis Rogozinski & Appenzeller, Mitteilungsbl. AfM 13(4): 152-160, ills., SEM-ills., 1989. Typus: *Rogozinski* 25/86 (ZSS). [Mexico: Coahuila] [RPS 40]

Mammillaria virginis Fittkau & Kladiwa, in Krainz, Die Kakt. part 46/47, C VIIIc, ill., 1971. Typus: *Fittkau* s.n. (ZSS). [Mexico: Guerrero] [RPS 22]

Mammillaria vivipara var. **alversonii** (Coulter) L. Benson, Cacti of Arizona, ed. 2, 118, 1950. Basionym: *Cactus radiosus* var. *alversonii*. [RPS -]

Mammillaria vivipara var. **chlorantha** (Engelmann) L. Benson, Cacti of Arizona, ed. 2, 117, 1950. Basionym: *Mammillaria chlorantha*. [RPS -]

Mammillaria voburnensis var. **collinsii** (Britton & Rose) Reppenhagen, Gattung Mammillaria, 175-176, 1987. Basionym: *Neomammillaria collinsii*. [The neotype designation by Reppenhagen, l.c. (*Reppenhagen* 958) is unnecessary, as the original type is extant at US (cited by D. R. Hunt in Bradleya 1: 125, 1983).] [RPS 38]

Mammillaria voburnensis var. **eichlamii** (Quehl) Reppenhagen, Gattung Mammillaria, 178-180, SEM-ills. 169-170 (p. 222), 1987. Basionym: *Mammillaria eichlamii*. [RPS 38]

Mammillaria voburnensis var. **quetzalcoatl** Reppenhagen, Gattung Mammillaria, 177-178, fig. 66 (p. 199), SEM-ills. 167-168 (p. 221), 1987. Typus: *Reppenhagen* 1504 (ZSS). [Mexico: Oaxaca] [RPS 38]

Mammillaria wohlschlageri Reppenhagen, Gattung Mammillaria, 49-50, fig. 7 (p. 184), SEM-ills. 33-34 (p. 205), 1987. Typus: *Reppenhagen* 2132 (ZSS). [Mexico: Zacatecas] [RPS 38]

Mammillaria wrightii fa. **wolfii** D. Hunt, Cact. Succ. J. Gr. Brit. 41(4): 97-98, 1979. Typus: *Lau et Schreier* in *Lau* 1042 (K). [First published invalidly by Schreier as *Mammillaria wrightii* var. *wolfii*.] [Mexico: Chihuahua] [RPS 30]

Mammillaria wrightii var. **viridiflora** (Britton & Rose) W. T. Marshall, Arizona's Cactuses, 104, 1950. Basionym: *Mammillaria wilcoxii* var. *viridiflora*. [RPS 2]

Mammillaria wrightii var. **wilcoxii** (Toumey *ex* Schumann) W. T. Marshall, Arizona's Cactuses, 104, 1950. Basionym: *Mammillaria* × *wilcoxii*. [RPS 2]

Mammillaria xaltianguensis Sanchez-Mejorada, Anales Inst. Biol. Mex. 44(Bot. Ser. 1): 29-32, ills., 1975. Typus: *Sanchez-Mejorada* 70-0802 (MEXU). [Mexico: Guerrero] [RPS 25]

Mammillaria xaltianguensis var. **aguilensis** Reppenhagen, Gattung Mammillaria, 106-107, fig. 27 (p. 189), SEM-ills. 85-86 (p. 211), 1987. Typus: *Reppenhagen* 1046 (ZSS). [Mexico: Michoacan] [RPS 38]

Mammillaria xochipilli Reppenhagen, Gattung Mammillaria, 167-169, fig. 63 (p. 198), SEM-ills. 157-158 (p. 220), 1987. Typus: *Reppenhagen* 1709 (ZSS). [Mexico: Hidalgo] [RPS 38]

Mammillaria zacatecasensis Shurly, Cact. Succ. J. Gr. Br. 22(3): 51, 58, ill., 1960. Typus: *Schwarz* s.n. (K 6620, ZSS). [See H. Krainz, Die Kakt. Lief. 17, 1961, for information on typification.] [Mexico: Zacatecas] [RPS 11]

Mammillaria zephyranthoides var. **heidiae** (Krainz) J. Lüthy, Kakt. and. Sukk. 38(1): 11, 1987. Basionym: *Mammillaria heidiae*. [RPS 38]

Mammillaria zubleri Reppenhagen, Gattung Mammillaria, 48-49, fig. 6 (p. 184), SEM-ills. 31-32 (p. 204), 1987. Typus: *Zubler* s.n. (ZSS). [Mexico: Tamaulipas] [RPS 38]

Mammillaria zuccariniana var. *ochroleuca* E. Kuhn, Inform.-brief ZAG Mammillarien 6(3): 31, 1980. *Nom. inval.* (Art. 36.1, 37.1). [Given as provisional name.] [RPS -]

Mammilloydia F. Buxbaum, Oesterr. Bot. Zeitschr. 98(1-2): 64-65, 1951. Typus: *Mammillaria candida* Scheidweiler. [RPS 2]

Mammilloydia candida (Scheidweiler) Buxbaum, Österr. Bot. Zeitschr. 98(1-2): 64-65, 1951. Basionym: *Mammillaria candida*. [RPS 2]

Mammilloydia candida ssp. **ortizrubiana** (Bravo) Buxbaum, in Krainz, Die Kakt. Lief. 55/56: C IVf, 1974. Basionym: *Neomammillaria ortizrubiana*. [Sphalm. 'ortiz-rubiona'. Dated Dec. 1973, published early in 1974.] [RPS 25]

Mammilloydia ortizrubiana (Bravo) Buxbaum, Österr. Bot. Zeitschr. 98(1-2): 64-65, 1951. Basionym: *Neomammillaria ortizrubiana*. [Sphalm. 'ortiz-rubiona'.] [RPS 2]

Marenopuntia Backeberg, Desert Pl. Life 22(3): 27, 1950. Typus: *Opuntia marenae*. [RPS 1]

Marenopuntia marenae (S. H. Parsons) Backeberg, Desert Pl. Life. 22(3): 27-28, 1950. Basionym: *Opuntia marenae*. [RPS 1]

Marginatocereus marginatus var. *oaxacensis* Backeberg, Descr. Cact. Nov. 3: 8, 1963. *Nom. inval.* (Art. 9.5). [Erroneously included as valid in RPS 14.] [RPS 14]

Maritimocereus Akers & Buining, Succulenta 1950(4): 6-9, 1950. Typus: *Maritimocereus gracilis*. [Sphalm. 'Maritinocereus' (corrected by Byles in Saguaroland Bull. 8(8): 91, 1954).] [RPS 1]

Maritimocereus gracilis Akers & Buining, Succulenta 1950: 50-52, 1950. [RPS 1]

Maritimocereus nanus Akers & Buining, Sukkulentenkunde 4: 44, 1951. *Nom. inval.* (Art. 36.1). [Sphalm. 'Maritinocereus nanus'.] [RPS 4]

Marniera Backeberg, Cact. Succ. J. (US) 22(5): 153, 1950. Typus: *Phyllocactus macropterus*. [RPS 1]

Marniera chrysocardium (Alexander) Backeberg, Die Cact. 2: 737, 1959. Basionym: *Epiphyllum chrysocardium*. [RPS 10]

Marniera macroptera (Lemaire) Backeberg, Cact. Succ. J. (US) 23(1): 17, 1951. Basionym: *Phyllocactus macroptera*. [RPS 3]

Marniera macroptera var. **kimnachii** (Bravo) Backeberg, Kakt.-Lex., 459, 1966. Basionym: *Epiphyllum crenatum* var. *kimnachii*. [RPS 17]

Marshallocereus Backeberg, Cact. Succ. J. (US) 22(5): 154, 1950. Typus: *Cereus aragonii.* [RPS 1]

Marshallocereus aragonii (F. A. C. Weber) Backeberg, Cact. Succ. J. (US) 23: 121, 1951. Basionym: *Cereus aragonii.* [RPS -]

Marshallocereus thurberi (Engelmann) Backeberg, Cact. Succ. J. (US) 23(4): 121, 1951. Basionym: *Cereus thurberi.* [The name given as basionym was itself a combination, and is here corrected.] [RPS 3]

Marshallocereus thurberi var. **littoralis** (K. Brandegee) Backeberg, Die Cact. 4: 2163, 1960. Basionym: *Cereus thurberi* var. *littoralis.* [RPS 11]

Matucana subgen. **Incaica** Ritter, Kakt. and. Sukk. 16: 229, 1965. Typus: *M. aureiflora* Ritter. [RPS 16]

Matucana aurantiaca (Vaupel) Buxbaum, in H. Krainz, Die Kakt. Lief. 54: C Vb, 1973. Basionym: *Echinocactus aurantiacus.* [RPS 24]

Matucana aureiflora Ritter, Kakt. and. Sukk. 16: 229-230, 1965. [RPS 16]

Matucana blancii Backeberg, Nation. Cact. Succ. J. 11(4): 70-71, 1956. [RPS 7]

Matucana blancii var. **nigriarmata** Backeberg, Nation. Cact. Succ. J. 11(4): 70-71, 1956. [RPS 7]

Matucana breviflora Rauh & Backeberg, in Backeberg, Descr. Cact. Nov. [1:] 18, 1957. Typus: *Rauh* K123 (1956) (ZSS). [Dated 1956, published 1957.] [Peru] [RPS 7]

Matucana brunnescens Rauh, Cactus (Paris) No. 50: 64, 1956. *Nom. inval.* (Art. 36.1). [Casually mentioned in the text. Again (invalidly) described by Backeberg in l.c. No. 53: 139, 1957 (cf. RPS 8).] [RPS 7]

Matucana calliantha Ritter, Kakt. Südamer. 4: 1490-1491, 1981. Typus: *Ritter* 1308 (U). [Peru: Amazonas] [RPS 32]

Matucana calliantha var. **prolifera** Ritter, Kakt. Südamer. 4: 1491, 1981. Typus: *Ritter* 1308a (U). [Peru: Cajamarca] [RPS 32]

Matucana calocephala Skarupke, Stachelpost 9(46): 99-101, 1973. *Nom. inval.* (Art. 37.1). [RPS -]

Matucana calvescens (Kimnach & P. C. Hutchison) Buxbaum, Rep. Pl. Succ. 24: 7, 1975. Basionym: *Borzicactus calvescens.* [RPS 24]

Matucana celendinensis Ritter, Succulenta 46(1): 4, ill., 1967. [Latin diagnosis published without type in l.c. 45(8): 118, 1966.] [RPS 17]

Matucana cereoides Rauh & Backeberg, in Backeberg, Descr. Cact. Nov. [1:] 19, 1957. Typus: *Rauh* K159 (1956) (ZSS). [Dated 1956, published 1957.] [Peru] [RPS 7]

Matucana comacephala Ritter, Succulenta 37(3): 92-93, 1958. Typus: *Ritter* 587 (ZSS). [See also F. Ritter, Kakt. Südamer. 4: 1492, 1981.] [Peru: Ancash] [RPS 9]

Matucana crinifera Ritter, Taxon 12(3): 125, 1963. Typus: *Ritter* 595 (U, ZSS). [Type cited for U l.c., 29.] [Peru: Ancash] [RPS 14]

Matucana currundayensis Ritter, Succulenta 37(12): 139-140, ill., 1958. Typus: *Ritter* 164 (ZSS). [Included as *nomen nudum* in Backeberg, Die Cact. 2: 1085, 1959 (cf. RPS 11).] [Peru: La Libertad] [RPS 9]

Matucana elongata Rauh & Backeberg, in Backeberg, Descr. Cact. Nov. [1:] 19, 1957. Typus: *Rauh* K53 (1956) (?, ZSS [seeds only]). [Dated 1956, published 1957.] [Peru] [RPS 7]

Matucana formosa Ritter, Taxon 12: 125, 1963. [RPS 14]

Matucana formosa var. **minor** Ritter, Taxon 12: 125, 1963. [RPS 14]

Matucana fruticosa Ritter, Succulenta 46(1): 3-4, ill., 1967. Typus: *Ritter* 1307 (ZSS [type status ?]). [No herbarium cited for the type.] [Latin diagnosis published without type in l.c. 45(8): 117-118, 1966.] [Peru: Cajamarca] [RPS 17]

Matucana hastifera Ritter, Kakt. Südamer. 4: 1496-1497, 1981. Typus: *Ritter* 1306 (). [Peru: Huari] [RPS 32]

Matucana haynei var. **elongata** (Backeberg) Ritter, Backeberg's Descr. & Erört. taxon. nomenkl. Fragen, 37, 1958. Basionym: *Matucana elongata.* [Combination repeated in Ritter, Kakt. Südamer. 4: 1497, 1981.] [RPS 15]

Matucana haynei var. **erectipetala** Rauh & Backeberg, in Backeberg, Descr. Cact. Nov. [1:] 18, 1957. [Dated 1956, published 1957.] [RPS 7]

Matucana haynei var. **gigantea** Ritter, Backeberg's Descr. & Erört. taxon. nomenkl. Fragen, 37, 1958. Basionym: *Matucana multicolor.* [RPS 15]

Matucana herzogiana Backeberg, Nation. Cact. Succ. J. 11(4): 70-71, 1956. [RPS 7]

Matucana herzogiana var. **perplexa** Backeberg, Nation. Cact. Succ. J. 11(4): 70-71, 1956. [RPS 7]

Matucana huagalensis (Donald & Lau) Bregman *et al.*, Succulenta 67(7/8): 155-157, ills., SEM-ills., 1988. Basionym: *Borzicactus huagalensis.* [RPS 39]

Matucana humboldtii (Humboldt, Bonpland & Kunth) Buxbaum, in Krainz, Die Kakt. Lief. 54: C Vb, 1973. Basionym: *Cactus humboldtii.* [RPS 24]

Matucana hystrix Rauh & Backeberg, in Backeberg, Descr. Cact. Nov. [1:] 19, 1957. Typus: *Rauh* K113 (1956) (ZSS). [Dated 1956, published 1957.] [Peru] [RPS 7]

Matucana hystrix var. **atrispina** Rauh & Backeberg, in Backeberg, Descr. Cact. Nov. [1:] 19, 1957. [Dated 1956, published 1957.] [RPS 7]

Matucana hystrix var. **umadeavoides** Rauh & Backeberg, in Backeberg, Descr. Cact. Nov. [1:] 19, 1957. [Dated 1956, published 1957.] [RPS 7]

Matucana icosagona (Humboldt, Bonpland & Kunth) Buxbaum, in Krainz, Die Kakt. 54, CVb, 1973. Basionym: *Cactus icosagonus.* [RPS 24]

Matucana intertexta Ritter, Taxon 12(3): 125, 1963. [RPS 14]

Matucana intertexta var. **celendinensis** (Ritter) Bregman *et al.*, Succulenta 67(5): 100-102, ills., SEM-ills., 1988. Basionym: *Matucana celendinensis.* [RPS 39]

Matucana krahnii (Donald) Bregman, Kakt. and. Sukk. 37(12): 253, ill., 1986. Basionym: *Borzicactus krahnii.* [RPS 37]

Matucana madisoniorum (P. C. Hutchison) Rowley, Rep. Pl. Succ. 22: 10, 1973. Basionym: *Borzicactus madisoniorum.* [RPS 22]

Matucana madisoniorum var. **pujupatii** (Donald & Lau) Rowley, Rep. Pl. Succ. 22: 10, 1971. Basionym: *Borzicactus madisoniorum* var. *pujupatii.* [RPS 22]

Matucana megalantha Ritter, Succulenta 46(1): 3, ill. (p. 1), 1967. [Latin diagnosis published without type in l.c. 45(8): 117, 1966.] [RPS 17]

Matucana mirabilis Buining, Sukkulentenkunde 7/8: 39-41, ills., 1963. Typus: *Akers* s.n. (ZSS [authentic material ?]). [Ex cult. A. Buining.] [Peru: Dept. Lima] [RPS 14]

Matucana multicolor Rauh & Backeberg, in Backeberg, Descr. Cact. Nov. [1:] 19, 1957. Typus: *Rauh* K115 (1956) (ZSS). [Dated 1956, published 1957.] [Peru] [RPS 7]

Matucana multicolor var. *breviflora* (Backeberg) Ritter, Kakt. Südamer. 4: 1500, 1981. *Nom. inval.* (Art. 33.2), based on *Matucana breviflora.* [RPS 32]

Matucana multicolor var. *hystrix* (Backeberg) Ritter, Kakt. Südamer. 4: 1499, 1981. *Nom. inval.* (Art. 33.2), based on *Matucana hystrix.* [RPS 32]

Matucana myriacantha (Vaupel) Buxbaum, in Krainz, Die Kakt. Lief. 54: C Vb, 1973. Basionym: *Echinocactus myriacanthus.* [RPS 24]

Matucana oreodoxa (Ritter) Slaba, Kaktusy 22(6): 130, 1986. Basionym: *Eomatucana oreodoxa.* [RPS 37]

Matucana pallarensis Ritter, Kakt. Südamer. 4: 1501, 1981. Typus: *Ritter* 1076 (U). [Peru: La Libertad] [RPS 32]

Matucana paucicostata Ritter, Taxon 12: 124, 1963. [RPS 14]

Matucana polzii Diers, Donald & Zecher, Kakt. and. Sukk. 37(6): 114-119, ills., SEM-ills., 1986. Typus: *Zecher* 762 (KOELN). [RPS 37]

Matucana pujupatii (Donald & Lau) Bregman, Willdenowia 17(1/2): 179, ill. (p. 178), 1988. Basionym: *Borzicactus madisoniorum* var. *pujupatii.* [RPS 39]

Matucana purpureoalba Ritter, Kakt. Südamer. 4: 1502-1503, 1981. Typus: *Ritter* 1063 (U). [Peru: La Libertad] [RPS 32]

Matucana ritteri Buining, Succulenta 38(1): 2-4, 1959. Typus: *Ritter* 299 (ZSS). [A taxon with the same name is mentioned in Backeberg, Die Cact. 2: 1084, 1959, with the authors "Krainz & Rupf", but without Latin diagnosis (cf. RPS 10).] [Peru: La Libertad] [RPS 10]

Matucana supertexta Ritter, Kakt. Südamer. 4: 1504, 1981. Typus: *Ritter* 690 (U). [Peru] [RPS 32]

Matucana tuberculata (Donald) Bregman et al., Succulenta 66(9): 175, ills. (pp. 176-177), 1987. Basionym: *Borzicactus tuberculatus.* [RPS 38]

Matucana tuberculosa Ritter, Kakt. Südamer. 4: 1505, 1981. Typus: *Ritter* 1073 (U). [Peru: La Libertad] [RPS 32]

Matucana variabilis Rauh & Backeberg, in Backeberg, Descr. Cact. Nov. [1:] 19, 1957. [Dated 1956, published 1957.] [RPS 7]

Matucana variabilis var. **fuscata** Rauh & Backeberg, in Backeberg, Descr. Cact. Nov. [1:] 20, 1957. Typus: *Rauh* K95 (1956) (ZSS [type status ?]). [Dated 1956, published 1957. Backeberg cites K95a as type of *M. variabili* var. *variabilis* and K95b as type of var. *fuscata*, for which Rauh cites K95 as type.] [Peru] [RPS 7]

Matucana weberbaueri fa. **flammea** (Donald) Bregman et al., Succulenta 68(6): 143, 1989. Basionym: *Borzicactus weberbaueri* var. *flammeus.* [RPS 40]

Matucana weberbaueri var. *blancii* (Backeberg) Rauh, Cactus (Paris) No. 50: 64, 1956. *Nom. inval.* (Art. 33.2), based on *Matucana blancii.* [Again mentioned casually (as new variety) in l.c. No. 53: 139, 1957 (RPS 8).] [RPS 7]

Matucana winteri Ritter, Kakt. Südamer. 4: 1506, 1981. Typus: *Ritter* 691 (U). [Peru: La Libertad] [RPS 32]

Matucana yanganucensis Rauh & Backeberg, in Backeberg, Descr. Cact. Nov. [1:] 18, 1957. Typus: *Rauh* K55b (1956) (ZSS). [Dated 1956, published 1957.] [Peru] [RPS 7]

Matucana yanganucensis var. **albispina** Rauh & Backeberg, in Backeberg, Descr. Cact. Nov. [1:] 18, 1957. Typus: *Rauh* K55 (1956) (ZSS). [Dated 1956, published 1957.] [Peru] [RPS 7]

Matucana yanganucensis var. **fuscispina** Rauh & Backeberg, in Backeberg, Descr. Cact. Nov. [1:] 18, 1957. [Dated 1956, published 1957.] [RPS 7]

Matucana yanganucensis var. **longistyla** Rauh, Beitr. Kenntn. Peruan. Kakt.veg., 352, 1958. [RPS 10]

Matucana yanganucensis var. **parviflora** Rauh & Backeberg, in Backeberg, Descr. Cact. Nov. [1:] 19, 1957. [Dated 1956, published 1957.] [RPS 7]

Matucana yanganucensis var. **salmonea** Rauh & Backeberg, in Backeberg, Descr. Cact. Nov. [1:] 18, 1957. Typus: *Rauh* K55d (1956) (ZSS). [Dated 1956, published 1957.] [Peru] [RPS 7]

Matucana yanganucensis var. **setosa** Ritter, Kakt. Südamer. 4: 1507-1508, 1981. Typus: *Ritter* 592a (U, ZSS). [Type cited for U l.c. 1: iii, 1979.] [Peru: Ancash] [RPS 32]

Matucana yanganucensis var. **suberecta** Rauh & Backeberg, in Backeberg, Descr. Cact. Nov. [1:] 18, 1957. Typus: *Rauh* K66 (1956) (ZSS). [Dated 1956, published 1957.] [Peru] [RPS 7]

Mediocactus hahnianus Backeberg, Descr. Cact. Nov. [1:] 10, 1957. Neotypus: *Rojas et West* 8499 (HNT, MO, US). [= UCBG 37.1164-1 = HBG 33559, grown from seed collected from this introduction in the Botanical Garden at Asuncion. Neotype designated by Kimnach in Cact. Succ. J. (US) 58(2): 61, 1987.] [Dated 1956 on the cover, published early in 1957 according to copy at Kew.] [Paraguay] [RPS 7]

Mediocactus hassleri (Schumann) Backeberg, Die Cact. 2: 798, 1959. Basionym: *Cereus hassleri*. [RPS 10]

Mediocactus lindmanii (F. A. C. Weber) Backeberg, Die Cact. 2: 798, 1959. Basionym: *Cereus lindmanii*. [RPS 10]

Mediocactus setaceus (Salm-Dyck) Borg, Cacti, ed. 2, 213, 1951. Basionym: *Cereus setaceus*. [RPS 9]

Mediolobivia auranitida var. **flaviflora** Backeberg, Descr. Cact. Nov. [1:] 31, 1957. [Dated 1956, published 1957.] [RPS 7]

Mediolobivia auranitida var. **gracilis** (Wessner) Backeberg, Die Cact. 3: 1522, 1959. Basionym: *Lobivia auranitida*. [RPS 10]

Mediolobivia aureiflora fa. **kesselringiana** (Cullmann) Köhler, Kakt. and. Sukk. 24: 206-207, 1973. Basionym: *Mediolobivia kesselringiana*. [RPS 24]

Mediolobivia aureiflora subvar. **leucolutea** Backeberg, Descr. Cact. Nov. [1:] 30, 1957. [Dated 1956, published 1957.] [RPS 7]

Mediolobivia aureiflora subvar. **lilacinostoma** Backeberg, Descr. Cact. Nov. [1:] 30, 1957. [Dated 1956, published 1957.] [RPS 7]

Mediolobivia aureiflora var. **boedekeriana** (Backeberg) Backeberg, Die Cact. 3: 1488, 1959. Basionym: *Mediolobivia boedekeriana*. [RPS 10]

Mediolobivia aureiflora var. **duursmaiana** (Backeberg) Backeberg, Die Cact. 3: 1492, 1959. Basionym: *Mediolobivia duursmaiana*. [RPS 10]

Mediolobivia aureiflora var. **rubelliflora** (Backeberg) Backeberg, Die Cact. 3: 1490, 1959. Basionym: *Mediolobivia rubelliflora*. [RPS 10]

Mediolobivia aureiflora var. **rubriflora** (Backeberg) Backeberg, Die Cact. 3: 1490, 1959. Basionym: *Mediolobivia rubriflora*. [RPS 10]

Mediolobivia aureiflora var. **sarothroides** (Werdermann) Backeberg, Die Cact. 3: 1491, 1959. Basionym: *Rebutia sarothroides*. [RPS 10]

Mediolobivia conoidea var. **columnaris** (Wessner) Backeberg, Die Cact. 3: 1498, 1959. Basionym: *Lobivia columnaris*. [RPS 10]

Mediolobivia costata var. *brachyantha* (Backeberg) Donald, Cactus (Paris) No. 40: 39, 1954. *Nom. inval*. (Art. 33.2), based on *Mediolobivia brachyantha*. [RPS 5]

Mediolobivia costata var. *eucaliptana* (Backeberg) Donald, Cactus (Paris) No. 40: 39, 1954. *Nom. inval*. (Art. 33.2), based on *Mediolobivia eucaliptana*. [RPS 5]

Mediolobivia euanthema var. **fricii** Backeberg, Descr. Cact. Nov. [1:] 30, 1957. [Dated 1956, published 1957.] [RPS 7]

Mediolobivia euanthema var. *pygmaea* (Backeberg) Donald, Cactus (Paris) No. 40: 39, 1954. *Nom. inval*. (Art. 33.2), based on *Mediolobivia pygmaea*. [RPS 5]

Mediolobivia euanthema var. *ritteri* (Wessner) Donald, Cactus (Paris) No. 40: 39, 1954. *Nom. inval*. (Art. 33.2), based on *Mediolobivia ritteri*. [RPS 5]

Mediolobivia famatimensis (Spegazzini) Pazout, Friciana 4(23): 21, 1964. *Nom. inval*. (Art. 33.2), based on *Echinocactus famatimensis*. [RPS 15]

Mediolobivia fuauxiana Backeberg, Descr. Cact. Nov. [1:] 31, 1957. [Dated 1956, published 1957.] [RPS 7]

Mediolobivia haagei var. *atrovirens* (Backeberg) Donald, Cactus (Paris) No. 40: 39, 1954. *Nom. inval*. (Art. 33.2), based on *Mediolobivia atrovirens*. [RPS 5]

Mediolobivia haagei var. *digitiformis* (Backeberg) Donald, Cactus (Paris) No. 40: 39, 1954. *Nom. inval*. (Art. 33.2), based on *Mediolobivia digitiformis*. [RPS 5]

Mediolobivia haagei var. **flavovirens** (Backeberg) Backeberg, Cact. Succ. J. (US) 23(3): 81, 1951. Basionym: *Lobivia neohaageana* var. *flavovirens*. [RPS 3]

Mediolobivia haagei var. *orurensis* (Backeberg) Donald, Cactus (Paris) No. 40: 39, 1954. *Nom. inval*. (Art. 33.2), based on *Lobivia orurensis*. [RPS 5]

Mediolobivia haagei var. *pectinata* (Backeberg) Donald, Cactus (Paris) No. 40: 39, 1954. *Nom. inval*. (Art. 33.2), based on *Mediolobivia pectinata*. [RPS 5]

Mediolobivia haefneriana Cullmann, Kakt. and. Sukk. 6(2): 119-120, 1955. [RPS 6]

Mediolobivia hirsutissima Cardenas, Cact. Succ. J. (US) 43: 244, 1971. [RPS 22]

Mediolobivia ithyacantha Cardenas, Cact. Succ. J. (US) 42: 35, 1970. [RPS 21]

Mediolobivia neopygmaea Backeberg, Descr. Cact. Nov. [1:] 30, 1957. [Dated 1956, published 1957.] [RPS 7]

Mediolobivia pectinata var. **atrovirens** (Backeberg) Backeberg, Die Cact. 3: 1506, 1959. Basionym: *Lobivia atrovirens*. [RPS 10]

Mediolobivia pectinata var. **digitiformis** (Backeberg) Backeberg, Die Cact. 3: 1506, 1959. Basionym: *Lobivia digitiformis*. [RPS 10]

Mediolobivia pectinata var. **neosteinmannii** Backeberg, Descr. Cact. Nov. [1:] 30, 1957. [Dated 1956, published 1957.] [RPS 7]

Mediolobivia pectinata var. **orurensis** (Backeberg) Backeberg, Die Cact. 3: 1505, 1959. Basionym: *Lobivia orurensis*. [RPS 10]

Mediolobivia pygmaea (R. E. Fries) Backeberg, Die Cact. 3: 1502, 1959. Basionym: *Echinopsis pygmaea*. [RPS 10]

Mediolobivia pygmaea var. **flavovirens** (Backeberg) Backeberg, Die Cact. 3: 1503, 1959. Basionym: *Lobivia neohaageana* var. *flavovirens*. [RPS 10]

Mediolobivia ritteri var. *pilifera* Fric *ex* Backeberg, Die Cact. 3: 1518, 1959. *Nom. inval*. (Art. 37.1). [Based on a *nomen nudum* of Fric. Erroneously included as valid in RPS 10.] [RPS 10]

Mediolobivia schmiedcheniana var. **einsteinii** Backeberg, Descr. Cact. Nov. [1:] 30, 1957. [Dated 1956, published 1957. Incorrectly given as invalid name by Eggli in Bradleya

3: 100, 1985.] [RPS 7]

Mediolobivia schmiedcheniana var. **karreri** Fric *ex* Backeberg, Descr. Cact. Nov. [1:] 30, 1957. [Dated 1956, published 1957. Incorrectly given as invalid name by Eggli in Bradleya 3: 100, 1985.] [RPS 7]

Mediolobivia schmiedcheniana var. **rubriviridis** Fric *ex* Backeberg, Descr. Cact. Nov. [1:] 30, 1957. [Dated 1956, published 1957. Incorrectly given as invalid name by Eggli in Bradleya 3: 100, 1985.] [RPS 7]

Mediolobivia schmiedcheniana var. **steineckei** Backeberg, Descr. Cact. Nov. [1:] 30, 1957. [Dated 1956, published 1957. Incorrectly given as invalid name by Eggli in Bradleya 3: 100, 1985.] [RPS 7]

Mediolobivia spiralisepala Schütz, Kaktusy 11: 51-53, 1975. [First used as a *nomen nudum* in the form of *Pygmaeolobivia spiralisepala* by Jajo.] [RPS 26]

Megalolobivia Y. Ito, Bull. Takarazuka Insectarium 71: 13-20, 1950. *Nom. inval.* (Art. 36.1). [RPS 3]

×**Meierara** P. V. Heath, Epiphytes 7(28): 91, 1983. [= *Aporocactus* × *Cryptocereus* × *Heliocereus*.] [RPS 34]

Melocactus ser. **Guitartii** Meszaros, Acta Bot. Acad. Sci. Hung. 24: 304, 1978. Typus: *Melocactus guitartii* Leon. [RPS 32]

Melocactus ser. **Harlowii** Meszaros, Acta Bot. Acad. Sci. Hung. 24: 304, 1978. Typus: *Melocactus harlowii* (Britton & Rose) Vaupel. [RPS 32]

Melocactus ser. **Matanzanus** Meszaros, Acta Bot. Acad. Sci. Hung. 24: 303, 1978. Typus: *Melocactus matanzanus* Leon. [RPS 32]

Melocactus acispinosus Buining & Brederoo, in Krainz, Die Kakt. Lfg. 62, CVId, 1975. [RPS 27]

Melocactus actinacanthus Areces Mallea, Ciencias ser. 10 Bot. (9): 4-11, ills., 1976. Typus: *Areces* 28367 (HAJB). [Cuba: Las Villas] [RPS 27]

Melocactus acunae var. **flavispinus** Meszaros, Acta Bot. Acad. Sci. Hung. 22: 138, 1977. [Sphalm. 'acunai'. Dated 1976.] [RPS 27]

Melocactus acunae var. **laguanaensis** Meszaros, Acta Bot. Acad. Sci. Hung. 22: 138, 1977. [Sphalm. 'acunai'. Dated 1976.] [RPS 27]

Melocactus × **albicephalus** Buining & Brederoo (*pro spec.*), in Krainz, Die Kakt. Lief. 52: C VId, 3 pp. of text, ills., 1973. Typus: *Horst et Uebelmann* HU 350 (1971) (U, ZSS). [= *M. glaucescens* × *M. ernestii*; hybrid status first suggested in 1981 and 1982, and confirmed by P. J. Braun in Bradleya 6: 94, 1988 and by Taylor in l.c. 9: 52, 1991.] [Brazil: Bahia] [RPS 24]

Melocactus amethystinus Buining & Brederoo, in Krainz, Die Kakt. Lief. 50-51: C VId, 4 pp. of text, ills., 1973. Typus: *Horst et Uebelmann* HU 270 (1968) (U). [Probably already published in 1972 ?] [Brazil: Bahia] [RPS 24]

Melocactus ammotrophus Buining & Brederoo *ex* Bercht & Brederoo, Succulenta 63(2): 33-38, ills., 1984. Typus: *Horst et Uebelmann* HU 353 (1971) (U, ZSS [fruits, seeds only]). [Brazil: Minas Gerais] [RPS 35]

Melocactus amstutziae Rauh & Backeberg, in Backeberg, Descr. Cact. Nov. [1:] 36, 1957. Typus: *Amstutz* 1 (ZSS). [Dated 1956, published 1957.] [Peru] [RPS 7]

Melocactus arcuatispinus Brederoo & Eerkens, Succulenta 62(5): 97-100, ills., 1983. Typus: *Horst et Uebelmann* HU 424 (1974) (U [missing], ZSS [fruits, seeds]). [Brazil: Bahia] [RPS 34]

Melocactus axiniphorus Buining & Brederoo, Succulenta 55(10): 193-197, ills., 1976. Typus: *Horst et Uebelmann* HU 450 (U, ZSS [fruits, seeds]). [Brazil: Bahia] [RPS 27]

Melocactus azulensis Buining & Brederoo, Kakt. and. Sukk. 28(7): 154-156, ills., 1977. Typus: *Horst et Uebelmann* HU 168 (U, ZSS). [Brazil: Minas Gerais] [RPS 28]

Melocactus azureus Buining & Brederoo, Kakt. and. Sukk. 22(6): 101-103, ills., 1971. Typus: *Horst et Uebelmann* HU 256 (1968) (U [missing], ZSS). [Brazil: Bahia] [RPS 22]

Melocactus azureus var. **krainzianus** (Buining & Brederoo) P. J. Braun, Bradleya 6: 94, 1988. Basionym: *Melocactus krainzianus*. [RPS 39]

Melocactus barbarensis Antesberger, Kakt. Orch.-Rundschau 13(2): 13-19, ills., 1988. Typus: *Antesberger* 13 (SZU). [Corrected to 'barbara' in RPS 39.] [Aruba] [RPS 39]

Melocactus bellavistensis Rauh & Backeberg, in Backeberg, Descr. Cact. Nov. [1:] 36, 1957. Typus: *Rauh* K137 (1954) (ZSS [syn]). [Dated 1956, published 1957.] [Peru] [RPS 7]

Melocactus bellavistensis fa. **minor** Rauh & Backeberg, in Backeberg, Descr. Cact. Nov. [1:] 36, 1957. [Dated 1956, published 1957.] [RPS 7]

Melocactus borhidii Meszaros, Acta Bot. Acad. Sci. Hung. 22: 135, 1977. [Dated 1976. Erroneously included under the mis-printed name *M. borthidii* in RPS 27.] [RPS 27]

Melocactus bozsingianus Antesberger, Kakt. Orch.-Rundschau 13(2-3): 17-23, ills., 1989. Typus: *Antesberger* 3 (SZU). [It is questionable whether the publication is "available" in the sense of ICBN 29.1; a final decision could affect the validity of the name.] [Netherlands Antilles: Aruba] [RPS 40]

Melocactus brederooianus Buining, Succulenta 51: 28-33, 1972. [RPS 23]

Melocactus canescens Ritter, Kakt. Südamer. 1: 134, 1979. [RPS 30]

Melocactus canescens var. **montealtoi** Ritter, Kakt. Südamer. 1: 134, 1979. [RPS 30]

Melocactus citrispinus H. & B. Antesberger, Salzburger Nachrichten 1989(30. August): 6, ill., 1989. *Nom. inval.* (Art. 36, 37, 29.4). [Published as provisional name in a newspaper.] [RPS 40]

Melocactus concinnus Buining & Brederoo, Kakt. and. Sukk. 23(1): 5-7, ills., 1972. Typus: *Horst et Uebelmann* HU 214 (1968) (U, ZSS). [Brazil: Bahia] [RPS 23]

Melocactus conoideus Buining & Brederoo, in Krainz, Die Kakt. Lief. 55-56: C VId, ills., 1974. Typus: *Horst et Uebelmann* HU 183 (1972) (U, ZSS). [Probably already published in 1973 ?] [Brazil: Bahia] [RPS 25]

Melocactus conquistaensis W. Uebelmann, Feld-Nummern, [unpaged], 1972. *Nom. inval.* (Art. 36.1, 37.1). [Published as provisional name.] [RPS -]

Melocactus coronatus (Lamarck) Backeberg, Die Cact. 4: 2569, 1960. Basionym: *Cactus coronatus*. [RPS 11]

Melocactus cremnophilus Buining & Brederoo, Cact. Succ. J. (US) 44: 3-5, 1972. [RPS 23]

Melocactus curvicornis Buining & Brederoo, Kakt. and. Sukk. 23: 33-35, 1972. [RPS 23]

Melocactus dawsonii Bravo, Cact. Suc. Mex. 10: 27-29, 1965. [RPS 16]

Melocactus deinacanthus Buining & Brederoo, Kakt. and. Sukk. 24(10): 217-219, ills., 1973. Typus: *Horst et Uebelmann* HU 153 (1971) (U, ZSS). [Brazil: Bahia] [RPS 24]

Melocactus deinacanthus fa. **mulequensis** (Buining & Brederoo) P. J. Braun, Bradleya 6: 94, 1988. Basionym: *Melocactus mulequensis*. [RPS 39]

Melocactus deinacanthus ssp. **florschuetzianus** (Buining & Brederoo) P. J. Braun, Bradleya 6: 94, 1988. Basionym: *Melocactus florschuetzianus*. [RPS 39]

Melocactus deinacanthus ssp. **longicarpus** (Buining & Brederoo) P. J. Braun, Bradleya 6: 94, 1988. Basionym: *Melocactus longicarpus*. [RPS 39]

Melocactus diersianus Buining & Brederoo, Kakt. and. Sukk. 26(8): 169-171, ills., 1975. Typus: *Horst et Uebelmann* HU 404 (1974) (U, ZSS [fruits, seeds]). [Brazil: Minas Gerais] [RPS 26]

Melocactus diersianus fa. **rubrispinus** (Ritter) P. J. Braun, Bradleya 6: 94, 1988. Basionym: *Melocactus rubrispinus*. [RPS 39]

Melocactus douradensis Hovens & Strecker, Succulenta 63(1): 1, 3-6, ills., 1984. Typus: *Hovens et al.* 81-172 (U [missing]). [Brazil: Bahia] [RPS 35]

Melocactus erythracanthus Buining & Brederoo, Cact. Succ. J. (US) 45(5): 223-226, ills., 1973. Typus: *Horst et Uebelmann* HU 220 (U, ZSS). [Brazil: Bahia] [RPS 24]

Melocactus estevesii P. J. Braun, Kakt. and. Sukk. 41(1): 6-10, ills., 1989. Typus: *Esteves Pereira* 157 (UFG 12385, B, K, ZSS). [Vol. for 1990, issue already distributed on Dec. 22, 1989.] [Brazil: Roraima] [RPS 40]

Melocactus evae Meszaros, Acta Bot. Acad. Sci. Hung. 22: 135, 1977. [Dated 1976.] [RPS 27]

Melocactus ferreophilus Buining & Brederoo, in Krainz, Die Kakt. 52, CVId, 2 pp. of text, ills., 1973. Typus: *Horst et Uebelmann* HU 217 (1967) (U, ZSS [fruits, seeds]). [Brazil: Bahia] [RPS 24]

Melocactus florschuetzianus Buining & Brederoo, Ashingtonia 2(2): 25-27, ills., 1975. Typus: *Horst et Uebelmann* HU 148 (1966) (U, ZSS). [Brazil: Minas Gerais] [RPS 26]

Melocactus fortalezensis Rauh & Backeberg, in Backeberg, Descr. Cact. Nov. [1:] 36, 1957. [Dated 1956, published 1957.] [RPS 7]

Melocactus giganteus Buining & Brederoo, Cact. Succ. J. (US) 45(5): 227-230, ills., 1973. Typus: *Horst et Uebelmann* HU 266 (1968) (U, ZSS [status ?]). [Brazil: Bahia] [RPS 24]

Melocactus glaucescens Buining & Brederoo, Cact. Succ. J. (US) 44(4): 159-161, ills., 1972. Typus: *Horst et Uebelmann* HU 219 (1967) (U [missing], ZSS [fruit, seeds]). [Brazil: Bahia] [RPS 23]

Melocactus glauxianus Brederoo & Bercht, Succulenta 63(3): 55-60, ills., 1984. Typus: *Horst et Uebelmann* HU 382 (U [missing]). [Brazil: Minas Gerais] [RPS 35]

Melocactus griseoloviridis Buining & Brederoo, Kakt. and. Sukk. 25(5): 98-100, ills., 1974. [Sphalm. 'grisoleoviridis'.] [RPS 25]

Melocactus guaricensis Croizat, Noved. Ci. Contr. Ocas. Mus. Hist. Nat. La Salle Caracas, ser. Bot., 1: 1, 1950. [RPS 9]

Melocactus harlowii var. *acunai* (León) Riha, Kaktusy 17(6): 124, 1981. *Nom. inval.* (Art. 33.2), based on *Melocactus acunae*. [RPS -]

Melocactus harlowii var. *nagyi* (Meszaro) Riha, Kaktusy 17(6): 124, 1981. *Nom. inval.* (Art. 33.2), based on *Melocactus nagyi*. [RPS -]

Melocactus helvolilanatus Buining & Brederoo, Succulenta 55: 261-265, 1976. [RPS 27]

Melocactus holguinensis Areces-Mallea, Ciencias ser. 10 Bot. (10): 3-12, ills., 1976. Typus: *Areces* 30911 (HAJB). [Cuba: Oriente] [RPS 27]

Melocactus huallancensis Rauh & Backeberg, in Rauh, Beitr. Kenntn. Peruan. Kakt.veg., 538, 1958. [RPS 10]

Melocactus inconcinnus Buining & Brederoo, Kakt. and. Sukk. 26: 193-195, 1975. [RPS 26]

Melocactus inconcinnus var. *brederooianus* (Buining) Ritter, Kakt. Südamer. 1: 137, 1979. Incorrect name (Art. 57.1), based on *Melocactus brederooianus*. [RPS 30]

Melocactus interpositus Ritter, Kakt. Südamer. 1: 140, 1979. [RPS 30]

Melocactus intortus var. **antonii** (Britton) Backeberg, Die Cact. 4: 2575, 1960. Basionym: *Cactus antonii*. [RPS 11]

Melocactus itabirabensis Uebelmann, Feld-Nummern, [unpaged], 1972. *Nom. inval.* (Art. 36.1, 37.1). [Published as provisional name.] [RPS -]

Melocactus jakusii Meszaros, Acta Acad. Sci. Hung. 22(1-2), 134, 1977. Typus: *Meszaros et Jakus* s.n. (HAC, ZSS [ex cult., type collection]). [Dated 1976.] [Cuba: Oriente] [RPS 27]

Melocactus krainzianus Buining & Brederoo, in Krainz, Die Kakt. Lief. 62: C VId, 4 pp. of text, 1975. Typus: *Horst et Uebelmann* HU 264 (1968) (U, ZSS [fruit, seeds]). [Illustrations included in the protologue are misidentified.] [Brazil: Bahia] [RPS 27]

Melocactus lanssensianus P. J. Braun, Succulenta 65(2): 25-30, 61-64, ills., 1986. Typus: *Horst et Uebelmann* HU 474 (ZSS). [Brazil: Pernambuco] [RPS 37]

Melocactus lensselinkianus Buining & Brederoo, Succulenta 53(4): 63-73, ills., 1974. Typus: *Horst et Uebelmann* HU 381

(U, ZSS [fruits, seeds]). [Brazil: Minas Gerais] [RPS 25]

Melocactus levitestatus Buining & Brederoo, Cact. Succ. J. (US) 45(6): 271-274. ills., 1973. Typus: *Horst et Uebelmann* HU 397 (U, ZSS [status ?]). [Brazil: Bahia] [RPS 24]

Melocactus loboguerreroi Cardenas, Cact. Suc. Mex. 12: 58-59, 67, 1967. Typus: *Cardenas* s.n. (LIL ?). [Erroneously given as invalid in RPS 18.] [Colombia: Dept. del Cauca] [RPS 18]

Melocactus longicarpus Buining & Brederoo, Cact. Succ. J. (US) 46(4): 191-194, ills., 1974. Typus: *Horst et Uebelmann* HU 149 (1966) (U, ZSS [seeds]). [Brazil: Minas Gerais] [RPS 25]

Melocactus longispinus Buining & Brederoo, Succulenta 56(6): 137-142, ills., 1977. Typus: *Horst et Uebelmann* HU 435 (1974) (U [missing], ZSS). [*Non Melocactus longispinus* Ritter *nom. nud.*] [Brazil: Bahia] [RPS 28]

Melocactus longispinus var. *barocensis* Uebelmann, Feld-Nummern, [unpaged], 1972. *Nom. inval.* (Art. 36.1, 37.1). [Published as provisional name.] [RPS -]

Melocactus macrodiscus var. **minor** Ritter, Kakt. Südamer. 1: 133, 1979. [RPS 30]

Melocactus margaritaceus Rizzini, Melocactus no Brasil 76-77, 1982. *Nom. inval.* (Art. 37.1). [RPS 33]

Melocactus margaritaceus var. *disciformis* Rizzini, Melocactus no Brasil 78, 1982. *Nom. inval.* (Art. 43.1). [RPS 33]

Melocactus margaritaceus var. *salvadoranus* Rizzini, Melocactus no Brasil 78, 1982. *Nom. inval.* (Art. 43.1). [RPS 33]

Melocactus maxonii var. **sanctae-rosae** L. D. Gómez, Brenesia 10/11: 127-133, 1977. [RPS 29]

Melocactus mazelianus Riha, Kakt. and. Sukk. 32(9): 214-217, 1981. [RPS 32]

Melocactus melocactoides fa. *capensis* Rizzini, Melocactus no Brasil, 81-82, 1982. *Nom. inval.* (Art. 37.1). [RPS 33]

Melocactus melocactoides fa. **civitatis** Rizzini, Melocactus no Brasil, 81, 1982. [RPS 33]

Melocactus melocactoides fa. **depressus** (Hooker) Rizzini, Melocactus no Brasil, 82, 84, 1982. Basionym: *Melocactus depressus*. [RPS 33]

Melocactus melocactoides fa. **exsanguis** Rizzini, Melocactus no Brasil, 83, 84, 1982. Typus: *M. C. Rizzini* s.n. (?). [Brazil] [RPS 33]

Melocactus melocactoides fa. *perspinosus* Rizzini, Melocactus no Brasil, 82, 1982. *Nom. inval.* (Art. 37.1). [RPS 33]

Melocactus melocactoides fa. *sanctaritae* Rizzini, Melocactus no Brasil, 82-84, 1982. *Nom. inval.* (Art. 37.1). [RPS 33]

Melocactus melocactoides var. **depressus** (Hooker) Rizzini, Melocactus no Brasil, 82, 1982. Basionym: *Melocactus depressus*. [RPS 33]

Melocactus melocactoides var. **itaipuassuensis** Rizzini, Melocactus no Brasil, 82, 84, 1982. Typus: *Rizzini et Rizzini* s.n. (?). [Brazil] [RPS 33]

Melocactus melocactoides var. *natalensis* Rizzini, Melocactus no Brasil, 82, 84, 1982. *Nom. inval.* (Art. 37.1). [RPS 33]

Melocactus melocactoides var. **violaceus** (Pfeiffer) Rizzini, Melocactus no Brasil 81, 1982. Basionym: *Melocactus violaceus*. [RPS 33]

Melocactus montanus Ritter, Kakt. Südamer. 1: 141, 1979. [RPS 30]

Melocactus mulequensis Buining & Brederoo, Succulenta 55(3): 46-51, ills., 1976. Typus: *Horst et Uebelmann* HU 122 (U, ZSS). [Brazil: Bahia] [RPS 27]

Melocactus nagyi Meszaros, Acta Bot. Acad. Sci. Hung. 22: 137, 1977. [Dated 1976. Sphalm. 'nagyii'.] [RPS 27]

Melocactus neomontanus van Heek & Hovens, Succulenta 63(4): 77-82, ills., 1984. Typus: *van Heek et Hovens* 81-135 (U [missing]). [Brazil: Bahia] [RPS 35]

Melocactus nitidus Ritter, Kakt. Südamer. 1: 139, 1979. [= *Melocactus longispinus* Ritter *nom. nud.* (*non M. longispinus* Buining & Brederoo).] [RPS 30]

Melocactus oaxacensis (Britton & Rose) Backeberg, Die Cact. 4: 2587, 1960. Basionym: *Cactus oaxacensis*. [RPS 11]

Melocactus onychacanthus Ritter, Succulenta 46(3): 40, ill., 1967. [First published invalidly (Art. 37.1) l.c. 45(8): 118, 1966 (given as valid in RPS 32).] [RPS 32]

Melocactus onychacanthus var. **albescens** Ritter, Kakt. Südamer. 4: 1308, 1981. Typus: *Ritter* 700b (U). [Peru: Cajamarca] [RPS 32]

Melocactus onychacanthus var. **conicus** Ritter, Kakt. Südamer. 4: 1308, 1981. Typus: *Ritter* 700a (U). [Peru: La Libertad] [RPS 32]

Melocactus oreas fa. **azulensis** (Buining et al.) P. J. Braun, Bradleya 6: 95, 1988. Basionym: *Melocactus azulensis*. [RPS 39]

Melocactus oreas fa. **erythracanthus** (Buining & Brederoo) P. J. Braun, Bradleya 6: 95, 1988. Basionym: *Melocactus erythracanthus*. [RPS 39]

Melocactus oreas ssp. **bahiensis** (Britton & Rose) Rizzini, Melocactus no Brasil, 52-53, 1982. Basionym: *Cactus bahiensis*. [RPS 33]

Melocactus oreas ssp. **cremnophilus** (Buining & Brederoo) P. J. Braun, Bradleya 6: 95, 1988. Basionym: *Melocactus cremnophilus*. [RPS 39]

Melocactus oreas ssp. **ernestii** (Vaupel) P. J. Braun, Bradleya 6: 95, 1988. Basionym: *Melocactus ernestii*. [RPS 39]

Melocactus oreas ssp. **rubrisaetosus** (Buining & Brederoo) P. J. Braun, Bradleya 6: 95, 1988. Basionym: *Melocactus rubrisaetosus*. [RPS 39]

Melocactus oreas var. **bahiensis** (Britton & Rose) Rizzini, Melocactus no Brasil, 52-53, 1982. Basionym: *Cactus bahiensis*. [RPS 33]

Melocactus oreas var. **longispinus** (Buining & Brederoo) P. J. Braun, Bradleya 6: 95, 1988. Basionym: *Melocactus longispinus*. [RPS 39]

Melocactus oreas var. *rubrisaetosus* (Buining & Brederoo) Rizzini, Melocactus no Brasil 52-

53, 1982. *Nom. inval.* (Art. 33.2), based on *Melocactus rubrisaetosus.* [RPS 33]

Melocactus oreas var. *submunitis* Rizzini, Melocactus no Brasil 54, 1982. *Nom. inval.* (Art. 37.1). [RPS 33]

Melocactus pachyacanthus Buining & Brederoo, Kakt. and. Sukk. 27(1): 1-3, ills., 1976. Typus: *Horst et Uebelmann* HU 407 (1972) (U, ZSS [seeds]). [Issue probably already distributed late in December 1975.] [Brazil: Bahia] [RPS 27]

Melocactus paucispinus Heimen & Paul, Kakt. and. Sukk. 34(10): 227-229, ills., 1983. Typus: *Heimen et al.* 81/149 (KOELN [Succulentarium]). [Brazil: Bahia] [RPS 34]

Melocactus peruvianus var. **amstutziae** (Rauh & Backeberg) Rauh & Backeberg, in Rauh, Beitr. Kenntn. peruan. Kakt.veg., 531, 1958. Basionym: *Melocactus amstutziae.* [RPS 10]

Melocactus peruvianus var. **canetensis** Rauh & Backeberg, in Rauh, Beitr. Kenntn. Peruan. Kakt.veg., 533, 1958. [RPS 10]

Melocactus peruvianus var. *lurinensis* Rauh & Backeberg, in Rauh, Beitr. Kenntn. Peruan. Kakt.veg., 533, 1958. *Nom. inval.* (Art. 37.1). [This taxon is based on 2 syntypes, *Rauh* K179 (1956) and *Rauh* K21b (1956). Erroneously included as valid in RPS 10.] [RPS 10]

Melocactus pruinosus var. **concinnus** (Buining & Brederoo) P. J. Braun, Bradleya 6: 95, 1988. Basionym: *Melocactus concinnus.* [RPS 39]

Melocactus radoczii Meszaros, Acta Bot. Acad. Sci. Hung. 22: 136, 1977. [Dated 1976.] [RPS 27]

Melocactus robustispinus Buining & Brederoo, Succulenta 56(5): 116-119, ills., 1977. Typus: *Horst et Uebelmann* HU 403 (U, ZSS [fruits, seeds]). [Brazil: Minas Gerais] [RPS 28]

Melocactus rubrisaetosus Buining & Brederoo, Succulenta 56(7): 161-167, ills., 1977. Typus: *Horst et Uebelmann* HU 137 (U, ZSS [seeds]). [Brazil: Bahia] [RPS 28]

Melocactus rubrispinus Ritter, Kakt. Südamer. 1: 135, 1979. [This name has also been used provisionally by W. Uebelmann in Feld-Nummern, [unpaged], 1972 for the taxon later described as *M. erythracanthus.*] [RPS 30]

Melocactus ruestii ssp. **cintalapensis** Elizondo, Cact. Suc. Mex. 31(3): 54-55, ill., 1986. Typus: *Elizondo* 189 (XAL). [Mexico] [RPS 37]

Melocactus ruestii ssp. **maxonii** (Rose) Elizondo, Cact. Suc. Mex. 31(3): 53, 1986. Basionym: *Cactus maxonii.* [RPS 37]

Melocactus ruestii ssp. **oaxacensis** (Britton & Rose) Elizondo, Cact. Suc. Mex. 31(3): 52, 1986. Basionym: *Cactus oaxacensis.* [RPS 37]

Melocactus ruestii ssp. **sanctae-rosae** (L. D. Gómez) Elizondo, Cact. Suc. Mex. 31(3): 54, 1986. Basionym: *Melocactus maxonii* var. *sanctae-rosae.* [RPS 37]

Melocactus saxicola Diers & E. Esteves Pereira, Kakt. and. Sukk. 35(9): 196-201, ills., 1984. Typus: *Esteves Pereira* 119 (KOELN [Succulentarium]). [Brazil] [RPS 35]

Melocactus schatzlii Till & Gruber, Kakt. and. Sukk. 33(4): 68-70, ills., 1982. Typus: *Gruber* 40 (WU). [RPS 33]

Melocactus schulzianus Buining & Brederoo, Succulenta 53: 86-92, 1974. [RPS 25]

Melocactus seabrasensis Uebelmann, Feld-Nummern, [unpaged], 1972. *Nom. inval.* (Art. 36.1, 37.1). [Published as provisional name.] [RPS -]

Melocactus securituberculatus Buining & Brederoo, Cact. Succ. J. (US) 48(1): 38-41, ills., 1976. Typus: *Horst et Uebelmann* HU 446 (1974) (U, ZSS). [Brazil: Bahia] [RPS 27]

Melocactus smithii (Alexander) Buining *ex* Rowley, Rep. Pl. Succ. 25: 11, 1976. Basionym: *Cactus smithii.* [First published invalidly (Art. 33.2) by Buining in Succulenta 53: 92, 1974.] [RPS 25]

Melocactus trujilloensis Rauh & Backeberg, in Backeberg, Descr. Cact. Nov. [1:] 36, 1957. Typus: *Schön* 4 (ZSS). [Dated 1956, published 1957.] [Peru] [RPS 7]

Melocactus trujilloensis var. **schoenii** Rauh & Backeberg, in Backeberg, Descr. Cact. Nov. [1:] 36, 1957. [Dated 1956, published 1957.] [RPS 7]

Melocactus uebelmannii P. J. Braun, Kakt. and. Sukk. 36(11): 232-237, ills., 1985. Typus: *Horst et Uebelmann* HU 528 (1982) (ZSS). [Brazil: Bahia] [RPS 36]

Melocactus unguispinus Rauh & Backeberg, in Backeberg, Descr. Cact. Nov. [1:] 36, 1957. [Dated 1956, published 1957.] [RPS 7]

Melocactus warasii Esteves Pereira & R. Bueneker, Bradea 2(30): 213, 1977. [RPS -]

Melocactus zehntneri (Britton & Rose) Backeberg, Die Cact. 4: 2575, 1960. Basionym: *Cactus zehntneri.* [RPS 11]

Melocactus zehntneri ssp. **canescens** (Ritter) P. J. Braun, Bradleya 6: 98, 1988. Basionym: *Melocactus canescens.* [RPS 39]

Melocactus zehntneri ssp. **robustispinus** (Buining & Brederoo) P. J. Braun, Bradleya 6: 98, 1988. Basionym: *Melocactus robustispinus.* [RPS 39]

Melocactus zehntneri var. *ananas* Rizzini, Melocactus no Brasil 64, 1982. *Nom. inval.* (Art. 37.1). [RPS 33]

Melocactus zehntneri var. **curvicornis** (Buining & Brederoo) P. J. Braun, Bradleya 6: 98, 1988. Basionym: *Melocactus curvicornis.* [RPS 39]

Melocactus zehntneri var. **viridis** Ritter, Kakt. Südamer. 1: 132, 1979. [RPS 30]

Mesechinopsis Y. Ito, Explan. Diagr. Austroechinocactinae, 72, 286, 1957. Typus: *Echinopsis hamatacantha.* [RPS 8]

Mesechinopsis ancistrophora (Spegazzini) Y. Ito, Explan. Diagr. Austroechinocactinae, 75, 1957. Basionym: *Echinopsis ancistrophora.* [RPS 8]

Mesechinopsis hamatacantha (Backeberg) Y. Ito, Explan. Diagr. Austroechinocactinae, 74, 1957. Basionym: *Echinopsis hamatacantha.* [RPS 8]

Mesechinopsis leucorhodantha (Backeberg) Y. Ito, Explan. Diagr. Austroechinocactinae, 76,

1957. Basionym: *Echinopsis leucorhodantha*. [RPS 8]

Mesechinopsis lobivioides (Backeberg) Y. Ito, Explan. Diagr. Austroechinocactinae, 75, 1957. Basionym: *Echinopsis lobivioides*. [RPS 8]

Mesechinopsis nakajimae Y. Ito, Explan. Diagr. Austroechinocactinae, 76, 286, 1957. [RPS 8]

Mesechinopsis pelecyrhachis (Backeberg) Y. Ito, Explan. Diagr. Austroechinocactinae, 76, 1957. Basionym: *Echinopsis pelecyrhachis*. [RPS 8]

Mesechinopsis polyancistra (Backeberg) Y. Ito, Explan. Diagr. Austroechinocactinae, 74, 1957. Basionym: *Echinopsis polyancistra*. [RPS 8]

Micranthocereus albicephalus (Buining & Brederoo) Ritter, Kakt. Südamer. 1: 108, 1979. Basionym: *Austrocephalocereus albicephalus*. [RPS 30]

Micranthocereus aureispinus Ritter, Kakt. Südamer. 1: 107, 1979. [RPS 30]

Micranthocereus auriazureus Buining & Brederoo, Cact. Succ. J. (US) 45(3): 120-123, ills., 1973. Typus: *Horst et Uebelmann* HU 346 (1971) (U, ZSS [status ?]). [Sphalm. 'auri-azureus'. The type is erroneously cited as *Horst* 348 in the original description; the correction is based on the field-number list of W. Uebelmann (Horst - Uebelmann numbers).] [Brazil: Minas Gerais] [RPS 24]

Micranthocereus densiflorus Buining & Brederoo, Cact. Succ. J. (US) 46(3): 113-116, ills., 1974. Typus: *Horst et Uebelmann* HU 221 (U, ZSS). [Brazil: Bahia] [RPS 25]

Micranthocereus dolichospermaticus (Buining & Brederoo) Ritter, Kakt. Südamer. 1: 108, 1979. Basionym: *Austrocephalocereus dolichospermaticus*. [RPS 30]

Micranthocereus estevesii (Buining & Brederoo) Ritter, Kakt. Südamer. 1: 108, 1979. Basionym: *Austrocephalocereus estevesii*. [RPS 30]

Micranthocereus flaviflorus Buining & Brederoo, Kakt. and. Sukk. 25: 25-27, 1974. [RPS 25]

Micranthocereus haematocarpus Ritter, Kakt. Südamer. 1: 105, 1979. [RPS 30]

Micranthocereus lehmannianus (Werdermann) Ritter, Kakt. and. Sukk. 19: 157, 1968. Basionym: *Cephalocereus lehmannianus*. [RPS 19]

Micranthocereus monteazulensis Ritter, Kakt. Südamer. 1: 105, 1979. [RPS 30]

Micranthocereus purpureus (Gürke) Ritter, Kakt. and. Sukk. 19: 157, 1968. Basionym: *Cephalocereus purpureus*. [RPS 19]

Micranthocereus ruficeps Ritter, Kakt. Südamer. 1: 106, 1979. [RPS 30]

Micranthocereus streckeri van Heek & van Criekinge, Kakt. and. Sukk. 37(5): 102-105, ills., 1986. Typus: *van Heek et al.* 85/250 (KOELN). [Brazil] [RPS 37]

Micranthocereus uilianus Brederoo & Bercht, Succulenta 63(8): 178-183, ills., 1984. Typus: *Horst et Uebelmann* HU 439 (U). [Brazil: Bahia] [RPS 35]

Micranthocereus violaciflorus Buining, Kakt. and. Sukk. 20(7): 129-130, ills., 1969. Typus: *Horst et Uebelmann* HU 275 (1968) (U [status ?], ZSS [status ?]). [Brazil: Minas Gerais] [RPS 20]

Micropuntia gracilicylindrica Wiegand & Backeberg, in Backeberg, Descr. Cact. Nov. [1:] 9, 1957. [Dated 1956, published 1957.] [RPS 7]

Micropuntia pygmaea Wiegand & Backeberg, in Backeberg, Descr. Cact. Nov. [1:] 9, 1957. [Dated 1956, published 1957.] [RPS 7]

Micropuntia tuberculosirhopalica Wiegand & Backeberg, in Backeberg, Descr. Cact. Nov. [1:] 9, 1957. [Dated 1956, published 1957.] [RPS 7]

Micropuntia wiegandii Backeberg, Descr. Cact. Nov. [1:] 9, 1957. [Dated 1956, published 1957.] [RPS 7]

Mila albisaetacens Rauh & Backeberg, in Backeberg, Descr. Cact. Nov. [1:] 11, 1957. [Dated 1956, published 1957.] [RPS 7]

Mila albo-areolata Akers, Fuaux Herb. Bull. 5: 2, 1953. [RPS 4]

Mila breviseta Rauh & Backeberg, in Backeberg, Descr. Cact. Nov. [1:] 12, 1957. [Dated 1956, published 1957.] [RPS 7]

Mila caespitosa fa. **albisaetacens** (Rauh & Backeberg) Donald, Ashingtonia 3(3-4): 62, 1978. Basionym: *Mila albisaetacens*. [RPS 29]

Mila caespitosa fa. **breviseta** (Rauh & Backeberg) Donald, Ashingtonia 3(3-4): 62, 1978. Basionym: *Mila breviseta*. [RPS 29]

Mila caespitosa fa. **cereoides** (Rauh & Backeberg) Donald, Ashingtonia 3(3-4): 61, 1978. Basionym: *Mila cereoides*. [RPS 29]

Mila caespitosa fa. **kubeana** (Backeberg & Werdermann) Donald, Ashingtonia 3(3-4): 62, 1978. Basionym: *Mila kubeana*. [RPS 29]

Mila caespitosa fa. **lurinensis** (Rauh & Backeberg) Donald, Ashingtonia 3(3-4): 62, 1978. Basionym: *Mila lurinensis*. [RPS 29]

Mila caespitosa fa. **sublanata** (Rauh & Backeberg) Donald, Ashingtonia 3(3-4): 61, 1978. Basionym: *Mila sublanata*. [RPS 29]

Mila caespitosa ssp. **nealeana** (Backeberg) Donald, Ashingtonia 3(3-4): 62, 1978. Basionym: *Mila nealeana*. [RPS 29]

Mila caespitosa var. **albo-areolata** (Akers) Donald, Ashingtonia 3(3-4): 62, 1978. Basionym: *Mila albo-areolata*. [Sphalm. 'alboareolata'.] [RPS 29]

Mila caespitosa var. **densiseta** (Rauh & Backeberg) Donald, Ashingtonia 3(3-4): 62, 1978. Basionym: *Mila densiseta*. [RPS 29]

Mila caespitosa var. **fortalezensis** (Rauh & Backeberg) Donald, Ashingtonia 3(3-4): 61, 1978. Basionym: *Mila fortalezensis*. [RPS 29]

Mila caespitosa var. **pugionifera** (Rauh & Backeberg) Donald, Ashingtonia 3(3-4): 61, 1978. Basionym: *Mila pugionifera*. [RPS 29]

Mila cereoides Rauh & Backeberg, in Backeberg, Descr. Cact. Nov. [1:] 11, 1957. Typus: *Rauh* K57 (1956) (ZSS). [Dated 1956, published 1957.] [Peru] [RPS 7]

Mila colorea Ritter, Kakt. Südamer. 4: 1341-

1342, 1981. Typus: *Ritter* 699 (U). [Peru: Ancash] [RPS 32]

Mila densiseta Rauh & Backeberg, in Backeberg, Descr. Cact. Nov. [1:] 11, 1957. [Dated 1956, published 1957.] [RPS 7]

Mila fortalezensis Rauh & Backeberg, in Backeberg, Descr. Cact. Nov. [1:] 11, 1957. Typus: *Rauh* K50 (1956) (ZSS). [Dated 1956, published 1957.] [Peru] [RPS 7]

Mila lurinensis Rauh & Backeberg, in Backeberg, Descr. Cact. Nov. [1:] 11, 1957. [Dated 1956, published 1957.] [RPS 7]

Mila nealeana fa. *senilis* Ritter, Kakt. Südamer. 4: 1341, 1981. *Nom. inval.* (Art. 37). [RPS 32]

Mila nealeana var. **tenuior** Rauh & Backeberg, in Backeberg, Descr. Cact. Nov. [1:] 11, 1957. [Dated 1956, published 1957.] [RPS 7]

Mila pugionifera Rauh & Backeberg, in Backeberg, Descr. Cact. Nov. [1:] 11, 1957. [Dated 1956, published 1957.] [RPS 7]

Mila sublanata Rauh & Backeberg, in Backeberg, Descr. Cact. Nov. [1:] 11, 1957. Typus: *Rauh* K57b (1956) (ZSS). [Dated 1956, published 1957.] [Peru] [RPS 7]

Mila sublanata var. **pallidior** Rauh & Backeberg, in Backeberg, Descr. Cact. Nov. [1:] 11, 1957. [Dated 1956, published 1957.] [RPS 7]

Miqueliopuntia Fric *ex* Ritter, Kakt. Südamer. 3: 869, 1980. Typus: *Opuntia miquelii.* [RPS 31]

Miqueliopuntia miquelii (Monville) Ritter, Kakt. Südamer. 3: 869-871, 1980. Basionym: *Opuntia miquelii.* [RPS 31]

Mirabella Ritter, Kakt. Südamer. 1: 108-110, 1979. Typus: *Acanthocereus albicaulis.* [RPS 30]

Mirabella albicaulis (Britton & Rose) Ritter, Kakt. Südamer. 1: 110, 1979. Basionym: *Acanthocereus albicaulis.* [RPS 30]

Mirabella minensis Ritter, Kakt. Südamer. 1: 111, 1979. [RPS 30]

Mitrocereus militaris (Audot) Buxbaum, Bot. Stud. 12: 54, 101, 1961. Basionym: *Cereus militaris.* [RPS 12]

Mitrocereus ruficeps (F. A. C. Weber) Backeberg, Die Cact. 4: 2244, 1960. Basionym: *Pilocereus ruficeps.* [RPS 11]

Monvillea adelmarii Rizzini & Mattos-F., Rev. Brasil Biol. 45(3): 301, 1985. Typus: *Adelmar F. Coimbra* s.n. (RB). [sphalm. 'adelmari'.] [Brazil] [RPS 36]

Monvillea alticostata Ritter, Kakt. Südamer. 1: 251-252, 1979. [RPS 30]

Monvillea apoloensis Cardenas, Cact. Succ. J. (US) 33: 74, 1961. [RPS 12]

Monvillea ballivianii Cardenas, Cactus (Paris) No. 64: 159, 1959. [RPS 10]

Monvillea calliantha Fuaux & Backeberg, Die Cact. 4: 2312-2313, 1960. *Nom. inval.* (Art. 37.1). [Erroneously included as valid in RPS 11.] [RPS 11]

Monvillea campinensis (Backeberg & Voll) Backeberg, Die Cact. 4: 2313, 1960. Basionym: *Pilocereus campinensis.* [RPS 11]

Monvillea chacoana Ritter, Kakt. Südamer. 2: 509-510, 1980. [RPS 31]

Monvillea ebenacantha Ritter, Kakt. Südamer. 2: 512-514, 1980. [RPS 31]

Monvillea euchlora (F. A. C. Weber) Backeberg, Die Cact. 4: 2301, 1960. Basionym: *Cereus euchlorus.* [RPS 11]

Monvillea jaenensis Rauh & Backeberg, in Backeberg, Descr. Cact. Nov. [1:] 33, 1957. [Dated 1956, published 1957. Repeated by Rauh in Sitz.Ber. Heidelberg. Akad. Wiss. Math.-Nat. Kl. 1. 506, 1958.] [RPS 7]

Monvillea jaenensis var. *columbiana* W. Rauh, Kakteen an ihren Standorten, 174, 1979. *Nom. inval.* (Art. 36.1, 37.1). [RPS 34]

Monvillea jaenensis var. **paucispina** Rauh & Backeberg, in Rauh, Beitr. Kenntn. Peruan. Kakt.veg., 507, 1958. [RPS 10]

Monvillea leucantha Ritter, Kakt. Südamer. 2: 511, 1980. Typus: *Ritter* 391 (U, ZSS). [Type cited for U l.c. 1: iii, 1979.] [Bolivia: Santa Cruz] [RPS 31]

Monvillea parapetiensis Ritter, Kakt. Südamer. 2: 510-511, 1980. [RPS 31]

Monvillea paxtoniana var. **borealis** Ritter, Kakt. Südamer. 1: 251, 1979. [RPS 30]

Monvillea piedadensis Ritter, Kakt. Südamer. 1: 114-115, 1979. [RPS 30]

Monvillea pugionifera Ritter, Kakt. Südamer. 4: 1305-1306, 1981. Typus: *Ritter* 1064 (U). [Peru: Cajamarca] [RPS 32]

Monvillea saddiana Rizzini & Mattos-F., Rev. Brasil Biol. 45(3): 304, 1985. Typus: *Mattos-F. et Saddi* 3648 (RB). [Brazil] [RPS 36]

Monvilleeae Ritter, Kakt. Südamer. 1: 230, 1979. Typus: *Monvillea.* [RPS 30]

Morangaya Rowley, Ashingtonia 1: 43-45, 1974. Typus: *Echinocereus pensilis.* [RPS 25]

Morangaya pensilis (K. Brandegee) Rowley, Ashingtonia 1(4): 44-45, ill. (p. 43), 1974. Basionym: *Cereus pensilis.* [RPS 25]

Morawetzia doelziana fa. **calva** (Backeberg) Ritter, Backeberg's Descr. & Erört. taxon. nomenkl. Fragen, 52, 1958. Basionym: *Morawetzia doelziana* var. *calva.* [Combination repeated in Ritter, Kakt. Südamer. 4: 1366-1367, 1981 (and reported as invalid in RPS 32).] [RPS 15]

Morawetzia doelziana var. **calva** Rauh & Backeberg, in Backeberg, Descr. Cact. Nov. [1:] 18, 1957. [Dated 1956, published 1957.] [RPS 7]

Morawetzia doelziana var. *fuscatispina* Backeberg, Kakt.-Lex., 281, 1966. *Nom. inval.* (Art. 37.1). [Erroneously included as valid in RPS 17.] [RPS 17]

Morawetzia sericata Ritter, Kakt. Südamer. 4: 1367-1368, 1981. Typus: *Ritter* 1309 (U). [Peru: Huancavelica] [RPS 32]

Morawetzia varicolor Knize, Biota 7: 254, 1969. Typus: *Knize* 345 (SMF 5972, ZSS [not found]). [= Soukup 6181] [Published either in Dec. 1968 or early in 1969. Sphalm. 'variicolor'.] [RPS 36]

× **Myrtgerocactus** Moran, Cact. Succ. J. (US) 34: 186, 1962. [Sphalm. 'Mytgerocactus'.] [= *Myrtillocactus* × *Bergerocactus.*] [RPS 13]

× **Myrtgerocactus lindsayi** Moran, Cact. Succ. J. (US) 34: 186, 1962. [Sphalm. 'Mytgerocactus'; = *Myrtillocactus cochal* × *Bergerocactus emoryi*.] [RPS 13]
× **Myrtillenocereus** Rowley, Nation. Cact. Succ. J. 37(3): 78, 1982. [= *Myrtillocactus* × *Stenocereus*.] [RPS 33]
Myrtillocactinae Buxbaum, Bot. Stud. 12: 93, 1961. Typus: *Myrtillocactus* Console. [RPS 12]
Myrtillocactus geometrizans fa. **hintonii** Fitz Maurice, Cact. Suc. Mex. 34(4): 89-91, ills., 1989. Typus: *Hinton* s.n. (MEXU, UNAM). [A monstrose form.] [Mexico: San Luis Potosi] [RPS 40]
Myrtillocactus geometrizans var. **grandiareolatus** (Bravo) Backeberg, Die Cact. 4: 2267, 1960. Basionym: *Myrtillocactus grandiareolatus*. [RPS 11]
× **Myrtocereus** Mottram, Contr. New Class. Cact. Fam., 59, 1990. [= *Myrtillocactus* × *Nyctocereus*.] [RPS 41]
Navajoa durispina Y. Ito, The Cactaceae, 473, 1981. *Nom. inval.* (Art. 37.1). [RPS -]
Navajoa fickeisenii Backeberg, Cact. Succ. J. Gr. Brit. 22: 49, 54, 1960. *Nom. inval.* (Art. 37.1). [RPS 11]
Neoabbottia grantiana (Britton) Buxbaum, Cact. Succ. J. (US) 38: 45, 1966. Basionym: *Leptocereus grantianus*. [RPS 17]
Neoabbottia paniculata var. **humbertii** Backeberg, Cact. Succ. J. (US) 27(2): 53, 1955. [RPS 6]
Neobesseya subgen. **Ortegocactus** (Alexander) Kladiwa, in Krainz, Die Kakt. Lief. 57: C VIIIc (April 1974), 1974. Basionym: *Ortegocactus*. [RPS 25]
Neobesseyae Buxbaum, Madroño 14(6): 198, 1958. Typus: *Neobesseya*. [Published as 'linea'. First mentioned (*nom. inval.*, Art. 36.1) in Österr. Bot. Zeitschr. 98(1/2): 75, 1951.] [RPS 9]
Neobinghamia Backeberg, Cact. Succ. J. (US) 22(5): 154, 1950. Typus: *Binghamia climaxantha*. [Emendation published in Backeberg, Descr. Cact. Nov. [1:] 5, 1957 (dated 1956) (cf. RPS 7).] [RPS 1]
Neobinghamia climaxantha var. **armata** Rauh & Backeberg, in Backeberg, Descr. Cact. Nov. [1:] 34, 1957. [Dated 1956, published 1957.] [RPS 7]
Neobinghamia climaxantha var. **lurinensis** Rauh & Backeberg, in Backeberg, Descr. Cact. Nov. [1:] 34, 1957. Typus: *Rauh* K182 (1956) (ZSS). [Dated 1956, published 1957.] [Peru] [RPS 7]
Neobinghamia climaxantha var. **subfusciflora** Rauh & Backeberg, in Backeberg, Descr. Cact. Nov. [1:] 34, 1957. Typus: *Rauh* K24 (1956) (ZSS). [Dated 1956, published 1957.] [Peru] [RPS 7]
Neobinghamia mirabilis Rauh & Backeberg, in Backeberg, Descr. Cact. Nov. [1:] 35, 1957. [Dated 1956, published 1957.] [RPS 7]
Neobinghamia multiareolata Rauh & Backeberg, in Backeberg, Descr. Cact. Nov. [1:] 34, 1957. Typus: *Rauh* K51 (1956) (ZSS). [Dated 1956, published 1957.] [Peru] [RPS 7]
Neobinghamia multiareolata var. **superba** Rauh & Backeberg, in Backeberg, Descr. Cact. Nov. [1:] 35, 1957. [Dated 1956, published 1957.] [RPS 7]
Neobinghamia villigera Rauh & Backeberg, in Backeberg, Descr. Cact. Nov. [1:] 35, 1957. [Dated 1956, published 1957.] [RPS 7]
Neobuxbaumia subgen. **Crassocereus** Backeberg, Descr. Cact. Nov. [1:] 6, 1957. Typus: *Neobuxbaumia polylopha* (De Candolle) Backeberg. [RPS 7]
Neobuxbaumia euphorbioides (Haworth) Buxbaum *ex* Bravo, Cact. Mex., ed. 2, 1: 658-659, 1978. Basionym: *Cereus euphorbioides*. [First published invalidly (Art. 33.2) by Buxbaum in Cactus (Paris) No. 40: 52, 1954 (cf. RPS 5).] [RPS 5]
Neobuxbaumia euphorbioides var. **olfersii** (Salm-Dyck) Bravo, Cact. Mex., ed. 2, 1: 660, 1978. Basionym: *Cereus olfersii*. [Erroneously accredited to Backeberg.] [RPS -]
Neobuxbaumia macrocephala (F. A. C. Weber) Dawson, Cact. Succ. J. (US) 24(6): 173, 1952. Basionym: *Cephalocereus macrocephalus*. [RPS 3]
Neobuxbaumia mezcalaensis var. **robusta** (Dawson) Backeberg, Die Cact. 4: 2203, 1960. Basionym: *Cephalocereus mezcalaensis* var. *robustus*. [First published as nude combination in Cact. Succ. J. (US) 23(4): 122, 1951 (and reported as new var. *robustus* (sic !) Backeberg in RPS 3).] [RPS 11]
Neobuxbaumia multiareolata (Dawson) Bravo, Scheinvar & Sanchez-Mejorada, Cact. Suc. Mex. 17: 120, 1972. Basionym: *Cephalocereus mezcalaensis* var. *multiareolatus*. [RPS 23]
Neobuxbaumia squamulosa Scheinvar & Sanchez-Mejorada, Cact. Suc. Mex. 35(1): 13-18, ills., SEM-ill., 1990. Typus: *Sanchez-Mejorada et al.* 4250 (MEXU). [Mexico: Michoacan] [RPS 41]
Neochilenia subgen. *Argyrochilenia* Y. Ito, Bull. Takarazuka Insectarium 71: 13-20, 1950. *Nom. inval.* (Art. 36.1). [RPS 3]
Neochilenia subgen. *Caphnochilenia* Y. Ito, Bull. Takarazuka Insectarium 71: 13-20, 1950. *Nom. inval.* (Art. 36.1). [RPS 3]
Neochilenia subgen. *Chrochilenia* Y. Ito, Bull. Takarazuka Insectarium 71: 13-20, 1950. *Nom. inval.* (Art. 36.1). [RPS 3]
Neochilenia subgen. *Melanochilenia* Y. Ito, Bull. Takarazuka Insectarium 71: 13-20, 1950. *Nom. inval.* (Art. 36.1). [RPS 3]
Neochilenia aerocarpa (Ritter) Backeberg, Die Cact. 3: 1826-1827, 1959. *Nom. inval.* (Art. 33.2). [RPS 10]
Neochilenia aerocarpa var. *fulva* (Ritter) Backeberg, Die Cact. 3: 1828, 1959. *Nom. inval.*, based on *Chileorebutia fulva, nom. inval.* (Art. 36.1). [RPS 10]
Neochilenia ambigua (Hildmann) Y. Ito, Explan. Diagr. Austroechinocactinae, 276, 1957. Basionym: *Echinocactus ambiguus*. [RPS 8]
Neochilenia andreaeana Backeberg, Kakt. and. Sukk. 10: 38, 1959. *Nom. inval.* (Art. 9.5,

37.1). [Erroneously included as valid in RPS 10.] [RPS 10]

Neochilenia aricensis (Ritter) Backeberg, Descr. Cact. Nov. 3: 9, 1963. Basionym: *Pyrrhocactus aricensis*. [RPS 14]

Neochilenia aspillagae (Söhrens) Y. Ito, Explan. Diagr. Austroechinocactinae, 277, 1957. Basionym: *Echinocactus aspillagai*. [Sphalm. 'aspillagai'.] [RPS 8]

Neochilenia atra Backeberg, Descr. Cact. Nov. 3: 8, 1963. *Nom. inval.* (Art. 9.5). [Erroneously included as valid in RPS 14.] [RPS 14]

Neochilenia calderana (Ritter) Backeberg, Descr. Cact. Nov. 3: 9, 1963. Basionym: *Pyrrhocactus calderanus*. [RPS 14]

Neochilenia carneoflora Kilian *ex* Backeberg, in Backeberg, Kakt.-Lex., 288, 1966. *Nom. inval.* (Art. 37.1). [Erroneously included as valid in RPS 17.] [RPS 17]

Neochilenia chilensis (Hildemann) Backeberg, Cact. Succ. J. (US) 23(3): 88, 1951. Basionym: *Chilenia chilensis*. [The name given as basionym was itself a combination, and is here corrected.] [RPS 3]

Neochilenia chorosensis (Ritter) Backeberg, Die Cact. 6: 3777, 1962. Basionym: *Pyrrhocactus chorosensis*. [RPS 13]

Neochilenia confinis (Ritter) Backeberg, Descr. Cact. Nov. 3: 9, 1963. Basionym: *Pyrrhocactus confinis*. [RPS 14]

Neochilenia deherdtiana Backeberg, Descr. Cact. Nov. 3: 8, 1963. *Nom. inval.* (Art. 9.5, 37.1). [RPS 14]

Neochilenia dimorpha (Ritter) Backeberg, Descr. Cact. Nov. 3: 9, 1963. Basionym: *Pyrrhocactus dimorphus*. [RPS 14]

Neochilenia duripulpa (Ritter) Backeberg, Descr. Cact. Nov. 3: 9, 1963. Basionym: *Chileorebutia duripulpa*. [RPS 14]

Neochilenia ebenacantha (Monville) Backeberg, Cact. Succ. J. (US) 23(3): 88, 1951. Basionym: *Echinocactus ebenacanthus*. [The name given as basionym is already a combination (here corrected).] [RPS 3]

Neochilenia ebenacantha var. **intermedia** (Labouret) Y. Ito, Explan. Diagr. Austroechinocactinae, 274, 1957. Basionym: *Echinocactus ebenacanthus* var. *intermedius*. [RPS 8]

Neochilenia ebenacantha var. **nova** (Schelle) Y. Ito, Explan. Diagr. Austroechinocactinae, 274, 1957. Basionym: *Echinocactus ebenacanthus* var. *novus*. [RPS 8]

Neochilenia eriocephala Backeberg, Die Cact. 3: 1830-1832, 1959. *Nom. inval.* (Art. 37.1). [Erroneously included as valid in RPS 10.] [RPS 10]

Neochilenia eriosyzoides (Ritter) Backeberg, Die Cact. 6: 3777, 1962. Basionym: *Horridocactus eriosyzoides*. [RPS 13]

Neochilenia eriosyzoides var. *roseiflora* Y. Ito, The Cactaceae, 457, 1981. *Nom. inval.* (Art. 37.1). [RPS -]

Neochilenia esmeraldana (Ritter) Backeberg, Descr. Cact. Nov. 3: 9, 1963. Basionym: *Chileorebutia esmeraldana*. [First mentioned as nude combination of manuscript name in Die Cact. 3: 1828, 1959 (cf. RPS 10).] [RPS 14]

Neochilenia esmeraldana var. *brevispina* Backeberg, Die Cact. 3: 1829, 1959. *Nom. inval.* (Art. 33.2, 37.1). [Published as provisional name.] [RPS 10]

Neochilenia flaviflora Y. Ito, The Cactaceae, 652, 1981. *Nom. inval.* (Art. 37.1). [RPS -]

Neochilenia floccosa (Ritter) Backeberg, Descr. Cact. Nov. 3: 9, 1963. Basionym: *Pyrrhocactus floccosus*. [RPS 14]

Neochilenia fobeana (Mieckley) Backeberg, Cact. Succ. J. (US) 23(3): 88, 1951. Basionym: *Neoporteria fobeana*. [The basionym given is already a combination.] [RPS 3]

Neochilenia fusca (Mühlenpfordt) Backeberg, Cact. Succ. J. (US) 23(3): 88, 1951. Basionym: *Echinocactus fuscus*. [The name given as basionym was itself a combination, and is here corrected.] [RPS 3]

Neochilenia glaucescens (Ritter) Backeberg, Descr. Cact. Nov. 3: 9, 1963. Basionym: *Pyrrhocactus glaucescens*. [RPS 14]

Neochilenia gracilis (Ritter) Backeberg, Descr. Cact. Nov. 3: 9, 1963. Basionym: *Pyrrhocactus gracilis*. [RPS 14]

Neochilenia hankeana var. *minor* Oehme *ex* Backeberg, Die Cact. 3: 1809, 1959. *Nom. inval.* (Art. 36.1, 37.1). [Given as "comb. nov." and based on *Echinocactus ebenacanthus* hort. var. *minor* hort. *non* Monville.] [RPS 10]

Neochilenia hankeana var. *taltalensis* Backeberg, Die Cact. 3: 1809, 1959. Based on *Ritter* 212. *Nom. inval.* (Art. 37.1). [Based on *Horridocactus taltalensis* Ritter *non* Hutchison (*nom. nud.*). Erroneously included as valid in RPS 10. This taxon is based on the same type as *Pyrrhocactus neohankeanus*.] [Chile] [RPS 10]

Neochilenia huascensis (Ritter) Backeberg, Descr. Cact. Nov. 3: 9, 1963. Basionym: *Pyrrhocactus huascensis*. [RPS 14]

Neochilenia hypogaea Ritter *ex* Backeberg, Die Cact. 3: 1830, 1959. *Nom. inval.* (Art. 36.1, 37.1). [Based on a manuscript name of Ritter, *Chileorebutia hypogaea*.] [RPS 10]

Neochilenia imitans Backeberg, Die Cact. 3: 1821, 1959. *Nom. inval.* (Art. 37.1). [Erroneously included as valid in RPS 10.] [RPS 10]

Neochilenia intermedia (Ritter) Backeberg, Descr. Cact. Nov. 3: 9, 1963. Basionym: *Pyrrhocactus intermedius*. [RPS 14]

Neochilenia iquiquensis (Ritter) Backeberg, Descr. Cact. Nov. 3: 9, 1963. Basionym: *Chileorebutia iquiquensis*. [RPS 14]

Neochilenia jussieui (Monville) Backeberg, Cact. Succ. J. (US) 23(3): 88, 1951. Basionym: *Echinocactus jussieui*. [The name given as basionym was itself a combination, and is here corrected.] [RPS 3]

Neochilenia krausii (Ritter) Backeberg, Die Cact. 3: 1832, 1959. *Nom. inval.* (Art. 33.2), based on *Chileorebutia krausii*, incorrect name (Art. 11.3). [RPS 10]

Neochilenia kunzei (Förster) Backeberg, Cact. Succ. J. (US) 23(3): 88, 1951. Basionym:

Echinocactus kunzei. [The name given as basionym was itself a combination, and is here corrected.] [RPS 3]

Neochilenia lembckei Backeberg, Die Cact. 3: 1822, 1959. *Nom. inval.* (Art. 37.1). [Erroneously included as valid in RPS 10.] [RPS 10]

Neochilenia malleolata (Ritter) Backeberg, Descr. Cact. Nov. 3: 9, 1963. Basionym: *Chileorebutia malleolata*. [RPS 14]

Neochilenia malleolata var. **solitaria** (Ritter) Backeberg, Descr. Cact. Nov. 3: 9, 1963. Basionym: *Chileorebutia malleolata* var. *solitaria*. [RPS 14]

Neochilenia mitis (Philippi) Backeberg, Die Cact. 3: 1818, 1959. Basionym: *Echinocactus mitis*. [RPS 10]

Neochilenia monte-amargensis Backeberg, Descr. Cact. Nov. 3: 8, 1963. *Nom. inval.* (Art. 9.5). [Erroneously included as valid in RPS 14.] [RPS 14]

Neochilenia napina (Philippi) Backeberg, Cact. Succ. J. (US) 23(3): 88, 1951. Basionym: *Echinocactus napinus*. [The name given as basionym is already a combination (here corrected).] [RPS 3]

Neochilenia napina var. *spinosior* Backeberg, Die Cact. 6: 3773, 1962. *Nom. inval.* (Art. 9.5). [Erroneously included as valid in RPS 13. First mentioned as new combination in l.c. 3: 1820, 1959 (cf. RPS 10).] [RPS 13]

Neochilenia neofusca Backeberg, Die Cact. 3: 1810, 1959. *Nom. inval.* (Art. 37.1). [Erroneously included as valid in RPS 10.] [RPS 10]

Neochilenia neoreichei Backeberg, Die Cact. 3: 1825, 1959. *Nom. inval.* (Art. 37.1). [Erroneously included as valid in RPS 10. Republished invalidly (Art. 37.1) by Ito, The Cactaceae, 461, 1981 [mention of a single collection, F. Ritter 501, but most probably the author has not seen any herbarium specimens which could serve as type; therefore, the name is still invalid].] [RPS 10]

Neochilenia nigricans (Link) Y. Ito, Explan. Diagr. Austroechinocactinae, 275, 1957. Basionym: *Echinopsis nigricans*. [RPS 8]

Neochilenia nigriscoparia Backeberg, Die Cact. 6: 3784, 1962. *Nom. inval.* (Art. 37.1). [Erroneously included as valid in RPS 13.] [RPS 13]

Neochilenia occulta (Philippi) Backeberg, Cact. Succ. J. (US) 23(3): 88, 1951. Basionym: *Chilenia occulta*. [The basionym given is already a combination.] [RPS 3]

Neochilenia odieri (Lemaire) Backeberg, Cact. Succ. J. (US) 23(3): 88, 1951. Basionym: *Echinocactus odieri*. [The name given as basionym is already a combination (here corrected).] [RPS 3]

Neochilenia odieri var. **magnifica** (Hildmann) Y. Ito, Explan. Diagr. Austroechinocactinae, 278, 1957. Basionym: *Echinocactus odieri* var. *magnificus*. [RPS 8]

Neochilenia odieri var. **mebbesii** (Hildmann) Y. Ito, Explan. Diagr. Austroechinocactinae, 278, 1957. Basionym: *Echinocactus odieri* var. *mebbesii*. [RPS 8]

Neochilenia odieri var. **spininigris** (Labouret) Y. Ito, Explan. Diagr. Austroechinocactinae, 278, 1957. Basionym: *Echinocactus odieri* var. *spininigris*. [RPS 8]

Neochilenia odoriflora (Ritter) Backeberg, Die Cact. 6: 3778, 1962. Basionym: *Pyrrhocactus odoriflorus*. [RPS 13]

Neochilenia paucicostata (Ritter) Backeberg, Die Cact. 6: 3780, 1962. Basionym: *Horridocactus paucicostatus*. [RPS 13]

Neochilenia paucicostata var. **viridis** (Ritter) Backeberg, Die Cact. 6: 3780, 1962. Basionym: *Horridocactus paucicostatus* var. *viridis*. [RPS 13]

Neochilenia pilispina (Ritter) Backeberg, Descr. Cact. Nov. 3: 9, 1963. Basionym: *Pyrrhocactus pilispinus*. [RPS 14]

Neochilenia pseudoreichei Lembcke & Backeberg, in Backeberg, Die Cact. 3: 1827, 1959. *Nom. inval.* (Art. 37.1). [Erroneously included as valid in RPS 10.] [RPS 10]

Neochilenia pulchella Ritter *ex* Backeberg, Die Cact. 6: 3781, 1962. *Nom. inval.*, based on *Horridocactus pulchellus*, *nom. inval.* (Art. 36.1, 37.1). [Based on a *nomen novum* by Ritter. Erroneously included as valid in RPS 13.] [RPS 13]

Neochilenia pygmaea (Ritter) Backeberg, Descr. Cact. Nov. 3: 9, 1963. Basionym: *Pyrrhocactus pygmaeus*. [RPS 14]

Neochilenia recondita (Ritter) Backeberg, Descr. Cact. Nov. 3: 9, 1963. Basionym: *Pyrrhocactus reconditus*. [RPS 14]

Neochilenia reichei (Schumann) Backeberg, Cact. Succ. J. (US) 23(4): 117, 1951. Basionym: *Echinocactus reichei*. [The name given as basionym was itself a combination, and is here corrected.] [RPS 3]

Neochilenia residua (Ritter & Buining) Backeberg, Descr. Cact. Nov. 3: 10, 1963. Basionym: *Pyrrhocactus residuus*. [RPS 14]

Neochilenia robusta (Ritter) Backeberg, Die Cact. 6: 3783, 1962. Basionym: *Pyrrhocactus robustus*. [RPS 13]

Neochilenia robusta var. **vegasana** (Ritter) Backeberg, Die Cact. 6: 3783, 1962. Basionym: *Pyrrhocactus robustus* var. *vegasanus*. [RPS 13]

Neochilenia rostrata (Jacobi) Backeberg, Cact. Succ. J. (US) 23(4): 117, 1951. Basionym: *Chilenia rostrata*. [The name given as basionym is already a combination.] [RPS 3]

Neochilenia rupicola (Ritter) Backeberg, Descr. Cact. Nov. 3: 10, 1963. Basionym: *Pyrrhocactus rupicolus*. [RPS 14]

Neochilenia saxifraga Backeberg, Die Cact. 6: 3777, 1962. *Nom. inval.* (Art. 36.1, 37.1). [Casually mentioned only.] [RPS -]

Neochilenia scoparia (Ritter) Backeberg, Descr. Cact. Nov. 3: 10, 1963. Basionym: *Pyrrhocactus scoparius*. [Repeated in Kakt.-Lex., 294, 1966.] [RPS 14]

Neochilenia setosiflora (Ritter) Backeberg, Descr. Cact. Nov. 3: 10, 1963. Basionym: *Pyrrhocactus setosiflorus*. [RPS 14]

Neochilenia setosiflora var. **intermedia** (Ritter) Backeberg, Descr. Cact. Nov. 3: 10, 1963. Basionym: *Pyrrhocactus setosiflorus* var. *intermedius*. [RPS 14]

Neochilenia simulans (Ritter) Backeberg, Descr. Cact. Nov. 3: 10, 1963. Basionym: *Pyrrhocactus simulans*. [RPS 14]

Neochilenia taltalensis (P. C. Hutchison) Backeberg, Die Cact. 3: 1812, 1959. Basionym: *Neoporteria taltalensis*. [RPS 10]

Neochilenia totoralensis (Ritter) Backeberg, Descr. Cact. Nov. 3: 10, 1963. Basionym: *Pyrrhocactus totoralensis*. [RPS 14]

Neochilenia transitensis (Ritter) Backeberg, Descr. Cact. Nov. 3: 10, 1963. Basionym: *Pyrrhocactus transitensis*. [RPS 14]

Neochilenia trapichensis Ritter *ex* Backeberg, Die Cact. 3: 1834, 1959. *Nom. inval.* (Art. 37.1). [RPS -]

Neochilenia wagenknechtii (Ritter) Backeberg, Die Cact. 6: 3783, 1962. Basionym: *Pyrrhocactus wagenknechtii*. [RPS 13]

Neodawsonia guiengolensis Bravo, Anales Inst. Biol. UNAM 27(1): 12, 1957. [RPS 8]

Neodawsonia nana Bravo, Anales Inst. Biol. UNAM 27(1): 15, 1957. [RPS 8]

Neodawsonia nizandensis Bravo & MacDougall, Anales Inst. Biol. UNAM 29: 82, 1959. [RPS 10]

Neodawsonia totolapensis Bravo & MacDougall, Anales Inst. Biol. UNAM 29: 74, 1959. [RPS 10]

Neodiscocactus Ito, The Cactaceae, 9, 531, 1981. *Nom. inval.* (Art. 32). [Based on *Discocactus* Pfeiffer 1837.] [RPS -]

Neodiscocactus boliviensis (Backeberg) Y. Ito, The Cactaceae, 531, 1981. *Nom. inval.* (Art. 33.2, 43.1), based on *Discocactus boliviensis*. [RPS -]

Neodiscocactus hartmannii (Schumann) Y. Ito, The Cactaceae, 532, 1981. *Nom. inval.* (Art. 33.2, 43.1), based on *Echinocactus hartmannii*. [RPS -]

Neodiscocactus heptacanthus (Rodriguez) Y. Ito, The Cactaceae, 532, 1981. *Nom. inval.* (Art. 33.2, 43.1), based on *Malacocarpus heptacanthus*. [RPS -]

Neoevansia lazaro-cardenasii J. Contreras et al., Cact. Suc. Mex. 25(3): 51-54, 1980. [RPS 31]

Neoevansia striata (Brandegee) Sanchez-Mejorada, Cact. Suc. Mex. 18: 22, 1973. Basionym: *Cereus striatus*. [RPS 24]

Neoevansia zopilotensis (Meyran) Sanchez-Mejorada, Cact. Suc. Mex. 18: 24, 1973. Basionym: *Wilcoxia zopilotensis*. [RPS 24]

Neolloydia sect. *Conoideae* Kladiwa & Fittkau, in Krainz, Die Kakt. C VIIIb (June 1971), 1971. *Nom. illeg.* (Art. 22). [Typus: *N. conoidea* Britton & Rose; = sect. *Neolloydia*.] [RPS 22]

Neolloydia sect. **Smithianae** Kladiwa & Fittkau, in Krainz, Die Kakt., C VIIIb (June 1971), 1971. Typus: *N. smithii*. [RPS 22]

Neolloydia conoidea var. **ceratites** (Quehl) Kladiwa & Fittkau, in Krainz, Die Kakt. Lief. 48-49: C VIIIb, 1971. Basionym: *Mammillaria ceratites*. [Combination repeated in l.c. Jan. 1972.] [RPS 22]

Neolloydia conoidea var. **grandiflora** (Otto) Kladiwa & Fittkau, in Krainz, Die Kakt. Lief. 48-49: C VIIIb, 1971. Basionym: *Mammillaria grandiflora*. [Combination repeated in l.c. Jan. 1972.] [RPS 22]

Neolloydia conoidea var. **matehualensis** (Backeberg) Kladiwa & Fittkau, in Krainz, Die Kakt. Lief. 48-49: C VIIIb, 1971. Basionym: *Neolloydia matehualensis*. [Combination repeated in l.c. Jan. 1972.] [RPS 22]

Neolloydia conoidea var. **texensis** (Britton & Rose) Kladiwa & Fittkau, in Krainz, Die Kakt. Lief. 48-49: C VIIIb, 1971. Basionym: *Neolloydia texensis*. [Combination repeated in l.c. Jan. 1972.] [RPS 22]

Neolloydia durangensis (Runge) L. Benson, Cact. Succ. J. (US) 46: 81, 1974. Basionym: *Echinocactus durangensis*. [RPS 25]

Neolloydia erectocentra (Coulter) L. Benson, The Cacti of Arizona ed. 3, 24, 191, 1969. Basionym: *Echinocactus erectocentrus*. [RPS 20]

Neolloydia erectocentra var. **acunensis** (W. T. Marshall) L. Benson, Cacti of Arizona, ed. 3, 25, 192, 1969. Basionym: *Echinocactus acunensis*. [RPS 20]

Neolloydia gautii L. Benson, Cact. Succ. J. (US) 46: 80, 1974. [RPS 25]

Neolloydia grandiflora var. **robusta** Riha, Kakt. and. Sukk. 32(2): 126-128, 1981. [RPS 32]

Neolloydia intertexta (Engelmann) L. Benson, Cact. Succ. J. (US) 41: 233, 1969. Basionym: *Echinocactus intertextus*. [RPS 20]

Neolloydia intertexta var. **dasyacantha** (Engelmann) L. Benson, Cact. Succ. J. (US) 41: 233, 1969. Basionym: *Echinocactus intertextus* var. *dasyacanthus*. [RPS 20]

Neolloydia johnsonii (Parry) L. Benson, The Cacti of Arizona ed. 3, 25, 192, 1969. Basionym: *Echinocactus johnsonii*. [RPS 20]

Neolloydia laui (Glass & Foster) E. F. Anderson, Bradleya 4: 22, 1986. Basionym: *Turbinicarpus laui*. [RPS 37]

Neolloydia lophophoroides (Werdermann) E. F. Anderson, Bradleya 4: 22, 1986. Basionym: *Thelocactus lophophoroides*. [RPS 37]

Neolloydia macdowellii (Rebut *ex* Quehl) Moore, Baileya 19: 166, 1975. Basionym: *Echinocactus macdowellii*. [Sphalm. 'mcdowellii'.] [RPS 26]

Neolloydia mandragora (Fric *ex* A. Berger) E. F. Anderson, Bradleya 4: 14, 1986. Basionym: *Echinocactus mandragora*. [RPS 37]

Neolloydia mariposensis (Hester) L. Benson, Cact. Succ. J. (US) 41: 188, 1969. Basionym: *Echinomastus mariposensis*. [RPS 20]

Neolloydia pseudomacrochele (Backeberg) E. F. Anderson, Bradleya 4: 22, 1986. Basionym: *Strombocactus pseudomacrochele*. [RPS 37]

Neolloydia pseudopectinata (Backeberg) E. F. Anderson, Bradleya 4: 15, 1986. Basionym: *Pelecyphora pseudopectinata*. [RPS 37]

Neolloydia schmiedickeana (Bödeker) E. F. Anderson, Bradleya 4: 19, 1986. Basionym: *Echinocactus schmiedickeanus*. [RPS 37]

Neolloydia schmiedickeana var. **dickisoniae** (Glass & Foster) E. F. Anderson, Bradleya 4:

20, 1986. Basionym: *Turbinicarpus schmiedickeanus* var. *dickisoniae*. [RPS 37]

Neolloydia schmiedickeana var. **flaviflora** (G. Frank & Lau) E. F. Anderson, Bradleya 4: 20, 1986. Basionym: *Turbinicarpus flaviflorus*. [RPS 37]

Neolloydia schmiedickeana var. **gracilis** (Glass & Foster) E. F. Anderson, Bradleya 4: 19, 1986. Basionym: *Turbinicarpus gracilis*. [RPS 37]

Neolloydia schmiedickeana var. **klinkeriana** (Backeberg & Jacobsen) E. F. Anderson, Bradleya 4: 20, 1986. Basionym: *Turbinicarpus klinkerianus*. [RPS 37]

Neolloydia schmiedickeana var. **macrochele** (Werdermann) E. F. Anderson, Bradleya 4: 20, 1986. Basionym: *Echinocactus macrochele*. [RPS 37]

Neolloydia schmiedickeana var. **schwarzii** (Shurly) E. F. Anderson, Bradleya 4: 20, 1986. Basionym: *Strombocactus schwarzii*. [RPS 37]

Neolloydia smithii (Mühlenpfordt) Kladiwa & Fittkau, in H. Krainz, Die Kakt. Lief. 46/47: C VIIIb (June 1971), 1971. Basionym: *Echinocactus smithii*. [RPS 22]

Neolloydia smithii cv. **Senilis** Rowley, Rep. Pl. Succ. 23: 8, 1974. [Based on *Neolloydia beguinii* var. *senilis* Borg 1937 (*nom. inval.* Art. 36.1).] [RPS 23]

Neolloydia smithii var. *beguinii* (F. A. C. Weber) Kladiwa & Fittkau, in Krainz, Die Kakt. Lief. 46/47: C VIIIb (June 1971), 1971. *Nom. illeg.* (Art. 63.1), based on *Echinocactus beguinii, nom. inval.* (Art. 63.1). [RPS 22]

Neolloydia subterranea (Backeberg) H. Moore, Baileya 19(4): 165, 1975. Basionym: *Echinocactus subterraneus*. [RPS 26]

Neolloydia subterranea var. **zaragosae** (Glass & Foster) E. F. Anderson, Bradleya 4: 14, 1986. Basionym: *Gymnocactus subterraneus* var. *zaragosae*. [RPS 37]

Neolloydia unguispina (Engelmann) L. Benson, Cact. Succ. J. (US) 46: 80, 1974. Basionym: *Echinocactus unguispinus*. [RPS 25]

Neolloydia valdeziana (H. Möller) E. F. Anderson, Bradleya 4: 14, 1986. Basionym: *Pelecyphora valdeziana*. [RPS 37]

Neolloydia viereckii var. **major** (Glass & Foster) E. F. Anderson, Bradleya 4: 15, 1986. Basionym: *Gymnocactus viereckii* var. *major*. [RPS 37]

Neolloydia warnockii L. Benson, Cact. Succ. J. (US) 41: 186-188, 1969. [RPS 20]

Neolobivia Y. Ito, Explan. Diagr. Austroechinocactinae, 54, 284, 1957. Typus: *Lobivia wrightiana*. [First invalidly published in Bull. Takarazuka Insectarium 71: 13-20, 1950 (Art. 36.1, cf. RPS 3).] [RPS 8]

Neolobivia divaricata Rittter, Kakt. Südamer. 4: 1337-1338, 1981. Typus: *Ritter* 696 (U). [Peru: Cuzco] [RPS 32]

Neolobivia echinata (Rausch) Ritter, Kakt. Südamer. 4: 1338, 1981. Basionym: *Lobivia echinata*. [RPS 32]

Neolobivia hertrichiana (Backeberg) Ritter, Kakt. Südamer. 4: 1334-1335, 1981. Basionym: *Lobivia hertrichiana*. [RPS 32]

Neolobivia incaica (Backeberg) Ritter, Kakt. Südamer. 4: 1335, 1981. Basionym: *Lobivia incaica*. [RPS 32]

Neolobivia kratochviliana (Backeberg) Y. Ito, Explan. Diagr. Austroechinocactinae, 57, 1957. Basionym: *Echinopsis kratochviliana*. [RPS 8]

Neolobivia minuta (Ritter) Ritter, Kakt. Südamer. 4: 1336, 1981. Basionym: *Lobivia minuta*. [RPS 32]

Neolobivia nakaii Y. Ito, Explan. Diagr. Austroechinocactinae, 57, 284, 1957. [RPS 8]

Neolobivia nakaii var. **albiflora** Y. Ito, Explan. Diagr. Austroechinocactinae, 58, 285, 1957. [RPS 8]

Neolobivia nakaii var. **albocinnabarina** Y. Ito, Explan. Diagr. Austroechinocactinae, 58, 285, 1957. [RPS 8]

Neolobivia nakaii var. **atrorosiflora** Y. Ito, Explan. Diagr. Austroechinocactinae, 58, 285, 1957. [RPS 8]

Neolobivia nakaii var. **atrorubriflora** Y. Ito, Explan. Diagr. Austroechinocactinae, 58, 285, 1957. [RPS 8]

Neolobivia nakaii var. **rosiflora** Y. Ito, Explan. Diagr. Austroechinocactinae, 58, 285, 1957. [RPS 8]

Neolobivia prolifera Ritter, Kakt. Südamer. 4: 1335-1356, 1981. Typus: *Ritter* 1459 (U). [Peru: Cuzco] [RPS 32]

Neolobivia ritteri (Wessner) Y. Ito, Explan. Diagr. Austroechinocactinae, 55, 1957. Basionym: *Lobivia ritteri*. [RPS 8]

Neolobivia segawae Y. Ito, Explan. Diagr. Austroechinocactinae, 56, 284, 1957. [RPS 8]

Neolobivia segawae var. **albiroseiflora** Y. Ito, Explan. Diagr. Austroechinocactinae, 56, 284, 1957. [Sphalm. 'albirosiflora'.] [RPS 8]

Neolobivia segawae var. **roseiflora** Y. Ito, Explan. Diagr. Austroechinocactinae, 56, 284, 1957. [Sphalm. 'rosiflora'.] [RPS 8]

Neolobivia segawae var. **rubriviolaciflora** Y. Ito, Explan. Diagr. Austroechinocactinae, 56, 284, 1957. [RPS 8]

Neolobivia vilcabambae (Ritter) Ritter, Kakt. Südamer. 4: 1337, 1981. Basionym: *Lobivia vilcabambae*. [RPS 32]

Neolobivia winteriana (Ritter) Ritter, Kakt. Südamer. 4: 1333-1334, 1981. Basionym: *Lobivia winteriana*. [RPS 32]

Neolobivia wrightiana (Backeberg) Y. Ito, Explan. Diagr. Austroechinocactinae, 55, 1957. Basionym: *Lobivia wrightiana*. [RPS 8]

Neolobivia xiphacantha Y. Ito, Explan. Diagr. Austroechinocactinae, 58, 285, 1957. [This name should probably have been published as *Echinopsis xiphacantha*, as various varieties are published (as new combinations) under *Echinopsis* (ue.).] [RPS 8]

Neoporteria subgen. **Pyrrhocactus** (A. Berger *ex* Backeberg) Donald & Rowley, Baileya 19: 164, 1975. Basionym: *Pyrrhocactus*. [RPS 26]

Neoporteria subgen. *Theloneoporteria* Y. Ito, Bull. Takarazuka Insectarium 71: 13-20,

1950. *Nom. inval.* (Art. 36.1). [RPS 3]
Neoporteria subgen. *Tropidoneoporteria* Y. Ito, Bull. Takarazuka Insectarium 71: 13-20, 1950. *Nom. inval.* (Art. 36.1). [RPS 3]
Neoporteria andreaeana (Backeberg) Donald & Rowley, Cact. Succ. J. Gr. Brit. 28: 55, 1966. *Nom. inval.*, based on *Neochilenia andreaeana, nom. inval.* (Art. 9.5, 37.1). [Erroneously included as valid in RPS 17.] [RPS 17]
Neoporteria aricensis (Ritter) Donald & Rowley, Cact. Succ. J. Gr. Brit. 28: 55, 1966. Basionym: *Pyrrhocactus aricensis.* [RPS 17]
Neoporteria aricensis var. *floribunda* (Backeberg) A. E. Hoffmann, Cact. Fl. Chil., 176, ill. (p. 177), 1989. *Nom. inval.*, based on *Reicheocactus floribundus, nom. inval.* (Art. 37.1). [RPS 40]
Neoporteria aricensis var. **saxifraga** (Ritter) A. E. Hoffmann, Cact. Fl. Chil., 176, ill. (p. 177), 1989. Basionym: *Pyrrhocactus saxifragus.* [RPS 40]
Neoporteria armata (Ritter) Krainz, Kat. ZSS ed. 2, 86, 1967. Basionym: *Pyrrhocactus armatus.* [RPS 18]
Neoporteria backebergii Donald & Rowley, Cact. Succ. J. Gr. Brit. 28: 55, 1966. *Nom. inval.* (Art. 9.5, 37.1). [*Nom. nov. pro Pyrrhocactus atrispinosus* Backeberg 1963 (*nom. inval.* Art. 9.5); *non Neoporteria atrispinosa* Backeberg. Erroneously included as valid in RPS 17.] [RPS 17]
Neoporteria bicolor (Akers & Buining) Donald & Rowley, Cact. Succ. J. Gr. Brit. 28: 55, 1966. Basionym: *Islaya bicolor.* [RPS 17]
Neoporteria bulbocalyx (Werdermann) Donald & Rowley, Cact. Succ. J. Gr. Brit. 28: 55, 1966. Basionym: *Echinocactus bulbocalyx.* [RPS 17]
Neoporteria calderana (Ritter) Donald & Rowley, Cact. Succ. J. Gr. Brit. 28: 55, 1966. Basionym: *Pyrrhocactus calderanus.* [RPS 17]
Neoporteria calderana fa. **gracilis** (Ritter) Donald & Rowley, Cact. Succ. J. Gr. Brit. 28: 55, 1966. Basionym: *Pyrrhocactus gracilis.* [RPS 17]
Neoporteria carrizalensis (Ritter) A. E. Hoffmann, Cact. Fl. Chil., 178, ill. (p. 179), 1989. Basionym: *Pyrrhocactus carrizalensis.* [Incorrectly used (Art. 57), see below.] [RPS 40]
Neoporteria carrizalensis var. *totoralensis* (Ritter) A. E. Hoffmann, Cact. Fl. Chil., 178, ill. (p. 179), 1989. Incorrect name (Art. 57), based on *Pyrrhocactus totoralensis.* [Incorrectly used name (Art. 57), as *P. totoralensis* has priority over *N. carrizalensis.*] [RPS 40]
Neoporteria castanea Ritter, Taxon 12(1): 34, 1963. Typus: *Ritter* 236 (U, ZSS). [Chile: Talca] [RPS 14]
Neoporteria castanea var. **tunensis** Ritter, Taxon 12: 34, 1963. [RPS 14]
Neoporteria catamarcensis (F. A. C. Weber) Donald & Rowley, Cact. Succ. J. Gr. Brit. 28: 55, 1966. Basionym: *Echinopsis catamarcensis.* [RPS 17]
Neoporteria centeteria (Lehmann) Rowley, Excelsa 12: 39, 1986. Basionym: *Echinocactus centeterius.* [RPS 37]
Neoporteria chilensis fa. **confinis** (Hildmann) Donald & Rowley, Cact. Succ. J. Gr. Brit. 28: 55, 1966. Basionym: *Echinocactus chilensis* var. *confinis.* [RPS 17]
Neoporteria chilensis var. **confinis** (Hildmann) Y. Ito, Explan. Diagr. Austroechinocactinae, 214, 1957. Basionym: *Echinocactus chilensis* var. *confinis.* [RPS 8]
Neoporteria chilensis var. *cylindrica* (Backeberg) Y. Ito, Explan. Diagr. Austroechinocactinae, 214, 1957. *Nom. inval.*, based on *Chilenia chilensis* var. *cylindrica, nom. inval.* (Art. 43.1). [RPS 8]
Neoporteria choapensis (Ritter) Donald & Rowley, Cact. Succ. J. Gr. Brit. 28: 55, 1966. Basionym: *Pyrrhocactus choapensis.* [RPS 17]
Neoporteria chorosensis (Ritter) Donald & Rowley, Cact. Succ. J. Gr. Brit. 28: 55, 1966. Basionym: *Pyrrhocactus chorosensis.* [RPS 17]
Neoporteria clavata var. *nigrihorrida* A. E. Hoffmann, Cact. Fl. Chil., 160, ill. (p. 162), 1989. *Nom. inval.* (Art. 33.2). [The basionym reference (*Chilenia nigrihorrida* Backeberg 1935) is incorrect as it refers to an invalid publication of the basionym which was only validated in 1938 as *C. nigrihorrida* Backeberg *ex* A. W. Hill.] [RPS 40]
Neoporteria clavata var. **parviflora** Ritter, Suculenta 1966: 131, 1966. [RPS 17]
Neoporteria clavata var. **procera** Ritter, Taxon 12: 34, 1963. [RPS 14]
Neoporteria coimasensis var. **robusta** (Ritter) Ritter, Kakt. Südamer. 3: 1041-1042, 1980. Basionym: *Neoporteria robusta.* [RPS 31]
Neoporteria confinis (Rittter) Donald & Rowley, Cact. Succ. J. Gt. Brit. 28: 55, 1966. Basionym: *Pyrrhocactus confinis.* [RPS 17]
Neoporteria crispa (Ritter) Donald & Rowley, Cact. Succ. J. Gr. Brit. 28: 55, 1966. Basionym: *Pyrrhocactus crispus.* [RPS 17]
Neoporteria curvispina (Bertero) Donald & Rowley, Cact. Succ. J. Gr. Brit. 28(4): 55, 1966. Basionym: *Cactus curvispinus.* [RPS 17]
Neoporteria curvispina fa. **albicans** (Hildmann) Donald & Rowley, Cact. Succ. J. Gr. Brit. 28: 56, 1966. Basionym: *Echinocactus geissei* var. *albicans.* [RPS 17]
Neoporteria curvispina fa. **descendens** (Ritter) Donald & Rowley, Cact. Succ. J. Gr. Brit. 28: 55, 1966. Basionym: *Horridocactus andicola* var. *descendens.* [RPS 17]
Neoporteria curvispina fa. **gracilis** (Ritter) Donald & Rowley, Cact. Succ. J. Gr. Brit. 28: 56, 1966. Basionym: *Pyrrhocactus lissocarpus* var. *gracilis.* [RPS 17]
Neoporteria curvispina fa. **minor** (Ritter) Donald & Rowley, Cact. Succ. J. Gr. Brit. 28: 55, 1966. Basionym: *Pyrrhocactus echinus* var. *minor.* [RPS 17]
Neoporteria curvispina fa. **mollensis** (Ritter) Donald & Rowley, Cact. Succ. J. Gr. Brit. 28: 55, 1966. Basionym: *Horridocactus andicola* var. *mollensis.* [RPS 17]

Neoporteria curvispina fa. **orientalis** (Ritter) Donald & Rowley, Cact. Succ. J. Gr. Brit. 28: 55, 1966. Basionym: *Pyrrhocactus aconcaguensis* var. *orientalis*. [RPS 17]

Neoporteria curvispina fa. **robusta** (Ritter) Donald & Rowley, Cact. Succ. J. Gr. Brit. 28: 55, 1966. Basionym: *Horridocactus andicola* var. *robustus*. [RPS 17]

Neoporteria curvispina fa. *subaequalis* (Backeberg) Donald & Rowley, Cact. Succ. J. Gr. Brit. 28: 56, 1966. *Nom. inval*., based on *Horridocactus kesselringianus* var. *subaequalis, nom. inval.* (Art. 37.1). [Erroneously included as valid in RPS 17.] [RPS 17]

Neoporteria curvispina var. **aconcaguensis** (Ritter) Donald & Rowley, Cact. Succ. J. Gr. Brit. 28: 55, 1966. Basionym: *Pyrrhocactus aconcaguensis*. [RPS 17]

Neoporteria curvispina var. **andicola** (Ritter) Donald & Rowley, Cact. Succ. J. Gr. Brit. 28: 55, 1966. Basionym: *Horridocactus andicola*. [RPS 17]

Neoporteria curvispina var. **carrizalensis** (Ritter) Donald & Rowley, Cact. Succ. J. Gr. Brit. 28: 55, 1966. Basionym: *Pyrrhocactus carrizalensis*. [RPS 17]

Neoporteria curvispina var. **echinus** (Ritter) Donald & Rowley, Cact. Succ. J. Gr. Brit. 28: 55, 1966. Basionym: *Pyrrhocactus echinus*. [RPS 17]

Neoporteria curvispina var. **engleri** (Ritter) A. E. Hoffmann, Cact. Fl. Chil., 182, ill. (p. 185), 1989. Basionym: *Horridocactus engleri*. [RPS 40]

Neoporteria curvispina var. **garaventae** (Ritter) Donald & Rowley, Cact. Succ. J. Gr. Brit. 28: 56, 1966. Basionym: *Horridocactus garaventae*. [Sphalm. 'garaventai'.] [RPS 17]

Neoporteria curvispina var. **geissei** (Poselger) Donald & Rowley, Cact. Succ. J. Gr. Brit. 28: 56, 1966. Basionym: *Echinocactus geissei*. [RPS 17]

Neoporteria curvispina var. **grandiflora** (Ritter) Donald & Rowley, Cact. Succ. J. Gr. Brit. 28: 56, 1966. Basionym: *Pyrrhocactus grandiflorus*. [RPS 17]

Neoporteria curvispina var. **heinrichiana** (Backeberg) Donald & Rowley, Cact. Succ. J. Gr. Brit. 28: 56, 1966. Basionym: *Horridocactus heinrichianus*. [RPS 17]

Neoporteria curvispina var. **kesselringiana** (Dölz) Donald & Rowley, Cact. Succ. J. Gr. Brit. 28: 56, 1966. Basionym: *Horridocactus kesselringianus*. [RPS 17]

Neoporteria curvispina var. **lissocarpa** (Ritter) Donald & Rowley, Cact. Succ. J. Gr. Brit. 28: 56, 1966. Basionym: *Pyrrhocactus lissocarpus*. [RPS 17]

Neoporteria curvispina var. **marksiana** (Ritter) A. E. Hoffmann, Cact. Fl. Chil., 182, 1989. Basionym: *Pyrrhocactus marksianus*. [RPS 40]

Neoporteria curvispina var. **transitensis** (Ritter) Donald & Rowley, Cact. Succ. J. Gr. Brit. 28: 56, 1966. Basionym: *Pyrrhocactus transitensis*. [RPS 17]

Neoporteria curvispina var. **vallenarensis** (Ritter) Donald & Rowley, Cact. Succ. J. Gr. Brit. 28: 56, 1966. Basionym: *Pyrrhocactus vallenarensis*. [RPS 17]

Neoporteria deherdtiana (Backeberg) Donald & Rowley, Cact. Succ. J. Gr. Brit. 28: 56, 1966. *Nom. inval*., based on *Neochilenia deherdtiana, nom. inval.* (Art. 9.5, 37.1). [Erroneously included as valid in RPS 17.] [RPS 17]

Neoporteria densispina Y. Ito, Explan. Diagr. Austroechinocactinae, 213, 293, 1957. [= *Chilenia densispina* Backeberg 1937 (*nom. nud.*).] [RPS 8]

Neoporteria dimorpha (Ritter) Donald & Rowley, Cact. Succ. J. Gr. Brit. 28: 56, 1966. Basionym: *Pyrrhocactus dimorphus*. [RPS 17]

Neoporteria dubia (Backeberg) Donald & Rowley, Cact. Succ. J. Gr. Brit. 28: 56, 1966. Basionym: *Pyrrhocactus dubius*. [RPS 17]

Neoporteria ebenacantha (Monville) Y. Ito, Cacti, 77, 1952. *Nom. inval.* (Art. 33.2). [See also Backeberg, Die Cact. 3: 1808-1809, 1959.] [RPS 4]

Neoporteria engleri (Ritter) Donald & Rowley, Cact. Succ. J. Gr. Brit. 28: 56, 1966. Basionym: *Horridocactus engleri*. [RPS 17]

Neoporteria engleri var. *krausii* (Backeberg) Donald & Rowley, Cact. Succ. J. Gr. Brit. 28: 56, 1966. *Nom. inval*., based on *Horridocactus engleri* var. *krausii, nom. inval.* (Art. 37.1). [Erroneously included as valid in RPS 17.] [RPS 17]

Neoporteria eriocephala (Backeberg) Donald & Rowley, Cact. Succ. J. Gr. Brit. 28: 56, 1966. *Nom. inval*., based on *Neochilenia eriocephala, nom. inval.* (Art. 37.1). [Erroneously included as valid in RPS 17.] [RPS 17]

Neoporteria eriocephala var. *glaucescens* (Ritter) Donald & Rowley, Cact. Succ. J. Gr. Brit. 28: 56, 1966. *Nom. inval.* (Art. 43.1), based on *Pyrrhocactus glaucescens*. [Erroneously included as valid in RPS 17.] [RPS 17]

Neoporteria eriosyzoides (Ritter) Donald & Rowley, Cact. Succ. J. Gr. Brit. 28: 56, 1966. Basionym: *Horridocactus eriosyzoides*. [RPS 17]

Neoporteria esmeraldana (Ritter) Donald & Rowley, Cact. Succ. J. Gr. Brit. 28: 56, 1966. Basionym: *Chileorebutia esmeraldana*. [RPS 17]

Neoporteria esmeraldana var. **malleolata** (Ritter) A. E. Hoffmann, Cact. Fl. Chil., 222, ill. (p. 223), 1989. Basionym: *Chileorebutia malleolata*. [RPS 40]

Neoporteria exsculpta (Otto) Borg, Cacti, ed. 2, 270, 1951. Basionym: *Echinocactus exsculptus*. [RPS 9]

Neoporteria gerocephala Y. Ito, Explan. Diagr. Austroechinocactinae, 211, 1957. [*Nom. nov. pro Echinocactus senilis* Philippi, etc. (cf. RPS 8).] [RPS 8]

Neoporteria hankeana (Förster) Donald & Rowley, Cact. Succ. J. Gr. Brit. 28: 56, 1966. Basionym: *Echinocactus hankeanus*. [RPS 17]

Neoporteria hankeana var. **minor** (Labouret)

Donald & Rowley, Cact. Succ. J. Gr. Brit. 28: 56, 1966. Basionym: *Echinocactus ebenacanthus* var. *minor*. [RPS 17]

Neoporteria hankeana var. *taltalensis* (Backeberg) Donald & Rowley, Cact. Succ. J. Gr. Brit. 28: 56, 1966. *Nom. inval.*, based on *Neochilenia hankeana* var. *taltalensis, nom. inval.* (Art. 37.1). [Erroneously included as valid in RPS 17.] [RPS 17]

Neoporteria heteracantha (Backeberg) Backeberg, Cact. Succ. J. (US) 23(4): 118, 1951. *Nom. inval.*, based on *Chilenia heteracantha, nom. inval.* (Art. 43.1). [RPS 3]

Neoporteria horrida (Remy *ex* Gay) D. Hunt, Bradleya 5: 93, 1987. Basionym: *Echinocactus horridus*. [RPS 38]

Neoporteria horrida var. **armata** (Ritter) A. E. Hoffmann, Cact. Fl. Chil., 190, ill. (p. 191), 1989. Basionym: *Pyrrhocactus armatus*. [RPS 40]

Neoporteria horrida var. **aspillagae** (Söhrens) A. E. Hoffmann, Cact. Fl. Chil., 190, ill. (p. 193), 1989. Basionym: *Echinocactus aspillagae*. [Sphalm. 'aspillagai'.] [RPS 40]

Neoporteria horrida var. **coliguayensis** (Ritter) A. E. Hoffmann, Cact. Fl. Chil., 190, 1989. Basionym: *Pyrrhocactus coliguayensis*. [Inadvertently published as 'Pyrrhocactus horridus var. coliguayensis' by a slip of the pen; the intention of the author is, however, clear, and the combination is therefore to be accepted as above.] [RPS 40]

Neoporteria horrida var. **limariensis** (Ritter) A. E. Hoffmann, Cact. Fl. Chil., 190, ill. (p. 191), 1989. Basionym: *Pyrrhocactus limariensis*. [RPS 40]

Neoporteria horrida var. **odoriflora** (Ritter) A. E. Hoffmann, Cact. Fl. Chil., 190, ill. (p. 193), 1989. Basionym: *Pyrrhocactus odoriflorus*. [RPS 40]

Neoporteria huascensis (Ritter) Donald & Rowley, Cact. Succ. J. Gr. Brit. 28: 56, 1966. Basionym: *Pyrrhocactus huascensis*. [RPS 17]

Neoporteria intermedia (Ritter) Donald & Rowley, Cact. Succ. J. Gr. Brit. 28: 56, 1966. Basionym: *Pyrrhocactus intermedius*. [RPS 17]

Neoporteria intermedia fa. *gracilis* (Ritter) A. E. Hoffmann, Cact. Fl. Chil., 194, ill. (p. 197), 1989. Incorrect name (Art. 57.1), based on *Pyrrhocactus gracilis*. [RPS 40]

Neoporteria intermedia fa. *pygmaea* (Ritter) A. E. Hoffmann, Cact. Fl. Chil., 194, ill. (p. 195), 1989. Incorrect name (Art. 57.1), based on *Pyrrhocactus pygmaeus*. [RPS 40]

Neoporteria intermedia var. *calderana* (Ritter) A. E. Hoffmann, Cact. Fl. Chil., 194, 1989. Incorrect name (Art. 57.1), based on *Pyrrhocactus calderanus*. [RPS 40]

Neoporteria intermedia var. *pilispina* (Ritter) A. E. Hoffmann, Cact. Fl. Chil., 194, ill. (p. 195), 1989. Incorrect name (Art. 57.1), based on *Pyrrhocactus pilispinus*. [RPS 40]

Neoporteria intermedia var. *pulchella* (Ritter) A. E. Hoffmann, Cact. Fl. Chil., 194, ill. (p. 197), 1989. Incorrect name (Art. 57.1), based on *Pyrrhocactus pulchellus*. [RPS 40]

Neoporteria intermedia var. *scoparia* (Ritter) A. E. Hoffmann, Cact. Fl. Chil., 194, 1989. Incorrect name (Art. 57.1), based on *Pyrrhocactus scoparius*. [RPS 40]

Neoporteria iquiquensis (Ritter) Donald & Rowley, Cact. Succ. J. Gr. Brit. 28: 56, 1966. Basionym: *Pyrrhocactus iquiquensis*. [RPS 17]

Neoporteria islayensis (Förster) Donald & Rowley, Cact. Succ. J. Gr. Brit. 28: 56, 1966. Basionym: *Echinocactus islayensis*. [RPS 17]

Neoporteria islayensis fa. **brevicylindrica** (Rauh & Backeberg) Donald & Rowley, Cact. Succ. J. Gr. Brit. 28: 56, 1966. Basionym: *Islaya brevicylindrica*. [RPS 17]

Neoporteria islayensis fa. **brevispina** (Rauh & Backeberg) Donald & Rowley, Cact. Succ. J. Gr. Brit. 28: 56, 1966. Basionym: *Islaya grandis* var. *brevispina*. [RPS 17]

Neoporteria islayensis fa. **grandiflorens** (Rauh & Backeberg) Donald & Rowley, Cact. Succ. J. Gr. Brit. 28: 56, 1966. Basionym: *Islaya grandiflorens*. [RPS 17]

Neoporteria islayensis fa. **minor** (Backeberg) Donald & Rowley, Cact. Succ. J. Gr. Brit. 28: 56, 1966. Basionym: *Islaya minor*. [RPS 17]

Neoporteria islayensis fa. **molendensis** (Vaupel) Donald & Rowley, Cact. Succ. J. Gr. Brit. 28: 56, 1966. Basionym: *Echinocactus molendensis*. [RPS 17]

Neoporteria islayensis fa. **spinosior** (Rauh & Backeberg) Donald & Rowley, Cact. Succ. J. Gr. Brit. 28: 56, 1966. Basionym: *Islaya grandiflorens* var. *spinosior*. [Combination probably invalid, as the citation given was not for the basionym, but for another heterotypic synonym.] [RPS 17]

Neoporteria islayensis var. **copiapoides** (Rauh & Backeberg) Donald & Rowley, Cact. Succ. J. Gr. Brit. 28: 56, 1966. Basionym: *Islaya copiapoides*. [RPS 17]

Neoporteria islayensis var. **divaricatiflora** (Ritter) Donald & Rowley, Cact. Succ. J. Gr. Brit. 28: 56, 1966. Basionym: *Islaya divaricatiflora*. [RPS 17]

Neoporteria islayensis var. **grandis** (Rauh & Backeberg) Donald & Rowley, Cact. Succ. J. Gr. Brit. 28: 56, 1966. Basionym: *Islaya grandis*. [RPS 17]

Neoporteria jussieui var. **chaniarensis** (Ritter) A. E. Hoffmann, Cact. Fl. Chil., 198, 1989. Basionym: *Pyrrhocactus chaniarensis*. [RPS 40]

Neoporteria jussieui var. **chorosensis** (Ritter) A. E. Hoffmann, Cact. Fl. Chil., 202, ill. (p. 203), 1989. Basionym: *Pyrrhocactus chorosensis*. [RPS 40]

Neoporteria jussieui var. **dimorpha** (Ritter) A. E. Hoffmann, Cact. Fl. Chil., 200, ill. (p. 201), 1989. Basionym: *Pyrrhocactus dimorphus*. [RPS 40]

Neoporteria jussieui var. **fobeana** (Mieckley) Donald & Rowley, Cact. Succ. J. Gr. Brit. 28: 57, 1966. Basionym: *Echinocactus fobeanus*. [RPS 17]

Neoporteria jussieui var. **huascensis** (Ritter) A. E. Hoffmann, Cact. Fl. Chil., 198, ill. (p. 199), 1989. Basionym: *Pyrrhocactus* *huascensis*. [RPS 40]

Neoporteria jussieui var. **setosiflora** (Ritter) A. E. Hoffmann, Cact. Fl. Chil., 200, ill. (p. 201), 1989. Basionym: *Pyrrhocactus setosiflorus*. [RPS 40]

Neoporteria jussieui var. **trapichensis** (Ritter) A. E. Hoffmann, Cact. Fl. Chil., 202, ill. (p. 203), 1989. Basionym: *Pyrrhocactus trapichensis*. [RPS 40]

Neoporteria jussieui var. **wagenknechtii** (Ritter) A. E. Hoffmann, Cact. Fl. Chil., 198, ill. (p. 199), 1989. Basionym: *Pyrrhocactus wagenknechtii*. [RPS 40]

Neoporteria krainziana (Ritter) Donald & Rowley, Cact. Succ. J. Gr. Brit. 28: 57, 1966. Basionym: *Islaya krainziana*. [RPS 17]

Neoporteria kunzei var. **confinis** (Ritter) A. E. Hoffmann, Cact. Fl. Chil., 204, ill. (p. 205), 1989. Basionym: *Pyrrhocactus confinis*. [Basionym reference given is slightly unclear but interpreted as typografical error.] [RPS 40]

Neoporteria laniceps Ritter, Taxon 12: 34, 1963. [Sphalm. 'planiceps', cf. Succulenta 1966(3): 36.] [RPS 14]

Neoporteria lindleyi (Förster) Donald & Rowley, Cact. Succ. J. Gr. Brit. 28: 57, 1966. Basionym: *Echinocactus lindleyi*. [RPS 17]

Neoporteria lindleyi fa. **curvispina** (Rauh & Backeberg) Donald & Rowley, Cact. Succ. J. Gr. Brit. 28: 57, 1966. Basionym: *Islaya paucispina* var. *curvispina*. [RPS 17]

Neoporteria litoralis Ritter, Succulenta 38(3): 28-29, 43, 1959. Typus: *Ritter* 219 (ZSS). [Chile] [RPS 10]

Neoporteria litoralis var. **intermedia** Ritter, Succulenta 45(9): 130-131, 1966. Typus: *Ritter* 224a (U, ZSS). [Chile] [RPS 17]

Neoporteria litoralis var. *robustispina* Y. Ito, The Cactaceae, 422, 1981. *Nom. inval.* (Art. 37.1). [Sphalm. 'littoralis'.] [RPS -]

Neoporteria marksiana (Ritter) Donald & Rowley, Cact. Succ. J. Gr. Brit. 28: 57, 1966. Basionym: *Pyrrhocactus marksianus*. [RPS 17]

Neoporteria marksiana var. **tunensis** (Ritter) Donald & Rowley, Cact. Succ. J. Gr. Brit. 28: 57, 1966. Basionym: *Pyrrhocactus marksianus* var. *tunensis*. [RPS 17]

Neoporteria megliolii (Rausch) Donald, Ashingtonia Species Cat. Cact. [part 12], unnumbered page, 1976. Basionym: *Pyrrhocactus megliolii*. [Distributed as centre-page insert with Ashingtonia 2(4).] [RPS 27]

Neoporteria melanacantha (Backeberg) Donald & Rowley, Cact. Succ. J. Gr. Brit. 28: 57, 1966. *Nom. inval.*, based on *Pyrrhocactus melanacanthus, nom. inval.* (Art. 9.5). [Erroneously included as valid in RPS 17.] [RPS 17]

Neoporteria microsperma Ritter, Succulenta 1963: 6, 1963. [RPS 14]

Neoporteria microsperma var. **graciana** Ritter, Kakt. Südamer. 3: 1036, 1980. [RPS 31]

Neoporteria microsperma var. **serenana** Ritter, Succulenta 1963: 6, 1963. [RPS 14]

Neoporteria monte-amargensis (Backeberg) Donald & Rowley, Cact. Succ. J. Gr. Brit. 28: 57, 1966. *Nom. inval.*, based on *Neochilenia monte-amargensis, nom. inval.* (Art. 9.5). [Erroneously included as valid in RPS 17.] [RPS 17]

Neoporteria multicolor Ritter, Taxon 12(1): 33-34, 1963. Typus: *Ritter* 243 (U ?, ZSS [status ?]). [Type cited for U in original description, for ZSS in Kakt. Südamer. 3: 1039-1041.] [Chile] [RPS 14]

Neoporteria napina fa. **glabrescens** (Ritter) Donald & Rowley, Cact. Succ. J. Gr. Brit. 28: 57, 1966. Basionym: *Chileorebutia glabrescens*. [RPS 17]

Neoporteria napina fa. **mebbesii** (Hildmann) Donald & Rowley, Cact. Succ. J. Gr. Brit. 28: 57, 1966. Basionym: *Echinocactus odieri* var. *mebbesii*. [RPS 17]

Neoporteria napina fa. **nuda** (Ritter) A. E. Hoffmann, Cact. Fl. Chil., 224, ill. (p. 225), 1989. Basionym: *Thelocephala nuda*. [RPS 40]

Neoporteria napina var. **aerocarpa** (Ritter) A. E. Hoffmann, Cact. Fl. Chil., 226, ill. (p. 229), 1989. Basionym: *Chileorebutia aerocarpa*. [RPS 40]

Neoporteria napina var. **duripulpa** (Ritter) A. E. Hoffmann, Cact. Fl. Chil., 224, ill. (p. 227), 1989. Basionym: *Chileorebutia duripulpa*. [RPS 40]

Neoporteria napina var. **fankhauseri** (Ritter) A. E. Hoffmann, Cact. Fl. Chil., 224, ill. (p. 227), 1989. Basionym: *Thelocephala fankhauseri*. [Sphalm. 'fankhauserii'.] [RPS 40]

Neoporteria napina var. **fulva** (Ritter) A. E. Hoffmann, Cact. Fl. Chil., 226, ill. (p. 229), 1989. Basionym: *Thelocephala fulva*. [The basionym cited is a *nomen nudum*; moreover, the basionym citation is incomplete. However, the validly published name *Thelocephala fulva* is cited in the synonymy and is taken to represent the basionym citation.] [RPS 40]

Neoporteria napina var. *lembckei* (Backeberg) A. E. Hoffmann, Cact. Fl. Chil., 224, ill. (p. 227), 1989. *Nom. inval.*, based on *Neochilenia lembckei, nom. inval.* (Art. 37.1). [RPS 40]

Neoporteria napina var. **mitis** (Philippi) Donald & Rowley, Cact. Succ. J. Gr. Brit. 28: 57, 1966. Basionym: *Echinocactus mitis*. [RPS 17]

Neoporteria napina var. *spinosior* (Backeberg) Donald & Rowley, Cact. Succ. J. Gr. Brit. 28: 57, 1966. *Nom. inval.*, based on *Neochilenia napina* var. *spinosior, nom. inval.* (Art. 9.5). [Erroneously included as valid in RPS 17. This combination was first used (as new variety) by Backeberg in Backeberg & Knuth, Kaktus-ABC, 260, 1935, but the name is invalid (Art. 36.1).] [RPS 17]

Neoporteria nidus fa. **densispina** (Y. Ito) Donald & Rowley, Cact. Succ. J. Gr. Brit. 28: 57, 1966. Basionym: *Neoporteria densispina*. [RPS 17]

Neoporteria nidus fa. **senilis** (Philippi) Donald & Rowley, Cact. Succ. J. Gr. Brit. 28: 57, 1966. Basionym: *Echinocactus senilis*. [RPS 17]

Neoporteria nidus var. **gerocephala** (Y. Ito) Ritter, Kakt. Südamer. 3: 1039, 1980. Basionym: *Neoporteria gerocephala*. [RPS 31]

Neoporteria nidus var. **matancillana** Ritter, Kakt. Südamer. 3: 1039, 1980. [RPS 31]

Neoporteria nidus var. **multicolor** (Ritter) A. E. Hoffmann, Cact. Fl. Chil., 164, ill. (p. 167), 1989. Basionym: *Neoporteria multicolor*. [RPS 40]

Neoporteria nigrihorrida var. *major* (Backeberg) Backeberg, Cact. Succ. J. (US) 23(4): 118, 1951. *Nom. inval.*, based on *Chilenia nigrihorrida* var. *major, nom. inval.* (Art. 43.1). [Erroneously included as valid in RPS 3.] [RPS 3]

Neoporteria nigrihorrida var. *minor* (Backeberg) Backeberg, Cact. Succ. J. (US) 23(4): 118, 1951. *Nom. inval.*, based on *Chilenia nigrihorrida* var. *minor, nom. inval.* (Art. 43.1). [RPS 3]

Neoporteria odieri var. *araneifera* Donald & Rowley, Cact. Succ. J. Gr. Brit. 28: 57, 1966. *Nom. inval.* (Art. 37.1). [Based on *Echinocactus odieri spinis nigris* Labouret, Monogr. Cact., 248, 1858.] [RPS 17]

Neoporteria odieri var. *krausii* (Ritter) A. E. Hoffmann, Cact. Fl. Chil., 230, ills. (p. 231), 1989. *Nom. inval.* (Art. 33.2), based on *Chileorebutia krausii*, incorrect name (Art. 11.3). [Sphalm. 'kraussii'. Basionym citation is incomplete (lacks information on supplement).] [RPS 40]

Neoporteria odieri var. **longirapa** (Ritter) A. E. Hoffmann, Cact. Fl. Chil., 230, ill. (p. 231), 1989. Basionym: *Thelocephala longirapa*. [RPS 40]

Neoporteria paucicostata (Ritter) Donald & Rowley, Cact. Succ. J. Gr. Brit. 28: 57, 1966. Basionym: *Horridocactus paucicostatus*. [RPS 17]

Neoporteria paucicostata var. **echinus** (Ritter) A. E. Hoffmann, Cact. Fl. Chil., 208, ill. (p. 211), 1989. Basionym: *Pyrrhocactus echinus*. [RPS 40]

Neoporteria paucicostata var. **floccosa** (Ritter) A. E. Hoffmann, Cact. Fl. Chil., 208, ill. (p. 211), 1989. Basionym: *Pyrrhocactus floccosus*. [RPS 40]

Neoporteria paucicostata var. **glaucescens** (Ritter) A. E. Hoffmann, Cact. Fl. Chil., 208, ill. (p. 211), 1989. Basionym: *Pyrrhocactus glaucescens*. [RPS 40]

Neoporteria paucicostata var. **neohankeana** (Ritter) A. E. Hoffmann, Cact. Fl. Chil., 208, ill. (p. 209), 1989. Basionym: *Pyrrhocactus neohankeanus*. [RPS 40]

Neoporteria paucicostata var. **viridis** (Ritter) Donald & Rowley, Cact. Succ. J. Gr. Brit. 28: 57, 1966. Basionym: *Horridocactus paucicostatus* var. *viridis*. [Combination repeated by A. E. Hoffmann, Cact. Fl. Chil., 208, 1989, but the basionym cited there is incorrect.] [RPS 40]

Neoporteria pilispina (Ritter) Donald & Rowley, Cact. Succ. J. Gr. Brit. 28: 57, 1966. Basionym: *Pyrrhocactus pilispinus*. [RPS 17]

Neoporteria pilispina fa. **pygmaea** (Ritter) Donald & Rowley, Cact. Succ. J. Gr. Brit. 28: 57, 1966. Basionym: *Pyrrhocactus pygmaeus*. [RPS 17]

Neoporteria planiceps Ritter, Taxon 12(1): 43, 1963. *Nom. inval.* (Art. 75). [Spelling error for *Neoporteria laniceps* Ritter, cf. Succulenta 1966(3): 36.] [RPS -]

Neoporteria polyrhaphis (Pfeiffer) Backeberg, Cact. Succ. J. (US) 23(4): 118, 1951. Basionym: *Bridgesia polyrhaphis*. [The basionym given is already a combination.] [RPS 3]

Neoporteria pseudochilensis (Backeberg) Y. Ito, Explan. Diagr. Austroechinocactinae, 216, 1957. *Nom. inval.*, based on *Chilenia pseudochilensis, nom. inval.* (Art. 43.1). [RPS 8]

Neoporteria rapifera Y. Ito, The Cactaceae, 421, ill., 1981. Typus: *Ritter* 714 (). [Ascribed to "Ritter & Ito". Based on the material cited of which a specimen may be at U but this was most probably not seen by Ito. Validity of name not assessed.] [RPS -]

Neoporteria recondita (Ritter) Donald & Rowley, Cact. Succ. J. Gr. Brit. 28: 57, 1966. Basionym: *Pyrrhocactus reconditus*. [RPS 17]

Neoporteria recondita var. **residua** (Ritter) A. E. Hoffmann, Cact. Fl. Chil., 212, ill. (p. 213), 1989. Basionym: *Pyrrhocactus residuus*. [RPS 40]

Neoporteria recondita var. **vexata** (Ritter) A. E. Hoffmann, Cact. Fl. Chil., 212, ill. (p. 213), 1989. Basionym: *Pyrrhocactus vexatus*. [RPS 40]

Neoporteria reichei fa. **aerocarpa** (Ritter) Donald & Rowley, Cact. Succ. J. Gr. Brit. 28: 57, 1966. Basionym: *Chileorebutia aerocarpa*. [RPS 17]

Neoporteria reichei fa. *carneoflora* (Kilian) Donald & Rowley, Cact. Succ. J. Gr. Brit. 28: 57, 1966. *Nom. inval.*, based on *Neochilenia carneoflora, nom. inval.* (Art. 37.1). [Erroneously included as valid in RPS17.] [RPS 17]

Neoporteria reichei fa. **duripulpa** (Ritter) Donald & Rowley, Cact. Succ. J. Gr. Brit. 28: 57, 1966. Basionym: *Chileorebutia duripulpa*. [RPS 17]

Neoporteria reichei fa. *floribunda* (Backeberg) Donald & Rowley, Cact. Succ. J. Gr. Brit. 28: 57, 1966. *Nom. inval.*, based on *Reicheocactus floribundus, nom. inval.* (Art. 37.1). [Erroneously included as valid in RPS 17.] [RPS 17]

Neoporteria reichei fa. *imitans* (Backeberg) Donald & Rowley, Cact. Succ. J. Gr. Brit. 28: 57, 1966. *Nom. inval.*, based on *Neochilenia imitans, nom. inval.* (Art. 37.1). [Erroneously included as valid in RPS 17.] [RPS 17]

Neoporteria reichei fa. **krausii** (Ritter) Donald & Rowley, Cact. Succ. J. Gr. Brit. 28: 57, 1966. Basionym: *Chileorebutia krausii*. [RPS 17]

Neoporteria reichei fa. *lembckei* (Backeberg) Donald & Rowley, Cact. Succ. J. Gr. Brit. 28: 57, 1966. *Nom. inval.*, based on *Neochilenia lembckei, nom. inval.* (Art. 37.1).

[Erroneously included as valid in RPS 17.] [RPS 17]

Neoporteria reichei fa. *neoreichei* (Backeberg) Donald & Rowley, Cact. Succ. J. Gr. Brit. 28: 57, 1966. *Nom. inval.*, based on *Neochilenia neoreichei, nom. inval.* (Art. 37.1). [Erroneously included as valid in RPS 17.] [RPS 17]

Neoporteria reichei fa. *pseudoreichei* (Lembcke & Backeberg) Donald & Rowley, Cact. Succ. J. Gr. Brit. 28: 57, 1966. *Nom. inval.*, based on *Neochilenia pseudoreichei, nom. inval.* (Art. 37.1). [Erroneously included as valid in RPS 17.] [RPS 17]

Neoporteria reichei fa. **solitaria** (Ritter) Donald & Rowley, Cact. Succ. J. Gr. Brit. 28: 57, 1966. Basionym: *Chileorebutia malleolata* var. *solitaria*. [RPS 17]

Neoporteria reichei var. **malleolata** (Ritter) Donald & Rowley, Cact. Succ. J. Gr. Brit. 28: 57, 1966. Basionym: *Chileorebutia malleolata*. [RPS 17]

Neoporteria residua (Ritter) Donald & Rowley, Cact. Succ. J. Gr. Brit. 28: 57, 1966. Basionym: *Pyrrhocactus residuus*. [RPS 17]

Neoporteria ritteri Donald & Rowley, Cact. Succ. J. Gr. Brit. 28: 57, 1966. [*Nom. nov. pro Pyrrhocactus wagenknechtii* Ritter (*non Neoporteria wagenknechtii* Ritter).] [RPS 17]

Neoporteria robusta Ritter, Taxon 12: 34, 1963. [RPS 14]

Neoporteria roseiflora Y. Ito, Cacti, 78, 1952. *Nom. inval.* (Art. 36.1). [Published as *nomen nudum*. Repeated by Y. Ito in Full Bloom of Cactus Flowers, 93, 1962 (given as new combination for *Thelocephala roseiflora* Y. Ito).] [RPS 4]

Neoporteria rupicola (Ritter) Donald & Rowley, Cact. Succ. J. Gr. Brit. 28: 57, 1966. Basionym: *Pyrrhocactus rupicolus*. [RPS 17]

Neoporteria sanjuanensis (Spegazzini) Donald & Rowley, Cact. Succ. J. Gr. Brit. 28: 57, 1966. Basionym: *Echinocactus sanjuanensis*. [RPS 17]

Neoporteria scoparia (Ritter) Donald & Rowley, Cact. Succ. J. Gr. Brit. 28: 58, 1966. Basionym: *Pyrrhocactus scoparius*. [RPS 17]

Neoporteria setiflora (Backeberg) Donald & Rowley, Cact. Succ. J. Gr. Brit. 28: 58, 1966. *Nom. inval.*, based on *Pyrrhocactus setiflorus, nom. inval.* (Art. 37.1). [Erroneously included as valid in RPS 17.] [RPS 17]

Neoporteria setosiflora (Ritter) Donald & Rowley, Cact. Succ. J. Gr. Brit. 28: 58, 1966. Basionym: *Pyrrhocactus setosiflorus*. [RPS 17]

Neoporteria setosiflora var. **intermedia** (Ritter) Donald & Rowley, Cact. Succ. J. Gr. Brit. 28: 58, 1966. Basionym: *Pyrrhocactus setosiflorus* var. *intermedius*. [RPS 17]

Neoporteria simulans (Ritter) Donald & Rowley, Cact. Succ. J. Gr. Brit. 28: 58, 1966. Basionym: *Pyrrhocactus simulans*. [RPS 17]

Neoporteria sociabilis Ritter, Succulenta 1963: 3, 1963. [RPS 14]

Neoporteria sociabilis var. **napina** Ritter, Succulenta 1963: 3, 1963. [RPS 14]

Neoporteria strausiana (Schumann) Donald & Rowley, Cact. Succ. J. Gr. Brit. 28: 58, 1966. Basionym: *Echinocactus strausianus*. [RPS 17]

Neoporteria subaiana (Backeberg) Donald & Rowley, Cact. Succ. J. Gr. Brit. 28: 58, 1966. *Nom. inval.*, based on *Pyrrhocactus subaianus, nom. inval.* (Art. 9.5, 37.1). [Erroneously included as valid in RPS 17.] [RPS 17]

Neoporteria subcylindrica (Backeberg) Backeberg, Cact. Succ. J. (US) 23(4): 119, 1951. *Nom. inval.*, based on *Chilenia subcylindrica, nom. inval.* (Art. 43.1). [RPS 3]

Neoporteria subgibbosa fa. **castanea** (Ritter) Donald & Rowley, Cact. Succ. J. Gr. Brit. 28: 58, 1966. Basionym: *Neoporteria castanea*. [RPS 17]

Neoporteria subgibbosa fa. **castaneoides** (Cels) Donald & Rowley, Cact. Succ. J. Gr. Brit. 28: 58, 1966. Basionym: *Echinocactus castaneoides*. [RPS 17]

Neoporteria subgibbosa fa. *heteracantha* (Backeberg) Donald & Rowley, Cact. Succ. J. Gr. Brit. 28: 58, 1966. *Nom. inval.*, based on *Chilenia heteracantha, nom. inval.* (Art. 43.1). [Erroneously included as valid in RPS 17.] [RPS 17]

Neoporteria subgibbosa fa. **litoralis** (Ritter) Donald & Rowley, Cact. Succ. J. Gr. Brit. 28: 58, 1966. Basionym: *Neoporteria litoralis*. [RPS 17]

Neoporteria subgibbosa fa. *major* (Backeberg) Donald & Rowley, Cact. Succ. J. Gr. Brit. 28: 58, 1966. *Nom. inval.*, based on *Chilenia nigrihorrida* var. *major, nom. inval.* (Art. 43.1). [Erroneously included as valid in RPS 17.] [RPS 17]

Neoporteria subgibbosa fa. *minor* (Backeberg) Donald & Rowley, Cact. Succ. J. Gr. Brit. 28: 58, 1966. *Nom. inval.*, based on *Chilenia nigrihorrida* var. *minor, nom. inval.* (Art. 43.1). [Erroneously included as valid in RPS 17.] [RPS 17]

Neoporteria subgibbosa fa. **serenana** (Ritter) Donald & Rose, Cact. Succ. J. Gr. Brit. 28: 58, 1966. Basionym: *Neoporteria microsperma* var. *serenana*. [RPS 17]

Neoporteria subgibbosa fa. *subcylindrica* (Backeberg) Donald & Rowley, Cact. Succ. J. Gr. Brit. 28: 58, 1966. *Nom. inval.*, based on *Chilenia subcylindrica, nom. inval.* (Art. 43.1). [Erroneously included as valid in RPS 17.] [RPS 17]

Neoporteria subgibbosa fa. **tunensis** (Ritter) Donald & Rowley, Cact. Succ. J. Gr. Brit. 28: 58, 1966. Basionym: *Neoporteria castanea* var. *tunensis*. [RPS 17]

Neoporteria subgibbosa var. **intermedia** (Ritter) A. E. Hoffmann, Cact. Fl. Chil., 168, 1989. Basionym: *Neoporteria litoralis* var. *intermedia*. [RPS 40]

Neoporteria subgibbosa var. **litoralis** (Ritter) A. E. Hoffmann, Cact. Fl. Chil., 168, ill. (p. 171), 1989. Basionym: *Neoporteria litoralis*. [RPS 40]

Neoporteria subgibbosa var. **mammillarioides** (Hooker) Donald & Rowley, Cact. Succ. J. Gr. Brit. 28: 58, 1966. Basionym: *Echinocactus mammillarioides*. [RPS 17]

Neoporteria subgibbosa var. **microsperma** (Ritter) Donald & Rowley, Cact. Succ. J. Gr. Brit. 28: 58, 1966. Basionym: *Neoporteria microsperma*. [RPS 17]

Neoporteria subgibbosa var. *nigrihorrida* (Backeberg) Donald & Rowley, Cact. Succ. J. Gr. Brit. 28: 58, 1966. *Nom. inval.*, based on *Chilenia nigrihorrida, nom. inval.* (Art. 43.1). [Erroneously included as valid in RPS 17.] [RPS 17]

Neoporteria subgibbosa var. **orientalis** Ritter, Succulenta 1966: 129, 131 ill., 1966. [RPS 17]

Neoporteria subgibbosa var. **robusta** (Ritter) A. E. Hoffmann, Cact. Fl. Chil., 168, ill. (p. 171), 1989. Basionym: *Neoporteria robusta*. [RPS 40]

Neoporteria subikii Riha, Kaktusy 23(6): 128-130, ills., 1987. Based on *Knize* 39. *Nom. inval.* (Art. 36.1). [Published as provisional name.] [Chile] [RPS 38]

Neoporteria taltalensis P. C. Hutchison, Cact. Succ. J. (US) 27(6): 181, 1955. [RPS 6]

Neoporteria taltalensis var. **transiens** (Ritter) A. E. Hoffmann, Cact. Fl. Chil., 216, ill. (p. 217), 1989. Basionym: *Pyrrhocactus transiens*. [RPS 40]

Neoporteria thiebautiana (Backeberg) Y. Ito, Explan. Diagr. Austroechinocactinae, 220, 1957. *Nom. inval.*, based on *Chilenia thiebautiana, nom. inval.* (Art. 43.1). [RPS 8]

Neoporteria totoralensis (Ritter) Donald & Rowley, Cact. Succ. J. Gr. Brit. 28: 58, 1966. Basionym: *Pyrrhocactus totoralensis*. [RPS 17]

Neoporteria tuberisulcata (Jacobi) Donald & Rowley, Cact. Succ. J. Gr. Brit. 28: 58, 1966. Basionym: *Echinocactus tuberisulcatus*. [RPS 17]

Neoporteria tuberisulcata var. **armata** (Ritter) Donald & Rowley, Cact. Succ. J. Gr. Brit. 28: 58, 1966. Basionym: *Pyrrhocactus armatus*. [RPS 17]

Neoporteria tuberisulcata var. **atroviridis** (Ritter) Donald & Rowley, Cact. Succ. J. Gr. Brit. 28: 58, 1966. Basionym: *Pyrrhocactus atroviridis*. [RPS 17]

Neoporteria tuberisulcata var. **cupreata** (Poselger) Donald & Rowley, Cact. Succ. J. Gr. Brit. 28: 58, 1966. Basionym: *Echinocactus cupreatus*. [RPS 17]

Neoporteria tuberisulcata var. **froehlichiana** (Schumann) Donald & Rowley, Cact. Succ. J. Gr. Brit. 28: 58, 1966. Basionym: *Echinocactus froehlichianus*. [RPS 17]

Neoporteria tuberisulcata var. **nigricans** (Linke) Donald & Rowley, Cact. Succ. J. Gr. Brit. 28: 58, 1966. Basionym: *Echinopsis nigricans*. [RPS 17]

Neoporteria tuberisulcata var. **robusta** (Ritter) Donald & Rowley, Cact. Succ. J. Gr. Brit. 28: 58, 1966. Basionym: *Pyrrhocactus robustus*. [RPS 17]

Neoporteria tuberisulcata var. **vegasana** (Ritter) Donald & Rowley, Cact. Succ. J. Gr. Brit. 28: 58, 1966. Basionym: *Pyrrhocactus robustus* var. *vegasanus*. [RPS 17]

Neoporteria umadeave (Fric) Donald & Rowley, Cact. Succ. J. Gr. Brit. 28: 58, 1966. Basionym: *Echinocactus umadeave*. [RPS 17]

Neoporteria umadeave var. *marayensis* (Backeberg) Donald & Rowley, Cact. Succ. J. Gr. Brit. 28: 58, 1966. *Nom. inval.*, based on *Pyrrhocactus umadeave* var. *marayesensis, nom. inval.* (Art. 9.5). [Erroneously included as valid in RPS 17.] [RPS 17]

Neoporteria vallenarensis Ritter, Kakt. Südamer. 3: 1032-1033, 1980. [RPS 31]

Neoporteria vallenarensis (Ritter) A. E. Hoffmann, Cact. Fl. Chil., 218, ill. (p. 219), 1989. *Nom. illeg.* (Art. 64.1), based on *Pyrrhocactus vallenarensis*. [*Non Neoporteria vallenarensis* Ritter 1980.] [RPS 40]

Neoporteria vallenarensis var. *atroviridis* (Ritter) A. E. Hoffmann, Cact. Fl. Chil., 218, ill. (p. 221), 1989. *Nom. inval.* (Art. 57), based on *Pyrrhocactus atroviridis*. [RPS 40]

Neoporteria vallenarensis var. **crispa** (Ritter) A. E. Hoffmann, Cact. Fl. Chil., 218, ill. (p. 221), 1989. Basionym: *Pyrrhocactus crispus*. [Incorrectly used name (Art. 57).] [RPS 40]

Neoporteria vallenarensis var. *domeykoensis* (Ritter) A. E. Hoffmann, Cact. Fl. Chil., 218, ill. (p. 221), 1989. *Nom. inval.* (Art. 57), based on *Pyrrhocactus eriosyzoides* var. *domeykoensis*. [RPS 40]

Neoporteria vallenarensis var. *transitensis* (Ritter) A. E. Hoffmann, Cact. Fl. Chil., 218, ill. (p. 219), 1989. *Nom. inval.* (Art. 57), based on *Pyrrhocactus transitensis*. [RPS 40]

Neoporteria villicumensis (Rausch) Donald, Ashingtonia Species Cat. Cact. [part 12], unnumbered page, 1976. Basionym: *Pyrrhocactus villicumensis*. [Distributed as centre-page insert with Ashingtonia 2(4).] [RPS 27]

Neoporteria villosa var. **atrispinosa** (Backeberg) Donald & Rowley, Cact. Succ. J. Gr. Brit. 28: 58, 1966. [RPS 17]

Neoporteria villosa var. **cephalophora** (Backeberg) Donald & Rowley, Cact. Succ. J. Gr. Brit. 28: 58, 1966. Basionym: *Chilenia cephalophora*. [RPS 17]

Neoporteria villosa var. **laniceps** (Ritter) A. E. Hoffmann, Cact. Fl. Chil., 172, ill. (p. 173), 1989. Basionym: *Neoporteria laniceps*. [RPS 40]

Neoporteria volliana (Backeberg) Donald & Rowley, Cact. Succ. J. Gr. Brit. 28: 58, 1966. Basionym: *Pyrrhocactus vollianus*. [RPS 17]

Neoporteria volliana var. **breviaristata** (Backeberg) Donald & Rowley, Cact. Succ. J. Gr. Brit. 28: 58, 1966. Basionym: *Pyrrhocactus vollianus* var. *breviaristatus*. [RPS 17]

Neoporteria wagenknechtii Ritter, Succulenta 1963: 5, 1963. [RPS 14]

Neoporteria wagenknechtii var. **microsperma** (Ritter) A. E. Hoffmann, Cact. Fl. Chil., 174,

1989. Basionym: *Neoporteria microsperma*. [RPS 40]

Neoporteria wagenknechtii var. **napina** Ritter, Succulenta 1963: 5, 1963. [RPS 14]

Neoporteria wagenknechtii var. **vallenarensis** (Ritter) A. E. Hoffmann, Cact. Fl. Chil., 174, ill. (p. 175), 1989. Basionym: *Neoporteria vallenarensis*. [RPS 40]

Neoporteria woutersiana (Backeberg) Donald & Rowley, Cact. Succ. J. Gr. Brit. 28: 58, 1966. *Nom. inval.*, based on *Delaetia woutersiana, nom. inval.* (Art. 37.1). [Erroneously included as valid in RPS17.] [RPS 17]

Neoraimondia arequipensis var. *aticensis* (Rauh & Backeberg) Rauh & Backeberg, in Rauh, Beitr. Kenntn. peruan. Kakt.veg., 263, 1958. *Nom. inval.* (Art. 33.2), based on *Neoraimondia aticensis*. [RPS 10]

Neoraimondia arequipensis var. **gigantea** (Backeberg) Ritter, Backeberg's Descr. & Erört. taxon. nomenkl. Fragen, [unpaged], 1958. Basionym: *Neoraimondia arequipensis*. [RPS 15]

Neoraimondia arequipensis var. **rhodantha** Rauh & Backeberg, in Backeberg, Descr. Cact. Nov. [1:] 13, 1957. [Dated 1956, published 1957.] [RPS 7]

Neoraimondia arequipensis var. **riomajensis** Rauh & Backeberg, in Backeberg, Descr. Cact. Nov. [1:] 13, 1957. [Dated 1956, published 1957.] [RPS 7]

Neoraimondia arequipensis var. **roseiflora** (Backeberg) Ritter, Backeberg's Descr. & Erört. taxon. nomenkl. Fragen, [unpaged], 1958. Basionym: *Neoraimondia roseiflora*. [RPS 15]

Neoraimondia aticensis Rauh & Backeberg, in Backeberg, Descr. Cact. Nov. [1:] 13, 1957. [Dated 1956, published 1957.] [RPS 7]

Neoraimondia gigantea var. **saniensis** Rauh & Backeberg, in Backeberg, Descr. Cact. Nov. [1:] 13, 1957. [Dated 1956, published 1957.] [RPS 7]

Neoraimondia herzogiana (Backeberg) Buxbaum, in Krainz, Kat. ZSS ed. 2, 89, 1967. Basionym: *Neocardenasia herzogiana*. [RPS 18]

Neoraimondia peruviana (Linné) Ritter, Kakt. Südamer. 4: 1267-1270, 1981. Basionym: *Cactus peruvianus*. [RPS 32]

Neotanahashia Y. Ito, Explan. Diagr. Austroechinocactinae, 113, 290, 1957. *Nom. illeg.* (Art. 63.1). [Sphalm. 'Netanahashia'.] [= *Reicheocactus* Backeberg.] [RPS 8]

Neotanahashia reichei (Backeberg) Y. Ito, Explan. Diagr. Austroechinocactinae, 114, 1957. Incorrect name (Art. 11.3), based on *Reicheocactus reichei*. [RPS 8]

Neowerdermannia peruviana Ritter, Kakt. Südamer. 4: 1338-1339, 1981. Typus: *Ritter* 191 (U, ZSS). [Peru: Moquegua] [RPS 32]

Neowerdermannia vorwerkii fa. **gielsdorfiana** (Backeberg) Krainz, Die Kakt. C VIf, 1969. Basionym: *Neowerdermannia vorwerkii* var. *gielsdorfiana*. [RPS 20]

Neowerdermannia vorwerkii var. *erectispina* Hoffmann & Backeberg, in Backeberg, Die Cact. 3: 1796, 1959. *Nom. inval.* (Art. 37.1). [Erroneously included as valid in RPS 10.] [RPS 10]

Neowerdermannia vorwerkii var. **gielsdorfiana** Backeberg, Cact. Succ. J. (US) 23(3): 86, 1951. [First mentioned (as *nomen nudum*) by Backeberg in Backeberg & Knuth, Kaktus ABC, 285, 1935.] [RPS 3]

Nopalea escuintlensis Matuda, Cact. Suc. Mex. 1(2): 43-44, 1956. [RPS 7]

Nopalea nuda Backeberg, Die Cact. 6: 3629, 1962. *Nom. inval.* (Art. 9.5). [Erroneously included as valid in RPS 13.] [RPS 13]

Nopaloxenia G. Herter, Cactus (Paris) No. 45: 204, 1955. *Nom. inval.* (Art. 75). [Presumed error for *Nopalxochia*.] [RPS 6]

× **Nopalxalis** Süpplie, Succulenta 67(6): 141, 1988. [= *Nopalxochia* × *Rhipsalis*.] [RPS 39]

× **Nopalxalis** cv. **Naranja** Süpplie, Succulenta 67(6): 141, 1988. [= (*Nopalxochia ackermannii* × *N. phyllanthoides*) × *Rhipsalis monacantha*.] [RPS 39]

Nopalxochia ackermannii cv. **Candida** (Alexander) Kimnach, Cact. Succ. J. (US) 53(2): 85, 1981. Basionym: *Epiphyllum ackermannii* fa. *candida*. [RPS -]

Nopalxochia ackermannii var. **conzattianum** (T. M. Macdougal) Kimnach, Cact. Succ. J. (US) 53(2): 85, 1981. Basionym: *Nopalxochia conzattianum*. [RPS 32]

Nopalxochia × **capelleana** (Roth) Rowley, in Backeberg, Die Cact. 2: 761, 1959. Basionym: *Phyllocactus capelleanus*. [RPS 10]

Nopalxochia horichii Kimnach, Cact. Succ. J. (US) 56(1): 4-8, ills., 1984. Typus: *Horich* s.n. (HNT, CR, F, US, ZSS). [= Huntington BG 36756.] [Costa Rica: San José] [RPS 35]

Normanbokea Kladiwa & Buxbaum, in Krainz, Die Kakt. C VIIIb (March 1969), 1969. Typus: *N. valdeziana*. [First mentioned as manuscript name in Krainz, Katalog ZSS ed. 2, 89, 1967 (*nomen nudum*).] [RPS 20]

Normanbokea pseudopectinata (Backeberg) Kladiwa & Buxbaum, in Krainz, Die Kakt. Lief. 40/41: C VIIIb, 1969. Basionym: *Pelecyphora pseudopectinata*. [RPS 20]

Normanbokea valdeziana (H. Möller) Kladiwa & Buxbaum, in Krainz, Die Kakt. Lief. 40/41: C VIIIb, 1969. Basionym: *Pelecyphora valdeziana*. [RPS 20]

Notocacteae Buxbaum, Madroño 14(6): 191, 1958. Typus: *Notocactus*. [Published as 'linea'.] [RPS 9]

Notocactus sect. *Anstrophiolatae* Havlicek, Kakt. Vilag 18(4): 80, 1989. *Nom. inval.* (Art. 36.1). [Based on *Notocactus alacriportanus*. Sphalm. for 'Astrophiolatae' ? Dated 1988.] [RPS 41]

Notocactus sect. *Campanulatiflorales* Havlicek, Kakt. Vilag 18(4): 76, 1989. *Nom. inval.* (Art. 36.1). [Based on *Notocactus mammulosus*. Dated 1988.] [RPS 41]

Notocactus sect. **Infundibuliflorales** Havlicek, Kakt. Vilag 18(4): 73, 1989. Typus: *Notocactus ottonis*. Dated 1988. [RPS 41]

Notocactus sect. *Interflorales* Havlicek, Kakt. Vilag 18(4): 77, 1989. *Nom. inval.* (Art. 36.1). [Based on *Notocactus rauschii*. Dated 1988. Used as provisional name since 1976.]

[RPS 41]

Notocactus sect. *Malacocarpoformia* Havlicek, Kakt. Vilag 18(4): 78, 1989. *Nom. inval.* (Art. 36.1). [Based on *Notocactus corynodes*. Dated 1988. Used as provisional name since 1976.] [RPS 41]

Notocactus sect. *Microseminiformia* Havlicek, Kakt. Vilag 18(4): 77, 1989. *Nom. inval.* (Art. 22.1, 36.1). [Based on *Notocactus schumannianus* (*lectotypus generis*). Dated 1988. Used as provisional name since 1976..] [RPS 41]

Notocactus sect. *Protoseminiformia* Havlicek, Kakt. Vilag 18(4): 78, 1989. *Nom. inval.* (Art. 36.1). [Based on *Notocactus graessneri*. Dated 1988. Used as provisional name since 1976.] [RPS 41]

Notocactus ser. *Brasilianae* Havlicek, Kakt. Vilag 18(4): 79, 1989. *Nom. inval.* (Art. 36.1). [Based on *Notocactus longispinus*. Dated 1988.] [RPS 41]

Notocactus ser. *Buxbaumianae* Havlicek, Kakt. Vilag 18(4): 80, 1989. *Nom. inval.* (Art. 36.1). [Based on *Notocactus alacriportanus*. Dated 1988.] [RPS 41]

Notocactus ser. *Cephalioideae* (Fric & Kreuzinger) Havlicek, Kakt. Vilag 18(4): 77, 1989. *Nom. inval.* (Art. 22). [Dated 1988. As this series contains all the taxa which can serve as lectotype for the genus, the name will be invalid.] [RPS 41]

Notocactus ser. *Corynodinae* Havlicek, Kakt. Vilag 18(4): 79, 1989. *Nom. inval.* (Art. 36.1). [Based on *Notocactus corynodes*. Dated 1988.] [RPS 41]

Notocactus ser. *Fricianae* Havlicek, Kakt. Vilag 18(4): 75, 1989. *Nom. inval.* (Art. 36.1). [Based on *Notocactus minimus*. Dated 1988. Used as provisional name since 1976.] [RPS 41]

Notocactus ser. *Graessnerianae* Havlicek, Kakt. Vilag 18(4): 78, 1989. *Nom. inval.* (Art. 36.1). [Based on *Notocactus graessneri*. Dated 1988.] [RPS 41]

Notocactus ser. *Herterianae* Havlicek, Kakt. Vilag 18(4): 75, 1989. *Nom. inval.* (Art. 36.1). [Based on *Notocactus herteri*. First used provisionally since 1976. Dated 1988.] [RPS 41]

Notocactus ser. *Krainzianae* Havlicek, Kakt. Vilag 18(4): 78, 1989. *Nom. inval.* (Art. 36.1). [Based on *Notocactus magnificus*. Dated 1988.] [RPS 41]

Notocactus ser. *Mammulosi* Fric & Kreuzinger *ex* Havlicek, Kakt. Vilag 18(4): 76, 1989. *Nom. inval.* (Art. 36.1). [Based on *Notocactus* group *Mammulosi* in Kreuzinger, Verzeichn. Amer. and. Sukk., 21, 1935. Dated 1988.] [RPS 41]

Notocactus ser. *Melchersianae* Havlicek, Kakt. Vilag 18(4): 77, 1989. *Nom. inval.* (Art. 36.1). [Based on *Notocactus mueller-melchersii*. Dated 1988. Used as provisional name since 1976.] [RPS 41]

Notocactus ser. **Paucispini** (Fric & Kreuzinger) Havlicek, Kakt. Vilag 18(4): 73, 1989. Basionym: *Notocactus* group *Paucispini*. [Notocactus ottonis. Dated 1988. First used as a provisional name since 1976. Validity of name not assessed] [RPS 41]

Notocactus ser. *Rauschianae* Havlicek, Kakt. Vilag 18(4): 77, 1989. *Nom. inval.* (Art. 36.1). [Based on *Notocactus rauschii*. Dated 1988. Used as provisional name since 1976.] [RPS 41]

Notocactus ser. *Scopanae* Havlicek, Kakt. Vilag 18(4): 74, 1989. *Nom. inval.* (Art. 36.1). [Based on *Notocactus scopa*. First used provisionally since 1976.] [RPS 41]

Notocactus ser. *Setacei* (Fric & Kreuzinger) Havlicek, Kakt. Vilag 18(4): 75, 1989. *Nom. inval.* (Art. 36.1). [Based on *Notocactus concinnus*. Dated 1988. Used as provisional name since 1976.] [RPS 41]

Notocactus ser. *Uebelmannianae* Havlicek, Kakt. Vilag 18(4): 75, 1989. *Nom. inval.* (Art. 36.1). [Based on *Notocactus uebelmannianus*. Dated 1988. Used as provisional name since 1976.] [RPS 41]

Notocactus ser. *Werdermannianae* Havlicek, Kakt. Vilag 18(4): 75, 1989. *Nom. inval.* (Art. 36.1). [Based on *Notocactus werdermannianus*. Dated 1988. Used as provisional name since 1976.] [RPS 41]

Notocactus subgen. *Acanthocephala* (Backeberg) Havlicek, Kakt. Vilag 18(4): 78, 1989. *Nom. illeg.* (Art. 63.1). [Dated 1988..] [RPS 41]

Notocactus subgen. **Brasilicactus** Buxbaum, in Krainz, Die Kakt. Lief. 35: C VIc (Jan. 1967), 1967. Typus: *Echinocactus graessneri* Schumann. Based on *Brasilicactus* Backeberg, *nom. illeg.* (Art. 63). [RPS 18]

Notocactus subgen. *Calocomocephalus* Y. Ito, Bull. Takarazuka Insectarium 71: 13-20, 1950. *Nom. inval.* (Art. 36.1). [RPS 3]

Notocactus subgen. *Eriocactus* (Backeberg) Buining *ex* Buxbaum, in Krainz, Die Kakt. Lief. 35: C VIc (Jan. 1967), 1967. *Nom. inval.* (Art. 22.1). [Combination first published (without basionym) by Buining in Succulenta 1957: 106..] [RPS 18]

Notocactus subgen. *Eriocephala* (Backeberg) Havlicek, Kakt. Vilag 18(4): 77, 1989. *Nom. inval.* (Art. 22). [Dated 1988..] [RPS 41]

Notocactus subgen. *Eunotocactus* Backeberg, Cact. Succ. J. (US) 12(5): 153, 1950. *Nom. inval.* (Art. 21.3). [Based on *Cactus ottonis* Lehmann.] [RPS 1]

Notocactus subgen. **Malacocarpus** (Schumann) Buxbaum, in Krainz, Die Kakt. Lief. 35: C VIc, 1967. Basionym: *Echinocactus* subgen. *Malacocarpus*. [RPS 18]

Notocactus subgen. *Mallocephalus* Y. Ito, Bull. Takarazuka Insectarium 71: 13-20, 1950. *Nom. inval.* (Art. 36.1). [RPS 3]

Notocactus subgen. **Neonotocactus** Backeberg, Cact. Succ. J. (US) 22(5): 153, 1950. Typus: *Echinocactus mammulosus* Lemaire. [RPS 1]

Notocactus subgen. *Notobrasilia* Havlicek, Kakt. Vilag 18(4): 79-80, 1989. *Nom. inval.* (Art. 36.1). [Based on *Notocactus alacriportanus*. Dated 1988.] [RPS 41]

Notocactus subgen. *Pogocephalus* Y. Ito, Bull. Takarazuka Insectarium 71: 13-20, 1950. *Nom. inval.* (Art. 36.1). [RPS 3]

Notocactus subgen. *Raphidocephalus* Y. Ito,

Bull. Takarazuka Insectarium 71: 13-20, 1950. *Nom. inval.* (Art. 36.1). [RPS 3]

Notocactus subser. *Arechavaletae* Havlicek, Kakt. Vilag 18(4): 73, 1989. *Nom. inval.* (Art. 36.1, 37.1). [Sphalm. 'Arechavaletai'. Dated 1988.] [RPS 41]

Notocactus subser. *Blaawianus* Havlicek, Kakt. Vilag 18(4): 76, 1989. *Nom. inval.* (Art. 36.1, 37.1). [Dated 1988.] [RPS 41]

Notocactus subser. *Concinnus* Havlicek, Kakt. Vilag 18(4): 75, 1989. *Nom. inval.* (Art. 36.1, 37.1). [RPS 41]

Notocactus subser. *Floricomus* Havlicek, Kakt. Vilag 18(4): 76, 1989. *Nom. inval.* (Art. 36.1, 37.1). [Dated 1988.] [RPS 41]

Notocactus subser. *Glaucinus* Havlicek, Kakt. Vilag 18(4): 74, 1989. *Nom. inval.* (Art. 36.1, 37.1). [Dated 1988.] [RPS 41]

Notocactus subser. *Linkii* Havlicek, Kakt. Vilag 18(4): 74, 1989. *Nom. inval.* (Art. 36.1, 37.1). [Dated 1988.] [RPS 41]

Notocactus subser. *Mammulosus* Havlicek, Kakt. Vilag 18(4): 76, 1989. *Nom. inval.* (Art. 36.1, 37.1). [Dated 1988.] [RPS 41]

Notocactus subser. *Muricatus* Havlicek, Kakt. Vilag 18(4): 74, 1989. *Nom. inval.* (Art. 36.1, 37.1). [Dated 1988.] [RPS 41]

Notocactus subser. *Ottonis* Havlicek, Kakt. Vilag 18(4): 73, 1989. *Nom. inval.* (Art. 36.1). [*Notocactus ottonis*. Dated 1988.] [RPS 41]

Notocactus subser. *Oxycostatus* Havlicek, Kakt. Vilag 18(4): 73, 1989. *Nom. inval.* (Art. 36.1, 37.1). [Dated 1988.] [RPS 41]

Notocactus subser. *Tabularis* Havlicek, Kakt. Vilag 18(4): 76, 1989. *Nom. inval.* (Art. 36.1, 37.1). [Dated 1988.] [RPS 41]

Notocactus acuatus (Link & Otto) Theunissen, Succulenta 60(6): 142, 1981. Basionym: *Echinocactus acuatus*. [RPS 32]

Notocactus acutus Ritter, Kakt. Südamer. 1: 169, 1979. [RPS 30]

Notocactus agnetae van Vliet, Succulenta 54: 6-11, 1975. [RPS 26]

Notocactus agnetae var. **aureispinus** van Vliet, Succulenta 54: 6-11, 1975. [RPS 26]

Notocactus agnetae var. **minor** van Vliet, Succulenta 54: 6-11, 1975. [RPS 26]

Notocactus alacriportanus (Backeberg & Voll) Buxbaum, in Krainz, Die Kakt. C VIc (Jan. 1967), 1967. Basionym: *Parodia alacriportana*. [RPS 28]

Notocactus allosiphon Marchesi, Bol. Soc. Argent. Bot. 14: 246, 1972. [RPS 24 gives the author as "Schlosser" and the citation as 3: 246-248.] [RPS 24]

Notocactus ampliocostatus (Ritter) Theunissen, Succulenta 60(6): 142, 1981. Basionym: *Eriocactus ampliocostatus*. [RPS 32]

Notocactus arachnitis Ritter, Succulenta 49: 108, 1970. [This is the spelling used in the original diagnosis; several authorities are of the opinion that it needs to be corrected to 'arachnites'.] [RPS 21]

Notocactus arachnitis var. **minor** Ritter, Succulenta 49: 108, 1970. [RPS 21]

Notocactus araneolarius (Reichenbach) Herter, Cactus (Paris) No. 42: 120, 1954. Basionym: *Echinocactus araneolarius*. [RPS 5]

Notocactus arechavaletae var. *alacriportanus* Ritter, Kakt. Südamer. 1: 165, 1979. *Nom. inval.* (Art. 37.1). [Erroneously regarded as valid in RPS, but based on two syntypes.] [RPS 30]

Notocactus arechavaletae var. **aureus** Ritter, Kakt. Südamer. 1: 166, 1979. Typus: *Ritter* 1396 (). [Brazil: Rio Grande do Sul] [RPS 30]

Notocactus arechavaletae var. **buenekeri** Ritter, Kakt. Südamer. 1: 166, 1979. Typus: *Ritter* 1027b (). [Brazil: Rio Grande do Sul] [RPS 30]

Notocactus arechavaletae var. **horstii** Ritter, Kakt. Südamer. 1: 166, 1979. Typus: *Ritter* 1027c (). [Brazil: Rio Grande do Sul] [RPS 30]

Notocactus arechavaletae var. **limiticola** Ritter, Kakt. Südamer. 1: 164, 1979. Typus: *Ritter et Horst* 1415 (). [Brazil: Rio Grande do Sul] [RPS 30]

Notocactus arechavaletae var. **nanus** Ritter, Kakt. Südamer. 1: 165, 1979. Typus: *Ritter* 1389 (). [Brazil: Rio Grande do Sul] [RPS 30]

Notocactus arechavaletae var. **rubescens** Ritter, Kakt. Südamer. 1: 166, 1979. Typus: *Ritter et Horst* 1027d (). [Brazil: Rio Grande do Sul] [RPS 30]

Notocactus arnostianus Lisal & Kolarik, Internoto 7(1): 6-8, ills., 1986. Typus: *Horst et Uebelmann* HU 338 (herb. Arb. Mus. Siles., Opava - Novy Dur, CSSR). [Individuals from the type collection have also been designated as type of *Notocactus ritterianus*.] [Brazil] [RPS 37]

Notocactus beltranii (Fric *ex* Fleischer & Schütz) Schäfer, Kakt./Sukk. 14: 36, 1980. Basionym: *Wigginsia beltranii*. [RPS 31]

Notocactus bezrucii (Fric) Schäfer, Kakt./Sukk. 14: 36, 1980. Basionym: *Malacocarpus bezrucii*. [RPS 31]

Notocactus bezrucii var. **centrispinus** (Fric) Theunissen, Succulenta 60(6): 140, 1981. Basionym: *Malacocarpus bezrucii* var. *centrispinus*. [RPS 32]

Notocactus bezrucii var. **corniger** (Fric) Theunissen, Succulenta 60(6): 140, 1981. Basionym: *Malacocarpus bezrucii* var. *corniger*. [RPS 32]

Notocactus blaauwianus van Vliet, Succulenta 55: 108-113, 1976. [RPS 27]

Notocactus blaauwianus var. **enormis** van Vliet, Succulenta 55: 113, 1976. [RPS 27]

Notocactus bommeljei van Vliet, Succulenta 47: 5-8, 1968. [*Nom. nov. pro Echinocactus muricatus* K. Schumann (1898), *non* Otto (1837).] [RPS 19]

Notocactus brederooianus Prestlé, Succulenta 64(4): 84-87, ills., SEM-ills., 1985. Typus: *Prestlé* 81 (U). [Brazil] [RPS 36]

Notocactus brevihamatus (Haage) Buxbaum, in Krainz, Die Kakt. C VIc, 1967. Basionym: *Parodia brevihamata*. [RPS 18]

Notocactus brevihamatus fa. *conjungens* (Ritter) Theunissen, Succulenta 60(6): 142, 1981. *Nom. inval.*, based on *Brasiliparodia brevihamata* fa. *conjungens, nom. inval.* (Art. 37.1). [RPS 32]

Notocactus brevihamatus var. *mollispinus* (Ritter) Theunissen, Succulenta 60(6): 142, 1981. *Nom. inval.*, based on *Brasiliparodia brevihamata* var. *mollispina, nom. inval.* (Art. 37.1). [RPS 32]

Notocactus buenekeri (Buining) Buxbaum, Kakt. and. Sukk. 17: 195, 1966. Basionym: *Parodia buenekeri.* [RPS 17]

Notocactus buenekeri fa. *conjungens* (Ritter) Theunissen, Succulenta 60(6): 142, 1981. *Nom. inval.*, based on *Brasiliparodia buenekeri* fa. *conjungens, nom. inval.* (Art. 37.1). [RPS 32]

Notocactus buenekeri var. *intermedius* (Ritter) Theunissen, Succulenta 60(6): 142, 1981. *Nom. inval.*, based on *Brasiliparodia buenekeri* var. *intermedia, nom. inval.* (Art. 37.1). [Sphalm. 'intermedia' (not corrected in RPS 32).] [RPS 32]

Notocactus buiningii Buxbaum, Kakt. and. Sukk. 19(12): 229-231, ills., 1968. Typus: *Horst et Uebelmann* HU 90 (U, ZSS [seeds only]). [Uruguay: Rivera] [RPS 19]

Notocactus campestrensis Ritter, Kakt. Südamer. 1: 177, 1979. [RPS 30]

Notocactus carambeiensis Buining & Brederoo, Kakt. and. Sukk. 24(1): 1-3, ills., 1973. Typus: *Horst et Uebelmann* HU 140a (1966) (U, ZSS). [Brazil: Paraná] [RPS 24]

Notocactus catarinensis (Ritter) Theunissen, Succulenta 60(6): 142, 1981. *Nom. inval.*, based on *Brasiliparodia catarinensis, nom. inval.* (Art. 37.1). [RPS 32]

Notocactus catarinensis Scheinvar, Fl. Ilust. Catarinense, Cact., 50, 1986. Typus: *Horst et Uebelmann* HU 40 (No actual specimen selected as holotype). [Published as a new combination for *Brasiliparodia catarinensis* Ritter (*nom. inval.*), but to be treated as a new taxon for nomenclatural reasons.] [Brazil: Santa Catarina] [RPS 37]

Notocactus claviceps (Ritter) Krainz, Kat. ZSS ed. 2, 89, 1967. Basionym: *Eriocactus claviceps.* [RPS 18]

Notocactus concinnus fa. **joadii** (Arechavaleta) Havlicek, Kakt. Vilag 18(4): 75, 1989. Basionym: *Echinocactus concinnus* var. *joadii.* [Dated 1988.] [RPS 41]

Notocactus concinnus var. **aceguensis** Gerloff, Internoto 10(4): 99-106, ills., SEM-ill., 1989. Typus: *Stockinger* 196 (B). [Brazil: Rio Grande do Sul] [RPS 40]

Notocactus concinnus var. **eremiticus** (Ritter) Gerloff, Internoto 9(4): 112, ills. (pp. 99-112), 1988. Basionym: *Notocactus eremiticus.* [RPS 39]

Notocactus concinnus var. **gibberulus** (Prestlé) Gerloff, Internoto 9(4): 112, ills. (pp. 99-112), 1988. Basionym: *Notocactus gibberulus.* [RPS 39]

Notocactus concinnus var. **joadii** (Arechavaleta) Y. Ito, Explan. Diagr. Austroechinocactinae, 243, 1957. Basionym: *Echinocactus joadii.* [The name given as basionym is probably an inexistant combination, which may invalidate the combination.] [RPS 8]

Notocactus concinnus var. **rubrigemmatus** (Abraham) Gerloff, Internoto 9(4): 112, ills. (pp. 99-112), 1988. Basionym: *Notocactus rubrigemmatus.* [RPS 39]

Notocactus corynodes (Otto *ex* Pfeiffer) Krainz, Kakt. and. Sukk. 17: 195, 1966. Basionym: *Echinocactus corynodes.* [RPS 17]

Notocactus courantii (Lemaire) Theunissen, Succulenta 60(6): 141, 1981. Basionym: *Echinocactus courantii.* [RPS 32]

Notocactus crassigibbus Ritter, Succulenta 49(7): 108, 1970. [RPS 21]

Notocactus cristatoides Ritter, Kakt. Südamer. 1: 190, 1979. [RPS 30]

Notocactus curvispinus Ritter, Kakt. Südamer. 1: 189, 1979. [RPS 30]

Notocactus darilhoensis Prestlé, Minimus 21(4): 71-73, ill., 1990. Based on *Prestlé* 354. *Nom. inval.* (Art. 36.1). [?] [RPS 41]

Notocactus discophalium Prestlé, Internoto 6(2): 60-62, ills., 1985. *Nom. inval.* (Art. 36.1). [RPS 36]

Notocactus eremiticus Ritter, Kakt. Südamer. 1: 180, 1979. [RPS 30]

Notocactus erinaceus (Haworth) Krainz, Kakt. and. Sukk. 17: 195, 1966. Basionym: *Cactus erinaceus.* [RPS 17]

Notocactus erinaceus subvar. **courantii** (Lemaire) Schäfer, Kakt./Sukk. 14: 47, 1980. Basionym: *Echinocactus courantii.* [RPS 31]

Notocactus erinaceus subvar. **depressus** (Spegazzini) Schäfer, Kakt./Sukk. 14: 47, 1980. Basionym: *Echinocactus acuatus* var. *depressus.* [RPS 31]

Notocactus erinaceus var. **spinosior** (Labouret) Rowley, Rep. Pl. Succ. 24: 8, 1975. Basionym: *Echinocactus tephracanthus* var. *spinosior.* [RPS 24]

Notocactus erinaceus var. **tephracanthus** (Link & Otto) Krainz, Die Kakt. 53, CVIc, 1973. Basionym: *Echinocactus tephracanthus.* [RPS 24]

Notocactus ernestii hort. *ex* Dupal, Minimus 21(4): 73-76, ills., 1990. *Nom. inval.* (Art. 36.1, 37.1). [Published as *nomen nudum* in synonymy.] [RPS -]

Notocactus erubescens (Osten *pro hybr.*) Marchesi, Bol. Soc. Argentina Bot. 14(3): 248, 1972. Basionym: *Echinocactus* × *erubescens.* [RPS 32]

Notocactus erythracanthus Schlosser & Brederoo *ex* Schlosser, Kakt. and. Sukk. 36(9): 186-189, ills., 1985. Typus: *Schlosser* 165 (MVM). [Uruguay] [RPS 36]

Notocactus erythranthus Theunissen, Kakt. / Sukk. 12(4): 85-86, 1977. *Nom. inval.* (Art. 36.1, 37.1). [Described as provisional name.] [RPS 28]

Notocactus eugeniae van Vliet, Succulenta 55: 24-27, 1976. [RPS 27]

Notocactus euvelenovskyi Fleischer & Schütz, Friciana 50: 24-25, ills. (p. 11, 24), 1976. *Nom. inval.* (Art. 36.1, 37.1). [Dated 1975, published 1976.] [RPS 31]

Notocactus ferrugineus H. Schlosser, Internoto 1982(2): 23-26, ills., 1982. Typus: *Schlosser* 211 (MVM). [RPS 33]

Notocactus floricomus var. **flavispinus** Backeberg, Cact. Succ. J. (US) 23(3): 85, 1951. [RPS 3]

Notocactus fricii (Arechavaleta) Krainz, Kakt. and. Sukk. 17: 195, 1966. Basionym: *Echinocactus fricii.* [RPS 17]

Notocactus fuscus Ritter, Kakt. Südamer. 1: 178, ill., 1979. [RPS 30]

Notocactus fuscus var. **longispinus** Ritter, Kakt. Südamer. 1: 179, 1979. [RPS 30]

Notocactus gerloffii Havlicek, Kakt. Vilag 18(4): 77, 1989. [*Nom. nov. pro Notocactus fuscus* var. *longispinus* Ritter 1979. Dated 1988.] [RPS 41]

Notocactus gibberulus Prestlé, Succulenta 65(6/7): 142-145, ills., SEM-ills., 1986. Typus: *Prestlé* 313 (U). [RPS 37]

Notocactus gladiatus (Link & Otto) Prestlé & Sida, Internoto 6(4): 121-127, ill., 1985. *Nom. inval.* (Art. 33.2, 34.1), based on *Echinocactus gladiatus.* [RPS 36]

Notocactus glaucinus Ritter, Kakt. Südamer. 1: 179, 1979. [RPS 30]

Notocactus glaucinus var. **depressus** Ritter, Kakt. Südamer. 1: 169, 1979. [RPS 30]

Notocactus glaucinus var. **gracilis** Ritter, Kakt. Südamer. 1: 168, 1979. [RPS 30]

Notocactus globularis Ritter, Kakt. Südamer. 1: 167, 1979. [RPS 30]

Notocactus graessneri fa. **microdasys** P. J. Braun, Kakt. and. Sukk. 37(3): 57, ill., 1986. Typus: *Horst et Uebelmann* HU 501 (ZSS). [RPS 37]

Notocactus graessneri var. **albiseta** (Cullmann) Krainz, Die Kakteen C VIa, 1960. Basionym: *Brasilicactus graessneri* var. *albisetus.* [RPS 11]

Notocactus graessneri var. *stellatus* hort. *ex* Havlicek, Kakt. Vilag 18(4): 78, 1989. *Nom. inval.* (Art. 36.1, 37.1). [Used as provisional name. Dated 1988.] [RPS 41]

Notocactus grandiensis Bergner, Internoto 10(2): 43-47, ills., map, 1989. *Nom. inval.* (Art. 36.1, 37.1). [RPS 40]

Notocactus grossei fa. **aureispinus** (Ritter) Theunissen *ex* Havlicek, Kakt. Vilag 18(4): 78, 1989. Basionym: *Eriocactus grossei* var. *aureispinus.* [The combination is ascribed to Theunissen. Dated 1988.] [RPS 41]

Notocactus grossei var. **aureispinus** (Ritter) Theunissen, Succulenta 60(6): 142, 1981. Basionym: *Eriocactus grossei* var. *aureispinus.* [RPS 32]

Notocactus gutierrezii Abraham, Kakt. and. Sukk. 39(7): 150-154, ills., 1988. Typus: *Abraham* 253 (KOELN). [Brazil: Rio Grande do Sul] [RPS 39]

Notocactus harmonianus Ritter, Kakt. Südamer. 1: 176, 1979. [RPS 30]

Notocactus herteri fa. **pseudoherteri** (Buining) Herm, Internoto 10(2): 39, ills. (pp. 35-38), 1989. Basionym: *Notocactus pseudoherteri.* [RPS 40]

Notocactus horstii Ritter, Succulenta 45: 3-4, 1966. [RPS 17]

Notocactus horstii var. **purpureiflorus** Ritter, Kakt. Südamer. 1: 184, 1979. [RPS 30]

Notocactus ibicuiensis Prestlé, Internoto 6(4): 99-103, ills., SEM-ills., 1985. Typus: *Stockinger* 116 (U). [Brazil] [RPS 36]

Notocactus incomptus Gerloff, Internoto 11(2/3): 37-41, ills., SEM-ill., 1990. Typus: *Horst et Uebelmann* HU 96 (1967) (ZSS). [Brazil: Rio Grande do Sul] [RPS 41]

Notocactus infernensis hort. *ex* Gerloff, Internoto 11(4): 155-157, ill., 1990. *Nom. inval.* (Art. 36.1, 37.1). [Published as provisional name.] [RPS 41]

Notocactus intricatus (Link & Otto) Herter, Cactus (Paris) No. 42: 120, 1954. Basionym: *Echinocactus intricatus.* [RPS 5]

Notocactus kovaricii (Fric) Krainz, Kakt. and. Sukk. 17: 195, 1966. Basionym: *Malacocarpus kovaricii.* [RPS 17]

Notocactus laetevirens Ritter, Kakt. Südamer. 1: 167, 1979. [RPS 30]

Notocactus langsdorfii (Lehmann) Krainz, Kakt. and. Sukk. 17: 195, 1966. Basionym: *Cactus langsdorfii.* [RPS 17]

Notocactus leninghausii fa. **apelii** (Heinrich) Krainz, Die Kakt. CVIc, 1968. Basionym: *Eriocephala leninghausii* fa. *apelii.* [RPS 19]

Notocactus leninghausii fa. **minor** (Ritter) Theunissen *ex* Havlicek, Kakt. Vilag 18(4): 78, 1989. Basionym: *Eriocactus leninghausii* var. *minor.* [The combination is ascribed to Theunissen.] [RPS 41]

Notocactus leninghausii var. **minor** (Ritter) Theunissen, Succulenta 60(6): 142, 1981. Basionym: *Eriocactus leninghausii* var. *minor.* [RPS 32]

Notocactus leprosorum (Ritter) Havlicek, Kaktusy 17(1): 8, 1981. Basionym: *Wigginsia leprosorum.* [Combination repeated by Theunissen in Succulenta 60(6): 141, 1981.] [RPS 32]

Notocactus leucocarpus (Arechavaleta) Schäfer, Kakt./Sukk. 14: 57, 1980. Basionym: *Echinocactus leucocarpus.* [RPS 31]

Notocactus linkii (Lehmann) Herter, Cactus (Paris) No. 42: 120, 1954. Basionym: *Cactus linkii.* [Repeated by Y. Ito in Explan. Diagr. Austroechinocactinae, 249, 1957 (cf. RPS 8).] [RPS 5]

Notocactus linkii fa. *multiflorus* (Fric *ex* Buining) Ritter, Kakt. Südamer. 1: 172, 1979. *Nom. inval.*, based on *Notocactus ottonis* var. *multiflorus, nom. inval.* (Art. 36.1). [Erroneously included as valid in RPS 30.] [RPS 30]

Notocactus linkii fa. *websterianus* hort. Debaecke, Internoto 9(2): 58-60, ills., 1988. *Nom. inval.* (Art. 36.1, 37.1). [Published as *nom. nud.*] [RPS 39]

Notocactus linkii var. **albispinus** Abraham, Succulenta 69(5): 117-120, ills., 1990. Typus: *Abraham* 281 (KOELN). [= *N. ottonis* var. *albispinus* Backeb., nom. inval. (Art. 37). First used by Gerloff & Schäfer in Internoto 10(1): 11-15, ills., 1989 (cf. RPS 40).] [Brazil: Rio Grande do Sul] [RPS 41]

Notocactus linkii var. **buenekeri** Ritter, Kakt. Südamer. 1: 172, 1979. [RPS 30]

Notocactus linkii var. **flavispinus** Buining & Brederoo, Kakt. and. Sukk. 29(2): 25-27, ills., 1978. Typus: *Horst et Uebelmann* HU 86 (U, ZSS). [Brazil: Rio Grande do Sul] [RPS 29]

Notocactus linkii var. **guaibensis** Ritter, Kakt. Südamer. 1: 172, 1979. [RPS 30]

Notocactus longispinus (Ritter) Havlicek, Kaktusy 17(1): 8, 1981. Basionym: *Wigginsia longispina.* [Combination repeated by Theunissen in Succulenta 60(6): 142, 1981.] [RPS 32]

Notocactus macambarensis Prestlé, Internoto 7(3): 67-72, ills., SEM-ills., 1986. Typus: *Prestlé* 425 (U). [RPS 37]

Notocactus macracanthus (Arechavaleta) Schäfer, Kakt. / Sukk. 14: 59, 1980. Basionym: *Echinocactus sellowii* var. *macracanthus.* [Sphalm. 'macrocanthus'.] [RPS 31]

Notocactus macrogonus (Arechavaleta) Schäfer, Kakt./Sukk. 14: 60, 1980. Basionym: *Echinocactus sellowii* var. *macrogonus.* [RPS 31]

Notocactus magnificus (Ritter) Krainz, Kakt. and. Sukk. 17: 195, 1966. Basionym: *Eriocactus magnificus.* [Combination repeated by N. P. Taylor in Cact. Succ. J. Gr. Brit. 42(1): 5, 1980, being of the opinion that the combination by Krainz was invalid under to Art. 33.2.] [RPS 17]

Notocactus mammulosus var. *albispinus* Oster, Internoto 11(2/3): 94, ill., 1990. *Nom. inval.* (Art. 36.1, 37.1). [Published as provisional name.] [RPS 41]

Notocactus mammulosus var. *arbolitoensis* Oster, Internoto 11(2/3): 94, ill., 1990. *Nom. inval.* (Art. 36.1, 37.1). [Published as provisional name.] [RPS 41]

Notocactus mammulosus var. *artinensis* Oster, Internoto 11(2/3): 94, ill., 1990. *Nom. inval.* (Art. 36.1, 37.1). [Published as provisional name.] [RPS 41]

Notocactus mammulosus var. **brasiliensis** Havlicek, Kaktusy 16(1): 5-7, 1980. Typus: *Anonymus* s.n. *in Havlicek* HAV VII/31 (Herb. Fac. Med. Univ. Carol., Plzen). [Incorrectly given as invalid in RPS 31.] [Uruguay: Salto] [RPS 31]

Notocactus mammulosus var. *gracilior* Oster, Internoto 11(2/3): 94, ill., 1990. *Nom. inval.* (Art. 36.1, 37.1). [Published as provisional name.] [RPS 41]

Notocactus mammulosus var. *gracilispinus* hort. *ex* Bergner, Internoto 11(4): 146, 1990. *Nom. inval.* (Art. 36.1, 37.1). [Published as nude name.] [RPS 41]

Notocactus mammulosus var. **hircinus** (Spegazzini) Y. Ito, Explan. Diagr. Austroechinocactinae, 245, 1957. Basionym: *Echinocactus mammulosus* var. *hircinus.* [RPS 8]

Notocactus mammulosus var. **minor** (Monville) Y. Ito, Explan. Diagr. Austroechinocactinae, 245, 1957. Basionym: *Echinocactus mammulosus* var. *minor.* [RPS 8]

Notocactus mammulosus var. *multiflorus* Oster, Internoto 11(2/3): 94, ill., 1990. *Nom. inval.* (Art. 36.1, 37.1). [Published as provisional name.] [RPS 41]

Notocactus mammulosus var. **pampeanus** (Spegazzini) Y. Ito, Explan. Diagr. Austroechinocactinae, 245, 1957. Basionym: *Echinocactus mammulosus* var. *pampeanus.* [Probably first published by Castellanos & Lelong ?] [RPS 8]

Notocactus mammulosus var. *paucicostatus* Oster, Internoto 11(2/3): 94, ill., 1990. *Nom. inval.* (Art. 36.1, 37.1). [Published as provisional name.] [RPS 41]

Notocactus mammulosus var. *rubrispinus* Oster, Internoto 11(2/3): 94, ill., 1990. *Nom. inval.* (Art. 36.1, 37.1). [Published as provisional name.] [RPS 41]

Notocactus mammulosus var. **spinosior** (Haage jr.) Y. Ito, Explan. Diagr. Austroechinocactinae, 245, 1957. Basionym: *Echinocactus mammulosus* var. *spinosior.* [RPS 8]

Notocactus mammulosus var. **submammulosus** (Spegazzini) Y. Ito, Explan. Diagr. Austroechinocactinae, 245, 1957. Basionym: *Echinocactus mammulosus* var. *submammulosus.* [RPS 8]

Notocactus megalanthus Schlosser & Brederoo, Kakt. and. Sukk. 32(7): 154-157, 1981. [RPS 32]

Notocactus megapotamicus var. **alacriportanus** Ritter, Kakt. Südamer. 1: 175, 1979. [RPS 30]

Notocactus megapotamicus var. **crucicentrus** Ritter, Kakt. Südamer. 1: 176, 1979. [RPS 30]

Notocactus megapotamicus var. **horstii** Ritter, Kakt. Südamer. 1: 175, 1979. [RPS 30]

Notocactus megapotamicus var. **vulgatus** Ritter, Kakt. Südamer. 1: 175, 1979. [RPS 30]

Notocactus memorialis Prestlé, Internoto 9(1): 3-7, ills., SEM-ills., 1988. Typus: *Schlosser* s.n. (U). [Uruguay] [RPS 39]

Notocactus meonacanthus Prestlé, Internoto 7(2): 35-39, ills., SEM-ills., 1986. Typus: *Prestlé* 318 (U). [RPS 37]

Notocactus miniatispinus (Ritter) Havlicek, Kakt. Vilag 18(4): 73, 1989. Basionym: *Notocactus securituberculatus* var. *miniatispinus.* [Dated 1988.] [RPS 41]

Notocactus minimus var. **tenuicylindricus** (Ritter) Havlicek, Kakt. Vilag 18(4): 75, 1989. Basionym: *Notocactus tenuicylindricus.* [Dated 1988.] [RPS 41]

Notocactus mueller-melchersii fa. **gracilispinus** (Krainz) Krainz, Die Kakt. C VIa, 1962. Basionym: *Notocactus mueller-melchersii* var. *gracilispinus.* [RPS 13]

Notocactus mueller-moelleri Fric *ex* Fleischer & Schütz, Friciana 51: 13, 29-30, 36, 45-46, 1976. [Name republished in Kakt. and. Sukk. 28(10): 234-235, 1977, and reported to be invalid in RPS 38 (as *nom. nud.*). Validity of 1976-publication not assessed. First mentioned by Fric in Kat. Kreuzinger, 1935, then by Buining (and attributed to Fric) as nude name in Succulenta 1957: 102-103 (cf. RPS 8), and again by Backeberg in Die Cact. 3: 1647, 1959 (cf. RPS 10).] [RPS 31]

Notocactus multicostatus Buining & Brederoo, in Krainz, Die Kakt. Lief. 55-56: C VIc, 1974. [Dated Dec. 1973, but according to most sources published only early in 1974.] [RPS 25]

Notocactus muricatus var. *flavifuscus* Schäfer, Succulenta 61(1): 6-7, 1982. *Nom. inval.*

(Art. 36.1). [RPS 33]

Notocactus neoarechavaletae Havlicek, Kakt. Vilag 18(4): 79, 1989. [Sphalm. 'neoarechavaletai'. Dated 1988. First used as nude name by Elsner in Succulenta 56: 143-145, 1977 (cf. RPS 28). Based on *Echinocactus acuatus* var. *arechavaletae* Spegazzini, *non Notocactus arechavaletae*.] [RPS 41]

Notocactus neoarechavaletae var. *kovaricii* (Fric *ex* A. Berger) Havlicek, Kaktusy 16(6): 136, 1980. *Nom. inval.* (Art. 43.1), based on *Malacocarpus kovaricii*. [Dated 1988, published 1989. Repeated in Kakt. Vilag 18(4): 79, 1989 (sphalm. 'kovarikii', *nom. inval.* Art. 33.2).] [RPS 31]

Notocactus neobuenekeri Ritter, Kakt. Südamer. 1: 181, 1979. [RPS 30]

Notocactus neohorstii Theunissen, Succulenta 60(6): 142, 1981. [*Nom. nov. pro Wigginsia horstii* Ritter 1979, *non Notocactus horstii* Ritter 1966.] [RPS 32]

Notocactus neohorstii var. **juvenaliformis** (Ritter) Theunissen, Succulenta 60(6): 142, 1981. Basionym: *Wigginsia horstii* var. *juvenaliformis*. [Combination repeated by Havlieck in Kaktusy / Sukulenty 2(4): 85, 1981.] [RPS 32]

Notocactus nigrispinus (Schumann) Buining, Succulenta 49: 179, 1970. *Nom. inval.* (Art. 33.2). [RPS 21]

Notocactus nivosus Prestlé, Internoto 5(2): 59-61, ill., 1984. *Nom. inval.* (Art. 36.1). [Published as provisional name.] [RPS 35]

Notocactus nivosus var. *albiflorus* Prestlé, Minimus 21(1): 9-13, 1990. Based on *Prestlé* 120. *Nom. inval.* (Art. 36, 37, 43). [Published as provisional name.] [Uruguay] [RPS 41]

Notocactus notabilis Dupal, Minimus 19(1): 11-13, ill., 1988. Based on *Schlosser* 218. *Nom. inval.* (Art. 36.1). [Published as provisional name.] [Uruguay] [RPS 39]

Notocactus olimarensis Prestlé, Internoto 6(3): 94-95, ill., 1985. Based on *Prestlé* 93. *Nom. inval.* (Art. 36.1). [Uruguay: Dept. Treinta-y-Tres] [RPS 36]

Notocactus orthacanthus (Link & Otto) van Vliet, Succulenta 49: 185, 1970. Basionym: *Echinocactus orthacanthus*. [RPS 21]

Notocactus ottoianus Y. Ito, Explan. Diagr. Austroechinocactinae, 247, 1957. *Nom. illeg.* (Art. -). [*Nom. nov. pro Cactus ottonis* Lehmann. The name seems to be invalid, as no reason for a new name is apparent.] [RPS 8]

Notocactus ottoianus var. *pallidior* (Monville) Y. Ito, Explan. Diagr. Austroechinocactinae, 249, 1957. *Nom. inval.* (Art. 43.1), based on *Echinocactus ottonis* var. *pallidior*. [Sphalm. 'paltidior'. The name may be invalid, if the combination is not clearly indicated (there is a question mark in this respect in RPS 8).] [RPS 8]

Notocactus ottoianus var. **schuldtii** (Kreuzinger) Y. Ito, Explan. Diagr. Austroechinocactinae, 248, 1957. Basionym: *Echinocactus ottonis* var. *schuldtii*. [RPS 8]

Notocactus ottoianus var. **spinosior** (Monville) Y. Ito, Explan. Diagr. Austroechinocactinae, 249, 1957. Basionym: *Echinocactus ottonis* var. *spinosior*. [RPS 8]

Notocactus ottoianus var. **tenuispinus** (Pfeiffer) Y. Ito, Explan. Diagr. Austroechinocactinae, 248, 1957. Basionym: *Echinocactus ottonis* var. *tenuispinus*. [RPS 8]

Notocactus ottonis fa. **elegans** (Backeberg & Voll) Havlicek, Kakt. Vilag 18(4): 73, 1989. Basionym: *Notocactus ottonis* var. *elegans*. [Dated 1988.] [RPS 41]

Notocactus ottonis var. **acutangularis** Ritter, Kakt. Südamer. 1: 163, 1979. [RPS 30]

Notocactus ottonis var. *albispinus* Backeberg, Die Cact. 6: 3756, 1962. *Nom. inval.* (Art. 37.1). [Erroneously included as valid in RPS 13.] [RPS 13]

Notocactus ottonis var. **brasiliensis** (Haage jr.) Y. Ito, Explan. Diagr. Austroechinocactinae, 249, 1957. Basionym: *Echinocactus ottonis* var. *brasiliensis*. [RPS 8]

Notocactus ottonis var. **elegans** Backeberg & Voll, Arq. Jard. Bot. Rio de Janeiro 9: 172, ill. (p. 171), 1950. [Volume for 1949, published 1950.] [RPS -]

Notocactus ottonis var. **globularis** (Ritter) Bergner, Internoto 9(3): 67-73, ills., SEM-ills., 1988. Basionym: *Notocactus globularis*. [RPS 39]

Notocactus ottonis var. **janousekianus** Papousek, Kaktusy 11: 123-126, 1975. [RPS 26]

Notocactus ottonis var. *linkii* (Lehmann) Y. Ito, Explan. Diagr. Austroechinocactinae, 249, 1957. *Nom. inval.* (Art. 33.2), based on *Cactus linkii*. [The combination is given as *Notocactus* var. *linkii*; the basionym citation refers to an inexistent name.] [RPS 8]

Notocactus ottonis var. **minor** (Förster) Y. Ito, Explan. Diagr. Austroechinocactinae, 249, 1957. Basionym: *Echinocactus ottonis* var. *minor*. [RPS 8]

Notocactus ottonis var. *multiflorus* Fric *ex* Buining, Succulenta 1957(9): 105-108, 1957. *Nom. inval.* (Art. 36.1). [Sphalm. 'Echinocactus ottonis'. Erroneously included as valid in RPS 8.] [RPS 8]

Notocactus ottonis var. *nigrispinus* Lück, Internoto 8(4): 124-125, ill., 1987. Based on *Rausch* 374. *Nom. inval.* (Art. 36.1). [Published as provisional name; the plant is also cultivated as *Notocactus nigrispinus*.] [Uruguay] [RPS 38]

Notocactus ottonis var. **paraguayensis** (Heese) Y. Ito, Explan. Diagr. Austroechinocactinae, 249, 1957. Basionym: *Echinocactus ottonis* var. *paraguayensis*. [RPS 8]

Notocactus ottonis var. **pfeifferi** (Monville) Y. Ito, Explan. Diagr. Austroechinocactinae, 249, 1957. Basionym: *Echinocactus ottonis* var. *pfeifferi*. [RPS 8]

Notocactus ottonis var. *stenogonus* Backeberg, Die Cact. 6: 3758, 1962. *Nom. inval.* (Art. 37.1). [Erroneously included as valid in RPS 13.] [RPS 13]

Notocactus ottonis var. **tenuispinus** (Link & Otto) Y. Ito, Cacti, 72, 1952. Basionym: *Echinocactus tenuispinus*. [RPS 4]

Notocactus ottonis var. **tortuosus** (Link & Otto)

Backeberg, Die Cact. 3: 1639-1640, 1959. Basionym: *Echinocactus tortuosus*. [Erroneously ascribed to Prestlé in RPS 35: 5, corrected in RPS 36: 4. Backeberg ascribed the combination to A. Berger. RPS 8 ascribes the combination to Y. Ito, Explan. Diagr. Austroechinocactinae, 249, 1957, but his combination is based on *Echinocactus ottonis* var. *tortuosus*, itself already a combination.] [RPS 35]

Notocactus ottonis var. **uruguayensis** (Arechavaleta) Y. Ito, Explan. Diagr. Austroechinocactinae, 249, 1957. Basionym: *Echinocactus ottonis* var. *uruguayensis*. [RPS 8]

Notocactus ottonis var. **vencluianus** Schütz, Kaktusy 65: 126, 1965. [RPS 16]

Notocactus ottonis var. **villa-velhensis** Backeberg & Voll, Arq. Jard. Bot. Rio de Janeiro 9: 174, ill. (p. 171), 1950. [Volume for 1949, published 1950.] [RPS -]

Notocactus oxycostatus Buining & Brederoo, in Krainz, Die Kakt. Lief. 53: C VIc, 4 pp. of text, ills., 1973. Typus: *Horst et Uebelmann* HU 299 (1968) (U, ZSS). [The publication date of this part (dated April 1973) is unclear and 1972 is reported repeatedly.] [Brazil: Rio Grande do Sul] [RPS 24]

Notocactus pampeanus var. **charruanus** (Arechavaleta) Y. Ito, Explan. Diagr. Austroechinocactinae, 246, 1957. Basionym: *Echinocactus pampeanus* var. *charruanus*. [RPS 8]

Notocactus pampeanus var. **rubellianus** (Arechavaleta) Y. Ito, Explan. Diagr. Austroechinocactinae, 246, 1957. Basionym: *Echinocactus pampeanus* var. *rubellianus*. [RPS 8]

Notocactus pampeanus var. **subplanus** (Arechavaleta) Y. Ito, Explan. Diagr. Austroechinocactinae, 246, 1957. Basionym: *Echinocactus pampeanus* var. *subplanus*. [RPS 8]

Notocactus pauciareolatus (Arechavaleta) Krainz, Kakt. and. Sukk. 17: 195, 1966. Basionym: *Echinocactus pauciareolatus*. [RPS 17]

Notocactus paulus Schlosser & Brederoo, Kakt. and. Sukk. 31(4): 114-116, 1980. [RPS 31]

Notocactus permutatus Ritter, Kakt. Südamer. 1: 188, 1979. [RPS 30]

Notocactus polyacanthus (Link & Otto) Theunissen, Succulenta 60(6): 141, 1981. Basionym: *Echinocactus polyacanthus*. [RPS 32]

Notocactus prolifer (Ritter) Theunissen, Succulenta 60(6): 142, 1981. Basionym: *Wigginsia prolifera*. [Sphalm. 'prolifera'.] [RPS 32]

Notocactus pseudoherteri Buining, Nation. Cact. Succ. J. 26(1): 2-3, 1971. Typus: *Horst et Uebelmann* HU 342 (). [Brazil] [RPS 22]

Notocactus pseudopulvinatus Gerloff, Internoto 8(4): 126-127, ill., 1987. Based on *van Vliet* 36. *Nom. inval.* (Art. 36.1). [Published as provisional name.] [Uruguay] [RPS 38]

Notocactus pulvinatus van Vliet, Succulenta 49: 50-52, 1970. [RPS 21]

Notocactus purpureus Ritter, Succulenta 49: 109, 1970. [RPS 21]

Notocactus rauschii van Vliet, Succulenta 48(1): 3-5, ills., 1969. Typus: *van Vliet* 34 (U, ZSS [status ?]). [Uruguay: Rivera] [RPS 20]

Notocactus rechensis Buining, Kakt. and. Sukk. 19: 23-24, 1968. [RPS 19]

Notocactus riosusannensis Prestlé, Internoto 6(1): 30-31, ill., 1985. *Nom. inval.* (Art. 36.1). [RPS 36]

Notocactus ritterianus Lisal & Kolarik, Internoto 7(1): 3-5, ills., 1986. Typus: *Horst et Uebelmann* HU 338 (herb. Arb. Mus. Siles., Opava - Novy Dur, CSSR). [Individuals from the type collection have also been designated as type of *Notocactus arnostianus*.] [Brazil] [RPS 37]

Notocactus roseiflorus Schlosser & Brederoo, Kakt. and. Sukk. 29(12): 273-277, 1978. [RPS 29]

Notocactus roseoluteus van Vliet, Succulenta 52(101): 108-113, 1973. [RPS 24]

Notocactus rubricostatus (Fric *ex* Fleischer & Schütz) Schäfer, Kakt./Sukk. 14: 86, 1980. Basionym: *Wigginsia rubricostata*. [RPS 31]

Notocactus rubrigemmatus Abraham, Kakt. and. Sukk. 39(2): 38-41, ills., 1988. Typus: *Schlosser* 156 (KOELN). [Uruguay: Rivera] [RPS 39]

Notocactus rubropedatus Ritter, Kakt. Südamer. 1: 189, 1979. [RPS 30]

Notocactus rudibuenekeri Abraham, Succulenta 67(6): 133-138, ills., 1988. Typus: *Abraham* 355 (KOELN [Succulentarium]). [Brazil: Rio Grande do Sul] [RPS 39]

Notocactus rutilans fa. **storianus** Pazout, Kaktusy 66: 125, 1966. [RPS 17]

Notocactus schaeferianus (Abraham & Theunissen) Havlicek, Internoto 9(2): 39, 1988. Basionym: *Wigginsia schaeferiana*. [RPS 39]

Notocactus schlosseri van Vliet, Succulenta 53: 10-13, 1974. [RPS 25]

Notocactus schumannianus ssp. *nigrispinus* (Schumann) T. Engel, Internoto 11(2/3): 54, 1990. *Nom. inval.* (Art. 33.2), based on *Echinocactus nigrispinus*. [RPS 41]

Notocactus schumannianus var. **ampliocostatus** (Ritter) Theunissen *ex* Havlicek, Kakt. Vilag 18(4): 77, 1989. Basionym: *Eriocactus ampliocostatus*. [The combination is ascribed to Theunissen. Dated 1988.] [RPS 41]

Notocactus schumannianus var. **nigrispinus** (Schumann) Y. Ito, Cacti, 73, 1952. Basionym: *Echinocactus nigrispinus*. [Basionym attributed to 'Haage. fil.'.] [RPS 4]

Notocactus scopa fa. *albilanatus* Jahn, Stachelpost 4(22): 75, 1969. *Nom. inval.* (Art. 37.1). [Sphalm. 'albilanata'. Erroneously included as valid in RPS 20.] [RPS 20]

Notocactus scopa fa. **candidus** (Pfeiffer) Krainz, Die Kakt. C VIa, 1961. Basionym: *Echinocactus scopa* var. *candidus*. [RPS 12]

Notocactus scopa fa. **daenikerianus** (Krainz) Krainz, Die Kakt. C VIa, 1961. Basionym:

Notocactus scopa var. *daenikerianus*. [RPS 12]

Notocactus scopa fa. *erythrinus* Havlicek, Internoto 9(4): 125-127, ills., 1988. *Nom. inval*. (Art. 36.1, 37.1). [Published as *nom. nud*.] [RPS 39]

Notocactus scopa fa. **glauserianus** (Krainz) Krainz, Die Kakt. C VIa, 1961. Basionym: *Notocactus scopa* var. *glauserianus*. [The entry in RPS 12 for this taxon is corrupted.] [RPS 12]

Notocactus scopa fa. *marchesii* Gerloff, Internoto 9(2): 40-42, ill., 1988. Based on *van Vliet* 73d. *Nom. inval*. (Art. 34.1 / 36.1). [Published as *nom. nud*. Previously mentioned by W.-R. Abraham at varietal status in an Internoto-'Rundbrief'.] [Uruguay] [RPS 39]

Notocactus scopa var. **albicans** (Arechavaleta) Hofacker, Internoto 11(2/3): 69, 1990. Basionym: *Echinocactus scopa* var. *albicans*. [RPS 41]

Notocactus scopa var. **candidus** (Pfeiffer) Y. Ito, Cacti, 73, 1952. Basionym: *Echinocactus scopa* var. *candidus*. [RPS 4]

Notocactus scopa var. **cobrensis** Gerloff, Internoto 11(1): 3-9, ills., 1990. Typus: *Horst et Uebelmann* HU 80 (1982) (ZSS). [Brazil: Rio Grande do Sul] [RPS -]

Notocactus scopa var. **machadoensis** Abraham, Succulenta 67(4): 81-84, ill., 1988. Typus: *Abraham* 361 (KOELN [Succulentarium]). [Detailed origin deposited with holotype.] [First published as a provisional name by Menges in Internoto 8(4): 109-111, ills., 1987.] [Brazil: Rio Grande do Sul] [RPS 39]

Notocactus scopa var. **marchesii** Abraham, Kakt. and. Sukk. 40(7): 174-176, ills., 1989. Typus: *Abraham* 34 (KOELN [Succulentarium]). [Uruguay: Dept. Treinta y Tres] [RPS 40]

Notocactus scopa var. *murielii* Hofacker, Internoto 10(2): 40, 1989. *Nom. inval*. (Art. 36.1, 37.1). [Published in synonymy.] [RPS -]

Notocactus scopa var. **ramosus** (Osten) Backeberg, Die Cact. 3: 1637, 1959. Basionym: *Notocactus scopa* fa. *ramosus*. [Brazil: Rio Grande do Sul] [RPS 10]

Notocactus scopa var. **xicoi** Abraham, Succulenta 67(5): 111-114, ills., 1988. Typus: *Abraham* 332 (KOELN [Succulentarium]). [Detailed habitat information deposited together with holotype.] [Brazil: Rio Grande do Sul] [RPS 39]

Notocactus scopa var. **xiphacanthus** Abraham, Succulenta 66(12): 256-260, ills., 1987. Typus: *Abraham* 156 (KOELN). [Uruguay: Dept. Lavalleja] [RPS 38]

Notocactus securituberculatus Ritter, Kakt. Südamer. 1: 169, 1979. [RPS 30]

Notocactus securituberculatus var. **miniatispinus** Ritter, Kakt. Südamer. 1: 169, 1979. [RPS 30]

Notocactus sellowii (Link & Otto) Theunissen, Succulenta 60(6): 141, 1981. Basionym: *Echinocactus sellowii*. [RPS 32]

Notocactus sellowii var. **macracanthus** (Arechavaleta) Theunissen, Succulenta 60(6): 141, 1981. Basionym: *Echinocactus sellowii* var. *macracanthus*. [RPS 32]

Notocactus sessiliflorus (Hooker) Krainz, Kakt. and. Sukk. 17: 195, 1966. Basionym: *Echinocactus sessiliflorus*. [RPS 17]

Notocactus sessiliflorus var. **martinii** (Labouret) Krainz, Kakt. and. Sukk. 17: 195, 1966. Basionym: *Malacocarpus martinii*. [RPS 17]

Notocactus sessiliflorus var. *stegmannii* (Backeberg) Havlicek, Kaktusy 15(5): 102-104, 1966. *Nom. inval*., based on *Malacocarpus stegmannii, nom. inval*. (Art. 37.1). [RPS 31 gives the publication date as 1979 and erroneously reports the name as valid.] [RPS 30]

Notocactus soldtianus van Vliet, Succulenta 54: 72-75, 1975. [RPS 26]

Notocactus spinibarbis Ritter, Kakt. Südamer. 1: 186, 1979. [RPS 30]

Notocactus stegmannii (Backeberg) Krainz, Kakt. and. Sukk. 17: 195, 1966. *Nom. inval*., based on *Malacocarpus stegmannii, nom. inval*. (Art. 37.1). [Sphalm. 'stegmanni'. Erroneously included as valid in RPS 17.] [RPS 17]

Notocactus stockingeri Prestlé, Succulenta 64(11): 225-229, ills., SEM-ills., 1985. Typus: *Stockinger* 141 (U). [Brazil: Rio Grande do Sul] [RPS 36]

Notocactus submammulosus var. **pampeanus** (Spegazzini) Backeberg, Die Cact. 3: 1649, 1959. Basionym: *Echinocactus pampeanus*. [RPS 10]

Notocactus succineus Ritter, Succulenta 49: 109, 1970. [sphalm. 'sucineus'.] [RPS 21]

Notocactus tabularis fa. *boomelianus* Prauser, Internoto 11(2/3): 86, 1990. *Nom. inval*. (Art. 36.1, 37.1). [Published as provisional name.] [RPS 41]

Notocactus tabularis fa. *setispinus* Prauser, Internoto 11(2/3): 86, 1990. *Nom. inval*. (Art. 36.1, 37.1). [Published as provisional name.] [RPS 41]

Notocactus tabularis fa. *splendens* Prauser, Internoto 11(2/3): 86, 1990. *Nom. inval*. (Art. 36.1, 37.1). [Published as provisional name.] [RPS 41]

Notocactus tabularis var. *brederooianus* (Prestlé) Prauser, Internoto 11(2/3): 87, 1990. *Nom. inval*. (Art. 34.1), based on *Notocactus brederooianus*. [RPS 41]

Notocactus tabularis var. *splendens* Prauser, Internoto 11(2/3): 84, 1990. *Nom. inval*. (Art. 36.1, 37.1). [Published as provisional name.] [RPS 41]

Notocactus tabularis var. *tolomban* Prauser, Internoto 11(2/3): 84, 1990. *Nom. inval*. (Art. 36.1, 37.1). [Published as provisional name.] [RPS 41]

Notocactus tenuicylindricus Ritter, Succulenta 49: 108, 1970. [RPS 21]

Notocactus tenuispinus (Link & Otto) Herter, Cactus (Paris) No. 44: 177, 1955. Basionym: *Echinocactus tenuispinus*. [RPS 6]

Notocactus tephracanthus (Link & Otto) Krainz, Kakt. and. Sukk. 17: 195, 1966. Basionym: *Echinocactus tephracanthus*.

[RPS 17]

Notocactus turbinatus (Arechavaleta) Krainz, Kakt. and. Sukk. 17: 195, 1966. Basionym: *Echinocactus sellowii* var. *turbinatus*. [RPS 17]

Notocactus uebelmannianus Buining, Kakt. and. Sukk. 19: 175-176, 1968. [RPS 19]

Notocactus uebelmannianus fa. **flaviflorus** Buining, Kakt. and. Sukk. 19: 176, 1968. [RPS 19]

Notocactus uebelmannianus var. *nilsonii* Königs, Internoto 11(4): 119-124, ills., 1990. *Nom. inval.* (Art. 37.1). [While the herbarium is given where the type is deposited, no actual collection is cited as being type or holotype.] [RPS 41]

Notocactus vanvlietii Rausch, Kakt. and. Sukk. 21(5): 89-90, ill., 1970. Typus: *Rausch* 376 (ZSS). [Type cited for W but placed in ZSS.] [Uruguay] [RPS 21]

Notocactus vanvlietii var. **gracilis** Rausch, Kakt. and. Sukk. 21(5): 90, 1970. Typus: *Rausch* 375 (W, ZSS). [Uruguay] [RPS 21]

Notocactus veenianus van Vliet, Succulenta 53: 171-173, 1974. [RPS 25]

Notocactus velenovskyi Fric *ex* Y. Ito, Explan. Diagr. Austroechinocactinae, 247, 294, 1957. [Sphalm. 'velennovskyi'. = *Notocactus velenovskyi* Fric 1921 (*nom. nud.*).] [RPS 8]

Notocactus villa-velhensis (Backeberg & Voll) Slaba, Kaktusy 20(1): 7, ill., 1984. Basionym: *Notocactus ottonis* var. *villa-velhensis*. [RPS 35]

Notocactus villa-velhensis var. *carambeiensis* (Buining & Brederoo) Gerloff, Internoto 11(4): 131, 1990. *Nom. inval.* (Art. 57.1), based on *Notocactus carambeiensis*. [RPS 41]

Notocactus vorwerkianus (Werdermann) Krainz, Kakt. and. Sukk. 17: 195, 1966. Basionym: *Echinocactus vorwerkianus*. [RPS 17]

Notocactus warasii (Ritter) Hewitt & Donald, Ashingtonia 1(11): Spec. Cat. Cact., 1975. Basionym: *Eriocactus warasii*. [RPS 30]

Notocactus winkleri van Vliet, Succulenta 54: 136-139, 1975. [RPS 26]

× **Nyctocephalocereus** Mottram, Contr. New Class. Cact. Fam., 62, 1990. [= *Nyctocereus* × *Cephalocereus*.] [RPS 41]

Nyctocerei Buxbaum (*pro linea*), Madroño 14(6): 181, 1958. Typus: *Nyctocereus*. [Repeated by Mottram, Contr. New Class. Cact. Fam., 6, 10, 1990, as *Echinocactinae* linea *Nyctocerei*, "comb. nov."] [RPS 41]

Nyctocereinae Buxbaum, Madroño 14(6): 181, 1958. Typus: *Nyctocereus*. [RPS 9]

Nyctocereus castellanosii Scheinvar, Anales Inst. Biol. UNAM, Ser. Bot. 47-53: 165-176, ills., SEM ills., 1984. Typus: *Scheinvar et Orozco* 2190 (MEXU). [Vol. for 1976-1982, published 1984.] [Mexico] [RPS 38]

Nyctocereus chontalensis Alexander, Cact. Succ. J. (US) 22(5): 132-133., 1950. [RPS 1]

Nyctocereus serpentinus var. **pietatis** Bravo, Cact. Suc. Mex. 17: 116-117, 1972. [RPS 23]

Oehmea F. Buxbaum, Sukkulentenkunde 4: 17, 1951. Typus: *Neomammillaria nelsonii*. [RPS 2]

Oehmea beneckei (Ehrenberg) Buxbaum, Sukkulentenkunde 4: 18, 1951. Basionym: *Mammillaria beneckei*. [RPS 2]

Oehmea nelsonii (Britton & Rose) Buxbaum, Sukkulentenkunde 4: 17, 1951. Basionym: *Mammillaria nelsonii*. [RPS 2]

Opuntia sect. *Aurantiacae* (Britton & Rose) Castellanos, Revista Fac. Ci. Agron. 6(2): 22, 1957. *Nom. inval.* (Art. 33.2). [No basionym indicated.] [RPS 9]

Opuntia sect. **Austrocylindropuntia** (Backeberg) Moran, Gentes Herbar. 8(4): 326, 1953. Basionym: *Austrocylindropuntia*. [RPS 4]

Opuntia sect. *Brasiliopuntia* (Schumann) Castellanos & Lelong, Revista Fac. Ci. Agron. 6(2): 22, 1957. *Nom. inval.* (Art. 33.2). [No basionym indicated.] [RPS 9]

Opuntia sect. *Corynopuntia* (Knuth) L. Benson, Cacti of Arizona, ed. 3, 63, 1969. *Nom. illeg.* (Art. 63.1). [Included as superfluous for sect. *Clavatae* Engelmann in RPS 20.] [RPS 20]

Opuntia sect. **Cylindropuntia** (Engelmann) Moran, Baileya 19: 166, 1975. Basionym: *Opuntia* subgen. *Cylindropuntia*. [RPS 26]

Opuntia sect. *Euplatyopuntia* Castellanos, Revista Fac. Ci. Agron. 6(2): 22, 1957. *Nom. inval.* (Art. 21.1). [RPS 9]

Opuntia sect. **Tephrocactus** (Lemaire) Moran, Baileya 19: 166, 1975. Basionym: *Tephrocactus*. [RPS 26]

Opuntia ser. **Nopaleae** Rowley, Rep. Pl. Succ. 22: 10, 1971. Typus: *Cactus cochenillifer* L. (= *Nopalea* Salm-Dyck). [RPS 22]

Opuntia ser. **Pseudogrusonia** Bravo, Cact. Suc. Mex. 17: 119, 1972. Typus: *Grusonia santamaria* Baxter. [RPS 23]

Opuntia subgen. *Chaffeyopuntia* Bravo, Cact. Suc. Mex. 7(1): 7, 1962. *Nom. inval.* (Art. 36.1, 37.1). [Given as "comb. nov".] [RPS -]

Opuntia subgen. **Consolea** L. Benson, Cact. Succ. J. (US) 46: 80, 1974. Typus: *O. brasiliensis* (Willdenow) Haworth. The status of the taxon (combination or new taxon) has not been assessed. [RPS 25]

Opuntia subgen. **Corynopuntia** (Knuth) Bravo, Cact. Suc. Mex. 17: 118, 1972. Basionym: *Corynopuntia*. [Typus: *Opuntia clavata*.] [RPS 23]

Opuntia subgen. **Grusonia** (Reiche) Bravo, Cact. Suc. Mex. 17: 118, 1972. Basionym: *Grusonia*. [Typus: *Cereus bradtianus*.] [RPS 23]

Opuntia subgen. **Marenopuntia** (Backeberg) Bravo, Cact. Suc. Mex. 17: 118, 1972. Basionym: *Marenopuntia*. [Sphalm. 'Marenopuntiae'.] [RPS 23]

Opuntia subsect. *Elatiores* (Britton & Rose) Castellanos, Revista Fac. Ci. Agron. 6(2): 24, 1957. *Nom. inval.* (Art. 33.2). [RPS 9]

Opuntia subsect. **Eriospermae** Castellanos, Revista Fac. Ci. Agron. 6(2): 22, 1957. Typus: not designated. Erroneously treated as invalid on account of lack of type in RPS 9. [RPS 9]

Opuntia subsect. *Streptacantha* (Britton & Rose) Castellanos, Revista Fac. Ci. Agron. 6(2): 24, 1957. *Nom. inval.* (Art. 33.2). [RPS 9]

Opuntia subsect. *Subinermis* (Engelmann)

Castellanos, Revista Fac. Ci. Agron. 6(2): 24, 1957. *Nom. inval.* (Art. 33.2). [RPS 9]

Opuntia subsect. *Sulphureae* (Britton & Rose) Castellanos, Revista Fac. Ci. Agron. 6(2): 22, 1957. *Nom. inval.* (Art. 33.2). [RPS 9]

Opuntia subsect. *Vulgares* (Engelmann) Castellanos, Revista Fac. Ci. Agron. 6(2): 24, 1957. *Nom. inval.* (Art. 33.2). [RPS 9]

Opuntia acanthocarpa var. **coloradensis** L. Benson, Cacti of Arizona, ed. 3, 20, 34, 37, 1969. [RPS 20]

Opuntia acanthocarpa var. **ganderi** (Wolf) L. Benson, Cact. Succ. J. (US) 41: 33, 1969. Basionym: *Opuntia acanthocarpa* ssp. *ganderi*. [RPS 20]

Opuntia acanthocarpa var. **major** (Engelmann & Bigelow) L. Benson, Cacti of Arizona, ed. 3, 20, 35, 1969. Basionym: *Opuntia echinocarpa* var. *major*. [RPS 20]

Opuntia acanthocarpa var. **ramosa** Peebles, in Benson, Cacti Arizona, 26, 1950. [RPS -]

Opuntia aciculata var. **orbiculata** Backeberg, Descr. Cact. Nov. [1:] 10, 1957. [Dated 1956, published 1957.] [RPS 7]

Opuntia aggeria Ralston & Hilsenbeck, Madroño 36: 226, 1989. Typus: *Anthony* 856 (MICH). [USA: Texas] [RPS 40]

Opuntia albisaetacens var. *robustior* Backeberg, Cactus (Paris) No. 75: 32, 1962. *Nom. inval.* (Art. 9.5). [Erroneously included as valid in RPS 13. Simultaneously published in Die Cact. 6: 3617.] [RPS 13]

Opuntia alexanderi var. **bruchii** (Spegazzini) Rowley, Nation. Cact. Succ. J. 13(1): 5, 1958. Basionym: *Opuntia bruchii*. [RPS 9]

Opuntia alexanderi var. **subsphaerica** (Backeberg) Rowley, Nation. Cact. Succ. J. 13(1): 5, 1958. Basionym: *Tephrocactus subsphaericus*. [RPS 9]

Opuntia alko-tuna Cardenas, Lilloa 23: 23, 1950. [RPS 4]

Opuntia arbuscula var. **congesta** (Griffiths) Rowley, Nation. Cact. Succ. J. 13(2): 25, 1958. Basionym: *Opuntia congesta*. [RPS 9]

Opuntia arcei Cardenas, Cact. Succ. J. (US) 28(4): 113-114, 1956. [RPS 7]

Opuntia armata Backeberg, Cactus (Paris) 38: 250, 1953. [RPS 4]

Opuntia armata var. **panellana** Backeberg, Descr. Cact. Nov. [1:] 9, 1957. [Dated 1956, published 1957.] [RPS 7]

Opuntia arrastradillo Backeberg, Cactus (Paris) 36: 181, 1953. [RPS 4]

Opuntia articulata (Pfeiffer) D. Hunt, Bradleya 5: 93, 1987. Basionym: *Cereus articulatus*. [syn. *O. articulata* Otto in Allg. Gartenzeit. 1: 367, 1833, *nom. nud.*] [RPS 38]

Opuntia articulata var. *oligacantha* (Spegazzini) Krainz, Kat. ZSS ed. 2, 91, 1967. *Nom. inval.* (Art. 43.1), based on *Opuntia diademata*. [RPS 18]

Opuntia articulata var. *syringacantha* (Pfeiffer) Krainz, Kat. ZSS ed. 2, 91, 1967. *Nom. inval.* (Art. 43.1), based on *Cereus syringacanthus*. [RPS 18]

Opuntia asplundii (Backeberg) Rowley, Nation. Cact. Succ. J. 13(1): 5, 1958. Basionym: *Tephrocactus asplundii*. [RPS 9]

Opuntia atroviridis fa. **longicylindrica** (Rauh & Backeberg) Krainz, Die Kakt. B, 1970. Basionym: *Tephrocactus atroviridis* var. *longicylindricus*. [RPS 21]

Opuntia atroviridis fa. **parviflora** (Rauh & Backeberg) Krainz, Die Kakt. B, 1970. Basionym: *Tephrocactus atroviridis* var. *parviflorus*. [RPS 21]

Opuntia atroviridis fa. **paucispina** (Rauh & Backeberg) Krainz, Die Kakt. B, 1970. Basionym: *Tephrocactus atroviridis* var. *paucispinus*. [RPS 21]

Opuntia aureispina (Brack & Heil) Pinkava & Parfitt, Sida 13: 128, 1988. Basionym: *Opuntia macrocentra* var. *aureispina*. [RPS 39]

Opuntia backebergii Rowley, Nation. Cact. Succ. J. 13(1): 5, 1958. Basionym: *Tephrocactus minor*. [*Nom. nov.*, *non Opuntia minor* C. Mueller 1858.] [RPS 9]

Opuntia bakeri J. E. Madsen, Fl. Ecuador 35: 47-48, 1989. Typus: *Madsen* 50430 (AAU, QCA). [Ecuador: Prov. Pichincha] [RPS 40]

Opuntia barkleyana (Daston) Rowley, Nation. Cact. Succ. J. 13(1): 5, 1958. Basionym: *Micropuntia barkleyana*. [RPS 9]

Opuntia basilaris ssp. **whitneyana** (Baxter) Munz, Aliso 4(1): 94, 1957. Basionym: *Opuntia whitneyana*. [RPS 8]

Opuntia basilaris var. **heilii** Welsh & Neese, Great Basin Naturalist 43(4): 700, 1983. Typus: *Neese* 5938 (BRY). [USA] [RPS 37]

Opuntia basilaris var. **woodburyi** W. Earle, Saguaroland Bull. 34(2): 15, 1980. [RPS 31]

Opuntia bensonii Sanchez-Mejorada, Cact. Suc. Mex. 17: 47-50, 1972. [RPS 23]

Opuntia berteri (Colla) A. E. Hoffmann, Cact. Fl. Chil., 244, ill. (p. 245), 1989. Basionym: *Cactus berteri*. [RPS 40]

Opuntia bispinosa Backeberg, Cactus (Paris) No. 75: 32, 1962. *Nom. inval.* (Art. 9.5). [Erroneously included as valid in RPS 13. Simultaneously published in Die Cact. 6: 3607.] [RPS 13]

Opuntia blancii (Backeberg) Rowley, Nation. Cact. Succ. J. 13(1): 5, 1958. Basionym: *Tephrocactus blancii*. [RPS 9]

Opuntia brachyclada ssp. **humistrata** Wiggins & Wolf, in: Le Roy, Ill. Flora Pacific States 3: ?, 1951. [RPS 2]

Opuntia brachyrhopalica (Daston) Rowley, Nation. Cact. Succ. J. 13(1): 5, 1958. Basionym: *Micropuntia brachyrhopalica*. [RPS 9]

Opuntia bradleyi Rowley, Nation. Cact. Succ. J. 13(1): 4, 1958. [*Nom. nov. pro Austrocylindropuntia intermedia* Rauh & Backeberg, *non* Opuntia intermedia Salm-Dyck *nec* Gray.] [RPS 9]

Opuntia chichensis (Cardenas) Rowley, Nation. Cact. Succ. J. 13(1): 5, 1958. Basionym: *Tephrocactus chichensis*. [RPS 9]

Opuntia chichensis var. **colchana** (Cardenas) Rowley, Nation. Cact. Succ. J. 13(1): 5, 1958. Basionym: *Tephrocactus chichensis* var. *colchanus*. [RPS 9]

Opuntia chilensis (Backeberg) Rowley, Nation. Cact. Succ. J. 13(1): 5, 1958. Basionym: *Tephrocactus chilensis*. [RPS 9]

Opuntia chisosensis (Alexander) Ferguson,

Cact. Succ. J. (US) 58(3): 124, ills., 1986. Basionym: *Opuntia lindheimeri* var. *chisosensis*. [RPS 37]

Opuntia chlorotica var. **gosseliniana** (F. A. C. Weber) Ferguson, Cact. Succ. J. (US) 60(4): 159, ill. (p. 156), 1988. Basionym: *Opuntia gosseliniana*. [RPS 39]

Opuntia chuquisacana Cardenas, Lilloa 23: 20, 1950. [RPS 4]

Opuntia clavarioides var. **ruiz-lealii** (Castellanos) Rowley, Nation. Cact. Succ. J. 13(2): 25, 1958. Basionym: *Opuntia ruiz-lealii*. [RPS 9]

Opuntia cochabambensis Cardenas, Rev. Agric. Cochabamba 9: 20-22, 1953. [RPS 5]

Opuntia colubrina Castellanos, Lilloa 27: 81-84, 1953. [Dated 1953, published 1955 according to RPS 6.] [RPS 6]

Opuntia compressa var. *allairei* (Griffiths) Weniger, Cacti of the Southwest, 210-211, 1970. *Nom. inval.* (Art. 33.2), based on *Opuntia allairei*. [Not dated.] [RPS -]

Opuntia compressa var. **ammophila** (Small) L. Benson, Cact. Succ. J. (US) 41: 124, 1969. Basionym: *Opuntia ammophila*. [RPS 20]

Opuntia compressa var. **austrina** (Small) L. Benson, Cact. Succ. J. (US) 41: 125, 1969. Basionym: *Opuntia austrina*. [RPS 20]

Opuntia compressa var. *fuscoatra* (Engelmann) Weniger, Cacti of the Southwest, 207-208, 1970. *Nom. inval.* (Art. 33.2), based on *Opuntia fuscoatra*. [Not dated. Sphalm. 'fusco-atra'.] [RPS -]

Opuntia compressa var. *grandiflora* (Engelmann) Weniger, Cacti of the Southwest, 209-210, 1970. *Nom. inval.* (Art. 33.2), based on *Opuntia grandiflora*. [Not dated.] [RPS -]

Opuntia compressa var. *humifusa* (Rafinesque) Weniger, Cacti of the Southwest, 202-203, 1970. *Nom. inval.* (Art. 33.2), based on *Opuntia humifusa*. [Not dated.] [RPS -]

Opuntia compressa var. *stenochila* (Engelmann) Weniger, Cacti of the Southwest, 211, 1970. *Nom. inval.* (Art. 33.2), based on *Opuntia stenochila*. [Not dated.] [RPS -]

Opuntia conoidea (Ritter *ex* Backeberg) Rowley, Nation. Cact. Succ. J. 13(2): 25, 1958. *Nom. inval.*, based on *Tephrocactus conoideus*, *nom. inval.* (Art. 37.1). [Repeated by A. E. Hoffmann in Cact. Fl. Chil., 246, ill. (p. 247), 1989, and reported as place of first publication in RPS 40, but treated as invalid (Art. 37.1).] [RPS 9]

Opuntia corotilla var. **aurantiaciflora** (Rauh & Backeberg) Rowley, Nation. Cact. Succ. J. 13(1): 5, 1958. Basionym: *Tephrocactus corotilla* var. *aurantiaciflorus*. [RPS 9]

Opuntia crassicylindrica (Rauh & Backeberg) Rowley, Nation. Cact. Succ. J. 13(1): 5, 1958. Basionym: *Tephrocactus crassicylindricus*. [RPS 9]

Opuntia crispicrinita (Rauh & Backeberg) Rowley, Nation. Cact. Succ. J. 13(1): 5, 1958. Basionym: *Tephrocactus crispicrinitus*. [RPS 9]

Opuntia crispicrinita subvar. **flavicoma** (Rauh & Backeberg) Rowley, Nation. Cact. Succ. J. 13(1): 5, 1958. Basionym: *Tephrocactus crispicrinitus* subvar. *flavicomus*. [RPS 9]

Opuntia crispicrinita var. **cylindracea** (Rauh & Backeberg) Rowley, Nation. Cact. Succ. J. 13(1): 5, 1958. Basionym: *Tephrocactus crispicrinitus* var. *cylindraceus*. [RPS 9]

Opuntia crispicrinita var. **tortispina** (Rauh & Backeberg) Rowley, Nation. Cact. Succ. J. 13(1): 5, 1958. Basionym: *Tephrocactus crispicrinitus* var. *tortispinus*. [RPS 9]

Opuntia curassavica var. **colombiana** Backeberg, Descr. Cact. Nov. [1.] 10, 1957. [Dated 1956, published 1957.] [RPS 7]

Opuntia cylindrarticulata (Rauh & Backeberg) Rowley, Nation. Cact. Succ. J. 13(1): 5, 1958. Basionym: *Tephrocactus cylindrarticulatus*. [RPS 9]

Opuntia cylindrolanata (Rauh & Backeberg) Rowley, Nation. Cact. Succ. J. 13(1): 5, 1958. Basionym: *Tephrocactus cylindrolanatus*. [RPS 9]

Opuntia densiaculeata (Backeberg) Rowley, Nation. Cact. Succ. J. 13(1): 4, 1958. Basionym: *Cylindropuntia densiaculeata*. [RPS 9]

Opuntia dillenii var. **reitzii** Scheinvar, Feddes Repert. 95(6-5): 277, ills., 1984. Typus: *Scheinvar* 2982 (MEXU, HBR). [Mexico] [RPS 35]

Opuntia dillenii var. **tehuantepecana** Bravo, Cact. Suc. Mex. 9: 55-56, 1964. [RPS 15]

Opuntia dimorpha var. **pseudorauppianus** (Backeberg) Rowley, Nation. Cact. Succ. J. 13(2): 25, 1958. Basionym: *Tephrocactus pseudorauppianus*. [RPS 9]

Opuntia duvalioides (Backeberg) Rowley, Nation. Cact. Succ. J. 13(1): 5, 1958. Basionym: *Tephrocactus duvalioides*. [RPS 9]

Opuntia duvalioides var. **albispina** (Backeberg) Rowley, Nation. Cact. Succ. J. 13(1): 5, 1958. Basionym: *Tephrocactus duvalioides* var. *albispinus*. [RPS 9]

Opuntia echinacea (Ritter) A. E. Hoffmann, Cact. Fl. Chil., 247, ill. (p. 248), 1989. Basionym: *Tephrocactus echinaceus*. [RPS 40]

Opuntia echinocarpa var. **wolfei** L. Benson, Cact. Succ. J. (US) 41: 33, 1969. [RPS 20]

Opuntia echios var. **barringtonensis** Dawson, Cact. Succ. J. (US) 34: 104, 1962. [RPS 13]

Opuntia echios var. **gigantea** (J. Howell) D. Porter, Madroño 25(1): 58, 1978. Basionym: *Opuntia echios* ssp. *gigantea*. [RPS 33]

Opuntia echios var. **inermis** Dawson, Cact. Succ. J. (US) 34: 103, 1962. [RPS 13]

Opuntia echios var. **prolifera** Dawson, Cact. Succ. J. (US) 34: 104, 1962. [RPS 13]

Opuntia echios var. **zacana** (Howell) Anderson & Walkington, Madroño 20: 256, 1970. Basionym: *Opuntia zacana*. [RPS 21]

Opuntia engelmannii var. *aciculata* (Griffiths) Weniger, Cacti of the Southwest, 178-179, 1970. *Nom. inval.* (Art. 33.2), based on *Opuntia aciculata*. [Not dated.] [RPS -]

Opuntia engelmannii var. *alta* (Griffiths) Weniger, Cacti of the Southwest, 175, 1970. *Nom. inval.* (Art. 33.2), based on *Opuntia alta*. [Not dated.] [RPS -]

Opuntia engelmannii var. *cacanapa* (Griffiths &

Hare) Weniger, Cacti of the Southwest, 177, 1970. *Nom. inval.* (Art. 33.2), based on *Opuntia cacanapa.* [Not dated.] [RPS -]

Opuntia engelmannii var. **flavispina** (L. Benson) Parfitt & Pinkava, Madroño 35: 348, 1989. Basionym: *Opuntia phaeacantha* var. *flavispina.* [RPS 40]

Opuntia engelmannii var. *flexispina* (Griffiths) Weniger, Cacti of the Southwest, 178, 1970. *Nom. inval.* (Art. 33.2), based on *Opuntia flexispina.* [Not dated.] [RPS -]

Opuntia engelmannii var. **flexospina** (Griffiths) Parfitt & Pinkava, Madroño 35: 348, 1989. Basionym: *Opuntia flexospina.* [RPS 40]

Opuntia engelmannii var. **lindheimeri** (Engelmann) Parfitt & Pinkava, Madroño 35: 346, 1989. Basionym: *Opuntia lindheimeri.* [RPS 40]

Opuntia engelmannii var. **linguiformis** (Griffiths) Parfitt & Pinkava, Madroño 35: 347, 1989. Basionym: *Opuntia linguiformis.* [First invalidly (Art. 33.2) proposed by Weniger in Cacti of the Southwest, 181, 1970 [not dated].] [RPS 40]

Opuntia engelmannii var. *subarmata* (Griffiths) Weniger, Cacti of the Southwest, 180-181, 1970. *Nom. inval.* (Art. 33.2), based on *Opuntia subarmata.* [RPS -]

Opuntia engelmannii var. *texana* (Griffiths) Weniger, Cacti of the Southwest, 174, 1970. *Nom. inval.* (Art. 33.2), based on *Opuntia texana.* [Not dated.] [RPS -]

Opuntia engelmannii var. **wootonii** (Griffiths) Anthony, Amer. Midlands Natural. 40: 249-250, 1956. Basionym: *Opuntia wootonii.* [RPS 7]

Opuntia erinacea var. **aurea** (Baxter) Welsh, Great Basin Naturalist 46(2): 255, 1986. Basionym: *Opuntia aurea.* [RPS 37]

Opuntia erinacea var. **columbiana** (Griffiths) L. Benson, Cact. Succ. J. (US) 41: 124, 1969. Basionym: *Opuntia columbiana.* [RPS 20]

Opuntia erinacea var. **juniperina** (Britton & Rose) W. T. Marshall, Arizona's Cactuses, ed. 2, 36, 1953. Basionym: *Opuntia juniperina.* [Concurrently published in Saguaroland Bull. 7(9): 36.] [RPS 4]

Opuntia erinacea var. **utahensis** (Engelmann) L. Benson, Cacti of Arizona, ed. 3, 20, 77, 80, 1969. Basionym: *Opuntia sphaerocarpa* var. *utahensis.* [RPS 20]

Opuntia estevesii P. J. Braun, Cact. Succ. J. (US) 62(4): 165-169, ills., 1990. Typus: *Esteves Pereira* 191 (UFG 12.377, B, ZSS). [Brazil: Bahia] [RPS 41]

Opuntia excelsa Sanchez-Mejorada, Cact. Suc. Mex. 17: 67-73, 1972. [RPS 23]

Opuntia ferocior (Backeberg) Rowley, Nation. Cact. Succ. J. 13(1): 5, 1958. Basionym: *Tephrocactus ferocior.* [RPS 9]

Opuntia ficus-indica fa. **reticulata** Backeberg, Descr. Cact. Nov. [1:] 10, 1957. [Dated 1956, published 1957.] [RPS 7]

Opuntia flexuosa (Backeberg) Rowley, Nation. Cact. Succ. J. 13(1): 5, 1958. Basionym: *Tephrocactus flexuosus.* [RPS 9]

Opuntia floccosa subvar. **aurescens** (Rauh & Backeberg) Rowley, Nation. Cact. Succ. J. 13(1): 5, 1958. Basionym: *Tephrocactus floccosus* subvar. *aurescens.* [RPS 9]

Opuntia floccosa var. **canispina** (Rauh & Backeberg) Rowley, Nation. Cact. Succ. J. 13(1): 5, 1958. Basionym: *Tephrocactus floccosus* var. *canispinus.* [RPS 9]

Opuntia floccosa var. **cardenasii** Rowley, Rep. Pl. Succ. 22: 11, 1973. Basionym: *Tephrocactus malyanus.* [RPS 22]

Opuntia floccosa var. **crassior** (Backeberg) Rowley, Nation. Cact. Succ. J. 13(1): 5, 1958. Basionym: *Tephrocactus floccosus* var. *crassior.* [RPS 9]

Opuntia floccosa var. **ovoides** (Rauh & Backeberg) Rowley, Nation. Cact. Succ. J. 13(1): 5, 1958. Basionym: *Tephrocactus floccosus* var. *ovoides.* [RPS 9]

Opuntia fragilis var. **denudata** Wiegand & Backeberg, in Backeberg, Descr. Cact. Nov. [1:] 10, 1957. [Dated 1956, published 1957.] [RPS 7]

Opuntia fragilis var. **parviconspicua** Backeberg, Descr. Cact. Nov. [1:] 10, 1957. [Dated 1956, published 1957.] [RPS 7]

Opuntia fulvicoma (Rauh & Backeberg) Rowley, Nation. Cact. Succ. J. 13(1): 5, 1958. Basionym: *Tephrocactus fulvicomus.* [RPS 9]

Opuntia fulvicoma var. **bicolor** (Rauh & Backeberg) Rowley, Nation. Cact. Succ. J. 13(1): 5, 1958. Basionym: *Tephrocactus fulvicomus* var. *bicolor.* [RPS 9]

Opuntia galapageia subvar. **barringtonensis** (Dawson) Backeberg, Kakt.-Lex., 317, 1966. Basionym: *Opuntia echios* var. *barringtonensis.* [RPS 17]

Opuntia galapageia subvar. **inermis** (Dawson) Backeberg, Kakt.-Lex., 317, 1966. Basionym: *Opuntia echios* var. *inermis.* [RPS 17]

Opuntia galapageia subvar. *orientalis* (Howell) Backeberg, Die Cact. 1: 562, 1958. *Nom. inval.* (Art. 33.2), based on *Opuntia megasperma* ssp. *orientalis.* [The name given as basionym was itself a combination, and is here corrected.] [RPS 9]

Opuntia galapageia subvar. **prolifera** (Dawson) Backeberg, Kakt.-Lex., 317, 1966. Basionym: *Opuntia echios* var. *prolifera.* [RPS 17]

Opuntia galapageia var. *brossetii* Backeberg, Kakt.-Lex., 317, 1966. *Nom. inval.* (Art. 37.1). [Erroneously included as valid in RPS 17.] [RPS 17]

Opuntia galapageia var. *echios* (Howell) Backeberg, Die Cact. 1: 561, 1958. *Nom. inval.*, based on *Opuntia echios* var. *typica, nom. inval.* (Art. 24.3). [RPS 9]

Opuntia galapageia var. *gigantea* (Howell) Backeberg, Die Cact. 1: 562, 1958. *Nom. inval.* (Art. 33.2), based on *Opuntia echios* ssp. *gigantea.* [The name given as basionym was itself a combination, and is here corrected.] [RPS 9]

Opuntia galapageia var. **helleri** (Schumann) Backeberg, Die Cact. 1: 562, 1958. Basionym: *Opuntia helleri.* [RPS 9]

Opuntia galapageia var. **insularis** (Steward) Backeberg, Die Cact. 1: 561, 1958. Basionym: *Opuntia insularis.* [RPS 9]

Opuntia galapageia var. **macrocarpa** Dawson, Cact. Succ. J. (US) 37: 141, 1965. [RPS 16]
Opuntia galapageia var. **myriacantha** (F. A. C. Weber) Backeberg, Die Cact. 1: 561, 1958. Basionym: *Opuntia myriacantha*. [RPS 9]
Opuntia galapageia var. **profusa** Anderson & Walkington, Madroño 20: 256, 1970. [RPS 21]
Opuntia galapageia var. *saxicola* (Howell) Backeberg, Die Cact. 1: 562, 1958. *Nom. inval.* (Art. 33.2), based on *Opuntia saxicola*. [RPS 9]
Opuntia galapageia var. *zacana* (Howell) Backeberg, Die Cact. 1: 562, 1958. *Nom. inval.* (Art. 33.2), based on *Opuntia zacana*. [RPS 9]
Opuntia glomerata var. **calva** (Lemaire) Rowley, Nation. Cact. Succ. J. 13(1): 5, 1958. Basionym: *Opuntia calva*. [RPS 9]
Opuntia glomerata var. **gracilior** (Salm-Dyck) Rowley, Nation. Cact. Succ. J. 13(1): 5, 1958. Basionym: *Opuntia platyacantha* var. *gracilior*. [RPS 9]
Opuntia glomerata var. **inermis** (Spegazzini) Rowley, Nation. Cact. Succ. J. 13(1): 5, 1958. Basionym: *Opuntia diademata* var. *inermis*. [RPS 9]
Opuntia glomerata var. **oligacantha** (Spegazzini) Rowley, Nation. Cact. Succ. J. 13(1): 5, 1958. Basionym: *Opuntia diademata* var. *oligacantha*. [RPS 9]
Opuntia glomerata var. **polyacantha** (Spegazzini) Rowley, Nation. Cact. Succ. J. 13(1): 5, 1958. Basionym: *Opuntia diademata* var. *polyacantha*. [RPS 9]
Opuntia gosseliniana var. **santa-rita** (Griffiths & Hare) L. Benson, Cacti of Arizona, ed. 2, 65, 1950. Basionym: *Opuntia chlorotica* var. *santa-rita*. [RPS -]
Opuntia gracilicylindrica (Wiegand & Backeberg) Rowley, Nation. Cact. Succ. J. 13(1): 5, 1958. Basionym: *Micropuntia gracilicylindrica*. [RPS 9]
Opuntia hamiltonii (Gates) Rowley, Nation. Cact. Succ. J. 13(2): 25, 1958. *Nom. inval.*, based on *Grusonia hamiltonii, nom. inval.* (Art. 36.1, 37.1). [First used as nude name by Gates in Marshall & Bock, Cact., 68, 1941 (Art. 36.1).] [RPS 9]
Opuntia heacockiae G. K. Arp, Sida 10(3): 207-208, ills., 1984. Typus: *Arp* 4841 (SMU, CSU, POM). [Sphalm. 'heacockae'.] [USA] [RPS 35]
Opuntia heliabravoana Scheinvar, Anales Inst. Biol. Mex. 45: 75-86, 1975. [Dated 1974, published 1975 according to Index Kewensis.] [RPS 27]
Opuntia heliae Matuda, Cact. Suc. Mex. 1(2): 23-24, 1955. [RPS 6]
Opuntia hirschii (Backeberg) Rowley, Nation. Cact. Succ. J. 13(1): 5, 1958. Basionym: *Tephrocactus hirschii*. [RPS 9]
Opuntia horstii Heinrich, in Backeberg, Descr. Cact. Nov. 3: 10, 1963. *Nom. inval.* (Art. 9.5). [Erroneously included as valid in RPS 14.] [RPS 14]
Opuntia hossei (Krainz & Grässner) Rowley, Nation. Cact. Succ. J. 13(1): 5, 1958. Basionym: *Tephrocactus hossei*. [Sphalm. 'hosseii'.] [RPS 9]
Opuntia huajuapensis Bravo, Anales Inst. Biol. UNAM 25(1-2): 483-484, ill., 1954. Typus: *Anonymus* s.n. (MEXU). [Mexico: Oaxaca] [RPS 6]
Opuntia humahuacana (Backeberg) Rowley, Nation. Cact. Succ. J. 13(1): 4, 1958. Basionym: *Cylindropuntia humahuacana*. [RPS 9]
Opuntia humifusa var. **ammophila** (Small) L. Benson, Cact. Succ. J. (US) 48(2): 59, 1976. Basionym: *Opuntia ammophila*. [RPS 27]
Opuntia humifusa var. **austrina** (Small) Dress, Baileya 19: 164-165, 1975. Basionym: *Opuntia austrina*. [RPS 26]
Opuntia hypogaea fa. **rossiana** (Heinrich & Backeberg) Krainz, Kat. ZSS ed. 2, 94, 1967. Basionym: *Tephrocactus pentlandii* var. *rossianus*. [RPS 18]
Opuntia hystricina var. *bensonii* Backeberg, Die Cact. 1: 609, 1958. *Nom. inval.* (Art. 36.1, 37.1). [*Nom. nov. pro Opuntia erinacea* var. *typica sensu* Benson 1950 (*sine descr. lat.*). Erroneously included as valid in RPS 9.] [RPS 9]
Opuntia hystricina var. **nicholii** (L. Benson) Backeberg, Die Cact. 1: 610, 1958. Basionym: *Opuntia nicholii*. [RPS 9]
Opuntia hystricina var. **ursina** (F. A. C. Weber) Backeberg, Die Cact. 1: 610, 608, 1958. Basionym: *Opuntia ursina*. [RPS 9]
Opuntia ignescens var. **steiniana** (Backeberg) Rowley, Nation. Cact. Succ. J. 13(1): 5, 1958. Basionym: *Tephrocactus ignescens* var. *steinianus*. [Sphalm. 'steinianus'.] [RPS 9]
Opuntia imbricata var. *arborescens* (Engelmann) Weniger, Cacti of the Southwest, 229-230, 1970. *Nom. inval.* (Art. 33.2), based on *Opuntia arborescens*. [Not dated.] [RPS -]
Opuntia imbricata var. **argentea** Anthony, Amer. Midland Natural. 55: 236, 1956. [RPS 7]
Opuntia imbricata var. **cardenche** (Griffiths) Bravo, Cact. Suc. Mex. 17: 119, 1972. Basionym: *Opuntia cardenche*. [RPS 23]
Opuntia imbricata var. **lloydii** (Rose) Bravo, Cact. Suc. Mex. 17: 119, 1972. Basionym: *Opuntia lloydii*. [RPS 23]
Opuntia imbricata var. *vexans* (Griffiths) Weniger, Cacti of the Southwest, 231-232, 1970. *Nom. inval.* (Art. 33.2), based on *Opuntia vexans*. [Not dated.] [RPS -]
Opuntia imbricata var. *viridiflora* (Britton & Rose) Weniger, Cacti of the Southwest, 231, 1970. *Nom. inval.* (Art. 33.2), based on *Opuntia viridiflora*. [Not dated.] [RPS -]
Opuntia inamoena fa. **spinigera** (Ritter) P. J. Braun & E. Esteves Pereira, Cact. Succ. J. (US) 61(6): 272, 1989. Basionym: *Platyopuntia inamoena* fa. *spinigera*. [RPS 40]
Opuntia inamoena var. **flaviflora** Backeberg, Descr. Cact. Nov. [1:] 10, 1957. [Dated 1956, published 1957.] [RPS 7]
Opuntia ipatiana Cardenas, Cactus (Paris) No. 34: 127-128, 1952. [RPS 3]
Opuntia jaliscana Bravo, Cact. Suc. Mex. 17:

119, 1972. [RPS 23]

Opuntia johnsonii Johnson, Cact. Succ. J. (US) 29(1): 22, 1957. *Nom. inval.* (Art. 36.1). [RPS 8]

Opuntia kelvinensis V. Grant & K.A. Grant, Evolution 25: 155, 1971. [RPS 25]

Opuntia laetevirens Backeberg, Cactus (Paris) No. 75: 32, 1962. *Nom. inval.* (Art. 9.5). [Erroneously included as valid in RPS 13. Simultaneously published in Die Cact. 6: 3617.] [RPS 13]

Opuntia lagopus subvar. **brachycarpa** (Rauh & Backeberg) Rowley, Nation. Cact. Succ. J. 13(1): 5, 1958. Basionym: *Tephrocactus lagopus* subvar. *brachycarpus*. [RPS 9]

Opuntia lagopus var. **aurea** (Rauh & Backeberg) Rowley, Nation. Cact. Succ. J. 13(1): 5, 1958. Basionym: *Tephrocactus lagopus* var. *aureus*. [RPS 9]

Opuntia lagopus var. **aureo-penicillata** (Rauh & Backeberg) Rowley, Nation. Cact. Succ. J. 13(1): 6, 1958. Basionym: *Tephrocactus lagopus* var. *aureo-penicillatus*. [RPS 9]

Opuntia lagopus var. **leucolagopus** (Rauh & Backeberg) Rowley, Nation. Cact. Succ. J. 13(1): 6, 1958. Basionym: *Tephrocactus lagopus* var. *leucolagopus*. [RPS 9]

Opuntia lagopus var. **pachyclada** (Rauh & Backeberg) Rowley, Nation. Cact. Succ. J. 13(1): 6, 1958. Basionym: *Tephrocactus lagopus* var. *pachycladus*. [RPS 9]

Opuntia leptocaulis var. **brittonii** (Ortega) Bravo, Cact. Suc. Mex. 17: 119, 1972. Basionym: *Opuntia brittonii*. [RPS 23]

Opuntia lilae Trujillo & Ponce, Ernstia 58-60: 1-7, ill., 1990. Typus: *Trujillo et Ponce* 18643 (MY). [Venezuela: Sucre] [RPS 41]

Opuntia lindheimeri var. **aciculata** (Griffiths) Bravo, Cact. Suc. Mex. 19: 47, 1974. Basionym: *Opuntia aciculata*. [RPS 25]

Opuntia lindheimeri var. **chisosensis** Anthony, Amer. Midland Natural. 55: 252-253, 1956. [RPS 7]

Opuntia lindheimeri var. **cuija** (Griffiths & Hare) L. Benson, Cact. Succ. J. (US) 41: 125, 1969. Basionym: *Opuntia engelmannii* var. *cuija*. [RPS 20]

Opuntia lindheimeri var. **ellisiana** (Griffiths) Hammer, Kulturpflanze 24: 268, 1976. Basionym: *Opuntia ellisiana*. [RPS 28]

Opuntia lindheimeri var. **lehmannii** L. Benson, Cact. Succ. J. (US) 41: 125, 1969. [RPS 20]

Opuntia lindheimeri var. **linguiformis** (Griffiths) L. Benson, Cact. Succ. J. (US) 41: 125, 1969. Basionym: *Opuntia linguiformis*. [RPS 20]

Opuntia lindheimeri var. **lucens** (Griffiths) Scheinvar, Phytologia 49(4): 320, 1981. Basionym: *Opuntia lucens*. [RPS 32]

Opuntia lindheimeri var. **subarmata** (Griffiths) Elizondo & Wehbe, Cact. Suc. Mex. 32(1): 17, 1987. Basionym: *Opuntia subarmata*. [First published invalidly l.c. 31(4): 99.] [RPS 38]

Opuntia lindheimeri var. **tricolor** (Griffiths) L. Benson, Cact. Succ. J. (US) 41: 125, 1969. Basionym: *Opuntia tricolor*. [RPS 20]

Opuntia littoralis var. **austrocalifornica** L. Benson & Walkington, Ann. Missouri Bot. Gard. 52: 269, 1965. [RPS 16]

Opuntia littoralis var. **martiniana** (L. Benson) L. Benson, Ann. Missouri Bot. Gard. 52: 270, 1965. Basionym: *Opuntia macrocentra* var. *martiniana*. [RPS 16]

Opuntia littoralis var. **piercei** (Fosberg) L. Benson & Walkington, Ann. Missouri Bot. Gard. 52: 270, 1965. Basionym: *Opuntia phaeacantha* var. *piercei*. [RPS 16]

Opuntia littoralis var. **vaseyi** (Coulter) L. Benson & Walkington, Ann. Missouri Bot. Gard. 52: 268, 1965. Basionym: *Opuntia mesacantha* var. *vaseyi*. [RPS 16]

Opuntia longispina var. *agglomerata* Backeberg, Cactus (Paris) No. 75: 32, 1962. *Nom. inval.* (Art. 9.5). [Erroneously included as valid in RPS 13. Simultaneously published in Die Cact. 6: 3609.] [RPS 13]

Opuntia longispina var. **brevispina** Backeberg, Cactus (Paris) No. 38: 250, 1953. [RPS 4]

Opuntia longispina var. **corrugata** (Salm-Dyck) Backeberg, Cactus (Paris) No. 38: 250, 1953. Basionym: *Opuntia corrugata*. [RPS 4]

Opuntia longispina var. **flavidispina** Backeberg, Cactus (Paris) No. 38: 250, 1953. [RPS 4]

Opuntia longispina var. **intermedia** Backeberg, Cactus (Paris) No. 38: 250, 1953. [RPS 4]

Opuntia lubrica var. **aurea** (Baxter) Backeberg, Die Cact. 1: 585, 1958. Basionym: *Opuntia aurea*. [RPS 9]

Opuntia macbridei var. **orbicularis** Rauh & Backeberg, in Backeberg, Descr. Cact. Nov. [1:] 10, 1957. [Dated 1956, published 1957.] [RPS 7]

Opuntia macrocentra var. **aureispina** Brack & Heil, Cact. Succ. J. (US) 60(1): 17, ill. (p. 18), 1988. Typus: *Anonymus* s.n. (SJNM 3777). [The type collection is *K. Heil 2191* according to Pinkava & Parfitt *sub O. aureispina*.] [USA: Texas] [RPS 39]

Opuntia macrocentra var. **martiniana** L. Benson, Cacti of Arizona, ed. 2, 64, 1950. [RPS 32]

Opuntia macrocentra var. **minor** Anthony, Amer. Midland Natural. 55: 244-245, fig. 21, 1956. [RPS 7]

Opuntia macrorhiza var. **potosina** Hernandez-Valencia, Acta Ci. Potosina 10(2): 155-168, 1988. Typus: *Hernandez-Valencia* 385 (SLPM). [Mexico] [RPS 41]

Opuntia macrorhiza var. **pottsii** (Salm-Dyck) L. Benson, Cacti of Arizona, ed. 3, 20, 89, 1969. Basionym: *Opuntia pottsii*. [RPS 20]

Opuntia mandragora (Backeberg) Rowley, Nation. Cact. Succ. J. 13(1): 6, 1958. Basionym: *Tephrocactus mandragora*. [RPS 9]

Opuntia marnieriana Backeberg, Cactus (Paris) 36: 181, 1953. [Sphalm. 'marnierana'.] [RPS 4]

Opuntia martiniana (L. Benson) Parfitt, Syst. Bot. 5(4): 416, 1981. Basionym: *Opuntia macrocentra* var. *martiniana*. [RPS 32]

Opuntia matudae Scheinvar, Phytologia 49(4): 324-328, 1981. [RPS 32]

Opuntia megasperma var. **mesophytica** Lundheim, Madroño 20: 254, 1970. [RPS 21]

Opuntia megasperma var. **orientalis** (J. Howell) D. Porter, Madroño 25(1): 58, 1978. Basionym: *Opuntia megasperma* ssp. *orientalis*. [RPS 33]

Opuntia microdasys subvar. *albiflora* Backeberg, Cactus (Paris) No. 75: 32, 1962. *Nom. inval.* (Art. 9.5). [Erroneously included as valid in RPS 13. Simultaneously published in Die Cact. 6: 3623.] [RPS 13]

Opuntia microdasys var. **albispina** Fobe *ex* Backeberg, Descr. Cact. Nov. [1:] 10, 1957. [Dated 1956, published 1957.] [RPS 7]

Opuntia minuscula (Backeberg) Rowley, Nation. Cact. Succ. J. 13(1): 6, 1958. Basionym: *Tephrocactus minusculus*. [RPS 9]

Opuntia minuscula var. **silvestris** (Backeberg) Krainz, Kat. ZSS ed. 2, 95, 1967. Basionym: *Tephrocactus silvestris*. [RPS 18]

Opuntia minuta (Backeberg) Castellanos, Lilloa 23: 12, 1950. Basionym: *Tephrocactus minutus*. [Repeated by Rowley in Nation. Cact. Succ. J. 13(1): 6, 1958 (cf. RPS 9).] [RPS 9]

Opuntia miquelii var. **filesii** (Backeberg) Rowley, Nation. Cact. Succ. J. 13(1): 4, 1958. Basionym: *Austrocylindropuntia miquelii* var. *filesii*. [RPS 9]

Opuntia mira (Rauh & Backeberg) Rowley, Nation. Cact. Succ. J. 13(1): 6, 1958. Basionym: *Tephrocactus mirus*. [RPS 9]

Opuntia mistiensis (Backeberg) Rowley, Nation. Cact. Succ. J. 13(1): 6, 1958. Basionym: *Tephrocactus mistiensis*. [RPS 9]

Opuntia muelleriana (Backeberg) Rowley, Nation. Cact. Succ. J. 13(1): 6, 1958. Basionym: *Tephrocactus muellerianus*. [RPS 9]

Opuntia multiareolata Backeberg, Cactus (Paris) No. 75: 32, 1962. *Nom. inval.* (Art. 9.5/37.1). [Erroneously included as valid in RPS 13. Simultaneously published in Die Cact. 6: 3614.] [RPS 13]

Opuntia nejapensis Bravo, Cact. Suc. Mex. 17: 115-116, 1972. [RPS 23]

Opuntia neoargentina (Backeberg) Rowley, Nation. Cact. Succ. J. 13(1): 4, 1958. Basionym: *Brasiliopuntia neoargentina*. [RPS 9]

Opuntia neochrysantha Bravo, Cact. Suc. Mex. 19: 19-22, 1974. [RPS 25]

Opuntia nicholii L. Benson, Cacti of Arizona, ed. 2, 48, 1950. [RPS 4]

Opuntia noodtiae (Backeberg & Jacobsen) Rowley, Nation. Cact. Succ. J. 13(1): 6, 1958. Basionym: *Tephrocactus noodtiae*. [RPS 9]

Opuntia nuda (Backeberg) Rowley, Rep. Pl. Succ. 22: 11, 1973. *Nom. inval.*, based on *Nopalea nuda, nom. inval.* (Art. 9.5). [Erroneously included as valid in RPS 22.] [RPS 22]

Opuntia obliqua Backeberg, Cactus (Paris) No. 75: 32, 1962. *Nom. inval.* (Art. 9.5). [Erroneously included as valid in RPS 13. Simultaneously published in Die Cact. 6: 3614.] [RPS 13]

Opuntia occidentalis var. **megacarpa** (Griffiths) Munz, Aliso 4(1): 94, 1957. Basionym: *Opuntia megacarpa*. [RPS 8]

Opuntia occidentalis var. **piercei** (Fosberg) Munz, Aliso 4(1): 94, 1957. Basionym: *Opuntia phaeacantha* var. *piercei*. [RPS 8]

Opuntia occidentalis var. **vaseyi** (Coulter) Munz, Aliso 4(1): 94, 1957. Basionym: *Opuntia mesacantha* var. *vaseyi*. [RPS 8]

Opuntia oricola Philbrick, Cact. Succ. J. (US) 36: 163-165, 1964. [RPS 15]

Opuntia orurensis Cardenas, Cact. Succ. J. (US) 28: 112-113, 1956. [RPS 7]

Opuntia paediophila Castellanos, Lilloa 23: 5-13, 1950. [Could be correctible to 'pediophila'.] [RPS 3]

Opuntia panellana (Backeberg) Backeberg, Die Cact. 6: 3619, 1962. Basionym: *Opuntia armata* var. *panellana*. [RPS 13]

Opuntia parryi var. **serpentina** (Engelmann) L. Benson, Cact. Succ. J. (US) 41: 33, 1969. Basionym: *Opuntia serpentina*. [RPS 20]

Opuntia pentlandii var. **fuauxiana** (Backeberg) Rowley, Nation. Cact. Succ. J. 13(1): 6, 1958. Basionym: *Tephrocactus pentlandii* var. *fuauxianus*. [RPS 9]

Opuntia pentlandii var. **rossiana** (Backeberg) Rowley, Nation. Cact. Succ. J. 13(1): 6, 1958. Basionym: *Tephrocactus pentlandii* var. *rossianus*. [RPS 9]

Opuntia phaeacantha var. **camanchica** (Engelmann & Bigelow) L. Benson, Cact. Succ. J. (US) 41: 125, 1969. Basionym: *Opuntia camanchica*. [Repeated (invalidly, Art. 33.2) by Wenigerin Cacti of the Southwest, 196, 1970 [not dated].] [RPS 20]

Opuntia phaeacantha var. **charlestonensis** (Clokey) Backeberg, Die Cact. 1: 508, 1958. Basionym: *Opuntia charlestonensis*. [RPS 9]

Opuntia phaeacantha var. **chihuahuensis** (Rose) Bravo, Cact. Suc. Mex. 19: 47, 1974. Basionym: *Opuntia chihuahuensis*. [RPS 25]

Opuntia phaeacantha var. **discata** (Griffiths) L. Benson & Walkington, Ann. Missouri Bot. Gard. 52: 265, 1965. Basionym: *Opuntia discata*. [RPS 17]

Opuntia phaeacantha var. **flavispina** L. Benson, Cact. Succ. J. (US) 46: 79, 1974. [RPS 25]

Opuntia phaeacantha var. **laevis** (Coulter) L. Benson, Cacti of Arizona, ed. 3, 21, 98, 1969. Basionym: *Opuntia laevis*. [RPS 20]

Opuntia phaeacantha var. **spinosibacca** (Anthony) L. Benson, Cact. Succ. J. (US) 41: 125, 1969. Basionym: *Opuntia spinosibacca*. [RPS 20]

Opuntia phaeacantha var. **superbospina** (Griffiths) L. Benson, Cact. Succ. J. (US) 46: 79, 1974. Basionym: *Opuntia superbospina*. [RPS 25]

Opuntia phaeacantha var. *tenuispina* (Engelmann) Weniger, Cacti of the Southwest, 197-198, 1970. *Nom. inval.* (Art. 33.2), based on *Opuntia tenuispina*. [Not dated.] [RPS -]

Opuntia phaeacantha var. **wootonii** (Griffiths) L. Benson, Cact. Succ. J. (US) 41: 125, 1969. Basionym: *Opuntia wootonii*. [RPS 20]

Opuntia picardoi Marnier-Lapostolle, Cactus (Paris) No. 66: 3-4, 1960. [Validity of name not assessed.] [RPS 11]

Opuntia pilifera var. **aurantisaeta** Backeberg,

Descr. Cact. Nov. [1:] 10, 1957. [Dated 1956, published 1957.] [RPS 7]

Opuntia platyacantha var. **angustispina** (Backeberg) Rowley, Nation. Cact. Succ. J. 13(1): 6, 1958. Basionym: *Tephrocactus platyacanthus* var. *angustispinus*. [RPS 9]

Opuntia platyacantha var. **neoplatyacantha** (Backeberg) Rowley, Nation. Cact. Succ. J. 13(1): 6, 1958. Basionym: *Tephrocactus platyacanthus* var. *neoplatyacanthus*. [RPS 9]

Opuntia poecilacantha Backeberg, Cactus (Paris) No. 75: 32, 1962. *Nom. inval.* (Art. 9.5). [Erroneously included as valid in RPS 13. Concurrently published in Die Cact. 6: 3615.] [RPS 13]

Opuntia polyacantha var. **juniperina** (Britton & Rose) L. Benson, The Cacti of Arizona ed. 3, 20, 72, 1969. Basionym: *Opuntia juniperina*. [RPS 20]

Opuntia polyacantha var. **rufispina** (Engelmann & Bigelow) L. Benson, Cacti of Arizona, ed. 3, 20, 71, 1969. Basionym: *Opuntia missouriensis* var. *rufispina*. [RPS 20]

Opuntia polyacantha var. **schweriniana** (Schumann) Backeberg, Die Cact. 1: 607, 1958. Basionym: *Opuntia schweriniana*. [RPS 9]

Opuntia posnanskyana Cardenas, Lilloa 23: 25, 1950. [RPS 4]

Opuntia pseudo-udonis (Rauh & Backeberg) Rowley, Nation. Cact. Succ. J. 13(1): 6, 1958. Basionym: *Tephrocactus pseudo-udonis*. [RPS 9]

Opuntia punta-caillan (Rauh & Backeberg) Rowley, Nation. Cact. Succ. J. 13(1): 6, 1958. Basionym: *Tephrocactus punta-caillan*. [Sphalm. 'puntia-caillan'.] [RPS 9]

Opuntia pygmaea (Wiegand & Backeberg) Rowley, Nation. Cact. Succ. J. 13(1): 5, 1958. Basionym: *Micropuntia pygmaea*. [RPS 9]

Opuntia pyrrhacantha var. **leucolutea** (Backeberg) Rowley, Nation. Cact. Succ. J. 13(1): 6, 1958. Basionym: *Tephrocactus pyrrhacanthus* var. *leucoluteus*. [RPS 9]

Opuntia rarissima (Backeberg) Rowley, Nation. Cact. Succ. J. 13(1): 6, 1958. Basionym: *Tephrocactus rarissimus*. [RPS 9]

Opuntia rauhii (Backeberg) Rowley, Nation. Cact. Succ. J. 13(1): 6, 1958. Basionym: *Tephrocactus rauhii*. [RPS 9]

Opuntia recondita var. **perrita** (Griffiths) Rowley, Nation. Cact. Succ. J. 13(2): 25, 1958. Basionym: *Opuntia perrita*. [RPS 9]

Opuntia riviereana Backeberg, Cactus (Paris) No. 75: 32, 1962. *Nom. inval.* (Art. 9.5). [Erroneously included as valid in RPS 13. Concurrently published in Die Cact. 6: 3628.] [RPS 13]

Opuntia roborensis Cardenas, Cact. Succ. J. (US) 42: 32, 1970. [RPS 21]

Opuntia robusta var. **guerrana** (Griffiths) Sanchez-Mejorada, Cact. Suc. Mex. 17: 119, 1972. Basionym: *Opuntia guerrana*. [RPS 23]

Opuntia robusta var. **larreyi** (F. A. C. Weber) Bravo, Cact. Suc. Mex. 17: 119, 1972. Basionym: *Opuntia larreyi*. [RPS 23]

Opuntia robusta var. **longiglochidiata** Backeberg, Descr. Cact. Nov. [1:] 10, 1957. [Dated 1956, published 1957.] [RPS 7]

Opuntia robusta var. **megalarthra** (Rose) Hammer, Kulturpflanze 24: 266, 1976. Basionym: *Opuntia megalarthra*. [RPS 28]

Opuntia rufida var. **tortiflora** Anthony, Amer. Midland Natural. 55: 240, 1956. [RPS 7]

Opuntia rzedowskii Scheinvar, Anales Inst. Biol. UUNAM, Ser. Bot. 47-53: 123-136, ills., SEM ills., 1984. *Scheinvar* 2216 (MEXU). [Vol. for 1976-1982, published 1984.] [Mexico] [RPS 38]

Opuntia salagria Castellanos, Lilloa 27: 85-89, 1953. [Dated 1953, published 1955 according to RPS 6.] [RPS 6]

Opuntia salmiana var. **albiflora** (Schumann) Rowley, Nation. Cact. Succ. J. 13(2): 25, 1958. Basionym: *Opuntia albiflora*. [RPS 9]

Opuntia salmiana var. **spegazzinii** (F. A. C. Weber) Rowley, Nation. Cact. Succ. J. 13(2): 25, 1958. Basionym: *Opuntia spegazzinii*. [RPS 9]

Opuntia sanguinea G. R. Proctor, J. Arnold Arb. 63(3): 239, 1982. Typus: *Proctor* 38043 (IJ). [RPS 33]

Opuntia santamaria (Baxter) Wiggins, in Shreve & Wiggins, Veg. Fl. Sonoran Desert 2: 966, 1964. Basionym: *Grusonia santamaria*. [Combination repeated by Bravo-H. in Cact. Suc. Mex. 17: 119, 1972.] [RPS 18]

Opuntia sarca Griffiths *ex* Scheinvar, Phytologia 49(4): 328-333, 1981. [RPS 32]

Opuntia scheinvariana Paniagua, Thesis Univ. San Carlos, Guatemala, 1980 (fide Cact. Suc. Mex. 28(3): 69, 1983), 1983. Typus: *Paniagua* 75 (USCG). [*Nom. inval.* (Art. 29.1) ?] [RPS 34]

Opuntia schottii var. **grahamii** (Engelmann) L. Benson, Cact. Succ. J. (US) 41: 124, 1969. Basionym: *Opuntia grahamii*. [RPS 20]

Opuntia shaferi var. **humahuacana** (Backeberg) Rowley, Cact. Succ. J. Gr. Brit. 44(1): 2, 1982. Basionym: *Cylindropuntia humahuacana*. [RPS 33]

Opuntia spectatissima (Daston) Rowley, Nation. Cact. Succ. J. 13(1) 5, 1958. Basionym: *Micropuntia spectatissima*. [RPS 9]

Opuntia sphaerica var. **rauppiana** (Schumann) Rowley, Nation. Cact. Succ. J. 13(2): 25, 1958. Basionym: *Opuntia rauppiana*. [RPS 9]

Opuntia sphaerica var. **unguispina** (Backeberg) Rowley, Nation. Cact. Succ. J. 13(2): 25, 1958. Basionym: *Opuntia unguispina*. [RPS 9]

Opuntia spinosibacca Anthony, Amer. Midland Natural. 55: 246-249, 1956. [RPS 7]

Opuntia stanlyi var. **peeblesiana** L. Benson, Cacti of Arizona, ed. 3, 20, 64, 1969. [RPS 20]

Opuntia steiniana (Backeberg) Rowley, Nation. Cact. Succ. J. 13(1): 4, 1958. Basionym: *Austrocylindropuntia steiniana*. [RPS 9]

Opuntia stenopetala var. **inermis** Bravo, Cact. Suc. Mex. 19: 27, 1974. [Sphalm. 'inerme'.]

[RPS 25]

Opuntia streptacantha var. **pachona** (Griffiths) Hammer, Kulturpflanze 24: 262, 1976. Basionym: *Opuntia pachona*. [RPS 28]

Opuntia stricta var. **dillenii** (Ker-Gawler) L. Benson, Cact. Succ. J. (US) 41: 126, 1969. Basionym: *Cactus dillenii*. [RPS 20]

Opuntia strigil var. **flexospina** (Griffiths) L. Benson, Cact. Succ. J. (US) 46: 79, 1974. Basionym: *Opuntia flexospina*. [RPS 25]

Opuntia subulata var. **exaltata** (A. Berger) Rowley, Rep. Pl. Succ. 23: 8, 1974. Basionym: *Opuntia exaltata*. [RPS 23]

Opuntia sulphurea var. **hildmannii** (Fric) Backeberg, Die Cact. 1: 413, 1958. Basionym: *Opuntia hildmannii*. [RPS 9]

Opuntia sulphurea var. **pampeana** (Spegazzini) Backeberg, Die Cact. 1: 414, 1958. Basionym: *Opuntia pampeana*. [RPS 9]

Opuntia tayapayensis Cardenas, Lilloa 23: 18, 1950. [RPS 4]

Opuntia tehuantepecana (Bravo) Bravo, Cact. Suc. Mex. 17: 119, 1972. Basionym: *Opuntia dillenii* var. *tehuantepecana*. [RPS 23]

Opuntia tephrocactoides (Rauh & Backeberg) Rowley, Nation. Cact. Succ. J. 13(1): 4, 1958. Basionym: *Austrocylindropuntia tephrocactoides*. [RPS 9]

Opuntia thurberi var. **alamosensis** (Britton & Rose) Bravo, Cact. Suc. Mex. 17: 119, 1972. Basionym: *Opuntia alamosensis*. [RPS 23]

Opuntia tilcarensis var. *rubellispina* Backeberg, Cactus (Paris) No. 75: 32, 1962. *Nom. inval.* (Art. 9.5). [Erroneously included as valid in RPS 13. Concurrently published in Die Cact. 6: 3617.] [RPS 13]

Opuntia tomentosa var. **hernandezii** (De Candolle) Bravo, Cact. Suc. Mex. 20: 96, 1975. Basionym: *Opuntia hernandezii*. [RPS 27]

Opuntia tomentosa var. **herrerae** Scheinvar, Phytologia 49(4): 332-338, 1981. [RPS 32]

Opuntia tomentosa var. **rileyi** (Gonzalez Ortega) Backeberg, Die Cact. 1: 542, 1958. Basionym: *Opuntia rileyi*. [RPS 9]

Opuntia tomentosa var. **spraguei** (Gonzalez Ortega) Backeberg, Die Cact. 1: 542, 1958. Basionym: *Opuntia spraguei*. [Sphalm. 'spranguei'.] [RPS 9]

Opuntia tortispina var. **cymochila** (Engelmann) Backeberg, Die Cact. 1: 483, 1958. Basionym: *Opuntia cymochila*. [RPS 9]

Opuntia tuberculosirhopalica (Wiegand & Backeberg) Rowley, Nation. Cact. Succ. J. 13(1): 5, 1958. Basionym: *Micropuntia tuberculosirhopalica*. [RPS 9]

Opuntia tunicata var. **chilensis** (Ritter) A. E. Hoffmann, Cact. Fl. Chil., 242, ill. (p. 242), 1989. Basionym: *Cylindropuntia tunicata* var. *chilensis*. [The basionym cited is incorrect; the correct basionym, however, is included in the synonymy.] [RPS 40]

Opuntia tunicata var. **davisii** (Engelmann & Bigelow) L. Benson, Cact. Succ. J. (US) 41: 124, 1969. Basionym: *Opuntia davisii*. [RPS 20]

Opuntia velutina var. **affinis** (Griffiths) Bravo, Cact. Suc. Mex. 17: 119, 1972. Basionym: *Opuntia affinis*. [RPS 23]

Opuntia velutina var. **macdougaliana** (Rose) Bravo, Cact. Suc. Mex. 17: 119, 1972. Basionym: *Opuntia macdougaliana*. [RPS 23]

Opuntia verschaffeltii var. **hypsophila** (Spegazzini) Rowley, Nation. Cact. Succ. J. 13(2): 25, 1958. Basionym: *Opuntia hypsophila*. [RPS 9]

Opuntia verschaffeltii var. **longispina** (Backeberg) Rowley, Nation. Cact. Succ. J. 13(1): 4, 1958. Basionym: *Austrocylindropuntia verschaffeltii* var. *longispina*. [RPS 9]

Opuntia vestita fa. *intermedia* (Backeberg) Krainz, Die Kakt. B, 1967. *Nom. inval.*, based on *Austrocylindropuntia vestita* var. *intermedia*, *nom. inval.* (Art. 9.5). [Erroneously included as valid in RPS 18.] [RPS 18]

Opuntia vestita fa. **maior** (Backeberg) Krainz, Die Kakt. B, 1967. Basionym: *Austrocylindropuntia vestita* var. *maior*. [RPS 18]

Opuntia vestita var. **chuquisacana** (Cardenas) Rowley, Nation. Cact. Succ. J. 13(2): 25, 1958. Basionym: *Opuntia chuquisacana*. [RPS 9]

Opuntia vestita var. **major** (Backeberg) Rowley, Nation. Cact. Succ. J. 13(1): 4, 1958. Basionym: *Austrocylindropuntia vestita* var. *major*. [RPS 9]

Opuntia violacea var. *castetteri* L. Benson, Cact. Succ. J. (US) 41: 125, 1969. Based on *Benson* 15433. *Nom. inval.* (Art. 43.1). [Erroneously included as valid in RPS 20.] [USA: Texas] [RPS 20]

Opuntia violacea var. *gosseliniana* (F. A. C. Weber) L. Benson, Cacti of Arizona, ed. 3, 21, 1969. *Nom. inval.* (Art. 43.1), based on *Opuntia gosseliniana*. [RPS -]

Opuntia violacea var. *gosseliniana* (F. A. C. Weber) L. Benson, Cacti of Arizona, ed. 3, 21, 92, 1969. *Nom. inval.* (Art. 43.1), based on *Opuntia gosseliniana*. [Erroneously included as valid in RPS 20.] [RPS 20]

Opuntia violacea var. *macrocentra* (Engelmann) L. Benson, Cacti of Arizona, ed. 3, 21, 1969. *Nom. inval.* (Art. 43.1), based on *Opuntia macrocentra*. [Erroneously included as valid in RPS 20.] [RPS 20]

Opuntia violacea var. *santa-rita* (Griffiths & Hare) L. Benson, Cacti of Arizona, ed. 3, 21, 92, 1969. *Nom. inval.* (Art. 43.1), based on *Opuntia chlorotica* var. *santa-rita*. [Erroneously included as valid in RPS 20.] [RPS 20]

Opuntia vulgaris var. **lemaireana** (Console) Backeberg, Die Cact. 1: 400, 1958. Basionym: *Opuntia lemaireana*. [RPS 9]

Opuntia weberi var. **setiger** (Backeberg) Rowley, Nation. Cact. Succ. J. 13(2): 25, 1958. Basionym: *Tephrocactus setiger*. [RPS 9]

Opuntia whipplei var. **enodis** Peebles, in Benson, The Cacti of Arizona, 31, 1950. [RPS -]

Opuntia whipplei var. **multigeniculata** (Clokey) L. Benson, Cacti of Arizona, ed. 3,

20, 38, 1969. Basionym: *Opuntia multigeniculata*. [RPS 20]

Opuntia whipplei var. **viridiflora** (Britton & Rose) L. Benson, Cact. Succ. J. (US) 46: 79, 1974. Basionym: *Opuntia viridiflora*. [RPS 25]

Opuntia wiegandii (Backeberg) Rowley, Nation. Cact. Succ. J. 13(1): 5, 1958. Basionym: *Micropuntia wiegandii*. [RPS 9]

Opuntia wigginsii L. Benson, The Cacti of Arizona ed. 3, 19, 32, 1969. [RPS 20]

Opuntia wilkeana (Backeberg) Rowley, Nation. Cact. Succ. J. 13(1): 6, 1958. Basionym: *Tephrocactus wilkeanus*. [RPS 9]

Opuntia woodsii Backeberg, Descr. Cact. Nov. [1:] 10, 1957. [Dated 1956, published 1957.] [RPS 7]

Opuntia yanganucensis (Rauh & Backeberg) Rowley, Nation. Cact. Succ. J. 13(1): 6, 1958. Basionym: *Tephrocactus yanganucensis*. [RPS 9]

Opuntia zehnderi (Rauh & Backeberg) Rowley, Nation. Cact. Succ. J. 13(1): 6, 1958. Basionym: *Tephrocactus zehnderi*. [RPS 9]

Oreocereus australis (Ritter) A. E. Hoffmann, Cact. Fl. Chile, 80, 1989. Basionym: *Arequipa australis*. [RPS 40]

Oreocereus celsianus var. **fossulatus** (Labouret) Krainz, Kat. ZSS ed. 2, 98, 1967. Basionym: *Pilocereus fossulatus*. [RPS 18]

Oreocereus celsianus var. **hendriksenianus** (Backeberg) Krainz, Kat. ZSS ed. 2, 98, 1967. Basionym: *Oreocereus hendriksenianus*. [RPS 18]

Oreocereus celsianus var. **maximus** (Backeberg) Krainz, Kat. ZSS ed. 2, 99, 1967. Basionym: *Oreocereus maximus*. [RPS 18]

Oreocereus celsianus var. **ritteri** (Cullmann) Krainz, Kat. ZSS ed. 2, 99, 1967. Basionym: *Oreocereus ritteri*. [RPS 18]

Oreocereus celsianus var. **trollii** (Kupper) Krainz, Kat. ZSS ed. 2, 99, 1967. Basionym: *Cereus trollii*. [RPS 18]

Oreocereus celsianus var. **varicolor** (Backeberg) Krainz, Kat. ZSS ed. 2, 99, 1967. Basionym: *Oreocereus varicolor*. [Sphalm. 'variicolor'.] [RPS 18]

Oreocereus doelzianus fa. **calva** (Rauh & Backeberg) Buxbaum, in Krainz, Die Kakteen Lief. 33, 1966. Basionym: *Morawetzia doelziana* var. *calva*. [RPS 17]

Oreocereus fossulatus var. **gracilior** (Schumann) Backeberg, Die Cact. 2: 1032, 1959. Basionym: *Pilocereus celsianus* var. *gracilior*. [RPS 10]

Oreocereus fossulatus var. *rubrispinus* Ritter, Kakt. Südamer. 2: 697-699, 1980. Based on *Ritter* 100ap.p.. *Nom. inval.* (Art. 43). [Erroneously included as valid in RPS 31.] [Bolivia: Murillo] [RPS 31]

Oreocereus hempelianus (Gürke) D. Hunt, Bradleya 5: 93, 1987. Basionym: *Echinopsis hempeliana*. [RPS 38]

Oreocereus hendriksenianus fa. **densilanatus** (Rauh & Backeberg) Krainz, Die Kakt. C Vb, 1963. Basionym: *Oreocereus hendriksenianus* var. *densilanatus*. [RPS 14]

Oreocereus hendriksenianus fa. **spinosissimus** (Rauh & Backeberg) Krainz, Die Kakt. C Vb, 1963. Basionym: *Oreocereus hendriksenianus* var. *spinosissimus*. [RPS 14]

Oreocereus hendriksenianus var. **densilanatus** Rauh & Backeberg, in Backeberg, Descr. Cact. Nov. [1:] 17, 1957. [Dated 1956, published 1957.] [RPS 7]

Oreocereus hendriksenianus var. **spinosissimus** Rauh & Backeberg, in Backeberg, Descr. Cact. Nov. [1:] 17, 1957. Typus: *Rauh* K38a (1954) (ZSS). [Dated 1956, published 1957.] [Peru] [RPS 7]

Oreocereus knizei Hewitt & Donald, Ashingtonia Species Cat. Cact. [part 9], unnumbered page, 1975. [*Nom. nov. pro Morawetzia varicolor* Knize (*non Oreocereus variicolor* Backeberg).] [RPS -]

Oreocereus leucotrichus (Philippi) Wagenknecht, Anales Acad. Chil. Ci. Nat. 20: 102, 1956. Basionym: *Echinocactus leucotrichus*. [The bibliographical citation may be wrong, and the name could be invalid (Art. 33.2).] [RPS 16]

Oreocereus piscoensis (Backeberg) Ritter, Kakt. Südamer. 4: 1365-1366, 1981. *Nom. inval.* (Art. 33.2), based on *Loxanthocereus piscoensis*. [RPS 32]

Oreocereus rettigii (Quehl) Buxbaum, in Krainz, Die Kakt. Lief. 55-56: C Vc, 1974. Basionym: *Echinocactus rettigii*. [Dated Dec. 1973, but published in 1974 according to all sources except Index Kewensis.] [RPS 25]

Oreocereus ritteri Cullmann, Kakt. and. Sukk. 9(7): 101-103, ills., 1958. Typus: *Ritter* 177a (ZSS [status ?]). [First mentioned l.c. (5): 71.] [Peru: Ayacucho] [RPS 9]

Oreocereus tacnaensis Ritter, Kakt. Südamer. 4: 1363-1365, 1981. Typus: *Ritter* 124 (U, ZSS). [Peru: Tacna] [RPS 32]

Oreocereus trollii var. **crassiniveus** (Backeberg) Backeberg, Die Cact. 2: 1038, 1959. Basionym: *Oreocereus crassiniveus*. [RPS 10]

Oreocereus trollii var. **tenuior** Backeberg, Descr. Cact. Nov. [1:] 17, 1957. [Dated 1956, published 1957.] [RPS 7]

Oreocereus varicolor Backeberg, Cact. Succ. J. (US) 23(1): 20, 1951. [Sphalm. 'variicolor'.] [RPS 3]

Oreocereus varicolor var. *tacnaensis* Ritter *ex* Backeberg, Die Cact. 6: 3696, 1962. *Nom. inval.* (Art. 36.1, 37.1). [Sphalm. 'variicolor'. Erroneously included as valid in RPS 13. Based on *Oreocereus tacnaensis* Ritter (at that time a *nom. nud.*). Erroneously treated as valid name by Eggli in Bradleya 3: 102, 1985.] [RPS 13]

Oroya baumannii Knize, Biota 7: 254, 1969. Typus: *Knize* 380 (ZSS [status ?]). [Sphalm. 'baumanii', corrected on account of the spelling of *Espostoa baumannii*.] [Peru: Lima] [RPS 36]

Oroya baumannii var. **rubrispina** Knize, Biota 7: 255, 1969. Typus: *Knize* 238a (ZSS [not found]). [sphalm. 'baumanii', corrected on account of the spelling of *Espostoa baumannii*.] [RPS 36]

Oroya borchersii (Bödeker) Backeberg, in Rauh, Beitr. Kenntn. peruan. Kakt.-veg., 487, 1958. Basionym: *Echinocactus borchersii.* [RPS -]

Oroya borchersii fa. **fuscata** (Rauh & Backeberg) Krainz, Die Kakt. C Vb, 1963. Basionym: *Oroya borchersii* var. *fuscata.* [RPS 14]

Oroya borchersii var. **fuscata** Rauh & Backeberg, in Backeberg, Descr. Cact. Nov. [1:] 32, 1957. [Dated 1956, published 1957.] [RPS 7]

Oroya depressa Backeberg *ex* Rauh, Cactus (Paris) No. 51: 96, 1956. *Nom. inval.* (Art. 36.1). [Validity of name not assessed, probably still a *nomen nudum*, as the name by Backeberg on which it is based. Reported as *nom. subnud.* in l.c. No. 53: 139, 1957, by RPS 8. See *O. neoperuviana* var. *depressa.*] [RPS 7]

Oroya gibbosa Ritter, Kakt. Südamer. 4: 1512-1513, 1981. Typus: *Ritter* 143a (U). [First mentioned (as *nomen nudum*) in Backeberg, Die Cact. 3: 1695, 1959 (cf. RPS 10).] [Peru: Apurimac] [RPS 32]

Oroya gibbosa var. *citriflora* Knize, Biota 7: 255, 1969. Based on *Knize* 386. *Nom. inval.* (Art. 43.1). [The name *Oroya gibbosa* was only validly described in 1980.] [RPS 36]

Oroya laxiareolata Rauh & Backeberg, in Backeberg, Descr. Cact. Nov. [1:] 32, 1957. [Dated 1956, published 1957.] [RPS 7]

Oroya laxiareolata var. *pluricentralis* Backeberg, Descr. Cact. Nov. 3: 10, 1963. *Nom. inval.* (Art. 9.5). [RPS 14]

Oroya neoperuviana var. **depressa** Rauh & Backeberg, in Backeberg, Descr. Cact. Nov. [1:] 32, 1957. Typus: *Rauh* K72 (1954) (ZSS). [Dated 1956, published 1957.] [Peru: Ayacucho] [RPS 7]

Oroya neoperuviana var. **ferruginea** Rauh & Backeberg, in Backeberg, Descr. Cact. Nov. [1:] 32, 1957. [Dated 1956, published 1957.] [RPS 7]

Oroya peruviana fa. *minima* Riha & Subik, Kaktusy 23(6): 142-144, ills., 1987. *Nom. inval.* (Art. 36.1). [Published as provisional name.] [RPS 38]

Oroya peruviana var. **baumannii** (Knize) Slaba, Kaktusy 21(1): 10, 1985. Basionym: *Oroya baumannii.* [RPS 36]

Oroya peruviana var. *citriflora* (Knize) Slaba, Kaktusy 21(1): 9, 1985. *Nom. inval.*, based on *Oroya gibbosa* var. *citriflora, nom. inval.* (Art. 43.1). [Erroneously included as valid in RPS 36: 8.] [RPS 36]

Oroya peruviana var. **conaikensis** Donald & Lau, Nat. Cact. Succ. J. 25: 33, 1970. [RPS 21]

Oroya peruviana var. **depressa** (Rauh & Backeberg) Rauh & Backeberg, in Rauh, Beitr. Kenntn. peruan. Kakt.veg., 478, 1958. Basionym: *Oroya neoperuviana* var. *depressa.* [Repeated by Slaba in Kaktusy 21(1): 9, 1985, and reported as correct place of citation in RPS 36.] [RPS 10]

Oroya peruviana var. **neoperuviana** (Backeberg) Slaba, Kaktusy 21(1): 7, 1985. Basionym: *Oroya neoperuviana.* [RPS 36]

Oroya peruviana var. *pluricentralis* (Backeberg) Ritter, Kakt. Südamer. 4: 1511, 1981. *Nom. inval.*, based on *Oroya laxiareolata* var. *pluricentralis, nom. inval.* (Art. 9.5). [RPS 32]

Oroya peruviana var. *tenuispina* Rauh, Beitr. Kenntn. peruan. Kakt.-veg., 487, ills., 1958. Based on *Rauh* K24(1954). *Nom. inval.* (Art. 37.1). [Based on sevaral syntypes.] [Peru] [RPS -]

Oroya subocculta Rauh & Backeberg, in Backeberg, Descr. Cact. Nov. [1:] 32, 1957. [Dated 1956, published 1957.] [RPS 7]

Oroya subocculta var. **albispina** Rauh & Backeberg, in Backeberg, Descr. Cact. Nov. [1:] 32, 1957. [Dated 1956, published 1957.] [RPS 7]

Oroya subocculta var. **fusca** Rauh & Backeberg, in Backeberg, Descr. Cact. Nov. [1:] 32, 1957. Typus: *Rauh* K2c (1956) (ZSS). [Dated 1956, published 1957.] [Peru] [RPS 7]

Oroya subocculta var. **laxiareolata** (Rauh & Backeberg) Slaba, Kaktusy 21(2): 34, 1985. Basionym: *Oroya laxiareolata.* [RPS 36]

Oroya subocculta var. *pluricentralis* (Backeberg) Slaba, Kaktusy 21(2): 34, 1985. *Nom. inval.*, based on *Oroya laxiareolata* var. *pluricentralis, nom. inval.* (Art. 9.5). [RPS 36]

Oroya subocculta var. *typica* Rauh & Backeberg, in Rauh, Beitr. Kenntn. peruan. Kakt.veg., 482, 1958. *Nom. inval.* (Art. 24.3). [RPS 10]

Ortegocactus Alexander, Cact. Succ. J. (US) 33: 39, 1961. Typus: *O. macdougalii.* [RPS 12]

Ortegocactus macdougalii Alexander, Cact. Succ. J. (US) 33: 39, 1961. [Concurrently described in Cact. Suc. Mex. 6: 25, 1961.] [RPS 12]

× **Pacherocactus** Rowley, Nation. Cact. Succ. J. 37(3): 78, 1982. [= *Pachycereus* × *Bergerocactus.*] [RPS 33]

× **Pacherocactus orcuttii** (K. Brandegee *pro sp.*) Rowley, Nation. Cact. Succ. J. 37(3): 78, 1982. Basionym: *Cereus orcuttii.* [= *Bergerocactus emoryi* × *Pachycereus pringlei.*] [RPS 33]

× *Pachgerocereus* Moran, Cact. Succ. J. (US) 34: 93, 1962. *Nom. inval.* (Art. H6.2). [Erroneously included as valid in RPS 13.] [= *Pachycereus* × *Bergerocactus*] [RPS 13]

× *Pachgerocereus orcuttii* (K. Brandegee *pro sp.*) Moran, Cact. Succ. J. (US) 34: 93, 1962. *Nom. inval.* (Art. 43.1, H6.2), based on *Cereus orcuttii.* [= *Pachycereus pringlei* × *Bergerocactus emoryi.* Erroneously included as valid in RPS 13.] [RPS 13]

Pachycereeae Buxbaum, Madroño 14(6): 186, 1958. Typus: *Pachycereus.* [Sphalm. 'Pachycereae'.] [RPS 9]

Pachycereinae Buxbaum, Bot. Stud. 12: 89, 1961. Typus: *Pachycereus.* [RPS 12]

Pachycereus subgen. **Lemaireocereus** (Britton & Rose) Bravo, Cact. Suc. Mex. 17: 119, 1972. Basionym: *Lemaireocereus.* [RPS 23]

Pachycereus gigas (Backeberg) Backeberg, Die Cact. 4: 2154, 1960. Basionym: *Pachycereus grandis* var. *gigas.* [RPS 11]

Pachycereus hollianus (F. A. C. Weber)

Buxbaum, Bot. Stud. 12: 19, 99, 1961. Basionym: *Cereus hollianus*. [RPS 12]

Pachycereus militaris (Audot) D. Hunt, Bradleya 5: 93, 1987. Basionym: *Cereus militaris*. [RPS 38]

Pachycereus schottii (Engelmann) D. Hunt, Bradleya 5: 93, 1987. Basionym: *Cereus schottii*. [RPS 38]

Pachycereus tehuantepecanus MacDougall, Cact. Suc. Mex. 1(4): 63-67, 1956. [RPS 7]

Pachycereus weberi (Coulter) Backeberg, Die Cact. 4: 2152, 1960. Basionym: *Cereus weberi*. [RPS 11]

Parodia sect. **Sulcatae** Weskamp, Gattung Parodia, 33, 1987. Typus: *P. formosa* Ritter. [RPS 38]

Parodia ser. **Austrospermae** F. Brandt, Kakt. Orch.-Rundschau 7(4): 55, 1982. Typus: *Parodia aureicentra* Backeberg. [RPS 33]

Parodia ser. **Brachyspermae** Buxbaum, in Krainz, Die Kakt. C VIc (Sept. 1966), 1966. Typus: *P. schwebsiana* Backeberg. [RPS 17]

Parodia ser. **Brasilispermae** F. Brandt, Kakt. Orch.-Rundschau 7(4): 57, 1982. Typus: *Parodia buenekeri* Buining. [RPS 33]

Parodia ser. **Calyptospermae** F. Brandt, Kakt. Orch.-Rundschau 7(4): 52, 1982. Typus: *Parodia ayopayana* Cardenas. First mentioned as nude name in l.c. 1975/76(2): 11, 1975, and in Frankfurter Kakt.-Freund 4(4): 8, 1977. [RPS 33]

Parodia ser. **Campestrae** F. Brandt, Cact. Succ. J. (US) 49(3): 117, 1977. Typus: *P. formosa* Ritter. First mentioned as provisional name in Kakt. Orch.-Rundschau 1975/76(2): 20, 1975, later mentioned as provisional name in Frankfurter Kakt.-Freund 5(1): 13, 1978. [RPS 28]

Parodia ser. *Dentispermae* F. Brandt, Frankfurter Kakt.-Freund 5(1): 12, 1978. *Nom. inval.* (Art. 36.1, 37.1). [Published as provisional name. First mentioned in Kakt. Orch.-Rundschau 1975/76(2): 19, 1975.] [RPS -]

Parodia ser. **Eriospermae** F. Brandt, Kakt. Orch.-Rundschau 7(4): 57-58, 1982. Typus: *Parodia chrysacanthion* (Schumann) Backeberg. [RPS 33]

Parodia ser. **Extremispermae** F. Brandt, Kakt. Orch.-Rundschau 7(4): 55, 1982. Typus: *Parodia tilcarensis* Backeberg. Sphalm. 'Extremuspermae'. [RPS 33]

Parodia ser. *Hamatacanthae* F. Brandt, Cact. Succ. J. (US) 49(3): 117, 1977. *Nom. inval.* (Art. 32.1, 22.1). [Based on *Parodia microsperma* (F. A. C. Weber) Spegazzini, *typus generis*. Repeated in Kakt. Orch.-Rundschau 7(4): 58, 1982 (and included in RPS 33 as new name). Again mentioned as provisional name in Frankfurter Kakt.-Freund 5(2): 13, 1978.] [RPS 28]

Parodia ser. **Intectispermae** F. Brandt, Kakt. Orch.-Rundschau 7(4): 52-53, 1982. Typus: *Parodia miguillensis* Cardenas. First mentioned as nude name in l.c. 1975/76(2): 11, 1975, and in Frankfurter Kakt.-Freund 4(4): 8, 1977 (spelled as *Intectospermae*). [RPS 33]

Parodia ser. **Longispermae** F. Brandt, Kakt. Orch.-Rundschau 7(4): 56, 1982. Typus: *Parodia taratensis* Cardenas. First mentioned as provisional name in l.c. 1975/76(2): 16, 1975, and in Frankfurter Kakt.-Freund 5(1): 9, 1978. [RPS 33]

Parodia ser. **Macranthae** Buxbaum, in Krainz, Die Kakt. C VIc (Sept. 1966), 1966. Typus: *P. maassii* A. Berger. [RPS 17]

Parodia ser. **Mesospermae** F. Brandt, Kakt. Orch.-Rundschau 7(4): 54, 1982. Typus: *Parodia andreaeoides* F. Brandt. [RPS 33]

Parodia ser. *Montanae* F. Brandt, Kakt. Orch.-Rundschau 1975/76(2): 21, 1975. *Nom. inval.* (Art. 36.1, 37.1). [Published as provisional name.] [RPS -]

Parodia ser. **Oblongispermae** Buxbaum, in Krainz, Die Kakt. C VIc (Sept. 1966), 1966. Typus: *P. comarapana* Cardenas. [RPS 17]

Parodia ser. **Obtextospermae** Buxbaum, in Krainz, Die Kakt. C VIc (Sept. 1966), 1966. Typus: *P. ayopayana* Cardenas. [RPS 17]

Parodia ser. **Ovispermae** F. Brandt, Kakt. Orch.-Rundschau 7(4): 54, 1982. Typus: *Parodia ocampoi* Cardenas. [RPS 33]

Parodia ser. *Praeprotoparodia* F. Brandt, Frankfurter Kakt.-Freund 4(4): 9, 1977. *Nom. inval.* (Art. 36.1, 37.1). [Published as provisional name. First mentioned as provisional name in Kakt. Orch.-Rundschau 1975/76(2): 12, 1975.] [RPS -]

Parodia ser. **Pseudonotospermae** F. Brandt, Kakt. Orch.-Rundschau 7(4): 56, 1982. Typus: *Parodia mairanana* Cardenas. First published as provisional name in l.c. 1975/76(2): 18, 1975, and in Frankfurter Kakt.-Freund 5(1): 12, 1978. [RPS 33]

Parodia subgen. **Brasilicactus** (Buxbaum) F. Brandt, Kakt. Orch.-Rundschau 7(4): 53, 66, 1982. Basionym: *Notocactus* subgen. *Brasilicactus*. [Sphalm. 'Brasilicactea'. Basionym given as *Brasilicactus* Backeberg, referring to an illegitimate generic name later republished legitimately at subgeneric rank; treated as bibliographical error and here corrected] [RPS 33]

Parodia subgen. *Eriocactus* (Backeberg) F. Brandt, Kakt. Orch.-Rundschau 7(4): 58, 61, 1982. *Nom. illeg.* (Art. 57.1). [Sphalm. 'Eriocactea'..] [RPS 33]

Parodia subgen. *Glochoparodia* Y. Ito, Bull. Takarazuka Insectarium 71: 13-20, 1950. *Nom. inval.* (Art. 36.1). [RPS 3]

Parodia subgen. *Haploparodia* Y. Ito, Bull. Takarazuka Insectarium 71: 13-20, 1950. *Nom. inval.* (Art. 36.1). [RPS 3]

Parodia subgen. **Protoparodia** Buxbaum, in Krainz, Die Kakt. C VIc (Sept. 1966), 1966. Typus: *P. maassii* A. Berger. [RPS 17]

Parodia subgen. *Raphidoparodia* Y. Ito, Bull. Takarazuka Insectarium 71: 13-20, 1950. *Nom. inval.* (Art. 36.1). [RPS 3]

Parodia agasta F. Brandt, Frankfurter Kakt.-Freund 3(4): 6-7, 1976. [First mentioned as provisional name in Kakt. Orch.-Rundschau 1975/76(2): 17, 1975, later again mentioned as provisional name in l.c. 5(1): 10, 1978.] [RPS 27]

Parodia aglaisma F. Brandt, Kakt. Orch. Rundschau 1975/76(4): 50-53, ills., 1976. [First mentioned as provisional name in l.c.

1975/76(2): 16, then again mentioned as provisional name in Frankfurter Kakt.-Freund. 5(1): 9, 1978.] [RPS 27]

Parodia alacriportana Backeberg & Voll, Arq. Jard. Bot. Rio de Janeiro 9: 166-169, ill. (p. 159), 1950. Typus: *Berger* s.n. (?RB). [Cited as "Viveiro 15.513" which may indicate a living plant.] [Volume for 1949, published 1950.] [Brazil: Rio Grande do Sul] [RPS -]

Parodia albo-fuscata F. Brandt, Kakt. Orch. Rundschau 1977(2): 22-25, 1977. [RPS 28]

Parodia allosiphon (Marchesi) N. P. Taylor, Bradleya 5: 93, 1987. Basionym: *Notocactus allosiphon*. [RPS 38]

Parodia ampliocostata (Ritter) F. Brandt, Kakt. Orch.-Rundschau 7(4): 61, 1982. Basionym: *Eriocactus ampliocostatus*. [RPS 33]

Parodia andreae F. Brandt, Stachelpost 8: 145-147, 1972. [RPS 23]

Parodia andreaeoides F. Brandt, Stachelpost 50: 38-40, 1974. [RPS 25]

Parodia applanata F. Brandt, Letzeb. Cacteefren 4(12): 1-6, ills., 1983. Based on *Lau* s.n.. *Nom. inval.* (Art. 9.5). [Based on *Parodia schwebsiana* var. *applanata* W. Hoffmann & Backeberg, *nom. inval.* (Art. 9.5). An emendated diagnosis is presented and a collection without number from Lau (1970) is selected as type. This could constitute a validation. The paper is republished in Kakt. Orch.-Rundschau 9(2): 28-30, 1984. The taxon is again mentioned as 'species nova' in l.c. 13(1): 1988.] [RPS 34]

Parodia argerichiana Weskamp, Kakt. and. Sukk. 36(1): 8-9, ill., 1985. Typus: *Herzog* 108 (B). [Argentina] [RPS 36]

Parodia atroviridis Backeberg, Descr. Cact. Nov. 3: 10, 1963. *Nom. inval.* (Art. 9.5). [Erroneously included as valid in RPS 14.] [RPS 14]

Parodia aureicentra var. **albifusca** Ritter, Cactus (Paris) No. 75: 24, 1962. [RPS 13]

Parodia aureicentra var. **lateritia** Backeberg, Descr. Cact. Nov. [1:] 31, 1957. [Dated 1956, published 1957.] [RPS 7]

Parodia aureicentra var. **omniaurea** Ritter, Cactus (Paris) No. 75: 24, 1962. [RPS 13]

Parodia aureispina var. **australis** F. Brandt, Kakt. and. Sukk. 20: 167-168, 1969. [RPS 20]

Parodia aureispina var. **elegans** Backeberg, Descr. Cact. Nov. [1:] 31, 1957. [Dated 1956, published 1957.] [RPS 7]

Parodia aureispina var. *erythrantha* Backeberg *ex* F. Brandt, Frankfurter Kakt.-Freund 5(3): 6, 1978. *Nom. inval.* (Art. 36.1, 37.1). [Given as new combination with the taxon *P. aureispina* var. *australis* F. Brandt as synonym. First mentioned as provisional name in Kakt. Orch.-Rundschau 1975/76(2): 23, 1975.] [RPS -]

Parodia aureispina var. **erythrostaminea** F. Brandt, Kakt. Orch.-Rundschau 1975/76(3): 42, 1976. [First mentioned in l.c. 1975/76(2): 23, 1975. Repeated (as *nom. inval.*, Art. 36.1, 37.1) in Frankfurter Kakt.-Freund 5(3): 6, 1978.] [RPS 27]

Parodia aureispina var. *rubriantha* F. Brandt, Kakt. Orch.-Rundschau 7(5): 77, 1982. *Nom. inval.*, based on *Parodia rubriflora, nom. inval.* (Art. 9.5). [RPS 33]

Parodia aureispina var. *rubriflora* (Backeberg) F. Brandt, Kakt. and. Sukk. 20: 166, 1969. *Nom. inval.*, based on *Parodia rubriflora, nom. inval.* (Art. 9.5). [Erroneously included as valid in RPS 20.] [RPS 20]

Parodia aureispina var. **scopaoides** (Backeberg) F. Brandt, Kakt. and. Sukk. 20: 165, 1969. Basionym: *Parodia scopaoides*. [RPS 20]

Parodia aureispina var. **vulgaris** F. Brandt, Kakt. and. Sukk. 20: 164-165, 1969. [Later republished (*nom. inval.*, Art. 9.5) in Kakt. Orch.-Rundschau 1975/76(3): 42, 1976] [RPS 20]

Parodia aurihamata Fric & Kreuzinger *ex* Y. Ito, Explan. Diagr. Austroechinocactinae, 270, 1957. *Nom. inval.*, based on *Microspermia aurihamata, nom. inval.* (Art. 36.1). [RPS 8]

Parodia ayopayana Cardenas, Cact. Succ. J. (US) 23(3): 98, 1951. [RPS 2]

Parodia ayopayana var. **elata** Ritter, Kakt. Südamer. 2: 542, 1980. [RPS 31]

Parodia backebergiana F. Brandt, Kakt. and. Sukk. 20: 111-112, 1969. [RPS 20]

Parodia bellavistana F. Brandt, Kakt. Orch. -Rundschau 7(2): 18-21, ills., 1982. *Brandt cult.* 90 (HEID). [RPS 33]

Parodia belliata F. Brandt, Kakt. Orch.-Rundschau 6(2): 23-26, 1981. [RPS 32]

Parodia bermejoensis F. Brandt, Kakt. Orch.-Rundschau 179(3): 32-35, 1979. [RPS 30]

Parodia betaniana Ritter, Kakt. Südamer. 2: 426-427, 1980. [RPS 31]

Parodia bilbaoensis Cardenas, Cact. Succ. J. (US) 38: 146, 1966. [RPS 17]

Parodia borealis Ritter, Taxon 13(3): 116-117, 1964. Typus: *Ritter* 120 (U, ZSS). [Type cited for U l.c. 12: 28, 1963.] [Bolivia: La Paz] [RPS 15]

Parodia brevihamata W. Haage *ex* Backeberg, Descr. Cact. Nov. [1:] 31, 1957. [Dated 1956, published 1957. Described without mention of a type; cf. correspondence between ZSS and Haage; based on cultivated material.] [RPS 7]

Parodia brevihamata var. *mollispina* (Ritter) F. Brandt, Gattung Parodia, 154, 1989. *Nom. inval.*, based on *Brasiliparodia brevihamata* var. *mollispina, nom. inval.* (Art. 37.1). [RPS 40]

Parodia buenekeri Buining, Succulenta 1962: 99, 1962. [RPS 13]

Parodia buenekeri var. *intermedia* (Ritter) F. Brandt, Gattung Parodia, 154, 1989. *Nom. inval.*, based on *Brasiliparodia buenekeri* var. *intermedia, nom. inval.* (Art. 37.1). [RPS 40]

Parodia buiningii (Buxbaum) N. P. Taylor, Bradleya 5: 93, 1987. Basionym: *Notocactus buiningii*. [RPS 38]

Parodia buxbaumiana F. Brandt, Kaktus 10: 81-83, 1975. [RPS 26]

Parodia caespitosa (Spegazzini) N. P. Taylor, Bradleya 5: 93, 1987. Basionym: *Echinocactus caespitosus*. [RPS 38]

Parodia caineana F. Brandt, Letzeb. Cacteefren 6(2): 18, 1985. Typus: *Anonymus* s.n. (HEID). [RPS 36]

Parodia callosa F. Brandt, Frankfurter Kakt.-Freund 4(4): 11, 1977. *Nom. inval.* (Art. 36.1, 37.1). [Published as provisional name. First mentioned as provisional name in Kakt. Orch.-Rundschau 1975/76(2): 14, 1975.] [RPS -]

Parodia camargensis Buining & Ritter, Succulenta 41(2): 18-21, ill., 1962. Typus: *Ritter* 86 (U, ZSS). [Bolivia: Sud-Cinti] [RPS 13]

Parodia camargensis var. **camblayana** Ritter, Succulenta 1962: 20, 1962. [RPS 13]

Parodia camargensis var. **castanea** Ritter, Succulenta 1962: 21, 1962. [RPS 13]

Parodia camargensis var. **prolifera** Ritter, Succulenta 1962: 21, 1962. [RPS 13]

Parodia camblayana (Ritter) F. Brandt, Kakt. Orch.-Rundschau 7(3): 40, 1982. *Nom. inval.* (Art. 33.2). [Erroneously included as valid in RPS 33: 10, corrected in RPS 37: 20. First mentioned (sphalm. 'camplayana') by Brandt as provisional name in l.c. 1975/76(2): 17, 1975, and again in Frankfurter Kakt.-Freund 5(1): 10, 1978.] [RPS 33]

Parodia camblayana var. *rubra* F. Brandt, Cactus 9(6): 121-122, 1977. *Nom. inval.* (Art. 43.1). [Erroneously included as valid in RPS 28.] [RPS 28]

Parodia camblayoides F. Brandt, Gattung Parodia, 125, ills., 1989. *Nom. inval.* (Art. 36.1, 37.1). [Published as *nom. prov.*] [RPS 40]

Parodia campestris F. Brandt, Kaktus 10: 61-64, 1975. [Sphalm. 'campestra'.] [RPS 26]

Parodia capillitaensis F. Brandt, Kaktusz Vilag 1977: 50-52, 1977. [RPS 28]

Parodia carapariana F. Brandt, Cact. Succ. J. (US) 49(3): 119-120, 1977. [First mentioned in Kakt. Orch.-Rundschau 1975/76(2): 20, 1975, and later again mentioned as provisional name in Frankfurter Kakt.-Freund 5(2): 12, 1978.] [RPS 28]

Parodia cardenasii Ritter, Succulenta 43: 58, 1964. [RPS 15]

Parodia cardenasii var. **major** F. Brandt, Kakt. Orch.-Rundschau 8(Sonderheft 3): 36, 1983. [Synonym: *Parodia chirimoyarana* F. Brandt.] [RPS 34]

Parodia carrerana Cardenas, Cactus 78: 93, 1963. [RPS 14]

Parodia castanea (Ritter) Ritter, Kakt. Südamer. 2: 519-522, 1980. Basionym: *Parodia camargensis* var. *castanea*. [RPS 31]

Parodia catamarcensis var. *rubriflora* Backeberg, Descr. Cact. Nov. 3: 10, 1963. *Nom. inval.* (Art. 9.5). [Sphalm. 'rubriflorens'.] [RPS 14]

Parodia catarinensis (Ritter) F. Brandt, Kakt. Orch.-Rundschau 7(4): 65, 1982. *Nom. inval.*, based on *Brasiliparodia catarinensis*, *nom. inval.* (Art. 37.1). [RPS 33]

Parodia catharosperma F. Brandt, Frankfurter Kakt.-Freund 4(4): 8, 1977. *Nom. inval.* (Art. 36.1, 37.1). [Published as provisional name. First mentioned as provisional name in Kakt. Orch.-Rundschau 1975/76(2): 11, 1975.] [RPS -]

Parodia chaetocarpa Ritter, Succulenta 43: 58, 1964. [RPS 15]

Parodia challamarcana F. Brandt, Stachelpost 8(37): 1-4, ills., 1972. [First mentioned as provisional name in Kakt. and. Sukk. 22: 12, ill., 1971. Probably still invalid (Art. 9.5), as the holotype is said to be "in coll. F. H. Brandt".] [RPS 22]

Parodia chirimoyarana F. Brandt, Cact. Succ. J. (US) 50(1): 16-17, 1978. Typus: *Anonymus in coll. F. Brandt* 45/a (HEID). [Bolivia: Prov. O'Connor] [RPS 29]

Parodia chlorocarpa Ritter, Kakt. Südamer. 2: 427, 1980. [RPS 31]

Parodia cintiensis Ritter, Succulenta 1962: 122, 1962. [RPS 13]

Parodia claviceps (Ritter) F. Brandt, Kakt. Orch.-Rundschau 7(4): 62, 1982. Basionym: *Eriocactus claviceps*. [RPS 33]

Parodia columnaris Cardenas, Cact. Succ. J. (US) 23(3): 95-95, 1951. [RPS 2]

Parodia columnaris var. **ochraceiflora** Ritter, Kakt. Südamer. 2: 539-540, 1980. [RPS 31]

Parodia comarapana Cardenas, Rev. Agric. Cochabamba 7(6): 24, 1951. [RPS 2]

Parodia comarapana var. **paucicostata** Ritter, Taxon 13: 117, 1964. [RPS 15]

Parodia commutans Ritter, Succulenta 43(2): 22-23, 1964. Typus: *Ritter* 729 (U, ZSS). [Bolivia: Sud Cinti] [RPS 15]

Parodia comosa Ritter, Cactus (Paris) 17(75): 21-22, 1962. Typus: *Ritter* 111 p.p. (U, ZSS). [Bolivia: La Paz] [RPS 13]

Parodia compressa Ritter, Cactus (Paris) No. 73/74: 9, 1962. [RPS 13]

Parodia concinna (Monville) N. P. Taylor, Bradleya 5: 93, 1987. Basionym: *Echinocactus concinnus*. [RPS 38]

Parodia cotacajensis F. Brandt, Kakt. Orch.-Rundschau 6(3): 55-58, 1981. [RPS 32]

Parodia crassigibba (Ritter) N. P. Taylor, Bradleya 5: 93, 1987. Basionym: *Notocactus crassigibbus*. [RPS 38]

Parodia crucinigricentra Fric *ex* Subik, Sukkulentenkunde 4: 31, 1951. *Nom. inval.* (Art. 32.1). [Based on *Microspermia crucinigricentra* Fric 1928 (*nom. nud.*) and given as new combination. Probably invalid. Concurrently published in Kakt. Listy 16(10): 164-166 (cf. RPS 3).] [RPS 2]

Parodia crucinigricentra var. *sibalii* Subik, Sukkulentenkunde 4: 32, 1951. *Nom. inval.* (Art. 43.1 ?). [Validity depends on the validity of the species. Concurrently published in Kakt. Listy 16(10): 167, 1951 (cf. RPS 3).] [RPS 2]

Parodia culpinensis F. Brandt, Stachelpost 9: 161-165, 1973. [RPS 24]

Parodia dextrohamata Backeberg, Descr. Cact. Nov. 3: 10, 1963. *Nom. inval.* (Art. 9.5). [Erroneously included as valid in RPS 14.] [RPS 14]

Parodia dextrohamata var. *stenopetala* Backeberg, Descr. Cact. Nov. 3: 10, 1963. *Nom. inval.* (Art. 43.1). [Erroneously included as valid in RPS 14.] [RPS 14]

Parodia dichroacantha F. Brandt & Weskamp, Kakt. and. Sukk. 18(5): 87-88, 1967. Typus: *Fechser* s.n. (B, ZSS [status ?]). [Argentina] [RPS -]

Parodia echinopsoides F. Brandt, Kaktus 11: 40-41, 1976. [First mentioned as provisional name in Kakt. Orch.-Rundschau 1975/76(2): 12, 1975, later again mentioned as nude name in Frankfurter Kakt.-Freund 4(4): 9, 1977.] [RPS 27]

Parodia echinus Ritter, Taxon 13: 117, 1964. [RPS 15]

Parodia elachisantha (F. A. C. Weber ex Roland-Gosselin) F. Brandt, Kakt. Orch.-Rundschau 7(4): 67, 1982. Basionym: *Echinocactus elachisanthus*. [RPS 33]

Parodia elachista Brandt, Kakt. Orch.-Rundschau 6(1): 1-4, 1981. [RPS 32]

Parodia elata F. Brandt, Cactus 8: 33-36, 1976. [First mentioned as provisional name in Kakt. Orch.-Rundschau 1975/76(2): 11, 1975, later again mentioned as *nomen nudum* in Frankfurter Kakt.-Freund 4(4): 8, 1977.] [RPS 27]

Parodia erinacea (Haworth) N. P. Taylor, Bradleya 5: 93, 1987. Basionym: *Cactus erinaceus*. [RPS 38]

Parodia erythrantha var. **thionantha** (Spegazzini) Backeberg, Kakt.-Lex., 342, 1966. Basionym: *Echinocactus microspermus* var. *thionanthus*. [RPS 17]

Parodia escayachensis (Vaupel) Backeberg, Die Cact. 3: 1612, 1959. Basionym: *Echinocactus escayachensis*. [Repeated in Kakt.-Lex., 342, 1966 (cf. RPS 17).] [RPS 10]

Parodia exiguita F. Brandt, Frankfurter Kakt.-Freund 5(2): 12, 1978. *Nom. inval.* (Art. 36.1, 37.1). [Published as provisional name. First mentioned as provisional name in Kakt. Orch.-Rundschau 1975/76(2): 21, 1975. Probably a spelling error for 'exquisita'?] [RPS -]

Parodia exquisita F. Brandt, Kakt. Orch.-Rundschau 2: 44-46, 1978. [RPS 29]

Parodia fechseri Backeberg, Descr. Cact. Nov. 3: 11, 1963. *Nom. inval.* (Art. 9.5). [Erroneously included as valid in RPS 14.] [RPS 14]

Parodia festiva F. Brandt, Frankfurter Kakt.-Freund 4(4): 11, 1977. *Nom. inval.* (Art. 36.1, 37.1). [Published as provisional name. First mentioned as provisional name in Kakt. Orch.-Rundschau 1975/76(2): 14, 1975.] [RPS -]

Parodia firmissima F. Brandt, Stachelpost 51: 65-68, 1974. [RPS 25]

Parodia formosa Ritter, Succulenta 43: 57, 1964. [RPS 15]

Parodia friciana F. Brandt, Stachelpost 9: 68-71, 1973. [RPS 24]

Parodia fulvispina Ritter, Cactus (Paris) 17(76): 54-55, 1962. Typus: *Ritter* 727 (U, ZSS). [Bolivia: Mendez] [RPS 13]

Parodia fulvispina var. **brevihamata** Ritter, Cactus 76: 55, 1962. [RPS 13]

Parodia fuscato-viridis Backeberg, Descr. Cact. Nov. 3: 11, 1963. *Nom. inval.* (Art. 9.5). [Erroneously included as valid in RPS 14.] [RPS 14]

Parodia gibbulosa Ritter, Kakt. Südamer. 2: 545-546, 1980. [RPS 31]

Parodia gibbulosoides F. Brandt, Stachelpost 7: 414-416, 1971. [RPS 22]

Parodia gigantea Fric *ex* Krainz, Sukkulentenkunde 6: 26-28, 1957. Typus: *Fric* s.n. (ZSS). [= *Microspermia gigantea* Fric (*nom. nud.*).] [Argentina] [RPS 8]

Parodia glischrocarpa Ritter, Kakt. Südamer. 2: 427-428, 1980. Typus: *Ritter* 923 (U, ZSS). [Type cited for U l.c. 1: iii, 1979.] [Argentina: Salta] [RPS 31]

Parodia gokrauseana Heinrich, Kakt. / Sukk. 1967: 25-26, ill., 1967. *Nom. inval.* (Art. 9.5). [Erroneously included as valid in RPS 18.] [RPS 18]

Parodia gracilis Ritter, Succulenta 43: 23, 1964. [RPS 15]

Parodia graessneri (Schumann) F. Brandt, Kakt. Orch.-Rundschau 7(4): 66, 1982. Basionym: *Echinocactus graessneri*. [RPS 33]

Parodia grossei (Schumann) F. Brandt, Kakt. Orch.-Rundschau 7(4): 62, 1982. Basionym: *Echinocactus grossei*. [RPS 33]

Parodia grossei var. **aureispina** (Ritter) F. Brandt, Gattung Parodia, 232, 1989. Basionym: *Eriocactus grossei* var. *aureispinus*. [RPS 40]

Parodia gummifera Backeberg & Voll, Arq. Jard. Bot. Rio de Janeiro 9: 169-170, ill. (p. 167), 1950. Typus: *Mello Barreto* s.n. (RB 65.045). [Volume for 1949, published 1950.] [Brazil: Minas Gerais] [RPS -]

Parodia gutekunstiana Backeberg, Die Cact. 3: 1604, 1959. *Nom. inval.* (Art. 37.1). [Erroneously included as valid in RPS 10.] [RPS 10]

Parodia haageana F. Brandt, Kakt. Orch.-Rundschau 1977(4): 53-55, 1977. [RPS 28]

Parodia haselbergii (F. Haage *ex* Rümpler) F. Brandt, Kakt. Orch.-Rundschau 7(4): 67, 1982. Basionym: *Echinocactus haselbergii*. [RPS 33]

Parodia hausteiniana Rausch, Kakt. and. Sukk. 21(3): 45, ill., 1970. Typus: *Rausch* 192 (ZSS). [Type cited for W but placed in ZSS.] [Bolivia] [RPS 21]

Parodia herteri (Werdermann) N. P. Taylor, Bradleya 5: 93, 1987. Basionym: *Echinocactus herteri*. [RPS 38]

Parodia herzogii Rausch, Kakt. and. Sukk. 32(2): 30-31, ill., 1981. Typus: *Rausch* 707a (ZSS). [Argentina: Salta] [RPS 32]

Parodia heteracantha Ritter *ex* Weskamp, Kakt. / Sukk. 21(4): 83-84, 116 (ill.), 1986. Typus: *Ritter* 926 (DR). [RPS 37]

Parodia horrida F. Brandt, Cactus 11(8): 113-115, 1979. [First mentioned (*nom. inval.*, Art. 36.1, 37.1) in Kakt. Orch.-Rundschau 1975/76(2): 25, 1975, and again in Frankfurter Kakt.-Freund 5(3): 8, 1978.] [RPS 30]

Parodia horstii (Ritter) N. P. Taylor, Bradleya 5: 93, 1987. Basionym: *Notocactus horstii*. [RPS 38]

Parodia hummeliana Lau & Weskamp, Kakt. and. Sukk. 29(10): 226-227, 1978. [RPS 29]

Parodia idiosa F. Brandt, Frankfurter Kakt.-Freund 3(2): 6-8, 1976. [First mentioned as provisional name in Kakt. Orch.-Rundschau 1975/76(2): 14, 1975, later again mentioned as provisional name in Frankfurter Kakt.-

Freund 4(4): 11, 1977.] [RPS 27]
Parodia ignorata F. Brandt, Stachelpost 8: 86-87, 1972. [RPS 23]
Parodia jujuyana Fric *ex* Subik, in Pazout, Valnicek & Subik, Kaktusy, 131, 1960. [RPS 12]
Parodia kilianana Backeberg, Descr. Cact. Nov. 3: 11, 1963. *Nom. inval.* (Art. 9.5). [Erroneously included as valid in RPS 14.] [RPS 14]
Parodia kilianana var. *dichroacantha* (F. Brandt & Weskamp) F. Brandt, Kakt. Orch.-Rundschau 1975/76(2): 25, 1975. *Nom. inval.* (Art. 43.1). [RPS 27]
Parodia knizei F. Brandt, Kakt. Orch.-Rundschau 9(1): 1-3, ill., 1984. Typus: *Knize* s.n. (HEID). [ex cult. F. Brandt acc. no. 95/a] [RPS 35]
Parodia koehresiana F. Brandt, Stachelpost 8: 113-115, 1972. [RPS 23]
Parodia krahnii Weskamp, Kakt. and. Sukk. 40(3): 58-60, ills., 1989. Typus: *Krahn* s.n. (WU). [ex cult. Weskamp] [First published as provisional name in Gattung Parodia, 569-570, fig. 93, 1987.] [Bolivia: Tarata] [RPS 40]
Parodia krasuckana F. Brandt, Kakt. and. Sukk. 23: 179-180, 1972. [RPS 23]
Parodia lamprospina F. Brandt, Frankfurter Kakt.-Freund 4(2): 6-7, 1977. [First mentioned as provisional name in Kakt. Orch.-Rundschau 1975/76(2): 18, 1975, later again mentioned as provisional name in l.c. 5(1): 11, 1978 (sphalm. 'lamprespina').] [RPS 28]
Parodia laui F. Brandt, Kakt. and. Sukk. 24: 244-245, 1973. [RPS 24]
Parodia legitima F. Brandt, Kaktusz Vilag 5: 5-8, 1975. [RPS 27]
Parodia leninghausii (F. Haage) F. Brandt, Kakt. Orch.-Rundschau 7(4): 61, 1982. Basionym: *Pilocereus leninghausii*. [RPS 33]
Parodia leninghausii var. **minor** (Ritter) F. Brandt, Gattung Parodia, 231, 1989. Basionym: *Eriocactus leninghausii* var. *minor*. [RPS 40]
Parodia liliputana (Werdermann) N. P. Taylor, Bradleya 5: 93, 1987. Basionym: *Blossfeldia liliputana*. [RPS 38]
Parodia lohaniana Lau & Weskamp, Kakt. and. Sukk. 30(6): 137-138, 1979. [RPS 30]
Parodia lohaniana var. *rubriflora* (Backeberg) Weskamp, Gattung Parodia, 421, 1987. *Nom. inval.*, based on *Parodia catamarcensis* var. *rubriflora, nom. inval.* (Art. 9.5). [Sphalm. 'rubriflorens'.] [RPS 38]
Parodia lychnosa F. Brandt, Kaktus 10: 42-43, 1975. [RPS 26]
Parodia maassii fa. **camblayana** (Ritter) Krainz, Kat. ZSS ed. 2, 101, 1967. Basionym: *Parodia camargensis* var. *camblayana*. [RPS 18]
Parodia maassii fa. **castanea** (Ritter) Krainz, Kat. ZSS ed. 2, 101, 1967. Basionym: *Parodia camargensis* var. *castanea*. [RPS 18]
Parodia maassii fa. **maxima** (Ritter) Krainz, Kat. ZSS ed. 2, 102, 1967. Basionym: *Parodia maxima*. [RPS 18]
Parodia maassii fa. **prolifera** (Ritter) Krainz, Kat. ZSS ed. 2, 101, 1967. Basionym: *Parodia camargensis* var. *prolifera*. [RPS 18]
Parodia maassii var. **albescens** Ritter, Succulenta 1963: 179, 1963. [RPS 14]
Parodia maassii var. **camargensis** (Buining & Ritter) Krainz, Kat. ZSS ed. 2, 101, 1967. Basionym: *Parodia camargensis*. [RPS 18]
Parodia maassii var. **carminatiflora** Ritter, Succulenta 1963: 179, 1963. [RPS 14]
Parodia maassii var. **commutans** (Ritter) Krainz, Kat. ZSS ed. 2, 101, 1967. Basionym: *Parodia commutans*. [RPS 18]
Parodia maassii var. **intermedia** Ritter, Succulenta 1963: 179, 1963. [RPS 14]
Parodia maassii var. **rectispina** Backeberg, Cact. Succ. J. (US) 23(3): 84, 1951. [RPS 3]
Parodia maassii var. **rubida** (Ritter) Krainz, Kat. ZSS ed. 2, 102, 1967. Basionym: *Parodia rubida*. [RPS 18]
Parodia maassii var. **shaferi** Ritter, Succulenta 1963: 179, 1963. [RPS 14]
Parodia maassii var. **subterranea** (Ritter) Krainz, Kat. ZSS ed. 2, 102, 1967. Basionym: *Parodia subterranea*. [RPS 18]
Parodia maassii var. **suprema** (Ritter) Krainz, Kat. ZSS ed. 2, 102, 1967. Basionym: *Parodia suprema*. [RPS 18]
Parodia macednosa F. Brandt, Cactus 9(3): 42-45, 1977. [First mentioned as provisional name in Kakt. Orch.-Rundschau 1975/76(2): 15, 1975, later again mentioned as provisional name in Frankfurter Kakt.-Freund. 5(1): 9, 1978.] [RPS 28]
Parodia macrancistra (Schumann) Y. Ito *ex* Weskamp, Gattung Parodia, 424-425, 1987. Basionym: *Echinocactus microspermus* var. *macrancistrus*. [Basionym erroneously cited as *Parodia microsperma* var. *macrancistra*, but with correct bibliographic citation. First invalidly published (incorrect basionym taxon given) by Y. Ito, Explan. Diagr. Austroechinocactinae, 270, 1957 (cf. RPS 8).] [RPS 38]
Parodia macrancistra var. *rigidispina* Fric & Kreuzinger *ex* Y. Ito, Explan. Diagr. Austroechinocactinae, 270, 1957. *Nom. inval.*, based on *Microspermia macrancistra* var. *rigidispina, nom. inval.* (Art. 36.1). [Invalid anyway at the time of publication, as the combination for the species had not been published validly.] [RPS 8]
Parodia magnifica (Ritter) F. Brandt, Kakt. Orch.-Rundschau 7(4): 62, 1982. Basionym: *Eriocactus magnificus*. [RPS 33]
Parodia mairanana Cardenas, Nation. Cact. Succ. J. 12(4): 84, 1957. [RPS 8]
Parodia mairanana var. *atra* Backeberg, Descr. Cact. Nov. 3: 11, 1963. *Nom. inval.* (Art. 9.5). [Erroneously included as valid in RPS 14.] [RPS 14]
Parodia malyana Rausch, Kakt. and. Sukk. 21(3): 45, 1970. Typus: *Rausch* 156 (W, ZSS). [First invalidly published (Art. 37.1) in l.c. 20(1): 8, ill., 1969.] [Argentina: Catamarca] [RPS 21]
Parodia malyana fa. *citriflora* Rausch, Kakt.

and. Sukk. 20: 8, 1969. *Nom. inval.* (Art. 37.1, 43.1). [RPS 20]

Parodia malyana ssp. *igneiflora* F. Brandt, Kakt. Orch.-Rundschau 13(1): 1, 1988. *Nom. inval.* (Art. 37.1). [Given under the heading "Combinationes Novae", but without basionym. At the same time, a Latin diagnosis is supplied and the name is designated as 'subspec. nova'. First invalidly published l.c. 6(4): 96, 1981 (sphalm. 'igneuiflora'), cf. RPS 32.] [RPS 39]

Parodia malyana var. *rubriflora* F. Brandt, Frankfurter Kakt.-Freund 6(2): 6-7, 1979. *Nom. inval.* (Art. 35). [RPS 30]

Parodia mammulosa (Lemaire) N. P. Taylor, Bradleya 5: 93, 1987. Basionym: *Echinocactus mammulosus*. [RPS 38]

Parodia matthesiana Heinrich, Kakt./Sukk. 1968: 37-38, 1968. *Nom. inval.* (Art. 9.5, 36.1). [Published as provisional name.] [RPS 19]

Parodia maxima Ritter, Succulenta 43(2): 23, 1964. Typus: *Ritter* 87 (U, ZSS). [Bolivia: Tarija] [RPS 15]

Parodia mendezana F. Brandt, Cactus 8: 93-96, 1976. [First mentioned as provisional name in Kakt. Orch.-Rundschau 1975/76(2): 17, 1975, later again mentioned as provisional name in Frankfurter Kakt.-Freund 5(1): 10, 1978.] [RPS 27]

Parodia mercedesiana Weskamp, Kakt. and. Sukk. 35(3): 56-57, ills., 1984. Typus: *Herzog* 116 (B). [RPS 35]

Parodia mesembrina F. Brandt, Kakt. Orch.-Rundschau 1977(3): 33-36, 1977. [First mentioned as provisional name in l.c. 1975/76(2): 21, 1975, later again mentioned as provisional new name in Frankfurter Kakt.-Freund 5(2): 13, 1978.] [RPS 28]

Parodia mesembrina var. **juanensis** Weskamp, Kakt. and. Sukk. 41(7): 122-123, ill., 1990. Typus: *Herzog* 88 (WU). [Argentina: San Juan] [RPS 41]

Parodia microsperma ssp. **horrida** (F. Brandt) Kiesling & Ferrari, Cact. Succ. J. (US) 62(4): 198, 1990. Basionym: *Parodia horrida*. [RPS 41]

Parodia microsperma var. **aurantiaca** F. Brandt, Kakt. Orch.-Rundschau 1977(2): 26, 1977. [First mentioned as provisional name in l.c. 1975/76(2): 21, 1975, later again mentioned as provisional new name in Frankfurter Kakt.-Freund 5(2): 13, 1978.] [RPS 28]

Parodia microsperma var. **brevispina** (Haage jr.) Y. Ito, Explan. Diagr. Austroechinocactinae, 267, 1957. Basionym: *Echinocactus microspermus* var. *brevispinus*. [RPS 8]

Parodia microsperma var. *brunispina* (Schelle) Y. Ito, Cacti, 76, 1952. *Nom. inval.* (Art. 32.1). [Based on *Echinocactus microspermus brunispina* Hort. (SChelle 1926), *nom. nud.*] [RPS 4]

Parodia microsperma var. *cafayatensis* Backeberg, Kakt.-Lex., 345, 1966. *Nom. inval.* (Art. 37.1). [Erroneously included as valid in RPS 17.] [RPS 17]

Parodia microsperma var. **elegans** (Haage jr.) Y. Ito, Explan. Diagr. Austroechinocactinae, 267, 1957. Basionym: *Echinocactus microspermus* var. *elegans*. [RPS 8]

Parodia microsperma var. **erythrantha** (Spegazzini) Weskamp, Rep. Pl. Succ. 24: 8, 1975. Basionym: *Echinocactus microspermus* var. *erythrantha*. [RPS 23]

Parodia microsperma var. *gigantea* Y. Ito, Cacti, 76, 1952. *Nom. inval.* (Art. 36.1). [Published as *nomen nudum*.] [RPS 4]

Parodia microsperma var. **microthele** (Backeberg) Krainz, Kat. ZSS ed. 2, 102, 1967. Basionym: *Parodia microthele*. [RPS 18]

Parodia microsperma var. **opulenta** F. Brandt, Kakt. Orch.-Rundschau 1975/76(5): 67, 1976. [First mentioned as provisional name in l.c. 1975/76(2): 21, 1975, later again mentioned as provisional new name in Frankfurter Kakt.-Freund 5(2): 13, 1978.] [RPS 27]

Parodia microsperma var. **rigidissima** Fric *ex* F. Brandt, Kakt. Orch.-Rundschau 1975/76(5): 68, 1976. [RPS 27]

Parodia microsperma var. *rubriflora* (Backeberg) Weskamp, Repert. Pl. Succ. 24: 8, 1975. *Nom. inval.*, based on *Parodia rubriflora, nom. inval.* (Art. 9.5). [RPS 24]

Parodia microsperma var. *tarija* F. Brandt, Frankfurter Kakt.-Freund 5(2): 13, 1978. *Nom. inval.* (Art. 36.1, 37.1). [First mentioned as provisional name in Kakt. Orch.-Rundschau 1975/76(2): 21, 1975.] [RPS -]

Parodia microsperma var. **thionantha** (Spegazzini) Y. Ito, Explan. Diagr. Austroechinocactinae, 266, 1957. Basionym: *Echinocactus microspermus* var. *thionanthus*. [First published by Y. Ito in Cacti, 76, 1952 (cf. RPS 4).] [RPS 8]

Parodia miguillensis Cardenas, Cact. Succ. J. (US) 33: 109, 1961. [Frequently mis-spelled 'miquillensis'.] [RPS 12]

Parodia minima F. Brandt, Frankfurter Kakt.-Freund 3(3): 6-7, 1976. [RPS 27]

Parodia minima Rausch, Succulenta 64(9): 177-178, ill., 1985. Based on *Rausch 757*. *Nom. illeg.* (Art. 64). [*non Parodia minima* F. Brandt 1976.] [RPS 36]

Parodia minuscula Rausch, Succulenta 64(11): 230, 1985. Typus: *Rausch* 757 (ZSS). [*Nom. nov. pro Parodia minima* Rausch *non* F. Brandt.] [RPS 36]

Parodia minuta Ritter, Kakt. Südamer. 2: 546, 1980. [RPS 31]

Parodia miranda F. Brandt, Kakt. Orch.-Rundschau 6(4): 93-96, 1981. [RPS 32]

Parodia mueller-melchersii (Fric *ex* Backeberg) N. P. Taylor, Bradleya 5: 93, 1987. Basionym: *Notocactus mueller-melchersii*. [RPS 38]

Parodia muhrii F. Brandt, Kakt. Orch.-Rundschau 1: 14-17, 1978. [RPS 29]

Parodia multicostata Ritter & Jelinek *ex* Ritter, Kakt. Südamer. 2: 529-530, 1980. Typus: *Ritter* 733 (U, ZSS). [Bolivia: Chuquisaca] [RPS 31]

Parodia mutabilis var. **carneospina** Backeberg, Cact. Succ. J. (US) 23(3): 84, 1951. [RPS 3]

Parodia mutabilis var. **elegans** Backeberg, Descr. Cact. Nov. [1:] 31, 1957. [Dated 1956, published 1957.] [RPS 7]

Parodia mutabilis var. **sanguiniflora** (Backeberg) F. Brandt, Kakt. Orch.-Rundschau 7(5): 77, 1982. Basionym: *Parodia sanguiniflora*. [First published invalidly (Art. 33.2) in Kakt. Orch.-Rundschau 1975/76(2): 23. Repeated by F. Brandt in Frankfurter Kakt.-Freund 5(3): 6, 1978 and later in Letzeb. Cacteefren 7(2): 21 (this citation is reported as correct in RPS 37).] [RPS 33]

Parodia mutabilis var. *scopaoides* (Backeberg) F. Brandt, Kakt. Orch.-Rundschau 1975/76(2): 23, 1976. *Nom. inval.* (Art. 33.2), based on *Parodia scopaoides*. [Repeated in Frankfurter Kakt.-Freund 5(3): 5, 1978.] [RPS 27]

Parodia nana Weskamp, Kakt. and. Sukk. 41(9): 186-187, ill., 1990. Typus: *Herzog* 133 (WU). [Argentina: Catamarca] [RPS 41]

Parodia neglecta F. Brandt, Kakt. and. Sukk. 24: 49-51, 1973. [RPS 24]

Parodia neglectoides F. Brandt, Stachelpost 9: 129-131, 1973. [RPS 24]

Parodia neohorstii (Theunissen) N. P. Taylor, Bradleya 5: 93, 1987. Basionym: *Notocactus neohorstii*. [RPS 38]

Parodia nigrispina (Schumann) F. Brandt, Kakt. Orch.-Rundschau 7(4): 61, 1982. Basionym: *Echinocactus nigrispinus*. [RPS 33]

Parodia obtusa Ritter, Succulenta 43(3): 44, 1964. [RPS 15]

Parodia obtusa ssp. *atochana* F. Brandt, Kakt. Orch.-Rundschau 13(1): 1, 1988. *Nom. inval.* (Art. 37.1). [See note above for *P. malyana* ssp. *igneiflora*.] [RPS 39]

Parodia obtusa var. *atochana* F. Brandt, Frankfurter Kakt.-Freund 4(3): 7, 1977. *Nom. inval.* (Art. 37.1). [RPS 28]

Parodia ocampoi Cardenas, Kakt. and. Sukk. 6(1): 101-103, 1955. [RPS 6]

Parodia occulta Ritter, Kakt. Südamer. 2: 528-529, 1980. [RPS 31]

Parodia otaviana Cardenas, Cactus (Paris) No. 78: 94, 1963. [RPS 14]

Parodia ottonis (Lehmann) N. P. Taylor, Bradleya 5: 93, 1987. Basionym: *Cactus ottonis*. [RPS 38]

Parodia ottonis var. **tortuosa** (Link & Otto) N. P. Taylor, Bradleya 5: 93, 1987. Basionym: *Echinocactus tortuosus*. [RPS 38]

Parodia otuyensis Ritter, Cactus (Paris) No. 76: 52, 1962. [RPS 13]

Parodia pachysa F. Brandt, Frankfurter Kakt.-Freund 5(4): 5-6, 1978. [RPS 29]

Parodia papagayana F. Brandt, Kaktus 11: 88-89, 1976. [First mentioned as provisional name in Kakt. Orch.-Rundschau 1975/76(2): 23, 1975, later published as new name in Frankfurter Kakt.-Freund 5(3): 5, 1978 (*nom. inval.*, Art. 36.1, 37.1).] [RPS 27]

Parodia parvula F. Brandt, Kaktus 10: 6-8, 1975. [RPS 26]

Parodia paucicostata (Ritter) Weskamp, Stachelpost 9(48): 189, 1973. *Nom. inval.* (Art. 34.1), based on *Parodia comarapana* var. *paucicostata*. [Combination ascribed to F. Brandt.] [RPS 25]

Parodia penicillata Fechser & Van der Steeg, Succulenta 1960: 77, 1960. [RPS 11]

Parodia penicillata var. *fulviceps* Backeberg, Kakt.-Lex., 459, 1966. *Nom. inval.* (Art. 37.1). [Erroneously included as valid in RPS 17.] [RPS 17]

Parodia penicillata var. *nivosa* Fechser, in Backeberg, Kakt.-Lex., 460, 1966. *Nom. inval.* (Art. 37.1). [Erroneously included as valid in RPS 17.] [RPS 17]

Parodia perplexa F. Brandt, Kakt. Orch.-Rundschau 6(5): 114-117, 1981. [RPS 32]

Parodia perplexa var. **cupreo-aurea** F. Brandt, Kakt. Orch.-Rundschau 9(S4): 18, ills., 1984. Typus: *Anonymus* s.n. (HEID). [ex cult. Brandt acc. no. 85/b.] [RPS 35]

Parodia piltziorum Weskamp, Kakt. and. Sukk. 31(7): 202-203, 1980. [RPS 31]

Parodia pluricentralis Backeberg *ex* F. Brandt, Stachelpost 7: 365-367, ills., 1971. *Nom. inval.* (Art. 9.5). [First mentioned (*nom. inval.* Art. 9.5, 36.1) by Backeberg in Kakt.-Lex. 346, 1966 (and erroneously included as valid in RPS 17).] [RPS 22]

Parodia pluricentralis var. *H-2* F. Brandt, Kakt. Orch.-Rundschau 1975/76(6): 85, 1976. *Nom. inval.* (Art. 43.1). [Erroneously included as valid in RPS 27. First mentioned as provisional name in l.c. 1975/76(2): 24, 1975, later repeated (invalidly) in Frankfurter Kakt.-Freund 5(3): 7, 1978.] [RPS 27]

Parodia pluricentralis var. *erythroflora* F. Brandt, Kakt. Orch.-Rundschau 1975/76(6): 84, 1976. *Nom. inval.* (Art. 43.1). [Erroneously included as valid in RPS 27. First mentioned as provisional name in l.c. 1975/76(2): 24, 1975, later repeated (invalidly) in Frankfurter Kakt.-Freund 5(3): 7, 1978.] [RPS 27]

Parodia pluricentralis var. *xanthoflora* F. Brandt, Kakt. Orch.-Rundschau 1975/76(6): 86, 1976. *Nom. inval.* (Art. 43.1). [Erroneously included as valid in RPS 27. First mentioned as provisional name in l.c. 1975/76(2): 24, 1975, later repeated (invalidly) in Frankfurter Kakt.-Freund 5(3): 7, 1978.] [RPS 27]

Parodia prestoensis F. Brandt, Kaktus 11: 54-55, 1976. [First mentioned as provisional name in Kakt. Orch.-Rundschau 1975/76(2): 18, 1975, and again mentioned as provisional name in Frankfurter Kakt.-Freund 5(1): 12, 1978.] [RPS 27]

Parodia procera Ritter, Taxon 13: 117, 1964. [Emendation by F. Brandt in Kakt. and. Sukk. 22: 203-206, 1971.] [RPS 15]

Parodia pseudoayopayana Cardenas, Cact. Succ. J. (US) 42: 187-188, 1970. [RPS 21]

Parodia pseudoprocera F. Brandt, Kakt. and. Sukk. 21 (7): 122-124, 1970. [RPS 21]

Parodia pseudoprocera ssp. *aurantiaciflora* F. Brandt, Kakt. Orch.-Rundschau 13(1): 1, 1988. *Nom. inval.* (Art. 37.1). [See note above for *P. malyana* ssp. *igneiflora*.] [RPS 39]

Parodia pseudoprocera var. *aurantiaciflora* F. Brandt, Letzebueger Cacteefren 3(2): 1-5, ills., 1982. *Nom. inval.* (Art. 37.1). [RPS 33]

Parodia pseudostuemeri Backeberg, Descr. Cact. Nov. 3: 11, 1963. *Nom. inval.* (Art. 9.5). [Erroneously included as valid in RPS 14.] [RPS 14]

Parodia pseudosubterranea F. Brandt, Kakt. Orch.-Rundschau 179(5): 65-68, 1979. [RPS 30]

Parodia pulchella F. Brandt, Kakt. Orch.-Rundschau 1975/76(2): 15, 1975. *Nom. inval.* (Art. 36.1, 37.1). [Published as provisional name, mentioned again in Frankfurter Kakt.-Freund 4(4): 12, 1977.] [RPS -]

Parodia punae Cardenas, Cact. Succ. J. (US) 42: 39, 1970. [RPS 21]

Parodia purpureo-aurea Ritter, Succulenta 43: 57, 1964. [RPS 15]

Parodia pusilla F. Brandt, Cact. Succ. J. (US) 49(3): 119-121, 1977. [First mentioned as provisional name in Kakt. Orch.-Rundschau 1975/76(2): 20, 1975, later again mentioned as provisional name in Frankfurter Kakt.-Freund 5(2): 12, 1978.] [RPS 28]

Parodia quechua F. Brandt, Kakt. Orch.-Rundschau 1977(1): 4-7, 1977. [First mentioned as provisional name in l.c. 1975/76(2): 14, 1975, later again mentioned as provisional name in Frankfurter Kakt.-Freund 4(4): 11, 1977.] [RPS 28]

Parodia rauschii Backeberg, Descr. Cact. Nov. 3: 11, 1963. *Nom. inval.* (Art. 9.5). [Erroneously included as valid in RPS 14.] [RPS 14]

Parodia rechensis (Buining) F. Brandt, Kakt. Orch.-Rundschau 7(4): 65, 1982. Basionym: *Notocactus rechensis*. [RPS 33]

Parodia rigida Backeberg, Descr. Cact. Nov. 3: 11, 1963. *Nom. inval.* (Art. 9.5). [Erroneously included as valid in RPS 14.] [RPS 14]

Parodia rigidissima (Fric) Y. Ito, Explan. Diagr. Austroechinocactinae, 270, 1957. Basionym: *Echinocactus rigidissimus*. [Sphalm. 'rigidissimac'.] [RPS 8]

Parodia rigidissima var. *rubriflora* Fric *ex* Y. Ito, Explan. Diagr. Austroechinocactinae, 270, 1957. *Nom. inval.*, based on *Microspermia rigidissima* var. *rubriflora, nom. inval.* (Art. 36.1). [RPS 8]

Parodia riograndensis F. Brandt, Kakt. Orch.-Rundschau 1975/76(5): 76-78, ills., 1976. [RPS 27]

Parodia riojensis Ritter & Weskamp, Gattung Parodia, 474-476, fig. 75, 1987. Typus: *Ritter* 917 (DR). [Argentina: Catamarca / La Rioja] [RPS 38]

Parodia ritteri Buining, Succulenta 389(2): 17-20, ill., 1959. Typus: *Ritter* 85 (U, ZSS). [Bolivia: Tarija] [RPS 10]

Parodia ritteri var. **cintiensis** (Ritter) Krainz, Kat. ZSS ed. 2, 102, 1967. Basionym: *Parodia cintiensis*. [RPS 18]

Parodia roseoalba Ritter, Succulenta 43: 23, 1964. [RPS 15]

Parodia roseoalba var. **australis** Ritter, Kakt. Südamer. 2: 523, 1980. [RPS 31]

Parodia rostrum-sperma F. Brandt, Stachelpost 9: 2-4, 1973. [RPS 24]

Parodia rubellihamata Backeberg, Descr. Cact. Nov. 3: 11, 1963. *Nom. inval.* (Art. 9.5). [Erroneously included as valid in RPS 14.] [RPS 14]

Parodia rubellihamata var. *aureiflora* Backeberg, Kakt.-Lex., 348, 1966. *Nom. inval.* (Art. 9.5, 43.1). [Erroneously included as valid in RPS 17.] [RPS 17]

Parodia rubida Ritter, Succulenta 43(3): 43-44, 1964. Typus: *Ritter* 725 (U, ZSS). [Bolivia: Sud Cinti] [RPS 15]

Parodia rubispina Kohler, Kakt. and. Sukk. 18: 212, 1967. *Nom. inval.* (Art. 37.1). [RPS 18]

Parodia rubriflora Backeberg, Descr. Cact. Nov. 3: 12, 1963. *Nom. inval.* (Art. 9.5). [Erroneously included as valid in RPS 14.] [RPS 14]

Parodia rubrihamata Fric *ex* Y. Ito, Explan. Diagr. Austroechinocactinae, 270, 1957. *Nom. inval.*, based on *Microspermia rubrihamata, nom. inval.* (Art. 36.1). [RPS 8]

Parodia rubristaminea Ritter, Kakt. Südamer. 2: 428, 1980. [RPS 31]

Parodia rutilans (Däniker & Krainz) N. P. Taylor, Bradleya 5: 93, 1987. Basionym: *Notocactus rutilans*. [RPS 38]

Parodia saint-pieana Backeberg, Descr. Cact. Nov. [1:] 31, 1957. [Dated 1956, published 1957.] [RPS 7]

Parodia salmonea F. Brandt, Kakt. and. Sukk. 24: 97-98, 1973. [RPS 24]

Parodia salmonea var. **carminata** F. Brandt, Kakt. and. Sukk. 24: 99, 1973. [RPS 24]

Parodia salmonea var. **lau-multicostata** F. Brandt, Kakt. Orch.-Rundschau 1975/76(1): 3-4, 1975. Typus: *Lau* s.n. (HEID). [ex cult. F. Brandt acc. no. 17/c.] [Publication repeated in l.c. 12(2): 16-17, 1987. The type has Brandt's accession no. 17/c.] [Bolivia: Chuquisaca] [RPS 38]

Parodia sanagasta (Fric) Y. Ito, Explan. Diagr. Austroechinocactinae, 270, 1957. Basionym: *Echinocactus sanagastus*. [RPS 8]

Parodia sanagasta var. *saltensis* F. Brandt, Kakt. Orch.-Rundschau 1975/76(2): 22, (3): 40, 1975. *Nom. inval.* (Art. 36.1, 37.1). [Published as provisional name. Mentioned again in Frankfurter Kakt.-Freund 5(3): 5, 1978.] [RPS 27]

Parodia sanagasta var. *viridior* Backeberg, Descr. Cact. Nov. 3: 12, 1963. *Nom. inval.* (Art. 9.5, 37.1). [Erroneously included as valid in RPS 14.] [RPS 14]

Parodia sanguiniflora var. **comata** Ritter, Kakt. Südamer. 2: 428-429, 1980. [RPS 31]

Parodia schumanniana (Nicolai) F. Brandt, Kakt. Orch.-Rundschau 7(4): 62, 1982. Basionym: *Echinocactus schumannianus*. [RPS 33]

Parodia schwebsiana fa. *applanata* (Hoffmann & Backeberg) Krainz, Die Kakt. C VIe, 1963. *Nom. inval.*, based on *Parodia schwebsiana* var. *applanata, nom. inval.* (Art. 9.5). [Erroneously included as valid in RPS 14.] [RPS 14]

Parodia schwebsiana fa. **salmonea** (Backeberg) Krainz, Die Kakt. C VIe, 1963. Basionym: *Parodia schwebsiana* var. *salmonea*. [RPS 14]

Parodia schwebsiana var. *applanata* W. Hoffmann & Backeberg, Die Cact. 3: 1598, 1959. *Nom. inval.* (Art. 9.5). [Erroneously included as valid in RPS 10.] [RPS 10]

Parodia schwebsiana var. **salmonea** Backeberg, Cact. Succ. J. (US) 23(3): 84, 1951. [RPS 3]

Parodia scopa (Sprengel) N. P. Taylor, Bradleya 5: 93, 1987. Basionym: *Cactus scopa*. [RPS 38]

Parodia scoparia Ritter, Kakt. Südamer. 2: 421-422, 1980. [RPS 31]

Parodia separata F. Brandt, Catus 8: 77-79, 1976. [First mentioned as provisional name in Kakt. Orch.-Rundschau 1975/76(2): 12, 1975, later again mentioned as nude name in Frankfurter Kakt.-Freund 4(4): 9, 1977.] [RPS 27]

Parodia setispina Ritter, Succulenta 43: 57, 1964. [RPS 15]

Parodia setosa Backeberg, Descr. Cact. Nov. 3: 12, 1963. *Nom. inval.* (Art. 9.5). [Erroneously included as valid in RPS 14.] [RPS 14]

Parodia sotomayorensis Ritter, Kakt. Südamer. 2: 530, 1980. [RPS 31]

Parodia spanisa F. Brandt, Hazai Kaktusz Tükör 1977: 22-24, 1977. [First mentioned as provisional name in Kakt. Orch.-Rundschau 1975/76(2): 25, 1975, later republished (as *nom. inval.*, Art. 36.1, 37.1) under the name *P. spaniosa* in Frankfurter Kakt.-Freund 5(3): 7, 1978.] [RPS 28]

Parodia spegazziniana F. Brandt, Stachelpost 7: 367-368, 1971. [RPS 22]

Parodia spegazziniana var. **aurea** F. Brandt, Kakt. Orch.-Rundschau 1975/76(6): 83, 1976. [First mentioned as provisional name in l.c. 1975/76(2): 23, 1975, later repeated (as *nom. inval.*, Art. 36.1, 37.1) in Frankfurter Kakt.-Freund 5(3): 9, 1978.] [RPS 27]

Parodia splendens Cardenas, Cact. Succ. J. (US) 33: 108, 1961. [RPS 12]

Parodia stereospina F. Brandt, Kaktusz Vilag 1977: 10-12, 1977. [First mentioned as provisional name in Kakt. Orch.-Rundschau 1975/76(2): 14, 1975, later repeated in Letzeb. Cacteefren 7(9): 203, 201 (ill.), 1986 (nom. inval., Art. 36). Validity of publication in Kakt. Vilag not assessed.] [RPS 28]

Parodia stuemeri var. *robustior* Backeberg, Descr. Cact. Nov. 3: 12, 1963. *Nom. inval.* (Art. 9.5). [Erroneously included as valid in RPS 14.] [RPS 14]

Parodia subterranea Ritter, Succulenta 43(3): 43, 1964. Typus: *Ritter* 731 (U, ZSS). [Bolivia: Sud Cinti] [RPS 15]

Parodia subtilihamata Ritter, Kakt. Südamer. 2: 537-538, 1980. [RPS 31]

Parodia succinea (Ritter) N. P. Taylor, Bradleya 5: 93, 1987. Basionym: *Notocactus succineus*. [RPS 38]

Parodia sucrensis F. Brandt, Kakt. Orch.-Rundschau 8(1): 1-4, ills., 1983. Typus: *Knize* s.n. (HEID). [Type ex cult. F. Brandt 93/a.] [RPS 34]

Parodia superba F. Brandt, Kakt. and. Sukk. 21: 15-16, 1970. [RPS 21]

Parodia suprema Ritter, Cactus (Paris) 17(76): 51-52, 1962. Typus: *Ritter* 912 (U, ZSS). [Bolivia: Tarija] [RPS 13]

Parodia tafiensis Backeberg, Descr. Cact. Nov. 3: 12, 1963. *Nom. inval.* (Art. 9.5). [Erroneously included as valid in RPS 14.] [RPS 14]

Parodia talaensis F. Brandt, Cactus 8: 57-60, 1976. [First mentioned as provisional name in Kakt. Orch.-Rundschau 1975/76(2): 22, 1975, later again published as new name in Frankfurter Kakt.-Freund 5(3): 5, 1978 (*nom. inval.*, Art. 36.1, 37.1).] [RPS 27]

Parodia tarabucina Cardenas, Cact. Succ. J. (US) 33: 108, 1961. [RPS 12]

Parodia taratensis Cardenas, Cact. Succ. J. (US) 36: 24, 1964. [RPS 15]

Parodia thieleana F. Brandt, Kakt. Orch.-Rundschau 1975/76(6): 80-82, ills., 1976. [RPS 27]

Parodia thionantha F. Brandt, Kakt. and. Sukk. 20: 156-157, 1969. [RPS 20]

Parodia tilcarensis var. **gigantea** (Fric *ex* Krainz) Backeberg, Die Cact. 3: 1603, 1959. Basionym: *Parodia gigantea*. [RPS 10]

Parodia tillii Weskamp, Succulenta 67(10): 210-211, ill. (p. 201), 1988. Typus: *Till* 96 (WU). [Bolivia: Dept. Chuquisaca / Santa Cruz] [RPS 39]

Parodia tojoensis F. Brandt, Kakt. Orch.-Rundschau 10(2): 13-15, ill., 1985. Typus: *Anonymus* s.n. (HEID). [The accession number of the type (Brandt 107/a) is also given for the type of *Weingartia ansaldoensis* !] [RPS 36]

Parodia tredecimcostata Ritter, Kakt. Südamer. 2: 538, 1980. [RPS 31]

Parodia tredecimcostata var. **aurea** Ritter, Kakt. Südamer. 2: 538-539, 1980. [RPS 31]

Parodia tredecimcostata var. **minor** Ritter, Kakt. Südamer. 2: 538, 1980. [RPS 31]

Parodia tuberculata Cardenas, Cact. Succ. J. (US) 23(3): 97-98, 1951. [RPS 2]

Parodia tuberculosi-costata Backeberg, Kakt.-Lex., 351, 1966. *Nom. inval.* (Art. 9.5). [Erroneously included as valid in RPS 17.] [RPS 17]

Parodia tuberculosi-costata var. *amblayana* F. Brandt, Frankfurter Kakt.-Freund 4(3): 7, 1977. *Nom. inval.* (Art. 43.1). [Erroneously included as valid in RPS 28.] [RPS 28]

Parodia tuberculosi-costata var. *cafayatensis* (Backeberg) Weskamp, Gattung Parodia, 520-521, 1987. *Nom. inval.*, based on *Parodia microsperma* var. *cafayatensis*, *nom. inval.* (Art. 37.1). [RPS 38]

Parodia tucumanensis Weskamp, Succulenta 69(6): 137-139, ill., 1990. Typus: *Lau* 471 (WU). [ex cult. Weskamp] [Argentina: Tucuman] [RPS 41]

Parodia uebelmanniana Ritter, Kakt. Südamer. 2: 425-426, 1980. [RPS 31]

Parodia uhligiana Backeberg, Descr. Cact. Nov. 3: 12, 1963. *Nom. inval.* (Art. 9.5). [Erroneously included as valid in RPS 14.] [RPS 14]

Parodia varicolor Ritter, Taxon 13: 117, 1964. [Sphalm. 'variicolor'.] [RPS 15]

Parodia varicolor var. **robustispina** Ritter, Taxon 13: 117, 1964. [Sphalm. 'variicolor'.]

[RPS 15]

Parodia wagneriana Weskamp, Kakt. / Sukk. 22(3): 70-71, ill. (p. 89), 1987. *Nom. inval.* (Art. 37.1). [Equals *Parodia pluricentralis* Backeberg *nom. inval.* (Art. 36.1 / 9.5).] [RPS 38]

Parodia warasii (Ritter) F. Brandt, Kakt. Orch.-Rundschau 7(4):62, 1982. Basionym: *Eriocactus warasii.* [Sphalm. 'warasi'.] [RPS 33]

Parodia weberiana F. Brandt, Kakt. and. Sukk. 20: 206-207, 1969. [RPS 20]

Parodia weberioides F. Brandt, Letzeb. Cacteefrenn 5(2): 29-34, ills., 1984. Typus: *Anonymus* s.n. (HEID). [ex cult. F. Brandt acc. no. 87/a.] [RPS 35]

Parodia werdermanniana (Herter) N. P. Taylor, Bradleya 5: 93, 1987. Basionym: *Notocactus werdermannianus.* [RPS 38]

Parodia weskampiana Krasucka & Spanowski, Kakt. / Sukk. 1968: 45-46, ills., 1968. [RPS 19]

Parodia xantho-camargensis F. Brandt, Kakt. Orch.-Rundschau 1975/76(2): 18, 1975. *Nom. inval.* (Art. 36.1, 37.1). [Published as provisional name. Again mentioned in Frankfurter Kakt.-Freund 5(1): 11, 1978.] [RPS -]

Parodia yamparaezi Cardenas, Cactus (Paris) No. 82: 43, 1964. [The epithet derives from a place name (not a personal name) and thus its termination presumably may not be corrected.] [RPS 15]

Parodia zaletaewana F. Brandt, Stachelpost 9: 30-33, 1973. [RPS 24]

Parodia zecheri Vasquez, Kakt. and. Sukk. 29(3): 49-50, 1978. [RPS 29]

Parodia zecheri ssp. *elachista* (F. Brandt) F. Brandt, Kakt. Orch.-Rundschau 13(1): 1, 1988. *Nom. inval.* (Art. 33.2), based on *Parodia elachista.* [RPS 39]

Parodia zecheri var. *elachista* (F. Brandt) F. Brandt, Kakt. Orch.-Rundschau 8(3): 54, 1983. *Nom. inval.* (Art. 33.2), based on *Parodia elachista.* [RPS 34]

Parodiae Mottram, Contr. New Class. Cact. Fam., 6, 12, 1990. Typus: *Parodia* Spegazzini. [Published at the rank of *linea.*] [RPS 41]

Parviopuntia Soulaire, Cactus (Paris) No. 46-47: 225-230, ills., 1955. *Nom. inval.* (Art. 36.1, 37.1). [RPS 6]

Parviopuntia boliviana (Salm-Dyck) Marnier-Lapostolle & Soulaire, Cactus (Paris) No. 48: 9, 1956. *Nom. inval.* (Art. 43.1), based on *Opuntia boliviana.* [RPS 7]

Parviopuntia chilensis (Backeberg) Marnier-Lapostolle & Soulaire, Cactus (Paris) No. 48: 10, ill., 1956. *Nom. inval.* (Art. 43.1), based on *Tephrocactus chilensis.* [RPS 7]

Parviopuntia corotilla (Schumann) Marnier-Lapostolle & Soulaire, Cactus (Paris) No. 48: 12, ill., 1956. *Nom. inval.* (Art. 43.1), based on *Opuntia corotilla.* [RPS 7]

Parviopuntia diademata (Lemaire) Marnier-Lapostoll & Soulaire, Cactus (Paris) No. 48: 9, ill., 1956. *Nom. inval.* (Art. 43.1), based on *Opuntia diademata.* [RPS 7]

Parviopuntia diademata var. *articulata* (Otto) Marnier-Lapostolle & Soulaire, Cactus (Paris) No. 48: 9, ill., 1956. *Nom. inval.* (Art. 43.1). [Basionym not given.] [RPS 7]

Parviopuntia diademata var. *calva* (Lemaire) Marnier-Lapostolle & Soulaire, Cactus (Paris) No. 48: 9, ill., 1956. *Nom. inval.* (Art. 43.1). [Basionym not given.] [RPS 7]

Parviopuntia diademata var. *inermis* (Spegazzini) Marnier-Lapostolle & Soulaire, Cactus (Paris) No. 48: 24, 1956. *Nom. inval.* (Art. 43.1), based on *Opuntia inermis.* [RPS 7]

Parviopuntia diademata var. *oligacantha* (Spegazzini) Marnier-Lapostolle & Soulaire, Cactus (Paris) No. 48: 24, 1956. *Nom. inval.* (Art. 43.1), based on *Opuntia oligacantha.* [RPS 7]

Parviopuntia diademata var. *ovata* (Pfeiffer) Marnier-Lapostolle & Soulaire, Cactus (Paris) No. 48: 24, 1956. *Nom. inval.* (Art. 43.1), based on *Opuntia ovata.* [RPS 7]

Parviopuntia diademata var. *papyracantha* (Philippi) Marnier-Lapostolle & Soulaire, Cactus (Paris) No. 48: 24, 1956. *Nom. inval.* (Art. 43.1), based on *Opuntia papyracantha.* [RPS 7]

Parviopuntia diademata var. *polyacantha* (Spegazzini) Marnier-Lapostolle & Soulaire, Cactus (Paris) No. 48: 11, ill., 1956. *Nom. inval.* (Art. 43.1). [Basionym not indicated.] [RPS 7]

Parviopuntia diademata var. *syringacantha* (Pfeiffer) Marnier-Lapostolle & Soulaire, Cactus (Paris) No. 48: 24, 1956. *Nom. inval.* (Art. 43.1), based on *Opuntia syringacantha.* [RPS 7]

Parviopuntia dimorpha (Förster) Marnier-Lapostolle & Soulaire, Cactus (Paris) No. 48: 24, 1956. *Nom. inval.* (Art. 43.1), based on *Opuntia dimorpha.* [RPS 7]

Parviopuntia duvalioides (Backeberg) Marnier-Lapostolle & Soulaire, Cactus (Paris) No. 48: 9, ill., 1956. *Nom. inval.* (Art. 43.1), based on *Tephrocactus duvalioides.* [RPS 7]

Parviopuntia ferocior (Backeberg) Marnier-Lapostolle & Soulaire, Cactus (Paris) No. 48: 12, ill., 1956. *Nom. inval.* (Art. 43.1), based on *Tephrocactus ferocior.* [Sphalm. 'ferosior'.] [RPS 7]

Parviopuntia glomerata (Haworth) Marnier-Lapostolle & Soulaire, Cactus (Paris) No. 48: 24, 1956. *Nom. inval.* (Art. 43.1), based on *Opuntia glomerata.* [RPS 7]

Parviopuntia glomerata var. *andicola* (Pfeiffer) Marnier-Lapostolle & Soulaire, Cactus (Paris) No. 48: 24, 1956. *Nom. inval.* (Art. 43.1), based on *Opuntia andicola.* [RPS 7]

Parviopuntia ignescens (Vaupel) Marnier-Lapostolle & Soulaire, Cactus (Paris) No. 48: 12, ill., 1956. *Nom. inval.* (Art. 43.1), based on *Opuntia ignescens.* [RPS 7]

Parviopuntia minor (Backeberg) Marnier-Lapostolle & Soulaire, Cactus (Paris) No. 48: 24, 1956. *Nom. inval.* (Art. 43.1). [Basionym not indicated.] [RPS 7]

Parviopuntia molinensis (Spegazzini) Marnier-Lapostolle & Soulaire, Cactus (Paris) No. 48: 24, 1956. *Nom. inval.* (Art. 43.1), based on *Opuntia molinensis.* [RPS 7]

Parviopuntia parviconspicua (Backeberg) Marnier-Lapostolle & Soulaire, Cactus (Paris) No. 48: 10, ill., 1956. *Nom. inval.* (Art. 43.1). [Basionym not indicated.] [RPS 7]

Parviopuntia pentlandii (Salm-Dyck) Marnier-Lapostolle & Soulaire, Cactus (Paris) No. 48: 12, ill., 1956. *Nom. inval.* (Art. 43.1), based on *Opuntia pentlandii.* [RPS 7]

Parviopuntia russellii (Britton & Rose) Marnier-Lapostolle & Soulaire, Cactus (Paris) No. 48: 24, 1956. *Nom. inval.* (Art. 43.1), based on *Opuntia russellii.* [RPS 7]

Parviopuntia tilcarensis (Backeberg) Marnier-Lapostolle & Soulaire, Cactus (Paris) No. 48: 11, ill., 1956. *Nom. inval.* (Art. 43.1), based on *Opuntia tilcarensis.* [RPS 7]

Pediocactus sect. **Navajoa** (Croizat) L. Benson, Cact. Succ. J. 34: 57, 1962. Basionym: *Navajoa.* [RPS 13]

Pediocactus sect. **Toumeya** (Britton & Rose) L. Benson, Cact. Succ. J. 34: 61, 1962. Basionym: *Toumeya.* [RPS 13]

Pediocactus bradyi L. Benson, Cact. Succ. J. (US) 34: 19, 1962. [See also Saguaroland Bull. 16: 40 1962.] [RPS 13]

Pediocactus bradyi var. *knowltonii* (L. Benson) Backeberg, Descr. Cact. Nov. 3: 12, 1963. *Nom. inval.* (Art. 33.2), based on *Pediocactus knowltonii.* [Erroneously included as valid in RPS 14.] [RPS 14]

Pediocactus despainii Welsh & Goodrich, Great Basin Naturalist 40: 83-86, 1980. [RPS 31]

Pediocactus glaucus (Schumann) Arp, Cact. Succ. J. (US) 44: 221, 1972. Basionym: *Echinocactus glaucus.* [RPS 23]

Pediocactus hermannii W. T. Marshall, Saguaroland Bull. 8(7): 78-81, 1954. [RPS 7]

Pediocactus knowltonii L. Benson, Cact. Succ. J. (US) 32: 193, 1960. [RPS 11]

Pediocactus mesae-verdae (Boissevain & Davidson) Arp, Cact. Succ. J. (US) 44: 222, 1972. Basionym: *Coloradoa mesae-verdae.* [RPS 23]

Pediocactus papyracanthus (Engelmann) L. Benson, Cact. Succ. J (US) 34: 61, 1962. Basionym: *Mammillaria papyracantha.* [RPS 13]

Pediocactus paradinei B. W. Benson, Cact. Succ. J. (US) 29(5): 136-137, 1957. [First mentioned as *Pediocactus sp.* by Marshall in Saguaroland Bull. 10: 89-91, 1956.] [RPS 8]

Pediocactus peeblesianus (Croizat) L. Benson, Cact. Succ. J. (US) 34: 58, 1962. Basionym: *Navajoa peeblesiana.* [RPS 13]

Pediocactus peeblesianus var. *fickeisenii* L. Benson, The Cacti of Arizona, ed. 3, 23-24, 186-187, 1969. *Nom. inval.*, based on *Navajoa fickeisenii, nom. inval.* (Art. 37.1). [Combination originally based on *Navajoa fickeisenii* Backeberg (*nom. inval.*, Art. 37.1), published in Cact. Succ. J. (US) 34: 59-60, 1962. The validation of the name is unsuccessful as two type specimens are cited (p. 24).] [RPS 20]

Pediocactus peeblesianus var. **maianus** L. Benson, The Cacti of Arizona, ed. 3, 24, 186-187, 1969. [RPS 20]

Pediocactus polyancistrus (Engelmann & Bigelow) Arp, Cact. Succ. J. (US) 44: 222, 1972. Basionym: *Echinocactus polyancistrus.* [RPS 23]

Pediocactus robustior (Coulter) Arp, Cact. Succ. J. (US) 44: 222, 1972. Basionym: *Echinocactus simpsonii* var. *robustior.* [RPS 23]

Pediocactus sileri (Engelmann) L. Benson, Cact. Succ. J. (US) 33: 53, 1961. Basionym: *Echinocactus sileri.* [RPS 12]

Pediocactus simpsonii var. *caespiticus* Backeberg, Die Cact. 5: 2846, 1961. *Nom. inval.* (Art. 37.1). [Erroneously included as valid in RPS 12.] [RPS 12]

Pediocactus simpsonii var. **hermannii** (W. T. Marshall) W. T. Marshall, Saguaroland Bull. 11: 78-80, 1957. Basionym: *Pediocactus hermannii.* [Repeated by Wiegand & Backeberg in Backeberg, Die Cact. 5: 2846, 1961 (cf. RPS 12, reported as correct place of combination).] [RPS 8]

Pediocactus simpsonii var. **indraianus** Hochstätter, Succulenta 69(9): 178-183, ills., 1990. Typus: *Hochstätter* 4 (HBG). [Spelling corrected to 'indranus' in l.c. 70(3): 55, 1991. Publication repeated in Kaktusblüte 1991: 22-25, ills., 1991.] [USA: Idaho] [RPS 41]

Pediocactus simpsonii var. **minor** (Engelmann) L. Benson, Cact. Succ. J. (US) 33: 51, 1961. Basionym: *Echinocactus simpsonii* var. *minor.* [F. Hochstätter (Succulenta 69: 170, 1990) gives two earlier publications for this combination, viz. Cockerell 1916 (which is unlikely having in mind the publication date of the genus) and Boissevain & Davidson 1940. Data not further assessed.] [RPS 12]

Pediocactus simpsonii var. **nigrispinus** Hochstätter, Succulenta 69(9): 178-183, ills., 1990. Typus: *Hochstätter* 15 (HBG). [Repeated in Kaktusblüte 1991: 18-22, ills., 1991.] [USA: Oregon] [RPS 41]

Pediocactus simpsonii var. **robustior** (Coulter) L. Benson, Cact. Succ. J. (US) 33: 52, 34: 19, 1961. Basionym: *Echinocactus simpsonii* var. *robustior.* [Emendation published by F. Hochstätter in Succulenta 69: 174-175, 1990, and repeated (and given as var. nov. !) in Kaktusblüte 1991: 14-18, ills., 1991.] [RPS 12]

Pediocactus whipplei (Engelmann & Bigelow) Arp, Cact. Succ. J. (US) 44: 222, 1972. Basionym: *Echinocactus whipplei.* [RPS 23]

Pediocactus winkleri Heil, Cact. Succ. J. (US) 51: 28, 1979. [RPS 30]

Pediocactus wrightiae (L. Benson) Arp, Cact. Succ. J. (US) 44: 222, 1972. Basionym: *Sclerocactus wrightiae.* [RPS 23]

Pelecyphora valdeziana var. **albiflora** Pazout, in Pazout, Valnicek & Subik, Kaktusy, 129, fig. 81, 1960. [Repeated by Pazout in Friciana 42: 9, 1967 (cf. RPS 18).] [RPS 12]

Peniocereus subgen. **Cullmannia** (Distefano) Buxbaum, in Krainz, Die Kakt. Lief. 62: C IIa, 1975. Basionym: *Cullmannia.* [RPS 27]

Peniocereus subgen. **Pseudoacanthocereus** Sanchez-Mejorada, Cact. Suc. Mex. 19: 38-39, 1974. Typus: *Cereus maculatus*

Weingart. *Non Pseudoacanthocereus* Ritter 1979. [RPS 25]

Peniocereus castellae Sanchez-Mejorada, Cact. Suc. Mex. 19: 12-14, 16-17, 1974. [RPS 25]

Peniocereus cuixmalensis Sanchez-Mejorada, Cact. Suc. Mec. 18: 91, 95, 1973. [RPS 24]

Peniocereus diguetii (F. A. C. Weber) Backeberg, Cact. Succ. J. (US) 23(4): 119, 1951. Basionym: *Neoevansia diguetii*. [The name given as basionym was itself a combination, and is here corrected.] [RPS 3]

Peniocereus fosterianus var. **multitepalus** Sanchez-Mejorada, Cact. Suc. Mex. 19: 48-55, 1974. [RPS 25]

Peniocereus fosterianus var. **nizandensis** Sanchez-Mejorada, Cact. Suc. Mex. 19: 48-55, 1974. [RPS 25]

Peniocereus haackeanus Backeberg, Descr. Cact. Nov. 3: 12, 1963. *Nom. inval.* (Art. 9.5). [Erroneously included as valid in RPS 14.] [RPS 14]

Peniocereus macdougallii var. *centrispinus* Backeberg, Die Cact. 6: 3843, 1962. *Nom. inval.* (Art. 37.1). [Erroneously included as valid in RPS 13.] [RPS 13]

Peniocereus maculatus (Weingart) Cutak, Cact. Succ. J. (US) 23(1-2): 75-77, 1951. Basionym: *Cereus maculatus*. [RPS 2]

Peniocereus marianus (Gentry) Sanchez-Mejorada, Cact. Suc. Mex. 7: 85-91, 102, 1962. Basionym: *Wilcoxia mariana*. [RPS 13]

Peniocereus marnierianus Backeberg, Cactus (Paris) No. 85: 103-108, 1965. *Nom. inval.* (Art. 37.1). [Erroneously included as valid in RPS 16 and 17. Publication repeated in Kakt.-Lex., 360, 1966.] [RPS 16]

Peniocereus occidentalis Bravo, Cact. Suc. Mex. 8: 79, 1963. *Nom. inval.* (Art. 37.1). [RPS 14]

Peniocereus serpentinus (Lagasca & Rodrigues) N. P. Taylor, Bradleya 5: 93, 1987. Basionym: *Cactus serpentinus*. [RPS 38]

Peniocereus tepalcatepecanus Sanchez-Mejorada, Cact. Suc. Mex. 19: 14-18, 1974. [RPS 25]

Peniocereus viperinus (F. A. C. Weber) Buxbaum, in Krainz, Die Kakt. Lfg. 62, CIIa, 1976. Basionym: *Cereus viperinus*. [RPS 27]

Peniocereus zopilotensis (Meyran) Buxbaum, in Krainz, Die Kakt. Lfg. 62, CIIa, 1976. Basionym: *Wilcoxia zopilotensis*. [RPS 27]

Pereskia subgen. **Neopereskia** Backeberg, Descr. Cact. Nov. [1:] 5, 1957. Typus: *Pereskia diaz-romeroana* Cardenas. The incorrect spelling 'Peireskia' is consistently used and here treated as an error to be corrected (including the subgeneric name 'Neopeireskia'). Dated 1956, published 1957. [RPS 7]

Pereskia aculeata fa. **rubescens** (Pfeiffer) Krainz, Die Kakt. Lief. 40/41: A (1. 3. 1969), 1969. Basionym: *Pereskia aculeata* var. *rubescens*. [RPS 20]

Pereskia aculeata var. **godseffiana** (hort.) Backeberg & Knuth *ex* Krainz, Die Kakt. Lief. 40/41: A (1. 3. 1969), 1969. Basionym: *Pereskia godseffiana*. [RPS -]

Pereskia antoniana (Backeberg) Rauh, Kakt. an ihren Standorten, 89, 1979. *Nom. inval.*, based on *Rhodocactus antonianus, nom. inval.* (Art. 9.5, 37.1). [RPS -]

Pereskia aureiflora Ritter, Kakt. Südamer. 1: 22, fig. 5, 1979. Typus: *Ritter* 1413 (U [?not deposited]). [cf. Leuenberger in Mem. New York Bot. Gard. 41: 83, 1986.] [Brazil: Minas Gerais] [RPS 30]

Pereskia corrugata Cutak, Cact. Succ. J. (US) 23(6): 173, figs. 95-96, 1951. Typus: *Cutak* s.n. (MO). [?] [RPS 2]

Pereskia diaz-romeroana Cardenas, Lilloa 23: 15, fig. 1, t. 1, 1950. Typus: *Cardenas* 4388 (US [lecto], US). [Bolivia: Cochabamba] [RPS 4]

Pereskia grandifolia var. **violacea** Leuenberger, Mem. New York Bot. Gard. 41: 116-118, ill., 1986. Typus: *F. C. F. da Silva* 89 (HRB, B). [Brazil] [RPS 37]

Pereskia higuerana Cardenas, Cactus (Paris) 19(80-81): 18, 1964. Typus: *Candia* s.n. (LIL [not located], US [not located]). [cf. Leuenberger, Mem. New York Bot. Gard. 41: 127, 1986.] [Bolivia: Vallegrande] [RPS 15]

Pereskia humboldtii var. **rauhii** (Backeberg) Leuenberger, Mem. New York Bot. Gard. 41: 69, 1986. Basionym: *Pereskia vargasii* var. *rauhii*. [RPS 37]

Pereskia quisqueyana Liogier (= H. Alain), Phytologia 43: 183, fig. 6, 1980. Typus: *Liogier* 27032 (UPR, NY, US). [Dominican Republic] [RPS -]

Pereskia saipinensis Cardenas, Cactus (Paris) 19(80-81): 17, ills., 1964. Typus: *Cardenas* 6122 (LIL [not located]). [Bolivia: Santa Cruz] [RPS 15]

Pereskia sparsiflora Ritter, Kakt. Südamer. 2: 482, fig. 330, 1980. Typus: *Ritter* 640 p.p. (US). [First mentioned by Backeberg, Die Cact. 6: 3573, 1962 (*nom. nud.*).] [Bolivia: Tarija] [RPS 13]

Pereskia stenantha Ritter, Kakt. Südamer. 1: 21, figs. 3-4, 1979. Typus: *Ritter* 1251 (U [not located]). [cf. Leuenberger in Mem. New York Bot. Gard. 41: 122-123, 1986.] [Brazil: Bahia] [RPS 30]

Pereskia vargasii H. Johnson, Cact. Succ. J. (US) 24(4): 115-116, 1952. Typus: *Vargas et Johnson* s.n. (not located). [Peru: Jaen] [RPS 3]

Pereskia vargasii var. **longispina** Rauh & Backeberg, in Backeberg, Descr. Cact. Nov. [1:] 6, 1957. Typus: *Rauh* K80 (lost). [cf. Leuenberger in Mem. New York Bot. Gard. 41: 69, 1986.] [Dated 1956, published 1957. Amplified description in Rauh, Beitr. Kenntn. Peruan. Kakt.veg., 190, 1958.] [Peru: Jaen] [RPS 7]

Pereskia vargasii var. **rauhii** Backeberg, Descr. Cact. Nov. [1:] 6, 1957. Typus: *Rauh et Hirsch* P2162 (not located). [Dated 1956, published 1957.] [Peru: Jaen] [RPS 7]

Peruvocereus multangularis (Willdenow) Akers, Cact. Succ. J. (US) 22(6): 174-175, 1950. Basionym: *Cactus multangularis*. [Sometimes spelled 'multiangularis'.] [RPS 2]

Pfeiffera erecta Ritter, Taxon 13(3): 116, 1964.

Typus: *Ritter* 883 (U, ZSS). [Type cited for U l.c. 12: 28, 1963.] [Bolivia: Santa Cruz] [RPS 15]

Pfeiffera gracilis Ritter, Kakt. Südamer. 2: 505, 1980. Typus: *Ritter* 882 (U, ZSS). [Type cited for U l.c. 1: iii, 1979.] [Bolivia] [RPS 31]

Pfeiffera ianthothele var. **boliviana** Ritter, Kakt. Südamer. 2: 503, 1980. [RPS 31]

Pfeiffera ianthothele var. **tarijensis** Ritter, Kakt. Südamer. 2: 503-504, 1980. Typus: *Ritter* 880 (U ?). [Bolivia: Dept. Tarija] [RPS 31]

Pfeiffera mataralensis Ritter, Taxon 13(3): 115-116, 1964. Typus: *Ritter* 363 (U, ZSS). [Type cited for U l.c. 12: 28, 1963.] [Bolivia: Florida] [RPS 15]

Pfeiffera mataralensis var. **floccosa** Ritter, Taxon 13: 116, 1964. [RPS 15]

Pfeiffera miyagawae Barthlott & Rauh, Cact. Succ. J. (US) 58(2): 63-65, ills., SEM ill., 1987. Typus: *Miyagawa* s.n. (HEID 32857, BONN, HNT, ZSS). [Bolivia: Cochabamba] [RPS 38]

Pfeiffera multigona Cardenas, Cactus (Paris) No. 82: 51, 1964. [RPS 15]

Pfeifferae Buxbaum, Madroño 14(6): 184, 1958. Typus: *Pfeiffera*. [Published as 'linea'.] [RPS 9]

Pfeifferinae Volgin, Feddes Repert. 97(9-10): 563, 1986. Typus: *Pfeiffera* Salm-Dyck. [RPS 37]

Phellosperma subgen. *Euphellosperma* Buxbaum, Österr. Bot. Zeitschr. 98(1-2): 92, 1951. *Nom. inval.* (Art. 21.3). [*Phellosperma tetrancistra* (Engelmann) Britton & Rose.] [RPS 2]

Phellosperma subgen. **Krainzia** (Backeberg) Buxbaum, Österr. Bot. Zeitschr. 98(1-2): 92, 1951. Basionym: *Krainzia*. [*Phellosperma longiflora* (Britton & Rose) Buxbaum] [RPS 2]

Phellosperma guelzowiana (Werdermann) Buxbaum, Österr. Bot. Zeitschr. 98(1-2): 92, 1951. Basionym: *Mammillaria guelzowiana*. [RPS 2]

Phellosperma longiflora (Britton & Rose) Buxbaum, Österr. Bot. Zeitschr. 98(1-2): 92, 1951. Basionym: *Neomammillaria longiflora*. [RPS 2]

Phellosperma pennispinosa (Krainz) Buxbaum, Österr. Bot. Zeitschr. 98(1-2): 91-92, 1951. Basionym: *Mammillaria pennispinosa*. [RPS 2]

Pilocanthus B. W. Benson & Backeberg, Kakt. and. Sukk. 8: 187-189, 1957. Typus: *Pilocanthus paradinei*. [RPS 8]

Pilocanthus paradinei (B. W. Benson) B. W. Benson & Backeberg, Kakt. and. Sukk. 8(12): 187-189, 1957. Basionym: *Pediocactus paradinei*. [RPS 8]

Pilocereus subgen. **Globicarpi** (Croizat) Croizat, Noved. Ci. Contr. Occas. Museo Nat. Hist. La Salle, Ser. Bot. 1: 4, 1950. Basionym: *Pilocereus* sect. *Globicarpi*. [RPS 5]

Pilocereus subgen. **Oblongicarpi** (Croizat) Croizat, Noved. Ci. Contr. Occas. Museo Nat. Hist. La Salle, Ser. Bot. 1: 4, 1950. Basionym: *Pilocereus* sect. *Oblongicarpi*. [RPS 5]

Pilocereus campinensis Backeberg & Voll, Arq. Jard. Bot. Rio de Janeiro 9: 161-164, ills., 1950. Typus: *Anonymus* s.n. (RB 65.045). [Volume for 1949, published 1950.] [Brazil: São Paulo] [RPS -]

Pilocereus floccosus Backeberg & Voll, Arq. Jard. Bot. Rio de Janeiro 9:151-155, ills. (pp. 150, 152), 1950. Based on *Brade* s.n.. *Nom. illeg.* (Art. 64.1). [*Non Pilocereus floccosus* Lemaire 1866 (new name provided by Byles & Rowley: *Pilosocereus floccosus*).] [Brazil: Minas Gerais] [RPS -]

Pilocereus mortensenii Croizat, Noved. Ci. Contr. Ocas. Mus. Hist. Nat. La Salle, ser. Bot. No. 1: 3-4, 1950. [Sphalm. 'mortenseni'.] [RPS 9]

Pilocereus wagenaarii Croizat, Noved. Ci. Contr. Ocas. Mus. Hist. Nat. La Salle, ser. Bot. No. 1: 2, 1950. [Sphalm. 'wagenaari'.] [RPS 9]

Pilocopiapoa Ritter, Kakt. and. Sukk. 12: 20-22, 1961. Typus: *P. solaris*. [First mentioned as nude name in Backeberg, Die Cact. 1: 56, 79, 1958 (cf. RPS 9) and in l.c. 3: 1895, 1959 (cf. RPS 10).] [RPS 12]

Pilocopiapoa solaris Ritter, Kakt. and. Sukk. 12: 20, 1961. [First mentioned as nude name in Backeberg, Die Cact. 3: 1896, 1959 (cf. RPS 10).] [RPS 12]

Pilopsis Y. Ito, Bull. Takarazuka Insectarium 71: 13-20, 1950. *Nom. inval.* (Art. 36.1). [RPS 3]

Pilosocereus Byles & G. Rowley, Cact. Succ. J. Gr. Brit. 19: 66, 1957. Typus: *Pilosocereus leucocephalus*. [*Nom. nov. pro Pilocereus* K. Schumann *emend.* Backeberg, *non Pilocereus* Lemaire 1839.] [RPS 8]

Pilosocereus subgen. **Floribunda** (Ritter) P. J. Braun, Bradleya 6: 90, 1988. Basionym: *Floribunda*. [RPS 39]

Pilosocereus subgen. **Lagenopsis** (Buxbaum) P. J. Braun, Bradleya 6: 89, 1988. Basionym: *Coleocephalocereus* subgen. *Lagenopsis*. [RPS 39]

Pilosocereus albisummus P. J. Braun & E. Esteves Pereira, Kakt. and. Sukk. 38(5): 126-131, ills., SEM-ills., 1987. Typus: *Esteves Pereira* 123 (ZSS, K). [Detailed locality information deposited together with holotype.] [Brazil: Minas Gerais] [RPS 38]

Pilosocereus alensis (F. A. C. Weber) Byles & Rowley, Cact. Succ. J. Gr. Brit. 19: 66, 1957. Basionym: *Pilocereus alensis*. [RPS 8]

Pilosocereus arenicola (Werdermann) Byles & Rowley, Cact. Succ. J. Gr. Brit. 19: 66, 1957. Basionym: *Pilocereus arenicola*. [RPS 8]

Pilosocereus arrabidae (Lemaire) Byles & Rowley, Cact. Succ. J. Gr. Brit. 19: 66, 1957. Basionym: *Pilocereus arrabidae*. [Sphalm. 'arribidae'.] [RPS 8]

Pilosocereus atroflavispinus Ritter, Kakt. Südamer. 1: 68, 1979. [RPS 30]

Pilosocereus aureispinus (Buining & Brederoo) Ritter, Kakt. Südamer. 1: 83, 1979. Basionym: *Coleocephalocereus aureispinus*. [RPS 30]

Pilosocereus aurilanatus Ritter, Kakt. Südamer. 1: 77, 1979. [RPS 30]

Pilosocereus aurisetus (Werdermann) Byles & Rowley, Cact. Succ. J. Gr. Brit. 19: 66, 1957. Basionym: *Pilocereus aurisetus.* [RPS 8]

Pilosocereus azureus Ritter, Kakt. Südamer. 1: 73, 1979. [RPS 30]

Pilosocereus backebergii (Weingart) Byles & Rowley, Cact. Succ. J. Gr. Brit. 19: 66, 1957. Basionym: *Cereus backebergii.* [RPS 8]

Pilosocereus bahamensis (Britton) Byles & Rowley, Cact. Succ. J. Gr. Brit. 19: 66, 1957. Basionym: *Cephalocereus bahamensis.* [RPS 8]

Pilosocereus barbadensis (Britton & Rose) Byles & Rowley, Cact. Succ. J. Gr. Brit. 19: 66, 1957. Basionym: *Cephalocereus barbadensis.* [RPS 8]

Pilosocereus bradei (Backeberg & Voll) Byles & Rowley, Cact. Succ. J. Gr. Brit. 19: 66, 1957. Basionym: *Pilocereus bradei.* [RPS 8]

Pilosocereus brasiliensis (Britton & Rose) Backeberg, Die Cact. 4: 2423, 1960. Basionym: *Cephalocereus brasiliensis.* [RPS 11]

Pilosocereus braunii E. Esteves Pereira, Kakt. and. Sukk. 38(5): 132, 1987. Typus: *Braun* 70 (ZSS). [Detailed locality information deposited together with the holotype.] [Additional detailed description in l.c. 40(1): 6-13, ills., SEM-ills., 1989.] [Brazil: Bahia] [RPS 38]

Pilosocereus brooksianus (Vaupel) Byles & Rowley, Cact. Succ. J. Gr. Brit. 19: 66, 1957. Basionym: *Cereus brooksianus.* [RPS 8]

Pilosocereus carolinensis Ritter, Kakt. Südamer. 1: 80, 1979. [RPS 30]

Pilosocereus carolinensis var. **robustispinus** Ritter, Kakt. Südamer. 1: 81, 1979. [RPS 30]

Pilosocereus catalani (Riccobono) Byles & Rowley, Cact. Succ. J. Gr. Brit. 19: 66, 1957. Basionym: *Pilocereus catalani.* [RPS 8]

Pilosocereus catingicola (Gürke) Byles & Rowley, Cact. Succ. J. Gr. Brit. 19: 66, 1957. Basionym: *Cereus catingicola.* [RPS 8]

Pilosocereus cenepequei Rizzini & Mattos-F., Rev. Brasil Biol. 46(2): 324, 1986. Typus: *Rizzini et Mattos-Filho* 39 (RB). [Brazil] [RPS 37]

Pilosocereus chrysacanthus (F. A. C. Weber) Byles & Rowley, Cact. Succ. J. Gr. Brit. 19: 66, 1957. Basionym: *Pilocereus chrysacanthus.* [RPS 8]

Pilosocereus chrysostele (Vaupel) Byles & Rowley, Cact. Succ. J. Gr. Brit. 19: 66, 1957. Basionym: *Cereus chrysostele.* [RPS 8]

Pilosocereus claroviridis (Backeberg) Byles & Rowley, Cact. Succ. J. Gr. Brit. 19: 66, 1957. Basionym: *Cereus claroviridis.* [RPS 8]

Pilosocereus coerulescens (Lemaire) Ritter, Kakt. Südamer. 1: 75, 1979. Basionym: *Pilocereus coerulescens.* [RPS 30]

Pilosocereus collinsii (Britton & Rose) Byles & Rowley, Cact. Succ. J. Gr. Brit. 19: 66, 1957. Basionym: *Cephalocereus collinsii.* [RPS 8]

Pilosocereus columbianus (Rose) Byles & Rowley, Cact. Succ. J. Gr. Brit. 19: 66, 1957. Basionym: *Cephalocereus columbianus.* [Sphalm. 'colombianus'.] [RPS 8]

Pilosocereus cometes (Scheidweiler) Byles & Rowley, Cact. Succ. J. Gr. Brit. 19: 66, 1957. Basionym: *Cereus cometes.* [RPS 8]

Pilosocereus cristalinensis P. J. Braun & E. Esteves Pereira, Kakt. and. Sukk. 38(5): 132, (7): 163-167, ills., SEM-ills., 1987. Typus: *Esteves Pereira* 73 (ZSS). [Brazil: Goiás] [RPS 38]

Pilosocereus cuyabensis (Backeberg) Byles & Rowley, Cact. Succ. J. Gr. Brit. 19: 66, 1957. Basionym: *Pilocereus cuyabensis.* [RPS 8]

Pilosocereus cyaneus Ritter, Kakt. Südamer. 4: 1516, 1981. [*Nom. nov. pro Pseudopilocereus azureus* Buining & Brederoo, *non Pilosocereus azureus* Ritter 1979.] [RPS 32]

Pilosocereus deeringii (Small) Byles & Rowley, Cact. Succ. J. Gr. Brit. 19: 66, 1957. Basionym: *Cephalocereus deeringii.* [RPS 8]

Pilosocereus densiareolatus Ritter, Kakt. Südamer. 1: 73, 1979. [RPS 30]

Pilosocereus diersianus (E. Esteves Pereira) P. J. Braun, Bradleya 6: 88, 1988. Basionym: *Pseudopilocereus diersianus.* [RPS 39]

Pilosocereus flavipulvinatus (Buining & Brederoo) Ritter, Kakt. Südamer. 2: 707, 1980. Basionym: *Pseudopilocereus flavipulvinatus.* [RPS 31]

Pilosocereus flavipulvinatus var. **carolinensis** (Ritter) Ritter, Kakt. Südamer. 2: 707, 1980. Basionym: *Pilosocereus carolinensis.* [RPS 31]

Pilosocereus flexibilispinus P. J. Braun & E. Esteves Pereira, Kakt. and. Sukk. 41(5): 82-88, ills., SEM-ills., 1990. Typus: *Esteves Pereira* 145 (UFG 12.364, ZSS). [Brazil: Tocantins] [RPS 41]

Pilosocereus floccosus Byles & Rowley, Cact. Succ. J. Gr. Brit. 19: 67, 1957. [*Nom. nov. pro Pilocereus floccosus* Backeberg & Voll 1949, *non P. floccosus* Lemaire 1866.] [RPS 8]

Pilosocereus fulvilanatus (Buining & Brederoo) Ritter, Kakt. Südamer. 1: 84, 1979. Basionym: *Pseudopilocereus fulvilanatus.* [RPS 30]

Pilosocereus gaturianensis Ritter, Kakt. Südamer. 1: 81, 1979. [RPS 30]

Pilosocereus gaumeri (Britton & Rose) Backeberg, Die Cact. 4: 2462, 1960. Basionym: *Cephalocereus gaumeri.* [RPS 11]

Pilosocereus gironensis Rauh & Backeberg *ex* Byles & Rowley, Cact. Succ. J. Gr. Brit. 19: 69, 1957. [First published (in advance of the generic name; Art. 43) in Backeberg, Descr. Cact. Nov. [1:] 34, 1957, and reported as valid in RPS 7.] [RPS 8]

Pilosocereus glaucescens (Labouret) Byles & Rowley, Cact. Succ. J. Gr. Brit. 19: 67, 1957. Basionym: *Pilocereus glaucescens.* [RPS 8]

Pilosocereus glaucochrous (Werdermann) Byles & Rowley, Cact. Succ. J. Gr. Brit. 19: 67, 1957. Basionym: *Pilocereus glaucochrous.* [RPS 8]

Pilosocereus gounellei (F. A. C. Weber) Byles & Rowley, Cact. Succ. J. Gr. Brit. 19: 67, 1957. Basionym: *Pilocereus gounellei.* [RPS 8]

Pilosocereus gounellei var. **zehntneri** (Britton

& Rose) Byles & Rowley, Cact. Succ. J. Gr. Brit. 19: 67, 1957. Basionym: *Cephalocereus zehntneri*. [RPS 8]

Pilosocereus gruberi Schatzl & Till, Kakt. and. Sukk. 33(1): 8-9, ills., 1982. Typus: *Gruber et Schatzl* s.n. (WU). [RPS 33]

Pilosocereus guerreronis (Backeberg) Byles & Rowley, Cact. Succ. J. Gr. Brit. 19: 67, 1957. Basionym: *Pilocereus guerreronis*. [RPS 8]

Pilosocereus hapalacanthus (Werdermann) Byles & Rowley, Cact. Succ. J. Gr. Brit. 19: 67, 1957. Basionym: *Pilocereus hapalacanthus*. [RPS 8]

Pilosocereus hermentianus (Monville) Byles & Rowley, Cact. Succ. J. Gr. Brit. 19: 67, 1957. Basionym: *Cereus hermentianus*. [RPS 8]

Pilosocereus juaruensis (Buining & Brederoo) P. J. Braun, Kakt. and. Sukk. 35(8): 181, 1984. Basionym: *Pseudopilocereus juaruensis*. [RPS 37]

Pilosocereus kanukuensis (Alexander) Leuenberger, Willdenowia 16: 506, 1987. Basionym: *Cephalocereus kanukuensis*. [Published in February. Combination repeated by P. J. Braun in Succulenta 66(5): 106-107, 1987 (publ. May).] [RPS 38]

Pilosocereus keyensis (Britton & Rose) Byles & Rowley, Cact. Succ. J. Gr. Brit. 19: 67, 1957. Basionym: *Cephalocereus keyensis*. [RPS 8]

Pilosocereus lanuginosus (Linné) Byles & Rowley, Cact. Succ. J. Gr. Brit. 19: 67, 1957. Basionym: *Cactus lanuginosus*. [RPS 8]

Pilosocereus leucocephalus (Poselger) Byles & Rowley, Cact. Succ. J. Gr. Brit. 19: 67, 1957. Basionym: *Pilocereus leucocephalus*. [RPS 8]

Pilosocereus lindaianus P. J. Braun & E. Esteves Pereira, Kakt. and. Sukk. 38(5): 132, 1987. Typus: *Esteves Pereira* 60 (ZSS). [Additional detailed description and illustrations in Kakt. and. Sukk. 39(10): 222-228, 1988.] [Brazil: Goiás] [RPS 38]

Pilosocereus luetzelburgii (Vaupel) Byles & Rowley, Cact. Succ. J. Gr. Brit. 19: 67, 1957. Basionym: *Cereus luetzelburgii*. [RPS 8]

Pilosocereus machrisii (Dawson) Backeberg, Die Cact. 4: 2419, 1960. Basionym: *Cephalocereus machrisii*. [RPS 11]

Pilosocereus magnificus (Buining & Brederoo) Ritter, Kakt. Südamer. 1: 72, 1979. Basionym: *Pseudopilocereus magnificus*. [RPS 30]

Pilosocereus maxonii (Rose) Byles & Rowley, Cact. Succ. J. Gr. Brit. 19: 67, 1957. Basionym: *Cephalocereus maxonii*. [RPS 8]

Pilosocereus millspaughii (Britton) Byles & Rowley, Cact. Succ. J. Gr. Brit. 19: 67, 1957. Basionym: *Cephalocereus millspaughii*. [RPS 8]

Pilosocereus minensis (Werdermann) Byles & Rowley, Cact. Succ. J. Gr. Brit. 19: 67, 1957. Basionym: *Cereus minensis*. [RPS 8]

Pilosocereus monoclonos (De Candolle) Byles & Rowley, Cact. Succ. J. Gr. Brit. 19: 67, 1957. Basionym: *Cereus monoclonos*. [RPS 8]

Pilosocereus moritzianus (Otto) Byles & Rowley, Cact. Succ. J. Gr. Brit. 19: 67, 1957. Basionym: *Cereus moritzianus*. [RPS 8]

Pilosocereus mortensenii (Croizat) Backeberg, Die Cact. 4: 2450, 1960. Basionym: *Pilocereus mortensenii*. [RPS 11]

Pilosocereus mucosiflorus (Buining & Brederoo) Ritter, Kakt. Südamer. 1: 84, 1979. Basionym: *Pseudopilocereus mucosiflorus*. [RPS 30]

Pilosocereus multicostatus Ritter, Kakt. Südamer. 1: 79, 1979. [RPS 30]

Pilosocereus nobilis (Haworth) Byles & Rowley, Cact. Succ. J. Gr. Brit. 19: 67, 1957. Basionym: *Cereus nobilis*. [RPS 8]

Pilosocereus oligolepis (Vaupel) Byles & Rowley, Cact. Succ. J. Gr. Brit. 19: 67, 1957. Basionym: *Cereus oligolepis*. [Sphalm. 'oligolepsis'.] [RPS 8]

Pilosocereus oreus Ritter, Kakt. Südamer. 1: 69, 1979. [RPS 30]

Pilosocereus pachycladus Ritter, Kakt. Südamer. 1: 70, 1979. [RPS 30]

Pilosocereus palmeri (Rose) Byles & Rowley, Cact. Succ. J. Gr. Brit. 19: 67, 1957. Basionym: *Cephalocereus palmeri*. [RPS 8]

Pilosocereus palmeri var. **victoriensis** (Vaupel) Backeberg, Kakt.-Lex. 367, 1966. Basionym: *Cereus victoriensis*. [RPS 17]

Pilosocereus paraguayensis Ritter, Kakt. Südamer. 1: 249, 1979. [RPS 30]

Pilosocereus parvus (Diers & E. Esteves Pereira) P. J. Braun, Bradleya 6: 88, 1988. Basionym: *Pseudopilocereus parvus*. [RPS 39]

Pilosocereus pentaedrophorus (Labouret) Byles & Rowley, Cact. Succ. J. Gr. Brit. 19: 67, 1957. Basionym: *Cereus pentaedrophorus*. [RPS 8]

Pilosocereus perlucens (Schumann) Byles & Rowley, Cact. Succ. J. Gr. Brit. 19: 67, 1957. Basionym: *Cereus perlucens*. [RPS 8]

Pilosocereus pernambucoensis Ritter, Kakt. Südamer. 1: 65, 1979. [RPS 30]

Pilosocereus pernambucoensis var. **caesius** Ritter, Kakt. Südamer. 1: 66, 1979. [RPS 30]

Pilosocereus pernambucoensis var. **montealtoi** Ritter, Kakt. Südamer. 1: 66, 1979. [RPS 30]

Pilosocereus piauhyensis (Gürke) Byles & Rowley, Cact. Succ. J. Gr. Brit. 19: 67, 1957. Basionym: *Cereus piauhyensis*. [RPS 8]

Pilosocereus pleurocarpus (Ritter) P. J. Braun, Bradleya 6: 88, 1988. Basionym: *Cipocereus pleurocarpus*. [RPS 39]

Pilosocereus polygonus (Lamarck) Byles & Rowley, Cact. Succ. J. Gr. Brit. 19: 67, 1957. Basionym: *Cactus polygonus*. [RPS 8]

Pilosocereus purpusii (Britton & Rose) Byles & Rowley, Cact. Succ. J. Gr. Brit. 19: 67, 1957. Basionym: *Cephalocereus purpusii*. [RPS 8]

Pilosocereus pusillibaccatus P. J. Braun & E. Esteves Pereira, Cact. Succ. J. (US) 58(6): 240-243, ills., 1986. Typus: *Esteves Pereira* 202 (ZSS). [Brazil] [RPS 37]

Pilosocereus pusilliflorus (Ritter) P. J. Braun, Bradleya 6: 90, 1988. Basionym: *Floribunda pusilliflora*. [RPS 39]

Pilosocereus quadricentralis (Dawson) Backeberg, Die Cact. 4: 2437, 1960. Basionym: *Cephalocereus quadricentralis*. [RPS 11]

Pilosocereus quadricostatus Ritter, Kakt.

Südamer. 1: 78, 1979. [RPS 30]
Pilosocereus robinii (Lemaire) Byles & Rowley, Cact. Succ. J. Gr. Brit. 19: 67, 1957. Basionym: *Pilocereus robinii*. [RPS 8]
Pilosocereus robustus Ritter, Kakt. Südamer. 1: 72, 1979. [*Non Cephalocereus robustus* Britton & Rose.] [RPS 30]
Pilosocereus rosae P. J. Braun, Kakt. and. Sukk. 35(8): 178-181, ills., 1984. Typus: *Horst et Uebelmann* HU 546 (1982) (ZSS). [Brazil: Minas Gerais] [RPS 35]
Pilosocereus royenii (Linné) Byles & Rowley, Cact. Succ. J. Gr. Brit. 19: 67, 1957. Basionym: *Cactus royenii*. [RPS 8]
Pilosocereus rupicola (Werdermann) Byles & Rowley, Cact. Succ. J. Gr. Brit. 19: 67, 1957. Basionym: *Pilocereus rupicola*. [RPS 8]
Pilosocereus ruschianus (Buining & Brederoo) P. J. Braun, Kakt. and. Sukk. 35(8): 181, 1984. Basionym: *Pseudopilocereus ruschianus*. [RPS 37]
Pilosocereus salvadorensis (Werdermann) Byles & Rowley, Cact. Succ. J. Gr. Brit. 19: 67, 1957. Basionym: *Pilocereus salvadorensis*. [RPS 8]
Pilosocereus sartorianus (Rose) Byles & Rowley, Cact. Succ. J. Gr. Brit. 19: 69, 1957. Basionym: *Cephalocereus sartorianus*. [RPS 8]
Pilosocereus saudadensis Ritter, Kakt. Südamer. 1: 82, 1979. [RPS 30]
Pilosocereus schoebelii P. J. Braun, Cact. Succ. J. (US) 58(4): 150-156, ills., SEM-ills., 1987. Typus: *Braun* 426 (ZSS, KOELN). [Brazil: Minas Gerais] [RPS 38]
Pilosocereus sergipensis (Werdermann) Byles & Rowley, Cact. Succ. J. Gr. Brit. 19: 69, 1957. Basionym: *Pilocereus sergipensis*. [RPS 8]
Pilosocereus splendidus Ritter, Kakt. Südamer. 1: 69, 1979. [RPS 30]
Pilosocereus sublanatus (Salm-Dyck) Byles & Rowley, Cact. Succ. J. Gr. Brit. 19: 69, 1957. Basionym: *Cereus sublanatus*. [RPS 8]
Pilosocereus × subsimilis Rizzini & Mattos-F., Rev. Brasil Biol. 46(2): 327, 1986. Typus: *Rizzini et Mattos-Filho* 41 (RB). [Recognized as hybrid by Taylor & Zappi in sched.] [Brazil] [RPS 37]
Pilosocereus superbus Ritter, Kakt. Südamer. 1: 67, 1979. [RPS 30]
Pilosocereus superbus var. **gacapaensis** Ritter, Kakt. Südamer. 1: 68, 1979. [RPS 30]
Pilosocereus superbus var. **lanosior** Ritter, Kakt. Südamer. 1: 68, 1979. [RPS 30]
Pilosocereus superbus var. **regius** Ritter, Kakt. Südamer. 1: 67, 1979. [RPS 30]
Pilosocereus superfloccosus (Buining & Brederoo) Ritter, Kakt. Südamer. 1: 84, 1979. Basionym: *Pseudopilocereus superfloccosus*. [RPS 30]
Pilosocereus supthutianus P. J. Braun, Kakt. and. Sukk. 36(5): 100-103, ills., (8): 170 [erratum], 1985. Typus: *Horst et Uebelmann* HU 547 (1982) (ZSS). [Brazil: Minas Gerais] [RPS 36]
Pilosocereus swartzii (Grisebach) Byles & Rowley, Cact. Succ. J. Gr. Brit. 19: 69, 1957. Basionym: *Cereus swartzii*. [RPS 8]
Pilosocereus tehuacanus (Weingart) Byles & Rowley, Cact. Succ. J. Gr. Brit. 19: 69, 1957. Basionym: *Pilocereus tehuacanus*. [RPS 8]
Pilosocereus tillianus Gruber & Schatzl, Kakt. and. Sukk. 33(8): 162-164, ills., 1982. Typus: *Gruber et Schatzl* 78 (WU, WU). [Venezuela: Merida] [RPS 33]
Pilosocereus tuberculatus (Werdermann) Byles & Rowley, Cact. Succ. J. Gr. Brit. 19: 69, 1957. Basionym: *Pilocereus tuberculatus*. [RPS 8]
Pilosocereus tuberculosus Rauh & Backeberg *ex* Byles & Rowley, Cact. Succ. J. Gr. Brit. 19: 69, 1957. Typus: *Rauh* K86a (1956) (ZSS). [First published (in advance of the generic name; Art. 43) in Backeberg, Descr. Cact. Nov. [1:] 34, 1957, and reported as valid in RPS 7.] [Peru] [RPS 8]
Pilosocereus tweedyanus (Britton & Rose) Byles & Rowley, Cact. Succ. J. Gr. Brit. 19: 69, 1957. Basionym: *Cephalocereus tweedyanus*. [RPS 8]
Pilosocereus ulei (Schumann) Byles & Rowley, Cact. Succ. J. Gr. Brit. 19: 69, 1957. Basionym: *Pilocereus ulei*. [(*Non Cephalocereus ulei* Gürke 1908.)] [RPS 8]
Pilosocereus urbanianus (Schumann) Byles & Rowley, Cact. Succ. J. Gr. Brit. 19: 69, 1957. Basionym: *Pilocereus urbanianus*. [RPS 8]
Pilosocereus vilaboensis (Diers & E. Esteves Pereira) P. J. Braun, Bradleya 6: 88, 1988. Basionym: *Pseudopilocereus vilaboensis*. [RPS 39]
Pilosocereus werdermannianus (Buining & Brederoo) Ritter, Kakt. Südamer. 1: 76, fig. 46 (p. 291), 1979. Basionym: *Pseudopilocereus werdermannianus*. [RPS 30]
Pilosocereus werdermannianus var. **densilanatus** Ritter, Kakt. Südamer. 1: 77, fig. 48-49 (pp. 291-292), 1979. Typus: *Ritter* 960 (U ?). [Brazil: Minas Gerais] [RPS 30]
Pilosocereus werdermannianus var. **diamantinensis** Ritter, Kakt. Südamer. 1: 77, 1979. [RPS 30]
Pilosocereus zehntneri (Britton & Rose) Ritter, Kakt. Südamer. 1: 74, 1979. Basionym: *Cephalocereus zehntneri*. [RPS 30]
Piptanthocereus aethiops (Haworth) Ritter, Kakt. Südamer. 2: 435, 1980. Basionym: *Cereus aethiops*. [RPS 31]
Piptanthocereus alacriportanus (Pfeiffer) Ritter, Kakt. Südamer. 1: 236, 1979. Basionym: *Cereus alacriportanus*. [RPS 30]
Piptanthocereus bageanus Ritter, Kakt. Südamer. 1: 238, 1979. [RPS 30]
Piptanthocereus cabralensis Ritter, Kakt. Südamer. 1: 235-236, 1979. [RPS 30]
Piptanthocereus calcirupicola Ritter, Kakt. Südamer. 1: 234, 1979. [RPS 30]
Piptanthocereus calcirupicola var. **pluricostatus** Ritter, Kakt. Südamer. 1: 235, 1979. [RPS 30]
Piptanthocereus cipoensis Ritter, Kakt. Südamer. 1: 236, 1979. [RPS 30]
Piptanthocereus colosseus Ritter, Kakt. Südamer. 2: 553-554, 1980. [RPS 31]
Piptanthocereus comarapanus (Cardenas) Ritter, Kakt. Südamer. 2: 555-556, 1980. Basionym: *Cereus comarapanus*. [RPS 31]

Piptanthocereus crassisepalus (Buining & Brederoo) Ritter, Kakt. Südamer. 1: 238, 1979. Basionym: *Cereus crassisepalus*. [RPS 30]
Piptanthocereus dayamii (Spegazzini) Ritter, Kakt. Südamer. 2: 554, 1980. Basionym: *Cereus dayamii*. [RPS 31]
Piptanthocereus forbesii var. **bolivianus** Ritter, Kakt. Südamer. 2: 554-555, 1980. [RPS 31]
Piptanthocereus goiasensis Ritter, Kakt. Südamer. 1: 234, 1979. [RPS 30]
Piptanthocereus huilunchu (Cardenas) Ritter, Kakt. Südamer. 2: 556, 1980. Basionym: *Cereus huilunchu*. [RPS 31]
Piptanthocereus lamprospermus (Schumann) Ritter, Kakt. Südamer. 1: 259, 1979. Basionym: *Cereus lamprospermus*. [RPS 30]
Piptanthocereus lanosus Ritter, Kakt. Südamer. 1: 259-260, 1979. [RPS 30]
Piptanthocereus lindenzweigianus (Gürke) Ritter, Kakt. Südamer. 1: 260, 1979. Basionym: *Cereus lindenzweigianus*. [RPS 30]
Piptanthocereus neonesioticus Ritter, Kakt. Südamer. 1: 237, 1979. [RPS 30]
Piptanthocereus neonesioticus var. **interior** Ritter, Kakt. Südamer. 1: 237-238, 1979. [RPS 30]
Piptanthocereus obtusus (Haworth) Ritter, Kakt. Südamer. 1: 231, 1979. *Nom. inval.* (Art. 33.2), based on *Cereus obtusus*. [RPS 30]
Piptanthocereus pachyrhizus (Schumann) Ritter, Kakt. Südamer. 1: 259, 1979. Basionym: *Cereus pachyrhizus*. [RPS 30]
Piptanthocereus sericifer Ritter, Kakt. Südamer. 1: 232, 1979. [RPS 30]
Piptanthocereus stenogonus (Schumann) Ritter, Kakt. Südamer. 1: 258, 1979. Basionym: *Cereus stenogonus*. [RPS 30]
Piptanthocereus uruguayensis Ritter, Kakt. Südamer. 1: 244, 1979. *Nom. inval.* (Art. 36.1). [See *Cereus uruguayanus* Kiesling.] [RPS 30]
Piptanthocereus xanthocarpus (Schumann) Ritter, Kakt. Südamer. 1: 258, 1979. Basionym: *Cereus xanthocarpus*. [RPS 30]
Platyopuntia albisaetacens (Backeberg) Ritter, Kakt. Südamer. 2: 501-502, 1980. Basionym: *Opuntia albisaetacens*. [RPS 31]
Platyopuntia apurimacensis Ritter, Kakt. Südamer. 4: 1257-1258, 1981. Typus: *Ritter* 1323 (U). [Peru: Apurimac] [RPS 32]
Platyopuntia atroglobosa (Backeberg) Ritter, Kakt. Südamer. 2: 502, 1980. *Nom. inval.*, based on *Tephrocactus atroglobosus, nom. inval.* (Art. 9.5). [Erroneously included as valid in RPS 31.] [RPS 31]
Platyopuntia brachyacantha Ritter, Kakt. Südamer. 2: 501, 1980. [RPS 31]
Platyopuntia brunneogemmia Ritter, Kakt. Südamer. 1: 34, 1979. [RPS 30]
Platyopuntia cardiosperma (Schumann) Ritter, Kakt. Südamer. 1: 246, 1979. *Nom. inval.* (Art. 33.2), based on *Opuntia cardiosperma*. [RPS 30]
Platyopuntia chilensis Ritter, Kakt. Südamer. 3: 889, 1980. [RPS 31]
Platyopuntia cognata Ritter, Kakt. Südamer. 1: 246, 1979. [RPS 30]
Platyopuntia conjungens Ritter, Kakt. Südamer. 2: 494-495, 1980. [RPS 31]
Platyopuntia cordobensis (Spegazzini) Ritter, Kakt. Südamer. 2: 403-404, 1980. Basionym: *Opuntia cordobensis*. [RPS 31]
Platyopuntia corrugata (Salm-Dyck) Ritter, Kakt. Südamer. 2: 410-411, 1980. Basionym: *Opuntia corrugata*. [RPS 31]
Platyopuntia discolor (Britton & Rose) Ritter, Kakt. Südamer. 2: 497, 1980. Basionym: *Opuntia discolor*. [RPS 31]
Platyopuntia dumetorum (A. Berger) Ritter, Kakt. Südamer. 1: 27, 1979. Basionym: *Opuntia dumetorum*. [RPS 30]
Platyopuntia ianthinantha Ritter, Kakt. Südamer. 2: 413, 1980. [RPS 31]
Platyopuntia inaequilateralis (A. Berger) Ritter, Kakt. Südamer. 4: 1256, 1981. Basionym: *Opuntia inaequilateralis*. [RPS 32]
Platyopuntia inaequilateralis var. **angustior** Ritter, Kakt. Südamer. 4: 1257, 1981. Typus: *Ritter* 1322 (U). [Type cited for U l.c. 1: iii, 1979.] [Peru: Huancavelica] [RPS 32]
Platyopuntia inamoena (Schumann) Ritter, Kakt. Südamer. 1: 32, 1979. Basionym: *Opuntia inamoena*. [RPS 30]
Platyopuntia inamoena fa. **spinigera** Ritter, Kakt. Südamer. 1: 32, 1979. Typus: *Ritter* 1252a (U). [Brazil: Bahia] [RPS 30]
Platyopuntia infesta Ritter, Kakt. Südamer. 4: 1258-1259, 1981. Typus: *Ritter* 309 (U). [Peru: Ancash] [RPS 32]
Platyopuntia interjecta Ritter, Kakt. Südamer. 2: 499, 1980. [RPS 31]
Platyopuntia kiska-loro (Spegazzini) Ritter, Kakt. Südamer. 2: 407, 1980. Basionym: *Opuntia kiska-loro*. [RPS 31]
Platyopuntia limitata Ritter, Kakt. Südamer. 1: 245, 1979. [RPS 30]
Platyopuntia microdisca (F. A. C. Weber) Ritter, Kakt. Südamer. 2: 408-410, 1980. Basionym: *Opuntia microdisca*. [RPS 31]
Platyopuntia nana (Kunth) Ritter, Kakt. Südamer. 2: 495-496, 1980. Basionym: *Cactus nanus*. [RPS 31]
Platyopuntia nigrispina (Schumann) Ritter, Kakt. Südamer. 2: 413-414, 1980. Basionym: *Opuntia nigrispina*. [RPS 31]
Platyopuntia orurensis (Cardenas) Ritter, Kakt. Südamer. 2: 502, 1980. Basionym: *Opuntia orurensis*. [RPS 31]
Platyopuntia pituitosa Ritter, Kakt. Südamer. 2: 404-405, 1980. [RPS 31]
Platyopuntia pyrrhantha Ritter, Kakt. Südamer. 2: 497-498, 1980. [RPS 31]
Platyopuntia quimilo (Schumann) Ritter, Kakt. Südamer. 2: 404, 1980. Basionym: *Opuntia quimilo*. [RPS 31]
Platyopuntia quitensis (F. A. C. Weber) Ritter, Kakt. Südamer. 4: 1255, 1981. Basionym: *Opuntia quitensis*. [RPS 32]
Platyopuntia retrorsa (Spegazzini) Ritter, Kakt. Südamer. 2: 496-497, 1980. Basionym: *Opuntia retrorsa*. [RPS 31]
Platyopuntia rubrogemmia Ritter, Kakt. Südamer. 1: 33, 1979. [RPS 30]
Platyopuntia salmiana (Palmentier *ex* Pfeiffer) Ritter, Kakt. Südamer. 2: 405-406, 1980.

Basionym: *Opuntia salmiana.* [RPS 31]
Platyopuntia saxatilis Ritter, Kakt. Südamer. 1: 32, 1979. Typus: *Ritter* 1035 (U). [Brazil: Minas Gerais] [RPS 30]
Platyopuntia soehrensii (Britton & Rose) Ritter, Kakt. Südamer. 2: 411-412, 1980. Basionym: *Opuntia soehrensii.* [RPS 31]
Platyopuntia soehrensii var. **grandiflora** Ritter, Kakt. Südamer. 2: 501, 1980. [RPS 31]
Platyopuntia soehrensii var. **tilcarensis** (Backeberg) Ritter, Kakt. Südamer. 2: 412, 1980. Basionym: *Opuntia tilcarensis.* [RPS 31]
Platyopuntia soehrensii var. **transiens** Ritter, Kakt. Südamer. 2: 388-389, 1980. [RPS 31]
Platyopuntia spinibarbis Ritter, Kakt. Südamer. 2: 499-500, 1980. Typus: *Ritter* 94 (U, ZSS [fragment]). [Type cited for U l.c. 1: iii, 1979.] [Bolivia: Tarija] [RPS 31]
Platyopuntia spinibarbis var. **grandiflora** Ritter, Kakt. Südamer. 2: 407-408, 1980. [RPS 31]
Platyopuntia sulphurea (G. Don *ex* Loudon) Ritter, Kakt. Südamer. 2: 407-408, 1980. Basionym: *Opuntia sulphurea.* [RPS 31]
Platyopuntia viridirubra Ritter, Kakt. Südamer. 1: 35, 1979. [RPS 30]
Platyopuntia vitelliniflora Ritter, Kakt. Südamer. 2: 498, 1980. [RPS 31]
Platyopuntia vulgaris (Miller) Ritter, Kakt. Südamer. 1: 35, 1979. Basionym: *Opuntia vulgaris.* [RPS 30]
Polaskia chende (Roland-Gosselin) Gibson & Horak, Ann. Missouri Bot. Gard. 65(4): 1006, 1979. Basionym: *Cereus chende.* [Dated 1978.] [RPS 30]
Polyanthocereus Backeberg, Kakt. and. Sukk. 4(3): 36, 1953. *Nom. inval.* (Art. 36.1, 37.1). [Name only.] [RPS 4]
Porfiria schwarzii (Fric) Crkal, Lovec Kaktusu 395, 1983. Basionym: *Haagea schwarzii.* [A spelling variant for *P. schwarztii.*] [RPS 34]
Praecereus F. Buxbaum, Beitr. Biol. Pfl. 44: 215-276, 1968. Typus: *Cephalocereus smithianus* Britton & Rose. [Concurrently published in Krainz, Die Kakt. Lief. 38-39: C IVb (July 1968). Name first mentioned as manuscript name in Krainz, Katalog ZSS ed. 2, 104, 1967.] [RPS 19]
Praecereus amazonicus (Schumann) Buxbaum, Beitr. Biol. Pfl. 44: 276, 1968. Basionym: *Cereus amazonicus.* [RPS 19]
Praecereus apoloensis (Cardenas) Buxbaum, Beitr. Biol. Pfl. 44: 275, 1968. Basionym: *Monvillea apoloensis.* [RPS 19]
Praecereus campinensis (Backeberg & Voll) Buxbaum, Beitr. Biol. Pfl. 44: 276, 1969. Basionym: *Pilocereus campinensis.* [RPS 19]
Praecereus jaenensis (Rauh & Backeberg) Buxbaum, Beitr. Biol. Pfl. 44: 276, 1968. Basionym: *Monvillea jaenensis.* [RPS 19]
Praecereus maritimus (Britton & Rose) Buxbaum, Beitr. Biol. Pfl. 44: 276, 1968. Basionym: *Monvillea maritima.* [RPS 19]
Praecereus smithianus (Britton & Rose) Buxbaum, Beitr. Biol. Pfl. 44: 273, 1968. Basionym: *Cephalocereus smithianus.* [Combination first mentioned (*comb. nud.*) in Krainz, Katalog ZSS ed. 2, 104, 1967.] [RPS 19]
Pseudoacanthocereus Ritter, Kakt. Südamer. 1: 47-48, 1979. Typus: *Acanthocereus brasiliensis.* [RPS 30]
Pseudoacanthocereus boreominarum Rizzini & Mattos-F., Rev. Brasil. Biol. 46(2): 327, 1986. Typus: *Rizzini et Mattos-Filho* 40 (RB). [Brazil] [RPS 37]
Pseudoacanthocereus brasiliensis (Britton & Rose) Ritter, Kakt. Südamer. 1: 47, 1979. Basionym: *Acanthocereus brasiliensis.* [RPS 30]
Pseudoechinocereus Krainz, Katalog ZSS ed. 2, 104, 1967. *Nom. inval.* (Art. 36.1, 37.1). [Described as *nomen nudum.*] [RPS -]
Pseudoechinocereus splendens Buining, Sukkulentenkunde 4: 45, 1951. *Nom. inval.* (Art. 36.1, 37.1). [Again mentioned (as *nomen nudum*) by Krainz in Katalog ZSS ed. 2, 104, 1967.] [RPS 4]
Pseudoespostoa melanostele var. **inermis** Backeberg, Cact. Succ. J. (US) 23(5): 149, 1951. [RPS 3]
Pseudoespostoa nana (Ritter) Backeberg, Kakt.-Lex. 371, 1966. Basionym: *Espostoa nana.* [RPS 17]
Pseudolobivia cv. **Geisenheim** Kilian, Stachelpost 4(24): 137, 1969. [RPS 20]
Pseudolobivia acanthoplegma Backeberg, Die Cact. 6: 3726, 1962. *Nom. inval.* (Art. 37.1). [Erroneously included as valid in RPS 13.] [RPS 13]
Pseudolobivia aurea (Britton & Rose) Backeberg, Cact. Succ. J. (US) 23(2): 49, 1951. Basionym: *Echinopsis aurea.* [The name given as basionym is already a combination, and is here corrected.] [RPS 3]
Pseudolobivia aurea var. **elegans** (Backeberg) Backeberg, Cact. Succ. J. (US) 23(2): 49, 1951. Basionym: *Lobivia aurea* var. *elegans.* [RPS 3]
Pseudolobivia aurea var. *fallax* (Oehme) Backeberg, Die Cact. 2: 1357, 1959. Incorrect name (Art. 56.1(b)), based on *Lobivia fallax.* [Erroneously included as valid in RPS 10.] [RPS 10]
Pseudolobivia aurea var. **grandiflora** Backeberg, Descr. Cact. Nov. [1:] 28, 1957. [Dated 1956, published 1957. Given as "comb. nov.", but with Latin diagnosis instead of a basionym (which could be *Lobivia aurea* var. *spinosior* Backeberg 1935 (*nom. inval.* Art. 36.1). The combination of this basionym is repeated in Die Cact. 2: 1357, 1959 (cf. RPS 10).] [RPS 7]
Pseudolobivia callichroma (Cardenas) Backeberg, Kakt.-Lex., 372, 1966. Basionym: *Echinopsis callichroma.* [RPS 17]
Pseudolobivia calorubra (Cardenas) Backeberg, Kakt.-Lex., 372, 1966. Basionym: *Echinopsis calorubra.* [RPS 17]
Pseudolobivia carmineoflora Hoffmann & Backeberg, in Backeberg, Die Cact. 2: 1355, 1959. *Nom. inval.* (Art. 37.1). [Erroneously included as valid in RPS 10.] [RPS 10]
Pseudolobivia ducis-pauli var. **rubriflora**

Schütz, Kakt. Listy. 16(6): 62-62, 1951. [RPS 3]

Pseudolobivia lecoriensis (Cardenas) Backeberg, Cact. Succ. J. (US) 35: 158-159, 1966. Basionym: *Echinopsis lecoriensis*. [RPS 17]

Pseudolobivia longispina var. *nigra* (Backeberg) Backeberg, Die Cact. 2: 1344, 1959. *Nom. inval.*, based on *Echinopsis nigra, nom. inval.* (Art. 36.1). [RPS 10]

Pseudolobivia longispina var. **rubriflora** (Schütz) Schütz, Friciana 1(7): 7, 1962. Basionym: *Pseudolobivia ducis-pauli* var. *rubriflora*. [Given as 'n. comb. subnud.' in RPS 13 - validity of name not further assessed.] [RPS 13]

Pseudolobivia luteiflora Backeberg, Kakt.-Lex., 373, 1966. *Nom. inval.* (Art. 9.5). [Erroneously included as valid in RPS 17.] [RPS 17]

Pseudolobivia obrepanda var. **fiebrigii** (Gürke) Backeberg, Die Cact. 2: 1350, 1959. Basionym: *Echinopsis fiebrigii*. [RPS 10]

Pseudolobivia orozasana (Ritter) Backeberg, Kakt.-Lex., 374, 1966. Basionym: *Echinopsis mamillosa* var. *orozasana*. [RPS 17]

Pseudolobivia pelecyrhachis var. **lobivioides** (Backeberg) Backeberg, Die Cact. 2: 1352, 1959. Basionym: *Echinopsis lobivioides*. [RPS 10]

Pseudolobivia rojasii (Cardenas) Backeberg, Die Cact. 2: 1351, 1959. Basionym: *Echinopsis rojasii*. [RPS 10]

Pseudolobivia rojasii var. **albiflora** (Cardenas) Backeberg, Die Cact. 2: 1351, 1959. Basionym: *Echinopsis rojasii* var. *albiflora*. [RPS 10]

Pseudolobivia toralapana (Cardenas) Backeberg, Kakt.-Lex., 374, 1966. Basionym: *Echinopsis toralapana*. [RPS 17]

Pseudolobivia torrecillasensis (Cardenas) Backeberg, Die Cact. 2: 1353, 1959. Basionym: *Echinopsis torrecillasensis*. [Repeated in Kakt.-Lex., 374, 1966 (cf. RPS 17, without basionym citation).] [RPS 10]

Pseudolobivia wilkeae Backeberg, Die Cact. 6: 3724-3725, ill., 1962. *Nom. inval.* (Art. 37.1). [Erroneously included as valid in RPS 13.] [RPS 13]

Pseudolobivia wilkeae var. *carminata* Backeberg, In Backeberg, Die Cact. 6: 3726, 1962. *Nom. inval.* (Art. 43.1). [Erroneously included as valid in RPS 13.] [RPS 13]

Pseudomammillaria F. Buxbaum, Österr. Bot. Zeitschr. 98(1-2): 84, 1951. Typus: *Mammillaria camptotricha* Dams. [RPS 2]

Pseudomammillaria albescens (Tiegel) Buxbaum, Österr. Bot. Zeitschr. 98(1-2): 84, 1951. Basionym: *Mammillaria albescens*. [RPS 2]

Pseudomammillaria camptotricha (Dams) Buxbaum, Österr. Bot. Zeitschr. 98(1-2): 84, 1951. Basionym: *Mammillaria camptotricha*. [RPS 2]

Pseudomammillaria decipiens (Scheidweiler) Buxbaum, Österr. Bot. Zeitschr. 98(1-2): 84, 1951. Basionym: *Mammillaria decipiens*. [RPS 2]

Pseudomammillaria kraehenbuehlii Krainz, Kakt. and. Sukk. 1970(12): Gesellschaftsnachrichten SKG, 4, 1970. Typus: *Krähenbühl* 153.69 (ZSS). [The number of the type is given as *Krähenbühl 153.68* in later publications.] [Mexico: Oaxaca] [RPS 21]

Pseudomitrocereus H. Bravo-H. & Buxbaum, in Buxbaum, Bot. Stud. 12: 48-54, 1961. Typus: *P. fulviceps*. [= *Mitrocereus* Backeberg 1942, *excl. typ.*] [RPS 12]

Pseudomitrocereus fulviceps (F. A. C. Weber) Bravo & Buxbaum, Bot. Stud. 12: 99, 1961. Basionym: *Pilocereus fulviceps*. [RPS 12]

Pseudonopalxochia Backeberg, Die Cact. 1: 52, 69, in clav, 1958. *Nom. inval.* (Art. 36.1). [RPS 9]

Pseudonopalxochia conzattianum (MacDougall) Backeberg, Die Cact. 2: 757, 1959. *Nom. inval.* (Art. 43.1), based on *Nopalxochia conzattianum*. [Erroneously included as valid in RPS 10.] [RPS 10]

Pseudopilocereus Buxbaum, Beitr. Biol. Pfl. 44: 215-276, 1968. Typus: *Pilocereus arrabidae* Lemaire. [Concurrently published in Krainz, Die Kakt. C IVb (July 1968). Name first mentioned as manuscript name in Krainz, Katalog ZSS ed. 2, 104, 1967 (including many nude combinations).] [RPS 19]

Pseudopilocereus arrabidae (Lemaire) Buxbaum, Beitr. Biol. Pfl. 44: 252, 1968. Basionym: *Pilocereus arrabidae*. [RPS 19]

Pseudopilocereus aurisetus (Werdermann) Buxbaum, Beitr. Biol. Pfl. 44: 252, 1968. Basionym: *Pilocereus aurisetus*. [RPS 19]

Pseudopilocereus azureus Buining & Brederoo, Kakt. and. Sukk. 26( ): 241-243, 1975. [If combined under *Pilosocereus*, the correct name is *Pilosocereus cyaneus* Ritter 1981 (*nom. nov., non Pilosocereus azureus* Ritter 1979).] [RPS 26]

Pseudopilocereus bradei (Backeberg & Voll) Buxbaum, Beitr. Biol. Pfl. 44: 252, 1968. Basionym: *Pilocereus bradei*. [RPS 19]

Pseudopilocereus catingicola (Gürke) Buxbaum, Beitr. Biol. Pfl. 44: 252, 1968. Basionym: *Cereus catingicola*. [RPS 19]

Pseudopilocereus chrysostele (Vaupel) Buxbaum, Beitr. Biol. Pfl. 44: 252, 1968. Basionym: *Cereus chrysostele*. [RPS 19]

Pseudopilocereus cuyabensis (Backeberg) Buxbaum, Beitr. Biol. Pfl. 44: 252, 1968. Basionym: *Pilocereus cuyabensis*. [RPS 19]

Pseudopilocereus diersianus E. Esteves Pereira, Kakt. and. Sukk. 32(4): 86-90, 1981. [RPS 32]

Pseudopilocereus flavipulvinatus Buining & Brederoo, Succulenta 58(6): 137-143, 1979. [RPS 30]

Pseudopilocereus floccosus (Byles & Rowley) Buxbaum, Beitr. Biol. Pfl. 44: 252, 1968. Basionym: *Pilosocereus floccosus*. [Erroneously based on *Pilocereus floccosus* Backeberg & Voll; treated as bibliographic error.] [RPS 19]

Pseudopilocereus fulvilanatus Buining & Brederoo, Kakt. and. Sukk. 24(7): 145-147, ills., 1973. Typus: *Horst et Uebelmann* HU 277 (1968) (U, ZSS). [Brazil: Minas Gerais]

[RPS -]

Pseudopilocereus glaucescens (Labouret) Buxbaum, Beitr. Biol. Pfl. 44: 252, 1968. Basionym: *Pilocereus glaucescens.* [RPS 19]

Pseudopilocereus glaucochrous (Werdermann) Buxbaum, Beitr. Biol. Pfl. 44: 252, 1968. Basionym: *Pilocereus glaucochrous.* [RPS 19]

Pseudopilocereus gounellei (F. A. C. Weber) Buxbaum, Beitr. Biol. Pfl. 44: 252, 1968. Basionym: *Pilocereus gounellei.* [RPS 19]

Pseudopilocereus hapalacanthus (Werdermann) Buxbaum, Beitr. Biol. Pfl. 44: 252, 1968. Basionym: *Pilocereus hapalacanthus.* [RPS 19]

Pseudopilocereus juaruensis Buining & Brederoo *ex* Theunissen et al., Kakt. and. Sukk. 29(7): 153-155, ills., 1978. Typus: *Horst et Uebelmann* HU 454 (U, ZSS). [Brazil: Mato Grosso] [RPS 29]

Pseudopilocereus luetzelburgii (Vaupel) Buxbaum, Beitr. Biol. Pfl. 44: 252, 1968. Basionym: *Cereus luetzelburgii.* [RPS 19]

Pseudopilocereus machrisii (Dawson) Buxbaum, Beitr. Biol. Pfl. 44: 252, 1968. Basionym: *Cephalocereus machrisii.* [Repeated by Buining & Brederoo in Succulenta 56: 90-94, 1977 (included as new name in RPS 28).] [RPS 19]

Pseudopilocereus magnificus Buining & Brederoo, Cact. Succ. J. (US) 44(2): 66-70, ills., 1972. Typus: *Horst et Uebelmann* HU 224 (1968) (U, ZSS). [Brazil: Minas Gerais] [RPS 23]

Pseudopilocereus mucosiflorus Buining & Brederoo, Kakt. and. Sukk. 28(9): 201-203, 1977. [RPS 28]

Pseudopilocereus nobilis (Haworth) Buxbaum, Beitr. Biol. Pfl. 44: 253, 1968. Basionym: *Cereus nobilis.* [RPS 19]

Pseudopilocereus oligolepis (Vaupel) Buxbaum, Beitr. Biol. Pfl. 44: 253, 1968. Basionym: *Cereus oligolepis.* [RPS 19]

Pseudopilocereus parvus Diers & E. Esteves Pereira, Kakt. and. Sukk. 33(5): 100-104, ills., 1982. Typus: *Esteves Pereira* 94 (KOELN [Succulentarium]). [RPS 33]

Pseudopilocereus pentaedrophorus (Labouret) Buxbaum, Beitr. Biol. Pfl. 44: 253, 1968. Basionym: *Cereus pentaedrophorus.* [RPS 19]

Pseudopilocereus piauhyensis (Gürke) Buxbaum, Beitr. Biol. Pfl. 44: 253, 1968. Basionym: *Cereus piauhyensis.* [RPS 19]

Pseudopilocereus rupicola (Werdermann) Buxbaum, Beitr. Biol. Pfl. 44: 253, 1968. Basionym: *Pilocereus rupicola.* [RPS 19]

Pseudopilocereus ruschianus Buining & Brederoo, Kakt. and. Sukk. 31(2): 33-35, 1980. [RPS 31]

Pseudopilocereus salvadorensis (Werdermann) Buxbaum, Beitr. Biol. Pfl. 44: 253, 1968. Basionym: *Pilocereus salvadorensis.* [RPS 19]

Pseudopilocereus sergipensis (Werdermann) Buxbaum, Beitr. Biol. Pfl. 44: 253, 1968. Basionym: *Pilocereus sergipensis.* [RPS 19]

Pseudopilocereus superfloccosus Buining & Brederoo, Cact. Succ. J. (US) 46(2): 60-63, 96, ills., 1974. Typus: *Horst et Uebelmann* HU 394 (1972) (U, ZSS). [Brazil: Bahia] [RPS 25]

Pseudopilocereus tuberculatus (Werdermann) Buxbaum, Beitr. Biol. Pfl. 44: 253, 1968. Basionym: *Pilocereus tuberculatus.* [RPS 19]

Pseudopilocereus ulei (Schumann) Buxbaum, Beitr. Biol. Pfl. 44: 253, 1968. Basionym: *Pilocereus ulei.* [RPS 19]

Pseudopilocereus vilaboensis Diers & E. Esteves Pereira, Kakt. and. Sukk. 34: 111, 1983. Typus: *Esteves Pereira* 20 (KOELN [Succulentarium]). [Brazil: Goiás] [RPS 34]

Pseudopilocereus werdermannianus Buining & Brederoo, Kakt. and. Sukk. 26(4): 74-77, ills., 1975. Typus: *Horst et Uebelmann* HU 227 (1972) (U, ZSS). [Brazil: Minas Gerais] [RPS 26]

Pseudorhipsalis harrisii (Gürke) Y. Ito, Cacti, 160, 1952. Basionym: *Rhipsalis harrisii.* [RPS 4]

Pseudoselenicereus Innes, Handbook Cacti & Succ., 31, 34, 1978. *Nom. inval.* (Art. 36.1, 37.1). [Sphalm. 'Pseudo-selenicereus'.] [RPS 29]

Pseudosolisia Ito, The Cactaceae, 477, 1981. *Nom. illeg.* (Art. 63.1). [Based on the type of *Normanbokea* Kladiwa & Buxbaum.] [RPS -]

Pseudosolisia valdeziana (Möller) Y. Ito, The Cactaceae, 477, 1981. *Nom. inval.* (Art. 33.2), based on *Pelecyphora valdeziana.* [Parenthetical author given as 'Bödeker'.] [RPS -]

Pseudosolisia valdeziana var. *albiflora* Y. Ito, The Cactaceae, 478, 1981. *Nom. inval.* (Art. 43.1). [RPS -]

Pseudozygocactus epiphylloides var. *bradei* (Campos-Porto & Castellanos) Backeberg, Die Cact. 2: 721, 1959. *Nom. inval.* (Art. 33.2), based on *Hariota epiphylloides* var. *bradei.* [Erroneously included as valid in RPS 10.] [RPS 10]

Pterocactus australis (F. A. C. Weber) Backeberg, Desert Pl. Life. 22(2): 17, 1950. Basionym: *Opuntia australis.* [RPS 3]

Pterocactus australis var. **arnoldianus** Backeberg, Cact. Succ. J. (US) 23(1): 14, 1951. [RPS 3]

Pterocactus gonjianii Kiesling, Cact. Succ. J. Gr. Brit. 44(3): 55-56, 1982. Typus: *Kiesling* 307 (LP). [RPS 33]

Pterocactus kuntzei fa. **lelongii** Ruiz Leal *ex* Kiesling, Cact. Succ. J. Gr. Brit. 44(3): 56, 1982. Typus: *Ruiz Leal* 12017 (RL). [RPS 33]

Pterocactus marenae (Parsons) Rowley, Nation. Cact. Succ. J. 13(1): 4, 1958. Basionym: *Opuntia marenae.* [RPS 9]

Pterocactus megliolii Kiesling, Bol. Soc. Argentina Bot. 14(1-2): 111-114, 1971. [RPS 29]

Pterocactus reticulatus Kiesling, Bol. Soc. Argent. Bot. 14(1-2): 144, 1971. Typus: *Kiesling* 56 (LP, SI). [Argentina: San Juan] [RPS 29]

Pterocactus skottsbergii (Britton & Rose)

Backeberg, Desert Pl. Life. 22(2): 17, 1950. Basionym: *Opuntia skottsbergii*. [RPS 3]

Pterocereinae Buxbaum, Bot. Stud. 12: 88, 1961. Typus: *Pterocereus*. [RPS 12]

Pterocereus MacDougall & Miranda, Ceiba 4(2): 135, 1954. Typus: *Pterocereus foetidus*. [RPS 5]

Pterocereus foetidus MacDougall & Miranda, Ceiba 4(2): 135, 1954. [RPS 5]

Pterocereus gaumeri (Britton & Rose) MacDougall & Miranda, Ceiba 4(2): 140, 1954. Basionym: *Pachycereus gaumeri*. [RPS 5]

Puna Kiesling, Hickenia 1(55): 289-290, ills., 1982. Typus: *P. clavarioides*. [RPS 33]

Puna clavarioides (Pfeiffer) Kiesling, Hickenia 1(55): 291-292, 1982. Basionym: *Opuntia clavarioides*. [Argentina] [RPS 33]

Puna subterranea (R.E. Fries) Kiesling, Hickenia 1(55): 293, 1982. Basionym: *Opuntia subterranea*. [Argentina] [RPS 33]

Pygmaeocereus Johnson & Backeberg, Nation. Cact. Succ. J. 12(4): 86, 1957. Typus: *Pygmaeocereus bylesianus*. [RPS 8]

Pygmaeocereus akersii Johnson *ex* Backeberg, Nation. Cact. Succ. J. 12(4): 87, 1957. *Nom. inval.* (Art. 36.1). [First mentioned by Johnson in 1955.] [RPS 8]

Pygmaeocereus bylesianus Andreae & Backeberg, Nation. Cact. Succ. J. 12(4): 86-87, 1957. [RPS 8]

Pygmaeocereus densiaculeatus Backeberg, Descr. Cact. Nov. 3: 12, 1963. *Nom. inval.* (Art. 9.5). [RPS 14]

Pygmaeocereus familiaris Ritter, Kakt. Südamer. 4: 1428-1429, fig. 1282, 1287-1288, 1205, 1981. Typus: *Ritter* 322 (U, ZSS). [Peru: Arequipa] [RPS 32]

Pygmaeocereus rowleyanus Backeberg, Die Cact. 6: 3718, 1962. *Nom. inval.* (Art. 9.5). [Erroneously included as valid in RPS 13.] [RPS 13]

Pygmaeocereus vespertinus Johnson, Kakt. and. Sukk. 20(5): 94, 1969. *Nom. inval.* (Art. 36.1, 37.1). [Casually mentioned in the text.] [RPS -]

Pyrrhocactus aconcaguensis Ritter, Succulenta 1960: 108, 1960. [RPS 11]

Pyrrhocactus aconcaguensis var. **orientalis** Ritter, Succulenta 1960: 108, 1960. [RPS 11]

Pyrrhocactus andicola (Ritter) Ritter, Succulenta 1959(10): 131, 1959. Basionym: *Horridocactus andicola*. [Sphalm. 'andicolus'.] [RPS 10]

Pyrrhocactus andicola var. **descendens** (Ritter) Ritter, Succulenta 1959(10): 131, 1959. Basionym: *Horridocactus andicola* var. *descendens*. [Sphalm. 'andicolus'.] [RPS 10]

Pyrrhocactus andicola var. **mollensis** (Ritter) Ritter, Succulenta 1959(10): 131, 1959. Basionym: *Horridocactus andicola* var. *mollensis*. [Sphalm. 'andicolus'.] [RPS 10]

Pyrrhocactus andicola var. **robustus** (Ritter) Ritter, Succulenta 1959(10): 131, 1959. Basionym: *Horridocactus andicola* var. *robustus*. [Sphalm. 'andicolus'.] [RPS 10]

Pyrrhocactus andreaeanus (Backeberg) Ritter, Succulenta 1959(10): 131, 1959. *Nom. inval.*, based on *Neochilenia andreaeana, nom. inval.* (Art. 9.5, 37.1). [Sphalm. 'andreaeana'. Erroneously included as valid in RPS 10.] [RPS 10]

Pyrrhocactus aricensis Ritter, Taxon 12(1): 32, 1963. Typus: *Ritter* 268 (ZSS). [Chile] [RPS 14]

Pyrrhocactus armatus Ritter, Succulenta 39(5): 49-51, ill., 1960. Typus: *Ritter* 449 (ZSS). [Chile] [RPS 11]

Pyrrhocactus aspillagae (Söhrens) Ritter, Succulenta 1959(10): 131, 1959. Basionym: *Echinocactus aspillagai*. [Sphalm. 'aspillagai'.] [RPS 10]

Pyrrhocactus atrospinosus Backeberg, Descr. Cact. Nov. 3: 13, 1963. *Nom. inval.* (Art. 9.5). [Erroneously included as valid in RPS 14.] [RPS 14]

Pyrrhocactus atroviridis Ritter, Succulenta 1960: 89, 1960. [RPS 11]

Pyrrhocactus bulbocalyx (Werdermann) Backeberg, Die Cact. 3: 1565, 1959. Basionym: *Echinocactus bulbocalyx*. [RPS 10]

Pyrrhocactus calderanus Ritter, Succulenta 1961: 13, 1961. [RPS 12]

Pyrrhocactus carrizalensis Ritter, Taxon 12(1): 33, 1963. Typus: *Ritter* 493 (U, ZSS). [Chile: Atacama] [RPS 14]

Pyrrhocactus chaniarensis Ritter, Kakt. Südamer. 3: 926-927, 1980. [RPS 31]

Pyrrhocactus chilensis (Hildmann) Ritter, Succulenta 1959(10): 131, 1959. Basionym: *Echinocactus chilensis*. [RPS 10]

Pyrrhocactus chilensis var. **albidiflorus** Ritter, Kakt. Südamer. 3: 927-928, 1980. [RPS 31]

Pyrrhocactus choapensis Ritter, Succulenta 30(12): 133-134, 1960. Typus: *Ritter* 238 (ZSS). [Chile] [RPS 11]

Pyrrhocactus chorosensis Ritter, Succulenta 1960: 121, 1960. [RPS 11]

Pyrrhocactus coliguayensis Ritter, Kakt. Südamer. 3: 929-930, 1980. [RPS 31]

Pyrrhocactus confinis Ritter, Succulenta 1961: 4, 1961. [RPS 12]

Pyrrhocactus crispus Ritter, Succulenta 1959(11): 137, 1959. [RPS 10]

Pyrrhocactus curvispinus var. *combarbalensis* Ritter, Kakt. Südamer. 3: 932, 1980. Based on *Ritter* 225c. *Nom. inval.* (Art. 36.1). [Published as provisional name.] [Chile] [RPS -]

Pyrrhocactus curvispinus var. *felipensis* Ritter, Kakt. Südamer. 3: 932, 1980. Based on *Ritter* 225b. *Nom. inval.* (Art. 36.1). [Published as provisional name.] [Chile] [RPS -]

Pyrrhocactus curvispinus var. *santiagensis* Ritter, Kakt. Südamer. 3: 932, 1980. Based on *Ritter* 225a. *Nom. inval.* (Art. 36.1). [Published as provisional name.] [Chile] [RPS -]

Pyrrhocactus deherdtianus (Backeberg) Kattermann, Cact. Succ. J. (US) 55(3): 123, ills., 1983. *Nom. inval.*, based on *Neochilenia deherdtiana, nom. inval.* (Art. 9.5, 37.1). [Epithet given as 'deherdtiana'.] [RPS 34]

Pyrrhocactus dimorphus Ritter, Succulenta

1962: 3, 1962. [RPS 13]

Pyrrhocactus echinus Ritter, Taxon 12(1): 33, 1963. Typus: *Ritter* 537 (U, ZSS). [Chile] [RPS 14]

Pyrrhocactus echinus var. **minor** Ritter, Taxon 12(1): 33, 1963. [RPS 14]

Pyrrhocactus engleri (Ritter) Ritter, Succulenta 1959(10): 131, 1959. Basionym: *Horridocactus engleri*. [RPS 10]

Pyrrhocactus eriosyzoides (Ritter) Ritter, Succulenta 1959(10): 131, 1959. Basionym: *Horridocactus eriosyzoides*. [RPS 10]

Pyrrhocactus eriosyzoides var. **domeykoensis** Ritter, Kakt. Südamer. 3: 938, 1980. [RPS 31]

Pyrrhocactus floccosus Ritter, Taxon 12: 32, 1963. [RPS 14]

Pyrrhocactus floccosus var. **minor** (Ritter) Ritter, Kakt. Südamer. 3: 939-940, 1980. Basionym: *Pyrrhocactus echinus* var. *minor*. [RPS 31]

Pyrrhocactus floribundus (Backeberg) Ritter, Kakt. Südamer. 3: 940-942, 1980. *Nom. inval.*, based on *Reicheocactus floribundus, nom. inval.* (Art. 37.1). [RPS 31]

Pyrrhocactus fuscus (Mühlenpfordt) Ritter, Succulenta 1959(10): 131, 1959. Basionym: *Echinocactus fuscus*. [RPS 10]

Pyrrhocactus garaventae (Ritter) Ritter, Succulenta 1959(10): 131, 1959. Basionym: *Horridocactus garaventae*. [Sphalm. 'garaventai'.] [RPS 10]

Pyrrhocactus glaucescens Ritter, Taxon 12: 33, 1963. [RPS 14]

Pyrrhocactus gracilis Ritter, Succulenta 1961: 129, 1961. [RPS 12]

Pyrrhocactus grandiflorus Ritter, Succulenta 1960: 41, 1960. [RPS 11]

Pyrrhocactus griseus Backeberg, Die Cact. 6: 3906, 1962. *Nom. inval.* (Art. 37.1). [RPS 13]

Pyrrhocactus heinrichianus (Backeberg) Ritter, Succulenta 1959(10): 131, 1959. Basionym: *Horridocactus heinrichianus*. [RPS 10]

Pyrrhocactus horridus var. **aconcaguensis** (Ritter) Ritter, Kakt. Südamer. 3: 948-949, 1980. Basionym: *Pyrrhocactus aconcaguensis*. [RPS 31]

Pyrrhocactus horridus var. **mutabilis** Ritter, Kakt. Südamer. 3: 946-947, 1980. [RPS 31]

Pyrrhocactus horridus var. **orientalis** (Ritter) Ritter, Kakt. Südamer. 3: 949, 1980. Basionym: *Pyrrhocactus aconcaguensis* var. *orientalis*. [RPS 31]

Pyrrhocactus horridus var. **robustus** (Ritter) Ritter, Kakt. Südamer. 3: 947-948, 1980. Basionym: *Pyrrhocactus robustus*. [RPS 31]

Pyrrhocactus huascensis Ritter, Succulenta 1961: 57, 1961. [RPS 12]

Pyrrhocactus intermedius Ritter, Taxon 12: 32, 1963. [RPS 14]

Pyrrhocactus iquiquensis Ritter, Taxon 12(1): 32, 1963. Typus: *Ritter* 201 (ZSS). [Chile] [RPS 14]

Pyrrhocactus jussieui (Monville) Ritter, Kakt. Südamer. 3: 951-952, 1980. Basionym: *Echinocactus jussieui*. [RPS 31]

Pyrrhocactus jussieui var. **australis** Ritter, Kakt. Südamer. 3: 952-953, 1980. [RPS 31]

Pyrrhocactus jussieui var. **spinosior** Ritter, Kakt. Südamer. 3: 952, 1980. [RPS 31]

Pyrrhocactus kesselringianus (Dölz) Ritter, Succulenta 1959(10): 131, 1959. Basionym: *Horridocactus kesselringianus*. [RPS 10]

Pyrrhocactus krausii Ritter, Kakt. Südamer. 3: 953-954, 1980. [RPS 31]

Pyrrhocactus kunzei (Förster) Y. Ito, Explan. Diagr. Austroechinocactinae, 228, 1957. Basionym: *Echinocactus kunzei*. [Sphalm. 'kuwzei'. Repeated by Ritter in Succulenta 1959(10): 130, 1959 (cf. RPS 10).] [RPS 8]

Pyrrhocactus limariensis Ritter, Succulenta 1960(12): 133, 1960. [Republished in Kakt. Südamer. 3: 956-957, 1980 (RPS 31).] [RPS 31]

Pyrrhocactus lissocarpus Ritter, Succulenta 1960: 17, 1960. [RPS 11]

Pyrrhocactus lissocarpus var. **gracilis** Ritter, Succulenta 1960: 17, 1960. [RPS 11]

Pyrrhocactus marksianus , Succulenta 39(1): 2-3, 1960. Typus: *Ritter* 234 (ZSS). [Chile] [RPS 11]

Pyrrhocactus marksianus Ritter, Succulenta 1960: 2, 1960. [RPS 11]

Pyrrhocactus marksianus var. **tunensis** Ritter, Succulenta 1960: 2, 1960. [RPS 11]

Pyrrhocactus megliolii Rausch, Kakt. and. Sukk. 25(10): 220-221, ill., 1974. Typus: *Rausch* 559 (ZSS). [Type cited for W but placed in ZSS.] [Argentina: San Juan] [RPS 25]

Pyrrhocactus melanacanthus Ritter *ex* Backeberg, Descr. Cact. Nov. 3: 13, 1963. *Nom. inval.* (Art. 9.5). [Erroneously included as valid in RPS 14.] [RPS 14]

Pyrrhocactus neohankeanus Ritter, Kakt. Südamer. 3: 959, 1980. [Incorrectly treated as *nom. illeg.* (Art. 67) in RPS 31.] [RPS 31]

Pyrrhocactus neohankeanus fa. *woutersianus* (Backeberg) Ritter, Kakt. Südamer. 3: 960, 1980. *Nom. inval.*, based on *Delaetia woutersiana, nom. inval.* (Art. 37.1). [RPS 31]

Pyrrhocactus neohankeanus var. **densispinus** Ritter, Kakt. Südamer. 3: 960, 1980. [Validity depends on assessment of questionable validity of *P. neohankeanus*.] [RPS 31]

Pyrrhocactus neohankeanus var. **elongatus** Ritter, Kakt. Südamer. 3: 960, 1980. [Validity depends on assessment of questionable validity of *P. neohankeanus*.] [RPS 31]

Pyrrhocactus neohankeanus var. **flaviflorus** Ritter, Kakt. Südamer. 3: 960, 1980. [Validity depends on assessment of questionable validity of *P. neohankeanus*.] [RPS 31]

Pyrrhocactus nigricans (Linke) Ritter, Succulenta 1959(10): 131, 1959. Basionym: *Echinopsis nigricans*. [RPS 10]

Pyrrhocactus occultus Ritter, Succulenta 1959(10): 131, 1959. *Nom. inval.* (Art. 36.1, 37.1). [Based on *Echinocactus occultus sensu* K. Schumann 1903 *non* Philippi.] [RPS 10]

Pyrrhocactus odoriflorus Ritter, Succulenta 1960: 116, 1960. [RPS 11]

Pyrrhocactus pachacoensis Rausch, Kakt. and. Sukk. 26(4): 73-74, ill., 1975. Typus: *Rausch* 556 (ZSS). [Type cited for W but placed in ZSS.] [Argentina: San Juan] [RPS 26]
Pyrrhocactus pamaensis Ritter, Kakt. Südamer. 3: 962-963, 1980. [RPS 31]
Pyrrhocactus paucicostatus (Ritter) Ritter, Succulenta 1959(10): 131, 1959. Basionym: *Horridocactus paucicostatus*. [RPS 10]
Pyrrhocactus paucicostatus var. **viridis** (Ritter) Ritter, Succulenta 1959(10): 131, 1959. Basionym: *Horridocactus paucicostatus* var. *viridis*. [RPS 10]
Pyrrhocactus pilispinus Ritter, Succulenta 41(4): 42-44, ills., 1962. Typus: *Ritter* 217 (ZSS). [Chile: Atacama] [RPS 13]
Pyrrhocactus platyacanthus Ritter, Kakt. Südamer. 2: 420, 1980. Typus: *Ritter* 448 (ZSS). [Argentina: Mendoza] [RPS 31]
Pyrrhocactus pulchellus Ritter, Kakt. Südamer. 3: 965-966, 1980. [RPS 31]
Pyrrhocactus pygmaeus Ritter, Taxon 12(1): 32, 1963. Typus: *Ritter* 519 (U, ZSS). [Chile: Antofagasta] [RPS 14]
Pyrrhocactus reconditus Ritter, Succulenta 41(3): 27-29, 1962. Typus: *Ritter* 204 (ZSS). [Chile: Antofagasta] [RPS 13]
Pyrrhocactus residuus Ritter, Taxon 12: 33, 1963. [RPS 14]
Pyrrhocactus robustus Ritter, Succulenta 39(6): 65-68, 1960. Typus: *Ritter* 239a (ZSS [status ?]). [No herbarium cited for type.] [Chile: Quillota] [RPS 11]
Pyrrhocactus robustus var. **vegasanus** Ritter, Succulenta 39(6): 65-68, 1960. Typus: *Ritter* 239 (ZSS [status ?]). [Type cited without herbarium.] [Chile: Quillota] [RPS 11]
Pyrrhocactus rupicolus Ritter, Taxon 12(1): 32, 1963. Typus: *Ritter* 213 (ZSS). [Chile] [RPS 14]
Pyrrhocactus sanjuanensis (Spegazzini) Ritter, Succulenta 1959(10): 130 (Oct.), 1959. Basionym: *Echinocactus sanjuanensis*. [Repeated by Backeberg in Die Cact. 3: 1573, 1959 (Nov.).] [RPS 10]
Pyrrhocactus saxifragus Ritter, Kakt. Südamer. 3: 970-971, 1980. Typus: *Ritter* 712 (U, ZSS). [Type cited for U l.c. 1: iii, 1979.] [Previously published as *nomen nudum* as *Chileorebutia saxifraga* Ritter and *Neochilenia saxifraga* Backeberg.] [Chile: Iquique] [RPS 31]
Pyrrhocactus scoparius Ritter, Succulenta 1962: 51, 1962. [RPS 13]
Pyrrhocactus setiflorus Backeberg, Descr. Cact. Nov. 3: 13, 1963. *Nom. inval.* (Art. 37.1). [*Nom. nov. pro Pyrrhocactus setosiflorus* Backeberg (*in synon.*) *non* Ritter.] [RPS 14]
Pyrrhocactus setosiflorus Ritter, Succulenta 1962: 70, 1962. [Concurrently published by Backeberg (as *nom. inval.*) in Die Cact. 6: 3745, 1962.] [RPS 13]
Pyrrhocactus setosiflorus var. **intermedius** Ritter, Succulenta 1962: 70, 1962. [RPS 13]
Pyrrhocactus simulans Ritter, Succulenta 1961: 35-36, 1961. [RPS 12]
Pyrrhocactus subaianus Backeberg, Die Cact. 3: 1575, 1959. *Nom. inval.* (Art. 9.5, 37.1). [Erroneously included as valid in RPS 10.] [RPS 10]
Pyrrhocactus taltalensis (P. C. Hutchison) Ritter, Succulenta 1959(10): 131, 1959. Basionym: *Neoporteria taltalensis*. [RPS 10]
Pyrrhocactus tenuis Ritter, Kakt. Südamer. 3: 976-977, 1980. [RPS 31]
Pyrrhocactus totoralensis Ritter, Succulenta 1961: 131, 1961. [RPS 12]
Pyrrhocactus transiens Ritter, Kakt. Südamer. 3: 977-978, 1980. [RPS 31]
Pyrrhocactus transitensis Ritter, Taxon 12: 33, 1963. [RPS 14]
Pyrrhocactus trapichensis Ritter, Kakt. Südamer. 3: 979-980, 1980. [RPS 31]
Pyrrhocactus truncatipetalus Ritter, Kakt. Südamer. 3: 980-981, 1980. [RPS 31]
Pyrrhocactus umadeave var. *marayesensis* Backeberg, Descr. Cact. Nov. 3: 13, 1963. *Nom. inval.* (Art. 9.5). [Erroneously included as valid in RPS 14.] [RPS 14]
Pyrrhocactus vallenarensis Ritter, Succulenta 1959(11): 157, 1959. [RPS 10]
Pyrrhocactus vexatus Ritter, Kakt. Südamer. 3: 981-982, 1980. [RPS 31]
Pyrrhocactus villicumensis Rausch, Kakt. and. Sukk. 25(12): 268, ill., 1974. Typus: *Rausch* 555 (ZSS). [Type cited for W but placed in ZSS.] [Argentina: San Juan] [RPS 25]
Pyrrhocactus vollianus Backeberg, Descr. Cact. Nov. [1:] 31, 1957. [Dated 1956, published 1957.] [RPS 7]
Pyrrhocactus vollianus var. **breviaristatus** Backeberg, Descr. Cact. Nov. [1:] 31, 1957. [Dated 1956, published 1957.] [RPS 7]
Pyrrhocactus wagenknechtii Ritter, Succulenta 1960: 82, 1960. [RPS 11]
Quiabentia chacoensis var. **jujuyensis** Backeberg, Descr. Cact. Nov. [1:] 6, 1957. [Dated 1956, published 1957.] [RPS 7]
Quiabentia pereziensis Backeberg, Descr. Cact. Nov. [1:] 6, 1957. [Dated 1956, published 1957.] [RPS 7]
Rathbunia neosonorensis Backeberg, Die Cact. 4: 2127, 1960. *Nom. inval.* (Art. 36.1/37.1). [Erroneously included as valid in RPS 11. Based on *Cereus sonorensis* Gürke.] [RPS 11]
× **Rathbunillocactus** Mottram, Contr. New Class. Cact. Fam., 72, 1990. [= *Rathbunia* × *Myrtillocactus*.] [RPS 41]
Rauhocereus Backeberg, Descr. Cact. Nov. [1:] 5, 1957. Typus: *Rauhocereus riosaniensis*. [Dated 1956, published 1957.] [RPS 7]
Rauhocereus riosaniensis Backeberg, Descr. Cact. Nov. [1:] 5, 20, 1957. [Dated 1956, published 1957.] [RPS 7]
Rauhocereus riosaniensis var. **jaenensis** Rauh *ex* Backeberg, Die Cact. 2: 1159, 1959. Typus: *Rauh* K75 (1956) (ZSS). [The same collection is also cited as type of *Rauhocereus riosaniensis* var. *riosaniensis* by Backeberg in Descr. Cact. Nov., 20, 1956 !] [First published as a *nomen nudum* by Rauh in Beitr. Kenntn. peruan. Kakt.veg., 367, 1958 (reported as valid in RPS 10).] [Peru] [RPS -]
Rebutia cv. **Alabaster** H. Kunzmann, Kakt. and. Sukk. 24: 161-162, ill., 1973. [= *Rebutia kariusiana* Wessner × *Rebutia* 'Stirnadels

Meisterstück'.] [RPS 24]

Rebutia sect. **Aylostera** (Spegazzini) Buining & Donald, Sukkulentenkunde 7/8: 98, 101, 1963. Basionym: *Aylostera*. [RPS 14]

Rebutia sect. **Cylindrorebutia** Buining & Donald, Sukkulentenkunde 7/8: 98, 101, 1963. Typus: *R. einsteinii* Fric. [RPS 14]

Rebutia sect. **Digitorebutia** Buining & Donald, Sukkulentenkunde 7/8: 98, 102, 1963. Typus: *R. haagei* Fric & Schelle. [RPS 14]

Rebutia sect. **Mediorebutia** Buining & Donald, Sukkulentenkunde 7/8: 98, 103, 1963. Typus: *R. calliantha* Bewerunge. [RPS 14]

Rebutia sect. **Setirebutia** Buining & Donald, Sukkulentenkunde 7/8: 98, 100, 1963. Typus: *R. aureiflora* Backeberg. [RPS 14]

Rebutia subgen. **Aylostera** (Spegazzini) Buining & Donald, Sukkulentenkunde 7/8: 98, 101, 1963. Basionym: *Aylostera*. [RPS 14]

Rebutia subgen. *Chinorebutia* Y. Ito, Bull. Takarazuka Insectarium 71: 13-20, 1950. *Nom. inval.* (Art. 36.1). [RPS 3]

Rebutia subgen. *Dichrorebutia* Y. Ito, Bull. Takarazuka Insectarium 71: 13-20, 1950. *Nom. inval.* (Art. 36.1). [RPS 3]

Rebutia albiareolata Ritter, Kakt. and. Sukk. 28(4): 78, 1977. Typus: *Ritter* 761 (U, ZSS). [Bolivia: Arce] [RPS 28]

Rebutia albiflora Ritter & Buin., Taxon 12: 29, 1963. [RPS 14]

Rebutia albipilosa Ritter, Taxon 12: 29, 1963. [RPS 14]

Rebutia albopectinata Rausch, Kakt. and. Sukk. 23(9): 236-237, 1972. Typus: *Rausch* 312 (ZSS). [Type cited for W but placed in ZSS.] [Bolivia] [RPS 23]

Rebutia almeyeri Heinrich, in Backeberg, Kakt.-Lex., 383, 1966. *Nom. inval.* (Art. 36.1, 37.1). [Based on cultivated material of unknown wild origin.] [RPS -]

Rebutia antarctica hort., Gard. Chron. 146(Aug. 1.): 47, 1959. *Nom. inval.* (Art. 36.1, 37.1). [RPS 10]

Rebutia archibuiningiana Ritter, Ashingtonia 3(1): 14-15, 1978. [RPS 29]

Rebutia arenacea Cardenas, Cact. Succ. J. (US) 23(3): 94-95, 1951. [RPS 2]

Rebutia auranitida (Wessner) Buining & Donald, Sukkulentenkunde 7/8: 101, 1963. Basionym: *Lobivia auranitida*. [RPS 14]

Rebutia auranitida fa. **gracilis** (Wessner) Buining & Donald, Sukkulentenkunde 7/8: 101, 1963. Basionym: *Lobivia auranitida* var. *gracilis*. [RPS 14]

Rebutia aureiflora fa. **rubelliflora** (Backeberg) Buining & Donald, Sukkulentenkunde 7/8: 100, 1963. Basionym: *Mediolobivia rubelliflora*. [RPS 14]

Rebutia aureiflora fa. **rubriflora** (Backeberg) Buining & Donald, Sukkulentenkunde 7/8: 100, 1963. Basionym: *Mediolobivia rubriflora*. [RPS 14]

Rebutia aureiflora fa. **sarothroides** (Werdermann) Buining & Donald, Sukkulentenkunde 7/8: 100, 1963. Basionym: *Rebutia sarothroides*. [RPS 14]

Rebutia aureiflora ssp. **elegans** (Backeberg) Donald, Ashingtonia 2(5): 90, 1976. Basionym: *Mediolobivia elegans*. [RPS 27]

Rebutia aureiflora var. **blossfeldii** (Werdermann) Donald, Ashingtonia 2(5): 88, 1976. Basionym: *Rebutia blossfeldii*. [RPS 27]

Rebutia aureiflora var. **elegans** (Backeberg) Buining & Donald, Sukkulentenkunde 7/8: 100, 1963. Basionym: *Mediolobivia elegans*. [RPS 14]

Rebutia aureiflora var. **sarothroides** (Werdermann) Donald, Ashingtonia 2(5): 90, 1976. Basionym: *Rebutia sarothroides*. [RPS 27]

Rebutia binnewaldiana Heinrich, in Backeberg, Kakt.-Lex., 383, 1966. *Nom. inval.* (Art. 36.1, 37.1). [Based on cultivated material from Bolivia (Huari Huari).] [RPS -]

Rebutia brachyantha (Wessner) Buining & Donald, Sukkulentenkunde 7/8: 102, 1963. Basionym: *Lobivia brachyantha*. [RPS 14]

Rebutia brachyantha Cardenas, Kakt. and. Sukk. 16: 74-75, 1965. *Nom. illeg.* (Art. 64.1). [*Non Rebutia brachyantha* (Wessner) Buining & Donald.] [RPS 16]

Rebutia brunescens Rausch, Kakt. and. Sukk. 23(9): 235-236, 1972. Typus: *Rausch* 480 (W, ZSS). [Bolivia] [RPS 23]

Rebutia brunneoradicata Ritter, Kakt. and. Sukk. 28(4): 77-78, 1977. [RPS 28]

Rebutia buiningiana Rausch, Kakt. and. Sukk. 23(4): 98-99, 1972. Typus: *Rausch* 511 (W, ZSS). [Argentina: Salta] [RPS 23]

Rebutia caineana Cardenas, Cact. Succ. J. (US) 38: 143-144, 1966. [RPS 17]

Rebutia cajasensis Ritter, Succulenta 56(3): 64, 1977. Typus: *Ritter* 1141 (U, ZSS). [Bolivia: Mendez] [RPS 28]

Rebutia calliantha fa. **breviseta** (Backeberg) Buining & Donald, Sukkulentenkunde 7/8: 104, 1963. Basionym: *Rebutia senilis* var. *breviseta*. [Republished in Cact. Succ. J. Gr. Brit. 27: 41, 1965 (cf. RPS 16).] [RPS 14]

Rebutia calliantha fa. **hyalacantha** (Backeberg) Buining & Don., Sukkulentenkunde 7/8: 103, 1963. Basionym: *Rebutia senilis* var. *hyalacantha*. [RPS 14]

Rebutia calliantha var. *beryllioides* Buining & Donald, Sukkulentenkunde 7/8: 103, 1963. *Nom. inval.* (Art. 37.1). [Erroneously included as valid in RPS 14.] [RPS 14]

Rebutia calliantha var. **kariusiana** (Wessner) Buining & Donald, Cact. Succ. J. Gt. Brit. 27: 41, 1965. Basionym: *Rebutia kariusiana*. [RPS 16]

Rebutia calliantha var. **krainziana** (Kesselring) Buining & Donald, Sukkulentenkunde 7/8: 103, 1963. Basionym: *Rebutia krainziana*. [RPS 14]

Rebutia camargoensis Rausch, Succulenta 55(3): 42, ill. (p. 41), 1976. Typus: *Rausch* 311 (ZSS). [Type cited for W but placed in ZSS.] [Bolivia] [RPS 27]

Rebutia canacruzensis Rausch, Kakt. and. Sukk. 27(3): 49-50, ill., 1976. Typus: *Rausch* 642 (ZSS). [Bolivia: Nor Cinti] [RPS 27]

Rebutia candiae Cardenas, Cact. Succ. J. (US) 33: 112, 1961. [RPS 12]

Rebutia canigueralii Cardenas, Cact. Succ. J. (US) 36: 26-27, 1964. [RPS 15]

Rebutia caracarensis Cardenas, Cact. Succ. J. (US) 42: 37-38, 1970. [RPS 21]
Rebutia carmeniana Rausch, Kakt. and. Sukk. 29(5): 105-106, ill., 1978. Typus: *Rausch* 690 (ZSS). [Argentina: Jujuy] [RPS 29]
Rebutia christinae Rausch, Kakt. and. Sukk. 26(7): 145-146, ill., 1975. Typus: *Rausch* 492a (ZSS). [Type cited for W but placed in ZSS.] [Argentina: Salta] [RPS 26]
Rebutia chrysacantha var. **elegans** (Backeberg) Backeberg, Die Cact. 3: 1539, 1959. Basionym: *Rebutia xanthocarpa* var. *elegans*. [RPS 10]
Rebutia chrysacantha var. *iseliniana* (Krainz) Donald, Cactus (Paris) No. 40: 39, 1954. *Nom. inval.* (Art. 33.2), based on *Rebutia senilis* var. *iseliniana*. [RPS 5]
Rebutia chrysacantha var. *kesselringiana* (Bewerunge) Donald, Cactus (Paris) No. 40: 39, 1954. *Nom. inval.* (Art. 33.2), based on *Rebutia senilis* var. *kesselringiana*. [RPS 5]
Rebutia cincinnata Rausch, Kakt. and. Sukk. 27(1): 4-5, ill., 1976. Typus: *Rausch* 300 (ZSS). [This issue was probably already distributed in late December 1975 (according to Index Kewensis).] [Bolivia: Potosi] [RPS 27]
Rebutia cintiensis Rausch, Verzeichnis Feldnummern W. Rausch, [unpaged], 1975. *Nom. inval.* (Art. 36.1, 37.1). [Published as provisional name, *non Rebutia cintiensis* Ritter 1977.] [RPS -]
Rebutia cintiensis Ritter, Ashingtonia 2(10): 206, 1978. [RPS 29]
Rebutia colorea Ritter, Kakt. and. Sukk. 28(4): 78, 1977. [RPS 28]
Rebutia corroana Cardenas, Cact. Succ. J. (US) 43: 244-245, 1971. [RPS 22]
Rebutia costata fa. **eucaliptana** (Backeberg) Buining & Donald, Sukkulentenkunde 7/8: 102, 1963. Basionym: *Lobivia eucaliptana*. [RPS 14]
Rebutia costata fa. *pilifera* Buining & Donald, Sukkulentenkunde 7/8: 102, 1963. *Nom. inval.* (Art. 37.1). [Erroneously included as valid in RPS 14. Publication repeated in Cact. Succ. J. Gr. Brit. 27: 40, 1965, again without type.] [RPS 14]
Rebutia cylindrica (Donald & Lau) Donald, Bradleya 5: 93, 1987. Basionym: *Sulcorebutia cylindrica*. [RPS 38]
Rebutia deminuta fa. **pseudominuscula** (Spegazzini) Buining & Donald, Sukkulentenkunde 7/8: 101, 1963. Basionym: *Echinocactus pseudominusculus*. [RPS 14]
Rebutia diersiana Rausch, Kakt. and. Sukk. 26(2): 25-26, ills., 1975. Typus: *Rausch* 631 (ZSS). [Type cited for W but placed in ZSS.] [Bolivia: Sud Cinti] [RPS 26]
Rebutia diersiana var. **atrovirens** Rausch, Kakt. and. Sukk. 26(2): 26, 1975. Typus: *Rausch* 633 (ZSS). [Type cited for W but placed in ZSS.] [Bolivia: Sud Cinti] [RPS 26]
Rebutia diersiana var. **minor** Rausch, Succulenta 58(11): 258-259, ill. (p. 257), 1979. Typus: *Rausch* 630 (ZSS). [Bolivia: Sud Cinti] [RPS 30]
Rebutia donaldiana Lau & Rowley, Ashingtonia 1: 76-78, 1974. [RPS 25]
Rebutia einsteinii fa. **conoidea** (Wessner) Donald, Ashingtonia 2(6): 112, 1976. Basionym: *Lobivia conoidea*. [RPS 27]
Rebutia einsteinii fa. **karreri** (Backeberg) Buining & Donald, Cact. Succ. J. Gr. Brit. 27: 39, 1965. Basionym: *Mediolobivia schmiedcheniana* var. *karreri*. [RPS 16]
Rebutia einsteinii fa. **rubriviridis** (Fric & Kreuzinger *ex* Backeberg) Donald, Ashingtonia 2(6): 112, 1976. Basionym: *Mediolobivia schmiedcheniana*. [RPS 27]
Rebutia einsteinii fa. **schmiedcheniana** (Köhler) Buining & Donald, Sukkulentenkunde 7/8: 101, 1963. Basionym: *Lobivia schmiedcheniana*. [RPS 14]
Rebutia einsteinii fa. **steineckei** (Fric & Kreuzinger *ex* Backeberg) Donald, Ashingtonia 2(6): 112, 1976. Basionym: *Mediolobivia schmiedcheniana* var. *steineckei*. [RPS 27]
Rebutia einsteinii var. **columnaris** (Wessner) Buining & Donald, Sukkulentenkunde 7/8: 101, 1963. Basionym: *Lobivia columnaris*. [RPS 14]
Rebutia einsteinii var. **conoidea** (Wessner) Buining & Donald, Sukkulentenkunde 7/8: 101, 1963. Basionym: *Lobivia conoidea*. [RPS 14]
Rebutia einsteinii var. **gonjianii** (Kiesling) Donald, Ashingtonia 1(7): 83, 1974. Basionym: *Rebutia gonjianii*. [RPS 25]
Rebutia einsteinii var. **karreri** (Fric *ex* Backeberg) Sida, Kaktusy 26(1): 19, 1990. Basionym: *Mediolobivia schmiedcheniana* var. *karreri*. [RPS 41]
Rebutia einsteinii var. **steineckei** (Backeberg) Buining & Donald, Sukkulentenkunde 7/8: 101, 1963. Basionym: *Mediolobivia schmiedcheniana* var. *steineckei*. [RPS 14]
Rebutia eos Rausch, Succulenta 51(1): 1-3, ill. (p. 1), 1972. Typus: *Rausch* 333 (W). [Argentina: Jujuy] [RPS 23]
Rebutia euanthema (Backeberg) Buining & Donald, Sukkulentenkunde 7/8: 102, 1963. Basionym: *Lobivia euanthema*. [RPS 14]
Rebutia euanthema fa. **fricii** (Backeberg) Buining & Donald, Sukkulentenkunde 7/8: 102, 1963. Basionym: *Mediolobivia euanthema* var. *fricii*. [RPS 14]
Rebutia euanthema fa. **neopygmaea** (Backeberg) Buining & Donald, Sukkulentnkunde 7/8: 102, 1963. Basionym: *Mediolobivia neopygmaea*. [RPS 14]
Rebutia euanthema fa. **oculata** (Werdermann) Buining & Donald, Sukkulentenkunde 7/8: 102, 1963. Basionym: *Rebutia oculata*. [RPS 14]
Rebutia eucaliptana (Backeberg) Ritter, Kakt. Südamer. 2: 602, 1980. Basionym: *Lobivia eucaliptana*. [RPS 31]
Rebutia fabrisii Rausch, Kakt. and. Sukk. 28(3): 52-53, ill., 1977. Typus: *Rausch* 688 (ZSS). [Argentina: Jujuy] [RPS 28]
Rebutia fabrisii var. **aureiflora** Rausch, Kakt. and. Sukk. 28(3): 53, 1977. Typus: *Rausch* 687 (ZSS). [Argentina: Jujuy] [RPS 28]
Rebutia fabrisii var. **nana** Rausch, Succulenta

64(5): 101-102, ills., 1985. Typus: *Rausch* 688a (ZSS). [Argentina: Jujuy] [RPS 36]

Rebutia fidaiana (Backeberg) D. Hunt, Bradleya 5: 94, 1987. Basionym: *Echinocactus fidaianus*. [RPS 38]

Rebutia fiebigiana Heinrich, in Backeberg, Kakt.-Lex., 383, 1966. *Nom. inval.* (Art. 36.1, 37.1). [Based on cultivated material from Bolivia (Huari Huari).] [RPS -]

Rebutia fiebrigii fa. **densiseta** Cullmann, Sukkulentenkunde 6: 25, 1957. [RPS 8]

Rebutia fiebrigii var. **densiseta** (Cullmann) Oeser, Kakt. and. Sukk. 27: 28-30, 1976. Basionym: *Rebutia fiebrigii*. [RPS 27]

Rebutia fiebrigii var. **vulpes** Ritter, Kakt. Südamer. 2: 619-620, 1980. [RPS 31]

Rebutia flavistyla Ritter, Ashingtonia 3(1): 14, 1978. [RPS 29]

Rebutia friedrichiana Rausch, Succulenta 55(6): 103, ill. (p. 101), 1976. Typus: *Rausch* 646 (ZSS). [Type cited for ZSS l.c. 56(2): 30, 1977.] [Bolivia: Sud Cinti] [RPS 27]

Rebutia froehlichiana Rausch, Succulenta 54(12): 225-226, 1975. Typus: *Rausch* 649 (ZSS). [Bolivia: Sud Cinti] [RPS 26]

Rebutia fulviseta Rausch, Kakt. and. Sukk. 21(2): 29, ill., 1970. Typus: *Rausch* 319 (W, ZSS). [Bolivia: Arce] [RPS 21]

Rebutia fusca Ritter, Kakt. and. Sukk. 28(4): 78, 1977. [RPS 28]

Rebutia glomeriseta Cardenas, Cact. Succ. J. (US) 23(3): 95, 1951. [RPS 2]

Rebutia glomerispina Cardenas, Cact. Succ. J. (US) 36: 40, 1964. [RPS 15]

Rebutia gonjianii Kiesling, Bol. Soc. Argent. Bot. 15(1): 132-135, 1973. [RPS 24]

Rebutia graciliflora Backeberg, Descr. Cact. Nov. 3: 13, 1963. *Nom. inval.* (Art. 9.5). [Erroneously included asd valid in RPS 14.] [RPS 14]

Rebutia gracilispina Ritter, Kakt. and. Sukk. 28(4): 76, 1977. [RPS 28]

Rebutia × *grandilacea* Pickoff, Cact. Succ. J. (US) 52(6): 285, 1980. *Nom. inval.* (Art. 40). [RPS 31]

Rebutia haagei var. **orurensis** (Backeberg) Sida, Kaktusy 26(1): 19, 1990. Basionym: *Lobivia orurensis*. [RPS 41]

Rebutia haseltonii Cardenas, Cact. & Succ. J. (US) 38: 143, 1966. [RPS 17]

Rebutia heliosa Rausch, Kakt. and. Sukk. 21(2): 30-31, ill., 1970. Typus: *Rausch* 314 (W, ZSS). [Bolivia: Tarija] [RPS 21]

Rebutia heliosa var. **cajasensis** Donald, Ashingtonia 3: 144-145, 1980. Typus: *Lau* 405 (K 41173, ZSS). [Bolivia: Tarija] [RPS 31]

Rebutia heliosa var. **condorensis** Donald, Ashingtonia 3: 143-144, 1980. Typus: *Lau* 401 (K, ZSS). [Bolivia: Tarija] [RPS 31]

Rebutia hoffmannii Diers & Rausch, Kakt. and. Sukk. 28(5): 105-106, ill., 1977. Typus: *Rausch* 521a (ZSS). [Argentina: Salta] [RPS 28]

Rebutia huasiensis Rausch, Kakt. and. Sukk. 28(2): 25-26, ill., 1977. Typus: *Rausch* 313 (ZSS). [Bolivia: Cinti] [RPS 28]

Rebutia hyalacantha (Backeberg) Backeberg, Die Cact. 3: 1551, 1959. *Nom. inval.* (Art. 57.1), based on *Rebutia senilis* var. *hyalacantha*. [Erroneously included as valid in RPS 10.] [RPS 10]

Rebutia inflexiseta Cardenas, Cact. Succ. J. (US) 42: 36-37, 1970. [RPS 21]

Rebutia iridescens Ritter, Kakt. and. Sukk. 28(4): 76, 1977. [*Non Rebutia iridescens* W. Rausch *nom. nud.*] [RPS 28]

Rebutia iscayachensis Rausch, Succulenta 56(1): 3, ill. (p. 1), 1977. Typus: *Rausch* 335B (ZSS). [Bolivia] [RPS 28]

Rebutia jujuyana Rausch, Kakt. and. Sukk. 24(7): 147-148, ill., 1973. Typus: *Rausch* 220 (ZSS). [Type cited for W but placed in ZSS.] [Argentina: Jujuy] [RPS 24]

Rebutia kariusiana Wessner, Kakt. and. Sukk. 14: 149, 1963. [RPS 14]

Rebutia kieslingii Rausch, Kakt. and. Sukk. 28(8): 177-178, ill., 1977. Typus: *Rausch* 694 (ZSS). [Argentina: Salta] [RPS 28]

Rebutia krainziana fa. *beryllioides* (Buining & Donald) Krainz, Kat. ZSS ed. 2, 107, 1967. *Nom. inval.*, based on *Rebutia calliantha* var. *beryllioides, nom. inval.* (Art. 37.1). [Erroneously included as valid in RPS 18.] [RPS 18]

Rebutia krainziana fa. **calliantha** (Bewerunge) Krainz & Haarmann, Kat. ZSS ed. 2, 107, 1967. Basionym: *Rebutia calliantha*. [RPS 18]

Rebutia krainziana var. **breviseta** (Backeberg) Donald, Nation. Cact. Succ. J. 12(1): 11, (2): 27-28, 1957. Basionym: *Rebutia senilis* var. *breviseta*. [RPS 8]

Rebutia krainziana var. **hyalacantha** (Backeberg) Buchheim, in Zander, Handwörterbuch Pfl.-namen ed. 10, 743, 1972. Basionym: *Rebutia senilis* var. *hyalacantha*. [RPS 23]

Rebutia krainziana var. **wessneriana** (Bewerunge) Krainz & Haarmann, Kat. ZSS ed. 2, 107, 1967. Basionym: *Rebutia wessneriana*. [RPS 18]

Rebutia kruegeri (Cardenas) Backeberg, Die Cact. 3: 1554, 1959. Basionym: *Aylostera kruegeri*. [RPS 10]

Rebutia kupperiana var. **spiniflora** Ritter, Kakt. and. Sukk. 28(4): 78, 1977. [RPS 28]

Rebutia lanosiflora Ritter, Kakt. and. Sukk. 28(4): 77, 1977. [RPS 28]

Rebutia leucanthema Rausch, Kakt. and. Sukk. 26(6): 125, ill., 1975. Typus: *Rausch* 305 (ZSS). [Type cited for W but placed in ZSS.] [Bolivia: Nor Cinti] [RPS 26]

Rebutia leucanthema var. **cocciniflora** Ritter, Succulenta 56: 63-64, 1977. [RPS 28]

Rebutia mamillosa Rausch, Succulenta 51(4): 69-70, ill., 1972. Typus: *Rausch* 302 (ZSS). [Type cited for W but placed in ZSS.] [Bolivia] [RPS 23]

Rebutia mamillosa var. **australis** Ritter, Kakt. and. Sukk. 28(4): 77, 1977. Typus: *Ritter* 341a (U, ZSS). [Bolivia: Mendez] [RPS 28]

Rebutia mamillosa var. **orientalis** Ritter, Kakt. and. Sukk. 28(4): 77, 1977. [RPS 28]

Rebutia margarethae Rausch, Kakt. and. Sukk. 23(1): 4-5, ill., 1972. Typus: *Rausch* 521 (W, ZSS). [Argentina: Salta] [RPS 32]

Rebutia marsoneri cv. **Spatula** Donald,

Ashingtonia 2(4): 71, 1976. [= *Rebutia marsoneri* var. *spatulata* hort.] [RPS 27]

Rebutia marsoneri fa. **sieperdaiana** (Buining) Buining & Donald, Sukkulentenkunde 7/8: 103, 1963. Basionym: *Rebutia sieperdaiana*. [RPS 14]

Rebutia marsoneri var. **sieperdaiana** (Buining) Donald, Nation. Cact. Succ. J. 12(1): 11, 1957. Basionym: *Rebutia sieperdaiana*. [First proposed by Donald in Cactus (Paris) No. 40: 39, 1954, as *comb. inval.* (Art. 33.2), cf. RPS 5.] [RPS 8]

Rebutia menesesii Cardenas, Cact. Succ. J. (US) 33: 113, 1961. [RPS 12]

Rebutia mentosa (Ritter) Donald, Bradleya 5: 93, 1987. Basionym: *Sulcorebutia mentosa*. [RPS 38]

Rebutia minuscula fa. *brunneoaurantiaca* Simon, Kakt. / Sukk. 1968(3): 47, 1968. *Nom. inval.* (Art. 36.1, 37.1). [Published as provisional name.] [RPS -]

Rebutia minuscula fa. *carminea* Buining *ex* Simon, Kakt. / Sukk. 1968(3): 47, 1968. *Nom. inval.* (Art. 33.2 / 36.1). [Published as nude combination, but no base name can be traced.] [RPS -]

Rebutia minuscula fa. *grandiflora* (Backeberg) Simon, Kakt. / Sukk. 1968(3): 47, 1968. *Nom. inval.* (Art. 33.2, 34.1), based on *Rebutia grandiflora*. [Published as provisional name.] [RPS -]

Rebutia minuscula fa. *intermedia* Donald *ex* Simon, Kakt. / Sukk. 1968(3): 47, 1968. *Nom. inval.* (Art. 36.1, 37.1). [Published as provisional name.] [RPS -]

Rebutia minuscula fa. **kariusiana** (Wessner) Donald, Ashingtonia 2: 43, 1975. Basionym: *Rebutia kariusiana*. [RPS 26]

Rebutia minuscula fa. **knuthiana** (Backeberg) Buining & Donald, Sukkulentenkunde 7/8: 99, 1963. Basionym: *Rebutia knuthiana*. [Repeated (*nom. inval.*, Art. 33.2) by Simon in Kakt. / Sukk. 1968(3): 47.] [RPS 14]

Rebutia minuscula fa. *rosiflora* Simon, Kakt. / Sukk. 1968(3): 47, 1968. *Nom. inval.* (Art. 36.1, 37.1). [Published as provisional name.] [RPS -]

Rebutia minuscula fa. **violaciflora** (Backeberg) Buining & Donald, Sukkulentenkunde 7/8: 99, 1963. Basionym: *Rebutia violaciflora*. [RPS 14]

Rebutia minuscula ssp. **grandiflora** (Backeberg) Donald, Ashingtonia 2: 43, 1975. Basionym: *Rebutia grandiflora*. [RPS 26]

Rebutia minuscula ssp. **violaciflora** (Backeberg) Donald, Ashingtonia 2: 43, 1975. Basionym: *Rebutia violaciflora*. [RPS 26]

Rebutia minuscula var. **grandiflora** (Backeberg) Krainz, Die Kakt. C Vc (July 1960), 1960. Basionym: *Rebutia grandiflora*. [RPS 11]

Rebutia minutissima Ritter, Kakt. and. Sukk. 28(4): 78, 1977. [RPS 28]

Rebutia mixta Ritter, Kakt. and. Sukk. 28(4): 76, 1977. [RPS 28]

Rebutia mixticolor Ritter, Kakt. and. Sukk. 28(4): 77, 1977. [RPS 28]

Rebutia mudanensis Rausch, Kakt. and. Sukk. 27(8): 169-170, ills., 1976. Typus: *Rausch* 689 (ZSS). [Argentina: Salta] [RPS 27]

Rebutia muscula Ritter & Thiele, Taxon 12(1): 29, 1963. Typus: *Ritter* 753 (U, ZSS). [Bolivia: O'Connor] [RPS 14]

Rebutia muscula var. **luteo-albida** F. Brandt, Kakt. Orch.-Rundschau 5(5): 61-64, 1980. [RPS 31]

Rebutia nazarenoensis (Rausch) B. Fearn & L. Pearcy, The Genus Rebutia, 71, 1981. Basionym: *Digitorebutia nazarenoensis*. [RPS 33]

Rebutia neocumingii (Backeberg) D. Hunt, Bradleya 5: 94, 1987. Basionym: *Weingartia neocumingii*. [RPS 38]

Rebutia neumanniana (Werdermann) D. Hunt, Bradleya 5: 94, 1987. Basionym: *Weingartia neumanniana*. [RPS 38]

Rebutia nitida Ritter, Ashingtonia 3(1): 14, 1978. [RPS 29]

Rebutia nogalesensis Ritter, Kakt. and. Sukk. 28(4): 78, 1977. [RPS 28]

Rebutia odontopetala Ritter, Kakt. and. Sukk. 28(4): 76, 1977. [RPS 28]

Rebutia orurensis (Backeberg) Ritter, Kakt. Südamer. 2: 594-595, 1980. Basionym: *Lobivia orurensis*. [RPS 31]

Rebutia padcayensis Rausch, Kakt. and. Sukk. 21(4): 65, ill., 1970. Typus: *Rausch* 322 (W, ZSS). [Bolivia] [RPS 21]

Rebutia pallida Rausch, Succulenta 56(10): 234, ill. (p. 233), 1977. Typus: *Rausch* 645 (ZSS). [Bolivia: Sud Cinti] [RPS 28]

Rebutia patericalyx Ritter, Kakt. and. Sukk. 28(4): 78, 1977. [RPS 28]

Rebutia pauciareolata Ritter, Kakt. and. Sukk. 28(4): 77, 1977. [RPS 28]

Rebutia paucicostata Ritter, Kakt. and. Sukk. 28(4): 77, 1977. [RPS 28]

Rebutia permutata Heinrich, in Backeberg, Descr. Cact. Nov. 3: 13, 1963. *Nom. inval.* (Art. 36.1, 37.1). [RPS 14]

Rebutia permutata fa. *gokrausei* Heinrich, in Backeberg, Descr. Cact. Nov. 3: 13, 1963. *Nom. inval.* (Art. 37.1, 43.1). [RPS 14]

Rebutia perplexa Donald, Ashingtonia 3(5-6): 150-151, ill., 1980. Typus: *Lau* 329a (K, ZSS). [Bolivia: Tarija] [RPS 31]

Rebutia poecilantha Ritter, Kakt. and. Sukk. 28(4): 77, 1977. [RPS 28]

Rebutia polymorpha Cardenas, Kakt. and. Sukk. 16: 115-116, 1965. [RPS 16]

Rebutia potosina Ritter, Kakt. and. Sukk. 28(4): 77, 1977. [RPS 28]

Rebutia prolifera Rausch, Verzeichnis Feldnummern W. Rausch, [unpaged], 1975. *Nom. inval.* (Art. 36.1, 37.1). [Published as provisional name.] [RPS -]

Rebutia prolifera var. *aureiflora* Rausch, Verzeichnis Feldnummern W. Rausch, [unpaged], 1975. *Nom. inval.* (Art. 36, 37, 43). [Published as provisional name.] [RPS -]

Rebutia pseudodeminuta fa. **albiseta** (Backeberg) Buining & Donald, Sukkulentenkunde 7/8: 101, 1963. Basionym: *Aylostera pseudodeminuta* var. *albiseta*. [RPS 14]

Rebutia pseudodeminuta fa. **grandiflora** (Backeberg) Buining & Donald, Sukkulentenkunde 7/8: 102, 1963. Basionym: *Aylostera pseudodeminuta* var. *grandiflora.* [RPS 14]

Rebutia pseudodeminuta fa. *rubrifilamentosa* Buining & Donald, Sukkulentenkunde 7/8: 102, 1963. *Nom. inval.* (Art. 37.1). [Erroneously included as valid in RPS 14. Also published in Cact. Succ. J. Gr. Brit. 27: 39, 1965, again without type.] [RPS 14]

Rebutia pseudodeminuta fa. **schneideriana** (Backeberg) Buining & Donald, Sukkulentenkunde 7/8: 102, 1963. Basionym: *Aylostera pseudodeminuta* var. *schneideriana.* [RPS 14]

Rebutia pulchella Rausch, Kakt. and. Sukk. 23(12): 340, ill., 1972. Typus: *Rausch* 320 (W [not found], ZSS). [Bolivia] [RPS 23]

Rebutia pulchra Cardenas, Cact. Succ. J. (US) 42: 38-39, 1970. [Sphalm. 'pulchera'.] [RPS 21]

Rebutia pulvinosa Ritter & Buining, Taxon 12: 29, 1963. [RPS 14]

Rebutia pygmaea fa. **atrovirens** (Backeberg) Buining & Donald, Sukkulentenkunde 7/8: 102, 1963. Basionym: *Lobivia atrovirens.* [RPS 14]

Rebutia pygmaea fa. **flavovirens** (Backeberg) Buining & Donald, Sukkulentenkunde 7/8: 102, 1961. Basionym: *Mediolobivia haagei* var. *flavovirens.* [RPS 14]

Rebutia pygmaea fa. **fuauxiana** (Backeberg) Buining & Donald, Sukkulentenkunde 7/8: 103, 1963. Basionym: *Mediolobivia fuauxiana.* [RPS 14]

Rebutia pygmaea fa. **haefneriana** (Cullmann) Buining & Donald, Sukkulentenkunde 7/8: 103, 1963. Basionym: *Mediolobivia haefneriana.* [RPS 14]

Rebutia pygmaea fa. **neosteinmannii** (Backeberg) Buining & Donald, Cact. Succ. J. Gr. Brit. 27: 40, 1965. Basionym: *Mediolobivia pectinata* var. *neosteinmannii.* [RPS 16]

Rebutia pygmaea var. **friedrichiana** (Rausch) Eriksson, Kaktus (Odense) 23(4): 81, ill., 1988. Basionym: *Rebutia friedrichiana.* [Given as 'sine comb.'.] [RPS 39]

Rebutia raulii Rausch, Kakt. and. Sukk. 31(6): 170-171, ill., 1980. Typus: *Rausch* 485 (ZSS). [Bolivia: Nor Cinti] [RPS 31]

Rebutia rauschii Zecher, Kakt. and. Sukk. 28(4): 73-74, ill., 1977. Typus: *Rausch* 297 (ZSS). [Bolivia: Potosi] [RPS 28]

Rebutia rauschii (G. Frank) D. Hunt, in Walters et al., Europ. Gard. Fl. 3: 245, 1989. *Nom. illeg.* (Art. 64.1), based on *Sulcorebutia rauschii.* [*Non Rebutia rauschii* Zecher.] [RPS -]

Rebutia ritteri (Wessner) Buining & Donald, Sukkulentenkunde 7/8: 103, 1963. Basionym: *Lobivia ritteri.* [RPS 14]

Rebutia ritteri fa. *hahniana* Buining & Donald, Sukkulentenkunde 7/8: 103, 1963. *Nom. inval.* (Art. 37.1). [Erroneously included as valid in RPS 14. Also published in Cact. Succ. J. Gr. Brit. 27: 40, 1965, again without type.] [RPS 14]

Rebutia ritteri fa. *peterseimii* Buining & Donald, Sukkulentenkunde 7/8: 103, 1963. *Nom. inval.* (Art. 37.1). [Erroneously included as valid in RPS 14. Also published in Cact. Succ. J. Gr. Brit. 27: 40, 1965, again without type.] [RPS 14]

Rebutia ritteri var. **nigricans** (Wessner) Buining & Donald, Sukkulentenkunde 7/8: 103, 1963. Basionym: *Lobivia nigricans.* [RPS 14]

Rebutia robustispina Ritter, Succulenta 56: 64-65, 1977. [RPS 28]

Rebutia robustispina var. **minor** Ritter, Succulenta 56: 65, 1977. [RPS 28]

Rebutia rosalbiflora Ritter, Kakt. and. Sukk. 28(4): 76, 1977. Typus: *Ritter* 1115 (U, ZSS). [Bolivia: Sud Chichas] [RPS 28]

Rebutia rosalbiflora var. **amblypetala** Ritter, Kakt. and. Sukk. 28(4): 76, 1977. [RPS 28]

Rebutia roseiflora Rausch, Verzeichnis Feldnummern W. Rausch, [unpaged], 1975. *Nom. inval.* (Art. 36.1, 37.1). [Published as provisional name.] [RPS -]

Rebutia rubiginosa Ritter, Taxon 12: 29, 1963. [RPS 14]

Rebutia rutiliflora Ritter, Kakt. and. Sukk. 28(4): 76, 1977. [RPS 28]

Rebutia salpingantha Ritter, Kakt. and. Sukk. 28(4): 77, 1977. [RPS 28]

Rebutia sanguinea Ritter, Succulenta 56: 65, 1977. [RPS 28]

Rebutia sanguinea var. **minor** Ritter, Succulenta 56: 65, 190, 1977. [RPS 28]

Rebutia schatzliana Rausch, Kakt. and. Sukk. 26(11): 244-245, ill., 1975. Typus: *Rausch* 640 (ZSS). [Bolivia: Nor Cinti] [RPS 26]

Rebutia senilis fa. *aurescens* (Backeberg) Simon, Kakt. / Sukk. 1968(3): 48, 1968. *Nom. inval.* (Art. 33.2). [Basionym name not traceable.] [RPS -]

Rebutia senilis fa. *blossfeldiana* Köhler *ex* Simon, Kakt. / Sukk. 1968(3): 48, 1968. *Nom. inval.* (Art. 36.1, 37.1). [Published as provisional name.] [RPS -]

Rebutia senilis fa. *breviseta* (Backeberg) Simon, Kakt. / Sukk. 1968(3): 47, 1968. *Nom. inval.* (Art. 33.2), based on *Rebutia senilis* var. *breviseta.* [Published as provisional name.] [RPS -]

Rebutia senilis fa. *cana* Backeberg *ex* Simon, Kakt. / Sukk. 1968(3): 48, 1968. *Nom. inval.* (Art. 36.1, 37.1). [Based on a catalogue name by Backeberg. Published as provisional name.] [RPS -]

Rebutia senilis fa. **chrysacantha** (Backeberg) Buining & Donald, Cact. Succ. J. Gr. Brit. 27: 38, 1965. Basionym: *Rebutia chrysacantha.* [Combination repeated by Simon (*comb. inval.*, Art. 33.2) in Kakt. / Sukk. 1968(3): 48.] [RPS 16]

Rebutia senilis fa. **elegans** (Backeberg) Buining & Donald, Sukkulentenkunde 7/8: 99, 1963. Basionym: *Rebutia xanthocarpa* var. *elegans.* [Republished in Cact. Succ. J. Gr. Brit. 27: 38, and again by Simon in Kakt. / Sukk. 1968(3): 48.] [RPS 14]

Rebutia senilis fa. *hyalacantha* Backeberg *ex* Simon, Kakt. / Sukk. 1968(3): 48, 1968. *Nom. inval.* (Art. 36.1, 37.1). [Published as provisional name and based on *Rebutia senilis* var. *hyalacantha*.] [RPS -]

Rebutia senilis fa. **iseliniana** (Krainz) Buining & Donald, Sukkulentenkunde 7/8: 99, 1963. Basionym: *Rebutia senilis* var. *iseliniana*. [Combination repeated (*comb. inval.*, Art. 33.2) by Simon in Kakt. / Sukk. 1968(3): 48.] [RPS 14]

Rebutia senilis fa. **kesselringiana** (Bewerunge) Buining & Donald, Sukkulentenkunde 7/8: 100, 1963. Basionym: *Rebutia senilis* var. *kesselringiana*. [Republished in Cact. Succ. J. Gr. Brit. 27: 38, and again repeated by Simon (*comb. inval.*, Art. 33.2) in Kakt. / Sukk. 1968(3): 48.] [RPS 14]

Rebutia senilis fa. **lilacino-rosea** (Backeberg) Buining & Donald, Sukkulentenkunde 7/8: 99, 1963. Basionym: *Rebutia senilis* var. *lilacino-rosea*. [Combination repeated (*comb. inval.*, Art. 33.2) by Simon in Kakt. / Sukk. 1968(3): 48.] [RPS 14]

Rebutia senilis fa. **schieliana** (Bewerunge) Donald, Ashingtonia 2: 50, 1975. Basionym: *Rebutia senilis* var. *schieliana*. [Combination first published by Simon (as *comb. inval.*, Art. 33.2) in Kakt. / Sukk. 1968(3): 48.] [RPS 26]

Rebutia senilis fa. **stuemeri** (Backeberg) Buining & Donald, Sukkulentenkunde 7/8: 99, 1963. Basionym: *Rebutia senilis* var. *stuemeri*. [Combination republished (*comb. inval.*, Art. 33.2) by Simon in Kakt. / Sukk. 1968(3): 47.] [RPS 14]

Rebutia senilis ssp. **chrysacantha** (Backeberg) Donald, Ashingtonia 2: 50, 1975. Basionym: *Rebutia chrysacantha*. [RPS 26]

Rebutia senilis var. *chrysacantha* (Backeberg) Buining & Donald *ex* Simon, Kakt. / Sukk. 1968(3): 48, 1968. *Nom. inval.* (Art. 33.2), based on *Rebutia chrysacantha*. [First invalidly published (Art. 33.2) by Donald in Cactus (Paris) No. 40: 39, 1954, and in Cact. Succ. J. Gr. Brit. 16: 44, 1954 (cf. RPS 5).] [RPS -]

Rebutia senilis var. *pallidior* Backeberg *ex* Donald, Cact. Succ. J. Gr. Brit. 16: 44, 1954. *Nom. inval.* (Art. 36.1). [Based on an unknown nude name of Backeberg. Concurrently published in Cactus (Paris) No. 40: 39, 1954.] [RPS 5]

Rebutia senilis var. **schieliana** Bewerunge, Kakt. and. Sukk. 8(7): 105-106, 1957. [RPS 8]

Rebutia senilis var. **sieperdaiana** (Buining) Backeberg, Die Cact. 3: 1546, 1959. Basionym: *Rebutia sieperdaiana*. [RPS 10]

Rebutia simoniana Rausch, Kakt. and. Sukk. 35(9): 204-205, ill., 1984. Typus: *Rausch* 739 (ZSS). [RPS 35]

Rebutia singularis Ritter, Ashingtonia 3(1): 12-13, 1978. [RPS 29]

Rebutia steinbachii var. *rosiflora* Backeberg, Cactus (Paris) No. 80/81: 5, 1964. *Nom. inval.* (Art. 37.1). [Erroneously included as valid in RPS 15.] [RPS 15]

Rebutia steinbachii var. *violaciflora* Backeberg, Cactus (Paris) No. 80/81: 6; l.c. No. 82: 52, 1964. *Nom. inval.* (Art. 37.1). [Sphalm. 'violacifera'. Erroneously included as valid in RPS 15.] [RPS 15]

Rebutia steinmannii var. **cincinnata** (Rausch) Ritter, Kakt. Südamer. 2: 602, 1980. Basionym: *Rebutia cincinnata*. [RPS 31]

Rebutia sumayana Rausch, Succulenta 65(4): 73-75, ill., 1986. Typus: *Rausch* 738 (ZSS). [Bolivia] [RPS 37]

Rebutia supthutiana Rausch, Kakt. and. Sukk. 27(8): 121-122, ill., 1976. Typus: *Rausch* 629 (ZSS). [Bolivia: Sud Cinti] [RPS 27]

Rebutia tamboensis Ritter, Ashingtonia 2(10): 207, ill., 1977. Typus: *Ritter* 1142 (U, ZSS). [Bolivia: O'Connor] [RPS 29]

Rebutia taratensis Cardenas, Cact. Succ. J. (US) 36: 26, 1964. [RPS 15]

Rebutia tarijensis Rausch, Kakt. and. Sukk. 26(9): 195-196, ill., 1975. Typus: *Rausch* 87 (ZSS). [Bolivia: Tarija] [RPS 26]

Rebutia tarvitaensis Ritter, Kakt. and. Sukk. 28(4): 78, 1977. [RPS 28]

Rebutia tiraquensis Cardenas, Cactus (Paris) No. 57: 257-259, 1957. [RPS 8]

Rebutia tiraquensis var. **longiseta** Cardenas, Cact. Succ. J. (US) 42: 186-188, 1970. [RPS 21]

Rebutia torquata Ritter & Buining, Succulenta 56: 63, 188, 1977. [RPS 28]

Rebutia totorensis Cardenas, Cactus (Paris) No. 57: 259-260, 1957. [RPS 8]

Rebutia tropaeoliptica Ritter, Kakt. and. Sukk. 28(4): 78, 1977. Typus: *Ritter* 1114 (U, ZSS [status ?]). [Sphalm. 'tropaeolipicta'.] [Bolivia: Sud Chichas] [RPS 28]

Rebutia tuberculato-chrysantha Cardenas, Cact. Succ. J. (US) 43: 246-247, 1971. [RPS 22]

Rebutia tuberosa Ritter, Taxon 12: 28, 1963. [RPS 14]

Rebutia tunariensis Cardenas, Cact. Succ. J. (US) 36: 38-40, 1964. [RPS 15]

Rebutia vallegrandensis Cardenas, Cact. Succ. J. (US) 42: 35-36, 1970. [RPS 21]

Rebutia villazonensis F. Brandt, Kakt. Orch.-Rundschau 8(2): 15-18, ills., 1983. Typus: *Anonymus ex cult. F. Brandt* 94/a (HEID). [?] [RPS 34]

Rebutia violaciflora var. **carminea** (Buining) Donald, Kakt. and. Sukk. 8(2): 24-7, 1957. Basionym: *Rebutia carminea*. [First proposed by Donald in Cactus (Paris) No. 40: 39, 1954, as *comb. inval.* (Art. 33.2), cf. RPS 5. Concurrently published in Nation. Cact. Succ. J. 12(1): 11, 1957.] [RPS 8]

Rebutia violaciflora var. **knuthiana** (Backeberg) Donald, Kakt. and. Sukk. 8(2): 24-27, 1957. Basionym: *Rebutia knuthiana*. [First proposed by Donald in Cactus (Paris) No. 40: 39, 1954, as *comb. inval.* (Art. 33.2), cf. RPS 5. Concurrently published in Nation. Cact. Succ. J. 12(1): 11, 1957.] [RPS 8]

Rebutia violascens Ritter, Kakt. and. Sukk. 28(4): 76, 1977. [RPS 28]

Rebutia vizcarrae Cardenas, Cact. Succ. J. (US) 42: 185-188, 1970. [RPS 21]

Rebutia vulpina Ritter, Succulenta 56: 66, 191, 1977. [RPS 28]

Rebutia wahliana Rausch, Succulenta 64(12): 257-258, ill., 1985. Typus: *Rausch* 654 (ZSS). [Bolivia: Tarija] [RPS 36]

Rebutia walteri Diers, Kakt. and. Sukk. 40(8): 186-190, ills., SEM-ills., 1989. Typus: *Hoffmann* 1960A (KOELN). [Argentina: Salta] [RPS 40]

Rebutia wessneriana cv. **Ruby** Donald, Ashingtonia 2(4): 71, 1976. [RPS 27]

Rebutia wessneriana cv. **Turbine** Donald, Ashingtonia 2(4): 71, 1976. [RPS 27]

Rebutia wessneriana fa. **calliantha** (Bewerunge) Buining & Donald, Ashingtonia 2(4): 70, 1976. Basionym: *Rebutia calliantha.* [RPS 27]

Rebutia wessneriana fa. *permutata* (Heinrich) Donald, Ashingtonia 2(4): 71, 1976. *Nom. inval.*, based on *Rebutia permutata, nom. inval.* (Art. 36.1, 37.1). [Erroneously included as valid in RPS 27.] [RPS 27]

Rebutia wessneriana ssp. *beryllioides* (Buining & Donald) Donald, Ashingtonia 2(4): 71, 1976. *Nom. inval.*, based on *Rebutia calliantha* var. *beryllioides, nom. inval.* (Art. 37.1). [Erroneously included as valid in RPS 27.] [RPS 27]

Rebutia wessneriana var. *beryllioides* (Buining & Donald) Buining & Donald, Succulenta 51: 222, 1972. *Nom. inval.*, based on *Rebutia calliantha* var. *beryllioides, nom. inval.* (Art. 37.1). [Erroneously included as valid in RPS 23.] [RPS 23]

Rebutia wessneriana var. **calliantha** (Bewerunge) Donald, Nation. Cact. Succ. J. 14: 5, 1959. Basionym: *Rebutia calliantha.* [Repeated by Krainz, Die Kakt. Lief. 14: C Vc, 1960 (cf. RPS 11).] [RPS 10]

Rebutia wessneriana var. **krainziana** (Kesselring) Buining & Donald, Succulenta 51: 225, 1972. Basionym: *Rebutia krainziana.* [RPS 23]

Rebutia wessneriana var. *permutata* (Heinrich) Buining & Donald, Succulenta 51: 225, 1972. *Nom. inval.*, based on *Rebutia permutata, nom. inval.* (Art. 36.1, 37.1). [Erroneously included as valid in RPS 23.] [RPS 23]

Rebutia xanthocarpa fa. **citricarpa** (Backeberg) Buining & Donald, Sukkulentenkunde 7/8: 100, 1963. Basionym: *Rebutia xanthocarpa* var. *citricarpa.* [Combination repeated by Simon (*comb. inval.*, Art. 33.2) in Kakt. / Sukk. 1968(3): 49.] [RPS 14]

Rebutia xanthocarpa fa. *citricarpa* Fric *ex* Simon, Kakt. / Sukk. 1968(3): 49, 1968. *Nom. inval.* (Art. 36, 37, 64). [*Non Rebutia xanthocarpa* fa. *citricarpa* (Backeberg) Simon.] [RPS -]

Rebutia xanthocarpa fa. *coerulescens* (Backeberg) Simon, Kakt. / Sukk. 1968(3): 49, 1968. *Nom. inval.* (Art. 33.2). [Basionym not traceable.] [RPS -]

Rebutia xanthocarpa fa. **dasyphrissa** (Werdermann) Buining & Donald, Sukkulentenkunde 7/8: 100, 1963. Basionym: *Rebutia dasyphrissa.* [Combination repeated by Simon (*comb. inval.*, Art. 33.2) in Kakt. / Sukk. 1968(3): 49.] [RPS 14]

Rebutia xanthocarpa fa. *elegans* Backeberg *ex* Simon, Kakt. / Sukk. 1968(3): 49, 1968. *Nom. inval.* (Art. 36.1, 37.1). [RPS -]

Rebutia xanthocarpa fa. *graciliflora* (Backeberg) Donald, Ashingtonia 2: 50, 1975. *Nom. inval.*, based on *Rebutia graciliflora, nom. inval.* (Art. 9.5). [RPS 26]

Rebutia xanthocarpa fa. *luteirosea* (Backeberg) Simon, Kakt. / Sukk. 1968(3): 49, 1968. *Nom. inval.* (Art. 33.2), based on *Rebutia xanthocarpa* var. *luteirosea.* [RPS -]

Rebutia xanthocarpa fa. **salmonea** (Backeberg) Buining & Donald, Sukkulentenkunde 7/8: 100, 1963. Basionym: *Rebutia xanthocarpa* var. *salmonea.* [Combination repeated by Simon (*comb. inval.*, Art. 33.2) in Kakt. / Sukk. 1968(3): 49.] [RPS 14]

Rebutia xanthocarpa fa. *salmonea* Fric *ex* Simon, Kakt. / Sukk. 1968(3): 49, 1968. *Nom. inval.* (Art. 36, 37, 64). [*Non Rebutia xanthocarpa* fa. *salmonea* (Backeberg) Simon.] [RPS -]

Rebutia xanthocarpa fa. **violaciflora** (Backeberg) Buining & Donald, Sukkulentenkunde 7/8: 100, 1963. Basionym: *Rebutia xanthocarpa* var. *violaciflora.* [RPS 14]

Rebutia xanthocarpa var. **citricarpa** Backeberg, Cact. Succ. J. (US) 23(3): 83, 1951. [RPS 3]

Rebutia xanthocarpa var. **coerulescens** Backeberg, Descr. Cact. Nov. [1:] 31, 1957. [Dated 1956, published 1957.] [RPS 7]

Rebutia xanthocarpa var. **elegans** Backeberg, Cact. Succ. J. (US) 23(3): 83, 1951. [First mentioned (*nom. inval.*, Art. 36.1) in Backeberg & Knuth, Kaktus-ABC, 279, 1935.] [RPS 3]

Rebutia xanthocarpa var. **luteirosea** Backeberg, Cact. Succ. J. (US) 23(3): 83, 1951. [RPS 3]

Rebutia xanthocarpa var. **salmonea** (Fric) Backeberg, Cact. Succ. J. (US) 23(3): 83, 1951. [RPS 3]

Rebutia xanthocarpa var. **violaciflora** Backeberg, Descr. Cact. Nov. [1:] 31, 1957. [Dated 1956, published 1957.] [RPS 7]

Rebutia yuquinensis Rausch, Kakt. and. Sukk. 31(10): 307, ill., 1980. Typus: *Rausch* 632 (ZSS). [Bolivia: Sud Cinti] [RPS 31]

Rebutia zecheri Rausch, Succulenta 56(2): 30, ill. (p. 29), 1977. Typus: *Rausch* 650 (ZSS). [Bolivia: Dept. Tarija] [RPS 28]

Rebutiae Mottram, Contr. New Class. Cact. Fam., 7, 11, 1990. Typus: *Rebutia.* [Validated by reference to *Rebutiinae* Donald 1955. Published at the rank of *linea.*] [RPS 41]

Rebutiinae Donald, Succulenta 1955(6): 84-86, 1955. Typus: *Rebutia.* [RPS 6]

Reicheocactus floribundus Backeberg, Die Cact. 6: 3802, 1962. *Nom. inval.* (Art. 37.1). [Erroneously included as valid in RPS 13.] [RPS 13]

Reicheocactus neoreichei (Backeberg) Backeberg, Die Cact. 6: 3801, 1962. *Nom. inval.*, based on *Neochilenia neoreichei, nom. inval.* (Art. 37.1). [Erroneously included as valid in RPS 13.] [RPS 13]

Rhipsales Buxbaum (*pro linea*), Madroño 14(6): 185, 1958. Typus: *Rhipsalis.* [Repeated by

Mottram, Contr. New Class. Cact. Fam., 7, 11, 1990 as *Cactinae* linea *Rhipsales* "comb. nov.".] [RPS 41]

Rhipsalidopsis gaertneri var. **tiburtii** (Backeberg & Voll) Moran, Gentes Herbar. 8(4): 342, 1953. Basionym: *Epiphyllopsis gaertneri* var. *tiburtii*. [RPS 4]

Rhipsalidopsis × **graeseri** (Werdermann) Moran, Gent. Herbar. 8(4): 342, 1953. Basionym: *Rhipsalis* × *graeseri*. [RPS 4]

Rhipsalidopsis rosea var. *remanens* Backeberg, Die Cact. 6: 3646, 1962. *Nom. inval.* (Art. 9.5, 37.1). [Erroneously included as valid in RPS 13.] [RPS 13]

Rhipsalis subgen. **Phyllarthrorhipsalis** Buxbaum, in Krainz, Die Kakt., C IVe (Oct. 1970), 1970. Typus: *R. pachyptera* Pfeiffer. [RPS 21]

Rhipsalis alboareolata Ritter, Kakt. Südamer. 1: 41, 1979. [RPS 30]

Rhipsalis baccifera ssp. **erythrocarpa** (Schumann) Barthlott, Bradleya 5: 100, 1987. Basionym: *Rhipsalis erythrocarpa*. [RPS 38]

Rhipsalis baccifera ssp. **horrida** (Baker) Barthlott, Bradleya 5: 100, 1987. Basionym: *Rhipsalis horrida*. [RPS 38]

Rhipsalis baccifera ssp. **mauritiana** (De Candolle) Barthlott, Bradleya 5: 100, 1987. Basionym: *Rhipsalis cassytha* var. *mauritiana*. [RPS 38]

Rhipsalis baccifera ssp. **rhodocarpa** (F. A. C. Weber) Süpplie, Rhipsalidinae, [27], 1990. Basionym: *Rhipsalis cassytha* var. *rhodocarpa*. [RPS 41]

Rhipsalis brevispina (Ritter) Kimnach, Cact. Succ. J. (US) 55(4): 181, 1983. *Nom. inval.*, based on *Acanthorhipsalis brevispina*, *nom. inval.* (Art. 37.1). [Erroneously included as valid in RPS 34.] [RPS 34]

Rhipsalis cassuthopsis Backeberg, Die Cact. 2: 660, 1959. Basionym: *Rhipsalis cassythoides*. [*Nom. nov.*, non *Rhipsalis cassythoides* G. Don 1834. Validity not assessed.] [RPS 10]

Rhipsalis clavellina Ritter, Kakt. Südamer. 1: 43, 1979. [RPS 30]

Rhipsalis coralloides Rauh, in Backeberg, Die Cact. 6: 3634, 1962. *Nom. inval.* (Art. 36.1, 37.1). [RPS 13]

Rhipsalis floccosa var. **gibberula** (F. A. C. Weber) Krainz, Kat. ZSS ed. 2, 109, 1967. Basionym: *Rhipsalis gibberula*. [RPS 18]

Rhipsalis flosculosa Ritter, Kakt. Südamer. 1: 42, 1979. [RPS 30]

Rhipsalis goebeliana hort. *ex* Backeberg, Descr. Cact. Nov. [1:] 10, 1957. [Dated 1956, published 1957.] [RPS 7]

Rhipsalis heptagona Rauh & Backeberg, in Backeberg, Descr. Cact. Nov. [1:] 10, 1957. [Dated 1956, published 1957.] [RPS 7]

Rhipsalis hohenauensis Ritter, Kakt. Südamer. 1: 248, 1979. [RPS 30]

Rhipsalis hylaea Ritter, Kakt. Südamer. 4: 1261-1262, 1981. Typus: *Ritter* 116 (U, ZSS). [Peru: Amazonas] [RPS 32]

Rhipsalis incachacana Cardenas, Cactus (Paris) No. 34: 125-126, 1952. [RPS 3]

Rhipsalis kirbergii Barthlott, Trop. Subtr. Pflanzenw. 10: 11-14, ills., 1974. Typus: *Rauh et Barthlott* 34364 (HEID, ZSS). [Ecuador: Manabi] [RPS 26]

Rhipsalis kirbergii var. **monticola** Barthlott, Trop. Subtr. Pflanzenw. 10: 15, 1974. [RPS 26]

Rhipsalis lumbricoides var. **leucorhaphis** (Schumann) Ritter, Kakt. Südamer. 1: 247, 1979. Basionym: *Rhipsalis leucorhaphis*. [RPS 30]

Rhipsalis monacantha var. **samaipatana** Cardenas, Nation. Cact. Succ. J. 12(4): 85, 1957. [RPS 8]

Rhipsalis monteazulensis Ritter, Kakt. Südamer. 1: 42, 1979. [RPS 30]

Rhipsalis occidentalis Barthlott & Rauh, Kakt. and. Sukk. 38(1): 16-19, ills., 1987. Typus: *Rauh et Barthlott* 35392 (HEID, BONN, HNT). [Peru: San Martin] [RPS 38]

Rhipsalis paranganiensis (Cardenas) Kimnach, Cact. Succ. J. (US) 55(4): 181, 1983. Basionym: *Acanthorhipsalis paranganiensis*. [RPS 34]

Rhipsalis quellebambensis Johnson, in Backeberg, Kakt.-Lex., 393, 1966. *Nom. inval.* (Art. 36.1, 37.1). [Based on cultivated material from Peru.] [RPS -]

Rhipsalis rauhiorum Barthlott, Trop. Subtr. Pflanzenw. 10: 15-21, ills., 1974. Typus: *Rauh et Barthlott* 35278 (HEID, ZSS). [Ecuador] [RPS 26]

Rhipsalis saxatilis (Friedrich & Redecker) Friedrich & Redecker, Rep. Pl. Succ. 25: 13, 1976. Basionym: *Lepismium saxatile*. [First published (without page reference, cf. Art. 33.2) in Kakt. and. Sukk. 25(8): 170, ill., 1974.] [RPS 25]

Rhodocactus antonianus Backeberg, Descr. Cact. Nov. 3: 13, 1963. *Nom. inval.* (Art. 9.5, 37.1). [RPS 14]

Rhodocactus conzattii (Britton & Rose) Backeberg, Die Cact. 1: 118, 1958. Basionym: *Pereskia conzattii*. [RPS 9]

Rhodocactus corrugatus (Cutak) Backeberg, Die Cact. 1: 118, 1958. Basionym: *Pereskia corrugata*. [RPS 9]

Rhodocactus higueranus (Cardenas) Backeberg, Kakt.-Lex., 396, 1966. Basionym: *Pereskia higuerana*. [RPS 17]

Rhodocactus sacharosa (Grisebach) Backeberg, Kakt.-Lex., 397, 1966. Basionym: *Pereskia sacharosa*. [RPS 17]

Rhodocactus saipinensis (Cardenas) Backeberg, Kakt.-Lex., 397, 1966. Basionym: *Pereskia saipinensis*. [RPS 17]

Rhodocactus tampicanus (F. A. C. Weber) Backeberg, Die Cact. 1: 115, 1958. Basionym: *Pereskia tampicana*. [RPS 9]

Ritterocereus chacalapensis Bravo & MacDougall, Anales Inst. Biol. UNAM 27: 311, 1957. [Concurrently published in Cact. Suc. Mex. 2(3): 49-52, 55, 1957.] [RPS 8]

Ritterocereus deficiens (Otto & Dietrich) Backeberg, Die Cact. 4: 2181, 1960. Basionym: *Cereus deficiens*. [First proposed as *comb. nud.* by Backeberg in Cact. Succ. J. (US) 23(4): 121, 1951 (RPS 3).] [RPS 11]

Ritterocereus eichlamii (Britton & Rose) Backeberg, Die Cact. 4: 2179, 1960.

Basionym: *Cereus laevigatus* var. *guatemalensis*. [First proposed as *comb. nud.* by Backeberg in Cact. Succ. J. (US) 23(4): 121, 1951.] [RPS 11]

Ritterocereus griseus (Haworth) Backeberg, Cact. Succ. J. (US) 23(4): 121, 1951. Basionym: *Cereus griseus*. [The name given as basionym was already a combination, and is here corrected.] [RPS 3]

Ritterocereus humilis (Britton & Rose) Backeberg, Cact. Succ. J. (US) 23(4): 121, 1951. Basionym: *Lemaireocereus humilis*. [RPS 3]

Ritterocereus laevigatus (Salm-Dyck) Backeberg, Die Cact. 4: 2178, 1960. Basionym: *Cereus laevigatus*. [RPS 11]

Ritterocereus montanus (Britton & Rose) Backeberg, Cact. Succ. J. (US) 23(4): 121, 1951. Basionym: *Lemaireocereus montanus*. [RPS 3]

Ritterocereus pruinosus (Otto) Backeberg, Die Cact. 4: 2183, 1960. Basionym: *Echinocactus pruinosus*. [First proposed as *comb. nud.* by Backeberg in Cact. Succ. J. (US) 23(4): 121 (RPS 3).] [RPS 11]

Ritterocereus queretaroensis (F. A. C. Weber) Backeberg, Die Cact. 4: 2184, 1960. Basionym: *Cereus queretaroensis*. [First proposed as *comb. nud.* by Backeberg in Cact. Succ. J. (US) 23(4): 121 (RPS 3).] [RPS 11]

Ritterocereus weberi (Coulter) Backeberg, Cact. Succ. J. (US) 23(4): 121, 1951. Basionym: *Cereus weberi*. [RPS 3]

Rodentiophila Ritter, in Backeberg, Die Cact. 1: 56, 78, in clav, 1958. *Nom. inval.* (Art. 36.1, 37.1). [Again mentioned in l.c. 3: 1799, 1959 (cf. RPS 10).] [Name republished (without type) by Ito, The Cactaceae, 429, 1981.] [RPS 9]

Rodentiophila atacamensis Ritter, in Backeberg, Die Cact. 3: 1799, 1959. *Nom. inval.* (Art. 36.1, 37.1). [First mentioned as provisional name in Katalog H. Winter, [unpaged], 1957. Republished invalidly (Art. 37.1, 43.1) by Ito, The Cactaceae, 429, 1981.] [RPS 10]

Rodentiophila megacarpa Ritter, Katalog H. Winter, [unpaged], 1957. *Nom. inval.* (Art. 36.1, 37.1). [Published as provisional name. Republished invalidly (Art. 37.1, 43.1) by Ito, The Cactaceae, 429, 1981.] [RPS -]

Rooksbya (Backeberg) Backeberg, Die Cact. 4: 2165, 1960. Basionym: *Carnegiea* subgen. *Rooksbya*. [RPS 11]

Rooksbya euphorbioides (Haworth) Backeberg, Die Cact. 4: 2167, 1960. Basionym: *Cereus euphorbioides*. [RPS 11]

Rooksbya euphorbioides var. **olfersii** (Salm-Dyck) Backeberg, Die Cact. 4: 2170, 1960. Basionym: *Cereus olfersii*. [RPS 11]

Roseocactus intermedius Backeberg & Kilian, Kakt. and. Sukk. 11: 149-152, 1960. Typus: *Schwarz* s.n. (ubi ?). [Treated as *nom. nud.* by Anderson, Kakt. and. Sukk. 16: 8, 1965.] [Mexico] [RPS 11]

Roseocactus kotschoubeyanus var. **macdowellii** (Backeberg) Backeberg, Die Cact. 5. 3075, 1961. Basionym: *Roseocactus kotschoubeyanus* ssp. *macdowellii*. [RPS 12]

Salpingolobivia Y. Ito, Explan. Diagr. Austroechinocactinae, 135, 290, 1957. Typus: *Echinopsis aurea*. [First invalidly published in Bull. Takarazuka Insectarium 71: 13-20, 1950.] [RPS 8]

Salpingolobivia andalgalensis (F. A. C. Weber) Y. Ito, Explan. Diagr. Austroechinocactinae, 141, 1957. Basionym: *Cereus andalgalensis*. [RPS 8]

Salpingolobivia aurea (Britton & Rose) Y. Ito, Explan. Diagr. Austroechinocactinae, 136, 1957. Basionym: *Echinopsis aurea*. [RPS 8]

Salpingolobivia aurea var. **aurantiaca** Y. Ito, Explan. Diagr. Austroechinocactinae, 138, 291, 1957. [RPS 8]

Salpingolobivia aurea var. **aureorubriflora** Y. Ito, Explan. Diagr. Austroechinocactinae, 137, 291, 1957. [RPS 8]

Salpingolobivia aurea var. **cinnabarina** Y. Ito, Explan. Diagr. Austroechinocactinae, 138, 291, 1957. [RPS 8]

Salpingolobivia aurea var. **elegans** (Backeberg) Y. Ito, Explan. Diagr. Austroechinocactinae, 137, 1957. Basionym: *Lobivia aurea* var. *elegans*. [RPS 8]

Salpingolobivia aurea var. *grandiflora* (Backeberg) Y. Ito, Explan. Diagr. Austroechinocactinae, 137, 1957. *Nom. inval.*, based on *Lobivia aurea* var. *grandiflora*, *nom. inval.* (Art. 36.1). [Erroneously included as valid in RPS 8.] [RPS 8]

Salpingolobivia aurea var. *robustior* (Backeberg) Y. Ito, Explan. Diagr. Austroechinocactinae, 137, 1957. *Nom. inval.*, based on *Lobivia aurea* var. *robustior*, *nom. inval.* (Art. 36.1). [Erroneously included as valid in RPS 8.] [RPS 8]

Salpingolobivia aurea var. **roseiflora** Y. Ito, Explan. Diagr. Austroechinocactinae, 138, 291, 1957. [RPS 8]

Salpingolobivia aurea var. **rubriflora** Y. Ito, Explan. Diagr. Austroechinocactinae, 137, 291, 1957. [RPS 8]

Salpingolobivia aurea var. **salmonea** Y. Ito, Explan. Diagr. Austroechinocactinae, 138, 291, 1957. [RPS 8]

Salpingolobivia cylindrica (Backeberg) Y. Ito, Explan. Diagr. Austroechinocactinae, 138, 1957. Basionym: *Lobivia cylindrica*. [RPS 8]

Salpingolobivia cylindrica var. **aureorubriflora** Y. Ito, Explan. Diagr. Austroechinocactinae, 139, 291, 1957. [RPS 8]

Salpingolobivia cylindrica var. **roseiflora** Y. Ito, Explan. Diagr. Austroechinocactinae, 139, 291, 1957. [RPS 8]

Salpingolobivia cylindrica var. **salmonea** Y. Ito, Explan. Diagr. Austroechinocactinae, 139, 291, 1957. [RPS 8]

Salpingolobivia densispina (Werdermann) Y. Ito, Explan. Diagr. Austroechinocactinae, 139, 1957. Basionym: *Echinopsis densispina*. [Basionym erroneously given as *Lobivia densispina*, treated as bibliographical error.] [RPS 8]

Salpingolobivia huascha (F. A. C. Weber) Y. Ito, Explan. Diagr. Austroechinocactinae,

141, 1957. Basionym: *Cereus huascha.* [RPS 8]
Salpingolobivia shaferi (Britton & Rose) Y. Ito, Explan. Diagr. Austroechinocactinae, 140, 1957. Basionym: *Lobivia shaferi.* [RPS 8]
Salpingolobivia spinosissima Y. Ito, Explan. Diagr. Austroechinocactinae, 140, 291, 1957. [RPS 8]
Salpingolobivia spinosissima var. **rubriflora** Y. Ito, Explan. Diagr. Austroechinocactinae, 140, 291, 1957. [RPS 8]
Samaipaticereus Cardenas, Cact. Succ. J. (US) 29: 141, 1952. Typus: *S. corroanus.* [RPS 3]
Samaipaticereus corroanus Cardenas, Cact. Succ. J. (US) 24(5): 141-143, 1952. [RPS 3]
Samaipaticereus inquisivensis Cardenas, Cactus (Paris) 12(57): 246-247, 1957. [RPS 8]
Schlumbergera subgen. **Zygocactus** (Schumann) Moran, Gentes Herbar. 8(4): 329, 1953. Basionym: *Zygocactus.* [RPS 4]
Schlumbergera × **buckleyi** (Buckley) Tjaden, Gard. Chron. 156: 421, 437, 444, 462, 468, 1964. Basionym: *Epiphyllum buckleyi.* [= *Schlumbergera truncata* × *S. russelliana*, the popular 'Christmas Cactus'. Combination repeated by D. Hunt in Kew Bull. 23: 259, 1969.] [RPS 15]
Schlumbergera × **exotica** Barthlott & Rauh, Kakt. and. Sukk. 28(12): 278, 1977. [= *Schlumbergera truncata* × *S. opuntioides.*] [RPS 28]
Schlumbergera obtusangula (Schumann) D. Hunt, Kew Bull. 23: 260, 1969. Basionym: *Cereus obtusangulus.* [RPS 20]
Schlumbergera opuntioides (Loefgren & Dusen) D. Hunt, Kew Bull. 23: 260, 1969. Basionym: *Epiphyllum opuntioides.* [RPS 20]
Schlumbergera orssichiana Barthlott & McMillan, Cact. Succ. J. (US) 50(1): 30-34, ills., 1978. Typus: *Anonymus* s.n. (HEID, ZSS). [Ex cult.] [Brazil] [RPS 29]
Schlumbergera × **reginae** McMillan & Orssich, Epiphytes 9(33): 8-9, ill., 1985. [= *S. orssichiana* × *S. truncata.*] [RPS 36]
Schlumbergera × **reginae** cv. **Bristol Queen** McMillan, Epiphytes 9(33): 8-9, ill., 1985. [RPS 36]
Schlumbergera truncata (Haworth) Moran, Gent. Herb. 8(4): 329 , 1953. Basionym: *Epiphyllum truncatum.* [RPS 4]
Schlumbergera truncata cv. **Lilac Beauty** Innes, Ashingtonia 1: 139, 143, 1975. [RPS 26]
Schlumbergera truncata var. **altensteinii** (Pfeiffer) Moran, Gent. Herbar. 8(4): 330, 1953. Basionym: *Epiphyllum altensteinii.* [RPS 4]
Schlumbergera truncata var. **delicata** (N. E. Brown) Moran, Gentes Herb. 8(4): 330, 1953. Basionym: *Epiphyllum delicatum.* [RPS 4]
Schlumbergera truncata var. *kautskyi* Horobin & McMillan, Epiphytes 14(56): 111-115, ill., 1990. *Nom. inval.* (Art. 37.1). [Attempt to validate as ssp. in l.c. 15(57), 1991.] [RPS 41]
Schlumbergerae Buxbaum, Madroño 14(6): 185, 1958. Typus: *Schlumbergera.* [Published as 'linea'.] [RPS 9]
Schlumbergerinae Volgin, Feddes Repert. 97(9-10): 563-564, 1986. Typus: *Schlumbergera* Lemaire. [RPS 37]
Sclerocactus blainei Welsh & Thorne, Great Basin Naturalist 45(3): 553, ill., 1985. Typus: *Welsh* 20580 (BRY). [USA] [RPS 36]
Sclerocactus contortus Heil, Cact. Succ. J. (US) 51(1): 25-27, 1979. [RPS 30]
Sclerocactus erectocentrus (J. Coulter) N. P. Taylor, Bradleya 5: 94, 1987. Basionym: *Echinocactus erectocentrus.* [RPS 38]
Sclerocactus glaucus (Schumann) L. Benson, Cact. Succ. J. (US) 38: 53-54, 1966. Basionym: *Echinocactus glaucus.* [RPS 17]
Sclerocactus intertextus (Engelmann) N. P. Taylor, Bradleya 5: 94, 1987. Basionym: *Echinocactus intertextus.* [RPS 38]
Sclerocactus intertextus var. **dasyacanthus** (Engelmann) N. P. Taylor, Bradleya 5: 94, 1987. Basionym: *Echinocactus intertextus* var. *dasyacanthus.* [RPS 38]
Sclerocactus johnsonii (Engelmann) N. P. Taylor, Bradleya 5: 94, 1987. Basionym: *Echinocactus johnsonii.* [RPS 38]
Sclerocactus mariposensis (Hester) N. P. Taylor, Bradleya 5: 94, 1987. Basionym: *Echinomastus mariposensis.* [RPS 38]
Sclerocactus mesae-verdae (Boissevain & Davidson) L. Benson, Cact. Succ. J. (US) 38: 54-55, 1966. Basionym: *Coloradoa mesae-verdae.* [RPS 17]
Sclerocactus papyracanthus (Engelmann) N. P. Taylor, Bradleya 5: 94, 1987. Basionym: *Mammillaria papyracantha.* [RPS 38]
Sclerocactus parviflorus var. **blessingiae** W. Earle, Saguaroland Bull. 34(3): 29, 1980. [RPS 31]
Sclerocactus parviflorus var. **intermedius** (Peebles) Woodruff & L. Benson, Cact. Succ. J. (US) 48(3): 133-134, 1976. Basionym: *Sclerocactus intermedius.* [RPS 27]
Sclerocactus pubispinus (Engelmann) L. Benson, Cact. Succ. J. (US) 38: 103-105, 1966. Basionym: *Echinocactus pubispinus.* [RPS 17]
Sclerocactus pubispinus var. **sileri** L. Benson, Cacti of Arizona, ed. 3, 23, 179, 1969. [RPS 20]
Sclerocactus pubispinus var. **spinosior** (Engelmann) Welsh, Great Basin Natural. 44(1): 67, 1984. Basionym: *Echinocactus whipplei* var. *spinosior.* [RPS 35]
Sclerocactus scheeri (Salm-Dyck) N. P. Taylor, Bradleya 5: 94, 1987. Basionym: *Echinocactus scheeri.* [RPS 38]
Sclerocactus schleseri Heil & Welsh, Great Basin Natural. 46(4): 677-678, 1986. *Heil* s.n. (BRY, NY). [Sphalm. 'schlesseri' (named for D. Schleser, cf. Cact. Succ. J. (US) 60: 35, 1988).] [RPS 38]
Sclerocactus spinosior (Engelmann) Woodruff & L. Benson, Cact. Succ. J. (US) 48(3): 131-132, 1976. Basionym: *Echinocactus whipplei* var. *spinosior.* [RPS 27]
Sclerocactus spinosior var. **blainei** (Welsh & Thorne) R. May, Cact. Succ. J. (US) 60(1):

35, 45, 1988. Basionym: *Sclerocactus blainei*. [RPS 39]

Sclerocactus spinosior var. **schleseri** (Heil & Welsh) R. May, Cact. Succ. J. (US) 60(1): 35, 45, 1988. Basionym: *Sclerocactus schleseri*. [RPS 39]

Sclerocactus terrae-canyonae Heil, Cact. Succ. J. (US) 51(1): 26-28, 1979. [RPS 30]

Sclerocactus uncinatus (Galeotti) N. P. Taylor, Bradleya 5: 94, 1987. Basionym: *Echinocactus uncinatus*. [RPS 38]

Sclerocactus uncinatus var. **crassihamatus** (F. A. C. Weber) N. P. Taylor, Bradleya 5: 94, 1987. Basionym: *Echinocactus crassihamatus*. [RPS 38]

Sclerocactus uncinatus var. **wrightii** (Engelmann) N. P. Taylor, Bradleya 5: 94, 1987. Basionym: *Echinocactus uncinatus* var. *wrightii*. [RPS 38]

Sclerocactus unguispinus (Engelmann) N. P. Taylor, Bradleya 5: 94, 1987. Basionym: *Echinocactus unguispinus*. [RPS 38]

Sclerocactus unguispinus var. **durangensis** (Runge) N. P. Taylor, Bradleya 5: 94, 1987. Basionym: *Echinocactus durangensis*. [RPS 38]

Sclerocactus warnockii (L. Benson) N. P. Taylor, Bradleya 5: 94, 1987. Basionym: *Neolloydia warnockii*. [RPS 38]

Sclerocactus wetlandicus Hochstätter, Succulenta 68(6): 123-126, ills., SEM-ills. (pp. 121, 124), 1989. Typus: *Hochstätter* 69.9.3 (HBG). [Publication repeated in Kaktusblüte 1990: 35-39, ills.] [USA: Utah] [RPS 40]

Sclerocactus whipplei var. **glaucus** (J. A. Purpus) Welsh, Great Basin Natural. 44(1): 68, 1984. Basionym: *Echinocactus glaucus*. [RPS 35]

Sclerocactus whipplei var. **heilii** Castetter, Pierce & Schwerin, Cact. Succ. J. (US) 48(2): 79-80, 1976. [RPS 27]

Sclerocactus whipplei var. **intermedius** (Peebles) L. Benson, Cact. Succ. J. (US) 38: 102, 1966. Basionym: *Sclerocactus intermedius*. [RPS 17]

Sclerocactus whipplei var. **reevesii** Castetter, Pierce & Schwerin, Cact. Succ. J. (US) 48(2): 80-82, 1976. [RPS 27]

Sclerocactus whipplei var. **roseus** (Clover & Jotter) L. Benson, Cact. Succ. J. (US) 38: 101, 1966. Basionym: *Sclerocactus parviflorus*. [RPS 17]

Sclerocactus wrightiae L. Benson, Cact. Succ. J. (US) 38: 55-57, 1966. [RPS 17]

× **Seleliocereus fulgidus** (Hooker) Rowley, Nation. Cact. Succ. J. 37(3): 79, 1982. Basionym: *Cereus fulgidus*. [RPS 33]

× *Seleniaporus* Rowley, Epiphytes 4(13): 13, 1972. *Nom. inval.* (Art. H6.2). [= *Selenicereus* × *Aporocactus*.] [RPS 31]

Selenicereus sect. **Cryptocereus** (Alexander) D. Hunt, Bradleya 7: 92-93, 1989. Basionym: *Cryptocereus*. [RPS 40]

Selenicereus sect. **Deamia** (Britton & Rose) D. Hunt, Bradleya 7: 93, 1989. Basionym: *Deamia*. [RPS 40]

Selenicereus sect. **Salmdyckia** D. Hunt, Bradleya 7: 91, 1989. Typus: *S. setaceus* (Salm-Dyck *ex* De Candolle) Werdermann; = *Mediocactus* Britton & Rose *quoad descr. tantum, excl. typ.* [RPS 40]

Selenicereus sect. **Strophocactus** (Britton & Rose) D. Hunt, Bradleya 7: 93, 1989. Basionym: *Strophocactus*. [RPS 40]

Selenicereus subgen. **Deamia** (Britton & Rose) Buxbaum, in Krainz, Die Kakt. Lief. 30: C IIa (June 1965), 1965. Basionym: *Deamia*. [RPS 28]

Selenicereus atropilosus Kimnach, Cact. Succ. J. (US) 50(6): 268-270, ills., 1978. Typus: *Boutin et Kimnach* 3190 (HNT, MEXU, US, ZSS). [Mexico: Jalisco] [RPS 29]

Selenicereus innesii Kimnach, Cact. Succ. J. (US) 54(1): 3-7, 1982. Typus: *Innes* s.n. (HNT). [West Indies: Windwards Islands] [RPS 33]

Selenicereus macdonaldiae var. **grusonianus** (Weingart) Backeberg, Die Cact. 2: 788, 1959. Basionym: *Cereus grusonianus*. [RPS 10]

Selenicereus mallisonii W. Beeson, Cact. Succ. J. Gr. Brit. 19(1): 14, 1957. *Nom. inval.* (Art. 36.1). [RPS 8]

Selenicereus megalanthus (Schumann) Moran, Gent. Herb. 8(4): 325, 1953. Basionym: *Cereus megalanthus*. [RPS 4]

Selenicereus mirandae Bravo, Cact. Suc. Mex. 12: 51-53, 65-66, 68, 1967. [RPS 18]

Selenicereus rizzinii Scheinvar, Rev. Bras. Biol. 34(2): 249-256, 1974. [RPS 29]

Selenicereus testudo (Karwinski) Buxbaum, in Krainz, Die Kakt. C IIa (June 1965), 1965. Basionym: *Cereus testudo*. [RPS 28]

Selenicereus tricae D. Hunt, Bradleya 7: 91, ills. (p. 90), 1989. Typus: *Hunt* 7076 (K). [ex cult. Hort. Kew 473-69.03879.] [Belize: El Cayo District] [RPS 40]

Selenicereus wittii (Schumann) Rowley, Excelsa 12: 36, 1986. Basionym: *Cereus wittii*. [RPS 37]

× **Seleniphylchia** Süpplie, Succulenta 67(12): 258-261, ills., 1988. [= (*Selenicereus* × *Epiphyllum*) × *Nopalxochia*.] [RPS 39]

× **Seleniphylchia** cv. **Nelson Mandela** Süpplie, Succulenta 67(12): 258-261, ills., 1988. [= (*Selenicereus grandiflorus* × *Epiphyllum crenatum*) × (*Nopalxochia ackermannii* × *N. phyllanthoides*)] [RPS 39]

× **Seleniphyllum** Rowley, in Backeberg, Die Cact. 6: 3557, 1962. [= *Selenicereus* × *Epiphyllum*. First mentioned as nude name in l.c. 2: 737, 755, 1959 (cf. RPS 10).] [RPS 13]

× **Seleniphyllum cooperi** Rowley, in Backeberg, Die Cact. 6: 3557, 1962. [= *Phyllocactus crenato-grandiflorus* Regel, Gartenflora 33: 357, 1884; = *Selenicereus grandiflorus* × *Epiphyllum crenatum*.] [RPS 13]

× **Seleniporocactus** Rowley, Nation. Cact. Succ. J. 37(3): 79, 1982. [= *Selenicereus* × *Aporocactus*.] [RPS 33]

× *Seleniporocactus* cv. *Grandiflorus Ruber* Rowley, Nation. Cact. Succ. J. 37(3): 79, 1982. *Nom. illeg.* (Art. ICNCP 27a). [Based on *Cereus grandiflorus fl. rubr. hort.* Focke 1881 (= *Aporocactus flagelliformis* × *Selenicereus grandiflorus*).] [RPS 33]

× **Selenirisia** Rowley, Nation. Cact. Succ. J.

37(3): 79, 1982. [= *Selenicereus* × *Harrisia*.] [RPS 33]

×**Selenochia** Rowley, Nation. Cact. Succ. J. 37(3): 79, 1982. [= *Selenicereus* × *Nopalxochia*.] [RPS 33]

Sericocactus Y. Ito, Explan. Diagr. Austroechinocactinae, 220, 293, 1957. Typus: *Echinocactus haselbergii*. [First invalidly published in Bull. Takarazuka Insectarium 71: 13-20, 1950 (Art. 36.1, cf. RPS 3).] [RPS 8]

Sericocactus haselbergii (Haage jr.) Y. Ito, Explan. Diagr. Austroechinocactinae, 223, 1957. Basionym: *Echinocactus haselbergii*. [RPS 8]

Seticereus chlorocarpus (Humboldt, Bonpland & Kunth) Backeberg, Die Cact. 2: 988, 1959. Basionym: *Cactus chlorocarpus*. [RPS 10]

Seticereus icosagonus var. **ferrugineus** (Backeberg) Backeberg, Kakt.-Lex., 406, 1966. Basionym: *Seticereus ferrugineus*. [The name given as basionym was itself a combination, and is here corrected and treated as bibliographical error.] [RPS 17]

Seticereus icosagonus var. **oehmeanus** (Backeberg) Backeberg, Die Cact. 2: 981, 1959. Basionym: *Seticereus oehmeanus*. [RPS 10]

Seticereus oehmeanus var. **ferrugineus** (Backeberg) Backeberg, Die Cact. 2: 982, 1959. Basionym: *Seticereus ferrugineus*. [RPS -]

Seticleistocactus Backeberg, Descr. Cact. Nov. 3: 13, 1963. Typus: *S. piraymirensis*. [RPS 14]

Seticleistocactus dependens (Cardenas) Backeberg, Kakt.-Lex., 406, 1966. Basionym: *Cleistocactus dependens*. [RPS 17]

Seticleistocactus piraymirensis (Cardenas) Backeberg, Descr. Cact. Nov. 3: 13, 1963. Basionym: *Cleistocactus piraymirensis*. [RPS 14]

Setiechinopsis mirabilis var. *gracilior* Backeberg, Descr. Cact. Nov. 3: 14, 1963. *Nom. inval.* (Art. 37.1). [RPS 14]

Siccobaccatus P. J. Braun & Esteves, Succulenta 69(1): 3-8, ills., 1990. Typus: *Austrocephalocereus dolichospermaticus*. [RPS 41]

Siccobaccatus dolichospermaticus (Buining & Brederoo) P. J. Braun & Esteves Pereira, Succulenta 69(1): 7, 1990. Basionym: *Austrocephalocereus dolichospermaticus*. [RPS 41]

Siccobaccatus estevesii (Buining & Brederoo) P. J. Braun & Esteves Pereira, Succulenta 69(1): 7, 1990. Basionym: *Austrocephalocereus estevesii*. [RPS 41]

Siccobaccatus estevesii ssp. **grandiflorus** (Diers & Esteves P.) P. J. Braun & Esteves Pereira, Succulenta 69(1): 7, 1990. Basionym: *Austrocephalocereus estevesii* ssp. *grandiflorus*. [RPS 41]

Siccobaccatus estevesii ssp. **insigniflorus** (Diers & Esteves P.) P. J. Braun & Esteves Pereira, Succulenta 69(1): 7, 1990. Basionym: *Austrocephalocereus estevesii* ssp. *insigniflorus*. [RPS 41]

×**Soehrenantha** Ito, Shaboten 92: 25, 1976. [= *Soehrensia* × *Cosmantha* (auct. ?) - the name may be invalid if the second generic name it is based on is not validly published. Publication repeated by Ito, The Cactaceae, 646, 1981.] [RPS 27]

×**Soehrenantha** cv. **Enyo-gyoku** Y. Ito, Shaboten 92: 25, 1976. [= *Soehrensia formosa* × *Cosmantha grandiflora*.] [RPS 27]

×**Soehrenantha** cv. **Hiran-Maru** Y. Ito, Shaboten 92: 25, 1976. [= *Soehrensia bruchii* × *Cosmantha grandiflora*.] [RPS 27]

×*Soehrenfuria* Ito, The Cactaceae, 649, 1981. *Nom. inval.* (Art. H6.2). [= *Soehrensia* × *Furiolobivia*.] [RPS -]

×**Soehrenlobivia** Ito, The Cactaceae, 650, 1981. [= *Soehrensia* × *Salpingolobivia*.] [RPS -]

Soehrensia allagantha Hirao & Y. Ito, The Cactaceae, 646-647, ill., 1981. *Nom. inval.* (Art. 37.1). [RPS -]

Soehrensia bruchii var. *albiflora* Y. Ito, The Cactaceae, 647-648, ill., 1981. *Nom. inval.* (Art. 37.1). [RPS -]

Soehrensia bruchii var. *aurantiflora* Y. Ito, The Cactaceae, 648-649, ill., 1981. *Nom. inval.* (Art. 37.1). [RPS -]

Soehrensia bruchii var. *aureorubriflora* Y. Ito, The Full Bloom of Cactus Flowers, 54, ill., 1962. *Nom. inval.* (Art. 37.1). [Included as valid in RPS 13.] [RPS 13]

Soehrensia bruchii var. *roseiflora* Y. Ito, The Cactaceae, 648, ill., 1981. *Nom. inval.* (Art. 37.1). [RPS -]

Soehrensia bruchii var. *roseorubriflora* Y. Ito, The Cactaceae, 648, ill., 1981. *Nom. inval.* (Art. 37.1). [RPS -]

Soehrensia formosa (Pfeiffer) Backeberg, Die Cact. 3: 1678, 1959. Basionym: *Echinocactus formosus*. [RPS 10]

Soehrensia formosa var. *maxima* Backeberg, Die Cact. 3: 1682, 1959. *Nom. inval.* (Art. 37.1). [Erroneously included as valid in RPS 10.] [RPS 10]

Soehrensia formosa var. *polycephala* Backeberg, Die Cact. 3: 1682, 1959. *Nom. inval.* (Art. 37.1). [Erroneously included as valid in RPS 10.] [RPS 10]

Soehrensia huascha var. *rosiflora* Y. Ito, The Full Bloom of Cactus Flowers, 51, ill., 1962. *Nom. inval.* (Art. 37.1). [Included as valid in RPS 13.] [RPS 13]

Soehrensia ingens Backeberg, Cact. Succ. J. (US) 23(3): 86, 1951. [Name ascribed to Britton & Rose.] [RPS 3]

Soehrensia korethroides (Werdermann) Backeberg, Cact. Succ. J. (US) 23(3): 86, 1951. Basionym: *Echinopsis korethroides*. [RPS 3]

Soehrensia oreopepon (Spegazzini) Backeberg, Die Cact. 3: 1674, 1959. Basionym: *Lobivia oreopepon*. [RPS 10]

Soehrensia smrziana (Backeberg) Backeberg, Die Cact. 3: 1677, 1959. Basionym: *Echinopsis smrziana*. [RPS 10]

Soehrensia uebelmanniana Lembcke & Backeberg, in Backeberg, Die Cact. 3: 1799-1800, 1925, 1959. *Nom. inval.* (Art. 37.1). [Erroneously included as valid in RPS 10.]

[RPS 10]

Stenocactus dichroacanthus var. **violaciflorus** (Quehl) Bravo, Cact. Suc. Mex. 27(1): 16, 1982. Basionym: *Echinocactus violaciflorus*. [RPS 33]

Stenocactus rectispinus Schmoll, Cact. Suc. Mex. 14: 65, 1969. [RPS -]

Stenocactus sulphureus (A. Dietrich) Bravo, Cact. Suc. Mex. 27(1): 16-17, 1982. Basionym: *Echinocactus sulphureus*. [RPS 33]

Stenocereinae Buxbaum, Bot. Stud. 12: 90, 1961. Typus: *Stenocereus*. [RPS 12]

Stenocereus subgen. *Hertrichocereiae* Bravo, Cact. Suc. Mex. 7(3): 57, 1962. *Nom. inval.* (Art. 36.1, 37.1). [Given as "comb. nov."..] [RPS -]

Stenocereus subgen. *Ritterocereiae* Bravo, Cact. Suc. Mex. 7(3): 57, 1962. *Nom. inval.* (Art. 36.1, 37.1). [Given as "comb. nov."..] [RPS -]

Stenocereus subgen. *Stenocereiae* Bravo, Cact. Suc. Mex. 7(3): 57, 1962. *Nom. inval.* (Art. 36.1, 37.1). [Given as "comb. nov.".] [RPS -]

Stenocereus alamosensis (J. Coulter) Gibson & Horak, Ann. Missouri Bot. Gard. 65(4): 1006, 1979. Basionym: *Cereus alamosensis*. [Dated 1978.] [RPS 30]

Stenocereus aragonii (F. A. C. Weber) Buxbaum, Bot. Stud. 12: 99, 1961. Basionym: *Cereus aragonii*. [RPS 12]

Stenocereus beneckei (Ehrenberg) Buxbaum, Bot. Stud. 12: 99, 1961. Basionym: *Cereus beneckei*. [RPS 12]

Stenocereus chrysocarpus Sanchez-Mejorada, Cact. Suc. Mex. 17: 95-98, 1972. [RPS 23]

Stenocereus deficiens (Otto & Dietrich) Buxbaum, Bot. Stud. 12: 100, 1961. Basionym: *Cereus deficiens*. [Repeated by Krainz in ZSS Kat. ed. 2, 112, 1967.] [RPS 12]

Stenocereus dumortieri (Scheidweiler) Buxbaum, Bot. Stud. 12: 100, 1961. Basionym: *Cereus dumortieri*. [RPS 12]

Stenocereus eruca (Brandegee) Gibson & Horak, Ann. Missouri Bot. Gard. 65(4): 1007, 1979. Basionym: *Cereus eruca*. [Dated 1978.] [RPS 30]

Stenocereus fricii Sanchez-Mejorada, Cact. Suc. Mex. 18: 89, 94, 1973. [RPS 24]

Stenocereus griseus (Haworth) Buxbaum, Bot. Stud. 12: 100, 1961. Basionym: *Cereus griseus*. [RPS 12]

Stenocereus gummosus (Brandegee) Gibson & Horak, Ann. Missouri Bot. Gard. 65(4): 1007, 1979. Basionym: *Cereus gummosus*. [Dated 1978.] [RPS 30]

Stenocereus hystrix (Haworth) Buxbaum, Bot. Stud. 12: 100, 1961. Basionym: *Cactus hystrix*. [RPS 12]

Stenocereus kerberi (Schumann) Gibson & Horak, Ann. Missouri Bot. Gard. 65(4): 1007, 1979. Basionym: *Cereus kerberi*. [Dated 1978.] [RPS 30]

Stenocereus laevigatus (Salm-Dyck) Buxbaum, Bot. Stud. 12: 100, 1961. Basionym: *Cereus laevigatus*. [RPS 12]

Stenocereus longispinus (Britton & Rose) Buxbaum, Bot. Stud. 12: 100, 1961. Basionym: *Lemaireocereus longispinus*. [RPS 12]

Stenocereus marginatus (De Candolle) Buxbaum, Bot. Stud. 12: 100, 1961. Basionym: *Cereus marginatus*. [RPS 12]

Stenocereus marginatus var. **gemmatus** (Zuccarini) Bravo, Cact. Suc. Mex. 19: 47, 1974. Basionym: *Cereus gemmatus*. [RPS 25]

Stenocereus martinezii (Gonzalez Ortega) Buxbaum, Bot. Stud. 12: 100, 1961. Basionym: *Lemaireocereus martinezii*. [Repeated by Krainz in ZSS Kat. ed. 2, 112, 1967; and by Bravo in Cact. Suc. Mex. 17: 119, 1972.] [RPS 12]

Stenocereus montanus (Britton & Rose) Buxbaum, Bot. Stud. 12: 101, 1961. Basionym: *Lemaireocereus montanus*. [RPS 12]

Stenocereus peruvianus (Miller) Kiesling, Darwiniana 24(1-4): 446, 1982. Basionym: *Cereus peruvianus*. [Based on *Cereus peruvianus* Miller *non Cactus peruvianus* Linné; the combination may not be valid since it is arguable that Miller's name cannot be treated as other than a combination for the Linnean *Cactus* basionym.] [RPS 33]

Stenocereus pruinosus (Otto) Buxbaum, Bot. Stud. 12: 101, 1961. Basionym: *Cereus pruinosus*. [RPS 12]

Stenocereus queretaroensis (F. A. C. Weber) Buxbaum, Bot. Stud. 12: 101, 1961. Basionym: *Cereus queretaroensis*. [RPS 12]

Stenocereus quevedonis (Gonzalez Ortega) Buxbaum, Bot. Stud. 12: 101, 1961. Basionym: *Lemaireocereus quevedonis*. [Repeated by Bravo in Cact. Suc. Mex. 17: 119, 1972.] [RPS 12]

Stenocereus standleyi (Gonzalez Ortega) Buxbaum, Bot. Stud. 12: 101, 1961. Basionym: *Lemaireocereus standleyi*. [RPS 12]

Stenocereus thurberi (Engelmann) Buxbaum, Bot. Stud. 12: 101, 1961. Basionym: *Cereus thurberi*. [RPS 12]

Stenocereus thurberi var. **littoralis** (Brandegee) Bravo, Cact. Suc. Mex. 17: 119-120, 1972. Basionym: *Cereus thurberi* var. *littoralis*. [RPS 23]

Stenocereus treleasii (Britton & Rose) Backeberg, Die Cact. 4: 2223, 1960. Basionym: *Lemaireocereus treleasii*. [Already proposed as *comb. nud* in Cact. Succ. J (US) 23(4): 120, 1951 (RPS 3) by Backeberg, giving Vaupel as original author.] [RPS 11]

Stenocereus weberi (Coulter) Buxbaum, Bot. Stud. 12: 101, 1961. Basionym: *Cereus weberi*. [RPS 12]

× *Stenomyrtillus* Rowley, Name that Succulent, 137, 1980. *Nom. inval.* (Art. H6.2). [= *Stenocereus* × *Myrtillocactus*.] [RPS 31]

Stetsonia coryne var. **procera** Ritter, Taxon 13: 116, 1964. [RPS 15]

Strombocacti Buxbaum, Madroño 14(6): 197, 1958. Typus: *Strombocactus*. [Published as 'linea'. First mentioned (*nom. inval.*, Art. 36.1) in Österr. Bot. Zeitschr. 98(1/2): 65, 1951.] [RPS 9]

Strombocactus denegrii (Fric) Rowley, Rep. Pl. Succ. 23: 9, 1974. Basionym: *Obregonia denegrii.* [RPS 23]
Strombocactus disciformis var. **seidelii** (Fric) Crkal, Lovec Kaktusu, 401, 1983. Basionym: *Strombocactus turbiniformis* var. *seidelii.* [RPS 34]
Strombocactus klinkerianus (Backeberg & Jacobsen) Buining, Succulenta 1951: 9, 1951. Basionym: *Turbinicarpus klinkerianus.* [RPS 9]
Strombocactus laui (Glass & Foster) Mays, Cactus 3(4): 84, 1979. Basionym: *Turbinicarpus laui.* [RPS 30]
Strombocactus polaskii (Backeberg) Hewitt, Ashingtonia Species Cat. Cact. [part 11], unnumbered page, 1975. *Nom. inval.*, based on *Turbinicarpus polaskii, nom. inval.* (Art. 37.1). [RPS -]
Strombocactus pseudomacrochele var. **krainzianus** (Frank) Rowley, Rep. Pl. Succ. 23: 10, 1974. Basionym: *Toumeya krainziana.* [RPS 23]
Strombocactus roseiflorus (Backeberg) Hewitt, Ashingtonia Species Cat. Cact. [part 11], unnumbered page, 1975. *Nom. inval.*, based on *Turbinicarpus roseiflorus, nom. inval.* (Art. 9.5). [RPS -]
Strombocactus schmiedickeanus var. **klinkerianus** (Backeberg & Jacobsen) Rowley, Rep. Pl. Succ. 23: 10, 1974. Basionym: *Turbinicarpus klinkerianus.* [Sphalm. 'klinkeranus'.] [RPS 23]
Submatucana Backeberg, Die Cact. 2: 1059, 1959. Typus: *Submatucana aurantiaca.* [First mentioned in a key by Backeberg, Die Cact. 1: 54, 73, 1958, as nude name (cf. RPS 9).] [RPS 10]
Submatucana aurantiaca (Vaupel) Backeberg, Die Cact. 2: 1061, 1959. Basionym: *Echinocactus aurantiacus.* [RPS 10]
Submatucana aureiflora (Ritter) Backeberg, Kakt.-Lex., 459, 1966. Basionym: *Matucana aureiflora.* [RPS 17]
Submatucana calvescens (Kimnach & P. C. Hutchison) Backeberg, Die Cact. 2: 1061, 1959. Basionym: *Borzicactus calvescens.* [RPS 10]
Submatucana currundayensis (Ritter) Backeberg, Die Cact. 6: 3702, 1962. *Nom. inval.* (Art. 33.2), based on *Matucana currundayensis.* [RPS 13]
Submatucana formosa (Ritter) Backeberg, Descr. Cact. Nov. 3: 14, 1963. Basionym: *Matucana formosa.* [RPS 14]
Submatucana formosa var. **minor** (Ritter) Backeberg, Descr. Cact. Nov. 3: 14, 1963. Basionym: *Matucana formosa* var. *minor.* [RPS 14]
Submatucana intertexta (Ritter) Backeberg, Descr. Cact. Nov. 3: 14, 1963. Basionym: *Matucana intertexta.* [RPS 14]
Submatucana madisoniorum (P. C. Hutchison) Backeberg, Kakt.-Lex., 412, 1966. Basionym: *Borzicactus madisoniorum.* [RPS 17]
Submatucana myriacantha (Vaupel) Backeberg, Die Cact. 2: 1063, 1959. Basionym: *Echinocactus myriacanthus.* [RPS 17]
Submatucana paucicostata (Ritter) Backeberg, Descr. Cact. Nov. 3: 14, 1963. Basionym: *Matucana paucicostata.* [Name first invalidly used (as new species) by Backeberg in Die Cact. 6: 3703, 1962 (*nom. nud.*).] [RPS 14]
Submatucana ritteri (Buining) Backeberg, Die Cact. 6: 3702, 1962. Basionym: *Matucana ritteri.* [RPS 13]
Subpilocereus atroviridis (Backeberg) Backeberg, Cact. Succ. J. (US) 23(4): 123, 1951. Basionym: *Cereus atroviridis.* [RPS 3]
Subpilocereus grenadensis (Britton & Rose) Backeberg, Die Cact. 4: 2383, 1960. Basionym: *Cereus grenadensis.* [RPS 11]
Subpilocereus horrispinus (Backeberg) Backeberg, Cact. Succ. J. (US) 23(4): 123, 1951. Basionym: *Cereus horrispinus.* [RPS 3]
Subpilocereus mortensenii (Croizat) Trujillo & Ponce, Ernstia 47: 28, 1988. Basionym: *Pilocereus mortensenii.* [RPS 39]
Subpilocereus remolinensis (Backeberg) Backeberg, Cact. Succ. J. (US) 23(4): 123, 1951. Basionym: *Cereus remolinensis.* [RPS 3]
Subpilocereus repandus (Linné) Backeberg, Cact. Succ. J. (US) 23(4): 123, 1951. Basionym: *Cereus repandus.* [RPS 3]
Subpilocereus repandus ssp. **micracanthus** (Wagenaar Hummelinck) Trujillo & Ponce, Ernstia 47: 28, 1988. Basionym: *Cereus margaritensis* var. *micracanthus.* [RPS 39]
Subpilocereus repandus var. *weberi* Backeberg, Die Cact. 4: 2385-2386, 1960. *Nom. inval.* (Art. 36.1, 37.1). [Erroneously included as valid in RPS 11.] [RPS 11]
Subpilocereus russelianus var. **margaritensis** (Johnston) Backeberg, Cact. Succ. J. (US) 23(4): 123, 1951. Basionym: *Pilocereus russelianus* ssp. *margaritensis.* [RPS 3]
Subpilocereus russelianus var. **micracanthus** (Wagenaar Hummelinck) Backeberg, Die Cact. 4: 2381, 1960. Basionym: *Cereus margaritensis* var. *micracanthus.* [RPS 11]
Subpilocereus wagenaari (Croizat) Backeberg, Cact. Succ. J. (US) 23(4): 123, 1951. Basionym: *Pilocereus wagenaarii.* [RPS 3]
Sulcorebutia Backeberg, Cact. Succ. J. Gr. Brit. 13(4): 96, 1951. Typus: *Rebutia steinbachii.* [RPS 2]
Sulcorebutia alba Rausch, Succulenta 50(5): 94-96, ill., 1971. Typus: *Rausch* 472 (W, ZSS). [Bolivia] [RPS 22]
Sulcorebutia albaoides (F. Brandt) Pilbeam, Sulcorebutia & Weingartia, 36, 1985. *Nom. inval.* (Art. 34.1a), based on *Weingartia albaoides.* [RPS 36]
Sulcorebutia albaoides var. *subfusca* (F. Brandt) Pilbeam, Sulcorebutia & Weingartia, 36, 1985. *Nom. inval.* (Art. 34.1a, 43.1), based on *Weingartia albaoides* var. *subfusca.* [RPS 36]
Sulcorebutia albissima (F. Brandt) Pilbeam, Sulcorebutia & Weingartia, 37, 1985. Basionym: *Weingartia albissima.* [RPS 36]
Sulcorebutia ambigua (Hildmann) F. Brandt,

Frankfurter Kakt.-Freund 3: 9, 1976. *Nom. inval.* (Art. 33), based on *Echinocactus ambiguus*. [The name given as basionym was itself a combination, and is here corrected and treated as bibliographical error.] [RPS 27]

Sulcorebutia arenacea (Cardenas) Ritter, Nation. Cact. Succ. J. 16: 81, 1961. Basionym: *Rebutia arenacea*. [RPS 12]

Sulcorebutia augustinii Hentzschel, Succulenta 68(7/8): 147-153, REM-ills., ill. (p. 145), 1989. Typus: *Swoboda* 152 (HBG). [Bolivia: Cochabamba] [RPS 40]

Sulcorebutia breviflora Backeberg, Kakt.-Lex., 414, 1966. [*Nom. nov. pro Rebutia brachyantha* Cardenas 1965 (*non Rebutia brachyantha* (Wessner) Buining & Donald 1963).] [RPS 17]

Sulcorebutia brevispina (F. Brandt) Pilbeam, Sulcorebutia & Weingartia, 41, 1985. *Nom. inval.* (Art. 34.1a), based on *Weingartia brevispina*. [RPS 36]

Sulcorebutia caineana (Cardenas) Donald, Cact. Succ. J. (US) 38: 143, 1971. Basionym: *Rebutia caineana*. [RPS 22]

Sulcorebutia callecallensis (F. Brandt) Pilbeam, Sulcorebutia & Weingartia, 41, 1985. *Nom. inval.* (Art. 34.1a), based on *Weingartia callecallensis*. [RPS 36]

Sulcorebutia candiae (Cardenas) Backeberg, Sukkulentenkunde 7/8: 104, 1963. Basionym: *Rebutia candiae*. [Repeated in Kakt.-Lex., 414, 1966.] [RPS 14]

Sulcorebutia canigueralii (Cardenas) Buining & Donald, Cact. Succ. J. Gt. Brit. 27: 57, 1965. Basionym: *Rebutia canigueralii*. [Repeated by Backeberg, Kakt.-Lex., 415, 1966.] [RPS 16]

Sulcorebutia caracarensis (Cardenas) Donald, Cact. Succ. J. (US) 43: 38, 1971. Basionym: *Rebutia caracarensis*. [RPS 22]

Sulcorebutia cardenasiana Vasquez, Kakt. and. Sukk. 26(3): 49, ill., 1975. Typus: *Vasquez* 544 (W, ZSS [status ?]). [Bolivia: Cochabamba] [RPS 26]

Sulcorebutia chilensis (Backeberg) F. Brandt, Frankfurter Kakt.-Freund 3: 9, 1976. *Nom. inval.* (Art. 33), based on *Neowerdermannia chilensis*. [The basionym given was itself already a combination (here corrected).] [RPS 27]

Sulcorebutia cintiensis (Cardenas) F. Brandt, Frankfurter Kakt.-Freund 3: 9, 1976. Basionym: *Weingartia cintiensis*. [RPS 27]

Sulcorebutia clavata (F. Brandt) Pilbeam, Sulcorebutia & Weingartia, 45, 1985. *Nom. inval.* (Art. 34.1a), based on *Weingartia clavata*. [RPS 36]

Sulcorebutia cochabambina Rausch, Succulenta 64(7-8): 152-153, ill., 1985. Typus: *Rausch* 275 (ZSS). [RPS 36]

Sulcorebutia corroana (Cardenas) Brederoo & Donald, Succulenta 52: 192, 1973. Basionym: *Rebutia corroana*. [RPS 24]

Sulcorebutia crispata Rausch, Kakt. and. Sukk. 21(6): 103, 1970. Typus: *Rausch* 288 (W 82-7534, ZSS). [Bolivia] [RPS 21]

Sulcorebutia croceareolata (F. Brandt) Pilbeam, Sulcorebutia & Weingartia, 47, 1985. *Nom. inval.* (Art. 34.1a), based on *Weingartia croceareolata*. [RPS 36]

Sulcorebutia cylindrica Donald & Lau, Ashingtonia 1(5): 56, 1974. [RPS 25]

Sulcorebutia cylindrica fa. *albiflora* Riha, Kaktusy 25(6): 134-136, ill., 1989. *Nom. inval.* (Art. 36.1, 37.1). [RPS 40]

Sulcorebutia erinacea (Ritter) F. Brandt, Frankfurter Kakt.-Freund 3: 9, 1976. Basionym: *Weingartia erinacea*. [RPS 27]

Sulcorebutia fidaiana (Backeberg) F. Brandt, Frankfurter Kakt.-Freund 3: 9, 1976. Incorrect name (Art. 11.3). [Incorrect name, including the type of the older genus *Weingartia* Werdermann.] [RPS 27]

Sulcorebutia fischeriana Augustin, Kakt. and. Sukk. 38(9): 210-216, ills., SEM-ills., 1987. Typus: *Swoboda* 79 (ZSS). [Bolivia: Chuquisaca] [RPS 38]

Sulcorebutia flavida (F. Brandt) Pilbeam, Sulcorebutia & Weingartia, 49, 1985. *Nom. inval.* (Art. 34.1a), based on *Weingartia flavida*. [RPS 36]

Sulcorebutia flavissima Rausch, Kakt. and. Sukk. 21(6): 105, 1970. Typus: *Rausch* 277 (W, ZSS). [Bolivia] [RPS 21]

Sulcorebutia formosa (F. Brandt) Pilbeam, Sulcorebutia & Weingartia, 50, 1985. *Nom. inval.* (Art. 34.1a), based on *Weingartia formosa*. [RPS 36]

Sulcorebutia frankiana Rausch, Kakt. and. Sukk. 21(6): 104-105, 1970. Typus: *Rausch* 290 (W, ZSS). [Bolivia: Sucre] [RPS 21]

Sulcorebutia frankiana var. **aureispina** Rausch, Succulenta 53(1): 3, 1974. Typus: *Rausch* 473 (ZSS). [Type cited for W but placed in ZSS.] [Bolivia: Sucre] [RPS 31]

Sulcorebutia glomeriseta (Cardenas) Ritter, Nation. Cact. Succ. J. 16: 81, 1961. Basionym: *Rebutia glomeriseta*. [RPS 12]

Sulcorebutia glomerispina (Cardenas) Buining & Donald, Cact. Succ. J. Gt. Brit. 27: 80, 1965. Basionym: *Rebutia glomerispina*. [Repeated by Backeberg, Kakt.-Lex., 415, 1966.] [RPS 16]

Sulcorebutia haseltonii (Cardenas) Donald, Cact. Succ. J (US) 43: 38, 1971. Basionym: *Rebutia haseltonii*. [RPS 22]

Sulcorebutia hediniana (Backeberg) F. Brandt, Frankfurter Kakt.-Freund 3: 9, 1976. Basionym: *Weingartia hediniana*. [RPS 27]

Sulcorebutia hoffmanniana (Backeberg) Backeberg, Kakt.-Lex., 415, 1966. *Nom. inval.*, based on *Lobivia hoffmanniana, nom. inval.* (Art. 9.5). [RPS 17]

Sulcorebutia inflexiseta (Cardenas) Donald, Cact. Succ. J. (US) 43: 38, 1971. Basionym: *Rebutia inflexiseta*. [RPS 22]

Sulcorebutia krahnii Rausch, Kakt. and. Sukk. 21: 104, 1970. Typus: *Rausch* 269 (W). [Bolivia: Santa Cruz] [RPS 21]

Sulcorebutia krugeri (Cardenas) Ritter, Nation. Cact. & Succ. J. 16: 81, 1961. Basionym: *Aylostera krugeri*. [Sphalm 'kruegeri'.] [RPS 12]

Sulcorebutia krugeri var. *hoffmanniana* (Backeberg) *hort. ex* Pilbeam, Sulcorebutia & Weingartia, 57, 1985. *Nom. inval.*, based on *Lobivia hoffmanniana, nom. inval.* (Art.

9.5). [Sphalm. 'kruegeri'. Repeated by Donald in Succulenta 65(10): 208, 1986 (cf. RPS 37).] [RPS 36]
Sulcorebutia lanata (Ritter) F. Brandt, Frankfurter Kakt.-Freund 3: 9, 1976. Basionym: *Weingartia lanata*. [RPS 27]
Sulcorebutia lecoriensis (Cardenas) F. Brandt, Frankfurter Kakt.-Freund 3: 9, 1976. Basionym: *Weingartia lecoriensis*. [RPS 27]
Sulcorebutia lepida Ritter, Nation. Cact. Succ. J. 17(1): 13, 1962. Typus: *Ritter* 369 (ZSS [status ?]). [No herbarium cited for type.] [Bolivia: Cochabamba] [RPS 13]
Sulcorebutia longigibba (Ritter) F. Brandt, Frankfurter Kakt.-Freund 3: 9, 1976. Basionym: *Weingartia longigibba*. [RPS 27]
Sulcorebutia losenickyana Rausch, Kakt. and. Sukk. 25(3): 49, ill., 1974. Typus: *Rausch* 477 (ZSS). [Type cited for W but placed in ZSS.] [Bolivia: Sucre] [RPS 25]
Sulcorebutia mariana Swoboda, Succulenta 68(1): 1, 3-8, ills., 1989. Typus: *Swoboda* 15 (HBG). [Bolivia: Cochabamba] [RPS 40]
Sulcorebutia markusii Rausch, Kakt. and. Sukk. 21(6): 103-104, ill., 1970. Typus: *Rausch* 195 (W, ZSS). [Bolivia] [RPS 21]
Sulcorebutia menesesii (Cardenas) Buining & Donald, Sukkulentenkunde 7/8: 104, 1963. Basionym: *Rebutia menesesii*. [Repeated by Backeberg, Kakt.-Lex., 415, 1966.] [RPS 14]
Sulcorebutia menesesii var. **kamiensis** Brederoo & Donald, Succulenta 65(8): 153, 155-158, (9): 190-193, (10): 207-209, ills., SEM-ills., 1986. Typus: *Lau* 974 (K). [Bolivia] [RPS 37]
Sulcorebutia menesesii var. **muschii** (Vasquez) Donald, Succulenta 65(10): 208, 1986. Basionym: *Sulcorebutia muschii*. [RPS 37]
Sulcorebutia mentosa Ritter, Succulenta 43: 102, 1964. [RPS 15]
Sulcorebutia mizquensis Rausch, Kakt. and. Sukk. 21(6): 102-103, ill, 1970. Typus: *Rausch* 194 (W, ZSS). [Bolivia] [RPS 21]
Sulcorebutia multispina (Ritter) F. Brandt, Frankfurter Kakt.-Freund 3: 9, 1976. Basionym: *Weingartia multispina*. [RPS 27]
Sulcorebutia muschii Vasquez, Succulenta 53: 43-44, 1974. Typus: *Vasquez* 562 (W). [Bolivia: Cochabamba] [RPS 25]
Sulcorebutia neocorroana F. Brandt, Frankfurter Kakt.-Freund 3: 9, 1976. *Nom. inval.* (Art. 36.1). [*Nom. nov. pro Weingartia corroana* (basionym *W. pulquinensis* var. *corroana*).] [RPS 27]
Sulcorebutia neocumingii (Backeberg) F. Brandt, Frankfurter Kakt.-Freund 3: 9, 1976. Basionym: *Weingartia neocumingii*. [RPS 27]
Sulcorebutia neumanniana (Backeberg) F. Brandt, Frankfurter Kakt.-Freund 3: 9, 1976. *Nom. inval.* (Art. 33). [Based on *Weingartia neumanniana* Werdermann (= *Echinocactus neumannianus* Backeberg *nom. illeg.*).] [RPS 27]
Sulcorebutia nigro-fuscata (F. Brandt) Pilbeam, Sulcorebutia & Weingartia, 66, 1985. *Nom. inval.* (Art. 34.1a), based on *Weingartia nigro-fuscata*. [RPS 36]
Sulcorebutia oenantha Rausch, Succulenta 50(6): 112-113, ill., 1971. Typus: *Rausch* 465 (W, ZSS). [Bolivia] [RPS 22]
Sulcorebutia oligacantha (F. Brandt) Pilbeam, Sulcorebutia & Weingartia, 67, 1985. *Nom. inval.* (Art. 34.1a), based on *Weingartia oligacantha*. [RPS 36]
Sulcorebutia pampagrandensis Rausch, Kakt. and. Sukk. 25(5): 97-98, ill., 1974. Typus: *Rausch* 466 (W [not found], ZSS). [Bolivia] [RPS 25]
Sulcorebutia perplexiflora (F. Brandt) Gertel, Kakt. and. Sukk. 36(3): 50, 1985. Basionym: *Weingartia perplexiflora*. [Simultaneously proposed invalidly (Art. 34.1a) by Pilbeam in Sulcorebutia & Weingartia, 68, 1985.] [RPS 36]
Sulcorebutia pilcomayensis (Cardenas) F. Brandt, Frankfurter Kakt.-Freund 3: 9, 1976. Basionym: *Weingartia pilcomayensis*. [Sphalm. 'pilcomayoensis'.] [RPS 27]
Sulcorebutia platygona (Cardenas) F. Brandt, Frankfurter Kakt.-Freund 3: 9, 1976. Basionym: *Weingartia platygona*. [RPS 27]
Sulcorebutia polymorpha (Cardenas) Backeberg, Kakt.-Lex., 416, 1966. Basionym: *Rebutia polymorpha*. [RPS 17]
Sulcorebutia pulchra (Cardenas) Donald, Cact. Succ. J. (US) 43: 39, 1971. Basionym: *Rebutia pulchra*. [Sphalm. 'pulchera'.] [RPS 22]
Sulcorebutia pulquinensis (Cardenas) F. Brandt, Frankfurter Kakt.-Freund 3: 9, 1976. Basionym: *Weingartia pulquinensis*. [RPS 27]
Sulcorebutia purpurea (Donald & Lau) Brederoo & Donald, Kakt. and. Sukk. 32(11): 273, 1981. Basionym: *Weingartia purpurea*. [RPS 32]
Sulcorebutia rauschii G. Frank, Kakt. and. Sukk. 20(1): 238-239, ill., 1969. Typus: *Rausch* 289 (W, ZSS). [Type cited for W. Rausch collection, but placed in W.] [Bolivia: Chuquisaca] [RPS 20]
Sulcorebutia riograndensis (Ritter) F. Brandt, Frankfurter Kakt.-Freund 3: 9, 1976. Basionym: *Weingartia riograndensis*. [RPS 27]
Sulcorebutia ritteri (F. Brandt) Ritter, Kakt. Südamer. 2: 645, 1980. Basionym: *Weingartia ritteri*. [RPS 31]
Sulcorebutia rubro-aurea (F. Brandt) Pilbeam, Sulcorebutia & Weingartia, 76, 1985. *Nom. inval.* (Art. 34.1a), based on *Weingartia rubro-aurea*. [RPS 36]
Sulcorebutia sanguineo-tarijensis (F. Brandt) Pilbeam, Sulcorebutia & Weingartia, 76, 1985. *Nom. inval.* (Art. 34.1a), based on *Weingartia sanguineo-tarijensis*. [RPS 36]
Sulcorebutia santiaginiensis Rausch, Kakt. and. Sukk. 30(10): 237, ill., 1979. Typus: *Rausch* 730 (ZSS). [Bolivia] [RPS 30]
Sulcorebutia steinbachii (Werdermann) Backeberg, Cact. Succ. J. Gr. Brit. 13(4): 96, 103, 1951. Basionym: *Rebutia steinbachii*. [RPS 2]
Sulcorebutia steinbachii var. **australis** Rausch, Succulenta 65(11): 240-241, ill., 1986. Typus: *Rausch* 729 (ZSS). [Bolivia] [RPS 37]

Sulcorebutia steinbachii var. *gracilior* Backeberg, Kakt.-Lex., 416, 1966. *Nom. inval.* (Art. 9.5). [Erroneously included as valid in RPS 17.] [RPS 17]

Sulcorebutia steinbachii var. **horrida** Rausch, Kakt. and. Sukk. 24(9): 193, 1973. Typus: *Rausch* 259 (ZSS). [Type cited for W but placed in ZSS.] [Bolivia: Cochabamba] [RPS 24]

Sulcorebutia steinbachii var. **polymorpha** (Cardenas) Pilbeam, Sulcorebutia & Weingartia, 80, 1985. Basionym: *Rebutia polymorpha*. [RPS 36]

Sulcorebutia steinbachii var. *rosiflora* Backeberg, Cactus (Paris) 19(80-81): 5, 1964. *Nom. inval.* (Art. 37.1). [RPS -]

Sulcorebutia steinbachii var. *violaciflora* Backeberg, Cactus (Paris) 19(80-81): 6, 1964. *Nom. inval.* (Art. 37.1). [Sphalm. 'violacifera', epithet corrected in Kakt.-Lex., 416, 1966.] [RPS -]

Sulcorebutia sucrensis (Ritter) F. Brandt, Frankfurter Kakt.-Freund 3: 9, 1976. Basionym: *Weingartia sucrensis*. [RPS 27]

Sulcorebutia swobodae Augustin, Kakt. and. Sukk. 35(6): 120-122, ills., 1984. Typus: *Swoboda* 27 (ZSS). [Bolivia: Cochambamba] [RPS 35]

Sulcorebutia tarabucoensis Rausch, Kakt. and. Sukk. 21(3): 45, 1970. Typus: *Rausch* 66 (W, ZSS). [First invalidly published (Art. 37.1) in l.c. 15: 92, 1964.] [Bolivia] [RPS 21]

Sulcorebutia taratensis (Cardenas) Buining & Donald, Cact. Succ. J. Gt. Brit. 27: 57, 1965. Basionym: *Rebutia taratensis*. [Repeated by Backeberg, Kakt.-Lex., 417, 1966.] [RPS 16]

Sulcorebutia taratensis var. **minima** Rausch, Kakt. and. Sukk. 21(3): 45, 1970. Typus: *Rausch* 196 (W 82-7533, ZSS). [First invalidly published (Art. 37.1) in l.c. 19: 112, 1968.] [Bolivia: Cochabamba] [RPS 21]

Sulcorebutia tarijensis Ritter, Ashingtonia 3(1): 13, 1978. [RPS 29]

Sulcorebutia tiraquensis (Cardenas) Ritter, Nation. Cact. Succ. J. 16: 81, 1961. Basionym: *Rebutia tiraquensis*. [Republished by Backeberg in Die Cact. 6: 3702, 1962] [RPS 12]

Sulcorebutia tiraquensis var. **aglaia** (F. Brandt) Sida, Kaktusy 26(1): 19, 1990. Basionym: *Weingartia aglaia*. [RPS 41]

Sulcorebutia tiraquensis var. *bicolorispina* Knize *ex* Pilbeam, Sulcorebutia & Weingartia, 85, 1985. *Nom. inval.* (Art. 37.1). [Based on *Weingartia aglaia* F. Brandt.] [RPS 36]

Sulcorebutia tiraquensis var. *electracantha* Backeberg, Descr. Cact. Nov. 3: 14, 1963. *Nom. inval.* (Art. 9.5). [Erroneously included as valid in RPS 14.] [RPS 14]

Sulcorebutia tiraquensis var. **longiseta** (Cardenas) Donald, Cact. Succ. J. (US) 43: 39, 1971. Basionym: *Rebutia tiraquensis* var. *longiseta*. [RPS 22]

Sulcorebutia torotorensis (Cardenas) F. Brandt, Frankfurter Kakt.-Freund 3: 9, 1976. Basionym: *Weingartia torotorensis*. [RPS 27]

Sulcorebutia totoralensis (F. Brandt) Pilbeam, Sulcorebutia & Weingartia, 90, 1985. *Nom. inval.* (Art. 34.1a), based on *Weingartia totoralensis*. [RPS 36]

Sulcorebutia totorensis (Cardenas) Ritter, Nation. Cact. Succ. J. 16: 81, 1961. Basionym: *Weingartia totorensis*. [Combination repeated by Brederoo & Donald in Kakt. and. Sukk. 32(11): 273, 1981 (cf. RPS 32).] [RPS 12]

Sulcorebutia totorensis var. **lepida** (Ritter) Pilbeam, Sulcorebutia & Weingartia, 91, 1985. Basionym: *Sulcorebutia lepida*. [Combining author given as 'Pilbeam *ex* Donald'.] [RPS 36]

Sulcorebutia tuberculato-chrysantha (Cardenas) Brederoo & Donald, Succulenta 52: 193, 1973. Basionym: *Rebutia tuberculato-chrysantha*. [RPS 24]

Sulcorebutia tunariensis (Cardenas) Buining & Donald, Cact. Succ. J. Gr. Brit. 27: 80, 1965. Basionym: *Rebutia tunariensis*. [Repeated by Backeberg, Kakt.-Lex., 417, 1966.] [RPS 16]

Sulcorebutia unguispina Rausch, Succulenta 64(6): 132-133, ill., 1985. Typus: *Rausch* 731 (ZSS). [Bolivia] [RPS 36]

Sulcorebutia vasqueziana Rausch, Kakt. and. Sukk. 21(6): 102, 1970. Typus: *Rausch* 284 (W, ZSS). [Bolivia: Sucre] [RPS 21]

Sulcorebutia vasqueziana var. **albispina** Rausch, Succulenta 52(12): 222, ill., 1973. Typus: *Rausch* 474 (W [not present], ZSS). [Bolivia: Sucre] [RPS 24]

Sulcorebutia verticillacantha Ritter, Nation. Cact. Succ. J. 17: 13, 1962. [as var. nova] [RPS 13]

Sulcorebutia verticillacantha fa. brevispina (F. Brandt) J. Pilbeam, Sulcorebutia & Weingartia, 97, 1985. Basionym: *Weingartia brevispina*. [Labelled as *forma nov.*, combining author given as 'Pilbeam *ex* Rausch'.] [RPS 36]

Sulcorebutia verticillacantha var. **albispina** (Rausch) Pilbeam, Sulcorebutia & Weingartia, 95, 1985. Basionym: *Sulcorebutia vasqueziana* var. *albispina*. [RPS 36]

Sulcorebutia verticillacantha var. **applanata** Donald & Krahn, Cact. Succ. J. Gr. Brit. 42(2): 37-38, 1980. [RPS 31]

Sulcorebutia verticillacantha var. **aureiflora** Rausch, Kakt. and. Sukk. 23(5): 123, ill., 1972. Typus: *Rausch* 479 (ZSS). [Type cited for W but placed in ZSS.] [*Weingartia rubro-aurea* F. Brandt 1984 is based on the same collection.] [Bolivia] [RPS 23]

Sulcorebutia verticillacantha var. **chatajillensis** Oeser & Brederoo, Kakt. and. Sukk. 35(10): 216-222, ills., 1984. Typus: *Anonymus* s.n. (HEID). [ex cult. Oeser acc. no. 568.] [RPS 35]

Sulcorebutia verticillacantha var. **cuprea** Rausch, Kakt. and. Sukk. 23(5): 125, 1972. Typus: *Rausch* 476 (ZSS). [Type cited for W but placed in ZSS.] [*Weingartia croceareolata* F. Brandt is based on the same type.] [Bolivia] [RPS 23]

Sulcorebutia verticillacantha var.

losenickyana (Rausch) Oeser, Kakt. and. Sukk. 35(10): 223, 1984. Basionym: *Sulcorebutia losenickyana*. [RPS 35]

Sulcorebutia verticillacantha var. **minima** (Rausch) Pilbeam, Sulcorebutia & Weingartia, 99, 1985. Basionym: *Sulcorebutia taratensis* var. *minima*. [RPS 36]

Sulcorebutia verticillacantha var. **ritteri** (F. Brandt) Donald & Krahn, Cact. Succ. J. Gr. Brit. 42(2): 38, 1980. Basionym: *Weingartia ritteri*. [RPS 31]

Sulcorebutia verticillacantha var. **verticosior** Ritter, Nation. Cact. Succ. J. 17: 13, 1962. [RPS 13]

Sulcorebutia vilcayensis (Cardenas) F. Brandt, Frankfurter Kakt.-Freund 3: 9, 1976. Basionym: *Weingartia vilcayensis*. [RPS 27]

Sulcorebutia vizcarrae (Cardenas) Donald, Cact. Succ. J. (US) 43: 40, 1971. Basionym: *Rebutia vizcarrae*. [RPS 22]

Sulcorebutia vizcarrae var. **laui** Brederoo & Donald, Succulenta 65(3): 52-55, (4): 89-93, (5): 106-108, ills., SEM-ills., 1986. Typus: *Lau* 324 (K). [RPS 37]

Sulcorebutia vorwerkii (Fric) F. Brandt, Frankfurter Kakt.-Freund 3: 9, 1976. *Nom. inval.* (Art. 33), based on *Neowerdermannia vorwerkii*. [The name cited as basionym was itself a combination, and is here corrected.] [RPS 27]

Sulcorebutia westii (P. C. Hutchison) F. Brandt, Frankfurter Kakt.-Freund 3: 9, 1976. Basionym: *Gymnocalycium westii*. [RPS 27]

Sulcorebutia xanthoantha Backeberg, Kakt.-Lex., 418, 1966. *Nom. inval.* (Art. 9.5, 37.1). [Erroneously included as valid in RPS 17.] [RPS 17]

Sulcorebutia zavaletae (Cardenas) Backeberg, Kakt.-Lex. 460, 1966. Basionym: *Aylostera zavaletae*. [RPS 17]

Tacinga atropurpurea var. **zehntnerioides** Backeberg, Descr. Cact. Nov. [1:] 10, 1957. [Dated 1956, published 1957.] [RPS 7]

Tacinga braunii E. Esteves Pereira, Kakt. and. Sukk. 40(6): 134-136, ills., 1989. Typus: *Braun* 864 (ZSS). [Brazil: Minas Gerais] [RPS 40]

Tephrocactus albiscoparius Backeberg, Cactus (Paris) No. 73/74: 5, 1962. *Nom. inval.* (Art. 9.5). [Erroneously included as valid in RPS 13. Concurrently described in Die Cact. 6: 3599, 1962.] [RPS 13]

Tephrocactus alboareolatus Ritter *ex* Backeberg, Descr. Cact. Nov. 3: 14, 1963. Typus: *Ritter* 184 (ZSS [type number]). [The description was based on a plant in the collection of Hegenbart (as no. 11); preserved material was available at the time of description.] [Peru: Arequipa] [RPS 14]

Tephrocactus alexanderi (Britton & Rose) Backeberg, Cactus (Paris) 8(38): 250, 1953. Basionym: *Opuntia alexanderi*. [RPS 4]

Tephrocactus alexanderi subvar. **brachyacanthus** (Spegazzini) Backeberg, Die Cact. 1: 294, 1958. Basionym: *Opuntia bruchii* var. *brachyacantha*. [RPS 9]

Tephrocactus alexanderi subvar. **macracanthus** (Spegazzini) Backeberg, Die Cact. 1: 293, 1958. Basionym: *Opuntia bruchii* var. *macracantha*. [RPS 9]

Tephrocactus alexanderi var. **bruchii** (Spegazzini) Backeberg, Cactus (Paris) 8(38): 250, 1953. Basionym: *Opuntia bruchii*. [RPS 4]

Tephrocactus alexanderi var. **subsphaericus** (Backeberg) Backeberg, Cactus (Paris) 8(38): 250, 1953. Basionym: *Tephrocactus subsphaericus*. [RPS 4]

Tephrocactus articulatus (Pfeiffer) Backeberg, Cactus (Paris) 8(38): 249, 1953. Basionym: *Cereus articulatus*. [The author abbreviation is normally given as "(Otto) Backeberg", on the assumption that *Opuntia articulata* Otto is the basionym, but this is an invalid nude name.] [RPS 4]

Tephrocactus articulatus fa. **syringacanthus** (Pfeiffer) Ritter, Kakt. Südamer. 2: 392-394, 1980. Basionym: *Cereus syringacanthus*. [RPS 31]

Tephrocactus articulatus var. **calvus** (Lemaire) Backeberg, Cactus (Paris) 8(38): 249, 1953. Basionym: *Opuntia calva*. [RPS 4]

Tephrocactus articulatus var. **diadematus** (Lemaire) Backeberg, Cactus (Paris) 8(38): 249, 1953. Basionym: *Opuntia diademata*. [RPS 4]

Tephrocactus articulatus var. **inermis** (Spegazzini) Backeberg, Cactus (Paris) 8(38): 249, 1953. Basionym: *Opuntia diademata* var. *inermis*. [RPS 4]

Tephrocactus articulatus var. **oligacanthus** (Spegazzini) Backeberg, Cactus (Paris) 8(38): 249, 1953. Basionym: *Opuntia diademata* var. *oligacantha*. [RPS 4]

Tephrocactus articulatus var. **ovatus** (Pfeiffer) Backeberg, Cactus (Paris) 8(38): 249, 1953. Basionym: *Cereus ovatus*. [RPS 4]

Tephrocactus articulatus var. **papyracanthus** (Philippi) Backeberg, Cactus (Paris) 8(38): 249, 1953. Basionym: *Opuntia papyracantha*. [RPS 4]

Tephrocactus articulatus var. **polyacanthus** (Spegazzini) Backeberg, Cactus (Paris) 8(38): 249, 1953. Basionym: *Opuntia diademata* var. *polyacantha*. [RPS 4]

Tephrocactus articulatus var. **syringacanthus** (Pfeiffer) Backeberg, Cactus (Paris) 8(38): 249, 1953. Basionym: *Cereus syringacanthus*. [RPS 4]

Tephrocactus asplundii Backeberg, Descr. Cact. Nov. [1:] 9, 1957. [Dated 1956, published 1957.] [RPS 7]

Tephrocactus atacamensis var. **chilensis** (Backeberg) Backeberg, Die Cact. 1: 340, 1958. Basionym: *Tephrocactus chilensis*. [RPS 9]

Tephrocactus atroglobosus Backeberg, Cactus (Paris) No. 73/74: 5, 1962. *Nom. inval.* (Art. 9.5). [Erroneously included as valid in RPS 13. Concurrently described in Die Cact. 6: 3905, 1962.] [RPS 13]

Tephrocactus atroviridis var. **longicylindricus** Rauh & Backeberg, in Rauh, Beitr. Kennt. Peruan. Kakt.veg., 212, 1958. [RPS 10]

Tephrocactus atroviridis var. **parviflorus** Rauh & Backeberg, in Rauh, Beitr. Kennt. Peruan. Kakt.veg., 213, 1958. [RPS 10]

Tephrocactus atroviridis var. **paucispinus** Rauh & Backeberg, in Rauh, Beitr. Kennt. Peruan. Kakt.veg., 214, 1958. [RPS 10]
Tephrocactus berteri (Colla) Ritter, Backeberg's Descr. & Erört. taxon. nomenkl. Fragen, 22, 1958. Basionym: *Echinocactus berteri*. [RPS 15]
Tephrocactus bicolor (Rauh & Backeberg) Rauh, Beitr. Kenntn. peruan. Kakt.veg., 223, 1958. Basionym: *Tephrocactus fulvicomus* var. *bicolor*. [RPS 10]
Tephrocactus blancii Backeberg, Descr. Cact. Nov. [1:] 8, 1957. [Dated 1956, published 1957.] [RPS 7]
Tephrocactus bolivianus (Salm-Dyck) Backeberg, Cact. Succ. J. (US) 23(1): 14, 1951. [RPS 3]
Tephrocactus camachoi (Espinosa) Backeberg, Die Cact. 1: 306, 1958. Basionym: *Opuntia camachoi*. [RPS 9]
Tephrocactus catacanthus Backeberg, Descr. Cact. Nov. 3: 14, 1963. *Nom. inval.* (Art. 9.5). [Erroneously included as valid in RPS 14.] [RPS 14]
Tephrocactus chichensis Cardenas, Nation. Cact. Succ. J. 7: 75-76, 1952. [RPS 3]
Tephrocactus chichensis var. **colchanus** Cardenas, Nation. Cact. Succ. J. 7: 75-76, 1952. [RPS 3]
Tephrocactus chilensis Backeberg, Cactus (Paris) 8(38): 250, 1953. [RPS 4]
Tephrocactus coloreus Ritter, in Backeberg, Descr. Cact. Nov. 3: 14, 1963. Typus: *Ritter* 513 (ubi ?). [The description was based on a plant in the collection of Hegenbart; preserved material was available at the time of description.] [?] [RPS 14]
Tephrocactus conoideus Ritter *ex* Backeberg, Die Cact. 1: 286, 1958. *Nom. inval.* (Art. 37.1). [Erroneously included as valid in RPS 9.] [RPS 9]
Tephrocactus corotilla (Schumann) Backeberg, Cactus (Paris) 8(38): 250, 1953. Basionym: *Opuntia corotilla*. [RPS 4]
Tephrocactus corotilla var. **aurantiaciflorus** Rauh & Backeberg, in Backeberg, Descr. Cact. Nov. [1:] 8, 1957. Typus: *Rauh* K147 (1956) (ZSS [status ?]). [Dated 1956, published 1957.] [Peru: Arequipa] [RPS 7]
Tephrocactus crassicylindricus Rauh & Backeberg, in Backeberg, Descr. Cact. Nov. [1:] 8, 1957. [Dated 1956, published 1957.] [RPS 7]
Tephrocactus crispicrinitus Rauh & Backeberg, in Backeberg, Descr. Cact. Nov. [1:] 7, 1957. Typus: *Rauh* K96a (1954) (ZSS). [Dated 1956, published 1957.] [Peru] [RPS 7]
Tephrocactus crispicrinitus subvar. **flavicomus** Rauh & Backeberg, in Backeberg, Descr. Cact. Nov. [1:] 7, 1957. [Dated 1956, published 1957.] [RPS 7]
Tephrocactus crispicrinitus var. **cylindraceus** Rauh & Backeberg, in Backeberg, Descr. Cact. Nov. [1:] 7, 1957. [Dated 1956, published 1957.] [RPS 7]
Tephrocactus crispicrinitus var. **tortispinus** Rauh & Backeberg, in Backeberg, Descr. Cact. Nov. [1:] 7, 1957. [Dated 1956, published 1957.] [RPS 7]
Tephrocactus curvispinus Backeberg, Descr. Cact. Nov. 3: 14, 1963. *Nom. inval.* (Art. 9.5). [Erroneously included as valid in RPS 14.] [RPS 14]
Tephrocactus cylindrarticulatus Cardenas, Nation. Cact. Succ. J. 7: 75-76, 1952. Typus: *Cardenas* 5004 (LIL, ZSS). [Bolivia: Nor Chichas] [RPS 3]
Tephrocactus cylindrolanatus Rauh & Backeberg, in Backeberg, Descr. Cact. Nov. [1:] 7, 1957. [Dated 1956, published 1957.] [RPS 7]
Tephrocactus dactyliferus (Vaupel) Backeberg, Die Cact. 1: 320, 1958. Basionym: *Opuntia dactylifera*. [RPS 9]
Tephrocactus dimorphus var. **pseudorauppianus** (Backeberg) Backeberg, Die Cact. 1: 301, 1958. Basionym: *Tephrocactus pseudorauppianus*. [RPS 9]
Tephrocactus duvalioides Backeberg, Cactus (Paris) 8(38): 250, 1953. [RPS 4]
Tephrocactus duvalioides var. **albispinus** Backeberg, Cactus (Paris) 8(38): 250, 1953. [RPS 4]
Tephrocactus echinaceus Ritter, Taxon 13(4): 145, 1964. Typus: *Ritter* 198 (U, ZSS [fragment]). [Type cited for U l.c. 12: 28, 1963.] [Chile] [RPS 15]
Tephrocactus ferocior Backeberg, Cactus (Paris) 8(38): 250, 1953. [RPS 4]
Tephrocactus flexispinus Backeberg, Descr. Cact. Nov. 3: 14, 1963. *Nom. inval.* (Art. 9.5). [Erroneously included as valid in RPS 14.] [RPS 14]
Tephrocactus floccosus fa. **denudata** (F. A. C. Weber) Ritter, Backeberg's Descr. & Erört. taxon. nomenkl. Fragen, [unpaged], 1958. Basionym: *Opuntia floccosa*. [RPS 15]
Tephrocactus floccosus fa. **rauhii** (Backeberg) Ritter, Backeberg's Descr. & Erört. taxon. nomenkl. Fragen, [unpaged], 1958. Basionym: *Tephrocactus rauhii*. [RPS 15]
Tephrocactus floccosus subvar. **aurescens** Rauh & Backeberg, in Backeberg, Descr. Cact. Nov. [1:] 7, 1957. [Dated 1956, published 1957.] [RPS 7]
Tephrocactus floccosus var. **aurescens** (Rauh & Backeberg) Rauh & Backeberg, in Rauh, Beitr. Kenntn. peruan. Kakt.veg., 203, 1958. Basionym: *Tephrocactus floccosus* subvar. *aurescens*. [RPS 10]
Tephrocactus floccosus var. **canispinus** Rauh & Backeberg, in Backeberg, Descr. Cact. Nov. [1:] 6, 1957. [Dated 1956, published 1957.] [RPS 7]
Tephrocactus floccosus var. *cardenasii* Marnier-Lapostolle, Cactus (Paris) No. 72: 137, 1961. *Nom. inval.* (Art. 37.1). [Erroneously included as valid in RPS 12.] [RPS 12]
Tephrocactus floccosus var. **crassior** Backeberg, Cactus (Paris) 8(38): 249, 1953. Neotypus: *Rauh* K26 (1954) (ZSS). [See Descr. Cact. Nov., 7, 1957 (dated 1956) for typification.] [Peru] [RPS 4]
Tephrocactus floccosus var. **denudatus** (F. A. C. Weber) Backeberg, Die Cact. 1: 235, 1958. Basionym: *Opuntia floccosa* var. *denudata*. [RPS 9]
Tephrocactus floccosus var. **lagopus**

(Schumann) Ritter, Backeberg's Descr. & Erört. taxon. nomenkl. Fragen, [unpaged], 1958. Basionym: *Opuntia lagopus*. [RPS 15]

Tephrocactus floccosus var. **ovoides** Rauh & Backeberg, in Backeberg, Descr. Cact. Nov. [1:] 7, 1957. [Dated 1956, published 1957.] [RPS 7]

Tephrocactus fulvicomus Rauh & Backeberg, in Backeberg, Descr. Cact. Nov. [1:] 9, 1957. Typus: *Rauh* K122 (1956) (ZSS [status ?]). [Dated 1956, published 1957.] [Peru] [RPS 7]

Tephrocactus fulvicomus var. **bicolor** Rauh & Backeberg, in Backeberg, Descr. Cact. Nov. [1:] 9, 1957. [Dated 1956, published 1957.] [RPS 7]

Tephrocactus glomeratus (Haworth) Backeberg, Cactus (Paris) 8(38): 249, 1953. Basionym: *Opuntia glomerata*. [RPS 4]

Tephrocactus glomeratus var. **andicola** (Pfeiffer) Backeberg, Cactus (Paris) 8(38): 249, 1953. Basionym: *Opuntia andicola*. [RPS 4]

Tephrocactus glomeratus var. *atratospinus* Backeberg, Descr. Cact. Nov. 3: 14, 1963. *Nom. inval.* (Art. 9.5). [Erroneously included as valid in RPS 14.] [RPS 14]

Tephrocactus glomeratus var. **fulvispinus** (Lemaire) Backeberg, Die Cact. 1: 283, 1958. Basionym: *Opuntia andicola* var. *fulvispina*. [RPS 9]

Tephrocactus glomeratus var. **gracilior** (Salm-Dyck) Backeberg, Cactus (Paris) 8(38): 249, 1953. Basionym: *Opuntia platyacantha* var. *gracilior*. [RPS 4]

Tephrocactus glomeratus var. *longispinus* Backeberg, Descr. Cact. Nov. 3: 14, 1963. *Nom. inval.* (Art. 9.5). [Erroneously included as valid in RPS 14.] [RPS 14]

Tephrocactus hegenbartianus Backeberg, Descr. Cact. Nov. 3: 15, 1963. Typus: *Ritter* 340 p.p. (ubi ?). [The description was based on a plant in the collection of Hegenbart; preserved material was available at the time of description.] [?] [RPS 14]

Tephrocactus heteromorphus (Philippi) Backeberg, Cactus (Paris) 8(38): 249, 1953. Basionym: *Opuntia heteromorpha*. [RPS 4]

Tephrocactus hirschii Backeberg, Descr. Cact. Nov. [1:] 8, 1957. [Dated 1956, published 1957.] [RPS 7]

Tephrocactus hossei Krainz & Gräser, Sukkulentenkunde 4: 29-30, 1951. [RPS 2]

Tephrocactus ignescens var. **steinianus** Backeberg, Descr. Cact. Nov. [1:] 9, 1957. [Dated 1956, published 1957.] [RPS 7]

Tephrocactus lagopus subvar. **brachycarpus** Rauh & Backeberg, in Backeberg, Descr. Cact. Nov. [1:] 7, 1957. [Dated 1956, published 1957.] [RPS 7]

Tephrocactus lagopus var. **aureo-penicillatus** Rauh & Backeberg, in Backeberg, Descr. Cact. Nov. [1:] 7, 1957. [Dated 1956, published 1957.] [RPS 7]

Tephrocactus lagopus var. **aureus** Rauh & Backeberg, in Backeberg, Descr. Cact. Nov. [1:] 7, 1957. [Dated 1956, published 1957.] [RPS 7]

Tephrocactus lagopus var. **leucolagopus** Rauh & Backeberg, in Backeberg, Descr. Cact. Nov. [1:] 7, 1957. Typus: *Rauh* K71 (1954) (ZSS). [Dated 1956, published 1957.] [Peru] [RPS 7]

Tephrocactus lagopus var. **pachycladus** Rauh & Backeberg, in Backeberg, Descr. Cact. Nov. [1:] 7, 1957. [Dated 1956, published 1957.] [RPS 7]

Tephrocactus leoninus (Rümpler) Backeberg, Die Cact. 6: 3596, 1962. *Nom. inval.* (Art. 33.2), based on *Opuntia leonina*. [Erroneously included as valid in RPS 13.] [RPS 13]

Tephrocactus longiarticulatus Backeberg, Descr. Cact. Nov. 3: 15, 1963. *Nom. inval.* (Art. 9.5). [Erroneously included as valid in RPS 14.] [RPS 14]

Tephrocactus malyanus Rausch, Kakt. and. Sukk. 22(3): 43-44, ill., 1971. Typus: *Rausch* 428 (W [not found], ZSS). [Peru] [RPS 22]

Tephrocactus mandragora Backeberg, Cactus (Paris) 8(38): 250, 1953. [RPS 4]

Tephrocactus melanacanthus Backeberg, Descr. Cact. Nov. 3: 15, 1963. *Nom. inval.* (Art. 9.5). [Erroneously included as valid in RPS 14.] [RPS 14]

Tephrocactus microclados Backeberg, Cactus (Paris) No. 75: 32, 1962. *Nom. inval.* (Art. 9.5, 37.1). [Erroneously included as valid in RPS 13. Concurrently described in Die Cact. 6: 3599, 1962.] [RPS 13]

Tephrocactus microsphaericus Backeberg, Kakt.-Lex., 426, 1966. *Nom. inval.* (Art. 9.5). [Erroneously included as valid in RPS 17.] [RPS 17]

Tephrocactus minor Backeberg, Cactus (Paris) 8(38): 250, 1953. [RPS 4]

Tephrocactus mirus Rauh & Backeberg, in Backeberg, Descr. Cact. Nov. [1:] 8, 1957. Typus: *Rauh* K111 (1956) (ZSS). [Dated 1956, published 1957.] [Peru] [RPS 7]

Tephrocactus molinensis (Spegazzini) Backeberg, Cactus (Paris) 8(38): 249, 1953. Basionym: *Opuntia molinensis*. [RPS 4]

Tephrocactus molinensis var. **denudatus** (F. A. C. Weber) Backeberg, Cactus (Paris) 8(38): 249, 1953. Basionym: *Opuntia floccosa* var. *denudata*. [RPS 4]

Tephrocactus muellerianus Backeberg, Descr. Cact. Nov. [1:] 8, 1957. [Dated 1956, published 1957.] [RPS 7]

Tephrocactus multiareolatus Ritter, Taxon: 13(4): 144, 1964. Typus: *Ritter* 275 (ZSS [not found], ZSS [seeds only]). [Peru: Arequipa] [RPS 15]

Tephrocactus neoglomeratus Y. Ito, The Cactaceae, 73, 1981. *Nom. inval.* (Art. 36.1, 37.1). [Published as *nom. nov.*, but it is unclear, which name is to be replaced.] [RPS -]

Tephrocactus neoglomeratus var. *andicolus* Y. Ito, The Cactaceae, 73, fig., 1981. *Nom. inval.* (Art. 43.1). [Given as *nom. nov.* for *Opuntia andicola* Pfeiffer 1837.] [RPS -]

Tephrocactus neoglomeratus var. *atratospinus* Y. Ito, The Cactaceae, 73, 1981. *Nom. inval.* (Art. 43.1). [Given as *nom. nov.* for *Tephrocactus glomeratus* var. *atrospinosus*

Backeberg 1963.] [RPS -]
Tephrocactus neoglomeratus var. *fulvispinus* Y. Ito, The Cactaceae, 73, 1981. *Nom. inval.* (Art. 43.1). [Given as *nom. nov.* for *Opuntia andicola* var. *fulvispina* Lemaire 1839.] [RPS -]
Tephrocactus neoglomeratus var. *gracilior* Y. Ito, The Cactaceae, 73, 1981. *Nom. inval.* (Art. 43.1). [Given as *nom. nov.* for *Opuntia platyacantha* var. *gracilior* Salm-Dyck 1850.] [RPS -]
Tephrocactus neoglomeratus var. *longispinus* Y. Ito, The Cactaceae, 73, 1981. *Nom. inval.* (Art. 43.1). [Given as *nom. nov.* for *Tephrocactus glomeratus* var. *longispinus* Backeberg 1963.] [RPS -]
Tephrocactus neuquensis (Borg) Backeberg, Cactus (Paris) 8(38): 250, 1953. Basionym: *Opuntia neuquensis*. [RPS 4]
Tephrocactus noodtiae Backeberg & Jacobsen, in Backeberg, Descr. Cact. Nov. [1:] 8, 1957. [Dated 1956, published 1957.] [RPS 7]
Tephrocactus ovatus (Pfeiffer) Ritter, Kakt. Südamer. 2: 395, 1980. Basionym: *Cereus ovatus*. [RPS 31]
Tephrocactus paediophilus (Castellanos) Ritter, Kakt. Südamer. 2: 395-396, 1980. Basionym: *Opuntia paediophila*. [See the basionym entry for notes on the spelling of the epithet.] [RPS 31]
Tephrocactus parvisetus Backeberg, Descr. Cact. Nov. 3: 15, 1963. *Nom. inval.* (Art. 9.5). [Erroneously included as valid in RPS 14.] [RPS 14]
Tephrocactus pentlandii var. *adpressus* Backeberg, Kakt.-Lex., 427, 1966. *Nom. inval.* (Art. 37.1). [Erroneously included as valid in RPS 17.] [RPS 17]
Tephrocactus pentlandii var. **fuauxianus** Backeberg, Cactus (Paris) 8(38): 250, 1953. [RPS 4]
Tephrocactus pentlandii var. **rossianus** Heinrich & Backeberg, Cactus (Paris) 8(38): 250, 1953. [RPS 4]
Tephrocactus platyacanthus var. **angustispinus** Backeberg, Descr. Cact. Nov. [1:] 8, 1957. [Dated 1956, published 1957.] [RPS 7]
Tephrocactus platyacanthus var. **deflexispinus** (Salm-Dyck) Backeberg, Cactus (Paris) 8(38): 249, 1953. Basionym: *Opuntia platyacantha* var. *monvillei*. [RPS 4]
Tephrocactus platyacanthus var. **monvillei** (Salm-Dyck) Backeberg, Cactus (Paris) 8(38): 249, 1953. Basionym: *Opuntia platyacantha* var. *monvillei*. [RPS 4]
Tephrocactus platyacanthus var. **neoplatyacanthus** Backeberg, Descr. Cact. Nov. [1:] 8, 1957. [Dated 1956, published 1957. *Nom. nov. pro Opuntia platyacantha* K. Schumann *non* Salm-Dyck.] [RPS 7]
Tephrocactus pseudo-udonis Rauh & Backeberg, in Backeberg, Descr. Cact. Nov. [1:] 7, 1957. [Dated 1956, published 1957.] [RPS 7]
Tephrocactus punta-caillan Rauh & Backeberg, in Backeberg, Descr. Cact. Nov. [1:] 8, 1957. [Dated 1956, published 1957.] [RPS 7]
Tephrocactus pyrrhacanthus (Schumann) Backeberg, Die Cact. 1: 336, 1958. Basionym: *Opuntia pyrrhacantha*. [RPS 9]
Tephrocactus pyrrhacanthus var. **leucoluteus** Backeberg, Descr. Cact. Nov. [1:] 9, 1957. [Dated 1956, published 1957.] [RPS 7]
Tephrocactus rauhii Backeberg, Cactus (Paris) No. 50: 62-63, 1956. [Repeated in Cact. Succ. J. Gr. Brit. 18(4): 103, 122, 1956; l.c. 19(1): 11, 1957; and in Descr. Cact. Nov. [1:] 7, 1956 [published 1957].] [RPS 7]
Tephrocactus reicheanus (Espinosa) Backeberg, Die Cact. 1: 284, 1958. Basionym: *Opuntia reicheana*. [RPS 9]
Tephrocactus russellii (Britton & Rose) Backeberg, Cactus (Paris) 8(38): 249, 1953. Basionym: *Opuntia russellii*. [RPS 4]
Tephrocactus sphaericus var. **glaucinus** Backeberg, Descr. Cact. Nov. 3: 15, 1963. Typus: *Ritter* 121a (ubi ?). [The description was based on a plant in the collection of Hegenbart; preserved material was available at the time of description.] [?] [RPS 14]
Tephrocactus sphaericus var. **rauppianus** (Schumann) Backeberg, Die Cact. 1: 298, 1958. Basionym: *Opuntia rauppiana*. [RPS 9]
Tephrocactus sphaericus var. **unguispinus** (Backeberg) Backeberg, Die Cact. 1: 297, 1958. Basionym: *Opuntia unguispina*. [RPS 9]
Tephrocactus staffordae (Bullock) Backeberg, Cactus (Paris) 8(38): 250, 1953. Basionym: *Opuntia staffordae*. [RPS 4]
Tephrocactus unguispinus Backeberg, Cact. Succ. J. (US) 23(1): 15, 1951. Basionym: *Opuntia unguispina*. [Name first mentioned in the original publication of the basionym.] [RPS 3]
Tephrocactus variflorus Backeberg, Cactus (Paris) No. 73/74: 5, 1962. *Nom. inval.* (Art. 37.1). [Sphalm. 'variiflorus'. Erroneously included as valid in RPS 13. Concurrently described in Die Cact. 6: 3594, 1962.] [RPS 13]
Tephrocactus virgultus Backeberg, Kakt.-Lex., 430, 1966. *Nom. inval.* (Art. 37.1). [Erroneously included as valid in RPS 17.] [RPS 17]
Tephrocactus weberi var. **deminutus** Rausch, Succulenta 65(12): 249, 251-252, ill., 1986. Typus: *Rausch* 241 (ZSS). [Argentina: Salta] [RPS 37]
Tephrocactus weberi var. **dispar** (Castellanos & Lelong) Backeberg, Cactus (Paris) 8(38): 249, 1953. Basionym: *Opuntia weberi* var. *dispar*. [RPS 4]
Tephrocactus weberi var. **setiger** (Backeberg) Backeberg, Die Cact. 1: 252, 1958. Basionym: *Tephrocactus setiger*. [RPS 9]
Tephrocactus yanganucensis Rauh & Backeberg, in Backeberg, Descr. Cact. Nov. [1:] 8, 1957. [Dated 1956, published 1957.] [RPS 7]
Tephrocactus zehnderi Rauh & Backeberg, in Backeberg, Descr. Cact. Nov. [1:] 9, 1957. Typus: *Rauh* K121 (1956) (ZSS [status ?]). [Dated 1956, published 1957.] [Peru] [RPS 7]
×**Thelobergia** Hirao, Col. Encycl. Cacti, 30, 1979. [= *Thelocactus* × *Leuchtenbergia*.]

[RPS 30]

Thelocacti Buxbum, Madroño 14(6): 196, 1958. Typus: *Thelocactus*. [Published as 'linea'. First mentioned (*nom. inval*., Art. 36.1) in Österr. Bot. Zeitschr. 98(1/2): 54, 1951.] [RPS 9]

Thelocactinae Buxbaum, Madroño 14(6): 196, 1958. Typus: *Thelocactus*. [RPS 9]

Thelocactus subgen. *Echinomastus* (Britton & Rose) Soulaire, Cactus (Paris) No. 46-47: 250, 1955. *Nom. inval*. (Art. 33.2). [RPS 6]

Thelocactus subgen. *Euthelocactus* (Schumann) Soulaire, Cactus (Paris) No. 46-47: 250, 1955. *Nom. inval*. (Art. 21.3). [RPS 6]

Thelocactus aguirreanus (Glass & Foster) Bravo, Cact. Succ. Mex. 25(3): 65, 1980. Basionym: *Gymnocactus aguirreanus*. [RPS 31]

Thelocactus beguinii N. P. Taylor, Bradleya 1: 113 in adnot., 1983. [*Nom. nov. pro Echinocactus beguinii* K. Schumann (*Gymnocactus*, *Neolloydia*, etc.), *nom. illeg*. (see E. F. Anderson in Bradleya 4: 17, 1986).] [RPS 34]

Thelocactus bicolor var. **bolaensis** (Schumann) Y. Ito, Cacti, 93, 1952. Basionym: *Echinocactus bicolor* var. *bolaensis*. [Spelling frequently changed to 'bolansis'.] [RPS 4]

Thelocactus bicolor var. **commodus** Haas, Kakt. and. Sukk. 39(4): 86-87, ills., 1988. Typus: *Haas* 212 (ZSS). [Mexico: Nuevo Leon] [RPS 39]

Thelocactus bicolor var. **pottsii** (Salm-Dyck) Backeberg, Die Cact. 5: 2809, 1961. Basionym: *Echinocactus bicolor* var. *pottsii*. [RPS 12]

Thelocactus bicolor var. **schottii** (Engelmann) Krainz, Die Kakt. C VIIIb, 1961. Basionym: *Echinocactus bicolor* var. *schottii*. [Repeated by Benson in Lundell et al., Fl. Texas 2: 291, 1969 (cf. RPS 21).] [RPS 12]

Thelocactus bicolor var. **schwarzii** (Backeberg) E. F. Anderson, Bradleya 5: 61, 1987. Basionym: *Thelocactus schwarzii*. [RPS 38]

Thelocactus bicolor var. *texensis* Backeberg, Die Cact. 6: 3872, 1962. *Nom. inval*. (Art. 37.1). [Erroneously included as valid in RPS 13.] [RPS 13]

Thelocactus bicolor var. **tricolor** (Schumann) Y. Ito, Cacti, 94, 1952. Basionym: *Echinocactus bicolor* var. *tricolor*. [RPS 4]

Thelocactus bicolor var. **wagnerianus** (A. Berger) Krainz, Die Kakt. C VIIIb, 1961. Basionym: *Echinocactus wagnerianus*. [RPS 12]

Thelocactus conothelos var. **argenteus** Glass & Foster, Cact. Succ. J. (US) 44(2): 47-48, ills., 1972. Typus: *Glass et Foster* 3176 (ZSS, MEXU, POM, ZSS). [Mexico: Nuevo Leon] [RPS 23]

Thelocactus conothelos var. **aurantiacus** Glass & Foster, Cact. Succ. J. (US) 44(2): 48, 50, ill., 1972. Typus: *Glass et Foster* 1383 (ZSS, MEXU, POM, ZSS). [Mexico: Nuevo Leon] [RPS 23]

Thelocactus conothelos var. **macdowellii** (Rebut *ex* Quehl) Glass & Foster, Cact. Succ. J. (US) 49(5): 220, 1977. Basionym: *Echinocactus macdowellii*. [RPS 28]

Thelocactus durangensis (Runge) Rowley, Excelsa 12: 31, 1986. Basionym: *Echinocactus durangensis*. [RPS 37]

Thelocactus flavidispinus (Backeberg) Backeberg, Cact. Succ. J. (US) 23(5): 150, 1951. Basionym: *Thelocactus bicolor* var. *flavidispinus*. [RPS 3]

Thelocactus goldii Bravo, Anales Inst. Biol. UNAM 26(1): 23-27, ills., 1955. Typus: *Sanchez Mejorada* s.n. (UNAM 3820). [English translation in Cact. Succ. J. (US) 27(5): 158, 1955.] [Mexico: Hidalgo] [RPS 6]

Thelocactus hertrichii (Weingart) Borg, Cacti, ed. 2, 347, 1951. Basionym: *Echinocactus hertrichii*. [RPS 9]

Thelocactus hexaedrophorus var. **fossulatus** (Scheidweiler) Backeberg, Die Cact. 5: 2800, 1961. Basionym: *Echinocactus fossulatus*. [RPS 12]

Thelocactus hexaedrophorus var. **lloydii** (Britton & Rose) Kladiwa & Fittkau, in Krainz, Die Kakt. Lief. 61: C VIIIb, 1975. Basionym: *Thelocactus lloydii*. [RPS 26]

Thelocactus hexaedrophorus var. **major** (Quehl) Y. Ito, Cacti, 94, 1952. Basionym: *Echinocactus hexaedrophorus* var. *major*. [RPS 4]

Thelocactus horripilus (Lemaire) Kladiwa & Fittkau, in Krainz, Die Kakt. Lief. 44-45: C VIIIb, 1970. Basionym: *Mammillaria horripila*. [Repeated in l.c. Lief. 61: C VIIIb (April 1975).] [RPS 21]

Thelocactus lausseri Riha & Busek, Kakt. and. Sukk. 37(8): 162-164, ills., SEM-ills., 1986. Typus: *Lausser* s.n. (PR 377 518). [First mentioned in a paper by A. Lausser in l.c. 37(6): 126-127, ills.] [Mexico] [RPS 37]

Thelocactus leucacanthus var. **ehrenbergii** (Pfeiffer) Bravo, Cact. Suc. Mex. 25(3): 65, 1980. Basionym: *Echinocactus ehrenbergii*. [RPS 31]

Thelocactus leucacanthus var. **porrectus** (Lemaire) Backeberg, Die Cact. 5: 2818, 1961. Basionym: *Echinocactus porrectus*. [RPS 12]

Thelocactus leucacanthus var. **sanchezmejoradae** (Meyran) Backeberg, Die Cact. 5: 2817, 1961. Basionym: *Thelocactus sanchezmejoradae*. [Sphalm. 'sanchezmejoradae'.] [RPS 12]

Thelocactus lophothele var. **nidulans** (Quehl) Kladiwa, in Krainz, Die Kakt. Lief. 61: C VIIIb, 1975. Basionym: *Echinocactus nidulans*. [RPS 26]

Thelocactus macdowellii (Rebut *ex* Quehl) C. Glass, Cact. Suc. Mex. 14: 4, 1969. Basionym: *Echinocactus macdowellii*. [Repeated by Kladiwa & Fittkau in Krainz, Die Kakt. Lief. 61: C VIIIb (April 1975).] [RPS 26]

Thelocactus macrochele fa. *polaskii* (Backeberg) Kladiwa, in Krainz, Die Kakt. Lfg. 62, CVIIIb, 1976. *Nom. inval*., based on *Turbinicarpus polaskii, nom. inval*. (Art. 37.1). [Erroneously included as valid in RPS 27.] [RPS 27]

Thelocactus macrochele var. **schwarzii**

(Shurly) Kladiwa, in Krainz, Die Kakt. Lfg. 62, CVIIIb, 1976. Basionym: *Strombocactus schwarzii.* [RPS 27]

Thelocactus matudae Sanchez-Mejorada & Lau, Cact. Suc. Mex. 23: 51-53, 1978. Typus: *Lau* s.n. (MEXU 227752). [Mexico: Nuevo Leon] [RPS 29]

Thelocactus pseudopectinatus (Backeberg) Anderson & Boke, Amer. J. Bot. 56: 326, 1969. Basionym: *Pelecyphora pseudopectinata.* [RPS 20]

Thelocactus rinconensis var. **nidulans** (Quehl) Glass & Foster, Cact. Succ. J. (US) 49(6): 245, 1977. Basionym: *Echinocactus nidulans.* [RPS 28]

Thelocactus rinconensis var. **phymatothele** (Poselger) Glass & Foster, Cact. Succ. J. (US) 49(6): 246, 1977. Basionym: *Echinocactus phymatothele.* [RPS 28]

Thelocactus sanchezmejoradae Meyran, Cact. Suc. Mex. 3(4): 77-78, 1958. [Sphalm. 'sanchezmejoradae'.] [RPS 9]

Thelocactus saussieri var. *longispinus* Y. Ito, The Cactaceae, 527-528, ill., 1981. *Nom. inval.* (Art. 37.1). [RPS -]

Thelocactus schottii (Engelmann) Kladiwa & Fittkau, in Krainz, Die Kakt. Lief. 61: C VIIIb, 1975. Basionym: *Echinocactus bicolor* var. *schottii.* [Incorrectly used name because *T. flavidispinus* Backeberg is cited in the synonymy (priority rule).] [RPS 26]

Thelocactus schwarzii Backeberg, Cact. Succ. J. Gr. Brit. 12(4): 81, ill. (p. 84), 1950. [Lecotypified by the illustration accompanying the original description (see E. F. Anderson in Bradleya 5: 61, 1987).] [RPS 1]

Thelocactus setispinus (Engelmann) E. F. Anderson, Bradleya 5: 59, 1987. Basionym: *Echinocactus setispinus.* [RPS 38]

Thelocactus smithii (Mühlenpfordt) Borg, Cacti, ed. 2, 346, 1951. Basionym: *Echinocactus smithii.* [RPS 9]

Thelocactus subterraneus var. **zaragosae** (Glass & Foster) Bravo, Cact. Suc. Mex. 25(3): 65, 1980. Basionym: *Gymnocactus subterraneus* var. *zaragosae.* [RPS 31]

Thelocactus tulensis var. **buekii** (Klein) E. F. Anderson, Bradleya 5: 65, 1987. Basionym: *Echinocactus buekii.* [RPS 38]

Thelocactus tulensis var. *longispinus* Y. Ito, The Cactaceae, 526, ill., 1981. *Nom. inval.* (Art. 37.1). [RPS -]

Thelocactus tulensis var. **matudae** (Sanchez-Mejorada & Lau) E. F. Anderson, Bradleya 5: 66, 1987. Basionym: *Thelocactus matudae.* [RPS 38]

Thelocactus unguispinus (Engelmann) Rowley, Excelsa 12: 30, 1986. Basionym: *Echinocactus unguispinus.* [RPS 37]

Thelocactus viereckii var. **major** (Glass & Foster) Bravo, Cact. Suc. Mex. 25(3): 65, 1980. Basionym: *Gymnocactus viereckii* var. *major.* [RPS 31]

Thelocephala Y. Ito, Explan. Diagr. Austroechinocactinae, 148, 292, 1957. Typus: *Neoporteria napina.* [RPS 8]

Thelocephala aerocarpa (Ritter) Ritter, Kakt. Südamer. 3: 1010-1011, 1980. *Nom. inval.*, based on *Chileorebutia aerocarpa, nom. inval.* (Art. 43.1). [RPS 31]

Thelocephala duripulpa (Ritter) Ritter, Kakt. Südamer. 3: 1010, 1980. Basionym: *Chileorebutia duripulpa.* [RPS 31]

Thelocephala esmeraldana (Ritter) Ritter, Kakt. Südamer. 3: 1021-1023, 1980. Basionym: *Chileorebutia esmeraldana.* [RPS 31]

Thelocephala fankhauseri Ritter, Kakt. Südamer. 3: 1002-1003, 1980. [RPS 31]

Thelocephala fulva Ritter, Kakt. Südamer. 3: 1011-1012, 1980. [Given as comb. nov., but in fact a sp. nov.] [RPS 31]

Thelocephala glabrescens (Ritter) Ritter, Kakt. Südamer. 3: 1003-1004, 1980. *Nom. inval.*, based on *Chileorebutia glabrescens, nom. inval.* (Art. 43.1). [RPS 31]

Thelocephala krausii (Ritter) Ritter, Kakt. Südamer. 3: 1017-1019, 1980. Basionym: *Chileorebutia krausii.* [RPS 31]

Thelocephala lembckei (Backeberg) Ritter, Kakt. Südamer. 3: 1005-1009, 1980. *Nom. inval.*, based on *Neochilenia lembckei, nom. inval.* (Art. 37.1). [RPS 31]

Thelocephala longirapa Ritter, Kakt. Südamer. 3: 1019, 1980. [RPS 31]

Thelocephala malleolata (Ritter) Ritter, Kakt. Südamer. 3: 1020, 1980. Basionym: *Chileorebutia malleolata.* [RPS 31]

Thelocephala malleolata var. **solitaria** (Ritter) Ritter, Kakt. Südamer. 3: 1020-1021, 1980. Basionym: *Chileorebutia malleolata* var. *solitaria.* [RPS 31]

Thelocephala napina (Philippi) Y. Ito, Explan. Diagr. Austroechinocactinae, 149, 1957. Basionym: *Echinocactus napinus.* [RPS 8]

Thelocephala napina var. *spinosior* Backeberg *ex* Y. Ito, Explan. Diagr. Austroechinocactinae, 149, 1957. *Nom. inval.* (Art. 36.1). [Based on *Neoporteria napina* var. *spinosior* Backeberg 1935, *nom. inval.* (Art. 36.1).] [RPS 8]

Thelocephala nuda Ritter, Kakt. Südamer. 3: 1004-1005, 1980. [RPS 31]

Thelocephala odieri (Salm-Dyck) Ritter, Kakt. Südamer. 3: 1012-1016, 1980. Basionym: *Echinocactus odieri.* [RPS 31]

Thelocephala reichei (Schumann) Ritter, Kakt. Südamer. 3: 1023-1024, 1980. Basionym: *Echinocactus reichei.* [RPS 31]

Thelocephala roseiflora Y. Ito, Explan. Diagr. Austroechinocactinae, 149, 292, 1957. [RPS 8]

Thelocephala tenebrica Ritter, Kakt. Südamer. 3: 1001-1002, 1980. Typus: *Ritter* 1092 (U, ZSS). [Type cited for U l.c. 1: iii, 1979.] [Chile: Aconcagua] [RPS 31]

Thrixanthocereus blossfeldiorum fa. *albidior* Backeberg, Die Cact. 6: 3863, 1962. *Nom. inval.* (Art. 37.1). [Erroneously included as valid in RPS 13; omitted from list published by Eggli in Bradleya 3: 97-102, 1985.] [RPS 13]

Thrixanthocereus blossfeldiorum fa. *paucicostata* Backeberg, Die Cact. 6: 3863, 1962. *Nom. inval.* (Art. 37.1). [Erroneously included as valid in RPS 13; omitted from list published by Eggli in Bradleya 3: 97-

102, 1985.] [RPS 13]
Thrixanthocereus blossfeldiorum fa. *typica* Backeberg, Die Cact. 6: 3863, 1962. *Nom. inval.* (Art. 24.3). [Erroneously included as valid in RPS 13; omitted from list published by Eggli in Bradleya 3: 97-102, 1985.] [RPS 13]
Thrixanthocereus cullmannianus Ritter, Kakt. and. Sukk. 12(8): 118-121, ills., 1961. Typus: *Ritter* 1065 (U, ZSS [seeds only]). [Peru: Cajamarca] [RPS 12]
Thrixanthocereus longispinus Ritter, Kakt. Südamer. 4: 1484-1485, 1981. [RPS 32]
Thrixanthocereus senilis Ritter, Kakt. and. Sukk. 12(6): 89-91, ills., 1961. Typus: *Ritter* 569 (U, ZSS). [First mentioned (without Latin diagnosis) by Backeberg in Die Cact. 4: 2477, 1960.] [Peru: Ancash] [RPS 12]
Toumeya fickeisenii (Backeberg) Kladiwa, Sukkulentenkunde 7/8: 46, 1963. *Nom. inval.*, based on *Navajoa fickeisenii, nom. inval.* (Art. 37.1). [Erroneously included as valid in RPS 14.] [RPS 14]
Toumeya klinkeriana (Backeberg & Jacobsen) Bravo & W. T. Marshall, Saguaroland Bull. 10(10): 116, 1956. Basionym: *Turbinicarpus klinkerianus.* [RPS 7]
Toumeya krainziana G. Frank, Kakt. and. Sukk. 11(11): 168-170, ills., 1960. Typus: *Wagner* s.n. (ZSS [status ?]). [Mexico] [RPS 11]
Toumeya pseudomacrochele var. **krainziana** (Frank) Kladiwa, in Krainz, Die Kakt. C VIIIb, 1966. Basionym: *Toumeya krainziana.* [Repeated by Krainz in ZSS Kat. ed. 2, 113, 1967.] [RPS 17]
Toumeya schmiedickeana var. **klinkeriana** (Backeberg & Jacobsen) Krainz, Die Kakt. C VIIIb (Sept. 1959), 1959. Basionym: *Turbinicarpus klinkerianus.* [RPS 10]
Toumeya schwarzii (Shurly) Bravo & W. T. Marshall, Saguaroland Bull. 11: 30-31, 1957. Basionym: *Strombocactus schwarzii.* [RPS 8]
Toumeya schwarzii var. *polaskii* (Backeberg) Kladiwa, Sukkulentenkunde 7/8: 46, 1963. *Nom. inval.*, based on *Turbinicarpus polaskii, nom. inval.* (Art. 37.1). [Erroneously included as valid in RPS 14.] [RPS 14]
Trichocereeae Buxbaum, Madroño 14(6): 189, 1958. Typus: *Trichocereus.* [Sphalm. 'Trichocereae'.] [Sphalm. 'Trichocereae'.] [RPS 9]
Trichocereinae Buxbaum, Madroño 14(6): 190, 1958. Typus: *Trichocereus.* [RPS 9]
Trichocereus cv. **Bernhard Kuderer** E. Kleiner, Kakt. and. Sukk. 31(4): 106-112, 1980. [RPS 31]
Trichocereus cv. **Elmar Marten** E. Kleiner, Kakt. and. Sukk. 31(4): 106-112, 1980. [RPS 31]
Trichocereus cv. **Franz Lang** E. Kleiner, Kakt. and. Sukk. 31(4): 106-112, 1980. [RPS 31]
Trichocereus cv. **Wilhelm Höch-Widmer** E. Kleiner, Kakt. and. Sukk. 31(4): 106-112, 1980. [RPS 31]
Trichocereus andalgalensis var. **auricolor** (Backeberg) Ritter, Kakt. Südamer. 2: 438, 1980. Basionym: *Trichocereus auricolor.* [RPS 31]
Trichocereus angelesiae Kiesling, Darwiniana 21(2-4): 314-315, 1978. [Sphalm. 'angelesii'.] [RPS 29]
Trichocereus antezanae Cardenas, Fuaux Herb. Bull. 5: 16, 1953. [RPS 4]
Trichocereus arboricola Kimnach, Cact. Succ. J. (US) 62(1): 3-5, ills., 1990. Typus: *Solomon* 11311 (HNT, MO, US). [Bolivia: Dept. Tarija] [RPS 41]
Trichocereus atacamensis var. **pasacana** (F. A. C. Weber *ex* Rümpler) Ritter, Kakt. Südamer. 2: 447, 1980. Basionym: *Pilocereus pasacana.* [RPS 31]
Trichocereus bruchii (Britton & Rose) Ritter, Kakt. Südamer. 2: 450, 1980. Basionym: *Lobivia bruchii.* [RPS 31]
Trichocereus bruchii var. **brevispinus** Ritter, Kakt. Südamer. 2: 450, 1980. [RPS 31]
Trichocereus cabrerae Kiesling, Hickenia 1: 30-31, 1976. [RPS 27]
Trichocereus callianthus Ritter, Kakt. Südamer. 2: 444, 1980. [RPS 31]
Trichocereus camarguensis Cardenas, Revista Agr. Cochabamba 11(8): 17, 1953. [RPS 4]
Trichocereus candicans fa. **rubriflorus** Ritter, Kakt. Südamer. 2: 440-443, 1980. [RPS 31]
Trichocereus candicans var. **courantii** (Schumann) Castellanos, Revista Fac. Ci. Agron. 6(2): 9, 1957. Basionym: *Cereus candicans* var. *courantii.* [RPS 9]
Trichocereus candicans var. **gladiatus** (Lemaire) A. Berger *ex* Castellanos, Revista Fac. Ci. Agron. 6(2): 9, 1957. [Combination first proposed (but not accepted) by A. Berger in Kakteen, 139, 1929. The combination by Castellanos is incorrectly given as superfluous in RPS 9.] [RPS 9]
Trichocereus candicans var. **nitens** (Salm-Dyck) Ritter, Kakt. Südamer. 2: 443, 1980. Basionym: *Cereus nitens.* [RPS 31]
Trichocereus candicans var. *roseoflorus* Backeberg, Descr. Cact. Nov. 3: 15, 1963. *Nom. inval.* (Art. 9.5, 37.1). [RPS 14]
Trichocereus candicans var. **tenuispinus** (Pfeiffer) Backeberg, Cactus (Paris) No. 41: 107, 1954. Basionym: *Cereus candicans* var. *tenuispinus.* [Repeated by Castellanos in Revista Fac. Ci. Agron. 6(2): 12, 1957 (cf. RPS 9).] [RPS 5]
Trichocereus catamarcensis Ritter, Kakt. Südamer. 2: 451-452, 1980. [RPS 31]
Trichocereus caulescens Ritter, Cactus (Paris) No. 87: 11, 1966. *Nom. inval.* (Art. 37.1). [RPS 17]
Trichocereus chalaensis Rauh & Backeberg, in Backeberg, Descr. Cact. Nov. [1:] 20, 1957. [Dated 1956, published 1957.] [RPS 7]
Trichocereus chilensis var. **australis** Ritter, Kakt. Südamer. 2: 451-452, 1980. [RPS 31]
Trichocereus chilensis var. **borealis** Ritter, Kakt. Südamer. 3: 1109, 1980. [RPS 31]
Trichocereus chilensis var. **conjungens** Ritter, Kakt. Südamer. 3: 1109, 1980. [RPS 31]
Trichocereus chuquisacanus Ritter, Cactus (Paris) No. 87: 12, 1966. *Nom. inval.* (Art. 37.1). [RPS 17]
Trichocereus clavatus Ritter, Kakt. Südamer. 2: 564, 1980. [RPS 31]
Trichocereus conaconensis Cardenas, Fuaux

Herb. Bull. 5: 24, 1953. [RPS 4]

Trichocereus courantii (Schumann) Backeberg, Cactus (Paris) No. 41: 107, 1954. Basionym: *Cereus candicans* var. *courantii*. [RPS 5]

Trichocereus crassicostatus Ritter, Cactus (Paris) No. 87: 13-14, 1966. *Nom. inval.* (Art. 37.1). [RPS 17]

Trichocereus cuzcoensis var. **knuthianus** (Backeberg) Ritter, Backeberg's Descr. & Erört. taxon. nomenkl. Fragen, [unpaged], 1958. Basionym: *Trichocereus knuthianus*. [RPS 15]

Trichocereus eremophilus Ritter, Kakt. Südamer. 2: 559, 1980. [RPS 31]

Trichocereus escayachensis Cardenas, Cact. Succ. J (US) 35: 157, 1963. [RPS 14]

Trichocereus fabrisii Kiesling, Hickenia 1: 29-30, 1976. [RPS 27]

Trichocereus formosus (Pfeiffer) Ritter, Kakt. Südamer. 2: 450, 1980. Basionym: *Echinocactus formosus*. [RPS 31]

Trichocereus fulvilanus Ritter, Kakt. and. Sukk. 13(10): 165-167, ill., 1962. Typus: *Ritter* 263 (U, ZSS). [Chile] [RPS 13]

Trichocereus glaucus Ritter, Kakt. and. Sukk. 13(11): 180-181, ill., 1962. Typus: *Ritter* 270 p.p. (U, ZSS). [Peru: Arequipa] [RPS 13]

Trichocereus glaucus fa. **pendens** Ritter, Kakt. and. Sukk. 13: 181, 1962. [RPS 13]

Trichocereus glaucus var. **pendens** (Ritter) Backeberg, Kakt.-Lex., 437, 1966. Basionym: *Trichocereus glaucus* fa. *pendens*. [RPS 17]

Trichocereus grandiflorus Backeberg, Kakt.-Lex., 438, 1966. *Nom. inval.* (Art. 37.1). [Erroneously included as valid in RPS 17.] [RPS 17]

Trichocereus grandis (Britton & Rose) Ritter, Kakt. Südamer. 2: 450, 1980. Basionym: *Lobivia grandis*. [RPS 31]

Trichocereus herzogianus Cardenas, Fuaux Herb. Bull. 5: 19, 1953. *Cardenas* 4826 (BOLV). [Bolivia: Prov. Loaiza] [RPS 4]

Trichocereus herzogianus var. **totorensis** Cardenas, Fuaux Herb. Bull. 5: 21, 1953. [RPS 4]

Trichocereus huascha var. **pecheretianus** (Backeberg) Kiesling, Darwiniana 21(2-4): 299, 1978. Basionym: *Helianthocereus pecheretianus*. [RPS 29]

Trichocereus ingens (Britton & Rose *ex* Backeberg) Ritter, Kakt. Südamer. 2: 451, 1980. Basionym: *Soehrensia ingens*. [RPS 31]

Trichocereus korethroides (Werdermann) Ritter, Kakt. Südamer. 2: 451, 1980. Basionym: *Echinopsis korethroides*. [RPS 31]

Trichocereus lobivioides Gräser & Ritter *ex* Ritter, Kakt. Südamer. 2: 444, 1980. [RPS 31]

Trichocereus manguinii Backeberg, Cactus (Paris) No. 35: 147, 1953. [RPS 4]

Trichocereus marzinzigianus Marzinzig, Stachelpost 10(49): 8-9, ills., 1974. *Nom. inval.* (Art. 36.1, 37.1). [RPS 25]

Trichocereus narvaecensis Cardenas, Fuaux Herb. Bull. 5: 25, 1953. [RPS 4]

Trichocereus neolamprochlorus Backeberg, Die Cact. 2: 1126, 1959. *Nom. inval.* (Art. 36.1). [*Nom. nov. pro Echinocereus lamprochlorus* sensu Rümpler, etc.] [RPS 10]

Trichocereus nigripilis (Philippi) Backeberg, Die Cact. 2: 1145, 1959. Basionym: *Cereus nigripilis*. [RPS 10]

Trichocereus orurensis Cardenas, Fuaux Herb. Bull. 5: 13, 1953. Typus: *Cardenas* 4824 (BOLV). [Bolivia: Prov. El Cercado] [RPS 4]

Trichocereus orurensis var. **albiflorus** Cardenas, Fuaux Herb. Bull. 5: 16, 1953. [RPS 4]

Trichocereus pachanoi fa. **peruvianus** (Britton & Rose) Ritter, Kakt. Südamer. 4: 1324, 1981. Basionym: *Trichocereus peruvianus*. [RPS 32]

Trichocereus pasacana var. *albicephalus* Y. Ito, Cacti, 143, 1952. *Nom. inval.* (Art. 36.1). [Published as *nomen nudum*.] [RPS 4]

Trichocereus poco var. **albiflorus** Cardenas, Fuaux Herb. Bull. 5: 12, 1953. [RPS 4]

Trichocereus poco var. **fricianus** Cardenas, Fuaux Herb. Bull. 5: 11, 1953. [RPS 4]

Trichocereus pseudocandicans (Backeberg) Kiesling, Hickenia 1: 32-34, 1976. *Nom. inval.*, based on *Helianthocereus pseudocandicans, nom. inval.* (Art. 37.1). [Erroneously included as valid in RPS 27.] [RPS 34]

Trichocereus puquiensis Rauh & Backeberg, in Backeberg, Descr. Cact. Nov. [1:] 20, 1957. [Dated 1956, published 1957.] [RPS 7]

Trichocereus quadratiumbonatus Ritter, Kakt. Südamer. 2: 566, 1980. [Erroneously given as 'quadriumbonatus' in RPS 31.] [RPS 31]

Trichocereus randallii Cardenas, Cact. Succ. J. (US) 1963: 158, 1963. [RPS 14]

Trichocereus riomizquensis Ritter, Kakt. Südamer. 2: 563-564, 1980. [RPS 31]

Trichocereus rowleyi (Friedrich) Kiesling, Darwiniana 21(2-4): 295-296, 1978. Basionym: *Echinopsis rowleyi*. [RPS 29]

Trichocereus rubinghianus Backeberg, Descr. Cact. Nov. 3: 15, 1963. *Nom. inval.* (Art. 9.5). [Erroneously included as valid in RPS 14.] [RPS 14]

Trichocereus santaensis Rauh & Backeberg, in Backeberg, Descr. Cact. Nov. [1:] 20, 1957. [Dated 1956, published 1957.] [RPS 7]

Trichocereus santiaguensis (Spegazzini) Backeberg, Die Cact. 2: 1107, 1959. Basionym: *Cereus santiaguensis*. [RPS 10]

Trichocereus schoenii Rauh & Backeberg, in Rauh, Beitr. Kennt. Peruan. Kakt.veg., 362, 1958. [RPS 10]

Trichocereus scopulicola Ritter, Cactus (Paris) No. 87: 14-15, 1966. *Nom. inval.* (Art. 37.1). [Erroneously included as valid in RPS 17.] [RPS 17]

Trichocereus serenanus Ritter, Kakt. and. Sukk. 16: 212, 1965. [Based in part on *Cereus nigripilis* Philippi. Validity of name not assessed.] [RPS 16]

Trichocereus skottsbergii Backeberg, Acta Hort. Gotob. 18: 146, 1951. [RPS 2]

Trichocereus skottsbergii var. **breviatus** Backeberg, Descr. Cact. Nov. [1:] 20, 1957. [Dated 1956, published 1957.] [RPS 7]

Trichocereus smrzianus (Backeberg)

Backeberg, Kakt.-Lex., 440, 1966. Basionym: *Echinopsis smrziana.* [RPS 17]

Trichocereus spachianus cv. **Hungaria** Konder, Kakt. and. Sukk. 18: 132-133, 1967. [= *Cereus hungaricus* Földi 1936.] [RPS 18]

Trichocereus spinibarbis (Otto *ex* Pfeiffer) Ritter, Kakt. and. Sukk. 16: 212, 1965. Basionym: *Cereus spinibarbis.* [RPS 16]

Trichocereus strigosus var. **flaviflorus** Ritter, Kakt. Südamer. 2: 440, 1980. Typus: *Ritter* 999 (U, ZSS). [Holotype cited for U l.c. 1: iii, 1979.] [Argentina: La Rioja] [RPS 31]

Trichocereus tacaquirensis (Vaupel) Cardenas, Revista Agric. (Cochabamba) 8: 17, 1953. Basionym: *Cereus tacaquirensis.* [Cited after Backeberg, Die Cact. 2: 1107.] [RPS -]

Trichocereus tacnaensis Ritter, Kakt. Südamer. 4: 1326, 1981. Typus: *Ritter* 155a (U). [Peru: Tacna] [RPS 32]

Trichocereus taquimbalensis Cardenas, Revista Agr. Cochabamba 11(8): 16, 1953. [RPS 4]

Trichocereus taquimbalensis var. **wilkeae** Backeberg, Descr. Cact. Nov. [1:] 20, 1957. [Dated 1956, published 1957.] [RPS 7]

Trichocereus taratensis Cardenas, Nation. Cact. Succ. J. 21: 14-15, 1966. [RPS 17]

Trichocereus tarijensis var. **densispinus** Ritter, Kakt. Südamer. 2: 449, 1980. [RPS 31]

Trichocereus tarijensis var. **orurensis** (Cardenas) Ritter, Kakt. Südamer. 2: 560, 1980. Basionym: *Trichocereus orurensis.* [RPS 31]

Trichocereus tarijensis var. **poco** (Backeberg) Ritter, Kakt. Südamer. 2: 560, 1980. Basionym: *Trichocereus poco.* [RPS 31]

Trichocereus tarijensis var. **totorillanus** Ritter, Cactus (Paris) 22(88): 26, ill., 1967. Typus: *Ritter* 851 (U). [Erroneously ommitted from all volumes of RPS.] [Bolivia: Prov. Mendez] [RPS -]

Trichocereus tarmaensis Rauh & Backeberg, in Backeberg, Descr. Cact. Nov. [1:] 20, 1957. [Dated 1956, published 1957.] [RPS 7]

Trichocereus tenuispinus Ritter, Cactus (Paris) 22(88): 27-28, ill., 1967. Typus: *Ritter* 616 (U). [Erroneously ommitted from all volumes of RPS. The collection number of the type is taken from Kakt. Südamer. 2: 564-565, 1980.] [Bolivia: Tarija] [RPS -]

Trichocereus tenuispinus var. **pajonalensis** Ritter, Cactus (Paris) 22(88): 27-28, ill., 1967. Typus: *Ritter* 866 (U). [Erroneously ommitted from all volumes of RPS.] [Bolivia: O'Connor] [RPS -]

Trichocereus terscheckii var. **montanus** Backeberg, Cact. Succ. J (US) 23(2): 45, 1951. [RPS 3]

Trichocereus terscheckioides Ritter, Kakt. Südamer. 2: 446, 1980. Based on *Ritter* 17. *Nom. inval.* (Art. 37.1). [Based on two syntypes, *Ritter* 17 and *Ritter* 993. Erroneously included as valid in RPS 31.] [Argentina: La Rioja] [RPS 31]

Trichocereus torataensis Ritter, Kakt. Südamer. 4: 1325-1326, 1981. Typus: *Ritter* 1467 (U). [Peru: Moquegua] [RPS 32]

Trichocereus totorensis (Cardenas) Ritter, Kakt. Südamer. 2: 562, 1980. Basionym: *Trichocereus herzogianus* var. *totorensis.* [RPS 31]

Trichocereus trichosus Cardenas, Cactus (Paris) No. 57: 249-251, 1957. [RPS 8]

Trichocereus tulhuayacensis Ochoa, Kakt. and. Sukk. 8: 106-107, 1957. Typus: *Ochoa* s.n. (ubi ?). [Repeated in Biota 2: 199-201, 1959 (cf. RPS 10).] [?] [RPS 8]

Trichocereus tunariensis Cardenas, Cactus (Paris) 14(64): 160, 1959. [RPS 10]

Trichocereus uebelmannianus (Lembcke & Backeberg) Ritter, Kakt. Südamer. 2: 451, 3: 1116-1117, 1980. *Nom. inval.*, based on *Soehrensia uebelmanniana, nom. inval.* (Art. 37.1). [Erroneously included as valid in RPS 31, and as a new and valid taxon by Eggli in Trop. subtrop. Pfl.welt 59: 120, 1987.] [RPS 31]

Trichocereus validus (Monville *ex* Salm-Dyck) Backeberg, Cactus (Paris) No. 46-47: 265-267, 1955. Basionym: *Echinopsis valida.* [RPS 6]

Trichocereus vasquezii Rausch, Kakt. and. Sukk. 25(9): 193-194, ill., 1974. Typus: *Rausch* 619 (W [not found], ZSS). [Bolivia] [RPS 25]

Trichocereus vatteri Kiesling, Hickenia 1: 31-32, 1976. [RPS 27]

Trichocereus volcanensis Ritter, Kakt. Südamer. 2: 448-449, 1980. Typus: *Ritter* 400 (U, ZSS). [Type cited for U l.c. 1: iii, 1979.] [Argentina: Jujuy] [RPS 31]

× *Trichoechinopsis imperialis* Hummel, in Backeberg, Die Cact. 2: 1283, 1959. *Nom. inval.* (Art. 36.1). [= *Echinopsis eyriesii* × *Trichocereus peruvianus vel aff.* ?] [RPS 10]

× **Trichopsis** Ito, The Cactaceae, 638, 1981. [= *Trichocereus* × *Echinopsis.*] [RPS -]

Turbinicarpus flaviflorus G. Frank & Lau, Kakt. and. Sukk. 30(1): 6-7, ill., 1979. Typus: *Lau* 1185 (ZSS). [Collection number not given in original publication, cited according to specimens at ZSS.] [Mexico: San Luis Potosi] [RPS 30]

Turbinicarpus gielsdorfianus (Werdermann) John & Riha, Kaktusy 19(1): 22, 1983. Basionym: *Echinocactus gielsdorfianus.* [Validation of a combination published already in Kaktusy 17(1): 17, but transgressing Art. 33.2.] [RPS 34]

Turbinicarpus gracilis Glass & Foster, Cact. Succ. J. (US) 48(4): 176-177, ill., 1976. Typus: *Glass et Foster* 3182 (ZSS). [Mexico: Nuevo Leon] [RPS 27]

Turbinicarpus horripilus (Lemaire) John & Riha, Kaktusy 19(1): 22, 1983. Basionym: *Mammillaria horripila.* [Validation of a combination published already in Kaktusy 17(1): 17, but transgressing Art. 33.2.] [RPS 34]

Turbinicarpus knuthianus (Bödeker) John & Riha, Kaktusy 19(1): 22, 1983. Basionym: *Echinocactus knuthianus.* [Validation of a combination published already in Kaktusy 17(1): 17, but transgressing Art. 33.2.] [RPS 34]

Turbinicarpus krainzianus (G. Frank) Backeberg, Die Cact. 5: 2890, 1961. Basionym: *Toumeya krainziana.* [RPS 12]

Turbinicarpus krainzianus fa. **minimus** G.

Frank, Succulenta 68(2): 40-41, (12): 272, ills., 1989. Typus: *Bonatz* s.n. (WU). [Sphalm. 'minima'.] [Mexico: Hidalgo] [RPS 40]

Turbinicarpus krainzianus var. **minimus** (G. Frank) Diers, Succulenta 69(12): 269, 1990. Basionym: *Turbinicarpus krainzianus* fa. *minimus*. [RPS 41]

Turbinicarpus laui Glass & Foster, Cact. Succ. J. (US) 43(3): 116-119, 1975. [RPS 27]

Turbinicarpus polaskii Backeberg, Die Cact. 5: 2883-2885, 1961. *Nom. inval.* (Art. 37.1). [RPS 12]

Turbinicarpus pseudomacrochele var. **krainzianus** (G. Frank) Glass & Foster, Cact. Succ. J. (US) 49(4): 173, 1977. Basionym: *Toumeya krainziana*. [RPS 28]

Turbinicarpus pseudopectinatus (Backeberg) Glass & Foster, Cact. Succ. J. (US) 49(4): 175, 1977. Basionym: *Pelecyphora pseudopectinata*. [RPS 28]

Turbinicarpus roseiflorus Backeberg, Descr. Cact. Nov. 3: 15, 1963. *Nom. inval.* (Art. 9.5). [Erroneously included as valid in RPS 14.] [RPS 14]

Turbinicarpus saueri (Bödeker) John & Riha, Kaktusy 19(1): 22, 1983. Basionym: *Echinocactus saueri*. [Validation of a combination published already in Kaktusy 17(1): 17, but transgressing Art. 33.2.] [RPS 34]

Turbinicarpus schmiedickeanus var. **dickisoniae** Glass & Foster, Cact. Succ. J. (US) 54(2): 74, 1982. Typus: *Dickison* s.n. (POM). [RPS 33]

Turbinicarpus schmiedickeanus var. **flaviflorus** (Frank & Lau) Glass & Foster, Cact. Succ. J. (US) 51(3): 123, ill., 1979. Basionym: *Turbinicarpus flaviflorus*. [RPS 30]

Turbinicarpus schmiedickeanus var. **gracilis** (Glass & Foster) Glass & Foster, Cact. Succ. J. (US) 49(4): 167, 1977. Basionym: *Turbinicarpus gracilis*. [RPS 28]

Turbinicarpus schmiedickeanus var. **klinkerianus** (Backeberg & Jacobsen) Glass & Foster, Cact. Succ. J. (US) 49(4): 168, 1977. Basionym: *Turbinicarpus klinkerianus*. [RPS 28]

Turbinicarpus schmiedickeanus var. **macrochele** (Werdermann) Glass & Foster, Cact. Succ. J. (US) 49(4): 168, 1977. Basionym: *Echinocactus macrochele*. [RPS 28]

Turbinicarpus schmiedickeanus var. **schwarzii** (Shurly) Glass & Foster, Cact. Succ. J. (US) 49(4): 169, 1977. Basionym: *Strombocactus schwarzii*. [RPS 28]

Turbinicarpus schwarzii (Shurly) Backeberg, Die Cact. 5: 2887, 1961. Basionym: *Strombocactus schwarzii*. [Already published as *comb. nud.* in Cact. Succ. J. (US) 23(5): 150, 1951 by Backeberg.] [RPS -]

Turbinicarpus swobodae L. Diers, Kakt. and. Sukk. 38(4): 86-91, ills., SEM-ills., 1987. Typus: *Swoboda* s.n. (KOELN). [Mexico: Nuevo Leon] [RPS 38]

Turbinicarpus valdezianus (H. Möller) Glass & Foster, Cact. Succ. J. (US) 49(4): 174, 1977. Basionym: *Pelecyphora valdeziana*. [RPS 28]

Turbinicarpus viereckii (Werdermann) John & Riha, Kaktusy 19(1): 22, 1983. Basionym: *Echinocactus viereckii*. [Validation of a combination published already in Kaktusy 17(1): 17, but transgressing Art. 33.2.] [RPS 34]

Turbinicarpus viereckii var. **major** (Glass & Foster) John & Riha, Kaktusy 19(1): 22, 1983. Basionym: *Gymnocactus viereckii* var. *major*. [Validation of a combination already published in Kaktusy 19(1): 22, but transgressing Art. 33.2.] [RPS 34]

Turbinicarpus ysabelae (K. Schlange) John & Riha, Kaktusy 19(1): 22, 1983. Basionym: *Thelocactus ysabelae*. [Validation of a combination already published in Kaktusy 17(1): 17, but transgressing Art. 33.2.] [RPS 34]

Turbinicarpus ysabelae var. **brevispinus** (K. Schlange) John & Riha, Kaktusy 19(1): 22, 1983. Basionym: *Thelocactus ysabelae* var. *brevispinus*. [Validation of a combination already published in Kaktusy 17(1): 17, but transgressing Art. 33.2.] [RPS 34]

Uebelmannia Buining, Succulenta (NL) 46: 159-163, 1967. Typus: *U. gummifera*. [RPS 18]

Uebelmannia buiningii Donald, Nation. Cact. Succ. J. 23(1): 2-3, ills., 1968. Typus: *Horst et Uebelmann* HU 141 (1966) (U, ZSS). [Brazil: Minas Gerais] [RPS 19]

Uebelmannia centeteria (Pfeiffer) Schnabel, Kakt. and. Sukk. 22: 164-167, 1971. Basionym: *Echinocactus centeterius*. [RPS 22]

Uebelmannia flavispina Buining & Brederoo, Succulenta 52(1): 9-10, ill., 1973. Typus: *Horst et Uebelmann* HU 361 (U, ZSS). [Brazil: Minas Gerais] [RPS 24]

Uebelmannia gummifera (Backeberg & Voll) Buining, Succulenta 46: 161, 1967. Basionym: *Parodia gummifera*. [RPS 18]

Uebelmannia gummifera fa. *cristata* Buining, Succulenta 48: 180-181, 1969. *Nom. inval.* (Art. 37.1). [RPS 20]

Uebelmannia meninensis Buining, Kakt. and. Sukk. 19: 151-152, 1968. [RPS 19]

Uebelmannia meninensis var. **rubra** Buining & Brederoo, in Krainz, Die Kakt. Lief. 55-56: C VIe, 1974. [RPS 25]

Uebelmannia pectinifera Buining, Nation. Cact. Succ. J. 22: 65, 86-87, 1967. [RPS 18]

Uebelmannia pectinifera var. **horrida** P. J. Braun, Kakt. and. Sukk. 35(12): 264-266, ills., 1984. Typus: *Horst et Uebelmann* HU 550 (1982) (ZSS). [Brazil: Minas Gerais] [RPS 35]

Uebelmannia pectinifera var. **multicostata** Buining & Brederoo, in Krainz, Die Kakt. Lief. 62: C VIe, ill., 1975. Typus: *Horst et Uebelmann* HU 362 (U, ZSS). [Brazil: Minas Gerais] [RPS 27]

Uebelmannia pectinifera var. **pseudopectinifera** Buining, Kakt. and. Sukk. 23(5): 125-126, ills., 1972. Typus: *Horst et Uebelmann* HU 280 (U, ZSS). [Brazil: Minas Gerais] [RPS 23]

Utahia peeblesiana (Croizat) Kladiwa, in Krainz, Die Kakt. C VIIIb, 1969. Basionym: *Navajoa peeblesiana.* [RPS 20]
Vatricania Backeberg, Cact. Succ. J. (US) 22(5): 154, 1950. Typus: *Cephalocereus guentheri.* [RPS 1]
Vatricania guentheri (Kupper) Backeberg, Cact. Succ. J. (US) 23: 149, 1951. Basionym: *Cephalocereus guentheri.* [RPS 9]
Weberbauerocereus albus Ritter, Kakt. and. Sukk. 13(6): 106-108, ill., 1962. Typus: *Ritter* 571 (U, ZSS). [Peru: Ancash] [RPS 13]
Weberbauerocereus cephalomacrostibas (Werdermann & Backeberg) Ritter, Kakt. Südamer. 4: 1353-1354, 1981. Basionym: *Cereus cephalomacrostibas.* [RPS 32]
Weberbauerocereus churinensis Ritter, Kakt. and. Sukk. 13: 133, 1962. [RPS 13]
Weberbauerocereus cuzcoensis Knize, Biota (Lima) 7(57): 256, 1968. [Publication date reported as 1969 in RPS 25, but publication is dated 1968.] [RPS 32]
Weberbauerocereus cuzcoensis var. **tenuiarboreus** Ritter, Kakt. Südamer. 4: 1358-1359, 1981. Typus: *Ritter* 662 (U). [Peru: Ayacucho] [RPS 32]
Weberbauerocereus fascicularis var. **horridispinus** (Rauh & Backeberg) Ritter, Backeberg's Descr. & Erört. taxon. nomenkl. Fragen, [unpaged], 1958. Basionym: *Weberbauerocereus horridispinus.* [RPS 15]
Weberbauerocereus horridispinus Rauh & Backeberg, in Backeberg, Descr. Cact. Nov. [1:] 27, 1957. Typus: *Rauh* K125 (1956) (ZSS [status ?]). [Dated 1956, published 1957.] [Peru] [RPS 7]
Weberbauerocereus johnsonii Ritter, Kakt. and. Sukk. 13(5): 72-74, ill., 1962. Typus: *Ritter* 570 (U, ZSS). [Peru: Cajamarca] [RPS 13]
Weberbauerocereus longicomus Ritter, Kakt. and. Sukk. 13: 117, 1962. [RPS 13]
Weberbauerocereus rauhii Backeberg, Descr. Cact. Nov. [1:] 27, 1957. Typus: *Rauh* K107 (1956) (ZSS). [Dated 1956, published 1957.] [Peru] [RPS 7]
Weberbauerocereus rauhii var. **laticornuus** Rauh, Beitr. Kennt. Peruan. Kakt.veg., 460, 1958. [Sphalm. 'laticornua'.] [RPS 10]
Weberbauerocereus seyboldianus Rauh & Backeberg, in Backeberg, Descr. Cact. Nov. [1:] 27, 1957. *Rauh* K148 (1956) (ubi ?). [Dated 1956, published 1957.] [Peru] [RPS 7]
Weberbauerocereus torataensis Ritter, Kakt. Südamer. 4: 1357, 1981. Typus: *Ritter* 194 (U, ZSS). [Peru: Moquegua] [RPS 32]
Weberbauerocereus weberbaueri var. **aureifuscus** Rauh & Backeberg, in Backeberg, Descr. Cact. Nov. [1:] 27, 1957. [Dated 1956, published 1957.] [RPS 7]
Weberbauerocereus weberbaueri var. **horribilis** Rauh & Backeberg, in Backeberg, Descr. Cact. Nov. [1:] 27, 1957. [Dated 1956, published 1957.] [RPS 7]
Weberbauerocereus weberbaueri var. **horridispinus** (Backeberg) Ritter, Kakt. Südamer. 4: 1356, 1981. Basionym: *Weberbauerocereus horridispinus.* [RPS 32]
Weberbauerocereus weberbaueri var. **humilior** Rauh & Backeberg, in Backeberg, Descr. Cact. Nov. [1:] 27, 1957. [Dated 1956, published 1957.] [RPS 7]
Weberbauerocereus winterianus Ritter, Kakt. and. Sukk. 13(4): 54, 1962. Typus: *Ritter* 165 (U, ZSS). [Peru: La Libertad] [RPS 13]
Weberbauerocereus winterianus var. **australis** Ritter, Kakt. and. Sukk. 13: 56, 1962. [RPS 13]
Weberocereus bradei (Britton & Rose) Rowley, Rep. Pl. Succ. 23(1972): 10, 1974. Basionym: *Eccremocactus bradei.* [Combination repeated by D. Hunt in Kew Mag. 2(4): 341, 1985.] [RPS 36]
Weberocereus glaber (Eichlam) Rowley, Nation. Cact. Succ. J. 37(2): 46, 1982. Basionym: *Cereus glaber.* [Combination repeated by D. Hunt in Kew Mag. 2(4): 341, 1985.] [RPS 33]
Weberocereus glaber var. **mirandae** (Bravo) Eliasson, Kaktus (Odense) 21(2): 44, 1986. Basionym: *Selenicereus mirandae.* [Erroneously ascribed to Kimnach in RPS 37: 7.] [RPS 37]
Weberocereus imitans (Kimnach & P. C. Hutchison) Buxbaum, Succulenta 57(6): 125, 1978. Basionym: *Werckleocereus imitans.* [Combination repeated by D. Hunt in Kew Mag. 2(4): 341, 1985.] [RPS 36]
Weberocereus rosei (Kimnach) Buxbaum, Succulenta 57(6): 125, 1978. Basionym: *Eccremocactus rosei.* [Repeated by Eliasson in Kaktus (Odense) 21(2): 44, 1986 (see RPS 29: 6). Combination also attributed to G. Rowley without data.] [RPS 37]
Weberocereus tonduzii (F. A. C. Weber) Rowley, Nation. Cact. Succ. J. 37(2): 46, 1982. Basionym: *Cereus tonduzii.* [Combination repeated by D. Hunt in Kew Mag. 2(4): 341, ill., 1985.] [RPS 33]
Weberocereus trichophorus Johnson & Kimnach, Cact. Succ. J. (US) 35: 203, 1963. [RPS 14]
Weingartia ser. **Nudispermae** F. Brandt, Kakt. Orch.-Rundschau 8(1): 6, 1983. Typus: *W. pulquinensis.* Epithet given as 'Nuduspermae', an orthographical error inadvertently not corrected in RPS 34. [RPS 34]
Weingartia ser. **Pilosocarpae** F. Brandt, Kakt. Orch.-Rundschau 8(1): 7, 1983. Typus: *W. haseltonii.* [RPS 34]
Weingartia ser. **Tectospermae** F. Brandt, Kakt. Orch.-Rundschau 8(1): 6-7, 1983. Typus: *W. aglaia.* [RPS 34]
Weingartia subgen. **Cumingia** F. Brandt, Kakt. Orch.-Rundschau 8(1): 5-6, 1983. Typus: *W. pulquinensis.* [RPS 34]
Weingartia subgen. *Spegazzinia* (Backeberg) F. Brandt, Kakt. Orch.-Rundschau 1977(5): 69, 1977. *Nom. inval.* (Art. 21). [*Spegazzinia* (*nom. inval.*, Art. 64) is based on the same type as *Weingartia.*] [RPS 28]
Weingartia subgen. **Sulcorebutia** (Backeberg) F. Brandt, Kakt. Orch.-Rundschau 1977(5): 69, 1977. Basionym: *Sulcorebutia.* [RPS 28]

Weingartia aglaia F. Brandt, Cactus 10(3): 54-56, 1978. [RPS 29]

Weingartia alba (Rausch) F. Brandt, Frankfurter Kakt.-Freund 5(2): 17, 1978. Basionym: *Sulcorebutia alba.* [RPS 29]

Weingartia albaoides F. Brandt, Letzeb. Cacteefren 4(8): 2-8, 1983. Typus: *Knize* 1266 (HEID). [Type ex cult. F. Brandt 102/a.] [Bolivia] [RPS 34]

Weingartia albaoides ssp. **subfusca** F. Brandt, Letzeb. Cacteefren 4(11): 1-7, ills., 1983. Typus: *Knize* 1255 (HEID). [Type ex cult. F. Brandt 102/b.] [Bolivia] [RPS 34]

Weingartia albissima F. Brandt, Kakt. Orch. Rundschau 5(1): 1-4, 1980. [RPS 31]

Weingartia ambigua (Hildmann) Backeberg, Cact. Succ. J. (US) 23(3): 86, 1951. Basionym: *Neoporteria ambigua.* [The basionym given is already a combination.] [RPS 3]

Weingartia ansaldoensis F. Brandt, Kakt. Orch.-Rundschau 10(3): 29-32, ills., 1985. Typus: *Anonymus* s.n. (HEID). [Ex cult. F. Brandt acc. no. 107/a (this number is also given for the type of *Parodia tojoensis* !).] [RPS 36]

Weingartia arenacea (Cardenas) F. Brandt, Kakt. Orch.-Rundschau 1979(5): 69, 1979. Basionym: *Rebutia arenacea.* [RPS 30]

Weingartia attenuata F. Brandt, Kakt. Orch.-Rundschau 10(1): 1-3, ill., 1985. Typus: *Knize* s.n. (HEID). [ex cult. Brandt acc. no. 105/a.] [RPS 36]

Weingartia aureispina (Rausch) F. Brandt, Frankfurter Kakt.-Freund 7(1): 113-115, 1980. Basionym: *Sulcorebutia frankiana* var. *aureispina.* [RPS 31]

Weingartia backebergiana F. Brandt, Kakt. Orch.-Rundschau 2(5): 68-71, 1977. Basionym: *Sulcorebutia steinbachii* var. *horrida.* [RPS 28]

Weingartia brachygraphisa F. Brandt, Kaktus 12(1): 12-14, 1977. [RPS 28]

Weingartia breviflora (Backeberg) F. Brandt, Frankfurter Kakt.-Freund 5(2): 17, 1978. Basionym: *Sulcorebutia breviflora.* [RPS 29]

Weingartia brevispina F. Brandt, Kakt. Orch.-Rundschau 5(2): 13-15, 1980. [RPS 31]

Weingartia buiningiana Ritter, Kakt. Südamer. 2: 659, 1980. [RPS 31]

Weingartia caineana (Cardenas) F. Brandt, Kakt. Orch.-Rundschau 5(1): 5, 1980. Basionym: *Rebutia caineana.* [RPS 31]

Weingartia candiae (Cardenas) F. Brandt, Kakt. Orch.-Rundschau 177(5): 69, 1977. Basionym: *Rebutia candiae.* [RPS 28]

Weingartia canigueralii (Cardenas) F. Brandt, Frankfurter Kakt.-Freund 5(2): 17, 1978. Basionym: *Rebutia canigueralii.* [RPS 29]

Weingartia caracarensis (Cardenas) F. Brandt, Kakt. Orch.-Rundschau 5(1): 5, 1980. Basionym: *Rebutia caracarensis.* [RPS 31]

Weingartia cardenasiana (Vasquez) F. Brandt, Frankfurter Kakt.-Freund 5(2): 17, 1978. Basionym: *Sulcorebutia cardenasiana.* [RPS 29]

Weingartia chilensis (Backeberg) Backeberg, Descr. Cact. Nov. 3: 15, 1963. Basionym: *Neowerdermannia chilensis.* [RPS 14]

Weingartia cintiensis Cardenas, Revista Agric. 10: 9-10, 1958. [RPS 9]

Weingartia clavata F. Brandt, Kakt. Orch.-Rundschau 1979(2): 16-19, 1979. [RPS 30]

Weingartia columnaris F. Brandt, Kakt. Orch.-Rundschau 11(2-3): 18, 1986. Typus: *Knize* s.n. (HEID). [ex cult. F. Brandt no. 109/a.] [RPS 37]

Weingartia corroana (Cardenas) Cardenas, Cactus (Paris) No 82: 49, 1964. Basionym: *Weingartia pulquinensis* var. *corroana.* [RPS 15]

Weingartia crispata (Rausch) F. Brandt, Frankfurter Kakt.-Freund 5(2): 18, 1978. Basionym: *Sulcorebutia crispata.* [RPS 29]

Weingartia croceareolata F. Brandt, Kakt. Orch.-Rundschau 9(4): 74-75, ill., 1984. Typus: *Rausch* 476 (HEID). [ex cult. F. Brandt acc. no. 98/a.] [Bolivia] [RPS 35]

Weingartia cylindrica (Donald) F. Brandt, Kakt. Orch.-Rundschau, Sonderheft S2: 40, 1978. Basionym: *Sulcorebutia cylindrica.* [RPS 29]

Weingartia electracantha (Backeberg) F. Brandt, Kakt. Orch.-Rundschau 7(5): 77, 1982. *Nom. inval.*, based on *Sulcorebutia electracantha, nom. inval.* (Art. 9.5). [RPS 33]

Weingartia erinacea Ritter, Cact. Succ. J. Gr. Brit. 23: 8, 1961. [RPS 12]

Weingartia erinacea var. **catarirensis** Ritter, Cact. Succ. J. Gt. Brit. 23: 9, 1961. [RPS 12]

Weingartia fidaiana ssp. **cintiensis** (Cardenas) Donald, Ashingtonia 3: 139, 1980. Basionym: *Weingartia cintiensis.* [RPS 31]

Weingartia flavissima (Rausch) F. Brandt, Kakt. Orch.-Rundschau 1979(5): 69, 1979. Basionym: *Sulcorebutia flavissima.* [RPS 30]

Weingartia formosa F. Brandt, Kakt. Orch.-Rundschau 1979(4): 46-49, 1979. [RPS 30]

Weingartia frankiana (Rausch) F. Brandt, Kakt. Orch.-Rundschau 5: 95, 1978. Basionym: *Sulcorebutia frankiana.* [RPS 29]

Weingartia glomeriseta (Cardenas) F. Brandt, Frankfurter Kakt.-Freund 4(3): 12-13, 1977. Basionym: *Rebutia glomeriseta.* [RPS 28]

Weingartia glomerispina (Cardenas) F. Brandt, Frankfurter Kakt.-Freund 5(2): 18, 1978. Basionym: *Rebutia glomerispina.* [RPS 29]

Weingartia gracilispina Ritter, Kakt. Südamer. 2: 658, 1980. [RPS 31]

Weingartia haseltonii (Cardenas) F. Brandt, Kakt. Orch.-Rundschau 5: 95, 1978. Basionym: *Rebutia haseltonii.* [RPS 29]

Weingartia hediniana Backeberg, Kakt. and. Sukk. 2: 2-3, 1950. [RPS 1]

Weingartia hoffmanniana F. Brandt, Letzeb. Cacteefrenn 5(11): 231-237, ills., 1984. Based on *Anonymus* s.n.. *Nom. inval.* (Art. 9.5). [Based on *Lobivia hoffmanniana* Backeberg, *nom. inval.* (Art. 9.5). First published by F. Brandt as a combination in Kakt. Orch.-Rundschau 5(1): 5, 1980 (RPS 31).] [RPS 35]

Weingartia inflexiseta (Cardenas) F. Brandt, Kakt. Orch.-Rundschau 5(1): 5, 1980. Basionym: *Rebutia inflexiseta.* [RPS 31]

Weingartia kargliana Rausch, Kakt. and. Sukk.

30(5): 105-106, ill., 1979. Typus: *Rausch* 677 (ZSS). [Bolivia] [RPS 30]

Weingartia knizei F. Brandt, Frankfurter Kakt.-Freund 4(3): 6-7, 1977. [RPS 28]

Weingartia krahnii (Rausch) F. Brandt, Kakt. Orch.-Rundschau, Sonderheft S2: 40, 1978. Basionym: *Sulcorebutia krahnii*. [RPS 29]

Weingartia krugeri (Cardenas) F. Brandt, Kakt. Orch.-Rundschau 177(5): 70, 1977. Basionym: *Aylostera krugeri*. [RPS 28]

Weingartia lanata Ritter, Nation. Cact. Succ. J. 16(1): 7-8, ill., 1961. Typus: *Ritter* 814 (U, ZSS). [Bolivia: Chuquisaca] [RPS 12]

Weingartia lanata fa. **platygona** (Cardenas) Donald, Ashingtonia 3: 129, 1980. Basionym: *Weingartia platygona*. [RPS 31]

Weingartia lanata ssp. **longigibba** (Ritter) Donald, Ashingtonia 3: 132, 138, 1980. Basionym: *Weingartia longigibba*. [RPS 31]

Weingartia lanata ssp. **pilcomayensis** (Cardenas) Donald, Ashingtonia 3: 129, 1980. Basionym: *Weingartia pilcomayensis*. [RPS 31]

Weingartia lanata ssp. **riograndensis** (Ritter) Donald, Ashingtonia 3: 132, 1980. Basionym: *Weingartia riograndensis*. [RPS 31]

Weingartia lecoriensis Cardenas, Cactus (Paris) No. 82: 47, 1964. [RPS 15]

Weingartia lepida (Ritter) F. Brandt, Kakt. Orch.-Rundschau 5(1): 5, 1980. Basionym: *Sulcorebutia lepida*. [RPS 31]

Weingartia longigibba Ritter, Cact. Succ. J. Gr. Brit. 16: 7, 1961. [RPS 12]

Weingartia losenickyana (Rausch) F. Brandt, Frankfurter Kakt.-Freund 5(2): 18, 1978. Basionym: *Sulcorebutia losenickyana*. [RPS 29]

Weingartia mairanana F. Brandt, Kakt. Orch.-Rundschau 8(3): 51-54, 1983. Typus: *Anonymus ex cult. F. Brandt* 101/a (HEID). [RPS 34]

Weingartia markusii (Rausch) F. Brandt, Frankfurter Kakt.-Freund 5(2): 18, 1978. Basionym: *Sulcorebutia markusii*. [RPS 29]

Weingartia mataralensis F. Brandt, Kakt. Orch.-Rundschau 9(5): 89-91, ill., 1984. Typus: *Anonymus* s.n. (HEID). [ex cult. F. Brandt acc. no. 104/a.] [RPS 35]

Weingartia menesesii (Cardenas) F. Brandt, Cact. Succ. J. (US) 33: 112, 1979. Basionym: *Rebutia menesesii*. [RPS 30]

Weingartia mentosa (Ritter) F. Brandt, Kakt. Orch.-Rundschau 1979(1): 6, 1979. Basionym: *Sulcorebutia mentosa*. [RPS 30]

Weingartia minima (Rausch) F. Brandt, Kakt. Orch.-Rundschau 11(1): 2, 1986. Incorrect name (Art. 57.1), based on *Sulcorebutia taratensis* var. *minima*. [Combination is incorrect as *Rebutia taratensis* Card. is given in the synonymy.] [RPS 37]

Weingartia miranda F. Brandt, Kakt. Orch.-Rundschau 10(4-5): 41-44, ill., 1986. Typus: *Knize* s.n. (HEID). [Ex cult. F. Brandt no. 108/a.] [Volume for 1985, but Kew copy is stamped 21. 2. 1986.] [RPS 37]

Weingartia mizquensis (Rausch) F. Brandt, Kakt. Orch.-Rundschau 1979(5): 69, 1979. Basionym: *Sulcorebutia mizquensis*. [RPS 30]

Weingartia mocharasensis Rausch, Verzeichnis Feldnummern W. Rausch, [unpaged], 1975. *Nom. inval.* (Art. 36.1, 37.1). [Published as provisional name.] [RPS -]

Weingartia multispina Ritter, Nation. Cact. Succ. J. 16(1): 7, ill., 1961. Typus: *Ritter* 372 (U, ZSS). [Bolivia: Cochabamba] [RPS 12]

Weingartia muschii (Vasquez) F. Brandt, Kakt. Orch.-Rundschau 1979(1): 6, 1979. Basionym: *Sulcorebutia muschii*. [RPS 30]

Weingartia neglecta F. Brandt, Kakt. Orch.-Rundschau 8(Sonderheft 3): 31-34, ills., 1983. Typus: *Knize* s.n. (). [Type ex cult. F. Brandt 103/a.] [Bolivia] [RPS 34]

Weingartia neocumingii Backeberg, Kakt. and. Sukk. 2(1): 2, 1950. [*Nom. nov. pro Echinocactus cumingii* Salm-Dyck 1849, *non* Hopffer 1843.] [RPS 1]

Weingartia neocumingii ssp. **pulquinensis** (Cardenas) Donald, Ashingtonia 3: 136, 1980. Basionym: *Weingartia pulquinensis*. [RPS 31]

Weingartia neocumingii ssp. **sucrensis** (Ritter) Donald, Ashingtonia 3: 137, 1980. Basionym: *Weingartia sucrensis*. [RPS 31]

Weingartia neocumingii var. **corroana** (Cardenas) Backeberg, Die Cact. 3: 1792, 1959. Basionym: *Weingartia pulquinensis* var. *corroana*. [Repeated by Donald in Ashingtonia 3: 136, 1980 (and reported as correct place of publication in RPS 31).] [RPS 10]

Weingartia neocumingii var. **hediniana** (Backeberg) Donald, Nation. Cact. Succ. J. 13(3): 56, 1958. Basionym: *Weingartia hediniana*. [RPS 9]

Weingartia neocumingii var. **koehresii** Oeser, Kakt. and. Sukk. 32(12): 286-289, 1981. [RPS 32]

Weingartia neocumingii var. **mairanensis** Donald, Ashingtonia 3: 137, 1980. [RPS 31]

Weingartia neocumingii var. **multispina** (Ritter) Donald, Ashingtonia 3: 136, 1980. Basionym: *Weingartia multispina*. [RPS 31]

Weingartia neocumingii var. **trollii** (Oeser) Donald, Ashingtonia 3: 137, 1980. Basionym: *Weingartia trollii*. [RPS 31]

Weingartia neumanniana var. *aurantia* Backeberg, Descr. Cact. Nov. 3: 15, 1963. *Nom. inval.* (Art. 9.5). [Erroneously included as valid in RPS 14. Based on *Ritter 50* according to F. Ritter.] [RPS 14]

Weingartia nigro-fuscata F. Brandt, Cactus 10(6): 113-116, 1978. [RPS 29]

Weingartia oenantha (Rausch) F. Brandt, Kakt. Orch.-Rundschau 1979(5): 69, 1979. Basionym: *Sulcorebutia oenantha*. [RPS 30]

Weingartia oligacantha F. Brandt, Kakt. Orch.-Rundschau 1979(1): 2-5, 1979. [RPS 30]

Weingartia pampagrandensis (Rausch) F. Brandt, Kakt. Orch.-Rundschau 1979(5): 69, 1979. Basionym: *Sulcorebutia pampagrandensis*. [RPS 30]

Weingartia pasopayana F. Brandt, Kakt. Orch.-Rundschau 9(3): 53-55, ill., 1984. Typus: *Lau* 387 (HEID). [ex cult. F. Brandt acc. no. 97/a.] [Bolivia] [RPS 35]

Weingartia perplexiflora F. Brandt, Letzeb.

Cacteefren 3(8): 4-5, 1982. Typus: *Rausch* 566 (HEID). [RPS 33]

Weingartia pilcomayensis Cardenas, Cactus (Paris) No. 82: 44-46, 1964. [RPS 15]

Weingartia platygona Cardenas, Cactus (Paris) No. 82: 50-51, 1964. [RPS 15]

Weingartia polymorpha (Cardenas) F. Brandt, Kakt. Orch.-Rundschau 1977(5): 70, 1977. Basionym: *Rebutia polymorpha*. [RPS 28]

Weingartia pulchra (Cardenas) F. Brandt, Kakt. Orch.-Rundschau 5: 95, 1978. Basionym: *Rebutia pulchra*. [RPS 29]

Weingartia pulquinensis Cardenas, Revista Agric. Cochabamba 7(6): 26-29, 1951. [RPS 2]

Weingartia pulquinensis var. **corroana** Cardenas, Revista Agric. Cochabamba 7(6): 26-29, 1951. [RPS 2]

Weingartia pulquinensis var. **mairananensis** (Donald) Pilbeam, Sulcorebutia & Weingartia, 124, 1985. Basionym: *Weingartia neocumingii* var. *mairananensis*. [RPS 36]

Weingartia purpurea Donald & Lau, Ashingtonia 1(5): 53, 1974. [RPS 25]

Weingartia pygmaea Ritter, Kakt. Südamer. 2: 652-653, 1980. [RPS 31]

Weingartia rauschii (Frank) F. Brandt, Kakt. Orch.-Rundschau 5: 95, 1978. Basionym: *Sulcorebutia rauschii*. [RPS 29]

Weingartia riograndensis Ritter, Cact. Succ. J. Gr. Brit. 23: 9, 1961. [RPS 12]

Weingartia ritteri F. Brandt, Kakt. Orch.-Rundschau 3: 75-77, 1978. [Sphalm. 'ritterii'.] [RPS 29]

Weingartia rubro-aurea F. Brandt, Kakt. Orch.-Rundschau 9(2): 36, ills., 1984. Typus: *Rausch* 479 (HEID). [Based on the same collection as *Sulcorebutia verticillacantha* var. *aureiflora* Rausch 1972.] [Bolivia] [RPS 35]

Weingartia saetosa F. Brandt, Kakt. Orch.-Rundschau 5(4): 41-44, 1980. [RPS 31]

Weingartia saipinensis F. Brandt, Kakt. Orch.-Rundschau 7(5): 71-74, 1982. Typus: *Brandt cult.* 100 (HEID). [RPS 33]

Weingartia sanguineo-tarijensis F. Brandt, Kakt. Orch.-Rundschau 5(3): 29-32, 1980. [RPS 31]

Weingartia saxatilis F. Brandt, Frankfurter Kakteen-Freund 8(1): 201-203, 1981. [Syn. *Sulcorebutia vasqueziana* var. *albispina* Rausch in Succulenta 52: 222, 1973.] [RPS 32]

Weingartia steinbachii (Werdermann) F. Brandt, Kakt. Orch.-Rundschau 1977(5): 69, 1977. Basionym: *Rebutia steinbachii*. [Repeated in Frankfurter Kakt.-Freund. 5(2): 17-18, 1978.] [RPS 28]

Weingartia sucrensis Ritter, Nation. Cact. Succ. J. 16: 8, 1961. [RPS 12]

Weingartia tarabucina F. Brandt, Letzeb. Cacteefren 6(4): 57, 1985. Typus: *Anonymus* s.n. (HEID). [ex cult. Brandt acc. no. 106/a.] [RPS 36]

Weingartia tarabucoensis (Rausch) F. Brandt, Kakt. Orch.-Rundschau 1979(5): 69, 1979. Basionym: *Sulcorebutia tarabucoensis*. [RPS 30]

Weingartia taratensis (Cardenas) F. Brandt, Kakt. Orch.-Rundschau 1977(5): 70, 1977. Basionym: *Rebutia taratensis*. [RPS 28]

Weingartia tarijensis F. Brandt, Kakt. Orch.-Rundschau 5: 92-94, 1978. [RPS 29]

Weingartia tiraquensis (Cardenas) F. Brandt, Kakt. Orch.-Rundschau 1977(5): 69, 1977. Basionym: *Rebutia tiraquensis*. [RPS 28]

Weingartia torotorensis Cardenas, Cact. Succ. J. (US) 43: 243, 1971. [RPS 22]

Weingartia totoralensis F. Brandt, Kakt. Orch. Rundschau 7(3): 32-35, ills., 1982. Typus: *Brandt cult.* 87/a (HEID). [RPS 33]

Weingartia totorensis (Cardenas) F. Brandt, Kakt. Orch.-Rundschau 1979(5): 69, 1979. Basionym: *Rebutia totorensis*. [RPS 30]

Weingartia trollii Oeser, Kakt. and. Sukk. 29(6): 129-131, 1978. [RPS 29]

Weingartia tuberculato-chrysantha (Cardenas) F. Brandt, Kakt. Orch.-Rundschau 5(1): 5, 1980. Basionym: *Rebutia tuberculato-chrysantha*. [RPS 31]

Weingartia tunariensis (Cardenas) F. Brandt, Kakt. Orch.-Rundschau 5: 95, 1978. Basionym: *Rebutia tunariensis*. [RPS 29]

Weingartia verticillacantha (Ritter) F. Brandt, Kakt. Orch.-Rundschau 1977(5): 70, 1977. Basionym: *Sulcorebutia verticillacantha*. [RPS 28]

Weingartia vilcayensis Cardenas, Cactus (Paris) No. 82: 46, 1964. [RPS 15]

Weingartia vizcarrae (Cardenas) F. Brandt, Kakt. Orch.-Rundschau 9(3): 55, 1984. Basionym: *Rebutia vizcarrae*. [RPS 35]

Weingartia vorwerkii (Fric) Backeberg, Descr. Cact. Nov. 3: 16, 1963. Basionym: *Neowerdermannia vorwerkii*. [RPS 14]

Weingartia vorwerkii var. *erectispina* (Hoffmann & Backeberg) Backeberg, Descr. Cact. Nov. 3: 16, 1963. *Nom. inval.*, based on *Neowerdermannia vorwerkii* var. *erectispina, nom. inval.* (Art. 37.1). [Erroneously included as valid in RPS 14.] [RPS 14]

Weingartia vorwerkii var. **gielsdorfiana** (Backeberg) Backeberg, Descr. Cact. Nov. 3: 16, 1963. Basionym: *Neowerdermannia vorwerkii* var. *gielsdorfiana*. [RPS 14]

Weingartia westii (P. C. Hutchison) Donald, Nation. Cact. Succ. J. 13(4): 67, 1958. Basionym: *Gymnocalycium westii*. [Repeated by Backeberg in Die Cact. 3: 1789, 1959 (and reported as correct place of citation in RPS 10).] [RPS 9]

Weingartia westii var. **lecoriensis** (Cardenas) Donald, Ashingtonia 3: 138, 1980. Basionym: *Weingartia lecoriensis*. [RPS 31]

Weingartia westii var. **vilcayensis** (Cardenas) Donald, Ashingtonia 3: 138, 1980. Basionym: *Weingartia vilcayensis*. [RPS 31]

Weingartia zavaletae (Cardenas) F. Brandt, Kakt. Orch.-Rundschau 5(1): 5, 1980. Basionym: *Aylostera zavaletae*. [RPS 31]

Werckleocereus glaber var. **mirandae** (Bravo) Kimnach, Cact. Succ. J. (US): 50(6): 270, 1978. Basionym: *Selenicereus mirandae*. [RPS 29]

Werckleocereus imitans Kimnach & P. C. Hutchison, Cact. Succ. J. (US) 28(5): 152-156, ill., 1956. Typus: *Lankester* s.n. (UC,

G, K, ZSS). [Costa Rica] [RPS 7]

Wigginsia D. Porter, Taxon 13: 210, 1964. Typus: *Echinocactus corynodes*. [*Nom. nov. pro Malacocarpus* Salm-Dyck 1850 *non Malacocarpus* Fischer & Meyer 1843.] [RPS 15]

Wigginsia acuata (Link & Otto) Ritter, Kakt. Südamer. 1: 197, 1979. Basionym: *Echinocactus acuatus*. [RPS 30]

Wigginsia arechavaletae (Spegazzini) D. Porter, Taxon: 13: 211, 1964. Basionym: *Echinocactus acuatus* var. *arechavaletae*. [RPS 15]

Wigginsia beltranii Fric *ex* Fleischer & Schütz, Friciana 50: 33-34, 36-37, 46-47, 1975. [Based on *Malacocarpus beltranii* Fric (*nom. nud.*). Publication date reported as 1976 in RPS 31.] [RPS 31]

Wigginsia bezrucii (Fric) Fleischer, Friciana 51: 17, 1976. Basionym: *Malacocarpus bezrucii*. [RPS 31]

Wigginsia bezrucii var. **centrispina** (Fric) Fleischer, Friciana 51: 33, 37, 1976. Basionym: *Malacocarpus bezrucii* var. *centrispinus*. [RPS 31]

Wigginsia bezrucii var. **corniger** (Fric) Fleischer, Friciana 51: 33, 37, 1976. Basionym: *Malacocarpus bezrucii* var. *corniger*. [RPS 31]

Wigginsia corynodes (Pfeiffer) D. Porter, Taxon 13: 211, 1964. Basionym: *Echinocactus corynodes*. [RPS 15]

Wigginsia courantii (Lemaire) Ritter, Kakt. Südamer. 1: 195, 1979. Basionym: *Echinocactus courantii*. [RPS 30]

Wigginsia erinacea (Haworth) D. Porter, Taxon 13: 210, 1964. Basionym: *Cactus erinaceus*. [RPS 15]

Wigginsia fricii (Arechavaleta) D. Porter, Taxon 13: 210, 1964. Basionym: *Echinocactus fricii*. [RPS 15]

Wigginsia horstii Ritter, Kakt. Südamer. 1: 199, 1979. [RPS 30]

Wigginsia horstii var. **juvenaliformis** Ritter, Kakt. Südamer. 1: 200, 1979. [RPS 30]

Wigginsia kovaricii (Fric) Backeberg, Kakt.-Lex., 233, 1966. Basionym: *Malacocarpus kovaricii*. [RPS 17]

Wigginsia langsdorfii (Lehmann) D. Porter, Taxon 13: 211, 1964. Basionym: *Cactus langsdorfii*. [RPS 15]

Wigginsia leprosorum Ritter, Kakt. Südamer. 1: 194-195, 1979. [RPS 30]

Wigginsia leucocarpa (Arechavaleta) D. Porter, Taxon 13: 211, 1964. Basionym: *Echinocactus leucocarpus*. [RPS 15]

Wigginsia longispina Ritter, Kakt. Südamer. 1: 198-199, 1979. [RPS 30]

Wigginsia macracantha (Arechavaleta) D. Porter, Taxon 13: 210, 1964. Basionym: *Echinocactus sellowii* var. *macracanthus*. [Sphalm. 'macrocantha'.] [RPS 15]

Wigginsia macrogona (Arechavaleta) D. Porter, Taxon 13: 210, 1964. Basionym: *Echinocactus sellowii* var. *macrogonus*. [RPS 15]

Wigginsia orthacantha (Link & Otto) Backeberg, Kakt.-Lex., 224, 1966. Basionym: *Malacocarpus orthacanthus*. [RPS 17]

Wigginsia polyacantha (Link & Otto) Ritter, Kakt. Südamer. 1: 193, 1979. Basionym: *Echinocactus polyacanthus*. [RPS 30]

Wigginsia prolifera Ritter, Kakt. Südamer. 1: 199, 1979. [RPS 30]

Wigginsia rubricostata Fric *ex* Fleischer & Schütz, Friciana 50: 37,47, 1975. [RPS -]

Wigginsia schaeferiana Abraham & Theunissen, Internoto 9(1): 21-25, ills., 1988. Typus: *Abraham* 163 (KOELN [Succulentarium]). [Additional data in l.c. (2): 52-53, ills.] [Uruguay] [RPS 39]

Wigginsia sellowii (Link & Otto) Ritter, Kakt. Südamer. 1: 196, 1979. Basionym: *Echinocactus sellowii*. [RPS 30]

Wigginsia sellowii var. **macracantha** (Arechavaleta) Ritter, Kakt. Südamer. 1: 197, 1979. Basionym: *Echinocactus sellowii* var. *macracanthus*. [RPS 30]

Wigginsia sessiliflora (Hooker) D. Porter, Taxon 13: 210, 1964. Basionym: *Echinocactus sessiliflorus*. [RPS 15]

Wigginsia sessiliflora var. **martinii** (Rümpler) D. Porter, Taxon 13: 210, 1964. Basionym: *Malacocarpus martinii*. [RPS 15]

Wigginsia sessiliflora var. *pauciareolata* (Arechavaleta) Backeberg, Kakt.-Lex., 224, 1966. *Nom. inval.* (Art. 33.2), based on *Malacocarpus pauciareolatus*. [RPS 17]

Wigginsia stegmannii (Backeberg) D. Porter, Taxon 13: 211, 1964. *Nom. inval.*, based on *Malacocarpus stegmannii, nom. inval.* (Art. 37.1). [Erroneously included as valid in RPS 15.] [RPS 15]

Wigginsia tephracantha (Link & Otto) D. Porter, Taxon 13: 210, 1964. Basionym: *Echinocactus tephracanthus*. [RPS 15]

Wigginsia tephracantha var. **courantii** (Salm-Dyck) Backeberg, Kakt.-Lex., 224, 1966. Basionym: *Malacocarpus tephracanthus* var. *courantii*. [RPS 17]

Wigginsia tephracantha var. **depressa** (Spegazzini) D. Porter, Taxon 13: 210, 1964. Basionym: *Echinocactus acuatus* var. *depressus*. [RPS 15]

Wigginsia turbinata (Arechavaleta) D. Porter, Taxon 13: 211, 1964. Basionym: *Echinocactus sellowii* var. *turbinatus*. [RPS 15]

Wigginsia vorwerkiana (Backeberg) D. Porter, Taxon 13: 210, 1964. Basionym: *Echinocactus vorwerkianus*. [RPS 15]

Wilcoxia albiflora Backeberg, Cactus (Paris) No. 33: 88 (= Suppl. Cactus No. 33: 16), ill., 1952. [Typified by cultivated material from the collection 'Les Cèdres'.] [RPS 3]

Wilcoxia kroenleinii A. Cartier, Succulentes 2(2): 2-3, 1980. [RPS 31]

Wilcoxia lazaro-cardenasii (Contreras & al.) Cartier, Maandbl. Liefhebb. Cact. Vetpl. 3(11): 164, 1990. *Nom. inval.* (Art. 33.2), based on *Neoevansia lazaro-cardenasii*. [RPS 41]

Wilcoxia nerispina Backeberg, Die Cact. 4: 2078, figs. 1960-1961, 1960. *Nom. inval.* (Art. 36.1, 37.1). [RPS -]

Wilcoxia schmollii var. *lanata* hort. *ex* Cartier, Maandbl. Liefhebb. Cact. Vetpl. 3(10): 146,

1990. *Nom. inval.* (Art. 36.1, 37.1). [Published as provisional name.] [RPS 41]

Wilcoxia tamaulipensis var. *brevispina* hort. *ex* Cartier, Maandbl. Liefhebb. Cact. Vetpl. 3(8): 123, 1990. *Nom. inval.* (Art. 36.1, 37.1). [Published as provisional name.] [RPS 41]

Wilcoxia tomentosa Bravo, Cact. Suc. Mex. 3(2): 28, 1958. [RPS 9]

Wilcoxia viperina var. **tomentosa** Bravo, Cact. Suc. Mex. 19: 47, 1974. [RPS 25]

Wilcoxia zopilotensis Meyran, Cact. Suc. Mex. 14: 51-54, 70-71, 1969. [RPS 20]

Wilmattea venezuelensis Croizat, Pittieria 4: 39-41, 1972. [RPS 24]

Winteria F. Ritter, Kakt. and. Sukk. 13: 4, 1962. *Nom. illeg.* (Art. 64). [*Nom. illeg., non Wintera* Murr. 1784 nec van Tiegh. 1900; subsequently replaced by *Hildewintera.*] [RPS 13]

Winteria aureispina Ritter, Kakt. and. Sukk. 13: 4, 1962. Incorrect name (Art. 11.3). [RPS 13]

Winterocereus Backeberg, Kakt.-Lex., 455, 1966. *Nom. illeg.* (Art. 63.1). [Erroneously included as legitimate in RPS 17..] [*Nom. nov. (illeg.) pro Winteria* Ritter 1962 (*non* Murray *nec* Forster) – *Hildewintera* Ritter 1966 has priority by 3 months.] [RPS 17]

Winterocereus aureispinus (Ritter) Backeberg, Kakt.-Lex., 455, 1966. Incorrect name, based on *Winteria aureispina,* incorrect name (Art. 11.3). [*Hildewintera aureispina* (Ritter) Ritter has priority by 3 months.] [RPS -]

Wittia himantoclada (Roland Gosselin) Woodson, Ann. Missouri Bot. Gard. 45(1): 88, 1958. Basionym: *Rhipsalis himantoclada.* [RPS 9]

Wittiocactus Rauschert, Taxon 31(3): 558, 1982. Typus: *Wittia amazonica* Schumann. [*Nom. nov. pro Wittia* Schumann 1903 *non* Pantocsek 1889.] [RPS 33]

Wittiocactus amazonicus (Schumann) Rauschert, Taxon 31(3): 559, 1982. Basionym: *Wittia amazonica.* [RPS 33]

Wittiocactus panamensis (Britton & Rose) Rauschert, Taxon 31(3): 559, 1982. Basionym: *Wittia panamensis.* [RPS 33]

Yungasocereus Ritter, Kakt. Südamer. 2: 668-669, 1980. Typus: *Samaipaticereus inquisiviensis.* [Name first mentioned in Backeberg, Die Cact. 2: 1359, 1959 (with *Y. microcarpus* as type; cf. RPS 10) and as manuscript name in Krainz, Katalog ZSS ed. 2, 117, 1967.] [RPS 31]

Yungasocereus inquisivensis (Cardenas) Ritter, Kakt. Südamer. 2: 669, 1980. Basionym: *Samaipaticereus inquisivensis.* [RPS 31]

Yungasocereus microcarpus Ritter *ex* Krainz, Katalog ZSS ed. 2, 117, 1967. Based on *Ritter* 332. *Nom. inval.* (Art. 36.1, 37.1). [Published as *nomen nudum.*] [Bolivia] [RPS -]

Zehntnerella chaetacantha Ritter, Kakt. Südamer. 1: 215, 1979. [RPS 30]

Zehntnerella chaetacantha var. **montealtoi** Ritter, Kakt. Südamer. 1: 215-216, 1979. [RPS 30]

Zehntnerella polygona Ritter, Kakt. Südamer. 1: 214, 1979. [RPS -]

Zygocactus microsphaericus (Schumann) Buxbaum, Kakt. and. Sukk. 8(9): 136, 1957. *Nom. inval.* (Art. 33.2), based on *Cereus microsphaericus.* [RPS 8]

Zygocactus truncatus var. **crenatus** Borg, Cacti, ed. 2, 420, 1951. [RPS 3]